NATIONAL ACADEMIES *Sciences Engineering Medicine*

NATIONAL ACADEMIES PRESS
Washington, DC

Carbon Utilization Infrastructure, Markets, and Research and Development

A Final Report

Committee on Carbon Utilization Infrastructure, Markets, Research and Development

Board on Energy and Environmental Systems

Division on Engineering and Physical Sciences

Board on Chemical Sciences and Technology

Division on Earth and Life Studies

Consensus Study Report

NATIONAL ACADEMIES PRESS 500 Fifth Street, NW Washington, DC 20001

This activity was supported by Contract DE-EP0000026/89303021FFE400026 between the National Academy of Sciences and the Department of Energy. Any opinions, findings, conclusions, or recommendations expressed in this publication do not necessarily reflect the views of any organization or agency that provided support for the project.

International Standard Book Number-13: 978-0-309-71775-5
International Standard Book Number-10: 0-309-71775-2
Digital Object Identifier: https://doi.org/10.17226/27732

This publication is available from the National Academies Press, 500 Fifth Street, NW, Keck 360, Washington, DC 20001; (800) 624-6242 or (202) 334-3313; http://www.nap.edu.

Printed in the United States of America.

Suggested citation: National Academies of Sciences, Engineering, and Medicine. 2024. *Carbon Utilization Infrastructure, Markets, and Research and Development: A Final Report*. Washington, DC: The National Academies Press. https://doi.org/10.17226/27732.

Consensus Study Reports published by the National Academies of Sciences, Engineering, and Medicine document the evidence-based consensus on the study's statement of task by an authoring committee of experts. Reports typically include findings, conclusions, and recommendations based on information gathered by the committee and the committee's deliberations. Each report has been subjected to a rigorous and independent peer-review process and it represents the position of the National Academies on the statement of task.

Proceedings published by the National Academies of Sciences, Engineering, and Medicine chronicle the presentations and discussions at a workshop, symposium, or other event convened by the National Academies. The statements and opinions contained in proceedings are those of the participants and are not endorsed by other participants, the planning committee, or the National Academies.

Rapid Expert Consultations published by the National Academies of Sciences, Engineering, and Medicine are authored by subject-matter experts on narrowly focused topics that can be supported by a body of evidence. The discussions contained in rapid expert consultations are considered those of the authors and do not contain policy recommendations. Rapid expert consultations are reviewed by the institution before release.

For information about other products and activities of the National Academies, please visit www.nationalacademies.org/about/whatwedo.

COMMITTEE ON CARBON UTILIZATION INFRASTRUCTURE, MARKETS, RESEARCH AND DEVELOPMENT

EMILY A. CARTER (NAS/NAE) (*Chair*), Princeton University and Princeton Plasma Physics Laboratory
SHOTA ATSUMI, University of California, Davis
MAKINI BYRON, Linde
JINGGUANG CHEN (NAE), Columbia University
STEPHEN COMELLO, Energy Futures Initiative Foundation
MAOHONG FAN, University of Wyoming and Georgia Institute of Technology
BENNY FREEMAN (NAE), The University of Texas at Austin
MATTHEW FRY, Great Plains Institute
SARAH M. JORDAAN, McGill University
HAROUN MAHGEREFTEH, University College London
AH-HYUNG (ALISSA) PARK, University of California, Los Angeles
JOSEPH B. POWELL (NAE), University of Houston
CLAUDIA A. RAMÍREZ RAMÍREZ, Delft University of Technology
VOLKER SICK, University of Michigan
SIMONE H. STEWART, National Wildlife Federation
JASON PATRICK TREMBLY, Ohio University
JENNY Y. YANG, University of California, Irvine
JOSHUA S. YUAN, Washington University in St. Louis

Study Staff

ELIZABETH ZEITLER, Study Director, Associate Board Director, Board on Energy and Environmental Systems (BEES)
CATHERINE WISE, Study Co-Director, Program Officer, BEES
KASIA KORNECKI, Senior Program Officer, BEES
LIANA VACCARI, Program Officer, Board on Chemical Sciences and Technology (BCST)
REBECCA PETRICH-DEBOER, Research Associate, BEES
JASMINE VICTORIA BRYANT, Research Assistant, BEES
KAIA RUSSELL, Senior Program Assistant, BEES
BRENT HEARD, Senior Program Officer, BEES
APARAJITA DATTA, Christine Mirzayan Science and Technology Policy Fellow, BEES
ISHITA KAMBOJ, Christine Mirzayan Science and Technology Policy Fellow, BCST
KAVITHA CHINTAM, Christine Mirzayan Science and Technology Policy Fellow, BEES

NOTE: See Appendix B, Disclosure of Unavoidable Conflicts of Interest.

BOARD ON CHEMICAL SCIENCES AND TECHNOLOGY

Reviewers

This Consensus Study Report was reviewed in draft form by individuals chosen for their diverse perspectives and technical expertise. The purpose of this independent review is to provide candid and critical comments that will assist the National Academies of Sciences, Engineering, and Medicine in making each published report as sound as possible and to ensure that it meets the institutional standards for quality, objectivity, evidence, and responsiveness to the study charge. The review comments and draft manuscript remain confidential to protect the integrity of the deliberative process.

We thank the following individuals for their review of this report:

JACK ANDREASEN, Breakthrough Energy
PATRICIA ANSEMS BANCROFT, The Dow Chemical Company
WESLEY BERNSKOETTER, University of Missouri-Columbia
HOLLY BUCK, University at Buffalo
GEOFFREY COATES (NAS), Cornell University
GREESHMA GADIKOTA, Cornell University
JAMES GLASS, Kinder Morgan, Inc.
MICHAEL GUARNIERI, National Renewable Energy Laboratory
DAVID JASSBY, University of California, Los Angeles
ALLAN KOLKER, U.S. Geological Survey
THOMAS MALLOUK (NAS), University of Pennsylvania
DANE McFARLANE, Horizon Climate Group
MARILENE PAVAN, LanzaTech Inc.
JOSHUA SCHAIDLE, National Renewable Energy Laboratory
CORINNE SCOWN, Lawrence Berkeley National Laboratory
DAVID SHOLL (NAE), Oak Ridge National Laboratory
VIJAY SWARUP, ExxonMobil
JAMES TOUR (NAE), Rice University
JOSH WICKS, Twelve

Although the reviewers listed above provided many constructive comments and suggestions, they were not asked to endorse the conclusions or recommendations of this report nor did they see the final draft before its release. The review of this report was overseen by **JOAN BRENNECKE (NAE),** The University of Texas at Austin, and **ANDREW BROWN, JR. (NAE),** Diamond Consulting. They were responsible for making certain that an independent examination of this report was carried out in accordance with the standards of the National Academies and that all review comments were carefully considered. Responsibility for the final content rests entirely with the authoring committee and the National Academies.

Contents

Preface

This second and final report of the National Academies of Sciences, Engineering, and Medicine's (the National Academies') congressionally mandated study on carbon utilization infrastructure, markets, and research and development should be read by anyone aspiring to help transition the world to a net-zero or net-negative carbon-emission civilization. This comprehensive report provides a sober look at some of the difficult tasks ahead of us. We must rise to the challenge, given the increasingly alarming changes to our climate. (Since the first report's publication in early 2023, we have witnessed many tragedies, including the destruction of Lahaina, Hawaii, and the perishing of nearly 100 of its citizens by wildfire; the unprecedented, widespread warming of the Atlantic Ocean; and the largest coral bleaching on record.)

This report focuses on an oft-neglected aspect of the transition to net-zero emissions: the carbon embedded in the essential products used in daily life, such as chemicals, plastics, and construction materials. While the energy system should be "decarbonized" by switching from fossil fuel to renewable or nuclear sources, zero-carbon alternatives cannot replace essential carbon-based products. Where carbon is crucial, we have two solutions. For short-lived products like chemicals or aviation fuel, we can stop using fossil carbon as a feedstock and instead develop processes that use recycled carbon. On the other hand, long-lived products can provide a place to store carbon for the long term, while meeting other market needs, such as for concrete, aggregates, or elemental carbon materials. We specifically must find *sustainable* ways to recycle and reuse carbon wastes, especially those causing the planet the most harm, to continue to produce many of the products needed for everyday life. This committee's mandate was to assess prospects for utilizing two particular carbon wastes in large abundance: carbon dioxide and coal waste. This final report expands far beyond, while building upon, the conclusions of the committee's first report, which spotlighted the current status, needs, and opportunities for CO_2 utilization market and infrastructure development.

In the following chapters, the committee examines market opportunities for carbon dioxide and coal waste utilization, the status of utilization technologies and their research, development, and demonstration needs, and the needs for and impacts of infrastructure for CO_2 utilization. The report also outlines the status and needs for life cycle, techno-economic, and equity assessments of CO_2 utilization systems, and the policy and regulatory frameworks needed for CO_2 utilization to contribute sustainably to a net-zero emissions future. With respect to specific processes and products, the report analyzes CO_2 mineralization, chemical, and biological conversion routes to make inorganic carbonate construction materials, elemental carbon materials, fuels, chemicals, and polymers. The report also examines coal waste utilization to produce long-lived carbon products and to extract critical minerals. This comprehensive review resulted in a research agenda for carbon utilization, discussed in individual chapters and in a final chapter along with Appendix E that organizes the information in multiple ways for multiple audiences.

Unlike the first report, which was completed in a relatively short time—under a year—to inform near-term infrastructure investment decisions, the committee developed this second report over roughly 15 months, holding throughout 2023 seven open information-gathering sessions with 41 speakers from multiple Department of Energy (DOE) offices, congressional authorizing committees, the Small Business Administration, companies, national laboratories, university researchers, and nonprofits. The committee held meetings for deliberation and worked independently and in small groups to produce the report's text, findings, recommendations, and research agenda. The committee has worked extraordinarily hard to produce this remarkably comprehensive analysis extending from basic research through societal impacts and everything in between. To cover that intellectual landscape appropriately, we added seven new members with complementary expertise to the committee that wrote the first report, to bring even more expertise on research, development, and demonstration needs and opportunities, coal waste utilization, life cycle issues, and societal considerations for carbon utilization. Together, I believe we have delivered a report of broad and lasting value to our sponsors within DOE and Congress, and for the nation and the world. As the chair, I want to extend my deepest gratitude to every committee member and to the National Academies staff for their indispensable contributions and steadfast commitment to our shared aspirations to preserve and sustain the planet for future generations.

Emily A. Carter, *Chair*
Committee on Carbon Utilization Infrastructure, Markets,
Research and Development

Summary

CARBON UTILIZATION AND THE TRANSITION TO NET-ZERO EMISSIONS

Carbon is an essential component of molecules and materials that are integral to life-sustaining atmospheric, geologic, biologic, and economic systems. Exploitation of fossil fuels has thrown the natural system of carbon flows out of balance, with ongoing and accelerating accumulation of waste carbon dioxide (CO_2) and other greenhouse gases (GHGs) in the atmosphere, causing global warming. Returning to a safer climate will require that atmospheric GHG concentrations be stabilized, and eventually lowered, primarily by ending the largest source of GHG flows into the atmosphere: CO_2 emissions from fossil fuel combustion. Some carbon-based systems that cannot be "decarbonized" with zero-carbon-emission substitutes will remain. For these systems, carbon will need to be managed, rather than eliminated. Examples include carbon-based molecules and materials associated with agriculture and consumer products; some fossil combustion emissions of CO_2 into the atmosphere that cannot or will not be ended; and removal of some CO_2 already in the atmosphere. Carbon management will be needed to establish a negative balance of GHG flows into and out of the atmosphere (i.e., net-negative emissions) during the decarbonization transition, and to maintain an even balance of GHG flows (i.e., net-zero emissions) once safe atmospheric concentrations of GHGs are reached.[1] In a net-zero future, CO_2 utilization—the conversion of CO_2 into marketable products—can operate at a global annual scale of multiple gigatonnes to provide an alternative, circular-carbon feedstock[2] for necessary carbon-based products and generate products that durably store carbon. Coal waste could also serve as a source of raw materials in a net-zero future, via conversion of carbon components to durable carbon-derived products or extraction of critical minerals and materials.

STUDY MANDATE

This study, sponsored by the Department of Energy (DOE), examines markets; infrastructure; and research, development, and demonstration (RD&D) needs for CO_2 and coal waste utilization in a net-zero emissions future, as requested by Congress in the Energy Act of 2020. The committee focused on regional and national market opportunities,

[1] For this report, net-zero emissions is the assumed final state of a safe climate system, although during and after the transition to net zero, periods of net-positive and net-negative emissions are both likely.

[2] A circular-carbon feedstock is a raw material that can participate in a circular carbon economy, where materials and energy are reused and recycled to prevent net emissions to the atmosphere.

infrastructure needs, and RD&D needs for technologies that transform CO_2 or coal waste[3] into products that will contribute to a net-zero emission future. It analyzed challenges in expanding infrastructure, mitigating environmental impacts, accessing capital, overcoming technical hurdles, and addressing geographic, community, and equity issues for carbon utilization. In a first report, the committee assessed the state of and opportunities to improve and expand on infrastructure for CO_2 utilization. The first report highlighted priority products that could be made from CO_2; discussed needs for enabling infrastructure; and overviewed policy, regulatory, and environmental justice considerations for utilization infrastructure.

For this second and final report from the study, the committee was tasked to identify potential market opportunities for CO_2 utilization; identify opportunities for federal support of small businesses; examine infrastructure for CO_2 utilization and the economic, climate, and environmental impacts of any well-integrated national CO_2 pipeline system applied for CO_2 utilization; assess current and emerging technologies and approaches for CO_2 utilization, identify their research needs, and develop a comprehensive research agenda to advance CO_2 utilization; and determine the feasibility of and opportunities for commercializing coal waste–derived products.

PRIORITY OPPORTUNITIES FOR CO_2- OR COAL WASTE–DERIVED PRODUCTS IN A NET-ZERO EMISSIONS FUTURE

CO_2 can be converted to products that could serve critical markets for carbon-based materials in a circular carbon economy, for durable carbon storage, or both. Priority product classes identified were fuels, construction materials, polymers, agrochemicals, chemicals and chemical intermediates, food and animal feed, and elemental carbon materials, at potential global scales of megatonnes to gigatonnes for individual products. Coal waste utilization market opportunities include long-lived products like construction materials and elemental carbon materials, as well as metals and minerals. The carbon feedstock (e.g., CO_2 from fossil or nonfossil sources, coal waste) and the product lifetime are important for assessing the climate impact and sustainability of different market opportunities. The committee considered product-specific market questions and examined factors that influence CO_2 utilization market development, including cost, competing feedstocks, technology and infrastructure development, supply chains, consumer demand, the regulatory environment, financial risks, and environmental and equity impacts. The committee recommends that DOE prioritize research on co-located capture and conversion, particularly for long-lived products that contribute to carbon sequestration (Recommendation 2-1); close information gaps on environmental, market, resource, and jobs impacts of CO_2 conversion (Recommendation 2-3); and support efforts to inform the public about carbon management (Recommendation 2-4).

LIFE CYCLE, TECHNO-ECONOMIC, AND SOCIETAL/EQUITY ASSESSMENTS OF CO_2 UTILIZATION PROCESSES, TECHNOLOGIES, AND SYSTEMS

CO_2 utilization is intimately tied to environmental, economic, and societal needs, and cannot be understood without the assessment of its life cycle, societal impact, and techno-economic status. The committee evaluated technology assessment approaches, focusing on capabilities, use, and critical issues for improvement. Techno-economic assessment (TEA) addresses economic questions such as cost or profitability of a developing technology or process, compared to competing solutions. Life cycle assessment (LCA) informs on the sustainability of technologies and processes by quantifying the environmental burdens from materials extraction through end of life. Critical issues for both TEA and LCA of CO_2 utilization technologies include transparency and consistency of system boundaries, incorporating CO_2 purity and source information, geographic and temporal relevance, and addressing uncertainty, especially for early-stage technologies. Equity assessments, including social LCA, seek to minimize negative outcomes and maximize positive outcomes from policies, programs, or processes, particularly for those facing inequality or disparities. In the federal government, equity assessment guidelines for the goal, scope, and system boundaries are still being developed, as are available data and localized information.

[3] Coal waste streams considered in this report are coal combustion residuals (fly ash, bottom ash, boiler slag, and flue gas desulfurization products), impoundment waste (coarse and fine refuse), and acid mine drainage (as a source of critical minerals).

TEA, LCA, and equity assessment can be applied throughout technology development, helping to inform design and RD&D priorities at early stages and informing deployment decisions at later stages. DOE's requirements for TEA and LCA are inconsistent, and there is a need to integrate TEA and LCA for decision support to avoid conflicting outcomes. Circular uses of carbon are poorly incorporated into existing methods for TEA, LCA, and equity assessments, and technical challenges need to be overcome to improve such methodologies. The committee recommends that federal research agencies make consistent TEA and LCA requirements for applied research for all but early-stage technologies (Recommendation 3-1) and that TEA and LCA be facilitated for mid-to-late-stage technologies via guidance and improved tools (Recommendation 3-3). DOE should require life cycle thinking for equity assessments to identify hotspots and integrate risk and societal assessments (Recommendation 3-5) and support research into assessment approaches that address circularity of CO_2-derived products and develop methods and tools for carbon traceability and custody (Recommendation 3-6).

POLICY AND REGULATORY FRAMEWORKS NEEDED FOR SUSTAINABLE CO_2 UTILIZATION

Realizing the climate change and market benefits of CO_2 utilization technology and infrastructure will require fundamental changes to current policy, economic, and regulatory structures. This report assumes a transition to a net-zero emissions future, which would require a price or limit on GHG emissions for technology competitiveness. CO_2 utilization has important equity and justice implications that need to be addressed during development and deployment.

Existing incentives stem from tax credits, permitting and regulatory frameworks, and federal legislation. The committee examined potential demand- and supply-side policy tools, such as procurement and deployment support, and noneconomic tools, such as common carrier status, clarity of standards and codes, and workforce development. It evaluated business development mechanisms, especially for small businesses. It emphasized that a better understanding and intentional focus on environmental justice is needed for CO_2 utilization to benefit impacted communities. In particular, environmental justice needs to be a focus during project development, infrastructure siting, and project selection. Principles of environmental justice are also useful in public discourse, community engagement, and policy development. Drivers, policies, and impacts for economically viable and sustainable carbon utilization are presented in Figure S-1.

The committee recommends that the General Services Administration and DOE develop upscaling opportunities for small businesses (Recommendation 4-4). DOE should work in partnership with community-centered councils and agencies to define, track, and measure impacts of CO_2 utilization projects and infrastructure, determine equity of distribution, and communicate outcomes (Recommendation 4-5). To better understand public perception of CO_2 utilization, nongovernmental organizations and research-conducting entities should identify gaps in knowledge about societal acceptance of or opposition to the sector (Recommendation 4-6). To improve community engagement, DOE should prioritize projects that incorporate meaningful community engagement frameworks into decision-making (Recommendation 4-7). To improve tangible benefits to surrounding communities, new CO_2 utilization infrastructure development should apply justice principles during the planning and design process (Recommendation 4-8).

CO_2 AND COAL WASTE UTILIZATION RESEARCH STATUS AND NEEDS

The committee described the status of RD&D efforts for CO_2 utilization, assessed current and emerging technologies and approaches, identified research needs, and developed a comprehensive research agenda to advance CO_2 utilization. This report updates a 2019 National Academies research agenda for gaseous carbon waste streams utilization (NASEM 2019). The committee identified research needs for coal waste utilization as part of examining commercialization feasibility and opportunities. The committee examined four approaches to CO_2 utilization RD&D: (1) mineralization of CO_2 into inorganic carbonates, (2) conversion of CO_2 into elemental carbon materials, and (3) chemical and (4) biological pathways for CO_2 conversion into organic chemicals and fuels. Coal waste utilization included use of carbon, rare earth elements, critical minerals, and other energy-relevant minerals. Chapters on each technology pathway describe existing and emerging products and processes, challenges, and RD&D opportunities. They conclude with overall RD&D needs and recommendations to address those needs. Figure S-2 summarizes the major feedstocks, processes, products, and applications for carbon utilization that are discussed in Chapters 5–9.

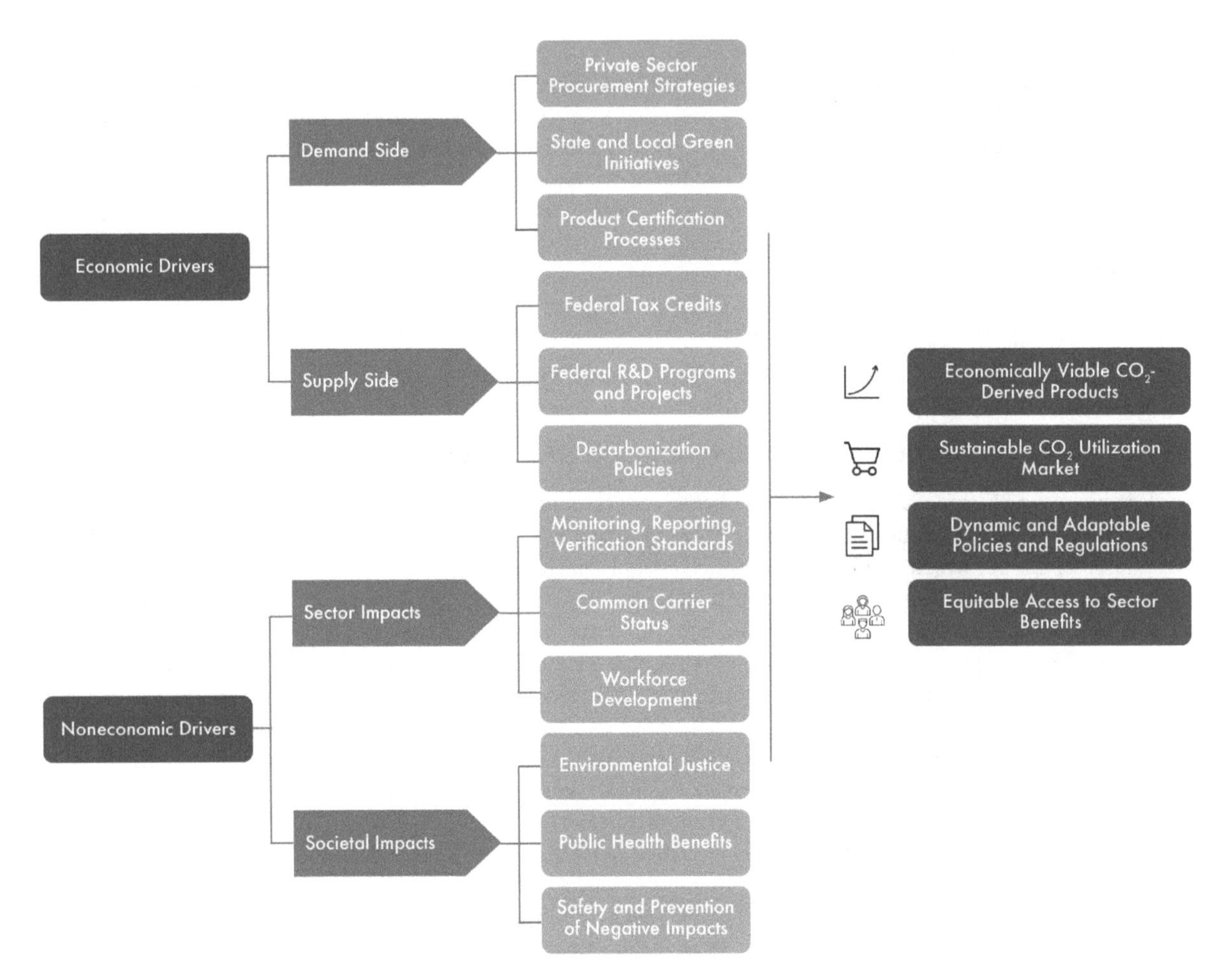

FIGURE S-1 Summary of policy and regulatory considerations and impacts for the emerging CO_2 utilization sector.
SOURCE: Icons from the Noun Project, https://thenounproject.com. CC BY 3.0.

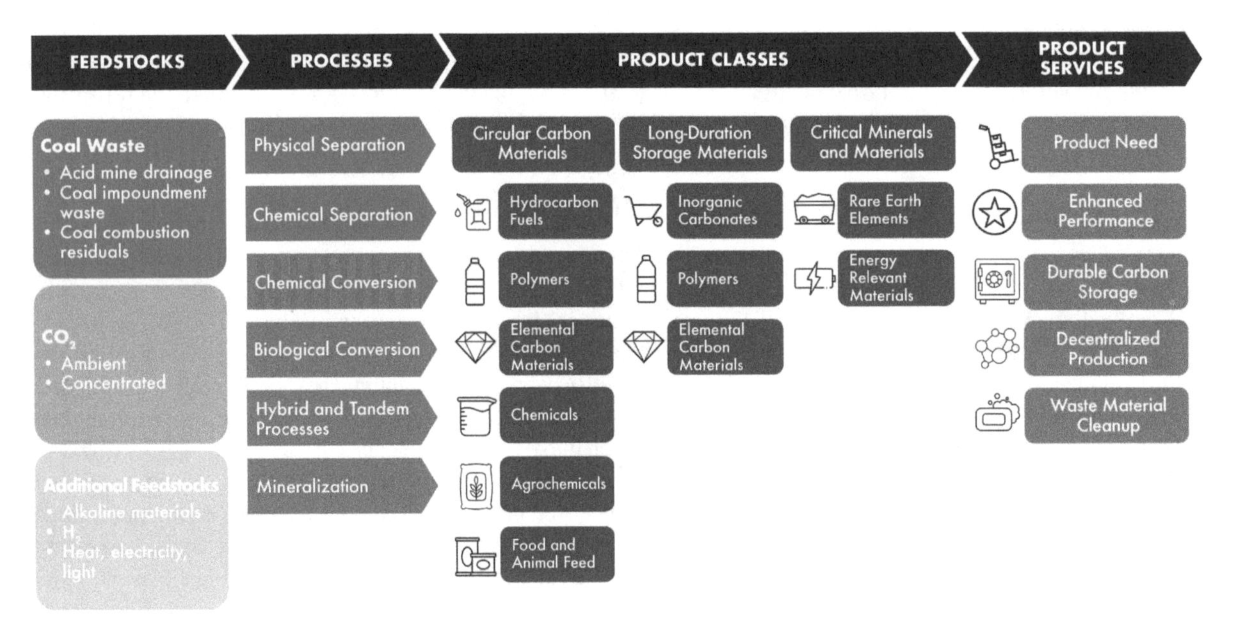

FIGURE S-2 Summary of the feedstocks, processes, product classes, and product services for CO_2 and coal waste utilization.
SOURCE: Icons from the Noun Project, https://thenounproject.com. CC BY 3.0.

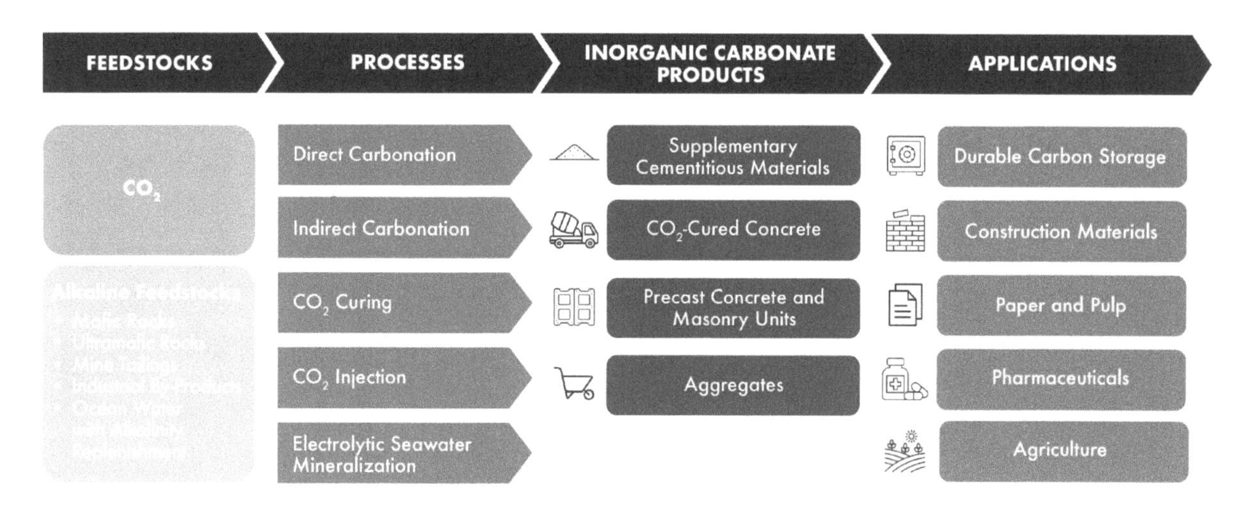

FIGURE S-3 Summary of the feedstock inputs, processes, products, and applications for mineral carbon utilization processes to form inorganic carbonates.
SOURCE: Icons from the Noun Project, https://thenounproject.com. CC BY 3.0.

Mineralization of CO_2 to Inorganic Carbonates

Reaction of CO_2 with calcium- and magnesium-bearing minerals or alkaline industrial wastes forms inorganic carbonates: thermodynamically stable, solid materials with durable carbon storage capability and tens of gigatonnes annual global market potential. Inorganic carbonates have applications as supplementary cementitious material, in concrete cured with CO_2, and as aggregates, all used primarily for construction materials. Technologies under development include carbonation of natural minerals or alkaline industrial wastes, enhanced carbon uptake by construction materials, electrolytic or biologically enhanced mineralization of brine and seawater, alternative cementitious materials and mineralization pathways (e.g., magnesium-based materials, pathways involving organic acids), and integrated processes. Current technical bottlenecks include large energy requirements for mining and processing of minerals, slow mineral dissolution and carbonation rates, managing feedstock impurities, and, for ocean-based processes, ensuring minimal environmental and ecosystem impacts. Testing and property validation of new materials for user and regulator acceptance are barriers to adoption. Figure S-3 presents the major features of mineral carbon utilization, including feedstocks, processes, products, and applications.

The committee identified mineralization RD&D needs and recommended actions in five areas. Cross-technology needs include evaluating and expanding mapping of alkaline resources; fundamental and translational research to improve energy efficiency, process efficiency, product selectivity, and scalability; and multimodal optimization of infrastructure to link feedstocks, mineralization sites, and product markets (Recommendation 5-1). Ocean-based CO_2 mineralization needs include understanding local environmental impacts; developing a protocol to assess and mitigate impacts from pH changes; and developing a testing platform for ocean-based concepts (Recommendation 5-2). Electrochemical CO_2 mineralization requires a full spectrum of RD&D activities, including catalyst development, cell design, membrane materials, and systems engineering/integration (Recommendation 5-3). Integration of carbon mineralization with metal recovery should be explored, including establishing university–industry–national laboratory collaborations for rapid scale up (Recommendation 5-4). Testing, standardization, and certification is required for construction materials produced from CO_2; materials discovery and characterization of new forms of mineral carbonates are required to enable new processes such as 3D-printed concrete (Recommendation 5-5).

Chemical Conversion to Elemental Carbon Materials

Elemental carbon materials are zero-, one-, two-, or three-dimensional (0D, 1D, 2D, or 3D) structures composed of carbon alone, and include products such as carbon dots, carbon nanotubes, graphene, carbon fibers, graphite, and carbon-carbon composites. These materials and their derivatives have elemental and bulk structures

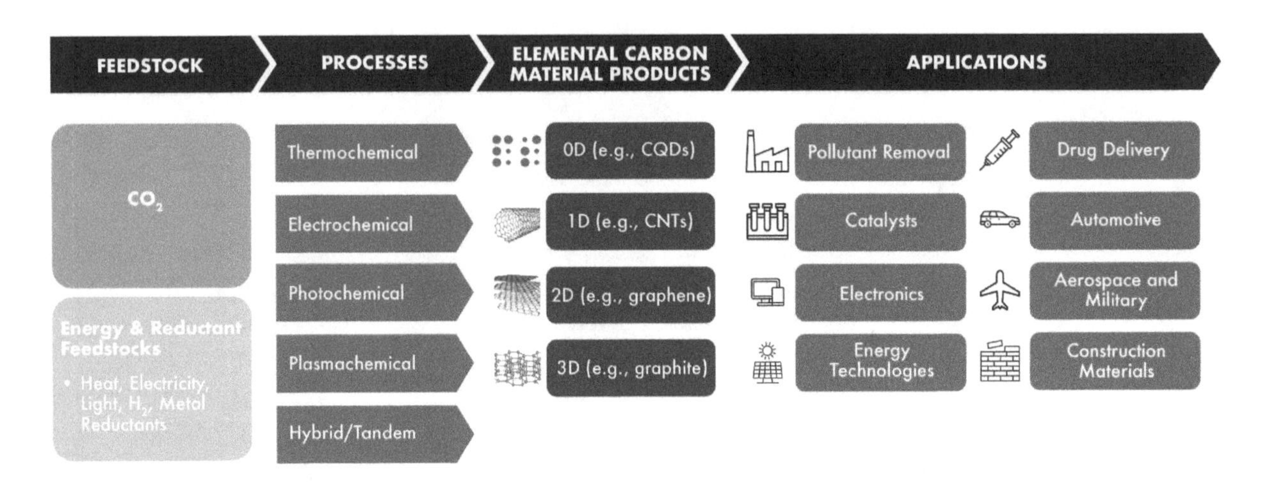

FIGURE S-4 Summary of the feedstock, processes, products, and applications for chemical conversion of CO_2 to form 0–3D elemental carbon materials.
SOURCE: Icons from the Noun Project, https://thenounproject.com. CC BY 3.0.

that yield properties like high conductivity, high mechanical strength, and active sites for catalysis. Markets for elemental carbon materials are growing as their novel structural and electronic properties are discovered and applications are found in the built environment, industry, health care, and environmental protection. Figure S-4 shows the major features of the input feedstock, processes, products, and applications associated with producing carbon materials from CO_2.

CO_2 can be reduced to elemental carbon via four major pathways: thermochemical, photochemical, electrochemical, and plasmachemical. Thermochemical reduction is the most mature but suffers from high energy requirements and low rates or deactivation through coking. Electrochemical reduction has relatively high selectivity and mild conditions, but slow rates owing to mass transfer, and high catalyst and separation costs. Photochemical and plasmachemical reduction have not been explored in depth. Directly derived products include fullerenes, hollow carbon spheres, carbon nanofiber and nanotubes, graphene, and graphite. Indirectly derived materials include carbon fiber and carbon-carbon composites.

Processes converting CO_2 to elemental carbon materials have several common challenges, including limited research to date, difficulty comparing across approaches, substantial energy requirements, and limited understanding of system stability and selectivity. Research needs include developing foundational knowledge across the four conversion types (Recommendation 6-1), discovery and development of catalysts and low-energy processes for morphologically selective production (Recommendation 6-2), discovery and development of catalysts that are active, morphologically selective, and robust for production of diverse elemental carbon materials from CO_2 (Recommendation 6-3), development of hybrid or tandem processes for CO_2 conversion to elemental carbon materials (Recommendation 6-4), and development of integrated CO_2 capture and conversion to elemental carbon materials (Recommendation 6-5).

Chemical Processes for CO_2 Conversion to Fuels, Chemicals, and Polymers

Chemical conversion of CO_2 can produce organic products, including fuels, chemicals and chemical intermediates, and polymers. Priority products include single-carbon compounds such as carbon monoxide, methanol, formic acid, urea, and methane; multicarbon compounds such as oxygenates, olefins, aromatics, and hydrocarbons; and polymers such as polycarbonates. Products can be generated via thermochemical, electrochemical, photochemical, plasmachemical, or integrated conversion routes, which each have their own challenges and RD&D opportunities. Figure S-5 shows the major feedstocks, processes, products, and applications for CO_2 conversion to fuels, chemicals, and polymers.

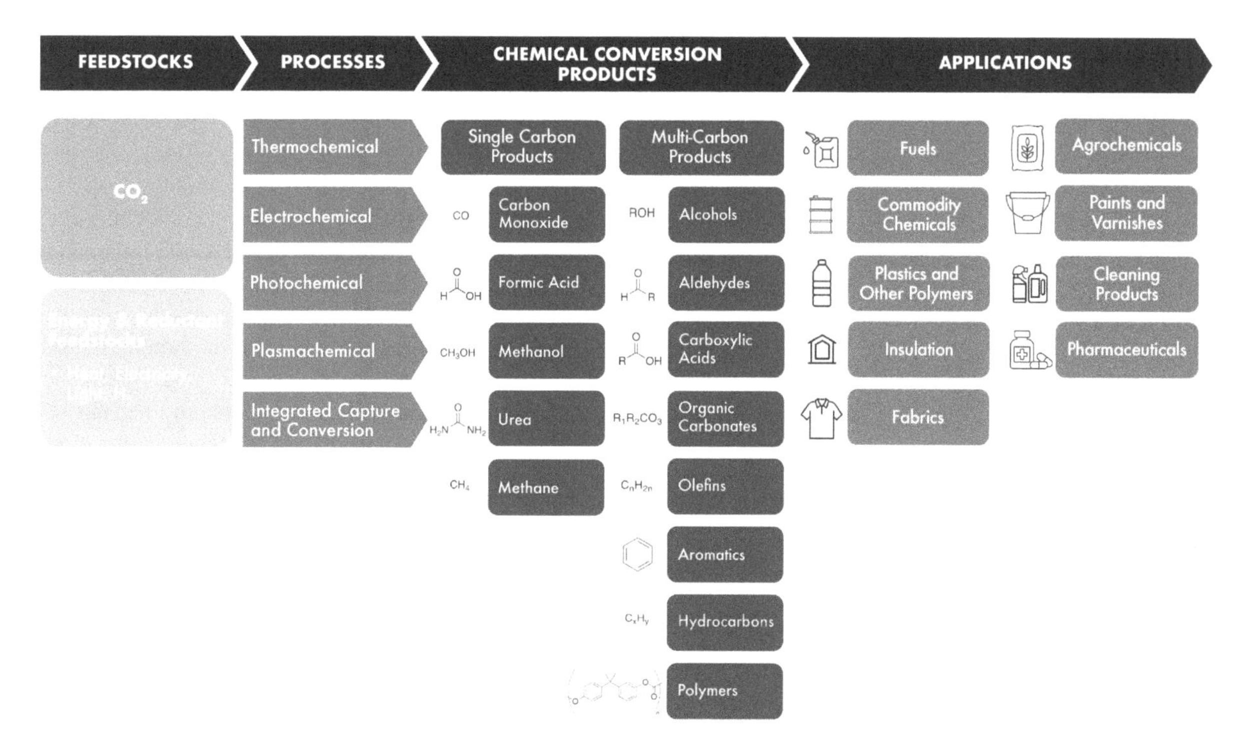

FIGURE S-5 Summary of the feedstocks, processes, products, and applications for chemical CO_2 conversion to make fuels, chemicals, and polymers.
SOURCE: Icons from the Noun Project, https://thenounproject.com. CC BY 3.0.

Challenges associated with high-temperature thermochemical CO_2 conversion include controlling catalyst selectivity and stability, and the need for carbon-neutral energy, hydrogen, or other reductants. Significant amounts of carbon-neutral energy also will be needed for the other conversion pathways—electrochemical, photo(electro)chemical, and plasmachemical. Additional challenges for electrochemical CO_2 conversion include long-term catalyst stability and robustness to impurities. Photo(electro)chemical and plasmachemical technologies require improved fundamental understanding of the steps from light absorption to reduction of CO_2 and of plasma-catalyst interactions, and improved reactor design and reaction engineering. Tandem catalysis and integrated capture and conversion of CO_2 could allow access to new products and improve energy efficiency, respectively.

The research agenda and recommendations offer guidance to address challenges with each chemical conversion pathway. Recommendations for thermochemical CO_2 conversion include RD&D on catalytic selectivity and stability, alternative reaction heating methods, production of low-carbon hydrogen and other reductants, and integration with renewable energy and energy storage (Recommendations 7-1 and 7-2). For electrochemical CO_2 conversion, RD&D needs include discovering and developing selective, active, and stable catalysts from abundant metals for diverse products; discovering and developing efficient, inexpensive, robust electrocatalysts for anodic reactions that enable CO_2 utilization; and developing membrane materials with improved properties, cost, and efficiency (Recommendation 7-3). Advancing photo(electro)chemical and plasmachemical CO_2 conversion requires gaining fundamental understanding of processes and interactions, materials discovery, and research to improve devices, reactor design, and reaction engineering (Recommendation 7-4). The committee recommends research on tandem catalysis to improve product portfolio options and integrated CO_2 capture and utilization to improve system efficiency (Recommendations 7-5 and 7-6). It also recommends design and development of catalysts for rapid, stereoselective polymerization with a broader class of monomers (Recommendation 7-7).

Biological CO_2 Conversion to Chemical, Fuel, and Polymer Products

Biological systems can convert CO_2 to fuels, chemicals, and polymers via photosynthetic, nonphotosynthetic, and hybrid (e.g., electro-bio and cell-free biochemical) pathways. This report considers direct conversion of CO_2 through autotrophic microorganisms, acetogenic microbes, or hybrid systems. Biological CO_2 conversion focuses on discovery and engineering of microbes or hybrid processes, rather than on products, as often multiple products are accessible with each system. Photosynthetic systems (i.e., microalgae and cyanobacteria) use light energy and water to fix CO_2 into products such as fuels, polymer precursors, and commodity chemicals. Nonphotosynthetic chemolithotrophic systems (e.g., acetogens) use the potential energy in inorganic compounds, such as H_2 or CO, to form biological reducing agents that can fix CO_2 under anaerobic conditions. Hybrid systems combine microorganism-based bioconversion with chemical catalysis (e.g., electro-, thermal-, plasma-, or photo-catalysis) or attempt biological conversions outside of the microbes themselves. Production of biopolymer precursors (e.g., butanediols, succinic acid, and isoprene) from CO_2 could enable a circular carbon economy for plastics and other polymeric materials. Figure S-6 shows the feedstocks, processes, products, and applications for direct biological conversion of CO_2.

Major challenges for photosynthetic production of chemicals from CO_2 include inefficient photosynthesis and CO_2 fixation; photorespiration of O_2 rather than utilization of CO_2, and cell shading inhibiting photosynthesis at high cell densities. Areas of opportunity include exploration of fast-growing cyanobacteria and eukaryotic algae, improved tools for genome and metabolic engineering, and development of large-scale cultivation strategies. Major challenges for nonphotosynthetic production include optimizing acetogen use of CO_2 and forming commercially advantageous products; multiple substances and complex physiochemical environments; and need for improving electron donors. Opportunities include developing photomixotrophic approaches that combine photosynthesis with chemolithotrophy, including in co-cultured conditions; enhancing acetogenic fermentation, including finding product targets beyond acetate; and discovering and scaling up processes. Major challenges for hybrid systems include providing suitable electron donors; discovering or engineering bioconversion systems capable of high-rate conversion to single-carbon compounds; and improving scalability, economic viability, and process integration to facilitate commercialization. Opportunities for hybrid systems include operating in ambient conditions, converting biocompatible two- and three-carbon chemically produced intermediates, and utilizing cell-free systems to avoid competing pathways while accelerating the discovery process. Understanding the carbon flux control

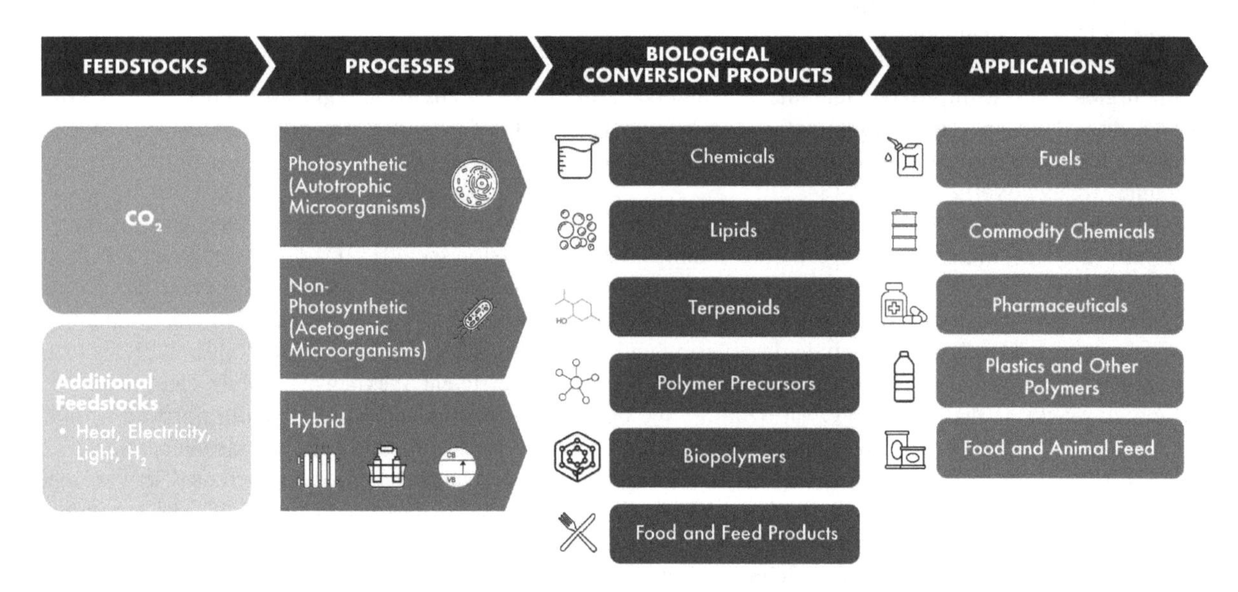

FIGURE S-6 Summary of the feedstocks, processes, products, and applications for direct biological conversion of CO_2.
SOURCE: Icons from the Noun Project, https://thenounproject.com. CC BY 3.0.

and bioenergetics of precursor production will help identify new pathways and improve productivity, conversion efficiency, and titer.

Biological CO_2 conversion requires both basic and applied research to discover and improve systems. Research needs include more sophisticated understanding of metabolism design principles; development of more efficient genetic manipulation tools; and better integration of carbon capture and conversion for photosynthetic systems. Experimental and computational approaches to enhance enzyme stability and efficiency can improve commercial viability by optimizing system efficiency, by-product titer, and productivity (Recommendation 8-2). Applied research needs for polymer precursors include improving reactor design and processes and integrating carbon capture technologies with biological reactors. For hybrid systems, improving the scalability of electrolytic technology remains a key hurdle. Additionally, electrocatalyst design should be explored to improve biocompatibility and develop microorganism and cell-free systems that efficiently produce target chemicals from catalysis-derived intermediates under conditions amenable to electrocatalysis (Recommendations 8-3 and 8-4). Reactor design improvements are also needed for hybrid systems, optimizing them to use specific intermediates and produce desired final products (Recommendation 8-5).

Coal Waste

Although production and consumption of coal have fallen substantially in the United States—trends that are expected to continue[4]—current and past use for power generation and industrial processes has resulted in voluminous waste material. Coal waste is both an environmental contaminant in need of remediation and a material containing potentially useful components. This report considers opportunities for beneficial reuse of coal wastes, including acid mine drainage (as a source of critical minerals), coal impoundment wastes, and coal combustion residuals, focusing especially on legacy waste streams. Wastes are generated during coal mining, preparation, and combustion, and are located predominantly in Appalachia and the Intermountain West.

Materials derived from acid mine drainage include pigments and critical minerals; from impoundment wastes include materials for use in construction, energy storage, and 3D printing, carbon fiber, and carbon foam; and from coal combustion residuals include materials for use in cement, concrete blocks, asphalt, drywall, and critical minerals. Coal waste contains hazardous components such as heavy metals and volatile organic compounds, necessitating risk assessments for materials with the potential to leach hazardous components, health assessment of occupational and user exposures, and product performance evaluation for applications in construction, manufacturing, and industry. Figure S-7 shows the major features of coal waste utilization to produce long-lived, solid carbon products and extract critical minerals and materials, including feedstocks, processes, products, and applications.

The committee identified RD&D needs for coal waste utilization in five areas. Facilitating the use of coal wastes to produce solid carbon products or critical minerals and materials will require evaluation and mapping of coal waste resources, development of strategies and infrastructure to link coal waste sites to markets, and improvements to physical and chemical methods for separating mineral matter from carbon in coal wastes (Recommendation 9-1). Improved transformation of coal waste requires applied research into production of long-lived solid carbon products; basic research to understand coal waste conversions; development of 3D printing media from coal wastes; performance evaluation of coal waste–derived materials in their desired application; establishment of standards to address environmental exposures and product safety; and data and tools to conduct LCA and TEA of coal waste utilization processes (Recommendations 9-2 and 9-3). Improved characterization and separations of coal wastes and its components are needed (Recommendation 9-4). Novel methods for extracting lithium, rare earth elements, and other energy-relevant critical materials from both solid and liquid waste streams need to be developed, as do techniques to separate individual elements from each other, especially separation of nickel from cobalt (Recommendation 9-5).

[4] In contrast, global production and consumption of coal are not in decline, so global markets for coal waste utilization technologies will likely exist even as U.S. coal waste volumes decrease.

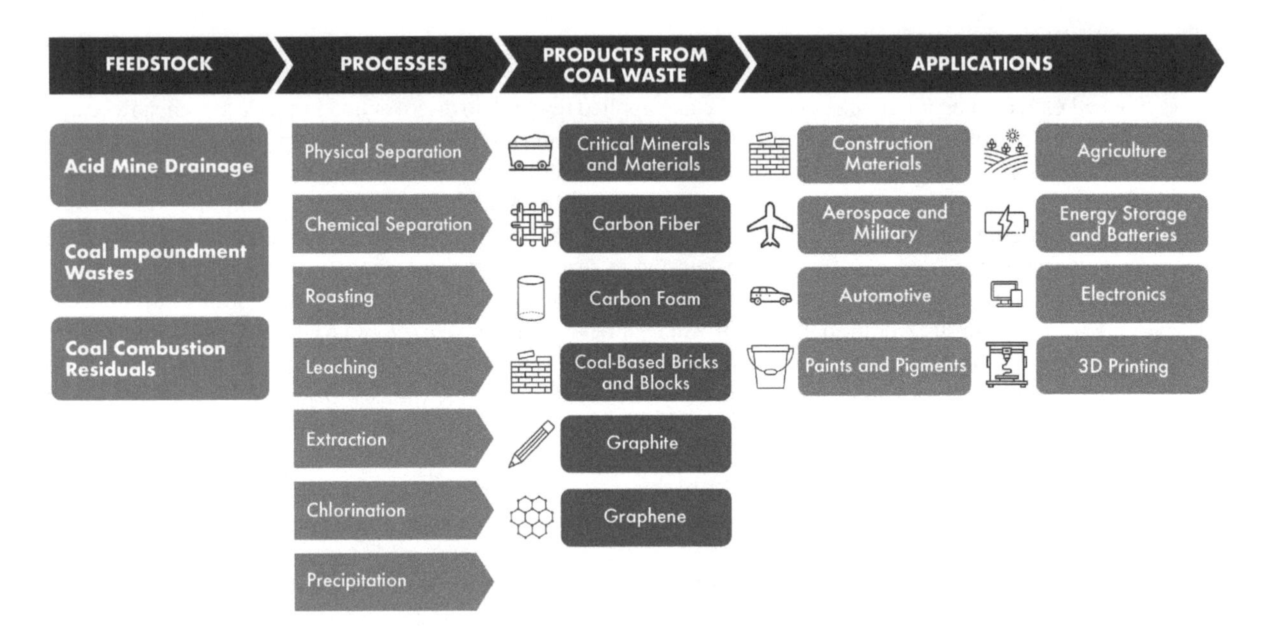

FIGURE S-7 Summary of the feedstocks, processes, products, and applications for coal waste utilization to produce long-lived, solid carbon products and extract critical minerals and materials.
SOURCE: Icons from the Noun Project, https://thenounproject.com. CC BY 3.0.

CO_2 UTILIZATION INFRASTRUCTURE

CO_2 utilization requires extensive infrastructure for CO_2 capture, purification, transportation, and conversion, and for enabling systems to provide hydrogen, electricity, water, CO_2 sequestration, and product transport. The first report of this committee (NASEM 2023) assessed the state of existing infrastructure for CO_2 transportation, use, and storage and identified priority opportunities for future development of such infrastructure. This report identifies opportunities and challenges for CO_2 utilization infrastructure planning at the regional or national scale and evaluates potential economic, climate, environmental, health, safety, justice, and societal impacts of CO_2 utilization infrastructure.

Existing infrastructure for CO_2 utilization includes 20 megatonnes (Mt) per year of CO_2 point source capture and a minimal amount of direct air capture, ~5,000 miles of CO_2 pipelines, ~400 gigawatts of carbon-free electricity, about half a Mt per year of low-carbon hydrogen capacity, and ~2 Mt per year of injection capacity for CO_2 storage. Regional and national assessments of infrastructure capacity indicate one or more orders of magnitude increased need CO_2 utilization and related systems. Extensive development and demonstration of CO_2 utilization infrastructure is under way, in part because of federal government investments authorized in the Infrastructure Investment and Jobs Act. Additionally, private developers are proposing CO_2 pipeline infrastructure, particularly in the Midwest associated with capture of ethanol and fertilizer plant emissions. The planned pipelines have faced opposition owing to health and safety concerns, and the potential to indirectly or directly enable continued fossil fuel use.

The need to access CO_2, other enabling inputs like H_2 and electricity, and product markets will generally result in a need to transport one or more of the inputs to or from a CO_2 utilization site. Multimodal, regional transportation can benefit from mathematical optimization models, which can address a variety of circumstances, such as distributed small- and medium-scale emitters, industrial clusters associated with large volumes of CO_2, and shared pipeline networks serving both CO_2 storage and utilization. In some cases, existing infrastructure can be retrofitted to accommodate CO_2 utilization, but technical, safety, and societal factors have to be considered. For instance, although some examples exist, there are significant economic and technical challenges with retrofitting liquid petroleum or natural gas pipelines to transport high pressure CO_2 or H_2, as they were originally designed for lower pressures and different material reactivities. The large infrastructure requirements across the value chain for CO_2 utilization could delay its growth if supporting infrastructure is slow to develop.

The committee examined economic, climate, environmental, health, safety, justice, and societal impacts of infrastructure for CO_2 utilization, especially a well-integrated regional or national CO_2 pipeline system applied for utilization. By enabling low-cost transportation of large volumes of CO_2, a pipeline system would likely incentivize greater build-out of carbon management technologies and infrastructure across the value chain. This would enable more CO_2 utilization, opening the CO_2 marketplace up to traditional market demand dynamics, and could be further bolstered by regulations and policies to support the dual use of CO_2 pipelines for both sequestration and utilization (Recommendation 10-3). Environmental impacts of CO_2 utilization are associated with energy used for capture, transport, and utilization; leaks of CO_2 or H_2; and land and water requirements. Health and safety implications of CO_2 pipeline systems include those associated with the value chain for utilization, especially CO_2 capture, such as increased or decreased facility pollutant emissions. CO_2 pipelines have significant risks that need to be addressed through proper planning, design, and public consultation in regulation, siting, construction, operation, and decommissioning. In particular, the Pipeline and Hazardous Materials Safety Administration needs to fund research into dispersion modeling and propagating brittle and ductile fractures, as well as realistic-scale test facilities (Recommendation 10-5) and to hold proactive, open-forum consultations with the public for updating its pipeline safety standards (Recommendation 10-1).

RESEARCH AGENDA

The committee developed a research agenda that identifies priority RD&D needs and recommended actions to be taken by government, industry, and academia to enable CO_2 and coal waste utilization in a net-zero future. Basic research, applied research, demonstrations, and enabling technology developments are needed across all CO_2 conversion and coal waste utilization pathways. Enabling needs were identified for markets, technology assessments, policy/equity, and infrastructure. Out of these research needs, three broad categories emerged—reaction-level understanding, systems-level understanding, and demonstration and deployment needs—further separated into 16 research themes. Figure S-8 illustrates the overlap in research themes among the different conversion pathways. Focus on reaction- and systems-level understanding will be most important for advancing chemical and biological CO_2 conversion, while mineralization and coal waste utilization require increased support for demonstration and deployment efforts. Research to support markets, technology assessments, policy/equity, and infrastructure includes identifying market opportunities for CO_2- and coal waste–derived products, developing tools to assess economic, societal, and environmental impacts of CO_2 utilization processes, understanding public perception of CO_2 utilization, and designing modeling and tools to support safe, efficient infrastructure development.

The full research agenda is presented in Chapter 11 (Table 11-1), supplemented by content in Appendix E. Research needs and recommended actions were developed in Chapters 2–10 based on the analysis of the status, barriers, and opportunities in each topical chapter, especially the technology pathway Chapters 5–9. All 71 research needs with their associated 35 recommendations are assembled in Table 11-1, classified by conversion pathway or enabling opportunity. Each research need identifies relevant funding agencies or other actors; specifies basic research, applied research, technology demonstration, or enabling technologies and processes; and denotes into which of the 16 research themes the research need falls. Each need indicates the relevant research area (mineralization, chemical, biological, coal waste utilization, LCA/TEA, markets, infrastructure, and societal impacts) and product class (construction materials, elemental carbon materials, chemicals, polymers, coal waste-derived carbon products, and critical material coal waste by-products), and whether the product is long- or short-lived. Finally, it notes the finding, recommendation, and/or chapter section associated with each research need to direct interested readers to more information on the topic. The research agenda indicates where DOE, other federal funding agencies, industry, and the research community can focus their efforts to improve CO_2 and coal waste utilization for a net-zero emissions future.

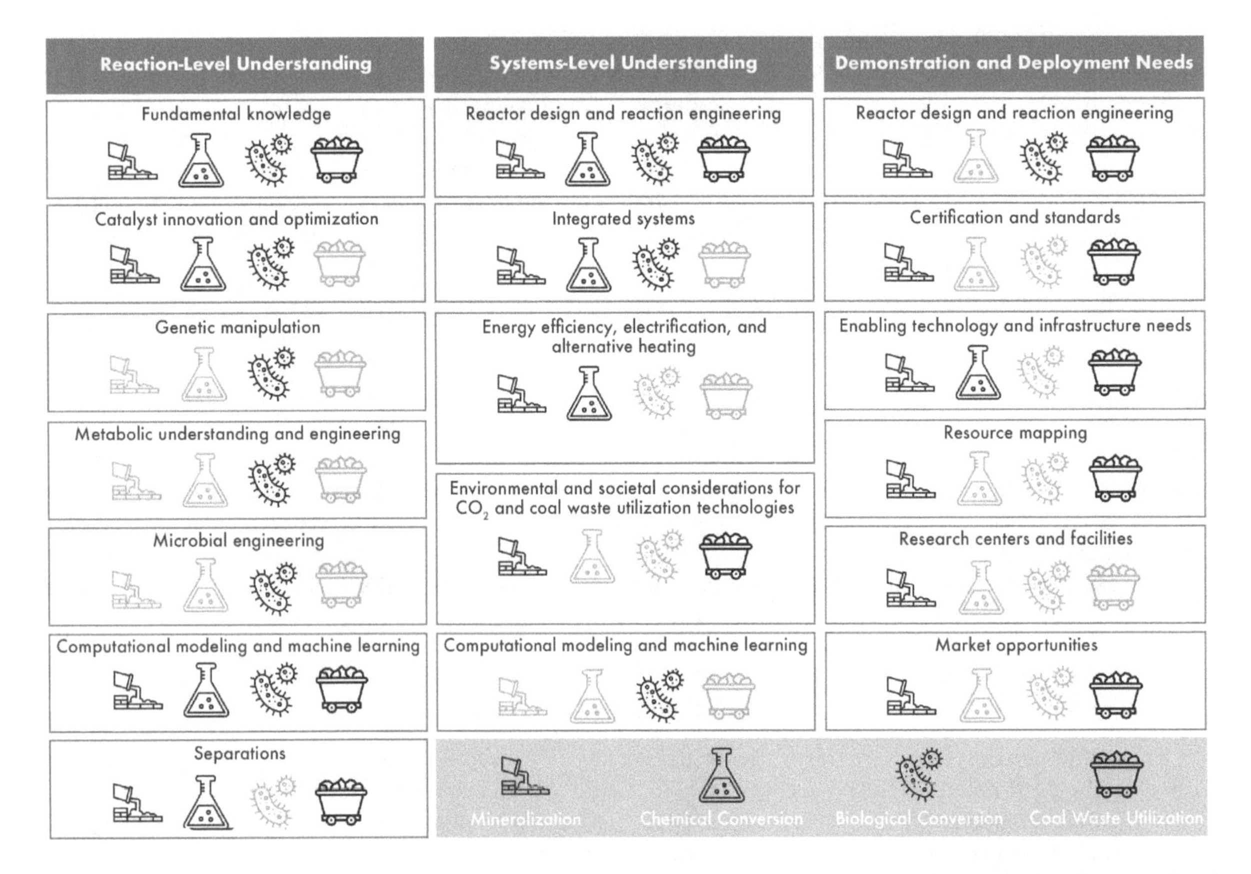

FIGURE S-8 Research themes for CO_2 and coal waste utilization RD&D needs, categorized by reaction- and systems-level understanding, and demonstration and deployment needs.
NOTE: Icons in black (see legend lower right) indicate which process(es)—mineralization, chemical conversion, biological conversion, and/or coal waste utilization—have RD&D needs in each theme.
SOURCE: Icons from the Noun Project, https://thenounproject.com. CC BY 3.0.

REFERENCES

NASEM (National Academies of Sciences, Engineering, and Medicine). 2019. *Gaseous Carbon Waste Streams Utilization: Status and Research Needs*. Washington, DC: The National Academies Press. https://doi.org/10.17226/25232.

NASEM. 2023. *Carbon Dioxide Utilization Markets and Infrastructure: Status and Opportunities: A First Report*. Washington, DC: The National Academies Press. https://doi.org/10.17226/26703.

1

Introduction

1.1 STUDY CONTEXT

To meet climate goals and limit the harmful effects of global warming, countries around the world are aiming to reach net-zero greenhouse gas (GHG) emissions across their economies by midcentury. Along this path toward net-zero emissions by 2050, the United States has set an intermediate goal of a 50–52 percent reduction in GHG emissions below 2005 levels by 2030 (DOS and EOP 2021), which aligns with the Paris Agreement target of limiting global warming to 1.5°C. Achieving net zero requires eliminating most emissions of carbon dioxide (CO_2) and other GHGs,[1] primarily via decarbonizing electricity generation, improving energy efficiency, and electrifying end uses (e.g., vehicles, buildings, industrial processes) where possible (DOS and EOP 2021; NASEM 2021). These actions, enabled by advances in low-carbon energy technologies and electrification, will significantly reduce the use of fossil fuels and resulting CO_2 emissions to the atmosphere. This report responds to a request from Congress to examine the role of carbon utilization in a net-zero emissions future.

While net GHG emissions to the atmosphere must end to achieve climate targets, carbon flows—particularly those related to embedded carbon in products—cannot be eliminated completely. As discussed in Chapter 2, global yearly materials flows are about 100 gigatonnes (Gt), including about 40 Gt of carbon-based materials, with about 15 Gt of that being fossil-derived carbon materials. Carbon-based products, including human-made chemicals, fuels, and materials, are central to global and national economies today, and many will remain important in a net-zero future. Historically, carbon-based products have been made from petroleum, natural gas, coal. The modern chemical industry was built to transform carbon-based molecules distilled from petroleum into a variety of products using inexpensive fossil fuel–derived heat. Most of these products are short-lived, and at end of life, become CO_2 via combustion or decay processes. When the carbon was fossil in origin, as is true for the vast majority of fuels, chemicals, and polymers produced today, then material end of life results in fossil CO_2 emissions to the atmosphere. To achieve net zero, these linear carbon flows from fossil feedstock to CO_2 in the atmosphere will need to shift to circular flows such that no new carbon enters the system, and instead any carbon emitted is

[1] The warming effects of CO_2 and other GHGs differ depending on their atmospheric lifetime and ability to absorb energy. These effects can be quantified and compared using Global Warming Potential (GWP), a measure of how much energy 1 ton of a GHG absorbs over a given amount of time (often 100 years) compared to the energy absorbed by 1 ton of CO_2 over the same amount of time (EPA 2024a).

captured and reused or stored. In a circular system, petroleum feedstocks will be largely unavailable[2] owing to their contribution to GHG emissions, and instead feedstocks will include biological material, recycled wastes, and CO_2. Using biological, recycled, or CO_2 feedstocks enable circularity by allowing carbon wastes, such as CO_2 from product degradation, to be incorporated into new products. The choice of sustainable carbon sources depends on many factors, including product composition and lifetime, feedstock cost, access to infrastructure, and regulatory and policy environment. To accommodate new feedstocks, the landscape of chemical and material products and processes is likely to change. Some carbon-based chemicals and materials will decline in use as zero-carbon alternatives become prominent (e.g., fuels replaced by electrification). Other carbon-based chemicals and materials are likely to increase in use because their relative value will increase (e.g., methanol or carbon monoxide as a more important intermediate for chemical synthesis, or carbon fibers as a replacement for higher-emitting materials in construction and manufacturing).

In addition to eliminating most sources of GHG emissions, long-term removal of CO_2 from the atmosphere will likely be needed to reach safe levels of GHGs for a stable climate. This removal could occur by geologic sequestration of CO_2 captured from the air or bodies of water or by incorporating captured CO_2 into long-lived products, especially those deployed at large (gigatonne) scales worldwide, such as concrete and aggregates. Such long-lived products additionally could contribute to emissions reductions by displacing heavily emitting processes like those used for producing conventional building materials.

The net-zero transition will require substantial amounts of critical minerals and materials to deploy clean energy technologies at scale. For example, lithium, cobalt, nickel, and graphite are used in batteries for electric vehicles and energy storage; silicon, copper, and silver are used in conventional photovoltaics; and copper, zinc, and rare earth elements (primarily neodymium) are used in wind turbines (IEA 2022). Currently, the United States imports the majority of minerals deemed "critical" by the U.S. Geological Survey (USGS): in 2022, imports comprised over 50 percent of demand for 43 critical minerals, with 12 of those being 100 percent imported (USGS 2023). Opportunities exist to extract some of these minerals from coal wastes, which could help to establish domestic supply chains while at the same time cleaning up legacy waste sites. The carbon constituents of coal waste can also be considered as a net-zero emissions feedstock for long-lived products, as explored in Chapter 9.

Expanding upon the committee's first report (NASEM 2023, summarized in Section 1.5), this report examines in greater depth the role of CO_2 utilization in a net-zero future, where CO_2 flows to the atmosphere are likely to be greatly reduced, and carbon wastes including CO_2 and coal waste streams will serve as feedstocks for carbon-based chemicals and materials, as well as for carbon storage in long-lived products. The report considers how chemicals and materials manufacturing could be adapted to take advantage of carbon wastes, particularly CO_2 and coal wastes, using low-carbon energy, and identifies circumstances in which CO_2 and coal wastes are advantaged feedstocks over biomass and other carbon wastes such as plastics. Specifically, it analyzes market opportunities, infrastructure requirements, and research, development, and demonstration (RD&D) needs for converting CO_2 into useful products, providing an update to the research agenda laid out in the 2019 National Academies' report *Gaseous Carbon Waste Streams Utilization: Status and Research Needs* (NASEM 2019). It also addresses the feasibility of deriving carbon materials and critical minerals from coal waste streams.

1.2 WHAT IS CO_2 UTILIZATION, AND HOW CAN IT CONTRIBUTE TO A NET-ZERO EMISSIONS FUTURE?

This study defines CO_2 utilization as the chemical transformation of CO_2 into a marketable product, which could include organic carbon-based fuels, chemicals, and materials (including polymers), inorganic carbonates, or elemental carbon materials. CO_2 conversion can occur via chemical, biological, or mineralization routes, with each having different requirements for energy, additional feedstocks, and infrastructure. CO_2 utilization processes

[2] In some limited applications, it may be technically or economically difficult to replace fossil carbon feedstocks with sustainable carbon feedstocks or to develop alternative non-carbon-based solutions in the near to medium term. As a result, during the transition, for some small-volume products, net zero may still be achieved by continued manufacture from fossil feedstocks accompanied by durable offsetting capture of CO_2 and geologic sequestration, or removals of CO_2 from the atmosphere. In the long term, these solutions are not viable because the production cost of oil and associated carbon costs will be prohibitively high.

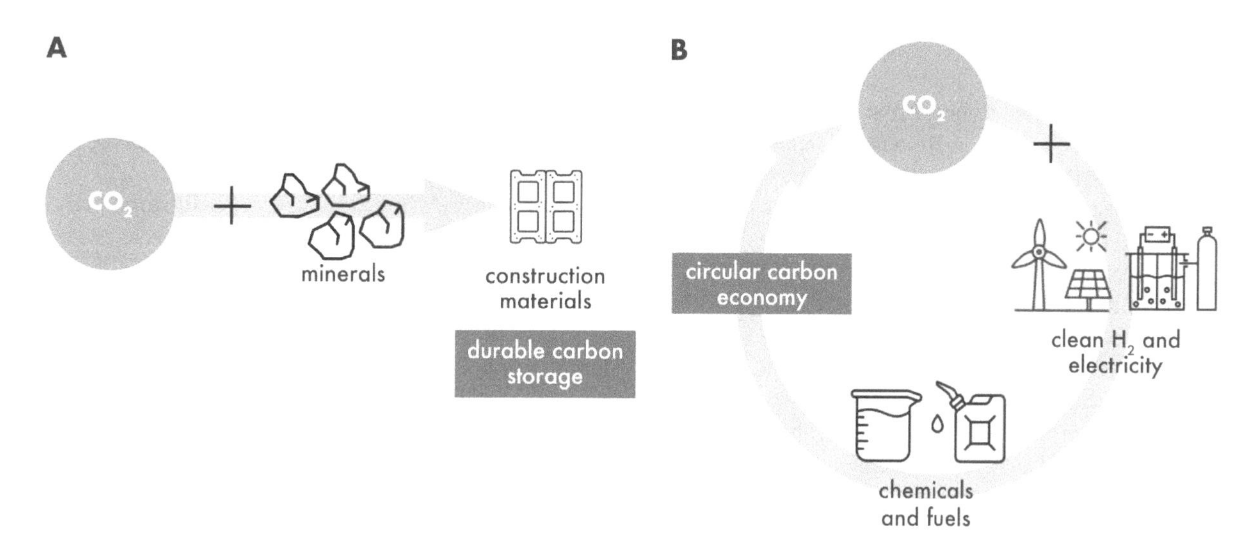

FIGURE 1-1 Schematic of the two primary roles of CO_2 utilization in a net-zero future, as an enabler of durable carbon storage in long-lived products (A) or a circular carbon economy via production of short-lived products (B).
NOTE: The committee defines clean hydrogen as having a GHG footprint of less than 2 kg CO_2 equivalent per kg H_2.
SOURCE: Icons from the Noun Project, https://thenounproject.com. CC BY 3.0.

span technology readiness levels, including a few large-scale, fully commercial activities (e.g., production of urea, salicylic acid, and organic carbonates); some pilot, demonstration, or limited-scale commercial facilities (e.g., production of methanol, carbon monoxide, and mineralized CO_2 products); and much research at the laboratory scale across all production routes and product classes. Priority products for CO_2 utilization are detailed in Chapter 2, and Chapters 5–8 cover RD&D needs for the various CO_2 utilization routes.

CO_2 utilization has two primary roles in a net-zero future: enabling circularity of carbon flows and carbon storage (Figure 1-1). Atmospheric, aquatic, or biogenic CO_2 is a feedstock for enabling a circular economy or net-zero-emissions approach to making carbon-based products,[3] alongside approaches that use biomass and other waste carbon feedstocks. CO_2 has several advantages and disadvantages relative to other sustainable carbon feedstocks for circularity of chemicals and materials. As a ubiquitous waste product of chemical combustion and decay, CO_2 is widely available. After gas stream purification, it is a uniform, nontoxic substance that can be cleanly inserted into chemical industry processes, as compared to mixed feedstocks of biomass and recycled waste products. CO_2 has lower land and, in many cases, water requirements than biomass cultivation, and it is easier to transport, although the most efficient transport is in pipelines, which can be challenging to plan and site. The primary disadvantage of CO_2 is its low energy, which makes chemical transformations very energy intensive; however, this is a necessary feature of a feedstock for circular carbon fuels. CO_2's single-carbon, oxidized chemical structure requires restructuring the chemical industry around processes that use more energy and hydrogen, and can build carbon-carbon bonds. (See Section 2.2.1 on factors that impact the ease of making products from CO_2.)

CO_2 utilization can enable the storage of CO_2-derived carbon in solid form in durable products, sequestering it from the atmosphere for climate-relevant timescales. In this role, CO_2 utilization is advantageous over alternatives like land- and ocean-based carbon dioxide removal and carbon capture and storage because it creates products with market value. However, the likely scale of CO_2-derived durable products is a few-gigatonnes annual global demand (Chapter 2 and NASEM 2023) compared to the projected tens of gigatonnes annual global removal required in the future (Pett-Ridge et al. 2023). Thus, geologic storage of CO_2 is preferred over durable CO_2-derived products for

[3] Zero-carbon energy carriers such as hydrogen and electricity are preferred over carbon-based fuels where feasible, owing to the energy efficiency of electricity use in motors and hydrogen use in fuel cells over combustion of fuels, the lower land and water requirements for electricity and hydrogen relative to biomass-based systems, and the avoided conversion energy and resource requirements for formation of carbon-based products.

meeting the full needs for carbon removal. CO_2 utilization is not likely to be a dominant source of carbon removal, nor of emissions mitigation for achieving net zero relative to other options for reducing emissions; however, it likely will contribute to product circularity and durable storage, where it has advantages. The extent of emissions reductions from CO_2 utilization could be impacted by factors such as societal acceptance of geologic carbon storage, ability to make long-lived products, and availability of renewable energy and hydrogen resources.

The use of CO_2 as a feedstock does not inherently reduce emissions relative to the use of fossil carbon feedstocks. To determine net-zero or net-negative product status, all emissions from the full product life cycle need to be considered, including upstream and downstream emissions associated with process, feedstock origin, energy use, product fate, co-product fate, and associated waste (see Chapter 3). For CO_2 utilization, it is particularly important to consider and match CO_2 source feedstocks and product sinks that can provide net-zero or net-negative emissions pathways. Specifically, long-lived products (lifetime >100 years[4]), such as concrete and aggregates, durably store carbon and can be produced sustainably with fossil or nonfossil CO_2. On the other hand, short-lived products (lifetime < 100 years), such as fuels, chemicals, and many plastics, store CO_2 from the atmosphere only temporarily and must be produced using atmospheric, aquatic, or biogenic sources of CO_2 to participate in a circular carbon economy.

Transitioning from today's heavily fossil fuel–dependent economy to a future economy with sustainable carbon feedstocks and net-zero or net-negative GHG emissions requires careful consideration of policies and technologies that promote emissions mitigation while ensuring that other societal needs and objectives are met. Current uses of fossil feedstocks do not include an internalized cost for the waste GHG products they emit, and as such, materials produced from fossil feedstocks are less expensive than they would be if their eventual fossil emissions to the atmosphere were priced. A future net-zero economy will require a new policy landscape to encourage net-zero or net-negative emissions technologies, and limit net-positive emissions technologies. Constraints on the use of fossil carbon will be stringent and may include preventing or pricing the incorporation of fossil-derived CO_2 into products that reemit CO_2 on a short timeframe (e.g., chemicals or fuels). These policies will also include economy-wide caps on emissions of fossil CO_2 and/or other GHGs or a price on carbon that is sufficiently high to greatly reduce fossil emissions. Emissions constraints would build on policies already being implemented to support technology development for a net-zero transition, such as tax credits for renewable energy production and sequestration or utilization of CO_2. While such forward-looking analysis is valuable to define and plan for an end goal, it overlooks complexities that will arise during the transition period as society decreases fossil fuel use. For example, in the near term when fossil emissions are being wound down, cost-effective emissions reductions may be achieved using fossil CO_2 as a feedstock for chemicals and fuels, even though this approach will not provide the net-zero emissions required in the long term. Collaborative transition planning by governments and the private sector will be needed to optimize investments and minimize stranded assets. This planning is particularly relevant for CO_2 utilization, where siting decisions have to consider fundamental shifts in product needs as well as matching CO_2 source with product lifetime. CO_2 utilization infrastructure planning is discussed further in Chapter 10, and more detail on policy options for CO_2 utilization can be found in Chapter 4.

1.3 WHAT IS COAL WASTE UTILIZATION AND HOW CAN IT CONTRIBUTE TO NATIONAL NEEDS?

In addition to CO_2 utilization, this report examines the feasibility of and opportunities for utilizing coal waste streams to produce carbon-based products and extract critical minerals and materials. Various waste streams arise from mining, processing, and using coal, including coal combustion residuals (fly ash, bottom ash, boiler slag, and flue gas desulfurization products), impoundment waste (coarse and fine refuse), and acid mine drainage. Coal combustion residuals are already used today as filler and raw materials in concrete, wallboard, asphalt, and other structural applications, as well as for soil modification (ACAA 2023; EPA 2023). Given the abundance, low cost, and high carbon density of coal waste streams, research efforts are targeting additional potential utilization opportunities in a wide spectrum of carbon-based products, from high-volume building materials and polymers to high-value graphite and carbon fiber (Stoffa 2023). Fly ash, wastes from coal mining and processing, and acid

[4] This committee's first report (NASEM 2023) chose a product lifetime of 100 years, in line with the United Nations Framework Convention on Climate Change, to differentiate between short- and long-lived products.

mine drainage are also potential sources of critical minerals and rare earth elements, and pilot-scale efforts are under way to develop separation and extraction processes (Kolker 2023). Using coal waste streams could enable environmental remediation, expand domestic supply chains for critical minerals and materials, and produce carbon-based products with improved performance and/or economics (Stoffa 2023). However, because these wastes contain nearly every element in the periodic table, separations can be challenging and pose safety and toxicity concerns (Kolker 2023; Stoffa 2023). Chapter 9 addresses opportunities and challenges with utilizing coal wastes.

1.4 CURRENT POLICY ENVIRONMENT FOR CO_2 UTILIZATION

Most uses of CO_2 within the scope of this study currently rely on subsidies or other incentives to be competitive with incumbent products, owing to various transient and persistent factors, including the unpriced costs of fossil hydrocarbon pollution from incumbent products, the small production scale of most CO_2 utilization products, the still-developing CO_2 utilization technologies, and the fundamental challenges of CO_2 conversion into organic products. A primary policy driver in the United States is the 45Q tax credit (IRA § 13104), which provides up to \$60/tCO_2 for utilization (or up to \$130/tCO_2 if paired with direct air capture [DAC]) (IRA 2022). The 45V tax credit for clean hydrogen production (IRA § 13204) could also benefit CO_2 utilization projects that require hydrogen as a feedstock, although the same facility cannot claim both the 45Q and 45V credits (IRA 2022). Additionally, the Carbon Dioxide Transportation Infrastructure Finance and Innovation (CIFIA) program, administered through the Department of Energy's (DOE's) Loan Programs Office with \$2.1 billion in appropriations between 2022 and 2026, finances common carrier transportation infrastructure to move CO_2 from the point of capture to the point of use or storage (DOE-LPO n.d.). The Utilization Procurement Grants (UPGrants) program, run by DOE's Office of Fossil Energy and Carbon Management and the National Energy Technology Laboratory, will facilitate procurement and use of CO_2-derived products by state and local governments and public utilities (NETL n.d.). Beyond government-funded initiatives, the voluntary carbon market and willingness of companies to pay a "green premium" for more sustainable products also support current CO_2 utilization efforts.

This study focuses on needs for a net-zero future, in which there will be significant constraints on the emission of fossil-derived CO_2 to the atmosphere, such as an explicit price on CO_2 or a limit on emissions. The committee considers opportunities and enabling environments for CO_2 utilization within this context, where there will be increased incentives for producing short-lived products in a circular carbon system, as well as for long-duration carbon storage, including in products (see Chapter 4). Applying constraints on CO_2 emissions to CO_2 utilization products or processes will require accounting for life cycle emissions of a CO_2 utilization process and the resulting product for compliance purposes. Products originally derived from a fossil feedstock could have end-of-life emissions associated with their use or degradation, and these emissions will need to be considered and managed. Emissions associated with enabling inputs such as electricity and hydrogen, transportation of CO_2 and CO_2-derived products, and any upstream or downstream processing may also impact the feasibility of certain CO_2 utilization pathways when total system emissions are constrained. Chapter 3 discusses life cycle assessment considerations for CO_2 utilization.

1.5 BRIEF REVIEW OF FIRST REPORT'S CONCLUSIONS

The committee's first report (NASEM 2023) examined the state of CO_2 capture, utilization, transportation, and storage infrastructure and analyzed opportunities for investment in CO_2 utilization infrastructure to serve a net-zero future. It found that most of the existing infrastructure, which was developed for enhanced oil recovery, does not align with opportunities for sustainable (i.e., net-zero) CO_2 utilization. Thus, to evaluate future infrastructure opportunities, the committee first considered factors that would influence the extent of CO_2 utilization in a net-zero economy. It found that the volume of CO_2 utilized will be driven by the market value of carbon-based products and competitiveness of CO_2 as a feedstock, as well as demand for services provided by carbon-based products; their relative cost compared to fossil-based products and other alternatives; availability of required inputs like clean hydrogen and clean electricity; and policy incentives and regulatory frameworks (Finding 3.10, NASEM 2023). As discussed above, another important consideration is the potential climate impact of the CO_2-derived product based on the CO_2 source, product lifetime, and any emissions associated with production, transportation, and use.

Considering these factors, the committee identified two priority near-term opportunities for CO_2 utilization infrastructure investment: (1) combining high-purity, low-cost CO_2 off-gas from bioethanol plants with clean hydrogen to make sustainable chemicals or fuels for heavy-duty transportation (e.g., shipping and aviation) and (2) mineralization using fossil or nonfossil CO_2 sources to generate mineral carbonates for construction materials, including concrete (Finding 6.1, NASEM 2023). It recommended strategies for infrastructure planning, such as co-locating CO_2 utilization with clean electricity, clean hydrogen, and other carbon management infrastructure; building in flexibility to connect CO_2 transport infrastructure to future utilization opportunities; and developing industrial clusters to manage large volumes of CO_2 without extensive pipeline networks and maintain jobs in regions with a large industrial presence (Recommendations 4.5, 6.2, 6.3, and 6.4, NASEM 2023). The committee also emphasized the importance of community engagement and equitable infrastructure development, recommending that regulatory agencies account for distributional impacts of CO_2 utilization projects, engage impacted communities early and throughout the project, and allow for alterations to project design and implementation (Recommendation 5.6, NASEM 2023). This report expands upon the findings and recommendations from the first report.

1.6 REPORT TASKING, SCOPE, AND KEY CONCEPTS

A congressional mandate in the Energy Act of 2020 required DOE to enter into an agreement with the National Academies of Sciences, Engineering, and Medicine to perform a study "to assess any barriers and opportunities relating to commercializing carbon, coal-derived carbon, and carbon dioxide in the United States" (U.S. Congress 2020, § 969A). DOE commissioned that this study be undertaken in two parts. In a first report, the committee would describe the current state of infrastructure for CO_2 transportation, use, and storage in the United States and identify priority opportunities for developing infrastructure to enable future CO_2 utilization processes and markets in a safe, cost-effective, and environmentally benign manner. The committee released this first report in December 2022 (NASEM 2023). A second, more comprehensive report (the present report) would provide additional detail on potential markets for products derived from CO_2; the economic, environmental, and climate impacts of CO_2 utilization infrastructure; RD&D needs to enable commercialization of CO_2 utilization technologies and processes; and opportunities for and feasibility of coal waste–derived carbon products and critical minerals. The full statement of task for the study is provided in the next section, followed by a description of the study scope and definitions of relevant concepts used in the report (Box 1-1).

1.6.1 Statement of Task

The National Academies of Sciences, Engineering, and Medicine will convene an ad hoc committee to assess infrastructure and research and development needs for carbon utilization, focused on a future where carbon wastes are fundamental participants in a circular carbon economy. In particular, the study will focus on regional and national market opportunities, infrastructure needs, and the research and development needs for technologies that can transform carbon dioxide and coal waste streams into products that will contribute to a future with zero net carbon emissions to the atmosphere. The committee will analyze challenges in expanding infrastructure, mitigating environmental impacts, accessing capital, overcoming technical hurdles, and addressing geographic, community, and equity issues for carbon utilization.

The committee will provide a first report that:

1. assesses the state of infrastructure for carbon dioxide transportation, use, and storage as of the date of the study, including pipelines, freight transportation, electric transmission, and commercial manufacturing facilities.
2. identifies priority opportunities for development, improvement, and expansion of infrastructure to enable future carbon utilization opportunities and market penetration. Such priority opportunities will consider how needs for carbon utilization infrastructure will interact with and capitalize on infrastructure developed for carbon capture and sequestration.

The committee will develop a second report that will evaluate the following:

1. Markets
   a. Identify potential markets, industries, or sectors that may benefit from greater access to commercial carbon dioxide to develop products that may contribute to a net-zero carbon future; identify the markets that are addressable with existing utilization technology and that still require research, development and demonstration;
   b. Determine the feasibility of, and opportunities for, the commercialization of coal waste–derived carbon products in commercial, industrial, defense, and agricultural settings; for medical, construction and energy applications; and for the production of critical minerals;
   c. Identify appropriate federal agencies with capabilities to support small business entities; and determine what assistance those federal agencies could provide to small business entities to further the development and commercial deployment of carbon dioxide–based products;
2. Infrastructure
   a. Building off the study's first report, assess infrastructure updates needed to enable safe and reliable carbon dioxide transportation, use, and storage for carbon utilization purposes. Assessment of infrastructure will consider how carbon utilization fits into larger carbon capture and sequestration infrastructure needs and opportunities;
   b. Describe the economic, climate, and environmental impacts of any well-integrated national carbon dioxide pipeline system as applied for carbon utilization purposes, including suggestions for policies that could: (i) improve the economic impact of the system; and (ii) mitigate climate and environmental impacts of the system;
3. Research, Development, and Demonstration
   a. Identify and assess the progress of emerging technologies and approaches for carbon utilization that may play an important role in a circular carbon economy, as relevant to markets determined in section 1a;
   b. Assess research efforts under way to address barriers to commercialization of carbon utilization technology, including basic, applied, engineering, and computational research efforts; and identify gaps in the research efforts;
   c. Update the 2019 National Academies' comprehensive research agenda on needs and opportunities for carbon utilization technology RD&D, focusing on needs and opportunities important to commercializing products that may contribute to a net-zero carbon future.

The first and second reports will provide guidance to infrastructure funders, planners, and developers and to research sponsors, as well as research communities in academia and industry, regarding key challenges needed to advance the infrastructure, market, science, and engineering required to enable carbon utilization relevant for a circular carbon economy.

1.6.2 Scope of Report

The committee determined the limits of its scope based on the congressional mandate for the study and the study's statement of task. The committee's definitions of relevant concepts and explanation of the study scope are outlined below.

1. **How is carbon utilization defined for the purposes of this report? What classes of carbon utilization are *not* in scope?**
   a. Carbon utilization is defined for the purposes of this report as the chemical transformation of concentrated CO_2 collected from the atmosphere, a body of water, or a waste gas stream into a carbon-containing product with market value.

b. In scope:
    i. Chemical, microbial, and mineralization transformations of CO_2.
c. Out of scope:
    i. Processes like enhanced oil recovery, fire suppression, and beverage carbonation that leave CO_2 untransformed.
    ii. CO_2 transformed into products via the growth of terrestrial plants and crops.

2. **What sources of carbon dioxide are considered in this report?**
    a. CO_2 captured and concentrated from the atmosphere through direct air capture.
    b. CO_2 captured from point sources before emission to the atmosphere, such as from power plants or industrial facilities.
    c. CO_2 dissolved from the atmosphere in natural or other bodies of water, where the dissolved CO_2 in those waters is used as feedstock for carbon utilization processes.

3. **What pathways to activate carbon dioxide are discussed in this report?**
    a. Chemical pathways to inorganic (mineral) products
    b. Chemical pathways to organic products
        i. Thermochemical
        ii. Electrochemical
        iii. Photochemical
        iv. Plasmachemical
        v. Hybrid pathways
    c. Biological pathways to organic products
        i. Photosynthetic
            1. Algae
            2. Cyanobacteria
        ii. Nonphotosynthetic
            1. Chemolithotrophs
            2. Bio-electrochemical processes

4. **To what extent is coal waste discussed in this report?**
    a. Utilization of coal waste streams, which was not discussed in the study's first report, is discussed herein.
    b. Coal waste includes coal combustion residuals (fly ash, bottom ash, boiler slag, and flue gas desulfurization products) and impoundment waste (coarse and fine refuse). Acid mine drainage also is considered as a potential source of critical minerals.
    c. Utilization of raw coal to produce mineral- or carbon-based products and selective mining of rare earth element-enriched portions of coal beds are out of scope for the study.

5. **What product classes are discussed in this report?**
    a. Inorganic carbonates
    b. Elemental carbon materials
    c. Fuels and commodity chemicals
    d. Polymer precursors and polymers
    e. Critical minerals and rare earth elements (only from coal waste)

6. **What assessments of carbon utilization products and processes does this report examine?**
    a. Life cycle assessments
    b. Techno-economic assessments
    c. Societal/equity assessments

7. **What CO_2 utilization infrastructure is considered in this report?**
   a. CO_2 utilization infrastructure systems, including capture, purification, transportation, utilization, and geologic storage.
   b. A U.S. regional- or national-scale pipeline system for carbon management, as applied to utilization.
   c. Enabling infrastructure for CO_2 utilization, including for
      i. Hydrogen
      ii. Electricity
      iii. Water
      iv. Energy storage
      v. Land use
      vi. Product transportation

BOX 1-1
Key Terms and Concepts for CO_2 Utilization

The following list provides definitions of key terms and concepts for CO_2 utilization that are used throughout this report.

- **Adoption readiness levels (ARLs)**—A framework complementary to technology readiness levels (see below) that assesses the readiness of a technology for commercialization and market uptake by evaluating risks related to value proposition, resource maturity, and license to operate. The ARL scale ranges from 1 to 9, with 1–3 classified as low readiness, 4–6 as medium readiness, and 7–9 as high readiness (Tian et al. 2023).
- **Carbon capture**—The process of separating and concentrating CO_2 from industrial or waste gas streams, ambient air, or bodies of water.
  - **Direct air capture (DAC)**—A technological process by which CO_2 is separated and concentrated from ambient air. DAC removes CO_2 from the atmosphere.
  - **Direct ocean capture (DOC) or capture from bodies of water**—A technological process by which CO_2 is separated and concentrated from the ocean or other bodies of water. DOC or capture from bodies of water indirectly results in CO_2 removal from the atmosphere.
  - **Point source capture**—A technological process by which CO_2 is separated and concentrated from waste gas streams at electric power plants, industrial facilities, and other sources of combustion or process emissions. Point source capture prevents CO_2 from being emitted to the atmosphere.
- **Carbon dioxide removal**—Technologies and processes that remove CO_2 from the atmosphere and bodies of water via durable storage in products or geological, terrestrial, or ocean reservoirs. The term is used to describe both engineered technologies, such as DAC, and enhancement of natural processes, such as soil carbon storage and enhanced weathering. It does not cover natural uptake of CO_2 without human intervention (Wilcox et al. 2021).
- **Carbon dioxide utilization (or conversion)**—The chemical or biological transformation of concentrated CO_2 collected from the atmosphere, a body of water, or an industrial or waste gas stream into a carbon-containing product with market value.
- **Circular carbon economy**—A system in which carbon, energy, and material flows are reduced, removed, recycled, and reused to achieve net-zero emissions (Williams et al. 2020).
  - **Carbon flow**—The movement of carbon in various forms among land, air, water, plants, living creatures, material products, and waste.
- **Community engagement**—A planned process through which members of a community—either based in a geographic location or formed around people of similar interest—work collaboratively with decision makers to address issues affecting their well-being. “It involves sharing information, building relationships and partnerships, and involving stakeholders in planning and making decisions with the goal of improving the outcomes of policies and programs” (CCI 2018).

continued

BOX 1-1 Continued

- **Critical material**—"Any non-fuel mineral, element, substance, or material that the Secretary of Energy determines: (i) has a high risk of supply chain disruption; and (ii) serves an essential function in one or more energy technologies, including technologies that produce, transmit, store, and conserve energy"; or a critical mineral, as defined by the Secretary of the Interior (Energy Act of 2020, § 7002).
- **Decarbonization**—Reducing or removing emissions of CO_2 and other GHGs throughout the economy, often by transitioning to zero-carbon processes. Some processes, such as the production of sustainable aviation fuels, are often referred to as decarbonization despite not removing carbon flows; a more appropriate term is "defossilization" (see below).
- **Defossilization**—Reducing or eliminating the use of fossil carbon throughout the economy, such that carbon-based products are made only from nonfossil feedstocks, and flows of carbon dioxide to and from the atmosphere do not include new fossil carbon.
- **Environmental justice**—The just treatment and meaningful involvement of all people—regardless of income, race, color, national origin, Tribal affiliation, or disability—in agency decision making and other federal activities related to climate change, the cumulative impacts of environmental and other burdens, and the legacy of racism or other structural or systemic barriers that disproportionately and adversely affect human health and the environment (EPA 2024b).
- **Feedstocks**—Material inputs to industrial processes to generate a product.
  - **Carbon waste streams**—Carbon-based gases or materials destined for disposal, either as emissions to the atmosphere or in a landfill, which could instead be reused in products in support of a circular carbon economy. Examples in scope for this report are CO_2 waste streams and coal-derived carbon wastes. Examples out of scope for this report include methane and biogas waste streams, plastic or other carbon-based product wastes, and bio-based wastes such as municipal, sanitary, and agricultural wastes.
  - **Fossil carbon**—The carbon in crude oil, coal, and natural gas; a nonrenewable source of carbon that formed from dead plant and animal matter under high temperature and pressure over millions of years (Renewable Carbon Initiative n.d.). Fossil carbon also includes any CO_2 or other waste carbon resulting from the use or decay of fossil-derived products, as well as CO_2 currently present in underground reservoirs. It is unclear how mineral carbonates, such as those that are decomposed to make cement, should be classified between fossil and nonfossil carbon.
  - **Nonfossil carbon**—Carbon derived from biogenic, atmospheric, or aquatic sources. It is unclear how mineral carbonates, such as those that are decomposed to make cement, should be classified between fossil and nonfossil carbon.
  - **Coal wastes**—Carbon and noncarbon waste streams that are generated throughout the coal supply chain, including acid mine drainage, coal impoundment wastes, and coal combustion residuals.
  - **Sustainable carbon feedstock**—A feedstock derived from nonfossil carbon that can support production of chemicals and materials in a net-zero-carbon economy without emissions from product degradation and decay.
- **Geologic carbon sequestration**—"The process of storing carbon dioxide in underground geologic formations. The CO_2 is usually pressurized until it becomes a liquid, and then it is injected into porous rock formations in geologic basins" (USGS n.d.).
- **Integrated system**—A system combining two or more CO_2 capture and/or utilization routes to produce any of the product classes within the scope of this report (inorganic carbonates, elemental carbon materials, chemicals, fuels, polymers). CO_2 conversion approaches could include mineralization, thermochemical, electrochemical, photo(electro)chemical, plasmachemical, or biological routes.
  - **Hybrid process**—A type of integrated system, typically used in the context of biological CO_2 utilization, involving the coupling of electro-, thermo-, photo-, or plasma-chemical conversion with bioconversion.
  - **Tandem process**—A type of integrated system where two or more CO_2 conversion routes are combined to occur in sequence. Tandem processes can occur on varying scales—for example, different conversions in separate coupled reactors, different conversions in the same reactor, or different conversions occurring at multiple sites on a single material.

BOX 1-1 Continued

- **Life cycle assessment**—An analysis of the environmental impacts, including but not limited to CO_2 flows, of a product, process, or system throughout its entire life cycle, from raw material extraction (cradle) to end of life (grave).
  - **Upstream emissions**—"Indirect emissions related to a reporting company's suppliers, from the purchased materials that flow into the company to the products and services the company utilizes" (Persefoni 2023).
  - **Downstream emissions**—"The emissions related to customers, from selling goods and services to their distribution, use, and end-of-life stages" (Persefoni 2023).
- **Linear carbon economy**—A system in which carbon, in the form of fossil fuels, is extracted and converted to valuable products and energy, which upon use or degradation emit CO_2 to the atmosphere without being reutilized (Williams et al. 2020).
- **Net-negative emissions**—The condition in which flows of CO_2 equivalents to the atmosphere are less than those removed from the atmosphere by technological or natural processes.
  - **Negative emissions**—A technology results in negative emissions if it removes physical emissions from the atmosphere, if the removed gases are stored out of the atmosphere in a manner intended to be permanent, if upstream and downstream GHG emissions associated with the removal and storage process, such as biomass origin, energy use, gas fate, and co-product fate, are comprehensively estimated and included in the emission balance, and if the total quantity of atmospheric GHGs removed and permanently stored is greater than the total quantity of GHGs emitted to the atmosphere. To fully understand climate impacts, evaluations of negative emissions technologies also need to estimate biogeophysical and potential nonlinear effects on Earth systems (Zickfeld et al. 2023).
  - **Net-negative emissions compatible**—A process that has the potential to result in net-negative emissions to the atmosphere over the course of its life cycle.
- **Net-positive emissions**—The condition in which flows of CO_2 equivalents to the atmosphere are greater than those removed from the atmosphere by technological or natural processes.
- **Net-zero carbon or net-zero GHG emissions**—The condition in which flows of CO_2 equivalents to and from the atmosphere are equal—that is, emissions of CO_2 and other GHGs are offset by removal of an equivalent amount through technological or natural processes.
  - **Net-zero emissions compatible**—A process that has the potential to result in net-zero emissions to the atmosphere over the course of its life cycle.
- **Product lifetime**—The amount of time between production and end of use or degradation of a product.
  - **Long-lived product or durable storage product (Track 1)**—In the current report, a product with a lifetime of more than 100 years, which stores CO_2 long enough to have a climate-relevant storage impact.
  - **Short-lived product or circular carbon product (Track 2)**—In the current report, a product with a lifetime of less than 100 years, which decomposes back to CO_2 in a short timespan, and which requires participation in a circular carbon economy for sustainability.
- **Public engagement**—The multifaceted ways in which people are involved in decisions about policies, programs, and services that impact issues of common importance and seek to solve shared problems.
- **Techno-economic assessment**—An integrated assessment of technical performance and economic feasibility that combines process modeling and engineering design with an economic evaluation to assess the (future) viability of a process, system configuration, or product.
- **Technology readiness level (TRL)**—"A type of measurement system used to assess the maturity level of a particular technology" (Manning 2023). This report uses the Department of Energy (DOE) TRL scale definitions, which range from TRL 1 "basic principles observed and reported" to TRL 9 "actual system operated over the full range of expected conditions." (see Table 1 of DOE 2015). Validation of components in the laboratory occurs at TRL 4, engineering or pilot-scale demonstration at TRL 6, and full-scale system demonstration at TRL 7.
- **Zero-carbon**—A product, technology, or process that does not require carbon flows for its operation, does not lead to emission of CO_2 to the atmosphere, and may not require any carbon-based materials at all. Often used to refer to products, technologies, or processes that can replace carbon-based products in a decarbonized economy, such as solar electricity generation replacing fossil fuel electricity generation.

1.7 OVERVIEW OF REPORT CHAPTERS AND CONTENT

This introductory chapter provides context and motivation for the study, describes the connection between this report and the committee's first report, and explains the study tasking and scope. Chapter 2 identifies market opportunities and requirements for carbon utilization in a net-zero future, considering the projected demand for carbon-based products and cases where CO_2 or coal waste is an advantageous feedstock. This is followed in Chapter 3 by a discussion of life cycle, techno-economic, and societal/equity assessments of CO_2 utilization processes, technologies, and systems. Chapter 4 assesses policy and regulatory frameworks needed to support sustainable CO_2 utilization; opportunities for small businesses to compete for CO_2 as a commodity; and environmental justice considerations when selecting, siting, and developing CO_2 utilization projects. Together, Chapters 2–4 lay the groundwork for understanding how and where CO_2 utilization could contribute to a net-zero future.

The next four chapters focus on RD&D needs for CO_2 utilization technologies and processes to generate inorganic carbonates (Chapter 5); elemental carbon materials (Chapter 6); and chemicals, fuels, and polymers via chemical routes (Chapter 7) or biological routes (Chapter 8). Chapter 9 examines the feasibility of and opportunities for deriving carbon-based materials and critical minerals from coal wastes. Chapter 10 discusses infrastructure needed to support CO_2 utilization, building on the first report's analysis with more detail on integrated infrastructure planning and the economic, environmental, health and safety, and environmental justice impacts of CO_2 utilization infrastructure development. Chapter 11 discusses the crosscutting research needs of CO_2 capture and purification and presents a research agenda for CO_2 and coal waste utilization, based on the committee's analyses in Chapters 5–9.

1.8 REFERENCES

ACAA (American Coal Ash Association). 2023. "About Coal Ash: What Are Coal Combustion Products?" ACAA. https://acaa-usa.org/about-coal-ash/what-are-ccps.

CCI (California Climate Investments). 2018. "Best Practices for Community Engagement and Building Successful Projects." https://ww2.arb.ca.gov/sites/default/files/auction-proceeds/cci-community-leadership-bestpractices.pdf.

DOE (Department of Energy). 2015. "Technology Readiness Assessment Guide." DOE G 413.3-4A. Washington, DC: Department of Energy.

DOE-LPO (Loan Programs Office). n.d. "Carbon Dioxide Transportation Infrastructure." https://www.energy.gov/lpo/carbon-dioxide-transportation-infrastructure.

DOS and EOP (Department of State and Executive Office of the President). 2021. "The Long-Term Strategy of the United States: Pathways to Net-Zero Greenhouse Gas Emissions by 2050." Washington, DC: Department of State and Executive Office of the President. https://www.whitehouse.gov/wp-content/uploads/2021/10/US-Long-Term-Strategy.pdf.

EPA (U.S. Environmental Protection Agency). 2023. "Frequent Questions about the Beneficial Use of Coal Ash." https://www.epa.gov/coalash/frequent-questions-about-beneficial-use-coal-ash.

EPA. 2024a. "Learn About Environmental Justice." February 6. https://www.epa.gov/environmentaljustice/learn-about-environmental-justice.

EPA. 2024b. "Understanding Global Warming Potentials." March 27. https://www.epa.gov/ghgemissions/understanding-global-warming-potentials.

IEA (International Energy Agency). 2022. "The Role of Critical Minerals in Clean Energy Transitions." World Energy Outlook Special Report. Paris, France: International Energy Agency. https://iea.blob.core.windows.net/assets/ffd2a83b-8c30-4e9d-980a-52b6d9a86fdc/TheRoleofCriticalMineralsinCleanEnergyTransitions.pdf.

IRA (Inflation Reduction Act). 2022. H.R. 5376—Inflation Reduction Act of 2022. Public Law 117-169. 117th Congress (2021–2022). https://www.congress.gov/bill/117th-congress/house-bill/5376.

Kolker, A. 2023. "Rare Earth Elements and Critical Minerals in Coal and Coal Byproducts." Presented at the Carbon Utilization Infrastructure, Markets, Research and Development Meeting #4. Virtually. June 28, 2023. https://www.nationalacademies.org/event/40093_06-2023_carbon-utilization-infrastructure-markets-research-and-development-meeting-4.

Manning, C.G. 2023. "Technology Readiness Levels." NASA. https://www.nasa.gov/directorates/somd/space-communications-navigation-program/technology-readiness-levels.

NASEM (National Academies of Sciences, Engineering, and Medicine). 2019. *Gaseous Carbon Waste Streams Utilization: Status and Research Needs*. Washington, DC: The National Academies Press. https://doi.org/10.17226/25232.

NASEM. 2021. *Accelerating Decarbonization of the U.S. Energy System*. Washington, DC: The National Academies Press. https://doi.org/10.17226/25932.

NASEM. 2023. *Carbon Dioxide Utilization Markets and Infrastructure: Status and Opportunities: A First Report*. Washington, DC: The National Academies Press. https://doi.org/10.17226/26703.

NETL (National Energy Technology Laboratory). n.d. "Utilization Procurement Grants (UPGrants)." https://www.netl.doe.gov/upgrants.

Persefoni. 2023. "Upstream vs Downstream: Breaking Down Scope 3." *Persefoni*, July 7. https://www.persefoni.com/learn/upstream-vs-downstream.

Pett-Ridge, J., H.Z. Ammad, A. Aui, M. Ashton, S.E. Baker, B. Basso, M. Bradford, et al. 2023. "Roads to Removal: Options for Carbon Dioxide Removal in the United States." LLNL-TR-852901. Lawrence Livermore National Laboratory. https://roads2removal.org.

Renewable Carbon Initiative. n.d. "Glossary." Renewable Carbon Initiative. https://renewable-carbon-initiative.com/renewable-carbon/glossary.

Stoffa, J. 2023. "Carbon Ore Processing Program." Presented at the Carbon Utilization Infrastructure, Markets, Research and Development Meeting #4. Virtually. June 28. https://www.nationalacademies.org/event/40093_06-2023_carbon-utilization-infrastructure-markets-research-and-development-meeting-4.

Tian, L., J. Mees, V. Chan, and W. Dean. 2023. "Commercial Adoption Readiness Assessment Tool (CARAT)." Department of Energy. https://www.energy.gov/sites/default/files/2023-03/Commercial%20Adoption%20Readiness%20Assessment%20Tool%20%28CARAT%29_030323.pdf.

U.S. Congress. 2020. "Division Z—Energy Act of 2020." H.R.133—Consolidated Appropriations Act, 2021. Public Law 116-260. 116th Congress (2019–2020). https://www.congress.gov/116/plaws/publ260/PLAW-116publ260.pdf.

USGS (U.S. Geological Survey). 2023. "Mineral Commodity Summaries 2023." Reston, VA: U.S. Geological Survey. https://pubs.usgs.gov/periodicals/mcs2023/mcs2023.pdf.

USGS. n.d. "What's the Difference Between Geologic and Biologic Carbon Sequestration?" https://www.usgs.gov/faqs/whats-difference-between-geologic-and-biologic-carbon-sequestration.

Wilcox, J., B. Kolosz, and J. Freeman. 2021. "Carbon Dioxide Removal Primer." https://cdrprimer.org.

Williams, E., A. Sieminski, and A. al Tuwaijri. 2020. "CCE Guide Overview." Riyadh, Saudi Arabia: King Abdullah Petroleum Studies and Research Center. https://www.cceguide.org/wp-content/uploads/2020/08/00-CCE-Guide-Overview.pdf.

Zickfeld, K., A.J. MacIsaac, J.G. Canadell, S. Fuss, R.B. Jackson, C.D. Jones, A. Lohila, et al. 2023. "Net-Zero Approaches Must Consider Earth System Impacts to Achieve Climate Goals." *Nature Climate Change* 13(12):1298–1305. https://doi.org/10.1038/s41558-023-01862-7.

2

Priority Opportunities for CO_2- or Coal Waste–Derived Products in a Net-Zero Emissions Future

2.1 CARBON FLOWS IN A NET-ZERO EMISSIONS FUTURE AND PATH TO CO_2-DERIVED PRODUCTS

Carbon-based materials derived from biogenic and fossil carbon currently play essential roles in our lives and the economy. Decomposition from decay or combustion of these materials leads to excess flows of CO_2 into the atmosphere, resulting in accumulating concentrations of CO_2, especially when the material was originally fossil derived. In a net-zero future, carbon flows will be in a global equilibrium so that CO_2 no longer accumulates in the atmosphere. Reducing emissions and CO_2 removal are needed to bring atmospheric CO_2 concentrations to levels that support stabilizing the global climate at acceptable conditions for human life, and then maintain those lower, stable concentrations of CO_2 in the atmosphere. Carbon flows in the economy, and the associated CO_2 emissions, will be greatly reduced by zero-carbon replacements for many products, especially fuels. However, carbon-based materials cannot be entirely eliminated because they (1) will continue to be part of natural and engineered biological and geological carbon cycles; (2) will continue to be necessary components of many products important in daily life; and (3) can be used to store carbon away from the atmosphere in durable products, or engineered and natural sequestration. CO_2 utilization can play a role in creating sustainable, circular, or net-zero-emissions; carbon-based systems for our future material needs; alongside other sustainable carbon feedstocks like biomass or recycled material. This chapter focuses on the market opportunities for CO_2 utilization in a net-zero future.[1] This report examines what carbon-based materials will be needed in a net-zero future, possible sources of sustainable carbon feedstocks for those materials, and what role CO_2 could play in supplying sustainable carbon.

Carbon-based biomass (24.6 gigatonnes [Gt]) and fossil hydrocarbons (15.1 Gt) represent nearly 40 percent of global resource flows today, with the remainder being minerals and ores (50.8 Gt, and 10.1 Gt, respectively) (de Wit et al. 2020, Figure 1). Carbon is not just an ingredient but is in fact the key element in such products as fuels, plastics, fertilizers, chemicals and chemical intermediates, and elemental carbon materials. Today, carbon-based chemical, fuel, and material products are dominantly manufactured with fossil carbon feedstocks,[2] so at the end of life, their consumption, disposal, or decay adds net-positive CO_2 emissions to the atmosphere. Using alternative feedstocks that enable circular carbon flows for carbon-based products is a key strategy for reducing

[1] While this chapter focuses on CO_2 utilization market opportunities, priority products from coal waste are also considered, especially critical minerals. The report covers coal waste utilization opportunities in detail in Chapter 9.

[2] Feedstocks are material inputs to industrial processes to generate a product.

fossil carbon emissions to the atmosphere. These feedstocks must be derived from sources or materials with low or zero life cycle greenhouse gas (GHG) emissions and integrated into industrial processes in a more sustainable way.[3] Examples of carbon feedstocks with lower life cycle emissions include biomass, recycled or waste carbon products such as plastics, captured CO and CO_2, biogas, and municipal solid waste.[4] Another important class of materials that can use CO_2 feedstocks is CO_2-derived mineral carbonates incorporated into construction materials, which do not traditionally incorporate CO_2, but where CO_2 can be incorporated as long-duration stored carbon.[5] During the transition to net-zero, an alternative to circular carbon feedstocks is the continued use of fossil feedstocks with compensatory capture and sequestration to prevent or remove an equivalent full life cycle amount of CO_2 emissions from the atmosphere. This report is tasked with examining a circular carbon future, and so this possibility of linear fossil production with offsetting is noted but not explored in depth. The report is also tasked to examine coal waste utilization opportunities, which are addressed in Section 2.2.3 and Chapter 9.

This chapter describes the market opportunities for products that will use captured CO_2 or coal waste as feedstocks to provide useful carbon-derived products in a net-zero future. Products fall into two classes: durable storage materials, with lifetimes greater than 100 years, and circular carbon materials, with lifetimes less than 100 years. The product lifetime distinction is important for understanding the two classes' climate impact. Durable storage materials will act as long-term sinks for carbon and could become instrumental in achieving an overall net-zero carbon future. Circular carbon materials will enable the sustainable cycling of nonfossil carbon in both natural and human-made systems, an essential aspect of moving from an extractive model of carbon mining and waste deposition into the atmosphere to a net-zero future with substantial climate and economic benefits.

There is no consensus on the stable need for carbon-based products in a net-zero future. Product volumes depend heavily on technology potential, the pace of transition, population and economic growth, available resources (CO_2, enabling, and competing), and policy choices based on priorities for decarbonization and other societal goals. Durable storage materials and circular carbon materials are distinct in their growth potential. Most durable storage materials have significant carbon utilization growth potential: they are currently not produced in large quantities (e.g., carbon fiber, nanotubes), have undeveloped yet significant potential for applications in new markets (e.g., carbon black in concrete, direct use of coal waste in construction materials), or have production method alternatives that incorporate CO_2 as a new ingredient, rather than a replacement for fossil carbon (e.g., concrete, aggregates). Because of their growth potential and the future need for carbon removal in a net-zero future, durable storage materials could result in both cost-effective removal of carbon from open environments and production of revenue-generating products at scales of Gt per year.

Short-lived, circular carbon materials to replace fossil-derived fuels and chemicals have a divergent growth trajectory, with some products expected to shrink and others expected to grow. In a net-zero future, some current hydrocarbon markets are expected to largely disappear and be replaced by zero-carbon solutions, notably electric power replacing fuels for ground transportation (NASEM 2023a). For example, the daily use of gasoline fuel in the United States is about 8 million barrels (EIA 2024). Within the fuels class, the production of aviation fuels will remain a large-volume need that could be met with CO_2 conversion (NASEM 2023b). Demand for other essential short-lived carbon products (e.g., chemicals and fertilizers) is expected to continue growing and can be integrated into a circular economy based on alternative carbon feedstocks, including CO_2. Within the chemicals class, this report distinguishes chemical intermediates from end products to emphasize their versatile role in the chemical industry. For example, ethylene or ethanol could be used as intermediates in the production of polymer material or aviation fuel. Figure 2-1 shows one estimate of (1) the embedded carbon in fuels for energy and transport, and in materials and chemicals in 2020 and 2050, and (2) a detailed description of the carbon embedded in chemicals

---

[3] Chapter 3 discusses life cycle assessment as applied to carbon utilization.

[4] It is also conceivable that some carbon-based products could be replaced in the future with materials that use silicon or sulfur as the backbone atom, but these options will not be addressed in this report (Barroso et al. 2019; Kausar et al. 2014; Petkowski et al. 2020).

[5] In this report, concrete and aggregates are considered carbon-derived materials, even though they traditionally do not use carbon as a feedstock. This helps to reduce the significant carbon burden created by life cycle emissions associated with construction materials (Park et al. 2024).

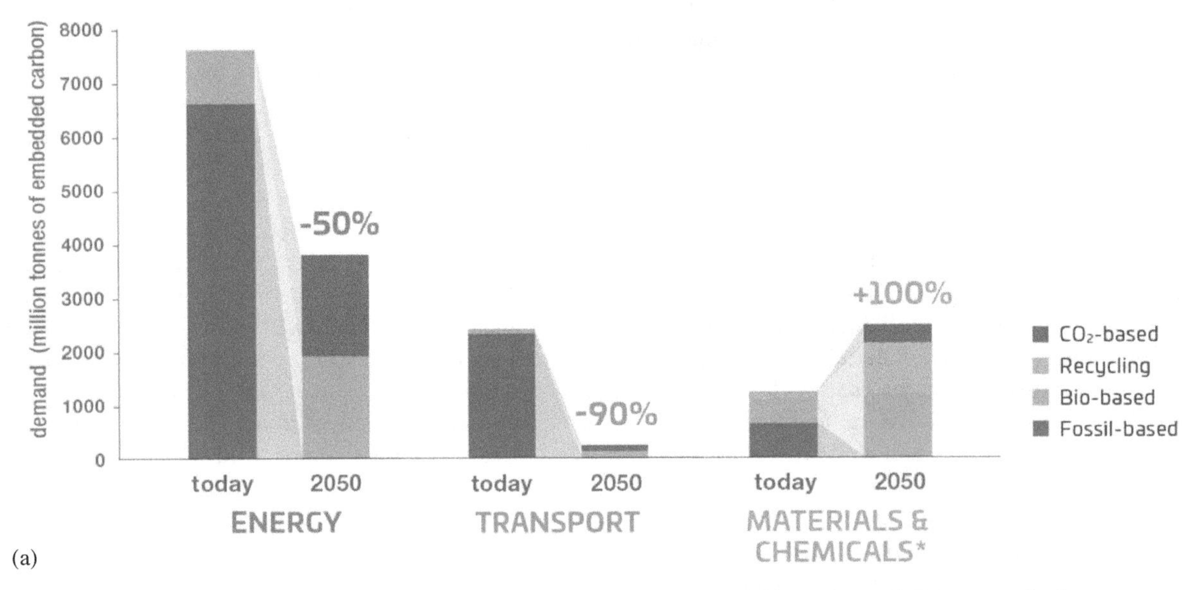

(a)

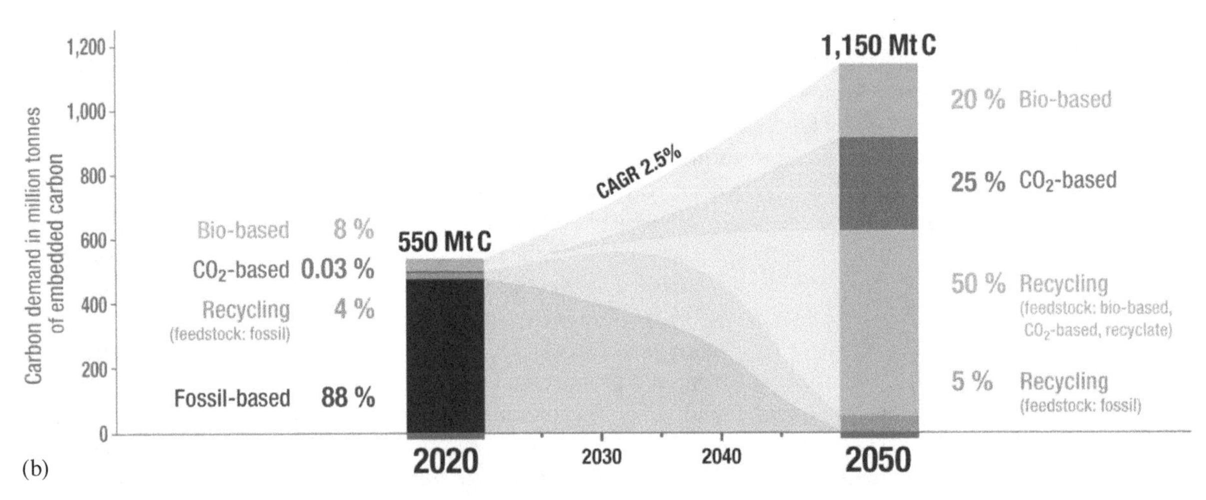

(b)

FIGURE 2-1 CO_2 utilization is projected to have a significant role in serving the demand for carbon-based materials globally in a defossilized future in 2050. Panel (a) shows the embedded carbon in fuels for energy and transport, and in materials and chemicals in 2020 and 2050, and panel (b) shows a detailed description of the carbon embedded in the subset of chemicals and materials, which is the output of the chemical industry: chemicals and derived materials (see footnote 7). The demand for fuels drops by 50 percent in the energy sector and 90 percent in the transport sector. Demand for chemicals and materials is projected to double by 2050. It illustrates projected growth for chemicals and derived materials of 2.5 percent compounded annual growth, leading to 1150 megatonnes (Mt) of carbon demanded in 2050 versus 550 Mt in 2020. Demand is served by bio-based, CO_2-derived, and recycled material, rather than fossil-based feedstocks. This figure does not include an assessment of carbon flows related to inorganic carbonate materials.
SOURCE: Kähler et al. (2023).

and derived materials globally.[6,7,8] The analysis indicates that fuel demand will drop by 50 percent in the energy sector and 90 percent in the transport sector. Demand for chemicals and materials is projected to double by 2050. When focused on the subset of materials and chemicals that includes chemicals and derived materials, especially polymers, the total carbon demand was 550 Mt annually in 2020, with 88 percent of that derived from fossil material, 8 percent derived from bio-based materials, and less than 5 percent being recycled or CO_2-derived (0.03 percent) (see Figure 2-1). In 2050, 25 percent of carbon demand for chemicals and derived materials is projected to be sourced from CO_2. Table 2-1 describes the committee's assessment of the priority products for a net-zero future. These include both durable storage and circular carbon materials.

The following sections describe existing markets and anticipated growth for three use cases for CO_2 or coal waste feedstocks: (1) incumbent products that could be replaced by products made from new carbon sources (e.g., sustainable aviation fuels, polymers, or chemicals and intermediates); (2) products that traditionally are not made from fossil carbon (e.g., concrete, aggregates) but that could incorporate CO_2 as a feedstock; and (3) products for which a current market is small but could grow substantially in a net-zero future (e.g., carbon fiber as a substitute for steel and aluminum). Specific market considerations for key product categories are analyzed in detail. Relevant factors for the market introduction of products from new carbon feedstocks are discussed, including access and availability to new feedstocks, suitable conversion technologies and infrastructure, consumer demand and acceptance, and regulatory environments. Sections on cost and financial risk are followed by a discussion of the need for techno-economic and life cycle assessments as well as analyses of the risk of unintended consequences. (Further material on these aspects is covered in Chapters 3 and 4 of this report.) The chapter then concludes with a summary of findings and recommendations.

2.2 MARKET OPPORTUNITIES FOR CO_2-DERIVED PRODUCTS

2.2.1 Factors That Impact Ease of Making Products from CO_2

In principle, all hydrocarbon fuels and chemicals, and many other materials, including inorganic carbonates, elemental carbon materials, and plastics, can be synthesized from CO_2. However, only some products and markets are likely to be attractive for investment in CO_2 conversion processes, relative to other sustainable carbon feedstock alternatives. The costs of producing specific products via various CO_2 utilization pathways versus competing pathways and feedstocks must be considered. Competing sustainable pathways include substituting the product with zero-carbon alternatives like electricity and hydrogen (most relevant for fuels) and manufacturing the product with other non-fossil carbon feedstocks like biomass or recycled carbon wastes. Uncertainties in future policy, market, and regulatory environments, as well as unknowns related to technological advancements, make it impossible to predict and compare future costs of producing specific products from different feedstocks. However, the physicochemical properties of CO_2 and potential CO_2-derived products provide some guidance on the ease of making different classes of products from CO_2 versus production from either incumbent net-positive emission feedstocks, or other net-zero emission feedstocks.

---

[6] Embedded carbon is the carbon present in the molecules or materials that constitute products. It differs from embodied carbon, which describes the life cycle carbon emissions associated with a product.

[7] As defined in Kähler et al. (2023), chemicals and derived materials are organic chemicals and polymers originating from the global chemical industry, including human-made fibers and rubber. This does not include chemicals derived from the heavy oil fraction (bitumen, lubricants, and paraffin waxes), nor does it include wood, pulp and paper, or natural textiles. The total estimated global demand for carbon embedded in chemicals and derived materials is approximately 550 megatonnes (Mt) per year, and 1200 Mt per year when the additional classes of materials and chemicals are included. None of the analyses in Kähler et al. (2023) include global embedded carbon in fuel products, such as gasoline, diesel, aviation fuel, natural gas, or coal.

[8] The future 2050 scenario for renewable carbon-based fuels, chemicals, and derived materials outlined in Kähler et al. (2023) assumes that demand for carbon-based fuels in the energy sector is reduced by 50 percent through use of electricity, hydrogen, and solar heating. Transportation carbon needs reduce by 90 percent due primarily to electrification and some hydrogen fuel. Demand increases by 100 percent, assuming a combined annual growth rate of 2.5 percent, for chemicals and derived materials. In this scenario, the shares of the renewable carbon sources for chemicals and derived materials are estimates based on ambitious rates of recycling (55 percent of embedded carbon), biomass limited by planting areas (20 percent), and the remainder of embedded carbon produced from CO_2 utilization (25 percent).

TABLE 2-1 Committee's Assessment of Priority Products from CO_2

Product Class	Example Priority Products	Competitors to CO_2-Derived Production in a Net-Zero Future	Current Global Production and Future Demand (gigatonnes [Gt] per year, year of estimate)[a]	Climate Benefits (lighter blue = lower climate benefit, darker blue = higher climate benefit)			Conversion Technology	Market Driver and Advantages of CO_2 Feedstock
				Durable carbon storage or circular carbon product	Amount of CO_2 used (tCO_2/tonne product)[a]	Global scale (Gt per year, estimated in 2050)[a]		
Fuels	Jet fuel Marine fuel Lipids	Biomass-derived carbon fuels Electrification Hydrogen Ammonia	*Jet fuel* 0.305 (2020) 3.07 (2050) *Marine fuel* 0.3 (2020)[b]	Circular	3–6 (Jet fuel)	3.07 (Jet fuel)	Chemical Biological	• Defossilization of needed product • Drop-in capability • Decentralized production opportunities
Inorganic Construction Materials	Concrete Aggregates	Incumbent construction materials (conventional concrete, aggregates, steel, aluminum, wood) Coal waste–derived products	*Concrete* 7 (2020) 32.3 (2050) *Aggregates* 45 (2020) 119 (2050)	Durable	.0.001–0.05 (Concrete) 0.087–0.44 (Aggregates)	32.3 (Concrete) 119 (Aggregates)	Mineralization	• CO_2 storage • Enhanced performance opportunities • Feedstock flexibility • Residual waste material use • Substitution for higher-emitting products
Polymers	Polycarbonates Polyurethanes Polylactic acid Polyhydroxy-alkanoate	Biomass-derived polymers Recycling	*Polycarbonates* 0.0015 (2007)[c] 0.024 (2020) *Polyurethanes* 0.06 (2050)	Circular or Durable	0.05–0.25 (Polyurethane)	0.06 (Polyurethane)	Chemical Biological	• Defossilization of needed product • Drop-in capability • CO_2 storage • Enhanced performance opportunities • Biodegradability (Polylactic acid and Polyhydroxyalkanoate) • Circularity
Agrochemicals Including Fertilizers	Urea	Biomass-derived agrochemicals	0.13 (2019)[d] 0.27 (2032)[e]	Circular	0.73[f]		Chemical	• Defossilization of needed product • Already commercialized CO_2 utilization (urea) • Scalability

Chemicals and Chemical Intermediates	Chemical products: CO Methanol Ethylene Formic acid Bioproducts: Butanediol Succinic acid Lactic acid	Biomass-derived chemicals Recycling	*Methanol* 0.110 (2022)[g] 0.432 (2050) *Ethylene* 0.168 (2020)[g] 0.25 (2050)[h] *Formic acid* 0.00078 (2020) 0.0140 (2050)	Circular	1.28–1.5 (Methanol) 0.49–0.96 (Formic acid)	0.432 (Methanol) 0.25 (Ethylene)[h] 0.0140 (Formic acid)	Chemical Biological	• Defossilization of needed product
Elemental Carbon Materials	Carbon black Carbon fiber Carbon nanotubes Graphene	Methane-derived elemental carbon materials Biomass-derived elemental carbon materials Coal waste	*Carbon black* 0.014 (2020) 0.07 (2050)	Circular or Durable	3.7–4.2 (Carbon black)	0.07 (Carbon black)	Chemical	• Defossilization of needed product • CO_2 storage • Substitution for higher-emitting products • Enhanced performance opportunities
Food and Animal Feed	Spent microbes	Low-impact animal and plant food production	*Animal feed* 0.337 (2020) 1.9 (2050)	Circular	0.5–0.7	1.9	Biological	• Defossilization of needed product • Enhanced performance opportunities

[a] Unless otherwise noted, data are from Sick et al. (2022b).
[b] From Statista Research Department (2023).
[c] From Neelis et al. (2007).
[d] From IEA (2019).
[e] From Chemanalyst (2023b).
[f] From Bazzanella and Ausfelder (2017).
[g] From CAETS (2023).
[h] From IEA (2018).

NOTE: The Climate Benefits column is color-coded, where light blue indicates low benefit, blue indicates medium benefit, and dark blue indicates high benefit.

SOURCES: Based on data from Bazzanella and Ausfelder (2017); CAETS (2023); Chemanalyst (2023b, 2023c); IEA (2019); Kähler et al. (2023); Mallapragada et al. (2023); Neelis et al. (2007); Sick et al. (2022b); Statista Research Department (2023).

TABLE 2-2 Standard Gibbs Free Energy to Convert CO_2 and Water to Reduced Products[a]

Product	Reaction	ΔG^0_{rxn} (kJ/mol)
Carbon monoxide	$CO_{2\,(g)} \rightarrow CO_{(g)} + 1/2O_{2\,(g)}$	257
Formic acid	$CO_{2\,(g)} + H_2O_{(l)} \rightarrow HCOOH_{(l)} + 1/2O_{2\,(g)}$	270
Methanol	$CO_{2\,(g)} + 2H_2O_{(l)} \rightarrow CH_3OH_{(l)} + 3/2O_{2\,(g)}$	703
Ethanol	$2CO_{2\,(g)} + 3H_2O_{(l)} \rightarrow C2H_5OH_{(l)} + 3O_{2\,(g)}$	1326
Ethylene	$2CO_{2\,(g)} + 2H_2O_{(l)} \rightarrow C2H_{4\,(g)} + 3O_{2\,(g)}$	1331
Ethane	$2CO_{2\,(g)} + 3H_2O_{(l)} \rightarrow C2H_{6\,(g)} + 7/2O_{2\,(g)}$	1468

[a] All thermodynamic quantities are calculated as described in Nitopi et al. (2019), using data from the NIST Chemistry Webbook (Linstrom and Mallard n.d.) and Lange's Handbook (Dean 1999).

CO_2 is a highly oxidized, single-carbon, and relatively unreactive feedstock. It can be transformed into products via low energy, non-reductive pathways where the carbon remains highly oxidized, such as into inorganic and organic carbonates, polycarbonates, urea, and carboxylic acids (Martín et al. 2015). CO_2 can also be converted into reduced carbon products, such as hydrocarbons and alcohols, via higher energy pathways (Shaw et al. 2024). Table 2-2 shows the energy required to form several reduced carbon products, with higher Gibbs free energy representing more thermodynamically difficult reactions (the related electrochemical reaction energetics are shown in Table 7-2) (Nitopi et al. 2019). All reactions are thermodynamically unfavorable (positive free energy) under standard conditions, which is to be expected from reductive transformations of CO_2 and water to form hydrocarbons and alcohols, and the most reduced and longer carbon chain products are more challenging thermodynamically.

The cost of making products depends both on the fundamental thermodynamics of conversion processes as well as reaction kinetics, and a variety of technology- and market-specific technical and economic factors. Kinetically, formation of single-carbon products is easier than multi-carbon products, which require challenging multi-step transformations. Improved catalysis, reaction design, and systems design can improve reaction kinetics (selectivity, rate, and yield). Sections 2.2.2 and 2.2.3 further describe the demand-side and supply-side market considerations for CO_2 utilization, and section 2.2.4 describes market considerations by product class. Chapter 7 examines the technology readiness (Figure 7-3) and compares scaling factors (Section 7.2.2.2) for processes to convert CO_2 to certain priority chemical intermediates and final products, providing an example of technical and economic factors to consider when selecting a conversion pathway for a particular product.

2.2.2 Demand for Products Derived from CO_2

The carbon-based product system will need to transform to one that can be net-zero-emitting while continuing to provide the remaining product services to the economy without relying on fossil carbon feedstocks. Future markets for short-lived, circular carbon products derived from CO_2 will be dependent on the demand for fuels, chemicals and chemical intermediates, and other such products; by the potential to supply such products from different sustainable feedstocks and will reflect restructuring of chemical markets based on competition with zero-carbon substitutes. Many of today's carbon-containing products, including most fuels, chemicals, and plastics, are derived from fossil carbon (petroleum, natural gas, and coal). Short-lived products emit their fossil carbon into the atmosphere during use or after disposal. In the future, a major class of these carbon-containing materials, hydrocarbon fuels for land- and sea-based transport, heating, and electricity generation, will largely disappear owing to improved zero-carbon options, and the limits placed on emissions that are likely to be required to achieve net-zero. Although projections for the rates of reduction in fossil fuel use vary, an example scenario is shown in Figure 2-2. The International Energy Agency's (IEA's) Net-Zero-Emissions Scenario projects a decline of fossil-carbon-derived fuels to 20 percent of all energy supply in 2050 (IEA 2021), from 80 percent in 2020.

The transition away from fossil fuels will follow different timelines in the developed world and developing regions where the demand for basic materials, electricity, and water, as well as the lack of infrastructure for net-zero options, might require the use of fossil carbon feedstocks to a larger extent and for longer times.

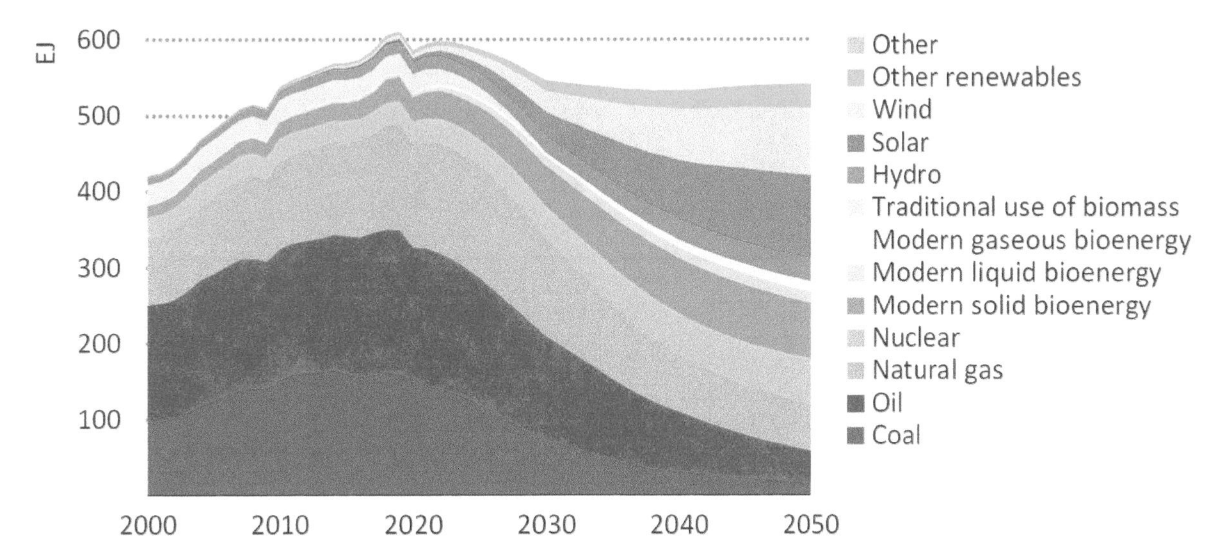

FIGURE 2-2 Total energy supply in the IEA's Net-Zero Emissions Scenario. Renewables and nuclear power displace most fossil fuel energy and the share of fossil fuels (coal, oil, and natural gas) falls from 80 percent of total energy supply in 2020 to just over 20 percent in 2050.
NOTE: EJ = exajoules.
SOURCE: IEA, 2021, *Net Zero by 2050: A Roadmap for the Global Energy Sector*, Paris: IEA, https://iea.blob.core.windows.net/assets/deebef5d-0c34-4539-9d0c-10b13d840027/NetZeroby2050-ARoadmapfortheGlobalEnergySector_CORR.pdf. CC BY 4.0.

In contrast to falling fuel demand, demand for the many nonfuel, carbon-based products is expected to grow, tracking with expected global economic development. For example, the IEA projected that demand for petrochemicals would represent nearly a third of demand growth for oil in 2030 and about half in 2050 (IEA 2018). To better understand the current chemical industry, Appendix I, Table I-1 describes the major fossil-derived chemical products, excluding fuels, by global volume in 2007, and their production methods. Although the data are from 2007, it describes a baseline of fossil chemical production, which in the future will need to evolve into an industry producing a related but not identical suite of products, with sustainable carbon feedstocks, and likely at larger volume overall, with projected increases in demand for chemicals production. The key question becomes how to source the required carbon to manufacture these products. In principle, all hydrocarbon fuels and chemicals can be synthesized from CO_2 and hydrogen. However, only some products and markets are likely to be attractive for investment in CO_2 conversion processes, relative to other sustainable carbon feedstock alternatives. The costs of producing specific products via various CO_2 utilization pathways versus competing pathways and feedstocks must be considered. Competing sustainable pathways include substituting the product with zero-carbon alternatives (most relevant for fuels), manufacturing the product with other nonfossil carbon feedstocks like biomass or recycled carbon wastes, or offsetting fossil carbon emissions from the product life cycle using negative emission technologies, such as capturing and geologically storing an equivalent amount of CO_2.

Preparing for the transition to non-fossil-sourced chemicals production needs to factor in the risks to growth in product demand, as it will play a crucial role in research investments, and planning and deploying new supply chains and infrastructure to provide raw material streams. The total addressable market estimates the demand for carbon-based products that could be satisfied by production from sustainable carbon feedstocks, including CO_2. Figure 2-3 illustrates examples of growth projections of the total addressable market for key product categories for which CO_2 utilization could be considered (Sick 2022b), based on a market analysis of historical published growth rates and industry leader expectations. Based on that, product-specific and constant compound annual growth rates were assumed to project the market demand.

Market penetration during the transition to net zero will depend critically on the cost of the new products compared to the incumbents, especially fossil-derived products. All hydrocarbon products derived from CO_2 require

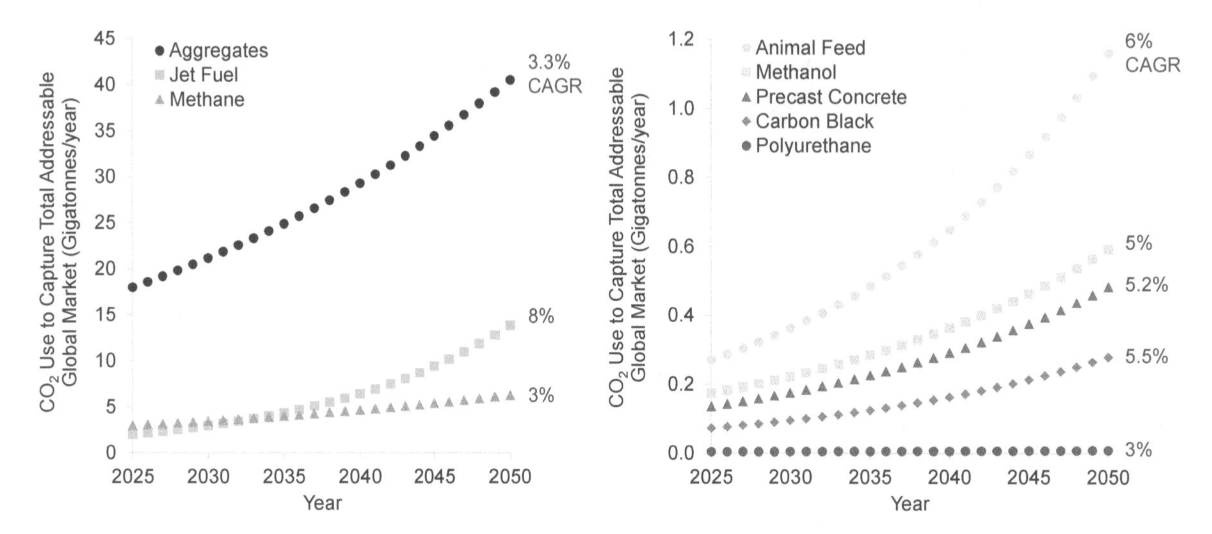

FIGURE 2-3 Projected CO_2 use to capture the total addressable global markets for key products that could be made with CO_2. Each projection uses a product-specific, constant compound annual growth rate (CAGR) to estimate the growth trajectory from 2025 to 2050 and assumes the average percent incorporation of CO_2 for the market class (i.e., CO_2-based aggregates assume 34 percent incorporation of CO_2 by mass, which may not be the rate achieved for all aggregate products). Detailed assumptions can be found on p. 158 of Sick (2022b). The left panel shows products that could use high volumes of CO_2, including durable storage aggregates and high-volume circular carbon uses, like jet fuel and methane. The right panel shows lower volume, primarily circular carbon products like chemicals, elemental carbon materials, and some high-volume durable storage materials that do not consume much CO_2 in their production, like precast concrete.
SOURCE: Adapted from Sick (2022b), https://dx.doi.org/10.7302/5825. CC BY 4.0.

the input of energy and hydrogen, either in molecular form or as water or another hydrogen donor. Most products also require the formation of carbon-carbon (C-C) bonds, which may need additional, capital-intensive reaction and separation steps. Fossil feedstocks already contain carbon in chemically reduced form ("hydrocarbons") and often also contain the desired C-C bonds. Given the large amounts of energy required and the capital intensity of the conversion processes, synthetic hydrocarbon products are therefore more expensive than those derived from petroleum or natural gas at their current prices. Any "green premium" can be an insurmountable barrier to (broad) market introduction. Procurement incentives, "buyers' clubs" such as the First Movers Coalition, and direct subsidies via tax rebates and other policy means will be important to kickstart the emerging industry. For CO_2 utilization to play a role in a future net-zero economy, the levelized cost of managing CO_2 via conversion to products compared to sequestration will have to be reduced, and the true (societal) cost of using fossil carbon needs to be incorporated into the economy as well (Black et al. 2023). Chapter 4 discusses policy needs for CO_2 utilization in greater depth.

2.2.3 Supply-Side Considerations for CO_2-Derived Products

Using CO_2 as a major feedstock for carbon-based products requires building up an entirely new industry, albeit one that can integrate substantial elements of current industries—for example, the construction material and chemical industries. The translation from invention to market-ready product gets increasingly expensive the closer the technology is to market introduction. Table 2-3 defines the technology readiness level (TRL) scale that describes the progress of a technology from research through development, and demonstration to operation. For technologies that achieve full commercial operation, times to market readiness usually are on the order of a decade, and any acceleration requires additional funding. The urgency to address climate change and secure access to sustainable carbon makes the long time-to-market a significant challenge.

TABLE 2-3 Definitions of Technology Readiness Levels

Level of Technology Development	Technology Readiness Level	TRL Definition	Description
System Operations	9	Actual system operated over the full range of expected conditions	The technology is in its final form and operated under the full range of operating conditions.
System Commissioning	8	Actual system completed and qualified through test and demonstration	The technology has been proven to work in its final form and under expected conditions.
	7	Full-scale, similar (prototypical) system demonstrated in relevant environment	Demonstration is shown of an actual system prototype in a relevant environment.
Technology Demonstration	6	Engineering/pilot-scale, similar (prototypical) system validation in relevant environment	Engineering-scale models or prototypes are tested in a relevant environment. This represents a major step up in a technology's demonstrated readiness.
Technology Development	5	Laboratory scale, similar system validation in relevant environment	The basic technological components are integrated so that the system configuration is similar to (matches) the final application in almost all respects.
	4	Component and/or system validation in laboratory environment	The basic technological components are integrated to establish that the pieces will work together. This is relatively "low fidelity" compared with the eventual system.
Research to Prove Feasibility	3	Analytical and experimental critical function and/or characteristic proof of concept	Active research and development (R&D) is initiated. This includes analytical studies and laboratory-scale studies to physically validate the analytical predictions of separate elements of the technology.
	2	Technology concept and/or application formulated	Practical applications can be invented. Applications are speculative, and there may be no proof or detailed analysis to support the assumptions.
Basic Technology Research	1	Basic principles observed and reported	Scientific research begins to be translated into applied R&D.

SOURCE: Adapted from DOE (2011).

CO_2 utilization could be implemented in several industries to manufacture products for a variety of applications. Table 2-4 collects the assessment of priority products from CO_2 utilization as examined in various studies, and their various applications in the economy. Some themes in priority products identified across studies include oxygenated chemicals like alcohols, aldehydes, and organic acids; chemical industry intermediates like CO, ethylene, and ethanol; chemicals with fuel applications like jet fuel, methanol, and gasoline; chemicals with organic carbonate groups, such as cyclic carbonates and polycarbonates and inorganic carbonates; and elemental carbon materials like carbon black and graphene. Most of these priority products follow the physicochemical trends identified in Section 2.2.1 that make them advantageous to synthesize from CO_2. Most have industry-facing applications like chemical intermediates and manufacturing inputs, while some have consumer-facing applications like fuels. Manufacturers of chemicals, fuels, polymers, and inorganic carbonates will find opportunities for CO_2 utilization.

The nascent CO_2 conversion industry is seeing increasing development, in response to expected demand for CO_2-derived products, existing market opportunities, and incentives. As shown in Figure 2-4, based on data from a global industry and literature survey, the number of developers working on technologies for CO_2 conversion to products has increased from 2016 to 2021, especially at lower TRL. Market-ready production capabilities were still very low. Another survey and analysis of self-reported data of developers in 2022 shows nearly a third of them operating at TRL 8 and 9 (Circular Carbon Network 2022).

Although the current petrochemical industry could be re-created with CO_2 as a feedstock, net-zero emissions requirements will entail shifts in supply and demand factors likely to result in a different composition of the chemical industry (IEA 2020). For supply-side factors, today's portfolio of chemicals in production and use

TABLE 2-4 Product Targets from CO_2 Utilization as Described in Selected Studies of Technical Potential

Chemicals and Materials	Product Application	Citations That Reference Priority Products
Acetic acid	Chemical intermediate Solvent	Huang et al. 2021
Alcohols	Solvent Detergent Fuel	Bazzanella and Ausfelder 2017
Aldehydes	Polymer Solvent Dye Cosmetics	Bazzanella and Ausfelder 2017
Carbon black	Filler for tires Pigment	Sick et al. 2022b
Carbon fiber	Replacements for steel and aluminum	Biniek et al. 2020
Carbon monoxide	Chemical intermediate	Grim et al. 2023 Sick et al. 2022b Huang et al. 2021 Biniek et al. 2020
Carbon nanotubes	Strengthener for concrete Optoelectronics Catalysis	Sick et al. 2022b
Cyclic carbonates	Solvent Battery electrolyte Intermediate for polymer synthesis	Bazzanella and Ausfelder 2017
Diesel/jet fuel/hydrocarbon fuels	Fuel	Sick et al. 2022b Huang et al. 2021 Biniek et al. 2020 IEA 2019
Dimethyl ether	Fuel additive LPG substitute	Huang et al. 2021 Bazzanella and Ausfelder 2017
Ethanol	Chemical intermediate Fuels	Grim et al. 2023 Huang et al. 2021 Biniek et al. 2020 IEA 2019
Ethylene	Chemical intermediate	Grim et al. 2023 Huang et al. 2021 Biniek et al. 2020
Formic acid	Preservative Adhesive Precursor Fuel cell substrate	Sick et al. 2022b Huang et al. 2021 Biniek et al. 2020 Bazzanella and Ausfelder 2017
Gasoline	Fuel	IEA 2019
Graphene	Electronics Batteries	Sick et al. 2022b
Inorganic carbonates	Cement Aggregate Concrete Soil stabilization Mineral filler	Sick et al. 2022b Biniek et al. 2020 IEA 2019 Bazzanella and Ausfelder 2017
Methane	Fuel	Sick et al. 2022b Huang et al. 2021 Biniek et al. 2020 IEA 2019

TABLE 2-4 Continued

Chemicals and Materials	Product Application	Citations That Reference Priority Products
Methanol	Acetic acid Ethylene, propylene Dimethyl ether Fuel Polymer precursor	Grim et al. 2023 Sick et al. 2022b Huang et al. 2021 Biniek et al. 2020 IEA 2019 Bazzanella and Ausfelder 2017
Organic acids	Surfactants Food industry Pharmaceutical industry	Bazzanella and Ausfelder 2017
Organic carbamates	Pesticide Polymer precursor Isocyanate precursor Agrochemicals Cosmetics Preservative	Bazzanella and Ausfelder 2017
Oxalic acid	Cleaning	Huang et al. 2021
Polycarbonate etherols	Polyurethane foams	Bazzanella and Ausfelder 2017
Polycarbonates	Polymer	Sick et al. 2022b Biniek et al. 2020 IEA 2019
Polyhydroxyalkanoate	Polymer	Sick et al. 2022b Biniek et al. 2020
Polyhydroxybutyrate	Polymer	Huang et al. 2021
Polypropylene carbonate	Packing foils/sheets	Bazzanella and Ausfelder 2017
Polyurethane	Polymer	Sick et al. 2022b Biniek et al. 2020 IEA 2019
Protein for animals	Animal feed	Sick et al. 2022b OECD-FAO Agriculture 2021
Protein for humans	Food	Sick et al. 2022b
Salicylic acid	Pharmaceuticals Cosmetics	Bazzanella and Ausfelder 2017
Urea	Fertilizer Resin	Bazzanella and Ausfelder 2017

NOTES: Bazzanella and Ausfelder (2017) evaluated the technologies, pathways, and abatement opportunities and challenges for the European chemical industry to be carbon neutral by 2050, including economic constraints, investments, and research and innovation requirements. Sick et al. (2022b) evaluated the utilization amount and market size for building materials, carbon additives, polymers, chemicals, food, and fuels between 2022 and 2050 in the context of the total addressable market for respective products. Additionally, they examined these products' development stages and developers in the market. Biniek et al. (2020) assessed current technologies and reviewed current developments for technology adoption and the economics of a range of use and storage scenarios. Grim et al. (2023) examined CO_2 conversion via low-temperature electrolysis and reported products that could most impact global emission levels, especially those that could serve as intermediate feedstock inputs to known, commercialized upgrading pathways for producing high-volume chemicals. Huang et al. (2021) examined direct (low- and high-temperature electrolysis, microbial electrosynthesis) and indirect (biological conversion, thermochemical conversion) pathways for production of 11 chemicals from CO_2, H_2, and electrical energy. The priority chemicals were identified by their near-term technical viability. IEA (2019) considered the near-term market potential for five categories of CO_2-derived services and products, including fuels, chemicals, building materials from minerals, building materials from waste, and CO_2 use to enhance the yields of biological processes, to scale them up to a market size of at least 10 $MtCO_2$/yr. OECD-FAO (2021) provided an assessment of the economic and social prospects and trends through 2030 for national, regional, and global agricultural commodity and fish markets with inputs from member countries of the Organisation for Economic Co-operation and Development and the Food and Agriculture Organization of the United Nations and commodity organizations, assuming no major changes in weather conditions or policies. The assessment highlighted that the implementation of climate smart production processes can mitigate the emissions impact of agriculture, especially in the livestock sector, discussed the prices, production, consumption and trade developments for biofuels, and the policies, regulations, and mandates for low-carbon agricultural practices, applications, and products in the member countries.

FIGURE 2-4 The increase in developers working on lower TRL maturity indicates a growing pipeline of opportunities to enter the market with new production capabilities from CO_2.
SOURCE: Sick et al. (2022b), https://dx.doi.org/10.7302/5825. CC BY 4.0.

stems from petroleum feedstock for carbon and focuses on oxidative conversions. Switching to carbon oxides as feedstock changes chemical pathways to reductive conversions, which changes the conversion processes and the intermediates and by-products involved. When using CO_2 as a feedstock, more chemicals likely will be produced via CO or alcohols, for example, versus when starting with hydrocarbons, where ethylene or aromatics are more common precursors. Such restructuring can have a significant impact on research and development (R&D) needs, as discussed in Chapters 5–8. Attention needs to be paid to which chemicals are likely to transition first, which might no longer be needed, and for which alternative carbon sources or noncarbon alternatives might be an option. The same will be true for inorganic carbonates (aggregates), concrete, and elemental carbon products. Decision criteria will include the cost of production compared to incumbents, specific demand-pull, and supply-push, especially via policy instruments. Estimates for those criteria, regional variations, and demand projections are inherently uncertain. As such, it is not surprising to find variations between studies that evaluate priority products for CO_2 utilization, as summarized in Table 2-4.

2.2.4 Markets for Materials from Coal Waste

Chapter 9 provides a deeper discussion of market opportunities for products made from coal waste, which offer the additional benefits of environmental and land remediation. Single- to double-digit growth rates of billion-dollar markets are projected for products from coal waste, including critical minerals and metals, pigments, direct use in construction materials, and coal waste–derived carbon materials (*Fortune Business Insights* 2023a, 2023b; Grand View Research 2022; SkyQuest Technology 2024; Stoffa 2023; Straits Research 2022).

Leveraging coal waste presents distinct challenges, primarily related to its fossil origin and the resulting potential for net-positive emissions, regional availability, and eventual diminishing supply. Any coal-derived product needs to be durable to avoid the introduction of new, fossil carbon into the atmosphere. The largest market value for durable products for beneficial coal waste reutilization include construction materials, energy storage materials/electronics, cement, and concrete (see Table 9-2). Coal ash is already a common additive in many concrete products. Globally, 70 to 90 million tons of coal impoundment waste are generated annually (Gassenheimer and Shaynak 2023), and several billion tons are stored in nearly 600 slurry impoundments across the United States (Environmental Integrity Project 2019). Although coal waste is localized and volume limited, existing transportation infrastructure for coal could be repurposed and leveraged to mitigate logistical hurdles. The value-added potential of coal waste utilization extends beyond carbon conversion or critical minerals extraction and encompasses environmental cleanup efforts, and local job creation or preservation in coal communities. Incentives driven by the growing demand for critical minerals could catalyze efforts to repurpose coal and simultaneously address local pollution concerns associated with coal waste piles, including fly ash cleanup (Granite et al. 2023).

2.2.5 Product-Specific Market Considerations

As outlined in Table 2-1, this section explores some of the main product classes targeted for CO_2 utilization (inorganic construction materials, fuels, polymers, chemicals and chemical intermediates, elemental carbon, and food and animal feed) and the specific market considerations for the future viability of each product class. Each section describes the incumbent production and use of the product class, why a transition in production is needed for a net-zero future, the implications of different sustainable carbon feedstocks, and the key market considerations for the product class.

2.2.5.1 Inorganic Carbonate Construction Materials

The inorganic carbonate construction materials product class includes cement, concrete, and aggregates used for constructing buildings and infrastructure—for example, roads, water systems, and so on. These materials are produced at a large scale and with low profit margins from raw materials such as ores, rocks, and wastes by mining, crushing, grinding, processing, and/or heating at high temperatures. Incumbent manufacturing processes do not employ CO_2 as a feedstock in concrete production and, in fact, emit large quantities of CO_2 through energy use and process emissions associated with the chemical transformations of the materials for cement production. Production and use of construction materials are distributed geographically, with limited long-distance transportation owing to their weight and volume. A net-zero future could result in reduced CO_2 emissions, consumption of waste materials including CO_2, and improved material properties (such as improved compressive strength).

Net-zero compatible alternatives to the production of inorganic construction materials include recycled materials, biogenic materials, and technologies that bypass the process-related CO_2 emissions inherent to cement production from carbonate minerals. Recycling is common in the construction industry, particularly in road infrastructure. Recycled materials can include inorganic materials, recycled plastic, and other wastes as fillers or formed into construction components. Biogenic materials include timber, laminated beams, and particle boards used in structural or other roles in buildings. Recycled or biogenic construction materials as replacements for inorganic building materials could have the advantages of reducing waste streams and resource depletion. However, based on competing needs in other parts of the economy and suitability for structural applications, the use of biogenic materials may be limited. Lower-emissions technologies are in development to produce cement and concrete. These novel approaches bypass the process-related emissions from the conversion of carbonate rocks to reactive calcium clinker, instead converting noncarbonate minerals to reactive calcium species. These technologies also often have reduced need for high-temperature reaction conditions, and thus can be electrified, providing further opportunities to reduce CO_2 emissions on a life cycle basis.

CO_2 utilization to produce inorganic construction materials includes direct reaction of dissolved CO_2 with minerals to form aggregates or powders; carbonation of alkaline industrial and demolition wastes to form components of concrete; exposing construction materials to CO_2 to enhance carbonation, including to cure concrete; and formation of alternative cementitious materials. (R&D status and needs for CO_2 utilization to produce inorganic carbonates are detailed in Chapter 5.) These processes are relatively well developed compared to other CO_2 utilization processes, could be rapidly deployed, and in some cases, result in cost advantages (Carey 2018;

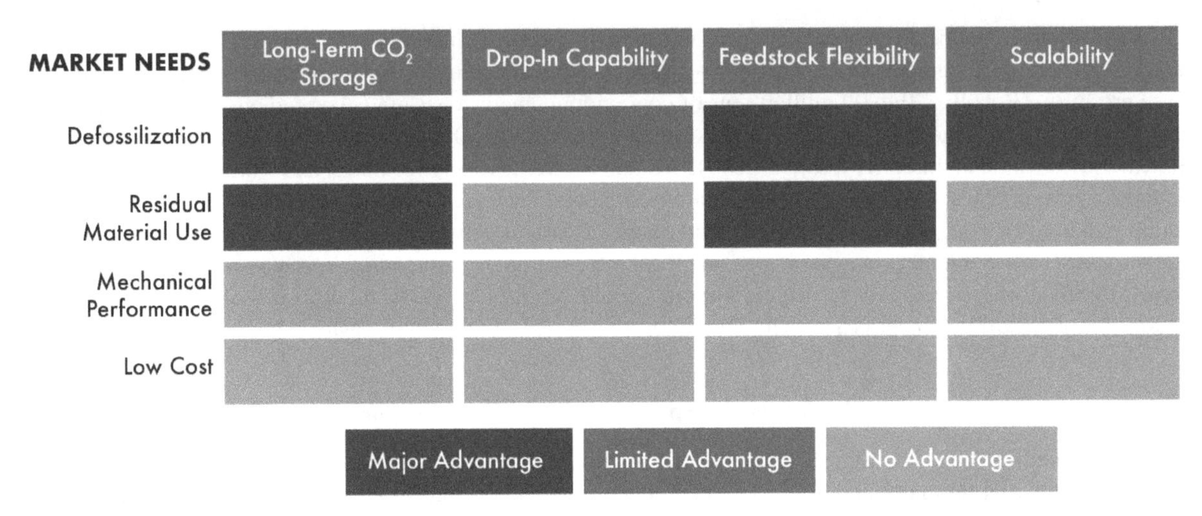

FIGURE 2-5 CO_2-derived aggregates have several advantages that could support market acceptance by meeting critical market needs. Leading needs are defossilization of the aggregates market, residual material use, mechanical performance, and low cost. Features of CO_2-derived aggregates that may meet those market needs are long-term CO_2 storage, drop-in capability, feedstock flexibility, and scalability. CO_2-derived aggregates are advantaged in the market by their ability to store CO_2 for the long term and use flexible feedstocks, leading to defossilization and residual material use. Given the large demand for aggregates, the scalability of new aggregate material production is a critical advantage of CO_2-derived aggregates.
SOURCE: Based on figures and material from Sick et al. (2022b), https://dx.doi.org/10.7302/5825. CC BY 4.0.

NASEM 2019; St. John 2024). The transformation produces solid carbonates, a stable, solid form of carbon that provides durable storage. Figure 2-5 maps the product-market fit via key market needs for the introduction of these new aggregates (listed as row headers: defossilization, residual material use, mechanical performance, and low cost), as well as features of CO_2-derived aggregates that may meet those market needs (listed as column headers: long-term CO_2 storage, drop-in capability, feedstock flexibility, and scalability), producing a heat map of areas with high potential for market pull. The market introduction of CO_2-based aggregates will be advantageous if the capabilities can successfully address the needs. As illustrated in the figure, CO_2-derived aggregates are advantaged in long-term CO_2 storage capability and feedstock flexibility to meet the market needs of defossilization and use of residual materials from construction and other industry sectors—for example, steel and coal wastes. Ensuring the required mechanical stability of new types of aggregate materials is a given expectation, and improved performance does not appear to create a market advantage. Equally, in a low-margin commodity market, low cost is expected.

Advantages of CO_2 utilization over incumbent concrete materials include improved properties, reduced material use, flexible feedstocks, reduced environmental impacts for a circular economy, and lower costs. For example, in precast concrete production, CO_2 curing accelerates the curing process, increases strength, reduces material needs, and can be cost-efficient. Also, technologies that mineralize CO_2 to limestone powders or carbonate fly ash can support novel three-dimensional (3D) concrete printing, which has the potential to provide environmental benefits, including less concrete waste and lower water use, faster construction, and lower costs (Yu et al. 2021; Zhu et al. 2021). The product–market fit for CO_2-cured concrete is summarized in Figure 2-6, showing a heat map of areas of high potential for market pull. Market inhibitors, such as local building codes and the cost and time required for testing and documentation can limit or prohibit the use of new materials, especially for small-scale producers, but are not fundamental inhibitors based on technical performance.

Key market questions for future viability of CO_2-derived inorganic carbonate building materials: Can new production technologies and reprocessing of waste materials overcome market inhibitors such as low profit margins, limited long-distance transportation of heavy, low-value commodities, and regulatory hurdles such as composition-based building codes? How can CO_2 availability accommodate the distributed nature of the construction industry?

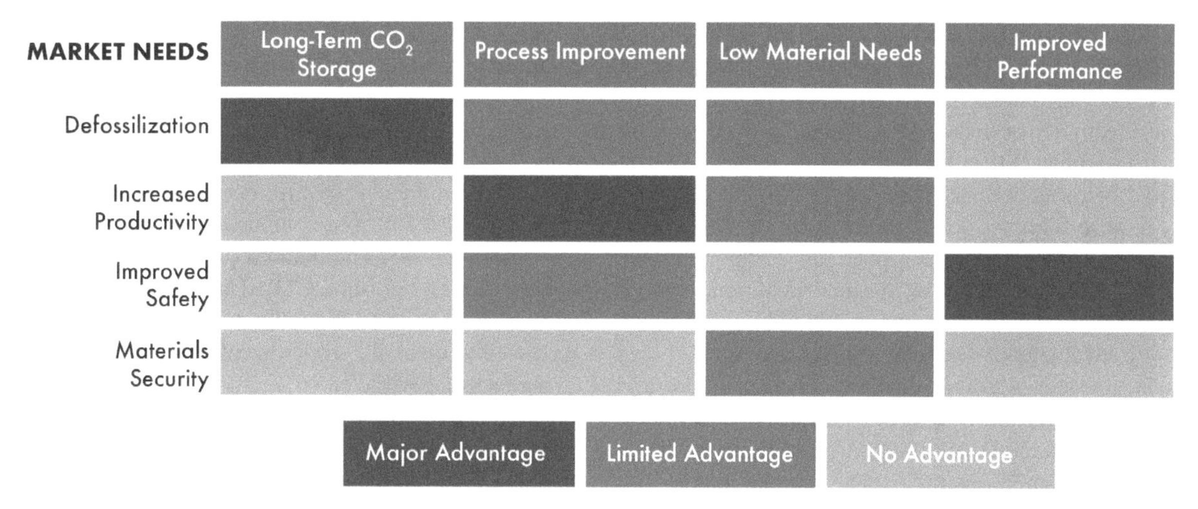

FIGURE 2-6 CO_2-cured concretes have several advantages that could support market acceptance by meeting critical market needs. Leading needs are defossilization of the concrete market, increased productivity, improved safety, and materials security. Features of CO_2-derived concretes that may meet those market needs are long-term CO_2 storage, process improvement, low material needs, and improved performance. CO_2-derived concretes are advantaged in the market by their ability to store CO_2 for the long term, improving processes and overall performance, leading to defossilization, increased productivity, and improved safety, respectively. These advantages can aid in the deployment of CO_2-cured concretes.
SOURCE: Based on figures and material from Sick et al. (2022b), https://dx.doi.org/10.7302/5825. CC BY 4.0.

2.2.5.2 Fuels

The global economy relies heavily on fossil fuels, with more than 80 percent of total energy from coal, oil, and natural gas (IEA 2019). Their combustion releases CO_2 and uses the energy in the fuel for electricity generation, vehicle propulsion, heating of buildings and industrial processes, and other energy needs. Fossil fuel production and combustion pollute the local and global environment, are harmful to human health, and are a major cause of climate change.

To eliminate the harms from fossil fuel production and use, most uses of fossil fuels must be replaced by zero-carbon alternatives in a net-zero future. Potential zero-carbon replacements for fossil fuel–powered systems include electric power generated from zero-carbon energy sources, hydrogen-powered systems including fuel cells, and energy efficiency measures to reduce the need for heat and power. Transitioning to zero-carbon electric power is more efficient than combustion, is less polluting, and leverages the existing power grid infrastructure. Drawbacks to electric power—for example, poor energy density and long recharging times for batteries—may make electricity unsuitable for some fuel substitution applications, especially in long-haul air and ocean transportation. Hydrogen fuel cell power is less efficient than using electric power directly, although it can have higher energy density. A major obstacle to hydrogen power is the requirement of new vehicle propulsion systems and hydrogen production, storage, and delivery infrastructure. For shipping, sustainably produced ammonia is being explored as an alternative zero-carbon fuel. However, the major concerns are safety, health issues, and the risk of highly elevated NO_x emissions from ammonia combustion (Bertagni et al. 2023).

Carbon-based alternatives to fossil fuel incumbents include biofuels such as ethanol, biodiesel, and jet fuel derived from bio-based sources. Bioethanol is already a major part of the transportation fuel system and can often be used in existing combustion, storage, and delivery systems with relatively minor modifications. Biofuels require significant land and water for crop production and result in pollution impacts from industrial agriculture, groundwater depletion, and fuel combustion. More details on the national prospects for the use of biomass resources can be found in the recently released DOE *2023 Billion-Ton Report* (DOE-BETO 2023).

Synthetic CO_2-derived liquid fuels can be produced by chemical and biological CO_2 utilization. (R&D status and needs for chemical and biological CO_2 utilization to fuels are detailed in Chapters 7 and 8, respectively.)

CO_2-derived fuels have similar advantages to biofuels, being energy-dense and in many cases, usable in the existing fuel combustion, storage, and delivery systems. CO_2-derived liquid fuel targets include methanol, ethanol, and jet fuel. They also have similar drawbacks to biofuels, including being less efficient and more expensive to produce than electricity or hydrogen, and leading to air pollution when combusted, although they are likely to have fewer land-use impacts than biofuels (Gabrielli et al. 2023). Synthetic aviation fuel will likely be the primary target for liquid fuel use, because of a lack of feasible technological alternatives, a greater need for energy density, the ability to absorb higher prices, and less concern about proximity to air pollution from combustion, with marine fuel (methanol) as an additional potentially important market. Aviation fuel may command a greater premium than marine fuel, depending on consumer willingness to pay (World Economic Forum 2023). The product–market fit for CO_2-derived fuels is summarized in Figure 2-7, showing a heat map of areas of high potential for market pull. A key advantage of synthetic aviation fuel is that it can be produced to meet the properties of current fossil-based fuels and used as a direct drop-in replacement, preserving all assets in the value chain, including aircraft. Distributed, co-located CO_2 capture and conversion plants could support scaling and increase supply stability, including for military needs (DoD 2023; U.S. Naval Research Laboratory 2012). Another competitor is offsetting fossil fuel combustion emissions with negative emissions technologies, which may have lower costs than replacing fossil fuel with bio-derived or synthetic CO_2-derived aviation fuel. However, CO_2-derived synthetic fuels offer more direct climate benefits than the purchase of negative emissions offsets and may be favored by future markets or regulatory structures.

Key market questions for future viability of CO_2-derived fuels: Where and when can synthetic fuels compete with direct electrification and other alternative fuels? How will utilization for short-lived products compete with sequestration for CO_2 sources? How will rereleased CO_2 be accounted for in market and regulatory monitoring, reporting, and verification schemes? What is the capacity to provide synthetic fuels in the context of competing demands for zero-carbon electricity and hydrogen?

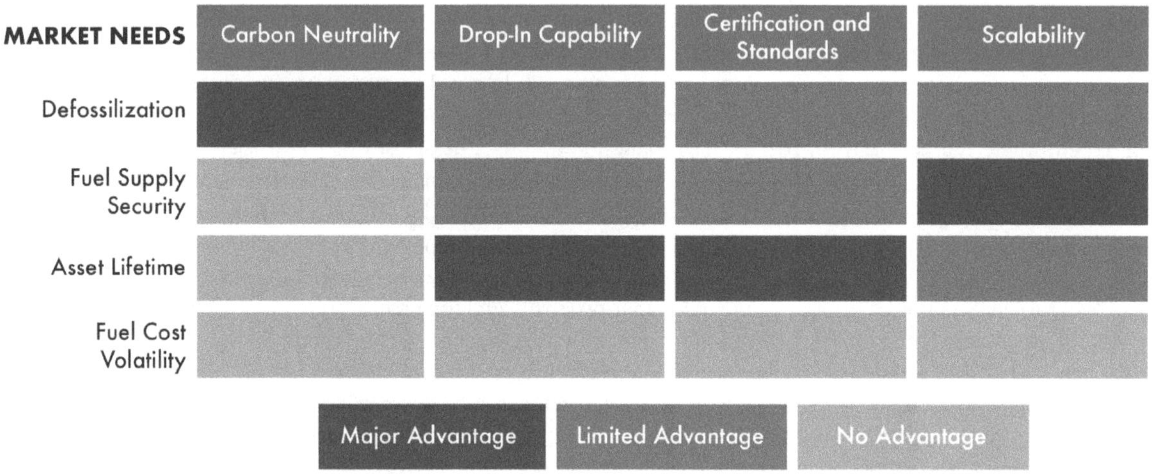

MARKET NEEDS
Carbon Neutrality
Drop-In Capability
Certification and Standards
Scalability
Defossilization
Fuel Supply Security
Asset Lifetime
Fuel Cost Volatility
Major Advantage
Limited Advantage
No Advantage

FIGURE 2-7 Synthetic, CO_2-derived aviation fuels have several advantages that could support market acceptance by meeting critical market needs. Leading needs are defossilization, fuel supply security, asset lifetime, and fuel cost volatility. Features of synthetic CO_2-derived aviation fuels that may meet those market needs are carbon neutrality, drop-in capability, certification and standards, and scalability. Synthetic CO_2-derived aviation fuels are advantaged by their ability to be carbon neutral, leading to defossilization. Given the demand for fuel supply security and asset lifetime in aviation, scalability and drop-in capability are critical advantages of synthetic CO_2-derived aviation fuels. Synthetic CO_2-derived aviation fuels can be produced to match the properties of incumbent fuels. This can preserve physical assets for fuel production, distribution, and use in airplanes and meet existing certifications and standards while meeting scalability and defossilization goals.
SOURCE: Based on figures and material from Sick et al. (2022b), https://dx.doi.org/10.7302/5825. CC BY 4.0.

2.2.5.3 Polymers

Polymers are currently predominantly synthesized from chemical intermediates derived from fossil carbon, with a smaller but significant market share derived from biomaterials or recycled polymer materials. Current methods of production and use result in CO_2 emissions, significant solid-waste streams, and local pollution.

Currently, the most important alternatives to fossil-derived polymers are biopolymers and polymers derived from recycled materials. In 2019, the polymer and plastics industry caused 3.4 percent of the global carbon emissions (OECD 2022). Furthermore, petrochemical plastics are notoriously recalcitrant to environmental degradation, causing substantial environmental hazards, including microplastics that impact human and wildlife health. Bioplastics, which encompass polymers made from biomass and polymers that biodegrade, represent a significant opportunity to reduce carbon emissions and other environmental hazards. Biopolymers like polyhydroxyalkanoate and polylactic acid provide local environmental benefits, as they are biodegradable or compostable and avoid lingering microplastics. Bio-based polymers produced from starch or sucrose derived from feedstocks like corn and sugarcane may confer an advantage over petrochemical plastics in reducing carbon emissions, although cultivation of corn and sugarcane comes with direct and indirect land use implications. Composting or recycling, described below, requires a value chain that includes infrastructure for separating and appropriate time and conditions for degradation.

Some polymers can be recycled either by mechanical or chemical recycling. Pure mechanical recycling via grinding or melting plastic products down to their base polymer requires high-quality, contaminant-free feedstock with uniform molecular composition. Mechanical recycling produces polymers with the same composition as its feedstock (Maureen 2023). Chemical or molecular recycling utilizes additional chemical inputs (solvents, enzymes) to break down recyclate into its constituent components (monomers, oligomers) to produce the same or different polymers (Luu 2023). Although often used synonymously with molecular recycling, chemical recycling sometimes refers to waste-to-energy processes, in which case CO_2 emissions are not minimized (Bell 2021). The viability of chemical or molecular recycling is limited by the complexity of plastic recyclate. Common additives like plasticizers and colors complicate the chemical recycling process owing to uncertainty or complexity of composition overwhelming existing molecular separation methods. Existing mechanical and chemical recycling methods can be energy-, water-, and land-intensive (Uekert et al. 2023). CO_2 emissions benefits of recycled versus virgin plastic manufacturing are circumstantial based on the composition, complexity, and quality of available feedstock.

CO_2 utilization to form polymers can proceed via the same intermediates as fossil fuel–derived polymers, but using CO_2-derived feedstocks, or via novel processes to incorporate CO_2 as a feedstock directly or via different intermediates. (R&D status and needs for chemical and biological CO_2 utilization to polymers are described in detail in Chapters 7 and 8, respectively.) Chemical intermediates such as ethylene, propylene, and aromatics used in current polymer production, to polyethylene, or polypropylene or polystyrene, can be generated from CO_2 via synthesis gas (Gao et al. 2017, 2020; Saeidi et al. 2021; Zhang et al. 2019). Direct utilization of CO_2 offers routes to other classes of polymers, such as polyurethanes made from CO_2-based polyols, polycarbonates, and polyhydroxyalkanoates (Afreen et al. 2021). These types of polymers or their building blocks could become key entry points for CO_2 use, with polyols already containing 20–40 percent CO_2 by weight. Limitations in thermal stability and mechanical properties of polycarbonates, polyols, and polyhydroxyalkanoates have restricted their widespread use (Ali et al. 2018; Capêto et al. 2024; Grignard et al. 2019; Styring et al. 2014). On the other hand, progress is being made to improve properties—for example, new synthesis methods have demonstrated polymers built from CO_2 that have flame-retardant properties (Ma et al. 2016). The product–market fit for CO_2-derived polymers is summarized in Figure 2-8, showing a heat map of areas of high potential for market pull. The market introduction of polymers made with CO_2 is facilitated not only by helping to defossilize the polymer industry but in particular by offering continued use of production facilities, improved recyclability, and the opportunity to provide entirely new performance characteristics.

Key market questions for future viability of CO_2-derived polymers: Can production costs be reduced, such as by co-location with CO_2 emitters? Can suitable CO_2-based polymers be made with favorable performance/cost balances? What is the competition for CO_2-derived versus bio-derived polymers and can biomass sourcing meet biopolymer demand? Can new, reductive synthesis methods that start with CO_2 be strategically used to design polymers with unique new purposes—for example, purpose-designed lifetimes?

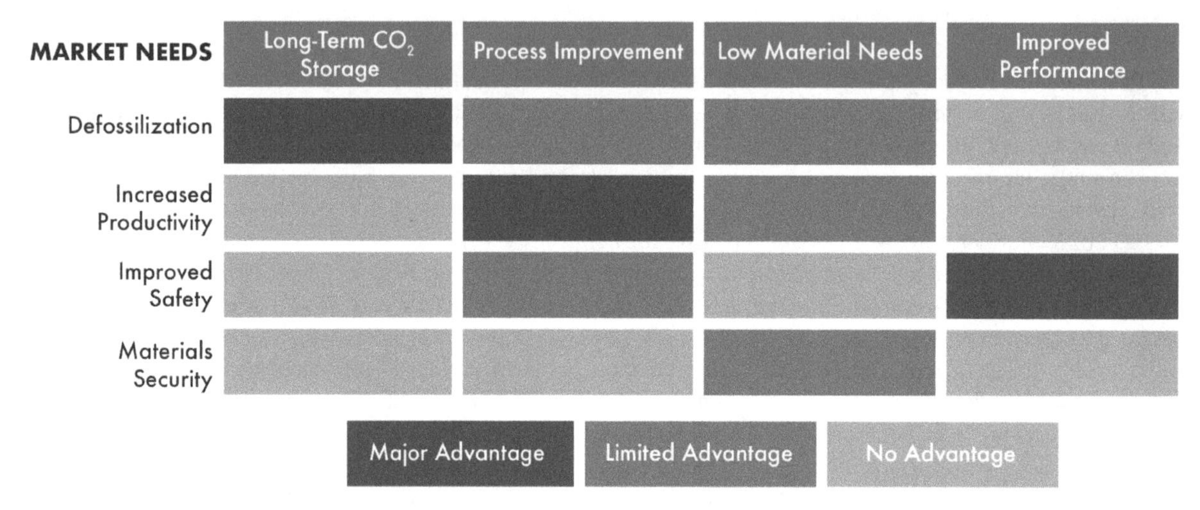

FIGURE 2-8 CO_2-derived polymers have several advantages that could support market acceptance by meeting critical market needs. Leading needs are defossilization, circularity, asset lifetime, and material performance. Features of CO_2-derived polymers that may meet those market needs are long-term CO_2 storage, drop-in capability, scalability, and replacement products. CO_2-derived polymers are advantaged by their ability to serve as replacement products delivering both material performance and circularity of the products. Given the large demand for polymers, scalability and drop-in capability are critical advantages of CO_2-derived polymers as well, provided properties and cost are favorable compared to incumbents to ensure timely market introduction.
SOURCE: Based on figures and material from Sick et al. (2022b), https://dx.doi.org/10.7302/5825. CC BY 4.0.

2.2.5.4 Chemicals and Chemical Intermediates

Current production of chemicals and chemical intermediates is almost entirely from fossil feedstocks of oil and gas and represents a small portion of fossil hydrocarbon use. The demand for chemical products is growing faster than the U.S. gross domestic product (GDP) and fuel demand. In a net-zero future, when fuel demand is likely to decrease dramatically, chemical demand will become a much more significant player in carbon-based product needs.

Alternatives to chemicals and intermediate production are primarily biobased materials. The efficiency and competitiveness of CO_2-derived chemicals compared to biomass-derived ones depend on factors like feedstock cost, energy requirements, and land/water use. Biomass is often more competitive for products requiring carbon-carbon bonds, which are often already present in bio-derived carbon feedstocks. Both CO_2- and biomass-derived materials are better suited to making oxygenated compounds, relative to fossil fuels. Biomass is used more easily for reduced compounds as compared to CO_2. Bio-derived compounds face higher water and land use implications than CO_2-derived materials (Gabrielli et al. 2023).

Carbon utilization is attractive for commodity chemical production to leverage existing infrastructure. Repurposing established facilities and processes offers a potentially cost-effective means to convert CO_2 to valuable products while simultaneously reducing GHG emissions. Chemical product targets include carbon monoxide, alcohols, light olefins, and carboxylic acids, both as final products and intermediates. The production of sustainable aviation fuels from CO_2 will result in many by-products that can enter the supply and production chains for chemicals in the same way that many chemicals we use today are by-products from reforming petroleum into gasoline, diesel, and kerosene fuels. Therefore, we may see some currently used chemicals and chemical intermediates disappear from markets while others enter.

Opportunities for broader market introduction will be higher the more downstream applications a product will have. This makes the drop-in replacement of entry-level chemicals and intermediates for a wider range of final products attractive. A key example is methanol, with high market needs as a base chemical and a potential new marine fuel. The product–market fit for CO_2-derived methanol is summarized in Figure 2-9, showing a heat map of areas of high potential for market pull. CO_2-derived methanol is a cost-effective drop-in replacement for its

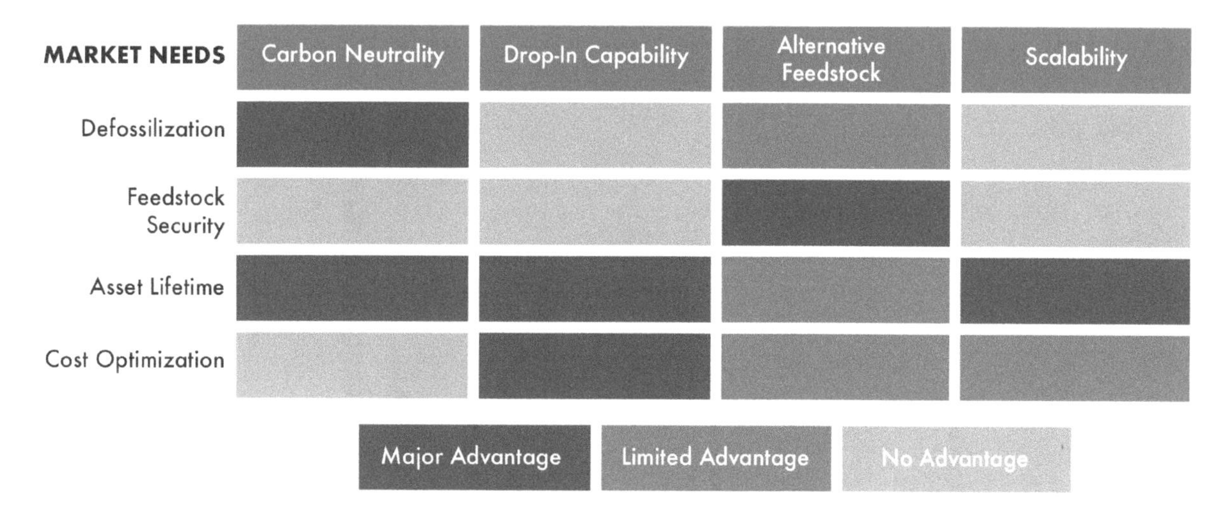

FIGURE 2-9 CO_2-derived methanol has several advantages that could support market acceptance by meeting critical market needs. Leading needs are defossilization, feedstock security, asset lifetime, and cost optimization. Features of CO_2-derived methanol that may meet those market needs are carbon neutrality, drop-in capability, alternative feedstock, and scalability. The carbon neutrality, drop-in capability, and scalability of CO_2-derived methanol especially aid in meeting the need for asset lifetime and cost optimization. Additionally, its carbon neutrality leads to defossilization and contributes to feedstock security by being an alternative option.
SOURCE: Based on figures and material from Sick et al. (2022b), https://dx.doi.org/10.7302/5825. CC BY 4.0.

chemically identical incumbent. The preservation of existing infrastructure in the chemical industry will be a key factor for adoption as geographically flexible feedstock availability and supply stability are increased.

Key market questions for future viability of CO_2-derived chemicals and chemical intermediates: Can CO_2-derived chemicals overcome the efficiency and competitiveness challenges presented by biomass-derived chemicals? Will future markets demand a price premium for a more sustainable product? Which opportunities exist for new products not yet available in this class? Will a new and different by-product stream from synthetic fuel production alter the chemical industry's well-established and global integrated supply chains and product mix?

2.2.5.5 Elemental Carbon Materials

Elemental carbon materials offer opportunities for long-term carbon storage, can potentially replace products made via high-carbon-emitting processes like steel production in some applications, and be used in high-value applications like electronics. Today, elemental carbon materials are primarily produced through combustion or pyrolysis of organic compounds (fossil sources) and synthesis through chemical vapor deposition techniques. Starting with biomass as a carbon source followed by subsequent combustion or pyrolysis could provide more sustainable pathways to elemental carbon products. These processes yield a range of materials, such as carbon black, graphite, graphene, and other carbon nanostructures, each with different properties and applications. Additionally, some of these materials can be produced through processes like electrochemical reduction or catalytic conversion, making them feasible CO_2 utilization targets. Particularly, graphene has many potential applications in energy, electronics, construction, and health care owing to its flexibility, lightness, and attractive mechanical and electronic properties.

Some elemental carbon products are likely to be used at lower volume, but in high-value applications, like electronics. Others could be deployed in very high-volume applications, with lower value, such as in construction materials. Small-volume, high-value markets may enable CO_2 utilization if buyers put a premium on CO_2-derived materials. Larger-volume, lower-value applications in the building industry present a significant market for material amendment or replacement. However, substituting carbon fibers and composites for steel and aluminum requires a significant industry shift and is more expensive, particularly for concrete. The product–market fit for elemental

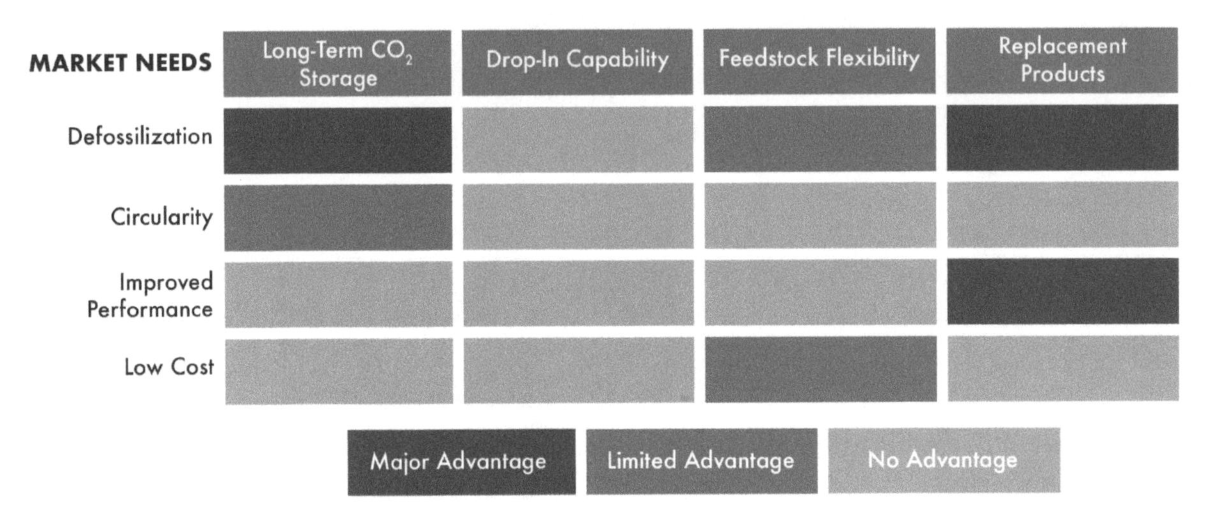

FIGURE 2-10 Elemental carbon has several advantages that could support market acceptance by meeting critical market needs. Leading needs are defossilization, circularity, improved performance, and low cost. Elemental carbon's ability to store CO_2 for the long term and be a replacement product can lead to defossilization of hard-to-abate sectors. Potentially superior mechanical and electronic properties will increase market interest.
SOURCE: Based on figures and material from Sick et al. (2022b), https://dx.doi.org/10.7302/5825. CC BY 4.0.

carbon materials is summarized in Figure 2-10, showing a heat map of areas of high potential for market pull. Key needs to address for successful market introduction are defossilization of target industries and providing suitable products to replace incumbents that suffer from a high carbon footprint—for example, aluminum and steel. While the production of carbon fibers, nanotubes, and graphene from CO_2 is still in its early stages, their value and potential to replace carbon-emission-intensive metals can lead to substantial growth. Conversion of CO_2 to carbon black could be pursued as a drop-in substitute for current production, but competition with incumbent producers will likely delay market penetration.

Key market questions for the future viability of CO_2-derived elemental carbon: Can new products overcome cost barriers and industry conservatism to replace carbon-emission-intensive metals like steel and aluminum in large-volume applications, particularly in the construction and automotive industry? What incentives may be needed? Can elemental carbon materials be recycled at the end of their use phase, which might be less than 100 years?

2.2.5.6 Food and Animal Feed

Current food and animal feed production is of biological origin, using plants and animals, and has sustainability challenges. While it is estimated that about 30 percent of produced food is wasted (NASEM 2023a), many lack access to enough food. The rising impacts of climate change also pose substantial risks to the food system through desertification and reduced land availability. Overfishing contributes to the loss of biodiversity and food resources from the oceans. Agricultural runoff pollutes waters and soils, leading to further ecosystem degradation. Animal agriculture (particularly the production of red meat) is especially resource-intensive and requires sustainability solutions in light of the growing global demand for animal protein. Alternatives need to be considered to ensure adequate nutrition for the world's human population and reduce the environmental burdens of food production.

The main alternatives to carbon dioxide utilization for food and animal feed are climate-smart agricultural methods. In addition to emissions reduction, these methods enhance agricultural resilience to climate-related risks, increase agricultural productivity, and improve financial returns for farmers (Kazimierczuk et al. 2023). Regenerative, digital, and controlled environment agriculture methods are among the most promising alternatives (Kazimierczuk et al. 2023). Regenerative methods focus on carbon sequestration through improved soil health and fertility, increasing water retention and percolation, reducing runoff, and strengthening system biodiversity and resilience (Elevitch et al. 2018). Digital methods integrate real-time or near-real-time feedback between sensors and equipment to make automated adjustments for emissions reduction and yield optimization. Controlled environment methods use indoor farming configurations like vertical farms, greenhouses, container farms, and integrated aquaponic systems to closely regulate the agricultural environment and reduce land and water usage (Goodman and Minner 2019).

CO_2 utilization can be leveraged in two ways in food production. First, increasing microbe-based production of drugs, food supplements, fuels, and chemicals leaves spent microbes as a waste material, which have high protein content and could be used directly as animal feed (LanzaTech 2023). This is analogous to other energy systems that use spent material as animal feed, such as ethanol production's coproduct of dry distillers grains, producing 38 million metric tons of feed for agricultural animals annually in 2018/19 (Olson and Capehart 2019). Department of Energy (DOE)-supported efforts on the algae-based conversion of CO_2 were recently summarized at the 2023 DOE's Office of Fossil Energy and Carbon Management/National Energy Technology Laboratory Carbon Management Research Project Review Meeting (NETL 2023).

Second, compounds derived from CO_2 conversion can be directly used for protein production via tissue engineering (e.g., cultivated meat or animal muscle cell cultures grown in reactors). Several such targeted commercialization activities are under way (Corbyn 2021; Mishra et al. 2020; Pander et al. 2020; Sillman et al. 2019). While market-ready production scales and acceptance are not expected until 2050 and beyond, consumer attitudes have been identified as a key issue in the market success of food replacements, especially alternative proteins (Van Loo et al. 2020). Competition for carbon-free electricity and hydrogen from other parts of the economy will be challenging for an emerging food production industry and is a key barrier for the industry. Additionally, regulatory barriers could challenge market entry.[9] The product–market fit for food and feeds is summarized in Figure 2-11, showing a heat map of areas of high potential for market pull.

Key market questions for the future viability of CO_2-derived food and animal feed: How can new products achieve Food and Drug Administration approval for human consumption? Will customers adopt "synthetic food"? What incentives may be needed? What are the techno-economic assessment (TEA) and life cycle assessment (LCA) considerations for cultivated protein products?

2.3 INFLUENCES ON CO_2 UTILIZATION MARKET DEVELOPMENTS

Potential revenue streams for CO_2-based products are trillions of dollars per year (Mason et al. 2023; NASEM 2023b), which could be an attractive driver to build up production capacity, depending on the unit economics per market. However, the successful market introduction of products made from new carbon feedstocks depends on a variety of factors, including feedstock availability and access, suitable conversion technologies and infrastructure, industrial participants in the value chain, consumer demand and acceptance, and regulatory environments. Furthermore, commercial success will depend on cost, cost-reduction strategies, financial risk management, and the ability to consistently meet demand, especially in commodity markets. TEA and LCA, including societal aspects, will be essential to understand environmental and equity risks and opportunities, and avoid unintended consequences. (See Chapter 3 for further details on LCA and TEA.)

Several studies project sizable opportunities for both climate benefits and economic potential for CO_2 as a carbon feedstock, especially conversion to long-lived products. Projections show that this could be possible at several Gt/year within decades (Biniek et al. 2020; Hepburn et al. 2019; IEA 2019; Jacobson and Lucas 2018; Sick 2018; Sick et al. 2022b).

Product adoption depends strongly on how fast market penetration proceeds, with timelines that stretch over decades. Figure 2-12 projects time needed to reach 10 percent market penetration and the time required to achieve a CO_2 utilization rate of 0.1 Gt per year for selected products. Given the urgent need to replace fossil carbon with

[9] As an example of regulatory inconsistency, the commercial sale of single-cell grown chicken meat for human consumption was recently allowed by regulators in the United States (Toeniskoetter 2022). In contrast, around the same time, the Italian government imposed a €60,000 fine for producing, selling, or importing laboratory-grown meats (Kirby 2023).

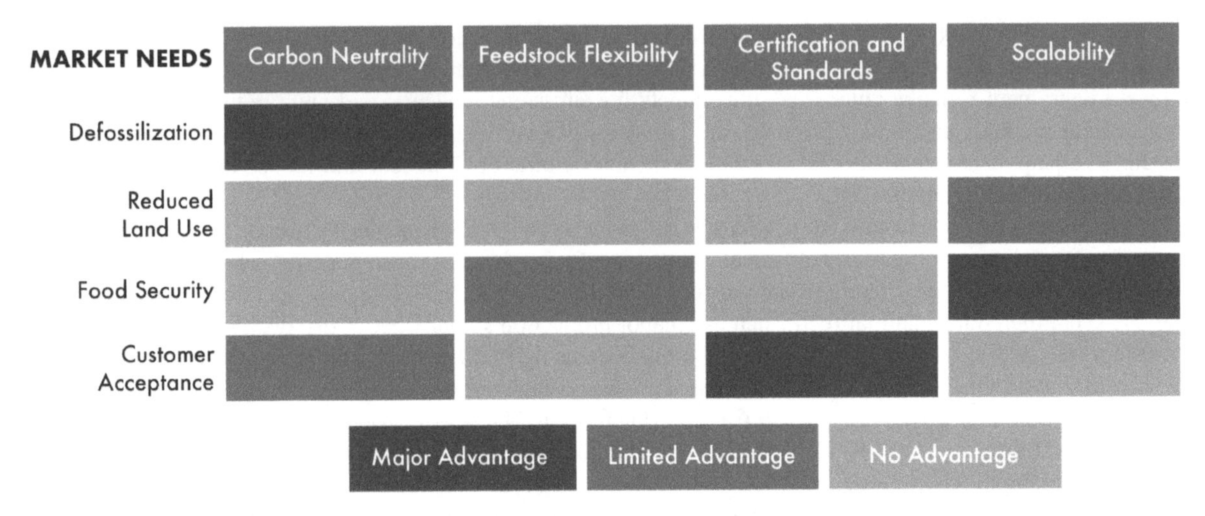

FIGURE 2-11 CO_2-derived food and animal feed has several advantages that could support market acceptance by meeting critical market needs. Primary market needs are defossilization, reduced land use, food security, and customer acceptance. CO_2-derived food and animal feed provides features in carbon neutrality, feedstock flexibility, certification and standards, and scalability. CO_2 as a feedstock to produce food and animal feed offers a chance to defossilize agricultural activities. If the right certification and standards are in place to ensure safety and to build consumer confidence, the scalability of suitable production technologies can significantly contribute to global food security.
SOURCE: Based on figures and material from Sick et al. (2022b), https://dx.doi.org/10.7302/5825. CC BY 4.0.

sustainable alternatives and produce durable stores of carbon, these low market uptake rates point to the need for rapid action to accelerate deployment. Comprehensive planning and evaluation are needed to ensure environmental benefits while also including economic and societal considerations (Newman et al. 2023).

Sections 2.3.1–2.3.8 detail important determinants of CO_2 market developments—namely cost, availability and access to feedstocks, technology and infrastructure, supply chains, consumer demand and acceptance, the regulatory environment, financial risks, and environmental and equity impacts. Establishing a CO_2 utilization industry would

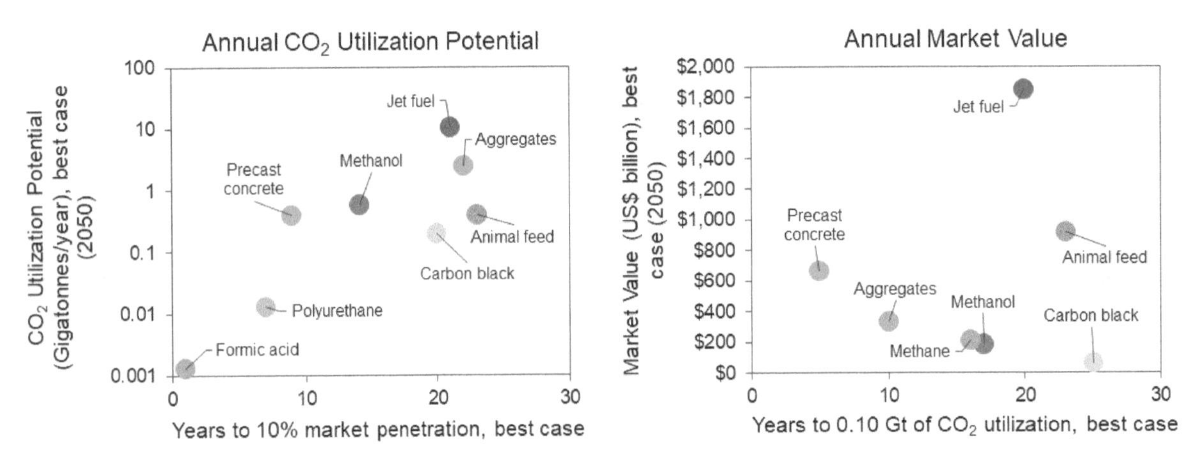

FIGURE 2-12 Differences in global market entry rates among products made from CO_2, showing the 2050 global best-case annual CO_2 utilization potential in Gt/year plotted versus years to 10 percent market penetration (left) and 2050 global best-case market value, in billion US$ plotted versus years to 100 Mt CO_2 utilization (right). A key competitive advantage for the United States is the large-scale availability of high-purity CO_2 from ethanol production that could be used to produce high-value sustainable aviation fuel.
SOURCE: Adapted from Sick et al. (2022b), https://dx.doi.org/10.7302/5825. CC BY 4.0.

benefit from a publicly available tracker that shows activity and progress with deployments, and the amounts of CO_2-based products that enter markets. This will also support tracking how much CO_2-based products contribute to reducing the carbon emissions burden.

2.3.1 Cost Factors for CO_2 Utilization Markets

The future use of CO_2 will be influenced by several cost factors that will determine the feasibility and scalability of carbon capture, utilization, and storage. One prevalent challenge is that, in many cases, the cost of producing CO_2-derived products exceeds that of incumbent alternatives. This cost disparity is driven by several factors, including the high upfront capital cost of CO_2 capture and transport, the energy expenditures required for possible purification and conversion processes, and the need to optimize and improve those processes (GAO 2022).

CO_2 utilization will be a highly capital-intensive endeavor to build the necessary production facilities or to retrofit some existing factories. The global cumulative investment in production facilities will be substantial for raw materials, labor, and construction (Sick et al. 2022a). CO_2 conversion facilities to form chemicals and fuels are especially capital-intensive, whereas facilities to produce aggregates and precast concrete, while more numerous, are less expensive to build to scale (Figure 2-13). For example, by 2050, meeting global aviation fuel demand with CO_2 utilization is estimated to require about 21,000 production facilities with annual capacities of 100 million liters of jet fuel each and estimated to cost $4.8 trillion (Sick et al. 2022b). Furthermore, for many CO_2-derived products, the dominant factor remains the cost of energy, typically electricity, which underscores the importance of energy efficiency and inexpensive, clean power generation in shaping the future of CO_2 utilization (Huang et al. 2021).

Capital and operating costs associated with CO_2 capture and transport are also important, especially as the industry evolves. The development of high-volume demand and compliance markets for some products will influence the trajectory of CO_2 capture costs for the market as a whole. For example, the aviation industry's quest for reduced carbon-intensity fuels could be a significant driver for increased capture volumes. Early opportunities for CO_2 capture may arise from existing processes like ethanol production, which have high-purity, proven technology,

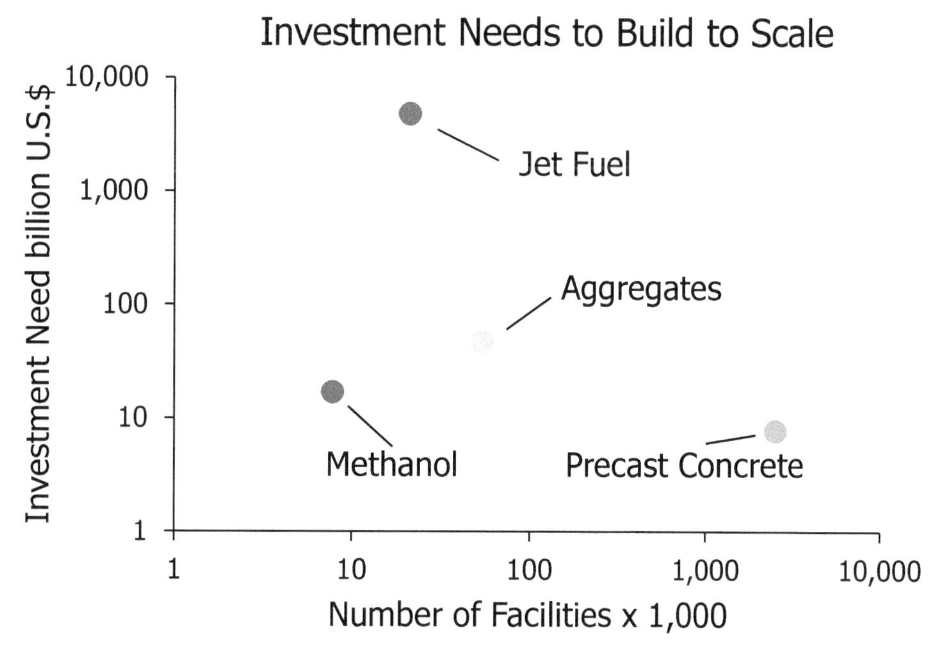

FIGURE 2-13 Estimated capital expenditures for facilities needed to provide full 2050 market demand with conversion of CO_2 to select products.
SOURCE: Based on data from Sick et al. (2022b), https://dx.doi.org/10.7302/5825. CC BY 4.0.

scale, and relatively low costs. The pace of capture process optimization, however, can become a critical cost driver. Local availability of sufficient sources of CO_2 at competitive cost will increase competitiveness of technologies by avoiding the need for expensive and potentially controversial transportation infrastructure.

2.3.2 CO_2 Supply Chains

The market for incumbent uses of CO_2 in the food and beverage industry and as a process gas globally reached approximately 236 million tonnes in 2022 and is projected to grow at a compound annual growth rate of 6.5 percent until 2035 to reach approximately 520 million tonnes (Chemanalyst 2023a). The addition of and shift toward CO_2 conversion to products could increase this market to Gt/year (IEA 2020; Sick et al. 2022b). Current CO_2 sources include ethanol, ammonia, and natural gas processing facilities, and future sources may include other industrial point sources and facilities drawing from ambient sources such as direct air capture (DAC) and direct ocean capture (DOC). The existing supply chain for the CO_2 industry lays a foundation for future CO_2 utilization-to-products in a net-zero market but needs to evolve to meet the challenges of tackling climate change. The CO_2 supply chain involves numerous aspects and actors, including carbon capture and separation from point or ambient sources, followed by purification, processing, and transportation to downstream applications and markets (DOE-EERE 2022). The success of CO_2 utilization-to-products depends on identifying best practices from existing supply chains and optimizing them to create long-term sustainability benefits.

Reliable availability and price stability of a feedstock are essential to build up downstream uptake. If competing demands for a feedstock exist, they may jeopardize companies, especially during the early scale-up phase when their needs are not at final capacity and when they are not yet established as a stable customer. The majority of CO_2 supply chains today have been developed for industrial applications that involve direct use of CO_2 without chemical conversion. This includes the food and beverage industry, which had the largest revenue share of the merchant CO_2 market in 2022, and enhanced oil recovery for depleted oil reserves. New applications for direct use of CO_2 are gaining prominence, such as the use of CO_2 in the medical sector as an inhalation gas in various surgeries (Grand View Research 2023) and CO_2-assisted enhanced metals recovery from spent lithium-ion batteries (Bertuol et al. 2016).

Chemical and biological conversions of CO_2 are not as prevalent in industry today, except in the manufacturing of urea for the fertilizer industry. However, as discussed throughout this chapter and projected in several market studies (Grand View Research 2023; Sick 2020), CO_2 as a carbon feedstock for products will quickly grow in relevance, albeit at different rates across the industry landscape. For a comprehensive view of these products, refer to Table 2-1. Incumbent direct-use applications of CO_2 will compete with emerging products for supply.

The industrial gas and the oil and gas industries historically have led investments in CO_2 supply chain development for merchant and enhanced oil recovery applications, respectively. The Oil and Gas Climate Initiative, representing 12 of the world's largest energy companies, is developing projects in regional, interconnected carbon capture, utilization, and storage supply chains at scale for industrial decarbonization (Oil and Gas Climate Initiative 2023). Alongside established companies, start-up companies will play a significant role in the carbon capture and utilization value chain, especially as new business models develop around "partial-chain" or specific components of supply (IEA 2023). In 2016, fewer than 200 entities were active in CO_2 utilization (Sick 2018). By 2022, that number increased to 274 (Circular Carbon Network 2022), indicating some growth but still at the very bottom of a typical S-curve for economic development.

By contrast, the number of companies that are pursuing CO_2 capture has risen more rapidly than those active in utilization. Government funding in the Infrastructure Investment and Jobs Act of 2021 (IIJA) and incentives provided in the Inflation Reduction Act of 2022 have spurred activity in both point source capture and DAC. For example, through the Carbon Capture Demonstration Projects Program, DOE has announced funding for three demonstration projects with the potential to capture nearly 8 million tons of CO_2 per year and is supporting front-end engineering design studies for an additional nine projects (CATF 2023). In support of DAC deployment, DOE has selected 2 of the 4 Regional Direct Air Capture Hubs authorized and appropriated in the IIJA and is funding feasibility and design studies for an additional 19 DAC projects (CATF 2023). All of these projects are structured as cost-share agreements with investment from both public and private partners. Additionally, carbon capture companies are being purchased by larger industrial entities. In 2023, Oxy, one of the largest oil producers in the United States, purchased all outstanding shares of the DAC company, Carbon Engineering Ltd., for approximately \$1.1 billion (Oxy 2023). The resulting entity is one of the two initially selected Regional Direct Air Capture Hubs. Efforts in DOC of CO_2 are impacted by the lack of emphasis on defining monitoring, reporting, and verification for DOC and policy support that instead favors point source and DAC.

Developments and investments in CO_2 transport have also gained momentum, although these efforts have faced legal, regulatory, and societal acceptance challenges (see Chapter 10). In 2023, ExxonMobil acquired pipeline operator Denbury for \$4.9 billion, making ExxonMobil the owner and operator of the largest CO_2 pipeline network in the United States (*AP News* 2023). Through the Carbon Dioxide Transportation Infrastructure Finance and Innovation Act, enacted as part of the IIJA, DOE's Loan Programs Office will support large-capacity, common carrier CO_2 transportation projects (DOE-LPO n.d.). As discussed further in Chapter 10, three major CO_2 pipelines that would traverse nearly 3600 miles were under development at the start of the committee's writing, although one of those projects, accounting for 1300 miles of pipeline, has since been canceled. Developers are also exploring the possibility of converting natural gas pipelines for CO_2 transport—for example, the Federal Energy Regulatory Commission approved Tallgrass Energy's Trailblazer Pipeline Company LLC to convert its existing 400-mile-long Trailblazer natural gas pipeline to a CO_2 transportation network (Ranevska 2023). However, as discussed in Section 10.3.2.2 of this report and Section 4.3.4 of the committee's first report (NASEM 2023b), there are significant challenges to performing such retrofits, and their feasibility has to be evaluated on a case-by-case basis. Liquefied natural gas carriers and shipping companies are expanding into CO_2 shipping (Northern Lights 2022), which is being explored as an alternative or complementary CO_2 transportation method in some cases. More discussion of CO_2 transport options, including multimodal transport, can be found in Chapter 10.

Developing a sustainable CO_2-to-products supply chain for a net-zero market requires established sources of CO_2 in significant quantities, coupled with long-term offtake agreements from CO_2 emitters, and pathways for economically sourcing clean hydrogen, electricity, and water where applicable. New business models and novel monetization strategies, such as carbon capture as a service, self-capture with third-party CO_2 offtake, CO_2 transport tolling fees, and voluntary carbon markets, will all depend on a coordinated rollout of infrastructure along the supply chain and defined regulations on long-term liability for "partial-chain" models. Stable consumer demand for CO_2-derived products will be driven by transparent definitions for carbon traceability and accounting and supportive government policies such as low-emission mandates and parity with incentives for CO_2 storage (Carbon Capture Coalition 2023). Issues of life cycle assessment are discussed in Chapter 3, and policy incentives and requirements are in discussed Chapter 4.

2.3.3 Availability and Access to Competing, Non-CO_2 Feedstocks

CO_2 utilization competes with alternative net-zero carbon emissions products that either allow for circularity or lead to durable carbon storage. Competing feedstock options include CO_2 (the focus of this report), biomass, and in some cases, replacement of the product with non-carbon-based alternative products or services. The choice and timescale for implementation of competing feedstocks will depend on infrastructure, feedstock source volume, consistent availability, price, and competition for the feedstock from other uses, and competition between different carbon feedstocks. A summary of non-CO_2 feedstock availability and readiness of conversion technologies is presented in Table I-2, in Appendix I.

Competing nonfossil carbon feedstocks from biological and recycled plastic streams have advantages and disadvantages relative to CO_2 feedstocks. Biomass and materials derived from biomass are already a feedstock for production of carbon-based materials, both as final products directly and as a feedstock to be converted to final products. However, biomass can cover only a fraction of all projected future carbon needs owing to its significant land use requirements and substantial competition for downstream use (food, animal feed) (Patrizio et al. 2021). Residual biomass and waste biomass, including municipal waste materials,[10] can offer local opportunities for carbon-based product manufacture, and avoid increasing cropped areas. However, lignocellulosic biomass

[10] Municipal solid waste can also be a source of critical minerals and other metals (Allegrini et al. 2013; Šyc et al. 2020).

utilization (constituting the majority of waste and residual biomass) still faces challenges, including developing more fermentable carbohydrate intermediates, lignin utilization pathways, and overcoming the mass transfer challenges caused by high solid loads. Furthermore, some types of biomass are geographically and seasonally constrained or present technological and economic challenges with transportation and processing (Energy Transitions Commission 2021). CO_2 conversion requires more energy, but typically less land use than biomass production.

Materials derived from recycled plastic will also compete with biomass and CO_2 as sources of carbon for chemicals synthesis (Gabrielli et al. 2023; Lange 2021). It is uncertain how much recycling of materials will compete with conversion of CO_2 as a carbon feedstock. To some extent, this is related to the uncertainty in material availability owing to multivariant routes for plastic materials (recycling, landfilling, incineration, loss as environmental pollution). As discussed in detail in Chapter 9, carbon-containing coal wastes can be used as a feedstock for durable storage materials, such as graphite, graphene, carbon fiber, carbon foam, and in concrete production.

In addition to competition from different carbon feedstocks, there is also competition for CO_2 from other processes. Sourcing CO_2 requires navigating a variety of complexities that range from entry-level maturity of some capture technologies, capture capacity, locations of sources and associated potential transportation requirements, needs for energy and other resources, and overall cost (Lebling et al. 2022; Mertens et al. 2023; Müller et al. 2020; NASEM 2019, 2022). Key federal legislation has favored CO_2 capture and geologic sequestration rather than utilization; owing to the federal incentives and other cost drivers, the bulk of current financial and deployment interest is still very focused on geologic sequestration.

2.3.4 CO_2 Utilization Technology and Infrastructure Development

Switching materials production to new carbon feedstocks for chemicals, elemental carbon materials, and inorganic carbonates often requires substantial investments in new equipment and/or infrastructure. Upfront capital investment as well as the ongoing operating costs will impact the choice of new feedstock. Petrochemical facilities will need to be retrofitted with large modifications to process CO_2 with substantial capital requirements. Once key entry-level chemicals or intermediates have been produced from CO_2—for example, methanol and ethylene—further upscaling is independent of the upstream process and can remain the same as is in production today, or in processes modified for sustainability, such as using renewable energy. Likewise, separation and reforming technology for fuel blends from Fischer-Tropsch synthesis can use installations from today's refineries or future sustainable ones. The production of aggregates and CO_2-cured concrete requires specialized new equipment that needs further technological development, investment, and a significant change to existing businesses. However, the distributed nature of concrete and aggregate production offers opportunities for local production, building on local CO_2 emission sources, thereby eliminating the need for increasingly controversial CO_2 transportation infrastructure (Adams 2023). Scale up of technologies from the laboratory to translated and tested real-world conditions will be a challenge because CO_2 utilization technologies are in the early stages of development, or although fully developed, are infrequently employed. The growth of engineering, procurement, and construction firms experienced with the CO_2 utilization industry will facilitate this design and manufacturing scale up. The choice of technology, particularly its efficiency and scalability, will need to be evaluated for each option. Chapter 10 provides additional details on infrastructure, and Chapter 4 discusses project impacts on host communities.

2.3.5 Consumer Demand and Acceptance

Consumer preferences and willingness to pay in a specific market can influence an investor's choice of carbon feedstock. An increase in procurement incentives for low-embodied carbon construction materials at local, regional, and federal levels is beginning to grow demand for materials, such as concrete and aggregates made with CO_2. Demand for CO_2-derived construction materials has been demonstrated but cannot be met owing to a lack of installed production capacity (Li et al. 2022; Roach 2023). For example, the San Francisco International Airport has developed standards and procured concrete that incorporates CO_2; however, none of the companies that they have worked with have been able to provide sufficient material to support new building construction, major renovations, and large infrastructure projects (Anthony Bernheim, personal communication, April 11, 2024). Similarly, demand for sustainable aviation fuel is increasing, especially in the aftermath of concerns about inadequate carbon offset programs (Astor 2022; Greenfield 2021; SDG Global Council on Future Fuels 2023) used for air travel that did not provide durable removal (West et al. 2020). Chapter 4 includes related content on the International Civil Aviation Organization (ICAO) standard for sustainable aviation fuel. As in the case of construction materials, a lack of production capacity is limiting the expansion of market introduction, along with other factors such as high cost and limited availability of carbon-free electricity and hydrogen.

Some consumers, especially institutional consumers, may be willing to pay a premium for a more sustainable product. However, others are concerned that using captured CO_2 in products perpetuates the use of fossil fuels, or they worry that CO_2 capture and conversion uses large amounts of energy (NASEM 2023a). Besides cost and moral hazard concerns, consumers might have concerns about the quality, safety, and health impacts of the products, or the manufacturing processes (Arning et al. 2021; Inwald et al. 2023; Lutzke and Árvai 2021; NASEM 2019; Wolske et al. 2019). Chapter 4, Section 4.4.1.2, "Strengthening Public Understanding of CO_2 Utilization through Engagement," discusses methods of addressing public concerns via engagement, especially through educational programming. Such engagement can increase the public's information about CO_2 utilization technologies and develop trust, accountability, and transparency between project designers, developers, the general public, and host communities. Related topics are also addressed in Chapters 3 and 9.

2.3.6 The Regulatory Environment

In the absence of a price or binding limit on carbon emissions, nonfossil alternatives in most cases will be more costly than fossil-derived products. Therefore, the market introduction and growth of different carbon feedstocks will depend critically on suitable policy support (Renewable Carbon Initiative 2022), public support for carbon pricing schemes and various means to deploy the revenue (Valencia et al. 2023), and location-specific pricing mechanisms or tax incentives and subsidies for carbon reduction efforts that can make the use of captured CO_2 or coal wastes more attractive (Thielges et al. 2022). These issues will be introduced here but are covered in more depth in Chapter 4.

The market need for net-zero carbon fuels offers an example of the regulatory environment considerations for CO_2-derived products. Liquid fuels are globally traded and used as commodity carriers of energy. Liquid fuels can be produced from coal, natural gas, petroleum, biomass, CO_2, or recycled waste materials such as plastics or paper. From a chemical and energy point of view, the most straightforward route to liquid fuels begins with petroleum, which contains stored energy in the form of hydrocarbons of appropriate size and composition to be easily refined into liquid fuels. Production of liquid fuels from petroleum is currently the least expensive option, given how the production and use of fossil fuels have developed and are supported (Black et al. 2023).

The alternative fossil and nonfossil feedstocks have chemical and energy disadvantages relative to petroleum. CO_2 has very low energy and requires H_2 or water and electricity to be converted to a fuel. It is also a one-carbon compound, and therefore requires the formation of carbon-carbon bonds to produce most liquid fuels. Most CO_2-to-fuel processes are at an early stage of technical development, and fewer still have been demonstrated or implemented at scale. Chapters 7 and 8 have more details about the status of CO_2-to-fuels processes. Despite these challenges, CO_2-to-fuel processes have advantages over petroleum in low- or net-zero-emissions fuel synthesis.

Owing to this combination of emissions advantages and chemical/energy disadvantages, CO_2-to-fuel conversion cannot be competitive with petroleum in the absence of broader policy support—that is, a price or limit on fossil carbon or fossil CO_2 emissions—and a build-out of production facilities. Other policy supports—such as subsidies, procurement mandates, or investment by early movers—can help accelerate market development. These will increase the overall societal costs for liquid fuels production, relative to other options. Other regulatory aspects, not related to the cost of production, include materials specifications, standards, and certifications—for example, property standards for kerosene or concrete materials. Chapter 4 details the regulations and their impact on CO_2 utilization opportunities.

2.3.7 Financial Risks

The cost factors described in Section 2.3.1 create many financial risks for producers and consumers. Fundamentally, a bankable business case is required for producers and finance providers to enter new markets and capture economic gains. For producers, the most significant source of financial risk will be insufficient or unstable demand, and thus an inability to generate sufficient revenue to cover capital and operating expenses. The demand uncertainty, while a full-scale market is being established, can challenge the economic viability of early carbon utilization projects. To mitigate this risk, producers operating in early-stage markets can seek grants from public or private sources to offset their capital costs, providing a crucial upfront financial boost to get projects off the ground. Additionally, buyers' clubs, like the First Movers Coalition, offer a way to pool demand and collectively ensure a price floor (World Economic Forum 2024). These coalitions help align carbon utilization projects within individual or collective marginal abatement curves, enabling better decision making based on the immediate need and availability of alternatives.

Consumers of CO_2-derived products also face financial risks, including insufficient or unreliable supply, higher costs, and concerns about product quality. To address these issues, consumers can consider pooled demand mechanisms, where multiple entities collaborate to bolster demand and mitigate supply risks. Government involvement in backstopping supply and providing procurement grants can enhance the resilience of supply chains and ensure a consistent flow of products. Additionally, the uncertainty stemming from unstable policy landscapes can be mitigated by engaging with regulators and advocating for policies that support the growth of CO_2-derived products, reducing financial risk for all stakeholders.

One of the most challenging financial risks stems from the commodity nature of many CO_2-derived products. Most incumbent producers of products that could be replaced by CO_2-derived alternatives are comfortable selling on global markets on future contracts and/or spot prices. Because of liquid markets, these producers can hedge the risk of being unable to offload their products by finding other buyers. In the current market, there is high illiquidity of CO_2-derived versions of products, and thus producers of CO_2-derived products rely on long-term contracts. This is a new way of doing business for buyers, who value the flexibility to acquire the best price as demand requires. Emerging approaches through financial engineering can accommodate this by having interested third parties purchase the "green premium" to offset the cost above the incumbent. However, these kinds of multilateral agreements could be cumbersome initially and will be heavily reliant on quick approval of IRS-sanctioned LCA models developed in consultation with DOE and the Department of the Treasury. Another alternative could be developing contractual vehicles that are analogous to physical or financial power purchase agreements used to encourage the development of solar and wind electricity.

2.3.8 Environmental and Equity Impacts

A switch of carbon feedstock from fossil sources to alternatives should offer CO_2 emissions benefits, but comprehensive LCAs that include broad and local societal factors must be conducted to understand the overall environmental impacts, as detailed in Chapter 3.

Although some production capability for CO_2- and coal waste–derived products could be based on adapted existing facilities, sourcing CO_2 and hydrogen, along with the associated required fossil-free electricity, will add demands on land, water, and potentially the host communities (Beswick et al. 2021; Chemnick et al. 2023; Qiu et al. 2022). It will be paramount to involve communities in the planning process early on to obtain buy-in and support (see Finding 5.9 and Recommendation 5.6 from the committee's first report; NASEM 2023b). Emergence of new technologies can falter if public opinion turns negative; in the broader context of carbon management, a substantial antagonistic attitude toward carbon capture, utilization, and storage technologies already exists (Arning et al. 2020; Bellamy and Raimi 2023). A key reason is a lack of familiarity in the public with the differences between—for example, CO_2 capture and subsequent sequestration versus utilization of CO_2 (Lutzke and Árvai 2021). In the context of using alternative carbon sources as product feedstock, consumer willingness to use such products is also not guaranteed (Engelmann et al. 2020; Lutzke and Árvai 2021). Suitable actions and policies are discussed in Chapter 4.

Environmental and economic impacts will depend strongly on the combination of the CO_2 source and the downstream fate—that is, the conversion process and the nature of the final product—as shown in Figure 2-14. The nature of the final product will have more global impact—for example, via emissions from use or decomposition—while the conversion process will have the most consequences locally, at or near the production site—for example, using limited local water supplies. Competition of CO_2 utilization for resources like hydrogen, CO_2, or clean electricity could lead

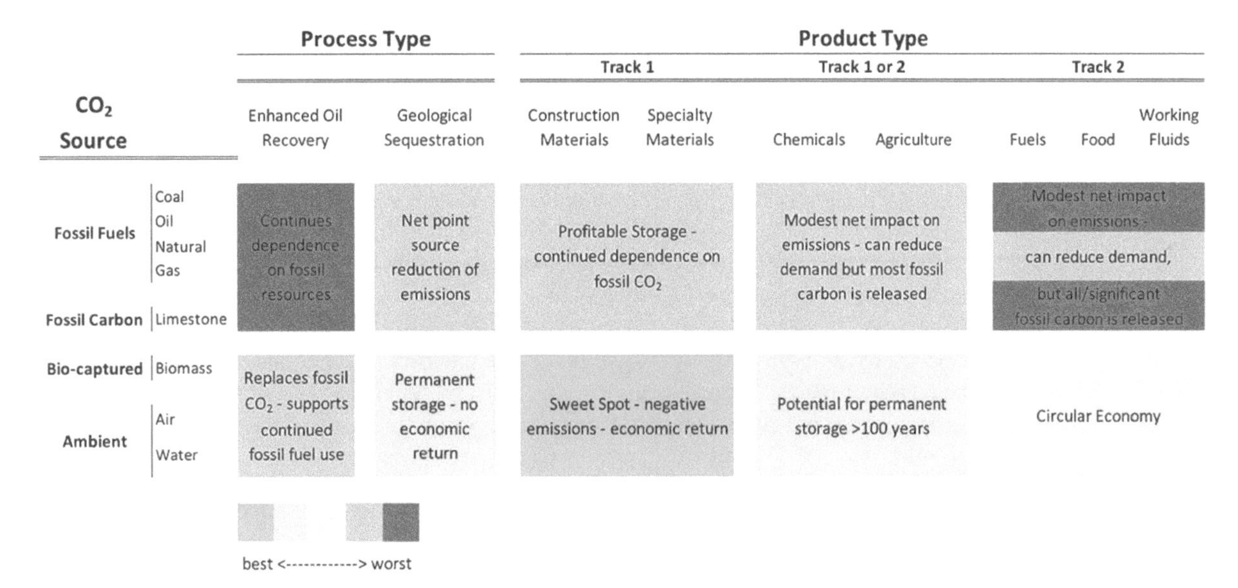

FIGURE 2-14 Illustration of the best cases for CO_2 utilization based on the relationship between sources and process or product type. Pathways are color-coded based on the combined outcomes of contribution to net-zero emissions and economic value (assuming a limit on emissions, but independent of economic incentives). Track 1 refers to durable storage products with lifetimes of >100 years, and Track 2 refers to circular carbon products with lifetimes of <100 years.
SOURCE: Mason et al. (2023), https://doi.org/10.3389/fclim.2023.1286588. CC BY 4.0.

to environmental or social impacts, such as renewable energy distorting local energy markets to the disadvantage of private electricity consumers (Ravikumar et al. 2020). Systems-level studies are needed to understand the broader impact on the environment, resource (re-)allocation and how those differentiate over location and time, jobs gains and/or losses, and other impacts (Faber and Sick 2022; SDG Global Council on Future Fuels 2023).

Current fossil-derived carbon products (e.g., fuels, polymers, and chemicals) are inexpensive, in part because the costs of pollution, particularly climate pollution, are externalized. A net-zero future is unlikely to be achieved via incentives alone, and so an economy-wide disincentive for emissions of CO_2, such as a cap or price on emissions, is likely to be required. The appropriate comparison to CO_2-derived products in a net-zero future is not the current cost of products, but instead the cost of products when emissions are implicitly or explicitly priced. (Another relevant comparison would be the cost of making products from unabated fossil fuels plus the socialized costs of unmitigated climate change, but this report does not focus on that present/future.) A price on emissions will change the prices of activities in the economy, including abatement/mitigation strategies for goods and services, and that may somewhat increase the cost of carbon-based consumer goods. These costs to certain stakeholders can be mitigated by other changes in the economy, such as subsidies to low-income households, that do not fundamentally change the incentives to limit emissions that a price on carbon is intended to provide. CO_2-based materials will compete against incumbents and alternative solutions that do not require carbon. As stated above, incumbent materials may become obsolete because non-carbon-based solutions provide the same function—for example, electricity powering ground vehicles instead of diesel and gasoline fuels. In some cases, non-carbon-based replacements will be less expensive than either fossil carbon-based incumbents or their sustainable carbon-based competitors.

2.4 CONCLUSIONS

CO_2 is a versatile resource used in diverse applications, including transforming captured CO_2 into valuable products. A world where CO_2 conversion to products is competitive may see a rise in circular economy practices, creating closed-loop systems where waste from one process becomes a resource for another. CO_2 conversion is necessary, in addition to biomass use and recycling of other carbon containing materials, to secure access to enough carbon once fossil carbon sources (petroleum, natural gas, coal) are no longer in use.

This report assumes a net-zero emissions future that will require a cap on, and eventual elimination of emissions and/or a cost for emitting fossil CO_2 as the basis for a viable introduction of CO_2 conversion to products. Without emissions prices or limits, it will be difficult for any such products to compete with fossil-based counterparts. This is also true for new carbon-use cases—for example, in industries like construction materials, where the added value of CO_2 use is largely the durable removal of carbon via long-lived storage products.

The cost of electricity significantly influences the overall cost structure of carbon utilization products, potentially affecting their competitiveness. As carbon pricing drives the economy toward net-zero, low-carbon electricity will become a crucial input for various carbon utilization processes, including DAC and DOC as sources of CO_2. The use of high-carbon electricity or hydrogen sources would negate emissions reductions benefits from carbon utilization. Currently, limited access to low-carbon electricity may slow investment and infrastructure development, but certain products, like aggregates and cured concrete, do not always require electricity for the conversion and are less impacted by its cost.

The success of the production of carbon-based materials from CO_2 or coal waste could also be linked to overall efforts to transition to a globally sustainable future as the detailed discussions in this chapter have demonstrated. Long-lived products will contribute toward a net-zero future as carbon sinks; short-lived products will be integrated in a circular carbon economy that runs without the need to add new fossil carbon. Additionally, entering the field of CO_2 conversion and use of coal waste for durable carbon products can open new markets. An example is the production of graphite and graphene materials, which both are critical materials for electrification but are either largely imported or not even available at scale.

This chapter has presented an overview of key factors that can enable and support the successful market introduction of CO_2-based products as well as competing or prohibiting factors. The overall economic and environmental benefits can be very large, but decisive and sustained action is required from the private and public sectors. All future scenarios will also rely on readily available and economically viable carbon feedstocks. Captured CO_2 may be advantageous in some instances, particularly when obtained from concentrated sources like ethanol plants. However, other scenarios could be better suited to alternative starting materials, such as coal waste, as covered in Section 2.2.3. Section 2.4.1 describes the committee's findings and recommendations on market opportunities and needs for CO_2 and coal waste utilization. Table 2-5 in Section 2.4.2 describes two research needs for CO_2 utilization market opportunities as identified by the committee, and a recommendation to address these needs.

2.4.1 Findings and Recommendations

The preceding discussions of market opportunities and influences led to the following findings and recommendations:

Finding 2-1: Large CO_2 removal opportunities—Potential market volumes for long-lived (durable) materials are very large and can lead to gigatonne carbon removal with coupled economic value. CO_2-derived construction materials and elemental carbon materials have the potential to be used by a variety of industries.

Finding 2-2: CO_2 conversion will be a key contributor to a circular carbon economy—CO_2 conversion to short-lived (circular) chemicals is required to source sufficient carbon for an overall circular carbon economy in the future. A key example is sustainable aviation fuel, for which biomass conversion alone can meet only a portion of demand. Short-lived carbon products decompose into CO_2, and thus the carbon can be used again only after separate capture.

Finding 2-3: Many products can be derived from CO_2 conversion—The key product categories in a future net-zero or net-negative economy include fuels, inorganic building materials, polymers, agrochemicals, chemicals and chemical intermediates, and elemental carbon materials.

Finding 2-4: Inhibiting price premiums for CO_2-based products—The levelized cost parity of carbon abatement favoring CO_2 conversion to most products over sequestration has not yet been reached. Although a net-zero future is assumed, market introduction and growth is often inhibited by substantial price premiums over incumbent materials. CO_2 utilization and storage must be compared based on their net benefits, including costs of mitigation, risk of storage, and durability of products, with nuanced trade-offs between the two options. Emerging "buyers' clubs" are beginning signs of market interest in a transition to CO_2-based manufacturing.

Finding 2-5: Substantial potential for co-benefits—CO_2 conversion to products, particularly co-located capture and conversion, can generate multiple societal benefits. They include the products made, CO_2 recycling or avoided emissions (e.g., at ethanol and cement factories), and potential negative emissions with direct air or direct ocean capture and conversion to long-lived products. For coal waste, combined benefits can be environmental remediation, access to critical minerals, and long-lived carbon products.

Recommendation 2-1: Prioritize co-located capture and conversion, especially for long-lived products that contribute to sequestration goals—In extension to Recommendation 6.1 in the committee's first report, the Department of Energy (DOE) should incentivize development work that produces high-volume valuable goods—for example, construction materials—as a means for carbon removal. DOE should consider prioritizing concerted research, development, and deployment efforts to integrate CO_2 capture and conversion into the portfolio of negative emissions strategies.

Finding 2-6: Infrastructure and supply chains are lacking—A future CO_2 and coal waste manufacturing industry can successfully emerge only when systems-level implications are fully and rigorously evaluated. Such evaluation includes implications for the sourcing of necessary raw materials and zero-carbon electricity and heat, as well as the impact of unintended consequences (e.g., excessive use of energy, loss of jobs, stranded assets). Despite these challenges, an opportunity exists to leverage existing chemical industry and construction materials industry infrastructure.

Finding 2-7: Public perception and understanding—There is an opportunity for DOE to increase carbon management education programs and public understanding.

Finding 2-8: The number of developers is still low—The number of emerging CO_2 conversion companies is growing, but slowly. While interest in CO_2 capture and conversion technologies is increasing and large amounts of capital are available to be deployed, clear policy signals toward a net-zero carbon future are needed for the investment community to expand engagement further.

Recommendation 2-2: Close information gaps—The Department of Energy should support systems-level studies to understand the broader impact of CO_2 conversion on the environment, markets, resource (re-)allocation, and jobs gains and/or losses. Related studies should be conducted to close information gaps to realize market opportunities for CO_2 conversion to (a) meet national needs for carbon products, (b) meet national targets for the transition to carbon neutrality, and (c) evaluate incentives and other policies for effectiveness.

Recommendation 2-3: Public engagement—Carbon management, including CO_2 utilization, is imperative for our future. Thus, the Department of Energy should support the creation and operation of efforts to educate the public about carbon management opportunities, needs, risks, and benefits.

Recommendation 2-4: Drive supporting policies—The Department of Energy should use science-based comparative system-level analysis to inform the creation of procurement incentives, carbon fees, and taxes that are needed to secure access to carbon in a nonfossil carbon future.

TABLE 2-5 Research Agenda for CO_2 Utilization Market Opportunities

Research, Development, and Demonstration Need	Funding Agencies or Other Actors	Basic, Applied, Demonstration, or Enabling	Research Area	Product Class	Long- or Short-Lived	Research Themes	Source
2-A. Understand broader impacts of CO_2 conversion on the environment, resource (re-)allocation, and jobs gains and/or losses.	DOE	Enabling	Societal Impacts	All	Long-lived Short-lived	Environmental and societal considerations for CO_2 and coal waste utilization technologies	Rec. 2-2
2-B. Understand broader impact of CO_2 conversion to (a) meet national needs for carbon products, (b) meet national targets for the transition to carbon neutrality, and (c) evaluate effectiveness of incentives and other policies.	DOE GSA State-level actors	Enabling	Markets Societal Impacts	All	Long-lived Short-lived	Market opportunities	Rec. 2-2

Recommendation 2-2: Close information gaps—The Department of Energy should support system-level studies to understand the broader impact of CO_2 conversion on the environment, markets, resource (re-)allocation, and jobs gains and/or losses. Related studies should be conducted to close information gaps to realize market opportunities for CO_2 conversion to (a) meet national needs for carbon products, (b) meet national targets for the transition to carbon neutrality, and (c) evaluate incentives and other policies for effectiveness.

NOTE: GSA = General Services Administration.

Recommendation 2-5: Certification and standards are needed—The U.S. Environmental Protection Agency, the National Institute of Standards and Technology, and the General Services Administration should develop processes for the certification, permitting, and approval of common CO_2-derived materials and coal waste using a uniform environmental product declaration to standardize and regulate the use of these products. The standards should consider both the life cycle impact and carbon intensity of products. ASTM International should develop building standards that support the use of CO_2-derived materials. These standards should include requirements for regulation at the local level.

Recommendation 2-6: Establish a tracker of reduced embodied carbon markets—To inform on progress made on defossilization efforts, the Department of Energy and the Department of Commerce should track regional, national, and international efforts to introduce CO_2-derived products and their market shares. This could include development of a CarbonStar program to label products based on their carbon intensity, as recommended in the committee's first report.

2.4.2 Research Agenda for Market Opportunities

Table 2-5 presents the committee's research agenda on market opportunities for CO_2 utilization technologies, including research needs (numbered by chapter), and related research agenda recommendations (a subset of research-related recommendations from the chapter). The table includes the relevant funding agencies or other actors; whether the need is for basic research, applied research, technology demonstration, or enabling technologies and processes for CO_2 utilization; the research theme(s) that the research need falls into; the relevant research area and product class covered by the research need; whether the relevant product(s) are long- or short-lived; and the source of the research need (chapter section, finding, or recommendation). The committee's full research agenda can be found in Chapter 11, Table 11-1.

2.5 REFERENCES

Adams, A. 2023. "Controversial Pipeline Canceled amid Safety Concerns, Regulatory Pushback." *Capitol News Illinois*. https://capitolnewsillinois.com/NEWS/controversial-pipeline-canceled-amid-safety-concerns-regulatory-pushback.

Afreen, R., S. Tyagi, G.P. Singh, and M. Singh. 2021. "Challenges and Perspectives of Polyhydroxyalkanoate Production from Microalgae/Cyanobacteria and Bacteria as Microbial Factories: An Assessment of Hybrid Biological System." *Frontiers in Bioengineering and Biotechnology* 9(February). https://doi.org/10.3389/fbioe.2021.624885.

Ali, M.H.M., H.A. Rahman, S.H. Amirnordin, and N.A. Khan. 2018. "Eco-Friendly Flame-Retardant Additives for Polyurethane Foams: A Short Review." *Key Engineering Materials* 791:19–28.

Allegrini, E., M.S. Holtze, and T. Fruergaard Astrup. 2013. "Metal Recovery from Municipal Solid Waste Incineration Bottom Ash (MSWIBA): State of the Art, Potential and Environmental Benefits." 3rd International Slag Valorisation Symposium. https://findit.dtu.dk/en/catalog/537f107b74bed2fd2100d7bb.

AP News. 2023. "Exxon Mobil Buys Denbury, Pipeline Company with Carbon Capture Expertise, for $5 Billion." *AP News*, July 13. https://apnews.com/article/exxon-mobil-denbury-carbon-capture-acquisition-e88462a294693e4139b24d6030ac3c2d.

Arning, K., J. Offermann-van Heek, A. Sternberg, A. Bardow, and M. Ziefle. 2020. "Risk-Benefit Perceptions and Public Acceptance of Carbon Capture and Utilization." *Environmental Innovation and Societal Transitions* 35(June):292–308. https://doi.org/10.1016/j.eist.2019.05.003.

Arning, K., J. Offermann-van Heek, and M. Ziefle. 2021. "What Drives Public Acceptance of Sustainable CO_2-Derived Building Materials? A Conjoint-Analysis of Eco-Benefits vs. Health Concerns." *Renewable and Sustainable Energy Reviews* 144(July):110873. https://doi.org/10.1016/j.rser.2021.110873.

Astor, M. 2022. "Do Airline Climate Offsets Really Work? Here's the Good News, and the Bad." *The New York Times*, May 18. https://www.nytimes.com/2022/05/18/climate/offset-carbon-footprint-air-travel.html.

Barroso, G., Q. Li, R.K. Bordia, and G. Motz. 2019. "Polymeric and Ceramic Silicon-Based Coatings—A Review." *Journal of Materials Chemistry A* 7(5):1936–1963. https://doi.org/10.1039/C8TA09054H.

Bazzanella, A.M., and F. Ausfelder. 2017. "Low Carbon Energy and Feedstock for the European Chemical Industry." *DECHEMA e.V.* https://dechema.de/dechema_media/Downloads/Positionspapiere/Technology_study_Low_carbon_energy_and_feedstock_for_the_European_chemical_industry.pdf.

Bell, L. 2021. "Plastic Waste Management Hazards: Waste-to-Energy, Chemical Recycling, and Plastic Fuels | IPEN." https://ipen.org/documents/plastic-waste-management-hazards-waste-energy-chemical-recycling-and-plastic-fuels.

Bellamy, R., and K.T. Raimi. 2023. "Communicating Carbon Removal." *Frontiers in Climate* 5. https://www.frontiersin.org/articles/10.3389/fclim.2023.1205388.

Bertagni, M.B., R.H. Socolow, J.M.P. Martirez, E.A. Carter, C. Greig, Y. Ju, T. Lieuwen, et al. 2023. "Minimizing the Impacts of the Ammonia Economy on the Nitrogen Cycle and Climate." *Proceedings of the National Academy of Sciences* 120(46):e2311728120. https://doi.org/10.1073/pnas.2311728120.

Bertuol, D.A., C.M. Machado, M.L. Silva, C.O. Calgaro, G.L. Dotto, and E.H. Tanabe. 2016. "Recovery of Cobalt from Spent Lithium-Ion Batteries Using Supercritical Carbon Dioxide Extraction." *Waste Management* 51(May):245–251. https://doi.org/10.1016/j.wasman.2016.03.009.

Beswick, R.R., A.M. Oliveira, and Y. Yan. 2021. "Does the Green Hydrogen Economy Have a Water Problem?" *ACS Energy Letters* 6(9):3167–3169. https://doi.org/10.1021/acsenergylett.1c01375.

Biniek, K., K. Henderson, M. Rogers, and G. Santoni. 2020. "Driving CO_2 Emissions to Zero (and Beyond) with Carbon Capture, Use, and Storage." McKinsey & Company.

Black, S., A.A. Liu, I. Parry, and N. Vernon. 2023. "IMF Fossil Fuel Subsidies Data: 2023 Update." Working Paper WP/23/169. IMF. https://www.imf.org/en/Publications/WP/Issues/2023/08/22/IMF-Fossil-Fuel-Subsidies-Data-2023-Update-537281.

CAETS (International Council of Academies of Engineering and Technological Sciences). 2023. "Towards Low-GHG Emissions from Energy Use in Selected Sectors." Washington, DC.

Capêto, A.P., M. Amorim, S. Sousa, J.R. Costa, B. Uribe, A.S. Guimarães, M. Pintado, and A.L.S. Oliveira. 2024. "Fire-Resistant Bio-Based Polyurethane Foams Designed with Two By-Products Derived from Sugarcane Fermentation Process." *Waste and Biomass Valorization* 15:2045–2059. https://doi.org/10.1007/s12649-023-02274-6.

Carbon Capture Coalition. 2023. "Carbon Capture Coalition Endorses the Bipartisan Captured Carbon Utilization Parity Act." February 28. https://carboncapturecoalition.org/carbon-capture-coalition-endorses-the-bipartisan-captured-carbon-utilization-parity-act.

Carey, P. 2018. "Discussion on TEA for Mature and Semi-Mature Technologies." Presentation to the committee. January 18. Washington, DC: National Academies of Sciences, Engineering, and Medicine.

CATF (Clean Air Task Force). 2023. "Two Years of IIJA: An Overview of Carbon Management Implementation to Date." https://www.catf.us/2023/12/two-years-iija-overview-carbon-management-implementation-date.

Chemanalyst. 2023a. "Carbon Dioxide (CO_2) Market Analysis: Industry Market Size, Plant Capacity, Production, Operating Efficiency, Demand and Supply, End-User Industries, Sales Channel, Regional Demand, Foreign Trade, Company Share, Manufacturing Process, 2015–2035." https://www.chemanalyst.com/industry-report/carbon-dioxide-market-630.

Chemanalyst. 2023b. "Methanol Market Size, Growth, Share, Analysis and Forecast, 2032." https://www.chemanalyst.com/industry-report/methanol-market-219.

Chemanalyst. 2023c. "Urea Market Size, Growth, Share, Analysis and Forecast, 2032." https://www.chemanalyst.com/industry-report/urea-market-666.

Chemnick, J. 2023. "The Carbon Removal Project That Puts Communities in the Driver's Seat." *E&E News Climatewire*. https://www.eenews.net/articles/the-carbon-removal-project-that-puts-communities-in-the-drivers-seat.

Circular Carbon Network. 2022. "Circular Carbon Market Report." https://circularcarbon.org/wp-content/uploads/2023/03/CCN-2022-MarketReport.pdf.

Corbyn, Z. 2021. "From Pollutant to Product: The Companies Making Stuff from CO2." *The Guardian*, December 5. https://www.theguardian.com/environment/2021/dec/05/carbon-dioxide-co2-capture-utilisation-products-vodka-jet-fuel-protein.

de Wit, M., J. Hoogzaad, and C. von Daniels. 2020. *The Circularity Gap Report*. Circle Economy. https://www.circularity-gap.world/2020

Dean, J.A., ed. 1999. *Lange's Handbook of Chemistry*. 15th ed. McGraw-Hill.

DoD (Department of Defense). 2023. "Department of Defense Operational Energy Strategy." F-42E610F. https://www.acq.osd.mil/eie/Downloads/OE/2023%20Operational%20Energy%20Strategy.pdf.

DOE (Department of Energy). 2011. "Technology Readiness Assessment Guide." DOE G 413.3-4A. Approved 9-15-2011. https://www.directives.doe.gov/directives-documents/400-series/0413.3-EGuide-04a-admchg1/@@images/file. Washington, DC: Department of Energy.

DOE-BETO (Bioenergy Technologies Office). 2023. "2023 Billion-Ton Report: An Assessment of U.S. Renewable Carbon Resources." https://www.energy.gov/eere/bioenergy/2023-billion-ton-report-assessment-us-renewable-carbon-resources.

DOE-EERE (Office of Energy Efficiency and Renewable Energy). 2022. "Carbon Capture, Transport, and Storage Assessment." https://www.energy.gov/sites/default/files/2022-02/Carbon%20Capture%20Supply%20Chain%20Report%20-%20Final.pdf.

DOE-LPO (Loans Program Office). n.d. "Carbon Dioxide Transportation Infrastructure." https://www.energy.gov/lpo/carbon-dioxide-transportation-infrastructure.

EIA (U.S. Energy Information Administration). 2024. "Weekly U.S. Product Supplied of Finished Motor Gasoline." https://www.eia.gov/dnav/pet/hist/LeafHandler.ashx?n=PET&s=WGFUPUS2&f=W.

Elevitch, C.R., D.N. Mazaroli, and D. Ragone. 2018. "Agroforestry Standards for Regenerative Agriculture." *Sustainability* 10(9):3337.

Energy Transitions Commission. 2021. "Bioresources Within a Net-Zero Emissions Economy: Making a Sustainable Approach Possible." Energy Transitions Commission. https://www.energy-transitions.org/publications/bioresources-within-a-net-zero-economy/#download-formwww.bio-based.eu/nova-papers.

Engelmann, L., K. Arning, A. Linzenich, and M. Ziefle. 2020. "Risk Assessment Regarding Perceived Toxicity and Acceptance of Carbon Dioxide-Based Fuel by Laypeople for Its Use in Road Traffic and Aviation." *Frontiers in Energy Research* 8. https://www.frontiersin.org/articles/10.3389/fenrg.2020.579814.

Environmental Integrity Project. 2019. "Coal's Poisonous Legacy Groundwater Contaminated by Coal Ash Across the U.S." https://environmentalintegrity.org/wp-content/uploads/2019/03/National-Coal-Ash-Report-Revised-7.11.19.pdf.

Faber, G., and V. Sick. 2022. "Identifying and Mitigating Greenwashing of Carbon Utilization Products." Global CO_2 Initiative. University of Michigan.

Fortune Business Insights. 2023a. "Cement Market Size, Share, Growth | Emerging Trends [2032]." *Fortune Business Insights*. https://www.fortunebusinessinsights.com/industry-reports/cement-market-101825.

Fortune Business Insights. 2023b. "Concrete Blocks and Bricks Market Size, Share | Growth [2029]." *Fortune Business Insights*. https://www.fortunebusinessinsights.com/concrete-blocks-and-bricks-market-103784.

Gabrielli, P., L. Rosa, M. Gazzani, R. Meys, A. Bardow, M. Mazzotti, and G. Sansavini. 2023. "Net-Zero Emissions Chemical Industry in a World of Limited Resources." *One Earth* 6(6):682–704. https://doi.org/10.1016/j.oneear.2023.05.006.

GAO (U.S. Government Accountability Office). 2022. "Decarbonization: Status, Challenges, and Policy Options for Carbon Capture, Utilization, and Storage." GAO-22-105274. https://www.gao.gov/products/gao-22-105274.

Gao, P., S. Li, X. Bu, S. Dang, Z. Liu, H. Wang, L. Zhong, et al. 2017. "Direct Conversion of CO_2 into Liquid Fuels with High Selectivity Over a Bifunctional Catalyst." *Nature Chemistry* 9(10):1019–1024. https://doi.org/10.1038/nchem.2794.

Gao, X., T. Atchimarungsri, Q. Ma, T.-S. Zhao, and N. Tsubaki. 2020. "Realizing Efficient Carbon Dioxide Hydrogenation to Liquid Hydrocarbons by Tandem Catalysis Design." *EnergyChem* 2(4):100038. https://doi.org/10.1016/j.enchem.2020.100038.

Gassenheimer, C., and C. Shaynak. 2023. "Coal Waste Recovery Presentation." Presentation to the committee. November 3. Washington, DC: National Academies of Sciences, Engineering, and Medicine.

Goodman, W., and J. Minner. 2019. "Will the Urban Agricultural Revolution Be Vertical and Soilless? A Case Study of Controlled Environment Agriculture in New York City." *Land Use Policy* 83:160–173.

Grand View Research. 2022. "Gypsum Board Market Size and Share Analysis Report, 2030." 978-1-68038-722–3. Grand View Research. https://www.grandviewresearch.com/industry-analysis/gypsum-board-market.

Grand View Research. 2023. "Carbon Dioxide Market Size, Share and Trends Analysis Report by Form (Solid, Liquid, Gas), by Source (Ethyl Alcohol, Ethylene Oxide), by Application (Food and Beverages, Oil and Gas, Medical), by Region, and Segment Forecasts, 2024–2030." https://www.grandviewresearch.com/industry-analysis/carbon-dioxide-market.

Granite, E., G. Bromhal, J. Wilcox, and M.A. Alvin. 2023. "Domestic Wastes and Byproducts: A Resource for Critical Material Supply Chains." *The Bridge* 53(3):59–66.

Greenfield, P. 2021. "Carbon Offsets Used by Major Airlines Based on Flawed System, Warn Experts." *The Guardian*, May 4. https://www.theguardian.com/environment/2021/may/04/carbon-offsets-used-by-major-airlines-based-on-flawed-system-warn-experts.

Grignard, B., S. Gennen, C. Jérôme, A.W. Kleij, and C. Detrembleur. 2019. "Advances in the Use of CO 2 as a Renewable Feedstock for the Synthesis of Polymers." *Chemical Society Reviews* 48(16):4466–4514. https://doi.org/10.1039/C9CS00047J.

Grim, R.G., J.R. Ferrell III, Z. Huang, L. Tao, and M.G. Resch. 2023. "The Feasibility of Direct CO_2 Conversion Technologies on Impacting Mid-Century Climate Goals." *Joule* 7(8):1684–1699. https://doi.org/10.1016/j.joule.2023.07.008.

Hepburn, C., E. Adlen, J. Beddington, E.A. Carter, S. Fuss, N. Mac Dowell, J.C. Minx, P. Smith, and C.K. Williams. 2019. "The Technological and Economic Prospects for CO_2 Utilization and Removal." *Nature* 575(7781):87–97. https://doi.org/10.1038/s41586-019-1681-6.

Huang, Z., R.G. Grim, J.A. Schaidle, and L. Tao. 2021. "The Economic Outlook for Converting CO_2 and Electrons to Molecules." *Energy and Environmental Science* 14(7):3664–3678. https://doi.org/10.1039/D0EE03525D.

IEA (International Energy Agency). 2018. "The Future of Petrochemicals." Paris: IEA. https://www.iea.org/reports/the-future-of-petrochemicals.

IEA. 2019. "Putting CO_2 to Use." 2019. Paris: IEA. https://www.iea.org/reports/putting-co2-to-use.

IEA. 2020. "CCUS in Clean Energy Transitions." Paris: IEA. https://www.iea.org/reports/ccus-in-clean-energy-transitions.

IEA. 2021. "Net Zero by 2050: A Roadmap for the Global Energy Sector." Paris: IEA. https://iea.blob.core.windows.net/assets/deebef5d-0c34-4539-9d0c-10b13d840027/NetZeroby2050-ARoadmapfortheGlobalEnergySector_CORR.pdf.

IEA. 2023. "How New Business Models Are Boosting Momentum on CCUS." Paris: IEA. https://www.iea.org/commentaries/how-new-business-models-are-boosting-momentum-on-ccus.

Inwald, J.F., W. Bruine de Bruin, M. Yaggi, and J. Árvai. 2023. "Public Concern About Water Safety, Weather, and Climate: Insights from the World Risk Poll." *Environmental Science and Technology* 57(5):2075–2083. https://doi.org/10.1021/acs.est.2c03964.

Jacobson, R., and M. Lucas. 2018. "Carbontech—a Trillion Dollar Opportunity." *Carbon180*. https://carbon180.medium.com/carbontech-a-trillion-dollar-opportunity-154a9c62cf1c.

Kähler, F., O. Porc, and M. Carus. 2023. "RCI Carbon Flows Report: Compilation of Supply and Demand of Fossil and Renewable Carbon on a Global and European Level." Renewable Carbon Initiative. https://renewable-carbon.eu/publications/product/the-renewable-carbon-initiatives-carbon-flows-report-pdf.

Kausar, A., S. Zulfiqar, and M.I. Sarwar. 2014. "Recent Developments in Sulfur-Containing Polymers." *Polymer Reviews* 54(2):185–267. https://doi.org/10.1080/15583724.2013.863209.

Kazimierczuk, K., S.E. Barrows, M.V. Olarte, and N.P. Qafoku. 2023. "Decarbonization of Agriculture: The Greenhouse Gas Impacts and Economics of Existing and Emerging Climate-Smart Practices." *ACS Engineering Au* 3(6):426–442.

Kirby, P. 2023. "Italy Bans Lab-Grown Meat in Nod to Farmers." *BBC News*, November 11. https://www.bbc.com/news/world-europe-67448116.

Lange, J.-P. 2021. "Towards Circular Carbo-Chemicals—the Metamorphosis of Petrochemicals." *Energy and Environmental Science* 14(8):4358–4376. https://doi.org/10.1039/D1EE00532D.

LanzaTech. 2023. "Biological CO_2 Fixation." Presentation to the committee. September 13. Washington, DC: National Academies of Sciences, Engineering, and Medicine.

Lebling, K., H. Leslie-Bole, and Z. Byrum. 2022. "6 Things to Know About Direct Air Capture." *World Resources Institute*. https://www.wri.org/insights/direct-air-capture-resource-considerations-and-costs-carbon-removal.

Li, N., L. Mo, and C. Unluer. 2022. "Emerging CO_2 Utilization Technologies for Construction Materials: A Review." *Journal of CO_2 Utilization* 65(November):102237. https://doi.org/10.1016/j.jcou.2022.102237.

Linstrom, P.J., and W.G. Mallard, eds. n.d. *NIST Chemistry WebBook*. NIST Standard Reference Database Number 69. Gaithersburg MD: National Institute of Standards and Technology. https://doi.org/10.18434/T4D303.

Lutzke, L., and J. Árvai. 2021. "Consumer Acceptance of Products from Carbon Capture and Utilization." *Climatic Change* 166(1):15. https://doi.org/10.1007/s10584-021-03110-3.

Luu, P. 2023. "What Is Chemical Recycling, Why Does It Have So Many Different Names, and Why Does It Matter?" *Closed Loop Partners*, August 15. https://www.closedlooppartners.com/what-is-chemical-recycling-why-does-it-have-so-many-different-names-and-why-does-it-matter.

Ma, K., Q. Bai, L. Zhang, and B. Liu. 2016. "Synthesis of Flame-Retarding Oligo(Carbonate-Ether) Diols via Double Metal Cyanide Complex-Catalyzed Copolymerization of PO and CO_2 Using Bisphenol A as a Chain Transfer Agent." *RSC Advances* 6(54):48405–48410. https://doi.org/10.1039/C6RA07325E.

Mallapragada, D.S., Y. Dvorkin, M.A. Modestino, D.V. Esposito, W.A. Smith, B.-M. Hodge, M.P. Harold, et al. 2023. "Decarbonization of the Chemical Industry Through Electrification: Barriers and Opportunities." *Joule* 7(1):23–41. https://doi.org/10.1016/j.joule.2022.12.008.

Martín, A.J., G.O. Larrazábal, and J. Pérez-Ramírez. 2015. "Towards Sustainable Fuels and Chemicals Through the Electrochemical Reduction of CO_2: Lessons from Water Electrolysis." *Green Chemistry* 17(12):5114-5130.

Mason, F., G. Stokes, S. Fancy, and V. Sick. 2023. "Implications of the Downstream Handling of Captured CO_2." *Frontiers in Climate* 5(September):1286588. https://doi.org/10.3389/fclim.2023.1286588.

Maureen, V. 2023. "Mechanical Recycling." *Plastic Smart Cities*, July 28. https://plasticsmartcities.org/mechanical-recycling.

Mertens, J., C. Breyer, K. Arning, A. Bardow, R. Belmans, A. Dibenedetto, S. Erkman, et al. 2023. "Carbon Capture and Utilization: More Than Hiding CO_2 for Some Time." *Joule* 7(3):442–449. https://doi.org/10.1016/j.joule.2023.01.005.

Mishra, A., J.N. Ntihuga, B. Molitor, and L.T. Angenent. 2020. "Power-to-Protein: Carbon Fixation with Renewable Electric Power to Feed the World." *Joule* 4(6):1142–1147. https://doi.org/10.1016/j.joule.2020.04.008.

Müller, L.J., A. Kätelhön, S. Bringezu, S. McCoy, S. Suh, R. Edwards, V. Sick, et al. 2020. "The Carbon Footprint of the Carbon Feedstock CO_2." *Energy and Environmental Science* 13(9):2979–2992. https://doi.org/10.1039/D0EE01530J.

NASEM (National Academies of Sciences, Engineering, and Medicine). 2019. *Gaseous Carbon Waste Streams Utilization: Status and Research Needs*. Washington, DC: The National Academies Press. https://doi.org/10.17226/25232.

NASEM. 2022. *A Research Strategy for Ocean-Based Carbon Dioxide Removal and Sequestration*. Washington, DC: The National Academies Press. https://doi.org/10.17226/26278.

NASEM. 2023a. *Accelerating Decarbonization in the United States: Technology, Policy, and Societal Dimensions*. Washington, DC: The National Academies Press. https://doi.org/10.17226/25931.

NASEM 2023b. *Carbon Dioxide Utilization Markets and Infrastructure: Status and Opportunities: A First Report*. Washington, DC: The National Academies Press. https://doi.org/10.17226/26703.

Neelis, M., M. Patel, K. Blok, W. Haije, and P. Bach. 2007. "Approximation of Theoretical Energy-Saving Potentials for the Petrochemical Industry Using Energy Balances for 68 Key Processes." *Energy* 32(7):1104–1123. https://doi.org/10.1016/j.energy.2006.08.005.

NETL (National Energy Technology Laboratory). 2023. "2023 FECM/NETL Carbon Management Research Project Review Meeting—Carbon Conversion—Proceedings." https://netl.doe.gov/23CM-CC-proceedings.

Newman, A.J.K., G.R.M. Dowson, E.G. Platt, H.J. Handford-Styring, and P. Styring. 2023. "Custodians of Carbon: Creating a Circular Carbon Economy." *Frontiers in Energy Research* 11. https://www.frontiersin.org/articles/10.3389/fenrg.2023.1124072.

Nitopi, S., E. Bertheussen, S.B. Scott, X. Liu, A.K. Engstfeld, S. Horch, B. Seger, et al. 2019. "Progress and Perspectives of Electrochemical CO_2 Reduction on Copper in Aqueous Electrolyte." *Chemistry Review* 119(12):7610–7672. https://doi.org/10.1021/acs.chemrev.8b00705.

Northern Lights. 2022. "Northern Lights Awards Ship Management Contract to "K" LINE." https://norlights.com/news/northern-lights-awards-ship-management-contract-to-k-line.

OECD (Organisation for Economic Co-operation and Development). 2022. *Global Plastics Outlook: Policy Scenarios to 2060*. Paris: OECD Publishing. https://doi.org/10.1787/aa1edf33-en.

OECD-FAO (Food and Agriculture Organization of the United Nations). 2021. "OECD-FAO Agricultural Outlook 2021–2030." OECD Publishing, Paris. https://doi.org/10.1787/19428846-en.

Oil and Gas Climate Initiative. 2023. "Building Towards Net Zero. Progress Report." https://3971732.fs1.hubspotusercontent-na1.net/hubfs/3971732/230804_OGCI_ProgressReport2023_V2.pdf.

Olson, D.W., and T. Capehart. 2019. "Dried Distillers Grains (DDGs) Have Emerged as a Key Ethanol Coproduct." *USDA Economic Research Service*. https://www.ers.usda.gov/amber-waves/2019/october/dried-distillers-grains-ddgs-have-emerged-as-a-key-ethanol-coproduct.

Oxy. 2023. "Occidental Enters into Agreement to Acquire Direct Air Capture Technology Innovator Carbon Engineering." https://www.oxy.com/news/news-releases/occidental-enters-into-agreement-to-acquire-direct-air-capture-technology-innovator-carbon-engineering.

Pander, B., Z. Mortimer, C. Woods, C. McGregor, A. Dempster, L. Thomas, J. Maliepaard, R. Mansfield, P. Rowe, and P. Krabben. 2020. "Hydrogen Oxidising Bacteria for Production of Single-cell Protein and Other Food and Feed Ingredients." *Engineering Biology* 4(2):21–24. https://doi.org/10.1049/enb.2020.0005.

Park, A.-H.A., J.M. Williams, J. Friedmann, D. Hanson, S. Kawashima, V. Sick, M.R. Taha, and J. Wilcox. 2024. "Challenges and Opportunities for the Built Environment in a Carbon-Constrained World for the Next 100 Years and Beyond." *Frontiers in Energy Research* 12:1388516. https://doi.org/10.3389/fenrg.2024.1388516.

Patrizio, P., M. Fajardy, M. Bui, and N. Mac Dowell. 2021. "CO2 Mitigation or Removal: The Optimal Uses of Biomass in Energy System Decarbonization." *IScience* 24(7):102765. https://doi.org/10.1016/j.isci.2021.102765.

Petkowski, J.J., W. Bains, and S. Seager. 2020. "On the Potential of Silicon as a Building Block for Life." *Life* 10(6). https://doi.org/10.3390/life10060084.

Qiu, Y., P. Lamers, V. Daioglou, N. McQueen, H.-S. de Boer, M. Harmsen, J. Wilcox, A. Bardow, and S. Suh. 2022. "Environmental Trade-Offs of Direct Air Capture Technologies in Climate Change Mitigation Toward 2100." *Nature Communications* 13(1):3635. https://doi.org/10.1038/s41467-022-31146-1.

Ranevska, S. 2023. "Trailblazer Pipeline Company Approved for CO_2 Transportation." *Carbon Herald*, November 1. https://carbonherald.com/trailblazer-pipeline-company-approved-for-co2-transportation.

Ravikumar, D., G. Keoleian, and S. Miller. 2020. "The Environmental Opportunity Cost of Using Renewable Energy for Carbon Capture and Utilization for Methanol Production." *Applied Energy* 279(December):115770. https://doi.org/10.1016/j.apenergy.2020.115770.

Renewable Carbon Initiative. 2022. "Renewable Carbon as a Guiding Principle for Sustainable Carbon Cycles." *Renewable Carbon Initiative*. https://renewable-carbon.eu/publications/product/renewable-carbon-as-a-guiding-principle-for-sustainable-carbon-cycles-pdf.

Roach, J. 2023. "Microsoft Lays Foundation for Green Building Materials of Tomorrow." *Microsoft News*. https://news.microsoft.com/source/features/sustainability/low-carbon-building-materials-for-datacenters.

Saeidi, S., S. Najari, V. Hessel, K. Wilson, F.J. Keil, P. Concepción, S.L. Suib, and A.E. Rodrigues. 2021. "Recent Advances in CO_2 Hydrogenation to Value-Added Products—Current Challenges and Future Directions." *Progress in Energy and Combustion Science* 85(July 1):100905. https://doi.org/10.1016/j.pecs.2021.100905.

SDG Global Council on Future Fuels. 2023. "Neither a Greenwasher nor a Greenhusher Be: A Guide to High Integrity Corporate Climate Action." Volans.

Shaw, W.J., M.K. Kidder, S.R. Bare, M. Delferro, J.R. Morris, F.M. Toma, S.D. Senanayake, et al. 2024. "A US perspective on Closing the Carbon Cycle to Defossilize Difficult-to-Electrify Segments of Our Economy." *Nature Reviews Chemistry* 8:376–400. https://doi.org/10.1038/s41570-024-00587-1.

Sick, V. 2018. "Global Roadmap Study for CO_2U Technologies." Global CO_2 Initiative. University of Michigan.

Sick, V. 2020. "Using CO_2 as an Industrial Feedstock Could Change the World. Here's How." *World Economic Forum*. https://www.weforum.org/agenda/2020/01/co2-as-industrial-feedstock.

Sick, V., G. Stokes, and F.C. Mason. 2022a. "CO_2 Utilization and Market Size Projection for CO_2-Treated Construction Materials." *Frontiers in Climate* 4(May):878756. https://doi.org/10.3389/fclim.2022.878756.

Sick, V., G. Stokes, and F.C. Mason. 2022b. "Implementing CO_2 Capture and Utilization at Scale and Speed." Presented at the Global CO_2 Initiative, University of Michigan. https://dx.doi.org/10.7302/5825.

Sillman, J., L. Nygren, H. Kahiluoto, V. Ruuskanen, A. Tamminen, C. Bajamundi, M. Nappa, et al. 2019. "Bacterial Protein for Food and Feed Generated via Renewable Energy and Direct Air Capture of CO2: Can It Reduce Land and Water Use?" *Global Food Security* 22(September):25–32. https://doi.org/10.1016/j.gfs.2019.09.007.

SkyQuest Technology. 2024. "Asphalt Market Share and Growth Analysis—Industry Forecast 2030." SkyQuest Technology. https://www.skyquestt.com/report/asphalt-market.

St. John, J. 2024. "US Steel Plant in Indiana to Host a $150M Carbon Capture Experiment." *Canary Media*, April 3. https://www.canarymedia.com/articles/carbon-capture/us-steel-plant-in-indiana-to-host-a-150m-carbon-capture-experiment.

Statista Research Department. 2023. "Annual Fuel Consumption by Ships Worldwide from 2019 to 2020, by Fuel Type." https://www.statista.com/statistics/1266963/amount-of-fuel-consumed-by-ships-worldwide-by-fuel-type.

Stoffa, J. 2023. "Carbon Conversion Program Overview." Presentation to the committee. June 28. Washington, DC: National Academies of Sciences, Engineering, and Medicine.

Straits Research. 2022. "Iron Oxide Pigments Market Size." *Straits Research*. https://straitsresearch.com/report/iron-oxide-pigments-market.

Styring, P., E.A. Quadrelli, and K. Armstrong, eds. 2014. "Carbon Dioxide Utilization: Closing the Carbon Cycle." Elsevier.

Šyc, M., F.G. Simon, J. Hykš, R. Braga, L. Biganzoli, G. Costa, V. Funari, and M. Grosso. 2020. "Metal Recovery from Incineration Bottom Ash: State-of-the-Art and Recent Developments." *Journal of Hazardous Materials* 393(July 5):122433. https://doi.org/10.1016/j.jhazmat.2020.122433.

Thielges, S., B. Olfe-Kräutlein, A. Rees, J. Jahn, V. Sick, and R. Quitzow. 2022. "Committed to Implementing CCU? A Comparison of the Policy Mix in the US and the EU." *Frontiers in Climate* 4. https://www.frontiersin.org/articles/10.3389/fclim.2022.943387.

Toeniskoetter, C. 2022. "Lab-Grown Meat Receives Clearance from F.D.A." *The New York Times*, November 17. https://www.nytimes.com/2022/11/17/climate/fda-lab-grown-cultivated-meat.html.

Uekert, T., A. Singh, J.S. DesVeaux, T. Ghosh, A. Bhatt, G. Yadav, S. Afzal, et al. 2023. "Technical, Economic, and Environmental Comparison of Closed-Loop Recycling Technologies for Common Plastics." *ACS Sustainable Chemistry and Engineering* 11(3):965–978. https://doi.org/10.1021/acssuschemeng.2c05497.

U.S. Naval Research Laboratory. 2012. "US Navy Research Aims to Produce Fuel from Sea Water." *Membrane Technology* 2012(10):10. https://doi.org/10.1016/S0958-2118(12)70212-4.

Valencia, F.M., C. Mohren, and A. Ramakrishnan. 2023. "Public Support for Carbon Pricing Policies and Different Revenue Recycling Options: A Systematic Review and Meta-Analysis of the Survey Literature." https://doi.org/10.21203/rs.3.rs-3528188/v1.

Van Loo, E.J., V. Caputo, and J.L. Lusk. 2020. "Consumer Preferences for Farm-Raised Meat, Lab-Grown Meat, and Plant-Based Meat Alternatives: Does Information or Brand Matter?" *Food Policy* 95(August):101931. https://doi.org/10.1016/j.foodpol.2020.101931.

West, T.A.P., J. Börner, E.O. Sills, and A. Kontoleon. 2020. "Overstated Carbon Emission Reductions from Voluntary REDD+ Projects in the Brazilian Amazon." *Proceedings of the National Academy of Sciences* 117(39):24188–24194. https://doi.org/10.1073/pnas.2004334117.

Wolske, K.S., K.T. Raimi, V. Campbell-Arvai, and P.S. Hart. 2019. "Public Support for Carbon Dioxide Removal Strategies: The Role of Tampering with Nature Perceptions." *Climatic Change* 152(3–4):345–361. https://doi.org/10.1007/s10584-019-02375-z.

World Economic Forum. 2023. "First Movers Coalition Aviation Commitment." https://www3.weforum.org/docs/WEF_First_Movers_Coalition_Aviation_Commitment_2023.pdf.

World Economic Forum. 2024. "First Movers Coalition." https://initiatives.weforum.org/first-movers-coalition/home.

Yu, K., W. McGee, T.Y. Ng, H. Zhu, and V.C. Li. 2021. "3D-Printable Engineered Cementitious Composites (3DP-ECC): Fresh and Hardened Properties." *Cement and Concrete Research* 143(May):106388. https://doi.org/10.1016/j.cemconres.2021.106388.

Zhang, X., A. Zhang, X. Jiang, J. Zhu, J. Liu, J. Li, G. Zhang, C. Song, and X. Guo. 2019a. "Utilization of CO_2 for Aromatics Production Over ZnO/ZrO_2-ZSM-5 Tandem Catalyst." *Journal of CO_2 Utilization* 29(January 1):140–145. https://doi.org/10.1016/j.jcou.2018.12.002.

Zhu, H., K. Yu, W. McGee, T.Y. Ng, and V.C. Li. 2021. "Limestone Calcined Clay Cement for Three-Dimensional-Printed Engineered Cementitious Composites." *ACI Materials Journal* 118(6). https://doi.org/10.14359/51733109.

3

Life Cycle, Techno-Economic, and Societal/ Equity Assessments of CO_2 Utilization Processes, Technologies, and Systems

3.1 INTRODUCTION

CO_2 utilization technologies and infrastructure will incur environmental impacts, potentially different from or differently distributed than those from existing technologies and infrastructure for chemicals and materials production. Developers, funders, and policy makers will consider environmental justice needs associated with their deployment and use. Thus, as CO_2 utilization technologies are being developed, decision-support tools are important to ascertain not just economic competitiveness but also the environmental sustainability and equity outcomes associated with their deployment. Such tools can quantify and inform the progress of emerging CO_2 utilization technologies and strategies for reaching net zero and a circular carbon economy. This chapter reviews methodologies and requirements to assess the contribution of CO_2 utilization to a net-zero emissions future—namely, techno-economic assessment (TEA), life cycle assessment (LCA), and societal/equity assessments.

TEA, LCA, and equity assessments provide vital information for decision makers considering the potential impacts of CO_2 utilization projects and related infrastructure, including implications on sustainability; they are broadly recognized as critical tools for evaluating existing and emerging technologies (Moni et al. 2019). TEA informs the economic viability of technologies, providing valuable information regarding the competitiveness of a product. LCA quantifies the environmental burdens associated with products, processes, or services from materials extraction through end of life. LCA and TEA can identify areas needing further attention and are an integral component of applied research and development (R&D) (Lettner and Hesser 2019), often used in a combined toolkit. Equity assessments seek to minimize unintended adverse outcomes while maximizing opportunities and positive outcomes from policies, programs, or processes, particularly for those facing inequity or disparities (Bradley et al. 2022b; Cremonese et al. 2022). Life cycle thinking can inform equity assessments to ensure that what appears to be a more benign choice does not result in unintended burden-shifting across the value chain. Social LCA (s-LCA) is a particular type of LCA aimed at establishing the human and societal impacts of a product's life cycle. These assessments are complementary: TEA and LCA were not designed to robustly examine questions of equity, and equity assessments do not capture the suite of impacts quantified in LCA and TEA. The Department of Energy (DOE) requires LCA and TEA to be completed as part of applied R&D programs (Skone et al. 2022) and is operationalizing environmental justice, energy justice, and equity into its CO_2 utilization R&D program in line with the Biden administration's

Justice40 initiative (Clark 2023; E.O. 14008).[1] Federal tools (Argonne National Laboratory's Greenhouse Gases, Regulated Emissions, and Energy Use in Technologies LCA models and DOE Office of Technology Transitions' Commercial Adoption Readiness Assessment Tool) and databases (the National Renewable Energy Laboratory Life Cycle Inventory and the National Energy Technology Laboratory Unit Process Library) exist to support these requirements. However, only certain programs within DOE require TEA and LCA estimates to be reported to program managers as part of the funding proposal and iteratively improved through project completion. This inconsistency can result in contradictory information and data gaps.

Carbon utilization technologies range in their technological development from early technology readiness level (TRL)[2] through pilot, demonstration, and full commercialization (IEA 2021; Sick et al. 2022). Because many CO_2 utilization technologies have yet to operate at commercial scale, LCA, TEA, and equity assessments are inherently challenging, and decision makers question how to gain useful insights into these earlier TRL technologies compared to those that are reaching demonstration and beyond (Cremonese et al. 2022; Goglio et al. 2020; Langhorst et al. 2022). Key issues facing LCA and TEA of early-stage technologies include paucity of data, incomplete understanding of scale up, a lack (in some cases) of comparator or proxy data, and uncertainty in how the technology will be deployed and in future market conditions. Recent studies (Langhorst et al. 2022; Newman et al. 2023, Figure 3) have shown demonstrable progress in managing these challenges by providing detailed guidance on how to streamline TEA-LCA, conduct analyses on emerging technologies, and interpret the results (Sick et al. 2023). Present efforts to improve global guidance have been limited aside from AssessCCUS, which includes collaborators from the United States, Canada, the United Kingdom, Germany, and Japan.

The committee's first report considered the full life cycle of CO_2 utilization processes when accounting for net CO_2 emissions from projects (NASEM 2023a). In doing so, the committee considered the durability of the carbon storage (product lifetime of greater or less than 100 years) alongside whether the carbon source was atmospheric, oceanic, biogenic, or fossil, noting that these factors determine whether CO_2 utilization is capable of being net-negative, net-zero, or net-positive in emissions. The committee also identified carbon accounting across the value chain as an important component of assessing sustainability for CO_2 utilization technologies, recommending that DOE "fund research to quantify the dynamic impact of CO_2-derived products, for example, their specific lifetime, on the CO_2 balance in the atmosphere" and that the United States "incorporate knowledge acquired from European projects and regulatory activities in addressing circular carbon economies and net-negative emissions" (Recommendation 3.1, NASEM 2023a). DOE also should support national laboratories, academia, and industry in performing TEA, LCA, and integrated systems analysis to identify CO_2 utilization approaches that are technologically feasible, sustainable, economically viable, and take into account relevant regulatory and policy frameworks, environmental justice impacts, and factors that may influence societal acceptance (Recommendation 6.1, NASEM 2023a).

This chapter reviews the current state of knowledge of LCA, TEA, and equity assessments for CO_2 utilization technology and infrastructure, providing results from recent studies as well as an overall synthesis of how these tools can inform decisions across TRLs, in concept, research, scale up, deployment, and diffusion. The chapter examines the challenges faced in recent studies, pointing to where methodological and data improvements are necessary to ensure that results can provide reliable decision-support. It assesses progress of CO_2 utilization technologies in meeting net-zero goals by reviewing current LCA and TEA results and identifying opportuni-

[1] The Justice40 initiative has the goal of ensuring that 40 percent of the overall benefits of certain federal investments go toward disadvantaged communities. Categories covered by Justice40 include climate change, clean energy and energy efficiency, clean transit, affordable and sustainable housing, training and workforce development, remediation and reduction of legacy pollution, and the development of critical clean water and wastewater infrastructure (E.O. 14008).

[2] TRL defines stages from basic research to maturity for full-scale market introduction. The stages begin with basic research, where principles are observed and reported, technology is conceived, and applications formulated. They proceed through research to prove feasibility, technology development, technology demonstration, and system commissioning, and end in system operations, where an actual system is operated in the final form over the full range of operating conditions. (See Table 2-3 in Chapter 2, which is modified from DOE [2011], for more detail about each TRL.)

ties from LCA, TEA, and equity assessments to improve decision-support for circularity, R&D investments, and infrastructure build-out.

3.1.1 Technology Assessment Approaches

Technology assessment aims to anticipate the current and future performance of a (novel) technology by integrating knowledge of the benefits and risks of emerging technologies, contributing to the formation of public and political opinion, and supplying effective, pragmatic, and sustainable options for decision making (UNCTAD 2022). In the case of CO_2 capture and utilization technologies, such assessments are far from straightforward, owing to both technology and market factors. Technology factors include the inherent uncertainties associated with low TRL; the time horizon of weeks to centuries in which environmental impacts are foreseen, depending on the lifetime of the final product and other factors; and the different impacts of interest, including global warming, and land and water use. Factors associated with the markets include the changing needs of society, such as shifts and upgrades in material end-use application and performance; the large and diverse number of actors in the value chains; the differences in scales at which technologies could be deployed (from niche to large markets); and the potential for CO_2 utilization technologies to disrupt the fossil-based incumbent. This heterogeneity strongly affects the future usage and performance of CO_2 utilization technologies. For example, producing sustainable aviation fuels from CO_2 requires large amounts of hydrogen and energy, both ideally sourced without CO_2 emissions. Furthermore, production facilities will be quite similar to petrochemical plants and refineries, and thus public opposition might become a factor in deployments.

Technology assessment transcends a purely technical and scientific evaluation, requiring a broad evaluation of the environmental, economic, and societal context and impacts of a technology. Holistic evaluation relies on three interconnected types of assessments:

- **Techno-Economic Assessment (TEA)** is an integrated assessment of the technical performance and economic feasibility of a technology. It combines process modeling and engineering design with an economic evaluation to assess the (future) viability of a process, system configuration, or product, also projecting if it will change the competitive landscape. As emerging technologies are often noncompetitive with incumbent technologies at their inception, understanding key drivers of performance, uncertainties associated with initial cost estimates, and potential pathways to improved performance are key results of early-stage TEAs (Buchner et al. 2018).
- **Environmental Life Cycle Assessment (e-LCA)** is an analytical tool that aims to provide insight into the environmental performance of technologies, processes, products, or services. It allows a comprehensive and systematic comparison of systems (e.g., a new technology versus a baseline condition) in order to identify potential shifting of environmental burdens across different environmental life cycle phases (e.g., raw material extraction, production, end use) or environmental compartments (e.g., soil, water, air). e-LCA is the most common form of LCA and often what is being referred to when "LCA" is used.
- **Equity and Social Impact Assessment** aims to support and advance societally equitable outcomes resulting from the deployment of a technology, product, or service. It provides information for decision makers about different forms of equity, such as intergenerational or temporal (i.e., equity across time), spatial or geographical (equity across space), racial or ethnic, socioeconomic, gender, cultural, and democratic (decision-making) equity (Parson and Mottee 2023). There are different approaches to conducting equity and societal impact assessment, including social life cycle assessment (s-LCA), which builds on e-LCA methodology and aims to quantify the impact of a product or process on society along the full life cycle. In contrast with e-LCA, the methodological development, application, and harmonization of s-LCA is still at a preliminary stage (McCord et al. 2023; Sala et al. 2015).

Box 3-1 defines terminology used in technology assessments.

BOX 3-1
Terminology

- **Prospective Versus Ex-Ante Technology Assessment:** *Prospective* assessment is the assessment of potential impacts as a consequence of deploying a technology in the future—for instance, the impacts of deploying a technology in 2050. *Ex-ante* assessment examines the impacts of a future technology—that is, it tries to identify the potential impacts of a technology that is still in the research and development (R&D) phase (the technology is not commercially available). Note that while ex-ante assessments are prospective in nature, not all prospective assessments are ex-ante.
- **Uncertainty and Sensitivity Analyses:** Uncertainty and sensitivity analyses are two approaches for evaluating models. Differences between the results of a model and observed values (in the real world) can result from either natural variability, known and unknown errors in the input data, the model parameters, or the model itself, owing to, for instance, lack of knowledge or (over)simplifications of reality. *Uncertainty analyses* aim to quantify the uncertainty in the outputs of a model that result from model assumptions and uncertainty in model input values. They reflect the uncertainty in the conclusions of the study. *Sensitivity analyses* try to identify how variation in input values relate to variation in output measures, with the most sensitive variables producing the most variation and hence the most significant opportunities for improvement of the model to reduce model uncertainty.
- **Environmental Impact Assessment (EIA) Versus Life Cycle Assessment (LCA) Versus Environmental Risk Assessment (ERA):** EIA and LCA are two approaches to assessing environmental impacts. *LCA* analyzes the environmental impacts of a product, process, or system throughout its entire life cycle, from raw material extraction (cradle) to end of life (grave). An *EIA* assesses the potential environmental impacts of proposed projects, policies, or plans at a *specific* location or area. It focuses on the potential impacts during the construction, operation, and decommissioning phases of the project (gate-to-gate)[a] and is generally conducted as part of the licensing of a project. A component of an EIA is an *Environmental Risk Assessment*, which assesses the nature and magnitude of *health* risks to humans and ecosystems at a local level. Approaches to combine ERA and LCA for technologies at low TRL (TRL 1 to 6) have started to appear in the literature (e.g., Tan et al. 2018; Van Harmelen et al. 2016). Although these studies do not focus on CO_2 utilization, they showcase that the combination of both approaches can provide valuable insights for technology developers and decision makers (Hauschild et al. 2022; Subramanian and Guinée 2021).
- **Attributional Versus Consequential LCA:** The two main types of LCA. *Consequential LCA* aims to assess the direct and indirect burdens that occur on the environment as a consequence of a possible decision (e.g., changes in the demand for a technology, process, or service). *Attributional LCA* focuses on providing information regarding how the impacts on the environment can be at-

Goal and scope definition is a common initial stage for both TEA and LCA. For integrated TEA-LCA, alignment in the goal, scope, data, and system elements provide consistency in results (Langhorst et al. 2022, Part E; Mahmud et al. 2021). Different goals lead to different comparisons, with varying data requirements and inventory creation efforts (inputs and outputs to TEA-LCA are compiled in inventories that document data for the analyses in line with the goal and scope definition). Conversely, the inventory also impacts the goal, especially if data are not available. Importantly, the assessment goal is specific to the individual study and the practitioner's perspective. Even when focusing on the same product system, the assessment goal can vary between studies, depending on factors such as the scope and size of the project, technological maturity, geographical region, and time horizon.

The *goal* for a TEA is to address techno-economic questions, such as the cost or potential profitability of a new technology, process, product, plant, or project. TEA is often carried out for a specific audience (e.g., assessment of a CO_2 utilization reaction concept for a funding agency, assessment of a CO_2 utilization plant concept for industry managers, assessment of CO_2 utilization technology options for policy makers). The goal informs the details needed to define the scope, and the goal and scope together then frame all subsequent work phases of the study. The TEA goal

tributed to a specific part of the life cycle of a technology process or service. The assessments answer different questions relevant to different types of stakeholders. Attributional LCAs are often used by, for example, technology developers or industrial decision makers, as such LCAs allow these actors to understand which parts of their value chains contribute the most to a given environmental impact. Consequential LCAs are often used by, for example, policy makers who are interested in the impacts of a decision, such as a policy to increase the share of a given product, at the macro level.

- **Static Versus Dynamic LCA:** Most LCAs are conducted in a static (or traditional) way, where all emissions are assumed to occur simultaneously, and consequently they are modeled as one pulse when calculating climate impacts. However, impacts can be dependent on the time of the emission. This aspect can be addressed by conducting Dynamic LCA. In this approach, the temporal profiles of emissions are included so that the result for each emission is a function of time in the estimation of climate impacts (Levasseur et al. 2010). Dynamic LCA is relatively less developed than traditional LCA and requires understanding of the changes over time regarding emission profiles both in the foreground and in the background (e.g., changes in electricity mix, infrastructure).
- **Footprint Analysis:**
  - **Ecological footprint:** Estimation of the resource consumption and waste assimilation requirements of land use (Dincer and Zamfirescu 2018).
  - **Product carbon footprint:** The effect of a product's greenhouse gas (GHG) emissions on Earth's climate (Ecochain 2023).
  - **Product environmental footprint:** Assessment of the environmental impacts of a product's chemical emissions and resource depletion (Ecochain 2023).
- **TEA and Life Cycle Costing (LCC):** TEA and LCC are two methodological frameworks to assess systematically the economic viability of a technology, with the main difference between them being the system boundaries used. While TEA typically adopts cradle-to-gate system boundaries, LCC covers all life cycle stages of a project, including cost during R&D and disposal phases. LCC increasingly includes environmental costs.
- **Hotspot Assessment and Analysis:** Hotspots are points, areas, or steps in the value chains of a product where significant environmental, cost, or equity problems may arise as a consequence of deploying a technology. Hotspot analysis is a methodological framework that incorporates information from various sources to identify hotspots and propose and prioritize actions to address impacts, often as a precursor to more detailed assessments (Barthel et al. 2015).

[a] In assessments, various terms are used as shorthand to describe the system boundaries of the assessment. For example, an assessment may include a "cradle-to-gate" boundary, meaning that it is associated with the activities taking place from materials extraction to the factory gate (i.e., before transportation to the consumer). Other terms include "cradle," from the origin of the item in question, or "grave" at its end of life (Bjørn et al. 2018).

also interacts with the subsequent work phase of data inventory creation (i.e., a listing of all relevant components, cost). The TEA *scope* describes what aspects of a product (or service) will be assessed and how a product (e.g., a plastic) or service (e.g., energy storage) will be compared to competing solutions. The first step of the TEA scope phase is to identify and describe the analyzed product system; the central elements are precise, quantitative descriptions of the function(s) of the new technology and selection of the comparison metrics in the form of functional units and reference flows that are needed to meet the functional unit. The next step is to specify the analyzed system elements and define system boundaries, followed by selecting benchmark systems for comparison. Last, the technological maturity of the product system will be used to select suitable assessment indicators. Since operational data are not available for technologies in development, one CO_2 utilization-specific challenge with the scoping phase is that many CO_2-derived products provide similar but nonidentical performance to benchmark products.

As noted by the International Organization for Standardization (ISO) 14044 standard for LCA, the goal and scope stages determine the intended application of the study, the reasons for carrying out the study, the intended audience of the study and whether the results are to be used in comparative assertions disclosed to public (European Committee for Standardization 2022a). These stages of LCA are carried out similarly to TEA, but with an environmental focus that includes identifying impacts of interest for the analysis.

The *system boundary* defines the limits of the product system and describes which system elements belong to it. For example, details of producing and delivering raw material streams—for example, CO_2, could be included in the assessment to analyze limitations or opportunities for improvements, or they could be treated as fixed input. More detail on transportation-specific considerations can be found in the committee's first report (NASEM 2023a). Material and energy flows crossing the system boundary are referred to as "input flows" and "output flows." A product system can have one or multiple input or output flows (e.g., coproducts or by-products, waste streams, various feedstocks, and various inputs for waste treatment); the latter often are referred to as multifunctional product systems or as having "multifunctionality." System boundaries can be defined for product systems and comparative benchmark systems and are derived from the assessment goal and product functions. System boundaries allow for a transparent and process-based comparison of the product and benchmark systems. They set the basis for reviewing what is included in a LCA or TEA study and for comparing different studies with each other. System boundaries must be consistent throughout the study.

The system boundaries, geographic and temporal representativeness, and choice of functional unit are key decisions that can impact LCA results of CO_2 utilization technologies (Hauschild et al. 2018). As for TEA, the system boundaries determine what elements will be included in the analysis (e.g., capture and transport of CO_2; inputs such as the required energy, hydrogen, and other raw materials; through plant decommissioning) (NASEM 2019).

3.1.2 Importance of Evaluating Carbon Utilization Systems and Individual Products and Processes in a Net-Zero Future

As discussed in Chapter 2, the economic potential for CO_2-derived products is substantial. The latest report of the Intergovernmental Panel on Climate Change (IPCC) indicated that "In order to reach net zero CO_2 emissions for the carbon needed in society (e.g., plastics, wood, aviation, fuels, solvents), it is important to close the use loops for carbon and carbon dioxide" (Bashmakov et al. 2022, p. 1163). The need for chemicals, fuels, and materials cannot be sustainably achieved without modifying their production, use, and disposal. CO_2 utilization technologies are a class of options to reduce dependence on nonrenewable resources while contributing to a net-zero future. However, the potential to remove CO_2 from the atmosphere or avoid emissions in the first place depends on the source of CO_2, the process and product that would be displaced, and the duration of storage (Mason et al. 2023). Schematically, Figure 3-1 summarizes the best-possible high-level outcomes of utilizing captured CO_2. These outcomes, which reflect the impact of deploying a CO_2 utilization technology on ambient concentrations of CO_2 in the atmosphere, are critical to decision making. The National Academies' report *Gaseous Carbon Waste Streams Utilization* summarized key considerations for performing LCA on different types of CO_2 utilization products (NASEM 2019, Table 8-2), a frame through which the technologies and products discussed in Chapters 5–8 can be viewed.

TEA, LCA, and equity assessments provide valuable feedback across TRLs as CO_2 utilization technologies move from proof-of-concept through prototype to eventual deployment. Owing to the wide range of outcomes, TEA and LCA can offer critical input to the design of early-stage technologies. The tools provide different types of decision-support at different stages of technological advancement (Figure 3-2). At early stages of technology readiness, results from LCA, TEA, and equity assessments are much more uncertain with much less information to understand actual operations and localized impacts. Recognizing these results as uncertain, they nonetheless can offer valuable insights into strategic investments in applied R&D to increase competitiveness or decrease negative impacts of the design. As technology advances toward deployment, additional spatial, temporal, and operational data will enable more accurate results for decision making. After substantial technical changes are made or more accurate data is obtained, repeated assessments will allow practitioners to obtain updated information and guidance for decisions on whether and how to proceed.

3.2 TECHNO-ECONOMIC ASSESSMENTS FOR CARBON UTILIZATION

TEAs are routinely conducted for CO_2 utilization technologies across TRLs, resulting in a wide range of published estimates. The committee reviewed recent TEAs and reviews of published literature to document the stage for the state of the art in the field. Early-stage technologies are still in development; therefore, such estimates

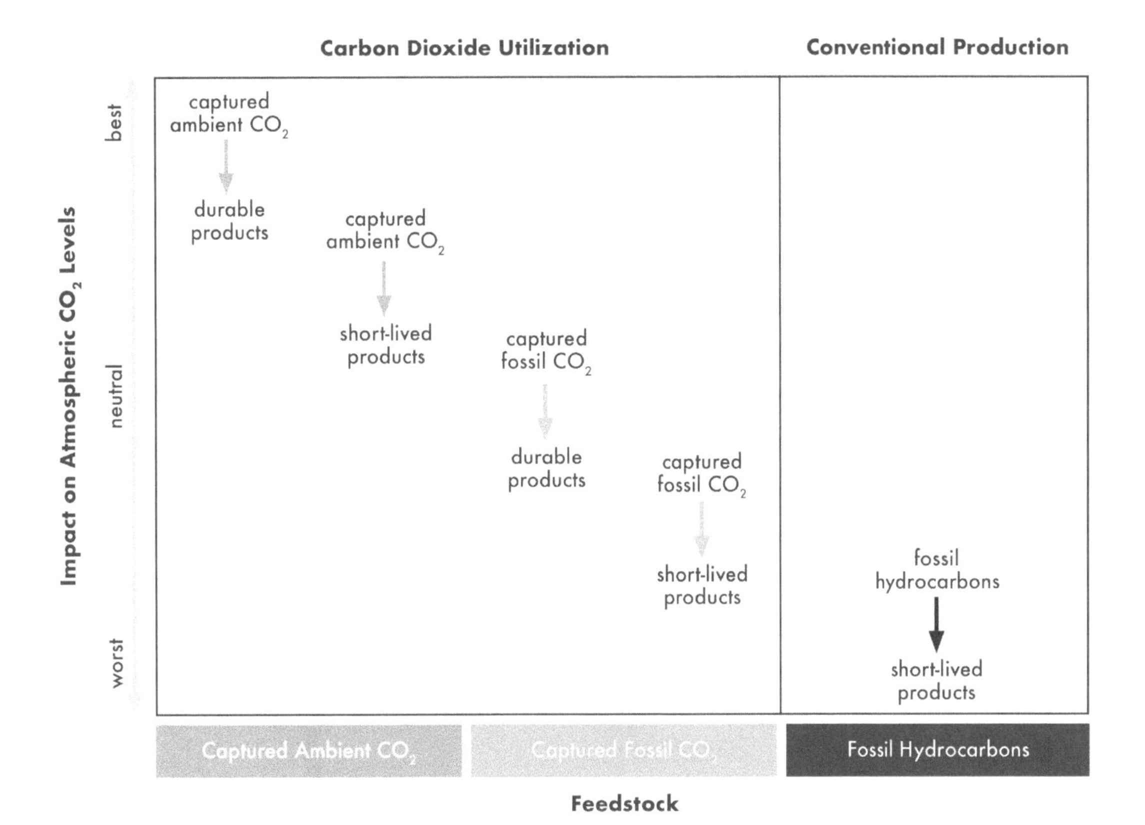

FIGURE 3-1 Impact of the product life cycle on ambient CO_2 levels is dependent on the CO_2 source and the lifetime of the CO_2-derived product.
NOTES: Assuming all other aspects of the life cycle emissions are equal, the best impact is associated with captured ambient CO_2 used as feedstock to make durable products. The worst CO_2 utilization option is use of captured fossil CO_2 as feedstock to make short-lived products. Worse than all carbon utilization options is conventional production of short-lived products directly from fossil hydrocarbons.

will change as technological advances are accomplished. Figure 3-3 shows the different components of a TEA for assessment of a future CO_2 utilization plant, illustrating how data and uncertainties in the performance of the plant's core technology (a CO_2 electrolyzer, which is at low TRL) impact the design and cost evaluation of the full plant (Vos et al. 2024). Although the figure is specific to low-temperature electrolyzers, the building blocks of the TEA are relevant for any technology. The representativeness of data inputs and levels of uncertainty in results for low-TRL technologies have implications for other process units within the system boundaries (e.g., the need to purify the streams). In turn, assumptions of technology performance prior to deployment influence expected material and equipment requirements, and subsequently the energy requirements, which are at the core of the cost estimations (e.g., levelized cost of production, payback time, or net present value).

Despite their inherent uncertainty, the results of early TEAs provide insights that can be used to shape the future development and deployment of a technology. For example, a review of production cost estimates for electrocatalytic conversion of CO_2 to different products showed that the technology is not generally competitive with fossil fuel alternatives, except possibly in a select few cases. At the time of this review, market prices for carbon monoxide and formate were \$0.18 and \$0.66/kg (Grim et al. 2020),[3] while production costs from CO_2 utilization technologies were estimated to be \$0.39 (0.18–0.64) and \$0.96 (0.10–2.63)/kg, respectively (Jordaan and Wang 2021, ranges in parentheses are minimum and maximum values).[4] Production cost estimates will evolve over time

[3] The carbon monoxide price is the average for 2014–2018, and the formate price is the average of 2014 and 2016.
[4] Study estimates refer to 2014–2022.

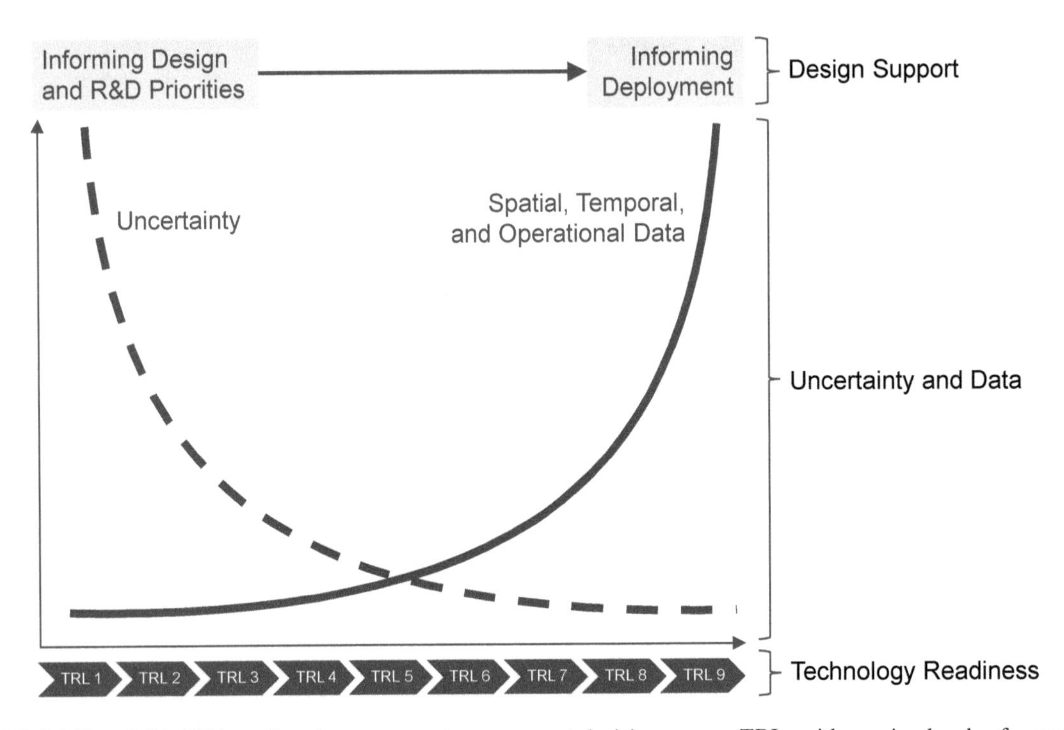

FIGURE 3-2 How LCA, TEA, and equity assessments can support decisions across TRLs with varying levels of uncertainty and data.
NOTE: The dashed line describes how uncertainty changes as TRLs increase and the solid line describes how available data on technologies change as TRLs increase.

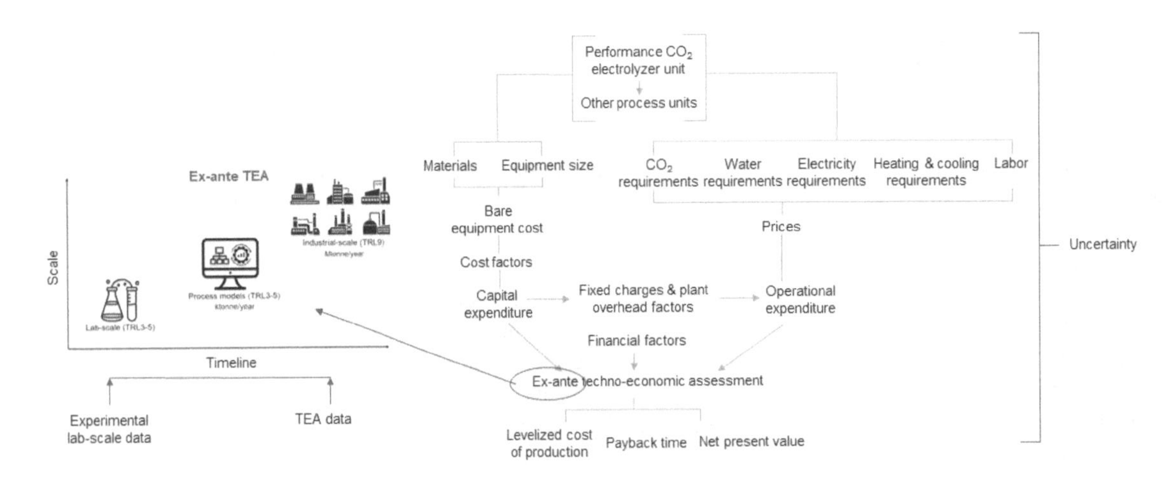

FIGURE 3-3 Visual overview of the methodological building blocks that comprise an ex-ante techno-economic assessment of a future low-temperature CO_2 electrolysis-based plant. For a low-TRL technology like a CO_2 electrolysis-based plant, data and uncertainties about the performance impact the design and cost evaluation.
SOURCE: Vos et al. (2024).

with technological advances and depend on the scope of the analysis, markets, prices, and local and geographical factors (Jordaan and Wang 2021). Such estimates provide value in showing the difference between published production costs and market prices, as well as by pointing to how variable estimates can be, depending on technologies, design, and assumptions. If the gap between production costs and market prices becomes narrow or nonexistent, a more comprehensive TEA can help identify the entry point for a first-of-a-kind (FOAK) plant. TEA estimates for FOAK plants will be less competitive than *n*th-of-a-kind (NOAK) plants, as further discussed in Section 3.2.3.3.

TEA provides critical information about how a CO_2 conversion technology can be competitive under specific technological and market conditions, informing both R&D and policy. For example, results can indicate how much the production costs must decrease to break even with present market prices (Ruttinger et al. 2022). A recent study completed consistent TEAs across five major electricity-driven CO_2 conversion technologies (low- and high-temperature electrolysis, microbial electrosynthesis, biological and thermochemical conversion with electrolytic hydrogen) and 11 distinct carbonaceous products, leveraging results to compare the minimum selling prices (i.e., required price to break even with production costs) to market prices (Huang et al. 2021; see Figure 3-4). Only one product, polyhydroxybutyrate, had a minimum selling price lower than market prices, but several other products show promise if technological advancements are realized (Figure 3-4). Such estimates will evolve over time as technology advances, as market conditions change, and if certain technologies are deployed in specific geographies. Low prices of renewable electricity and a price on carbon can play a substantial role in encouraging deployment as technologies approach competitiveness.

Owing to the evolving nature of these technologies and their estimated costs, the committee developed additional guidance on conducting TEAs across TRLs to produce valuable results for informing applied R&D through deployment.

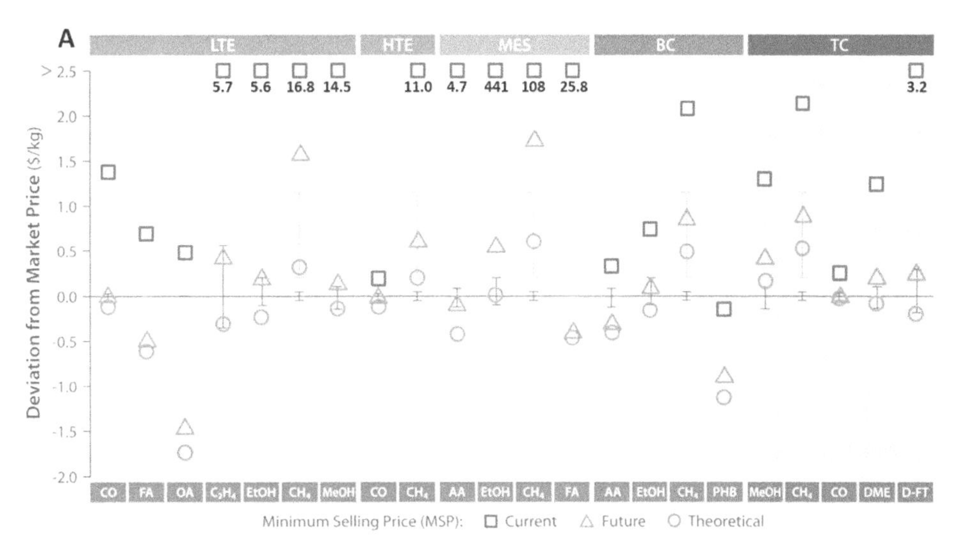

FIGURE 3-4 Minimum selling prices (i.e., required price to break even with production costs) across 11 carbonaceous products from CO_2 utilization—carbon monoxide (CO), formic acid (FA), oxalic acid (OA), ethylene (C_2H_4), ethanol (EtOH), methane (CH_4), methanol (MeOH), acetic acid (AA), polyhydroxybutyrate (PHB), dimethyl ether (DME), and Fischer–Tropsch liquids (FT)—produced by different reduction pathways—low-temperature electrolysis (LTE), high-temperature electrolysis (HTE), microbial electrosynthesis (MES), biological conversion (BC), or thermochemical conversion (TC). The current scenario includes a $40/tonne price on CO_2 and $0.068/kWh electricity prices. The future scenario assumes technological advancement with $20/tonne CO_2 and $0.03/kWh. The theoretical scenario assumes operations achieve thermodynamic limitations and/or best-case assumptions with prices on CO_2 and electricity as $0/tonne and $0.02/kWh, respectively.
SOURCE: Used with permission of the Royal Society of Chemistry from Z. Huang, R.G. Grim, J.A. Schaidle, and L. Tao, 2021, "The Economic Outlook for Converting CO_2 and Electrons to Molecules," *Energy & Environmental Science* 14(7):3664–3678; permission conveyed through Copyright Clearance Center, Inc.

3.2.1 What TEA Can and Cannot Do in Assessing CO_2 Utilization Technologies

TEAs usually are performed with a focus on the production phase and from a producer's point of view, but they can be expanded to include up- or downstream aspects—for example, the use and disposal phases of a product, as is done in LCC. Doing so might be useful if such aspects have an outsized impact on cost, which may be especially salient for CO_2 utilization technologies where accounting for the fate of the carbon in the product is important.

While related, assessment and decision making are separate efforts. A TEA is based generally on information from process design, and the results can be used as feedback or recommendations for improved design. However, TEA does not include technical development activities, such as chemical process design, but it builds on and feeds back into them. The studies are context-specific with respect to factors such as location, time horizon, or access to information and thus require specific assumptions in order to be meaningful. TEA can support project-specific decision making in both technological and economic contexts, such as specific R&D work or investment decisions. Reliably applying TEA results in a generalized context, such as for global policy making, can be quite challenging and therefore requires exercising considerable caution. At its core, TEA is about single-actor costs and how to minimize them, with some consideration of how this relates to the market of the product or function.

3.2.2 Use of TEA for Technologies at Different TRLs

Appropriate use of TEA results to direct R&D and deployment can be challenging for decision makers in industry and policy when indicators (e.g., net present value) are estimated using inconsistent methods or applied outside the intended context. Objective comparative analysis may not be possible without consistent and systematic methodology (Zimmermann and Schomäcker 2017). To resolve these challenges, the Global CO_2 Initiative led the development of community-directed harmonized guidelines for TEA and LCA specifically for CO_2 utilization (Langhorst et al. 2022). Opinions differ as to the usefulness of conducting TEA for TRLs below 3, and thus a TRL-dependent approach to TEA will be important (Buchner et al. 2018). Assessments for low TRL necessarily will be very limiting, driven by the lack of sufficiently accurate data, technology details, and information about potential applications, and can provide only high-level insights. Although cost estimation for TRL 1 is not advisable, a qualitative assessment could be conducted and used to recommend technology pathways. At TRL 2, limited quantitative assessments will be possible on mass, energy, and value efficiencies for general guidance. At higher TRLs, progressively more information will be available to allow the inclusion of more accurate data for additional elements that impact the cost of goods sold. Applying TEA results conducted at the earliest TRLs may limit innovation in basic science.

3.2.3 Critical Issues for CO_2 Utilization TEA

The level of detail available in a TEA depends on the TRL of the technology. Limitations in data availability and/or their respective accuracy are a key issue at low TRL; often relevant information has not been obtained in early-stage laboratory experiments (Buchner et al. 2018). Furthermore, projecting the equipment needs from bench-scale operation to scale up production processes to pilot-plant and full-scale deployment sizes is challenging. Capital expenditure estimation will be sensitive to the kind of equipment used and the estimation methods employed—for example, the *AACE International Cost Estimate Classification System*.[5] The greater the specificity in output information required, the higher the assessment effort in terms of time, complexity, detail, and work needed to gather data. In many cases, a technology assessment will include a multicriteria decision analysis in the interpretation phase (Chauvy et al. 2020; McCord et al. 2021), necessary for examining trade-offs and helping to prioritize the selection of options. Multiple-attribute decision making is useful for choosing between a set of specific and preselected variables that determine alternative solutions for the purpose of ideally narrowing down to one choice. By contrast, multiple-objective decision making is useful for a set of variables that will produce an infinite number of solutions—the Pareto group (with solutions on the Pareto frontier being considered optimal; Marler and Arora 2004).

[5] See AACE International, "Homepage," https://web.aacei.org, accessed August 5, 2024.

3.2.3.1 Transparency and Consistency Challenges with System Boundaries

The system boundaries in TEA and LCA are often inconsistent; most TEAs are gate-to-gate studies, while LCAs have broader system boundaries. This discrepancy has implications for the analysis of results. Differences in system boundaries can be a significant determinant of estimated CO_2 avoidance costs. To illustrate this point, Tanzer et al. (2023) assessed the impact of system boundaries in a biological carbon capture and storage (CCS) case study. The study used biochar to replace coal as clinker kiln fuel in a cement plant with CCS and included the uptake of CO_2 by concrete over time. System boundary choice varied net CO_2 estimates from −660 to +16 kg CO_2(eq)/t cement, and aligning boundaries shrank the avoidance cost range from 48–321 to 157–193 €/t CO_2(eq) (Tanzer et al. 2023).

3.2.3.2 CO_2 Purity

The CO_2 stream will contain multiple source-dependent impurities that may negatively impact the performance and costs of CO_2 utilization technologies. As indicated in Table 3-1, electrocatalytic and thermocatalytic routes are very sensitive to impurities. Impurities can impact electroreduction in several ways—for instance, they can compete with CO_2 for electrons or adsorb on or react with the catalyst surface and deleteriously modify its properties (Harman and Wang 2022). For conventional thermocatalytic processes, a major challenge is the presence of impurities such as H_2S, NH_3, carbonyl sulfide and alkali halides, which can result in catalyst deactivation owing to carbon deposition, sintering, pore blockage, and sulfur poisoning (Pattnaik et al. 2022). More on CO_2 stream purification can be found in Chapter 11, Section 11.1.2, noting that transport has less stringent requirements for purity than conversion. Appendix H contains further tables from the committee's first report with information on typical CO_2 stream impurities from different sources and CO_2 purity requirements for different transportation modes.

The cost to purify CO_2 streams depends on several factors, including process efficiency, operating conditions, energy requirements, safety considerations, and whether there is experience with similar applications in the industry. As such, impurities can significantly alter the business case. Vos et al. (2023) conducted a TEA of the pretreatment units needed to purify CO_2 from a bioethanol plant then used in a solid-oxide electrolysis unit to produce syngas; compositions and tolerance assumptions are shown in Table 3-2. The TEA indicates high capital expenditure (Capex) (almost 3 million euros in bare equipment costs) and energy costs, the former driven by units removing sulfur and alcohols and the latter mostly driven by the cryogenic distillation step used to remove noncondensable gases.

TABLE 3-1 Overview of Impurities of Concern by CO_2 Utilization Route

CO_2 Utilization Route	Required Purity	Impurities of Concern
Mineralization	Low	Most processes can work directly with flue gas if desired.
Biological conversion (anaerobic)	Low to medium	High tolerance to impurities except for oxygen.
Thermochemical conversion	High to very high	Heavy metals, sulfur, nitrogen, and carbon can poison the catalyst.
Electrochemical conversion	Very high	Heavy metals and sulfur (SO_2, H_2S, COS) at ppm levels can damage the electrochemical reactor.
CO_2 with food-grade purity[a]	Very high	Carbon monoxide, hydrocarbons, and metals (CO_2: >99 percent; H_2O: <2 ppm; CO: <10 ppm; C_xH_y: <50 ppmv; oil: <10 ppmw; and passing tests for acidity and red substances).

[a] Although use of CO_2 in the food and beverage industry is out-of-scope for this report, these purity levels have been included as a point of comparison, because many CO_2 utilization studies assume that the input CO_2 is food-grade purity. Purity levels are taken from EIGA (2018).

NOTE: ppm = parts-per-million; ppmv = parts-per-million by volume; ppmw = parts-per-million by weight.

SOURCE: Modified from NASEM (2023a).

TABLE 3-2 Composition of the CO_2 Stream by Component from a Bioethanol Plant, Electrolytic Reactor Degradation Type, and Tolerance Limits for Solid-Oxide Electrolysis

Component	ppm (except CO_2, wt%)	Degradation Type	Tolerable Amount	Unit Used
CO_2	90%[a]	C-poisoning		
CO	1	C-poisoning		
H_2O	5			
CH_4	3	C-poisoning		
SO_x	1	S-poisoning	<2	ppm
NO_x	1			
O_2	100		<5	%
H_2S	1	S-poisoning	~0.05	ppm
N_2	98,768			
Ar	—		~100	µmol/mol
Heavy metals	—	Metals	—	ppb level
Cl	—	Cl-poisoning	<5	ppm
Alcohols	3–950			
Other hydrocarbons	1	C-poisoning	~2	µmol/mol
Aromatics (benzene, toluene, xylene)	3			
Carbonyl sulfide	1	S-poisoning	<2	ppm
Dimethylsulfide	1	S-poisoning	<2	ppm
Ethers	1			

[a] All amounts in column 2 are in ppm, except for CO_2, which is given in percent.
SOURCE: Adapted from Vos et al. (2023).

To date, most experimental studies of CO_2 utilization technologies have used near-pure CO_2. There is a lack of understanding of the potential impact of impurities or partially converted feedstocks in technology performance and of mitigation strategies that go beyond developing more efficient pretreatment of streams. Such strategies can include developing more resilient or more easily regenerated catalysts and/or new approaches/reactions where impurities can act as promoters or useful reactants (Harman and Wang 2022). When industrial configurations are considered, potential impurities created in the conversion process are important to the downstream separations associated with recycle streams (Sarswat et al. 2022).

3.2.3.3 Scalability

Scaling a new technology to cost-competitive, industrial-scale production capacity requires upscaling from bench-top experiments to a full-size plant. The plant setup has to be scaled up physically to allow industrial-scale production volumes to meet market demands. Matching the capacity of the CO_2 source and the utilization plant would avoid unnecessary CO_2 transportation. This scale up is not simply making everything physically larger; it usually requires significant changes in plant design to achieve the necessary mass flows, temperature and pressure conditions, and more.[6] Scale up also requires integration of utilities and supporting processes such as heating and cooling to maximize the energy efficiency of an operation, often not restricted to the operation of one manufacturing process, but more frequently for the operation of a collection of manufacturing processes on a site

[6] Some processes do not follow rules of scale, such as those dependent on surface area for photo-driven reactions. Although the photo-driven reactions may not follow economies of scale, the downstream processes required to purify the products to market requirements tend to do so.

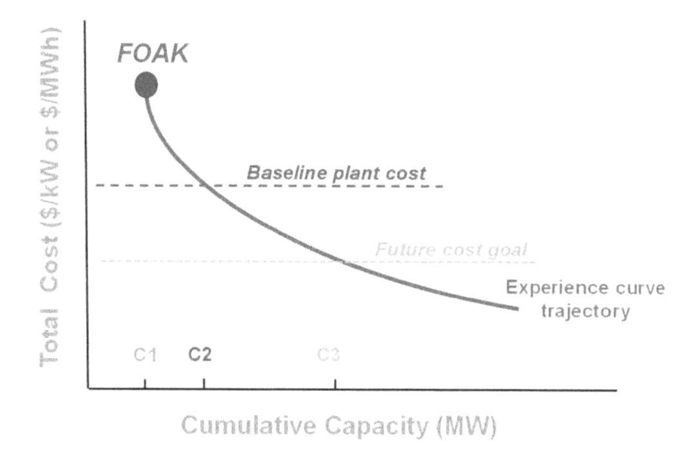

FIGURE 3-5 Illustrative cost trajectory of an advanced technology from first-of-its-kind to mature plant.
SOURCE: Roussanaly et al. (2021), https://doi.org/10.2172/1779820. CC BY 4.0.

to reuse heat. The resulting FOAK plants are more costly than NOAK plants, as indicated in Figures 3-5 and 3-6, owing to overdesign of equipment, redundancy of equipment to ensure that the plant will operate as desired, and design size limits or nonstandard material use in early-stage equipment.

Learning curves provide information about the speed at which costs would decrease in relation to the cumulative installed capacity. The FOAK cost and installed capacity, as well as learning rate are needed to estimate the shape of the learning curve (Figure 3-5). At the FOAK stage, learning rates are not yet established, and therefore data from other technologies are often used. For example, IEAGHG (2023) uses chlor-alkali and polymer electrolyte membrane (PEM)

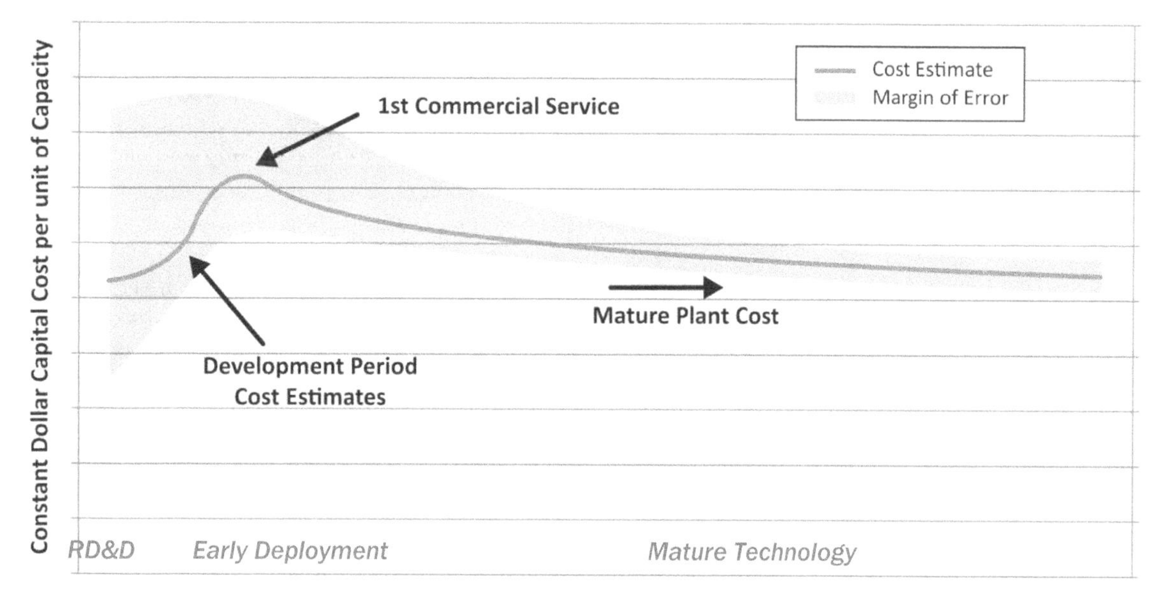

FIGURE 3-6 Typical costs trajectory and uncertainties of a novel technology.
SOURCE: Roussanaly et al. (2021), https://doi.org/10.2172/1779820. CC BY 4.0.

fuel cells to estimate learning for low- temperature CO_2 electroconversion and a solid oxide fuel cell for high-temperature CO_2 electroconversion. Incorporating learning rates allows Capex data from a FOAK plant to be used to derive cost projections for a NOAK plant (Greig et al. 2014; Rubin 2014, 2016; Van der Spek 2017).

Full-scale NOAK plants require process improvements and cost reductions both in terms of Capex to build the plants and operational expenses. Once multiple plants are in operation with the same or similar technologies, learning curve strategies for process improvements and cost reductions can be employed.

3.2.3.4 Geographic and Temporal Relevance

Local conditions can have a significant impact on the feasibility and viability of a new technology. Overall supply chain factors—for example, raw material availability, cost, transportation modes, and product distribution channels—are highly context-specific. An International Energy Agency (IEA) study found significant differences in the median projected costs of electricity generation technologies among regions and stressed that technology competitiveness depends on national and local conditions (IEA 2020b). To account for such differences, location factors can be used to convert construction costs of industrial plants among countries. Although these factors reflect average differences and values may differ within countries, they provide more realistic comparisons of TEAs across regions. As an example, the location factors published by Intratec for 2018 indicated that building a plant in China would be 16 percent less expensive than building a plant in the United States, while building the same plant in Germany would be 7 percent more expensive (Intratec Solutions 2023). These differences are not owing to exchange rates but rather reflect differences in labor costs, infrastructure availability, and so on.

Therefore, factoring regionality into assessments enables TEA to address the impact of local policy support or inhibitions and to quantify competing demands for the same resources (Jiang et al. 2020; Kähler et al. 2023; Ravikumar et al. 2020; Thielges et al. 2022). Alternative carbon sources also might experience strong seasonal variations in availability and cost, especially if derived from biomass, or might require critical raw materials. TEA uses commodity prices of chemicals and materials, which can be highly volatile; therefore, the year used for the analysis significantly affects the result. The price of steel, for example, quadrupled between 2020 and 2021 and remains at least 50 percent higher than in 2020 (Trading Economics 2024).

3.2.3.5 Addressing Uncertainty

Models used in TEA and LCA describe mathematical relationships between input variables and the desired output, which is any result or indicator of interest for a reference base case and additional scenarios that are crucial for the subsequent decision. Uncertainty and sensitivity analyses are key elements in assessments like TEA and LCA and need to be examined to put results in quantitative context. Uncertainty and sensitivity analyses are distinctly different in how they are conducted and what they mean, but are often confused with each other (see Box 3-1).

Uncertainty analyses determine the uncertainty associated with the model output, which in TEA can be a calculated profitability indicator (e.g., net present value or individual rate of return). The overall uncertainty is determined from the propagation of errors in input data as well as uncertainties in the model that describes the technology itself. It can even depend on the context in which the assessment is conducted—for example, the assumed market and or the environmental conditions of operations may result in uncertainty in estimates for net present value. Typically, each input variable has uncertainty associated with its value, which can be expressed with probability distributions that show the likelihood of the variable having a certain value. These distributions are then used as model input for a simulation that determines the uncertainty in the output. For multiple input variables with known or assumed probability distributions, the combined influence of all input variables on the output of the assessment can be determined rigorously with methods such as Monte Carlo simulations, which provide statistical (not physical) uncertainties associated with the model (Figure 3-7).

Sensitivity analyses (SAs) determine how sensitive the model output is to variations in one or more input variables. Uncertainty analyses and SAs are complementary, as SAs reveal how any uncertainty within the output is constructed and identify the key input variables that contribute most to the uncertainty. Figure 3-8 shows how a systematic variation of input variable values is used to determine their significance. SAs are especially powerful

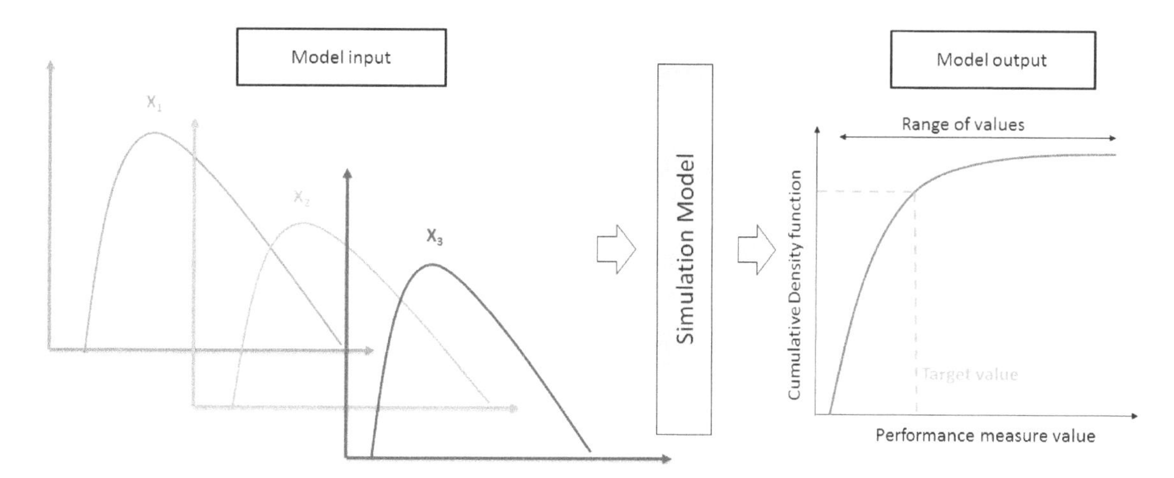

FIGURE 3-7 An assessment model can use probability functions of uncertainties for the input variables.
NOTE: These functions are the input for a simulation model that predicts the values of the selected indicators, including their uncertainty.
SOURCE: Langhorst et al. (2022), https://doi.org/10.7302/4190. CC-BY-ND 4.0.

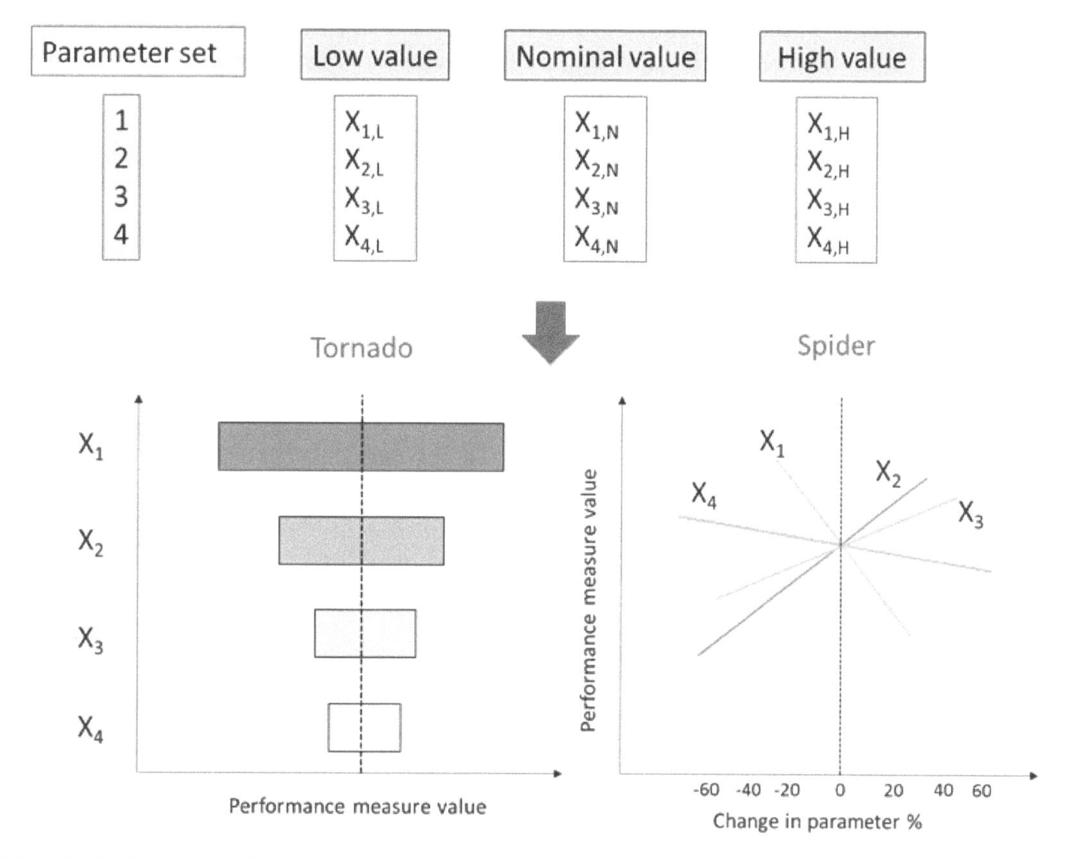

FIGURE 3-8 Visual representation of deterministic sensitivity analysis to determine the influence of each input variable.
NOTES: When parameters are uncertain, they can be varied from the assumed baseline to range from low to high values to demonstrate how the model results are impacted. Tornado and spider plots are useful ways to visualize the sensitivity of the model to each individual parameter.
SOURCE: Langhorst et al. (2022), https://doi.org/10.7302/4190. CC-BY-ND 4.0.

in providing insights into which variables have the highest impact on the output—for example, cost—and where resources should be allocated to develop improvements or alternatives. Likewise, SAs can identify variables that have minute impact on the output—that is, low sensitivity—and can be deprioritized in future work, improving the technology. Methods to conduct SAs include one-at-a-time sensitivity analysis, one- and multiple-way sensitivity analysis, scatterplot analysis, variance-based methods, and density-based methods.

Given the variety of techniques available, guidelines for selecting an uncertainty analysis method for TEAs are gaining traction in literature. An example in Appendix J (Roussanaly et al. 2021) shows a decision tree recommending the type of uncertainty analysis based on purpose—"what if" or "what will." "What if" uncertainty analyses address diagnostic questions and provide insights into changes in output resulting from changes in inputs. "What will" analyses are prognostic in nature and focus on understanding the conditions under which a result may be obtained with a certain probability (Roussanaly et al. 2021; Rubin 2019; Saltelli et al. 2008). Once a "what if" or "what will" analysis is chosen, an SA can then be performed. Expert elicitation approaches can be used to complement SAs. For instance, pedigree matrices[7] can be used to evaluate the knowledge base of a model or data (Edelen and Ingwersen 2016; Fernández-Dacosta et al. 2017; Pinto et al. 2024); an example of how to use these matrices in ex-ante TEA and LCA of a CO_2 to polyols plants is presented in Fernández-Dacosta et al. (2017).

3.2.3.6 Competing Technologies

TEA provides insights into the economic performance of a technology, and, in most cases, is discussed against a reference, also called a counterfactual case. For example, economic results from comparing a CO_2-derived product with a fossil-based product will be quite different from those comparing a CO_2-derived product to a bio-based product or a waste-based product (Singh et al. 2023). Complicating the interpretation of such comparisons, environmental impacts will be substantially different—land use may be a challenge for certain bio-based products—pointing to the importance of complementing TEA with LCA results. For CO_2 utilization technologies at early TRL, selecting the most likely competitor technology is not straightforward. Figure 3-9 shows a selection of 69 process routes to produce ethylene from alternative carbon sources (e.g., biomass, CO_2, waste). The variety of sources and process options to form just one product indicates that for early-stage TEA, reference routes beyond the fossil-based counterpart also need to be investigated to obtain robust insights into when a CO_2 utilization route performs better or worse under different scenarios. Furthermore, such analyses must be done on an equal basis, such that the system boundaries, temporal and geographic assumptions, and level of detail used to carry out the TEAs are as similar as possible so that the results can be compared fairly. If the TEA of the CO_2 technology is carried out for a given date in the future (e.g., 2040), potential changes in the reference technology (e.g., decline in cost owing to learning) must be accounted for.

3.3 LIFE CYCLE ASSESSMENTS FOR CARBON UTILIZATION

3.3.1 Summary of Recent LCA Reviews and Publications

The committee examined systematic reviews and recently published LCAs to determine the state of the art being applied to CO_2 utilization. Such systematic reviews are limited and cannot address the entirety of products and technologies; however, they do point to important conclusions about the state of LCA of CO_2 utilization. First, studies conducting LCAs of CO_2 utilization for the same product can examine many different types of technologies, and these technologies may incur different environmental impacts. Table 3-3 (in this chapter) and Tables J-1 and J-2 (in Appendix J) show compiled LCA results for CO_2 emissions released to produce methanol, dimethyl ether (DME), and dimethyl carbonate (DMC) from Garcia-Garcia et al. (2021), demonstrating the wide variety of technologies and processes that have been examined. Second, the system boundaries of published studies are highly variable and may not be consistent—for example, studies may include only a part of the supply chain, or they may examine impacts from materials extraction to the product as it leaves the production facilities

[7] A pedigree matrix analyzes the strengths and weaknesses of available information or knowledge base on an ordinal scale—for example, 1–5, low–high (Fernández-Dacosta et al. 2017).

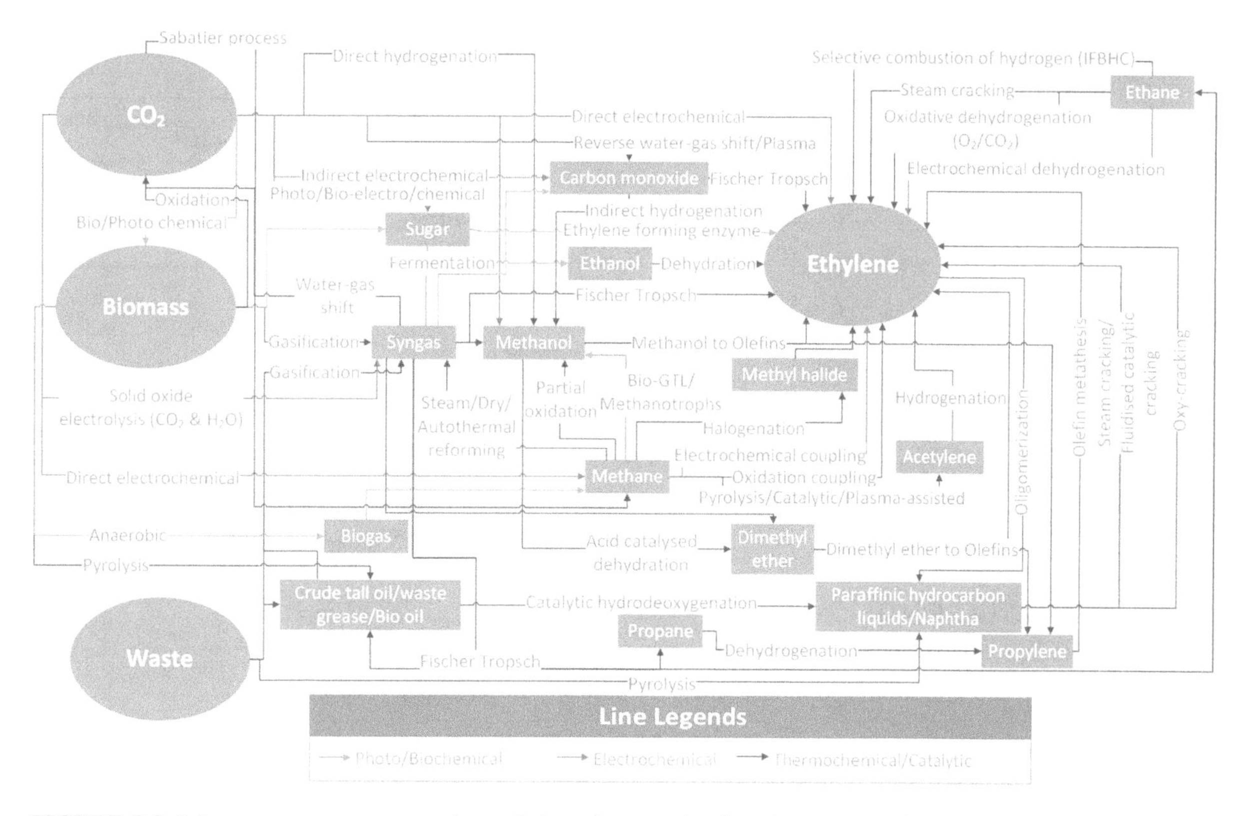

FIGURE 3-9 Select process routes to produce ethylene from nonfossil carbon sources: CO_2, biomass, and waste carbon materials.
SOURCE: Manalal et al. (n.d.).

(cradle-to-gate). Differences in results can be attributed to these variations in system boundaries or assumptions (e.g., source of hydrogen and energy) as well as whether feedstock-related emissions are considered. Results for the production of methane from CO_2 via the Sabatier reaction and a nickel-based catalyst with cradle-to-gate system boundaries showed large variations in methods (e.g., allocation and application of emissions credit) and energy source for electrolysis (e.g., specific renewable technologies and grid mixes), resulting in a wide range of results from -1.99 to $+16.7$ kg_{CO_2} $kg_{CH_4}^{-1}$ and from -0.04 to $+0.3$ kg_{CO_2} $MJ_{CH_4}^{-1}$, where the latter metric is relative to the lower heating value of methane in megajoules. Methane produced directly from fossil sources had emissions of 0.54 kg_{CO_2} $kg_{CH_4}^{-1}$ and 0.15 kg_{CO_2} $MJ_{CH_4}^{-1}$ respectively. Third, LCA results often focus on CO_2 emissions alone rather than a more comprehensive set of environmental impacts. The latter is recommended by ISO 14040 (European Committee for Standardization 2022a). As a result, important environmental impacts may be overlooked, and such studies can make conclusions only about climate-related impacts, not the overall sustainability of the products.

A meta-analysis of LCA studies examining a more comprehensive set of impacts highlighted the importance of including impacts beyond CO_2 alone: no CO_2 utilization-based chemical production alternative performed better in all impact categories than the conventional production technology (Thonemann 2020). For example, while formic acid production from CO_2 did not yield the lowest global warming potential (GWP), it showed significantly lower environmental impacts in most other impact categories. Depending on the source of heat and electricity, the global warming impact for formic acid could be reduced by as much as 95 percent compared to conventional production. Use of wind power resulted in lower environmental impacts across most impact categories compared to grid mixes that include fossil fuels. Given the challenges in completing consistent and comparable LCAs of CO_2 utilization technologies, the committee compiled additional guidance on LCA for CO_2 utilization, clarifying how to conduct these assessments and how decision makers can make use of results across TRLs.

TABLE 3-3 Compiled LCA Results for CO_2 Emissions Using Different System Boundaries, Assumptions, and Processes for Methanol Production from CO_2

Technology/Process	System Boundaries	CO_2 Emissions ($t_{CO_2\,eq}\ t^{-1}_{methanol}$ unless otherwise indicated)
Steam reforming or partial oxidation[a]	Cradle-to-gate	0.68–1.08
Bi-reforming (mix of dry/steam reforming)	Emissions from flue gases, steam and electricity generation, hydrogen and oxygen production and the natural gas supply chain were considered	1.768
Electrochemical methanol production	Gate-to-gate	$1.74 \times 10^{-6} t_{CO_2\,eq}\ MJ^{-1}$
Electrochemical reduction of CO_2 with water	Cradle-to-gate; utilities (electricity, heat, and water) are included; distribution is excluded	949
Hydrogenation of CO_2	Cradle-to-gate	−0.87–6.27[b]
	Process-related emissions only	0.123
Hydrogenation of CO_2 using solar energy	Cradle-to-gate	Normalized results against reference (100%); 278% larger GWP when using conventional fuels and −253% when using solar energy
Hydrogenation of CO_2, $Cu/ZnO/Al_2O_3$ catalyst	Cradle-to-gate	1.21–1.44
	Emissions from flue gases, steam and electricity generation, hydrogen and oxygen production, and the natural gas supply chain were considered	0.657–2.983[b]
	Gate-to-gate, excluding hydrogen generation and carbon capture	0.13
	Gate-to-gate, including CO_2 capture and conversion, hydrogen production, infrastructure; excluding transport, storage, and recovery and reuse of catalyst	0.226
Reduction of CO_2 to CO, water gas shift and methanol synthesis	Cradle-to-gate; construction, fuel production, and disassembly of the production plant are included; utilities (heat and electricity) are included	−1.70–1.87[b]
Tri-reforming	Emissions from flue gases, steam and electricity generation, hydrogen and oxygen production and the natural gas supply chain were considered	1.726–1.763[b]

[a] Standard production (non-CO_2 utilization) processes for comparison.
[b] Range contingent on hydrogen and electricity sources and other assumptions.
NOTE: Bi-reforming = mix of conventional steam (water) and dry (CO_2) reforming of methane to form synthesis gas that can be directly transformed into methanol; Tri-reforming = bi-reforming + partial oxidation of methane simultaneously.
SOURCE: Adapted from Garcia-Garcia (2021).

3.3.2 What LCA Can and Cannot Do in Assessing CO_2 Utilization Technologies

3.3.2.1 Dependencies

The accuracy and certainty of LCA results depend on the representativeness of the input data, which will be more uncertain for earlier-stage technologies. Recognizing uncertainty in the results informs R&D and design for early TRL technologies. For later-stage technologies, specific details about operation and siting are more available. Results for a CO_2 utilization technology deployed in one location and time horizon likely cannot be directly applied to a different geographical and temporal context, as data inputs may not be representative (e.g., electricity and heat generation, hydrogen production, and infrastructure requirements may be subject to large variability). For technologies that will be deployed in the future, present LCA results may not capture environmental changes

TABLE 3-4 Environmental Impact Categories, Scale of Impact, and Environmental Releases Used to Estimate the Impacts

Impact Category	Scale	Examples of Environmental Releases[a] from Life Cycle Inventory Data
Global Warming	Global	Carbon dioxide (CO_2), nitrous oxide (N_2O), methane (CH_4), chlorofluorocarbons (CFCs), hydrochlorofluorocarbons (HCFCs), methyl bromide (CH_3Br)
Stratospheric Ozone Depletion	Global	Chlorofluorocarbons (CFCs), hydrochlorofluorocarbons (HCFCs), halons, methyl bromide (CH_3Br)
Acidification	Regional, Local	Sulfur oxides (SO_x), nitric oxides (NO_x), hydrochloric acid (HCl), hydrofluoric acid (HF), ammonia (NH_3)
Eutrophication	Local	Phosphate (PO_4), nitrogen oxide (NO), nitrogen dioxide (NO_2), nitrates (NO_3^-), ammonia (NH_3)
Photochemical Smog	Local	Nonmethane hydrocarbon (NMHC)
Terrestrial Toxicity	Local	Toxic chemicals with a reported lethal concentration to rodents
Aquatic Toxicity	Local	Toxic chemicals with a reported lethal concentration to fish
Human Health	Global, Regional, Local	Total releases to air, water, and soil
Resource Depletion	Global, Regional, Local	Quantity of minerals used; quantity of fossil fuels used
Land Use	Global, Regional, Local	Quantity disposed of in a landfill or other land modifications
Water Use	Regional, Local	Water used or consumed

[a] The environmental releases are estimated in the inventory analysis, and they can contribute to specific impact categories.
SOURCE: Matthews et al. (2014), www.lcatextbook.com. CC BY-SA 4.0.

necessary to best represent future impacts. This challenge is particularly salient for impacts that require local or regional data, particularly when they change over time, as noted in Table 3-4.

Characterization factors transform the data inventories into impacts in the impact assessment stage of LCA. For example, GWP is a characterization factor used to translate the mass of each greenhouse gas (GHG) emission (e.g., kilogram of CO_2, CH_4, N_2O) into a common unit of mass of carbon dioxide equivalent (e.g., kilogram of CO_2 equivalent). Even global impact categories, such as climate change, are subject to important temporal factors specific to LCA of CO_2 utilization technologies. ISO 14040 notes the following temporal factors, each of which are relevant for CO_2 utilization: time horizon, discounting, temporal resolution of the inventory, time-dependent characterization, dynamic weighting, and time-dependent normalization (European Committee for Standardization 2022a; Lueddeckens et al. 2020). In a static LCA, for example, all emissions or avoided emissions are assumed to occur in 1 year, and a 100-year GWP is applied to estimate results. However, emissions inventories from CO_2 utilization technologies will occur at different times; thus, the emissions release or avoidance would imply a 100-year time horizon applied from the time of the event if using GWP 100. A dynamic LCA can provide more accurate results. Additionally, LCAs of presently deployed and future CO_2 utilization technologies will face different challenges in the uncertainty of impacts. Impact assessment models and the associated characterization factors can rapidly become outdated, so the accuracy of LCA results is tied to the data vintage associated with the most recent update. For example, a recent consensus was reached on how to characterize impacts of water consumption in terms of available water remaining per unit of surface in a given watershed relative to the world average, after demands from humans and ecosystems have been met (Boulay et al. 2018). While a forward-looking model is being developed to quantify future impacts of water consumption (Baustert et al. 2022), CO_2 utilization technologies to be deployed in the future may not be well represented for impact categories that are subject to large variability in environmental conditions over time (e.g., impacts related to nitrogen and other environmental loads). Understanding potential life cycle impacts is a critical aspect of examining the sustainability outcomes of CO_2 utilization technologies;

however, quantifying uncertainties in the results can inform decision makers about potential trade-offs and critical areas that warrant detailed investigation as more representative data become available (e.g., as facilities are sited).

3.3.2.2 Key Uncertainties in Quantifying and Interpreting Impacts

LCA seeks to quantify specific categories of environmental impacts based on a cause-effect chain linked to environmental releases from different processes. Impacts are quantified either as midpoint indicators or endpoint indicators. Midpoint impacts are considered intermediate indicators of environmental impacts that quantify changes in the environment caused by emissions or resource use but do not reflect the full consequence (e.g., GHG emissions expressed as CO_2 equivalents) (Kowalczyk et al. 2023). Endpoint indicators show much more aggregated environmental impacts, generally represented in three categories of damages: (1) effect on human health, (2) biodiversity, and (3) resource scarcity (Hauschild and Huibregts 2015). While endpoint impacts can more easily be interpreted by broad audiences, they are subject to greater uncertainty (Hauschild and Huijbregts 2015). Specific to CO_2 utilization, uncertainty will be further challenged by scarcity in life cycle inventory data (i.e., the data inputs and outputs for different life cycle stages), which will translate into greater uncertainties in impact assessment (Figure 3-10).

3.3.2.3 Consequential LCA

There is an increasing need to use LCAs to quantify impacts of system-level change—for example, if a CO_2-derived product has the potential to capture a substantial portion of a market, understanding the environmental impacts associated with the change in supply is valuable. Attributional LCA determines the share of global environmental burdens associated with a product by quantifying the environmentally relevant physical flows to and from the life cycle of the product (Ekvall 2019). Consequential LCA, on the other hand, estimates how a product affects global environmental burdens by determining how environmentally relevant physical flows

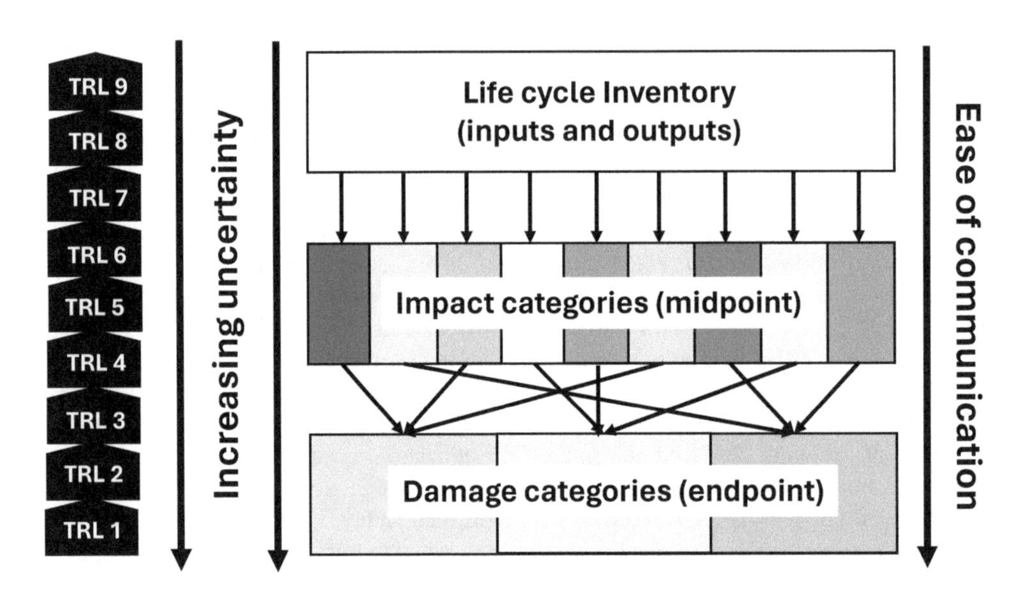

FIGURE 3-10 Key uncertainties in data and impacts for LCA of CO_2 utilization.
NOTES: As TRL increases, uncertainty in data and impacts decreases. True for LCA of any product, the uncertainty and ease of interpretation both increase when results are reported as endpoint indicators (damage categories) compared to midpoint indicators (impact categories). The shading illustrates that there are different impact and damage categories that LCAs may include.
SOURCE: Adapted from Burger Mansilha et al. (2019). The publisher for this copyrighted material is Mary Ann Liebert, Inc.

change in response to decisions, making it especially relevant for policy makers. Consequential LCA is particularly relevant for CO_2 utilization if product systems can be impacted by large-scale deployment—for example, gigatonne-scale carbon mineralization—as there is increasing potential for displacement of more carbon intensive products (NASEM 2019).

3.3.3 Use of LCA for Approaches and Technologies at Different TRLs

The need for LCA of early-stage, emerging technologies is widely recognized and has become a standard expectation in DOE's applied R&D projects on technologies across TRLs (Moni et al. 2020). Results generally show reductions in climate impacts compared to baseline technologies (Garcia-Garcia et al. 2021). Early-stage LCA can inform R&D, particularly through an iterative analysis that can be considered part of the R&D process (Lettner and Hesser 2019), incorporating knowledge from LCA into the design of the technology.

LCA results using databases representing present conditions, unit processes, and sectors are unlikely to be fully representative of the conditions where and when low TRL technologies may be deployed in the future. LCAs may overestimate impacts, as early-stage, emerging technologies are likely to have higher energy and material requirements; however, auxiliary requirements may be overlooked (Müller et al. 2020). LCAs of early-stage technologies typically use process simulation, manual calculations (e.g., stoichiometric equations), and proxies (Langhorst et al. 2022; Tsoy et al. 2020), and results depend on the benchmark data derived from these methods. LCA results for early-stage technologies are subject to higher levels of uncertainty compared to those for later-stage technologies. Despite this, LCAs of early-stage technologies are very useful in identifying critical pain points that have large influence on impact categories. This information could be used to prioritize and focus research.

Pilot and demonstration projects can provide useful baseline data representing present conditions; however, full scale up and deployment may be subject to different environmental and market conditions. Fully commercialized technologies can be characterized in the most representative sense, but CO_2 utilization technologies tend to be earlier stage. The impacts embodied in the required materials and the operational efficiencies will be project dependent on characteristics and infrastructure such as plant size, CO_2 transport, compression, and storage. The more the pilot or demonstration project resembles the proposed project or operations, the more representative the LCA's environmental impacts will be.

3.3.4 Critical Issues for CO_2 Utilization LCA

Present ISO LCA standards require updates to be more suitable to carbon capture and circular systems (Pinto et al. 2024), which the committee extends here to CO_2 utilization. Recently, the National Energy Technology Laboratory (NETL) developed specific guidance for CO_2 utilization (Skone et al. 2022)—in part based on NASEM (2019)—with additional directions on software, data, and tools to complete LCA in accordance with ISO for applicants to funding programs (Sick et al. 2020). Additional key challenges are described in the following sections.

3.3.4.1 System Boundaries and Choice of Functional Units

Decisions made in the goal and scope definition, where the system boundaries and functional units are chosen, affect LCA results for CO_2 utilization, and there is a lack of consensus on what to include within the system boundaries. Some experts recommend that CO_2 be considered as a regular feedstock with its own production emissions (von der Assen et al. 2013); however, the system boundaries may include the source of the CO_2 emissions (i.e., the emissions reductions). As with TEA, transparency in the choice of system boundaries provides clarity on whether different LCA results can be compared.

The functional unit is the denominator in a life cycle result, the basis for reference for the product system that represents the quantifiable function of the product under investigation (European Committee for Standardization 2022b). It plays a major role in determining the results for LCA, which is particularly important for CO_2 utilization with its potential impact across many different products. For example, one LCA compared CO_2 utilization technologies based on the treatment of 1 kg of CO_2 (Thonemann and Pizzol 2019). Because the marginal suppliers were those supplying CO_2 and H_2 rather than the products that may be displaced, the results are less relevant to the end products of each CO_2 utilization technology. Other studies recommended comparing LCA results for CO_2 utilization based on the products they are displacing, in terms of the analysis and functional units selected (Garcia-Garcia et al. 2021). A deeper integration of LCA with supply chain analysis for CO_2 utilization can enable an even more comprehensive understanding of these elements. NETL distinguishes between cradle-to-gate and cradle-to-grave, where the latter accounts for the end use of the product, and the former stops after the product is made (Skone et al. 2022).

3.3.4.2 Source of CO_2

The source of CO_2 (e.g., point sources versus direct air capture [DAC]) is an important consideration for the LCA of CO_2 utilization technologies (Müller et al. 2020). The life cycle carbon emissions of CO_2 capture have been reported to vary from negative to positive, meaning the CO_2 source is an important assumption. The challenge is thus determining the most environmentally beneficial sources, especially because the purity of CO_2 streams varies widely, generally from 5 to 35 percent by concentration but close to pure CO_2 in some cases (von der Assen et al. 2016). Some research suggests prioritizing point sources, as they result in the highest emissions benefits, noting that even low-purity CO_2 streams from cement production provide higher benefits than DAC (von der Assen et al. 2016). Published studies consider CO_2 differently: treating it as a negative emissions stream, an available pure CO_2 stream, or a co-product where emissions are split between the CO_2 and other products (Müller et al. 2020). Treating the CO_2 as a negative emission in LCA risks double counting emissions benefits (Lenzen 2008). DOE has recognized the challenge in overlooking the CO_2 source and assuming a negative emission: emissions benefits are attractive to both source and sink. Systems expansion can eliminate the risk of double counting by including the CO_2 source in the LCA—such an expansion is recommended by several studies (Cooney et al. 2022; Müller et al. 2020; Singh et al. 2023).

3.3.4.3 Duration of Carbon Storage

Durability of carbon abatement is a crucial factor for determining the life cycle impacts of CO_2-derived products. For example, synthetic fuels made from captured CO_2 are recombusted typically within 1 year, which rereleases CO_2 to the atmosphere. Therefore, synthetic fuels made from fossil CO_2 do not provide a net reduction in fossil-CO_2 emissions, except for global economic scenarios where production of synthetic fuel decreases oil and gas production owing to a drop in demand. The latter pathway readily occurs in the current global economy, which is 80 percent fossil-based, but will diminish in future net-zero scenarios in which energy increasingly is replaced by renewable or low-carbon nuclear sources.

Carbon-based products such as polyethylene, the dominant plastic, have a theoretical sequestration life of more than 1000 years in a landfill. However, 80 percent of waste is incinerated in the European Union, and 20 percent is incinerated in the United States, numbers that are expected to grow over concerns about land use and methane emissions. Thus, the European Union considers average sequestration life of polyethylene to be less than 100 years. de Kleijne et al. (2022) conclude that CO_2-based production of polyethylene is not compatible with the Paris Accord. Application of LCA to CO_2 utilization technologies is often challenged by making assumptions about the duration of storage, as actual duration is unknown. Carbon storage duration is also an important consideration for uncertainty and/or sensitivity analyses. Research is needed both to better understand storage duration through studies of the actual end of life of products and to better represent waste management in LCA and TEA of CO_2 utilization technologies.

3.3.4.4 Geographical and Temporal Representativeness

Geographical and temporal representativeness are widely recognized as critical data quality indicators for the accuracy of LCA results (Müller et al. 2016), creating challenges for early-stage technologies that have yet to be deployed. The results from one LCA in one region should not be applied to another region, as the environmental impacts often depend on local conditions. Characterization factors in typical LCAs have been identified as representative of large regions, but unsuitable at more granular scales such as counties (Pinto et al. 2024).

While LCA of earlier-stage technologies can provide useful information about potential impacts of deployment, these results should not be assumed as facilities are sited; rather, new LCAs need to be completed as relevant information becomes available. Once sites are selected, more detailed LCAs can support environmentally superior procurement decisions.

Temporal representativeness hinges on questions of key importance to the operation of the technology under examination. For example, use of static GWPs in LCA, as is typical, means that changes in emissions are considered as one pulse at the beginning of the assessment, rather than gradual emissions changes over the lifetime of operations (Langhorst et al. 2022; Levasseur et al. 2010, 2012). The selection of time horizon may overlook important emissions if not aligned with the duration of time the carbon is stored in the product (which is already uncertain). CO_2 storage duration is an important factor in the selection of time horizon; if the time horizon is too short, important emissions may be omitted from the analysis (von der Assen et al. 2013).

Whether emissions are reduced or avoided is also a critical question in LCA (Finkbeiner and Bach 2021). Typically, a service or pathway is replaced by an alternative, and it is important that system boundaries show the global cradle-to-grave impact of competing options (Finkbeiner and Bach 2021). LCAs often stop at intermediate boundaries (e.g., vehicle tank to tailpipe and not considering upstream supply chain) and underreport emissions relative to competing options. The value of the option varies with scenario (e.g., electric vehicles powered by coal plants today versus by renewable energy tomorrow), such that scenarios must be clearly stated. Avoiding petrochemical products through displacement with CO_2-derived products may realize emissions reductions in certain scenarios (Ruttinger et al. 2022).

3.3.4.5 Comprehensiveness of Impacts

The ISO standard recommends that LCAs include a comprehensive set of environmental impacts—for example, energy use, GHG emissions, and water consumption (Royal Society of Chemistry n.d.). LCA results for earlier-stage technologies may have limited accuracy for specific impact categories that depend on geographical representativeness. Estimates for the impacts of technologies and services can vary widely relative to the database used in assessment, as they may be too generic if the technology is earlier stage or if the characterization factors do not represent the environmental conditions at sites where facilities are being deployed. As technologies near deployment, geographical representativeness becomes increasingly important for interpreting the results for impacted communities. Complex trade-offs on weighting factors among categories may depend on local stakeholder values and drive substantial complexity that goes beyond results provided by LCA.

3.3.4.6 Addressing Uncertainty

Owing to the emergent nature of most CO_2 utilization technologies, uncertainty analyses are of particular importance. Uncertainty for early-stage technology necessitates an analysis of how well data inventories and impacts represent the time at which the technologies will be deployed and the regions where facilities may be sited (if it is possible to discern). Reasonableness checks against published literature results can support the identification of uncertainties. The methods outlined in Section 3.2.3.5 on addressing uncertainty in TEA also can be applied to address uncertainty in LCA results. For example, specific parameters may be identified as uncertain and impactful to cost or environment during the design stage (e.g., process simulation), which can be tested in sensitivity analyses informed by known operational performance and/or experimental results. Sensitivity analysis also can be informed by parameters found to be uncertain in experimental results for early-stage technologies. LCA-related assumptions (e.g., allocation) must be examined if they are used in the analysis (von der Assen et al. 2014). As discussed above, the duration of carbon storage is often unknown, yielding uncertainty in the long-term climate benefits of the technology under investigation. Scenarios for different fates can be tested to quantify how results may be impacted owing to specific outcomes. Break-even analysis[8] can play an important role in determining the potential extent of environmental risks in the face of uncertainty.

[8] An environmental break-even point refers to the point at which the environmental benefits of a product are reversed compared to another when varying a parameter in a parametric analysis (Messagie et al. 2012).

3.4 ASSESSING SOCIETAL IMPACT, PERCEPTION, AND EQUITY

As discussed further in Chapter 4, maintaining or developing social license is an essential component of technology development, particularly for technologies associated with products and industries that have been connected to past and ongoing societal harms. Developing an industry around CO_2 utilization will require proactive engagement with equity and justice concerns. These social considerations cannot be limited only to the site and direct operations of the carbon management project. Meaningfully addressing equity and justice will require a comprehensive look at the processes and products engaged by a project. LCA, and particularly s-LCA, can provide a structured framework for incorporating assessments of equity and societal impacts into the evaluation of a process or product.[9] s-LCA has been applied to fossil and nonfossil alternatives; yet less research has been completed that is focused specifically on CO_2 utilization (Ekener et al. 2018; Fortier et al. 2019; Iribarren et al. 2022). While s-LCA has been more directly applied, there are documented limitations of LCA and decision making in addressing equity to date (Bozeman et al. 2022).

3.4.1 What Equity-Specific Tools for Assessment Can and Cannot Do in Assessing the CO_2 Utilization Value Chain

3.4.1.1 Equity Evaluation

Assessments of a product or technology seek to minimize uncertainty about the quantification of positive and negative impacts and form an interpretation basis for decision makers to consider when determining whether to proceed with a project. s-LCA is a means of integrating assessments of equity, justice, and other issues of social license into the assessment of a technology. Specifically, s-LCAs connect the methodological approaches of LCA to quantify a technology's or a system's environmental, societal, and equity impacts (Bouillass et al. 2021; Bozeman et al. 2022; Sala et al. 2015; UNEP 2020). Figure 3-11 compares (traditional) e-LCA and s-LCA throughout the standard phases of an LCA.

While the standardized phases of LCA provide a template for creating a comprehensive impact assessment across the life cycle of a technology or process, LCA and equity/justice efforts are not meaningfully linked in relevant policy but instead are treated as separate considerations. For example, with the 45Q carbon utilization tax credit, DOE engages LCA for environmental assessment, but not for societal implications. Instead, DOE addresses societal implications for project funding by requiring Community Benefits Plans as a means of attaining Justice40 Initiative outcomes (see Chapter 4 for more detail), which does not connect a proposed project's equity and justice implications to the relevant product and process life cycle.

An s-LCA looks at the direct effect of evaluated products on stakeholders and the indirect impact of stakeholders on a product's socioeconomic processes (Yang et al. 2020). These stakeholders can be grouped into five categories—workers/employees, local community, society, consumers, and value chain actors—to discuss related subcategories (i.e., fair salary, health and safety, and social benefits) (Toniolo et al. 2020), with particular indicators and requirements detailed in Table 3-5. These indicators and requirements inform an s-LCA approach to understanding and accounting for equity/justice impacts, but there are other frameworks grounded in other disciplines and fields of practice that may be appropriate to employ. For example, frameworks from economics suggest including nonmarginal changes, modeling approaches that provide information on the environmental burdens that occur, because of changes in demand (Almeida et al. 2020), using frameworks based on theories of resource allocation and competitive equilibrium for public goods (Foley 1966), and a life cycle cost (LCC) analysis framework that analyzes the cost-effectiveness of an investment over its economic lifetime (Norris 2001). Social psychologists advocate for frameworks based on equity theory, wherein individuals are the most satisfied when they experience a relatively equal exchange of resources, rather than being greatly overbenefited or underbenefited (Van Dijk and Wilke 1993),

[9] The committee's first report connected the broad impacts of carbon capture and utilization to cost-benefit analyses (CBAs), noting that the distributional effects of a project can be integrated with the framework of a CBA (although there may be additional normative criteria employed to reflect societal views) (NASEM 2023a). LCA provides a full life cycle framework for accounting and has established practices for reporting and comparing different impact categories stemming from the same product or process—providing an additional analytical approach for integrating social and equity considerations into an analysis.

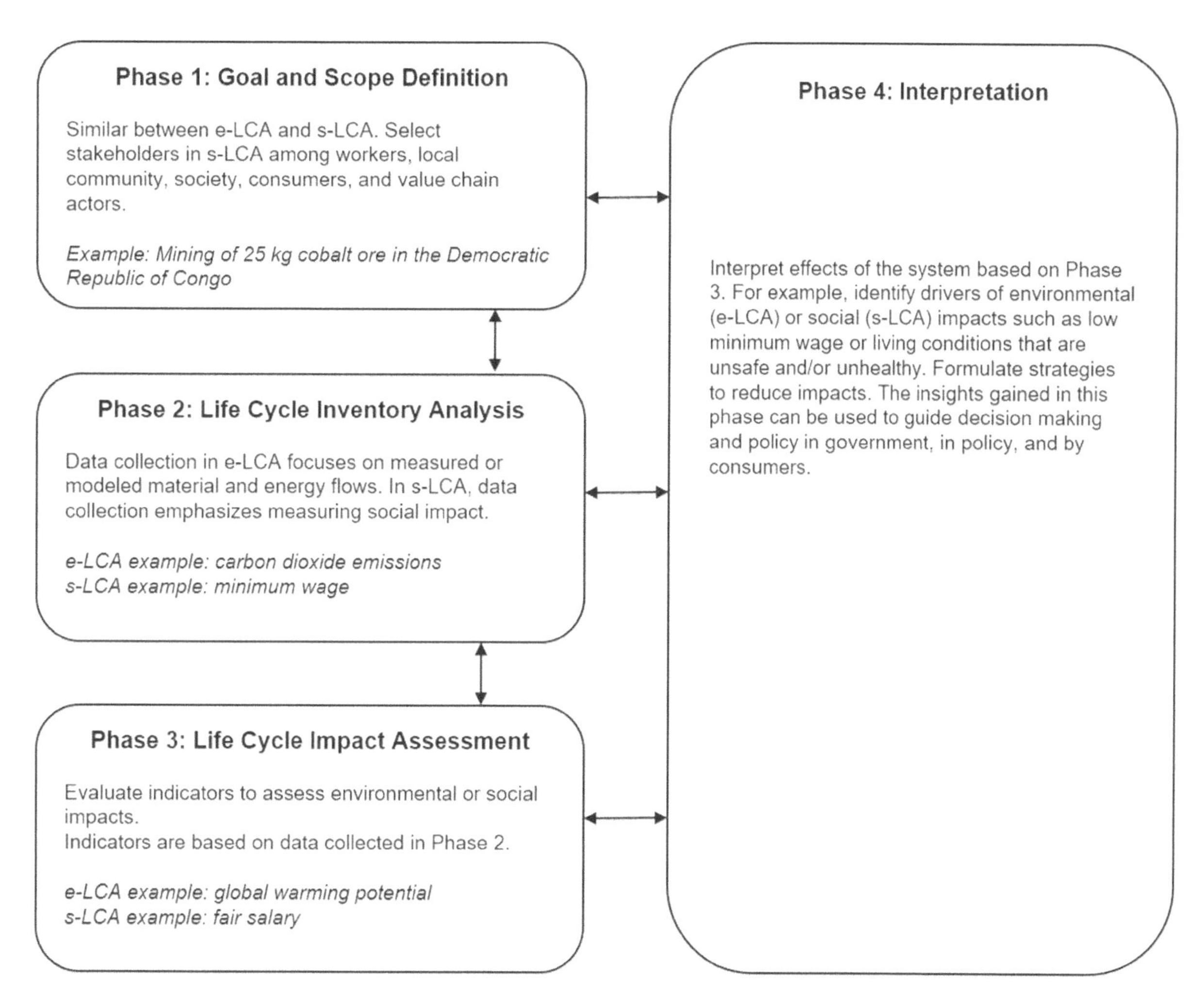

FIGURE 3-11 The four phases of life cycle assessment with illustrative examples to show the similarities and differences between environmental and social life cycle assessments (e-LCA and s-LCA).
NOTE: Both e-LCA and s-LCA are iterative processes with interactions between phases.
SOURCE: Modified from Bamana et al. (2021), https://doi.org/10.1016/j.oneear.2021.11.007. CC-BY-ND 4.0.

while public health, political science, communications, and public policy scholars recommend including frameworks based on an individual's core beliefs and values, or that of psychological constraint, along with the institutional framework of U.S. politics (Converse 2006; Druckman 2014; Feldman 1988, 2003).

3.4.1.2 Equity Assessment Tools and Uncertainties

No standardized s-LCA tools or methods are of similar quality to life cycle costing or e-LCAs (Reijnders 2022). Proposed methodologies give different weight to the role of local stakeholders and the need of a common social theory (Toniolo et al. 2020). Further challenging this issue, LCA and other decision-making tools have limited sociodemographic data, reducing the ability of practitioners to address important equity issues (Bozeman et al. 2022). Few studies have actively sought to improve s-LCA to include questions of distributional justice (Fortier et al. 2019), and the indicators presently used in s-LCA have been noted as insufficient to support the advancement of equity and justice (Bozeman et al. 2022; Greer 2023; Zeug et al. 2023). This limitation in s-LCA indicators points to the importance

TABLE 3-5 Stakeholder Categories, Indicators, and Corresponding Requirements

Stakeholder	Indicator	Requirement
Workers	Child labor	The absence of children working in the system.
	Fair salary	The salary should be no less than the minimum wage.
	Working hours	The average number of working hours should be limited to 8 hours/day and 48 hours/week.
	Forced labor	The abolition of forced labor.
	Discrimination/equal opportunity	The prevention of discrimination and the promotion of equal opportunities.
	Health and safety	The guarantee of worker's health and safety.
	Social benefits/social security	The suggestion of more than two social benefits provided by the organization.
	Freedom of association and collective bargaining	The presence of unions for representation and collective action; organization takes measures to promote the right to organize and right to collective bargaining where restricted by legislation.
Consumers	Health and safety	The guarantee of consumers' health and safety.
	Consumer privacy	The protection of consumers' right to privacy.
	Feedback mechanism	The presence of consumers' feedback mechanism.
	End-of-life responsibility	Information on end-of-life options or recalls policy for consumers.
Local Community	Local employment	The minimum percentage of local labor should be no less than 50 percent.
	Access to material resources	The sustainable utilization of natural resources and the recycling of used material.
	Access to immaterial resources (e.g., workers)	The promoting of community service.
	Delocalization and migration	The absence of forced resettlement caused by the system.
	Safe and healthy living conditions	The guarantee of safe and healthy surrounding communities.
	Respect of Indigenous rights	The protection of Indigenous rights.
	Community engagement	The consideration of the environment, health, or welfare of a community.
Society	Contribution to economic development	The promotion of economic contribution to society.
	Public commitments to sustainability issues	The promise or agreement related to the development of the system.
	Technology development	The development of efficient and environmentally friendly technologies.
	Corruption	The prevention of corruption of the system.
	Distribution of public good in indicators for society	Equitable and efficient distribution of and access to public goods to requirements.
Value Chain	Fair competition	The grantee of fair competition and the prevention of antitrust legislation or monopoly practices.
	Promoting social responsibility	The improvement of social responsibility contributed by the whole value chain of the investigated system.
	Supplier relationships	The cooperation between the supplies and the investigated system should be facilitated stably.
	Respect of intellectual property rights	The protection of the intellectual property rights by the involved actors within the value chain.

SOURCES: Adapted from Aparcana and Salhofer (2013); Arcese et al. (2013); Dreyer et al. (2010); Edvinsson et al. (2013); Feldman (2003); Foley (1966); Macombe and Loeillet (2014); Mattioda et al. (2017); Rajan et al. (2013); Toniolo et al. (2020); UNEP (2009); Van Dijk and Wilke (1993).

of advancing s-LCA methodology (Fortier et al. 2019) as well as combining s-LCA with more advanced tools that were developed specifically for equity. Community engagement tactics, for example, can help determine the weight of specific indicators in the s-LCA framework to further understand local outcomes.

Parameterizing social aspects generally requires judgment and cannot be determined by mathematics alone, so weighing what is critically important and considering community input in doing so is necessary to determine what metrics should be included and measured. The 2020 update to the United Nations Environment Programme and the Society for Environmental Toxicology and Chemistry Life Cycle Initiative s-LCA guidelines includes two approaches to data analysis: using generic databases or collecting and applying site specific information (UNEP 2020). For the former, determining localized issues of equity will be challenging. Equity assessments necessitate the collection of expert information including from impacted communities (Bradley et al. 2022a), meaning that generic s-LCA databases are unlikely to be sufficient in overcoming impacts. The guidelines emphasize the importance of site visits and working with relevant organizations for implementing s-LCA where site-specific information is used (UNEP 2020). Approaches that engage local communities hold promise for better integration with equity assessments but have yet to be implemented and will be challenging prior to proposing sites for facilities. Community engagement can provide insights to support equity beyond impact assessments and siting by including construction, operation, and decommissioning (Elmallah and Rand 2022). Participatory approaches that include residents of historically underserved communities also can improve the inclusion of minoritized[10] people in infrastructure site selection (Hasala et al. 2020).

Appropriately implementing s-LCA requires understanding its capabilities as well as the elements outside its scope. Specifically, s-LCA can inform improvements to a product or system but cannot provide solutions for sustainable consumption or living or ensure equitable outcomes (UNEP 2009). While some s-LCA procedures use public engagement to address communities' concerns and embed procedural justice principles into development processes (Bozeman et al. 2022; Fortier et al. 2019; Grubert 2023), incorporating public engagement and other participatory processes sometimes creates challenges related to societal acceptance and trust. The use of geospatial tools with social data layers, where possible, can support more equitable siting and inform aspects of the LCA—for example, resource consumption—to avoid increasing burdens or impacts on marginalized communities. Examples of such geospatial tools include EPA's EJScreen and the Council on Environmental Quality's (CEQ's) list of tools and resources (CEQ n.d.; EPA 2024).[11] Further discussion of community engagement and environmental justice can be found in Chapter 4.

3.4.2 Use of S-LCA for Approaches and Technologies at Different TRLs

s-LCA is still under development and cannot provide comprehensive assessments for early-stage technologies. Including the potential effect of technology deployment on a local community will be difficult because even technical and economic indicators—for example, specific types of equipment, amounts of water needed, or costs of production—are highly uncertain. At best, the output of assessments can be used to begin informing the public on expected performance, likely land uses, noise, and so on. There is a risk that communities will react negatively because accurate, definitive information is lacking, potentially being misunderstood as attempts to evade the community's critical concerns and needs (e.g., see Chailleux 2020 and Offermann-van Heek et al. 2018, 2020; see also Section 4.4.1.2 for more about how to strengthen public understanding of CO_2 utilization through

[10] The word *minoritized* describes the dynamic with which the status of "minority" is imposed upon certain groups through the use of power and systems, not just the statistical status of being a minority. The term demonstrates that there are intentional power structures that have resulted in certain populations experiencing discrimination/disenfranchisement (e.g., see Benitez 2011, Stewart 2013, and Wright-Mair 2023).

[11] Many federal tools explicitly exclude race as a factor in establishing impacts on marginalized communities, a practice criticized by many equity and justice advocates (e.g., see Sadasivam 2023 and WHEJAC 2022). From a policy perspective, the race-neutral criteria will often survive legal pushback. For example, race was considered as a factor for the CEQ mapping tool. However, race was ultimately excluded owing to efforts to make the tool more robust and durable against efforts to either (1) dispose of the tool or (2) use its "focus" on race to delegitimize it to the public (Frank 2023; NASEM 2023b). Frank (2023) notes that criteria such as poverty levels and environmental risks "capture the breadth of ways that racial discrimination is felt in society." However, equity and justice advocates still stress that policy "without explicit focus on race will not ultimately prioritize disadvantaged communities" (NASEM 2023b, p. 96). Policies and programs that acknowledge these shortcomings and incorporate supplemental equity and justice elements—such as meaningful community engagement or an explicit focus on environmental justice—could be beneficial for equity assessments. See Chapter 4 for more information about how to incorporate concepts of environmental justice into policy for the emerging CO_2 industry.

public engagement). s-LCA frameworks have not been designed to evaluate the societal impacts of early TRL technologies (i.e., technologies that may be deployed in a *future* sociotechnical system). Predicting how societal systems will develop is difficult, and because they are very context-specific (including geographic and temporal relevance), averages can be deceptive, as conditions may change drastically. Overall, there are ongoing discussions and coordination efforts on s-LCA with a focus on CO_2 conversion (McCord et al. 2023).

While not currently directed to s-LCA, Miller and Keoleian (2015) laid out a useful framework for transformative technologies that identified factors considered intrinsic, indirect, or external to the technology. Societal indicators could be incorporated into the indirect and external categories, delineating between what will be true for a given technology (intrinsic), and what may be true circumstantially (indirect and external).

3.4.3 S-LCA Goal, Scope, and System Boundaries

Following ISO standards and other guidelines, LCAs are not merely carbon accounting assessments but also provide information along multiple impact categories, including GHG emissions, water use, particulate matter emissions, eutrophication potential, and others. Given that results from LCAs are already structured to accommodate and present impacts in different categories, a congruent format exists for societal, equity, and justice indicators in the typical results reporting presentation. Toniolo et al. (2020) define two approaches to s-LCA: performance reference point methods (considering living and working conditions of workers at different life cycle phases) and impact pathways (considering the societal impacts using characterization models with indicators like those seen in e-LCAs).

The scope and level of data collection for different impact analyses and tools are detailed in Figure 3-12. The figure shows that s-LCA has both a significant system scope (the entirety or most of the product life cycle)

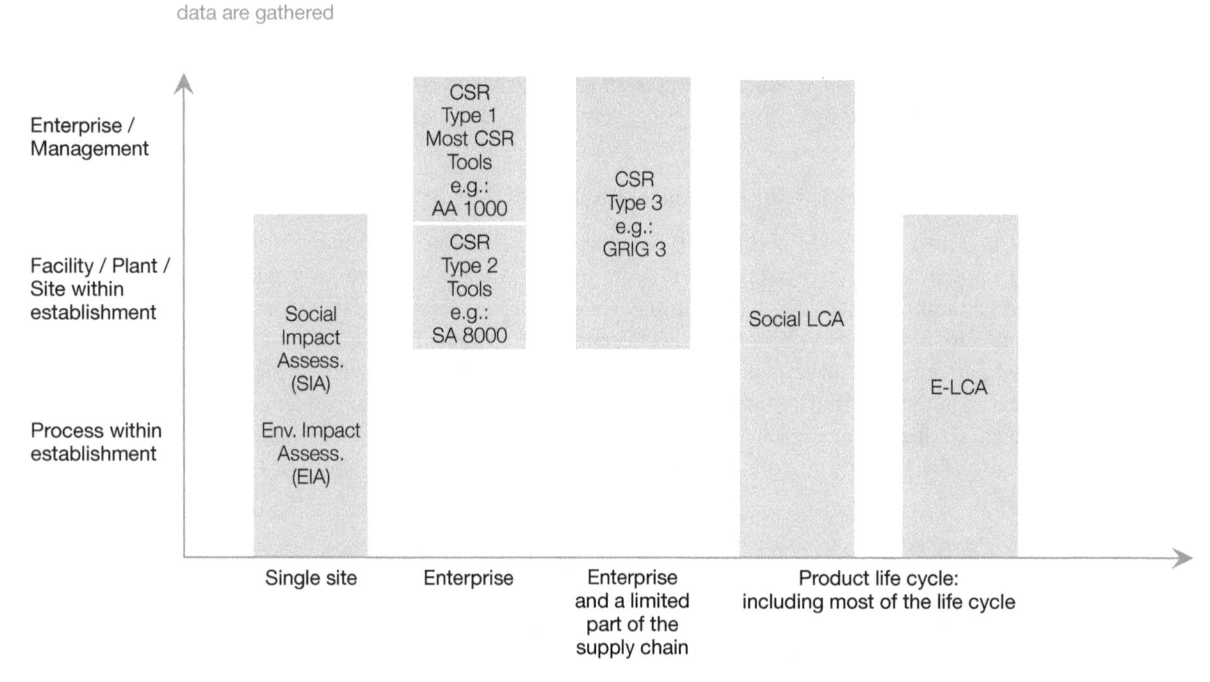

FIGURE 3-12 Various types of impact analysis (SIA, EIA, CSR, Social LCA, and E-LCA) are compared, showing the relevant system scope (horizontal axis) and level on which data are collected (vertical axis). The analyses have different goals and therefore are not necessarily interchangeable.
NOTE: CSR = Corporate Social Responsibility.
SOURCE: United Nations Environment Programme (2009). *Guidelines for Social Life Cycle Assessment of Products.* Figure 4 – Scope of CSR and Impact Assessment Techniques of Enterprises and Their Product. Paris.

as well as data-gathering requirements from the process, facility/plant/site, and enterprise/management, in contrast to other analyses and tools that have more limited scopes and levels of data collection.

One database to support s-LCA is the Product Social Impact Life Cycle Assessment (PSILCA) database (PSILCA n.d.), which allows a practitioner to review industrial sectors and societal indicators and can be used to assess the societal impacts of products along life cycles (Ciroth and Eisfeldt 2016).

The end use of a CO_2-derived product is an important consideration in building public trust and acceptance of the CO_2 utilization sector. Products that contribute to a circular economy by reducing waste and enhancing sustainability are more likely to find public acceptance owing to their role in climate change mitigation, as discussed in Chapter 4. Climate change mitigation and the role of the end product in a sustainable value chain both embody principles and considerations around equity and justice—as communities currently overburdened by waste and pollution are unlikely to support systems that would set them up to bear the burdens of a new CO_2-derived product. As social and technical LCAs are completed, environmental and social justice principles support decision making that minimizes harm and does not exacerbate historical inequities.

Not all social and community considerations may fit into the traditional LCA framework, but life cycle thinking can help identify potential impacts at every aspect of the value chain (Clark 2023). When retrofitting a facility with carbon capture equipment or building a DAC facility, research into the surrounding community and its historical relationship to existing infrastructure, infrastructure development, and energy and water resources provides decision-support information critical to improving outcomes (Zuniga-Teran et al. 2021). The same can be said for the community near where the energy for a project is being provided or developed, as well as where the parts or materials for the final site will be manufactured. While environmental justice is a potential outcome of projects, the historical injustices in a particular community require careful consideration. Such injustices can be best understood through just practices that include community engagement and restorative justice (see Chapter 4). The global nature of product development, including the supply chains that may provide construction materials and the markets through which a CO_2-derived product might be distributed, also require a global climate justice application to ensure that procedures to reduce harm in one locality do not result in externalities that might create or exacerbate harms elsewhere.

To further develop the technical and localized societal aspects of an LCA, developers and researchers can consider the boundaries of what might comprise a community. For example, an LCA and a Refined LCA, detailing the initially proposed DAC plan and the final phase hub plan, respectively, are separately required from a Community Benefits Plan for the DAC Hubs Funding Opportunity Announcement (DOE 2023). The goal of the LCAs is to demonstrate robust accounting of full life cycle emissions. Opportunities to discuss the project's relevance to the surrounding communities across the value chain are included in the portion of the LCA on impacts and discussion of potential co-benefits. If the captured CO_2 will be used for a product, the applicant must submit an LCA following the guidance document outlined by NETL (DOE 2023). This is a potential opportunity for a more robust discussion about incorporating social and equity considerations into LCA frameworks, especially when they are mandated by the federal government to receive funding. In the NETL guidance document, consideration of impacts from the product is limited to GWP and certain environmental impacts such as acidification, eutrophication, particulate matter emissions, and water consumption (Skone 2019). A narrow focus on traditional LCA impact categories will overlook potential societal concerns; considering s-LCA and equity considerations as part of funding decisions could play a substantial role in enacting environmental justice while moving toward tenets of a circular economy.

Use of resources like environmental product declarations (European Committee for Standardization 2020) conceivably could be adjusted and put in place to create LCA-type "scorecards" for certain products. Specific elements by which to judge the products across their lifetime (including disposal) would have to be agreed upon to further standardize and characterize their contribution to equity- and justice-related concerns and/or a circular economy. This is especially relevant if an incumbent product exists and the CO_2-derived version might have additional societal contributions that could either avoid or remedy injustices created by the existence or waste-stream of a particular product.

3.5 HANDLING CIRCULAR ECONOMY SYSTEMS

A key aspect when discussing CO_2 utilization technologies is their ability to achieve circular carbon chains. Circularity, in this context, refers to the transformation of atmospheric or biogenic CO_2 into a product, which would then rerelease the CO_2 at the end of its lifetime, making the flow of carbon circular. Such circularity could enable compatibility of CO_2 utilization technologies with climate neutral targets, only if the CO_2 utilization technologies themselves resulted in no net life cycle emissions. Without lifetime extensions of products (i.e., sufficient duration of carbon storage), CO_2 utilization technologies using fossil emissions have already been questioned regarding their compatibility with climate goals as they may risk carbon lock-in (de Kleijne et al. 2022). If fossil CO_2 were embedded in products and looped through multiple uses over time, the storage of carbon in products could theoretically be extended to significant periods of time (e.g., >100 years). With such extended product lifetimes, technologies that use fossil CO_2 could conceivably be compatible with climate targets under certain conditions (e.g., the fossil carbon is not leaked into the atmosphere) (Malins et al. 2023; Ramírez 2022; Stegmann et al. 2022). Such use of carbon would be an inherent component, for example, with the bold systems changes required to improve the environmental sustainability of the plastics industry (Vidal et al. 2024). This strategy is rather controversial, as it will require stakeholders (producers, users, policy makers) to keep strict custody of the carbon through multiple value chains over long periods independently of the country where the products are produced and used. The controversy takes root not only in questions of how to develop practical and implementable tracking systems but also in noting that net-zero energy systems prioritize moving away from dependence on fossil fuels (e.g., ISO 2022).

Few studies systematically assess the potential and impacts of such a strategy, partly owing to limitations of existing methodologies to address this issue. On the one hand, there are limitations to the use of current LCA methodologies to assess the environmental performance of circular value chains that go beyond one or two loops. Such limitations are unsurprising, as LCA was designed to assess linear value chains. Methodological developments in LCA are being conducted to address this limitation, including the life cycle gaps framework that aims to measure the system losses—the so-called life cycle gaps—between an ideal closed system and the status quo (Dieterle et al. 2018). On the other hand, circular economy metrics are nascent and have been mostly developed to measure or simulate circularity at the macro and meso level (e.g., at the regional, national, or global level) rather than the product level. Further challenging the development of metrics has been the historical lack of standardization; the circular economy and related performance measurement have recently undergone standardization by ISO (2024). Multiple efforts are under way to develop product-level circularity indicators. Three examples are the Circularity Economy Index (Di Maio and Rem 2015), which focuses on recycling process efficiency; the Material Circularity Indicator (Ellen MacArthur Foundation and Granta 2015), which focuses on measuring the use of virgin material and resultant waste; and product-level circularity (Linder et al. 2017), which focuses on the ratio between recirculated and total economic product value. None of these indicators have been tested in potential products of CO_2 utilization.

In addition to resolving these methodological challenges, there are concerns with the feasibility of the option owing to challenges of traceability and access to reliable data, which are further compounded by the large number of datasets that would be required. Technically it is, for instance, possible to trace fossil (or biogenic) carbon flows using ^{14}C methods (Palstra and Meijer 2010). This however would require that carbon flows are characterized, verified, and monitored throughout the different value chains over time, making this option difficult and likely to expensive to implement. One possibility that has been explored is the integration of blockchain technology and LCA (BC-LCA); however, the energy requirements warrant consideration. Frameworks proposed by Zhang et al. (2020) and Shou and Domenech (2022) focus on identifying key elements for BC-LCA, including mapping of value chains, identification of tracking methods, and data collection and validation protocols. The cases investigated in the literature examine products like leather for fashion and have yet to be applied to CO_2 utilization technologies.

3.6 CONCLUSIONS

3.6.1 Findings and Recommendations

Finding 3-1: Appropriate use of techno-economic and life cycle assessments—Techno-economic and life cycle assessments (TEAs and LCAs) provide insights into the economic and environmental performance of a technology, and in most cases, are discussed against a reference technology. TEA and LCA are most impactful at Technology Readiness Level (TRL) 3 and above and can inform deployment, but they also can be useful to understand limitations at earlier TRL levels and inform applied research and development (R&D). The accuracy and certainty of the results depend on the representativeness of the input data, which will be more uncertain for earlier-stage technologies; therefore, it is important to perform TEA and LCA iteratively as R&D progresses and incorporate the insights generated from TEA and LCA into the design of the technology.

Finding 3-2: Requirements for use of techno-economic and life cycle assessments—Requirements for use of techno-economic and life cycle assessments are not consistent across DOE applied research, development, and deployment programs. Contradictory information and data gaps will persist globally without efforts to develop broadly accepted, international guidance.

Finding 3-3: Integration of techno-economic and life cycle assessments—Integration of techno-economic and life cycle assessments (TEAs and LCAs) can provide insightful decision-support, allowing practitioners to simultaneously assess viability and evaluate potential negative impacts to avoid potential conflicting outcomes of a TEA and an LCA where the best-case scenarios for both are achieved for separate conditions.

Recommendation 3-1: Requirements of techno-economic and life cycle assessments for CO_2 utilization—The Department of Energy applied offices and other relevant funding agencies such as the Department of Defense and the National Science Foundation Directorate for Technology, Innovation and Partnerships should develop and maintain consistent requirements for techno-economic and life cycle assessments for technologies at Technology Readiness Level (TRL) 3+ as part of funded applied research and development projects to support increasing economic competitiveness and decreasing negative impacts on the environment. Requirements should inform the basis for collaboration through international agencies and organizations such as the International Energy Agency in improving global guidance. Life cycle and techno-economic assessments (LCAs and TEAs) are not recommended requirements for funding of early-stage TRL projects, as they could constrain innovation. CO_2 utilization assessments developed and maintained by relevant agencies for TRL 3+ should

a. **Report uncertainties and assumptions in the data input and model choice, as well as in the results.**
b. **Periodically reevaluate assumptions and revisit assessments as the technology advances. If the data are no longer relevant, a new TEA/LCA should be completed.**
c. **Justify the reference against which the technology is compared and assess the technology against more than one option that includes not only a fossil-based reference but also alternative carbon sources.**
d. **Integrate TEA and LCA to inform deployment by providing consistent economic and environmental guidance.**

Finding 3-4: CO_2 purity for techno-economic and life cycle assessments—CO_2 purity can play a significant role in the techno-economic and life cycle assessments (TEAs and LCAs) of CO_2 utilization technologies. The quality and cost of purification has often been overlooked in the CO_2 utilization literature. Further understanding of the implications of CO_2 purity for TEAs and LCAs, as well as for the scalability of the technology, is needed to steer technology deployment.

Recommendation 3-2: Research needs for CO_2 purity in techno-economic and life cycle assessments—The Department of Energy (DOE) and other relevant funding agencies should fund projects that examine the robustness of CO_2 utilization technologies to different CO_2 purities, as well as fund further research and development of CO_2 purification technologies. Insights from these projects should be disseminated to the larger community by DOE. DOE should require awardees of applied research and development funding for CO_2 utilization technologies to perform techno-economic and life cycle assessments that explicitly address the purity requirements of the CO_2 streams.

Finding 3-5: Performing techno-economic and life cycle assessments—Conducting techno-economic and life cycle assessments (TEAs and LCAs) is difficult and prone to mistakes if not performed by expert assessors. In general, public availability and selection of input data can be a serious concern. Additionally, International Organization for Standardization (ISO) LCA standards provide clear guidelines for neither CO_2 utilization technologies nor for the associated infrastructure across life cycle stages, which is further complicated by the implications of CO_2 being alternately considered as waste, product, or by-product. There is a lack of skilled experts and accessible tools to guide researchers and developers, to provide input data for assessments and guide the correct interpretation of results and their uncertainties.

Recommendation 3-3: Facilitating techno-economic and life cycle assessments—The Department of Energy (DOE) should facilitate the execution of techno-economic and life cycle assessments (TEAs and LCAs) for >5 Technology Readiness Level (TRL) technologies by

a. **Providing data preparation guidance for novice users of DOE LCA tools so that the data can be more easily analyzed.**
b. **Developing open-source software that guides users through LCA and TEA decision trees, similar to graphical user interface–supported interview-based software tools for tax reporting purposes.**
c. **Further supporting the development of user guidelines for an integrated approach to TEA-LCA, including how results can support decisions based on TRL (e.g., guiding research and development for earlier stage and supporting more sustainable procurement decisions as construction commences).**
d. **Continuing support for the development of databases by DOE and its national laboratories to ensure that information is readily available to complete robust TEAs and LCAs for CO_2 utilization technology and the associated infrastructure across life cycle stages.**

Finding 3-6: Non-CO_2-emissions impacts—A meta-analysis of life cycle assessments performed on CO_2-derived chemicals points to the importance of including impacts beyond CO_2 alone: no CO_2-utilization-based chemical production alternative available at this time performed better in all impact categories than the conventional production technology, and furthermore, production from CO_2 may not yield the lowest global warming potential while still achieving significantly reduced environmental impacts.

Finding 3-7: Social life cycle assessment tools—Social life cycle assessment (s-LCA) is a means for integrating aspects of equity and other issues of social license into the assessment of a technology; however, there are no standardized s-LCA tools or methods that are of similar quality to life cycle costing or environmental life cycle assessments. s-LCA is particularly challenging for technologies that are at low technology readiness level. s-LCA is valuable but is not a way of quantitatively measuring social license or justice.

Recommendation 3-4: Non-CO_2-emissions impacts within life cycle assessments—The Department of Energy and other relevant funding agencies such as the U.S. Environmental Protection Agency and the National Institute of Standards and Technology should support research into improving evaluation of non-CO_2-emissions impacts within life cycle assessments (LCAs) of CO_2 utilization technologies, including

a. **Evaluating the appropriate but differentiated applications for global and local impact categories, as the latter generally involves data and information with high spatial and temporal granularity (e.g., processes versus facilities, technology readiness level of various components of the technology).**
b. **Evaluating appropriate applications of social LCA (s-LCA) and further developing s-LCA tools and their potential integration with environmental LCA and techno-economic assessments.**

Finding 3-8: Use of social life cycle assessments—Potential community impacts are often site-specific, and social life cycle assessment (s-LCA) can be better applied nearer deployment. Using s-LCA to evaluate local impacts does not replace the need for life cycle assessments for determination of broader environmental impacts.

Recommendation 3-5: Life cycle thinking for equity assessments—It is challenging to evaluate equity within life cycle assessment because methods are underdeveloped; therefore, the Department of Energy (DOE) should prioritize assessment of supply chains through principles of life cycle thinking to enable equity assessments that extend beyond the physical borders of the project site. DOE should require life cycle thinking for equity assessments when project sites are being considered, in order to identify hotspots and integrate risk and societal assessments. Relevant agencies, such as DOE and the U.S. Environmental Protection Agency, should evaluate life cycle assessment tools for their applicability to equity assessments and environmental justice, based on technology readiness level, time to deployment, and challenges and opportunities for selecting the sites of facilities.

Finding 3-9: Circularity of carbon products—The use of circularity strategies that keep carbon in products through multiple cycles of use and recycling, are starting to be considered as a way to significantly extend the storage duration of carbon, including fossil carbon. There is, however, a lack of understanding of the technical, economic, environmental, social, and policy implications of such strategies. This includes assessing technological performance of products over multiple cycles; evaluation of potential leakage over time; need for fresh raw materials; life cycle impacts beyond global warming potential; design of new value chains; business cases and policy and regulatory mechanisms that can drive circularity of carbon products; monitoring, verification, and reporting; and social acceptance.

Recommendation 3-6: Implications of circularity on carbon storage—The Department of Energy and the National Institute of Standards and Technology should support research that examines the feasibility and impacts of extending the duration of carbon storage through circularity strategies of short-lived products. This includes

a. **Building on state-of-the-art life cycle assessment approaches that are able to address circularity of CO_2-derived products over time.**
b. **Development of approaches and tools that allow the traceability and custody of carbon across value chains over time, including mapping of value chains, identification of cost-efficient tracking methods, and data collection and validation protocols.**

3.6.2 Research Agenda for LCA-TEA Use with CO_2 Utilization Technologies

Table 3-6 presents the committee's research agenda for LCA and TEA of CO_2 utilization technologies, including research needs (numbered by chapter), and related research agenda recommendations (a subset of research-related recommendations from the chapter). The table includes the relevant funding agencies or other actors; whether the need is for basic research, applied research, technology demonstration, or enabling technologies and processes for CO_2 utilization; the research themes) that the research need falls into; the relevant research area and product class covered by the research need; whether the relevant product(s) are long- or short-lived; and the source of the research need (chapter section, finding, or recommendation). The committee's full research agenda can be found in Chapter 11.

TABLE 3-6 Research Agenda for Life Cycle and Techno-Economic Assessment of CO_2 Utilization Technologies

Research, Development, and Demonstration Need	Funding Agencies or Other Actors	Basic, Applied, Demonstration, or Enabling	Research Area	Product Class	Long- or Short-Lived	Research Themes	Source
3-A. Understanding the impact of fluctuations in CO_2 purity in the life cycle and techno-economic assessment of CO_2 utilization technologies.	DOE-EERE DOE-FECM	Enabling	LCA/ TEA	All	Long-lived Short-lived	Environmental and societal considerations for CO_2 and coal waste utilization technologies	Fin. 3-4 Rec. 3-2
3-B. Development of improved CO_2 purification technologies that are more flexible, modular, and less energy-intensive.	DOE-EERE DOE-FECM DOE-BES	Basic Applied	Chemical	All	Long-lived Short-lived	Separations	Fin. 3-4 Rec. 3-2 Sec. 11.1.2
Recommendation 3-2: Research needs for CO_2 purity in techno-economic and life cycle assessments—The Department of Energy (DOE) and other relevant funding agencies should fund projects that examine the robustness of CO_2 utilization technologies to different CO_2 purities as well as fund further research and development of CO_2 purification technologies. Insights from these projects should be disseminated to the larger community by DOE. DOE should require awardees of applied research and development funding for CO_2 utilization technologies to perform techno-economic and life cycle assessments that explicitly address the purity requirements of the CO_2 streams.							
3-C. Understanding of non-CO_2-emissions impacts of CO_2 utilization technologies within life cycle assessments (e.g., impacts on chemical toxicity, water requirements, and air quality of carbon mineralization at the gigatonne scale).	DOE-EERE DOE-FECM EPA USGS	Enabling	LCA/ TEA	All	Long-lived Short-lived	Environmental and societal considerations for CO_2 and coal waste utilization technologies	Fin. 3-6 Rec. 3-4
Recommendation 3-4: Non-CO_2-emissions impacts within life cycle assessments—The Department of Energy and other relevant funding agencies such as the U.S. Environmental Protection Agency and the National Institute of Standards and Technology should support research into improving evaluation of non-CO_2-emissions impacts within life cycle assessments (LCAs) of CO_2 utilization technologies, including **a. Evaluating the appropriate but differentiated applications for global and local impact categories, as the latter generally involves data and information with high spatial and temporal granularity (e.g., processes versus facilities, technology readiness level of various components of the technology).** **b. Evaluating appropriate applications of social LCA (s-LCA) and further developing s-LCA tools and their potential integration with environmental LCA and techno-economic assessments.**							
3-D. Development of life cycle assessment approaches that can address circularity of CO_2-derived products over time.	DOE-FECM National Laboratories NIST	Enabling	LCA/ TEA	Chemicals Polymers	Short-lived	Environmental and societal considerations for CO_2 and coal waste utilization technologies	Rec. 3-6
3-E. Understanding the flows of carbon through product life cycles to enable a circular carbon system, including identifying leakage potential from circular systems, the fate of products under different end of life conditions, and how processes and demand may evolve through multiple cycles of use and reuse.	DOE-FECM National Laboratories NIST	Enabling	LCA/ TEA	All	Long-lived Short-lived	Environmental and societal considerations for CO_2 and coal waste utilization technologies	Fin. 3-9 Rec. 3-6
3-F. Development of approaches and tools to trace carbon across value chains over time, including mapping of value chains, identification of tracking methods, and data collection and validation protocols.	DOE-FECM NIST	Enabling	LCA/ TEA	All	Long-lived Short-lived	Environmental and societal considerations for CO_2 and coal waste utilization technologies	Rec. 3-6
Recommendation 3-6: Implications of circularity on carbon storage—The Department of Energy and the National Institute of Standards and Technology should support research that examines the feasibility and impacts of extending the duration of carbon storage through circularity strategies of short-lived products. This includes: **a. Further development of life cycle assessment approaches that are able to address circularity of CO_2 based products over time** **b. Development of approaches and tools that allow the traceability and custody of carbon across value chains over time, including mapping of value chains, identification of tracking methods, and data collection and validation protocols.**							

NOTE: ARPA-E = Advanced Research Projects Agency–Energy; BES = Basic Energy Sciences; DoD = Department of Defense; DOE = Department of Energy; DOT = Department of Transportation; EERE = Office of Energy Efficiency and Renewable Energy; EPA = U.S. Environmental Protection Agency; FECM = Office of Fossil Energy and Carbon Management; FHWA = Federal Highway Administration; NSF = National Science Foundation; OSMRE = Office of Surface Mining Reclamation and Enforcement; OST-R = Office of the Assistant Secretary for Research and Technology; USGS = U.S. Geological Survey.

3.7 REFERENCES

Almeida, D.T.L., C. Charbuillet, C. Heslouin, A. Lebert, and N. Perry. 2020. "Economic Models Used in Consequential Life Cycle Assessment: A Literature Review. " *Procedia CIRP* 90:187–191. https://doi.org/10.1016/j.procir.2020.01.057.

Aparcana, S., and S. Salhofer. 2013. "Development of a Social Impact Assessment Methodology for Recycling Systems in Low-Income Countries." *International Journal of Life Cycle Assessment* 18(5):1106–1115. https://doi.org/10.1007/s11367-013-0546-8.

Arcese, G., M. Lucchetti, and R. Merli. 2013. "Social Life Cycle Assessment as a Management Tool: Methodology for Application in Tourism." *Sustainability* 5(8):3275–3287. https://doi.org/10.3390/su5083275.

Bamana, G., J.D. Miller, S.L. Young, and J.B. Dunn. 2021. "Addressing the Social Life Cycle Inventory Analysis Data Gap: Insights from a Case Study of Cobalt Mining in the Democratic Republic of the Congo." *One Earth* 4:1704–1714. https://doi.org/10.1016/j.oneear.2021.11.007.

Barthel, M., J.A. Fava, C.A. Harnanan, P. Strothmann, S. Khan, and S. Miller. 2015. "Hotspots Analysis: Providing the Focus for Action." In *Life Cycle Management. LCA Compendium—The Complete World of Life Cycle Assessment*, G. Sonnemann and M. Margni, eds. Dordrecht: Springer. https://doi.org/10.1007/978-94-017-7221-1_12.

Bashmakov, I.A., L.J. Nilsson, A. Acquaye, C. Bataille, J.M. Cullen, S. de la Rue du Can, M. Fischedick, Y. Geng, and K. Tanaka. 2022. "Industry." In *IPCC, 2022: Climate Change 2022: Mitigation of Climate Change. Contribution of Working Group III to the Sixth Assessment Report of the Intergovernmental Panel on Climate Change*, P.R. Shukla, J. Skea, R. Slade, A. Al Khourdajie, R. van Diemen, D. McCollum, M. Pathak, S. Some, P. Vyas, R. Fradera, M. Belkacemi, A. Hasija, G. Lisboa, S. Luz, and J. Malley, eds. Cambridge, UK: Cambridge University Press. https://doi.org/10.1017/9781009157926.013.

Baustert, P., E. Igos, T. Schaubroeck, L. Chion, A. Mendoza Beltran, E. Stehfest, D. van Vuuren, H. Biemans, and E. Benetto. 2022. "Integration of Future Water Scarcity and Electricity Supply into Prospective LCA: Application to the Assessment of Water Desalination for the Steel Industry." *Journal of Industrial Ecology* 26(4):1182–1194. https://onlinelibrary.wiley.com/doi/10.1111/jiec.13272.

Benitez, M. 2011. "Resituating Culture Centers Within a Social Justice Framework: Is There Room for Examining Whiteness?" In *Culture Centers in Higher Education: Perspectives on Identity, Theory, and Practice*, L.D. Patton, ed. New York: Routledge. https://doi.org/10.4324/9781003443971.

Bjørn, A., M. Owsianiak, A. Laurent, S.I. Olsen, A. Corona, and M.Z. Hauschild. 2018. "Scope Definition." *Life Cycle Assessment: Theory and Practice* 75–116.

Bouillass, G., I. Blanc, and P. Perez-Lopez. 2021. "Step-By-Step Social Life Cycle Assessment Framework: A Participatory Approach for the Identification and Prioritization of Impact Subcategories Applied to Mobility Scenarios." *The International Journal of Life Cycle Assessment* 26:2408–2435. https://link.springer.com/article/10.1007/s11367-021-01988-w.

Boulay, A.-M., J. Bare, L. Benini, M. Berger, M.J. Lathuillière, A. Manzardo, M. Margni, et al. 2018. "The WULCA Consensus Characterization Model for Water Scarcity Footprints: Assessing Impacts of Water Consumption Based on Available Water Remaining (AWARE)." *The International Journal of Life Cycle Assessment* 23(2):368–378. https://doi.org/10.1007/s11367-017-1333-8.

Bozeman III, J.F., E. Nobler, and D. Nock. 2022. "A Path Toward Systemic Equity in Life Cycle Assessment and Decision-Making: Standardizing Sociodemographic Data Practices." *Environmental Engineering Science* 39(9):759–769. https://doi.org/10.1089/ees.2021.0375.

Bradley, K., K. Aguillard, A. Benton, L. Erickson, S. Martinez, and B. McGill. 2022a. "Conducting Intensive Equity Assessments of Existing Programs, Policies, and Processes." Office of the Assistant Secretary for Planning and Evaluation, Department of Health and Human Services.

Bradley, K., K. Aguillard, A. Benton, L. Erickson, S. Martinez, and B. McGill. 2022b. "Tips for Conducting Equity Assessments." Office of the Assistant Secretary for Planning and Evaluation, Department of Health and Human Services. https://aspe.hhs.gov/sites/default/files/documents/b1b55dafaf47467313466932de31a6d5/Tips-Conduct-Equity-Assessmnts.pdf.

Buchner, G.A., A.W. Zimmermann, A.E. Hohgräve, and R. Schomäcker. 2018. "Techno-Economic Assessment Framework for the Chemical Industry—Based on Technology Readiness Levels." *Industrial and Engineering Chemistry Research* 57(25):8502–8517. https://doi.org/10.1021/acs.iecr.8b01248.

Burger Mansilha, M., M. Brondani, F.A. Farret, L. Cantorski da Rosa, and R. Hoffmann. 2019. "Life Cycle Assessment of Electrical Distribution Transformers: Comparative Study Between Aluminum and Copper Coils." *Environmental Engineering Science* 36(1):114–135.

CEQ (Council on Environmental Quality). n.d. "GHG Tools and Resources." https://ceq.doe.gov/guidance/ghg-tools-and-resources.html.

Chailleux, S. 2020. "Making the Subsurface Political: How Enhanced Oil Recovery Techniques Reshaped the Energy Transition." *Environment and Planning C: Politics and Space* 38(4):733–750. https://doi.org/10.1177/2399654419884077.

Chauvy, R., R. Lepore, P. Fortemps, and G. De Weireld. 2020. "Comparison of Multi-Criteria Decision-Analysis Methods for Selecting Carbon Dioxide Utilization Products." *Sustainable Production and Consumption* 24(October):194–210. https://doi.org/10.1016/j.spc.2020.07.002.

Ciroth, A., and F. Eisfeldt. 2016. "PSILCA—A Product Social Impact Life Cycle Assessment Database." https://www.openlca.org/wp-content/uploads/2016/08/PSILCA_documentation_v1.1.pdf.

Clark, C. 2023. "Perspective from the DOE's Justice40 Initiative." Presentation to the committee. May 3. Washington, DC: National Academies of Sciences, Engineering, and Medicine.

Converse, P.E. 2006. "The Nature of Belief Systems in Mass Publics (1964)." *Critical Review* 18:1–3, 1–74. https://doi.org/10.1080/08913810608443650.

Cooney, G., J. Benitez, U. Lee, and M. Wang. 2022. "Clarification to Recent Publication—Incremental Approach for the Life-Cycle Greenhouse Gas Analysis of Carbon Capture and Utilization." Technical Publications. Lemont, IL: Argonne National Laboratory. https://greet.es.anl.gov/files/ccu_lca_memo.

Cremonese, L., T. Strunge, B. Olfe-Kräutlein, T. Strung, H. Naims, A. Zimmerman, T. Langhorst, et al. 2022. "Making Sense of Techno-Economic and Life Cycle Assessment Studies for CO2 Utilization." Global CO_2 Initiative. University of Michigan.

de Kleijne, K., S.V. Hanssen, L. van Dinteren, M.A.J. Huijbregts, R. van Zelm, and H. de Coninck. 2022. "Limits to Paris Compatibility of CO_2 Capture and Utilization." *One Earth* 5(2):168–185. https://doi.org/10.1016/j.oneear.2022.01.006.

Di Maio, F., and P.C. Rem. 2015. "A Robust Indicator for Promoting Circular Economy Through Recycling." *Journal of Environmental Protection* 6(10):1095–1104. https://doi.org/10.4236/jep.2015.610096.

Dieterle, M., P. Schäfer, and T. Viere. 2018. "Life Cycle Gaps: Interpreting LCA Results with a Circular Economy Mindset." *Procedia CIRP* 69:764–768.

Dincer, I., and C. Zamfirescu. 2018. "Sustainability Dimensions of Energy." Pp. 102–151 in *Comprehensive Energy Systems*, I. Dincer, ed. Amsterdam, Elsevier. https://www.sciencedirect.com/science/article/abs/pii/B9780128095973001048.

DOE (Department of Energy). 2011. "Technology Readiness Assessment Guide." DOE G 413.3-4A. Approved 9-15-2011. https://www.directives.doe.gov/directives-documents/400-series/0413.3-EGuide-04a-admchg1/@@images/file. Washington, DC: Department of Energy.

DOE. 2023. "Funding Notice: Bipartisan Infrastructure Law: Regional Direct Air Capture Hubs." DE-FOA-0002735. https://www.energy.gov/fecm/funding-notice-bipartisan-infrastructure-law-regional-direct-air-capture-hubs.

Dreyer, L.C., M.Z. Hauschild, and J. Schierbeck. 2010 "Characterisation of Social Impacts in LCA. Part 2: Implementation in Six Company Case Studies." *The International Journal of Life Cycle Assessment* 15:385–402.

Druckman, J.N. 2014. "Pathologies of Studying Public Opinion, Political Communication, and Democratic Responsiveness." *Political Communication* 31(3):467–492.

Ecochain. 2023. "CO_2 footprint vs. Environmental Footprint of Products—The Main Differences." https://ecochain.com/blog/product-co2-footprint-vs-environmental-footprint.

Edelen, A., and W. Ingwersen. 2016. "Guidance on Data Quality Assessment for Life Cycle Inventory Data." EPA/600/R-16/096. U.S. Environmental Protection Agency. https://cfpub.epa.gov/si/si_public_file_download.cfm?p_download_id=528687&Lab=NRMRL.

Edvinsson, S., E.H. Lundevaller, and G. Malmberg. 2013. "Do Unequal Societies Cause Death Among the Elderly? A Study of the Health Effects of Inequality in Swedish Municipalities in 2006." *Global Health Action* 6:1–9. https://doi.org/10.3402/gha.v6i0.19116.

EIGA. 2018. "Minimum Specifications for Food Gas Applications." EIGA Doc 126/20, revision of Doc 126/18. https://www.eiga.eu/ct_documents/doc126-pdf.

Ekener, E., J. Hansson, and M. Gustavsson. 2018. "Addressing Positive Impacts in Social LCA—Discussing Current and New Approaches Exemplified by the Case of Vehicle Fuels." *The International Journal of Life Cycle Assessment* 23(3):556–568. https://doi.org/10.1007/s11367-016-1058-0.

Ekvall, T. 2019. "Attributional and Consequential Life Cycle Assessment." In *Sustainability Assessment at the 21st Century*, M.J. Bastante-Ceca, J.L. Fuentes-Bargues, L.H. Florin-Constantin Mihai, and C. Iatu, eds. IntechOpen. https://doi.org/10.5772/intechopen.78105.

Ellen Macarthur Foundation and Granta. 2015. "Circularity Indicators: An Approach to Measuring Circularity." Ellen Macarthur Foundation. https://emf.thirdlight.com/link/3jtevhlkbukz-9of4s4/@/preview/1?o.

Elmallah, S., and J. Rand. 2022. "After the Leases Are Signed, It's a Done Deal: Exploring Procedural Injustices for Utility-Scale Wind Energy Planning in the United States." *Energy Research and Social Science* 89(July):102549. https://doi.org/10.1016/j.erss.2022.102549.

E.O. (Executive Order) 14008. January 27, 2021. "Executive Order on Tackling the Climate Crisis at Home and Abroad." https://www.whitehouse.gov/briefing-room/presidential-actions/2021/01/27/executive-order-on-tackling-the-climate-crisis-at-home-and-abroad.

EPA (U.S. Environmental Protection Agency). 2024. "EJScreen: Environmental Justice Screening and Mapping Tool." https://www.epa.gov/ejscreen.

European Committee for Standardization. 2020. "ISO 14025—Environmental Labels and Declarations—Type III Environmental Declarations—Principles and Procedures 13.020.50." Berlin: Beuth Verlag GmbH. https://www.iso.org/standard/38131.html.

European Committee for Standardization. 2022a. "ISO 14040—Environmental Management—Life Cycle Assessment—Principles and Framework 13.020.10." Berlin: Beuth Verlag GmbH. https://www.iso.org/standard/37456.html.

European Committee for Standardization. 2022b. "ISO 14044—Environmental Management—Life Cycle Assessment—Requirements and Guidelines. 13.020.10." Berlin: Beuth Verlag GmbH. https://www.iso.org/standard/38498.html.

Feldman, S. 1988. "Structure and Consistency in Public Opinion: The Role of Core Beliefs and Values." *American Journal of Political Science* 32(2):416–440. https://doi.org/10.2307/2111130.

Feldman, S. 2003. "Values, Ideology, and the Structure of Political Attitudes." Pp. 477–508 in *Oxford Handbook of Political Psychology*, D.O. Sears, L. Huddy, and R. Jervis, eds. Oxford University Press.

Fernández-Dacosta, C., M. van der Spek, C.R. Hung, G.D. Oregionni, R. Skagestad, P. Parihar, D.T. Gokak, A.H. Strømman, and A. Ramirez. 2017. "Prospective Techno-Economic and Environmental Assessment of Carbon Capture at a Refinery and CO_2 Utilisation in Polyol Synthesis." *Journal of CO_2 Utilization* 21(October 1):405–422. https://doi.org/10.1016/j.jcou.2017.08.005.

Finkbeiner, M., and V. Bach. 2021. "Life Cycle Assessment of Decarbonization Options—Towards Scientifically Robust Carbon Neutrality." *International Journal of Life Cycle Assessment* 26:635–639. https://doi.org/10.1007/s11367-021-01902-4.

Foley, D.K. 1966. "Resource Allocation and the Public Sector." Yale University.

Fortier, M.-O.P., L. Teron, T.G. Reames, D.T. Munardy, and B.M. Sullivan. 2019. "Introduction to Evaluating Energy Justice Across the Life Cycle: A Social Life Cycle Assessment Approach." *Applied Energy* 236:211–219. https://doi.org/10.1016/j.apenergy.2018.11.022.

Frank, T. 2023. "How the White House Found EJ Areas Without Using Race." *E&E News*. https://www.eenews.net/articles/how-the-white-house-found-ej-areas-without-using-race.

Garcia-Garcia, G., M. Cruz Fernandez, K. Armstrong, S. Woolass, and P. Styring. 2021. "Analytical Review of Life-Cycle Environmental Impacts of Carbon Capture and Utilization Technologies." *ChemSusChem* 14(4):995–1015. https://chemistry-europe.onlinelibrary.wiley.com/doi/10.1002/cssc.202002126.

Goglio, P., A.G. Williams, N. Balta-Ozkan, N.R.P. Harris, P. Williamson, D. Huisingh, Z. Zhang, and M. Tavoni. 2020. "Advances and Challenges of Life Cycle Assessment (LCA) of Greenhouse Gas Removal Technologies to Fight Climate Changes." *Journal of Cleaner Production* 244:118896. https://doi.org/10.1016/j.jclepro.2019.118896.

Greer, F. 2023. "Leveraging Environmental Assessment and Environmental Justice to Deliver Equitable, Decarbonized Built Infrastructure." *Environmental Research: Infrastructure and Sustainability* 3(4):040401. https://doi.org/10.1088/2634-4505/ad084b.

Greig, C., A. Garnett, J. Oesch, and S. Smart. 2014. "Guidelines for Scoping and Estimating Early Mover CCS Projects." University Queensland, Brisbane. https://anlecrd.com.au/projects/guidelines-for-scoping-estimating-early-mover-ccs-projects.

Grim, R.G., Z. Huang, M.T. Guarnieri, J.R. Ferrell, L. Tao, and J.A. Schaidle. 2020. "Transforming the Carbon Economy: Challenges and Opportunities in the Convergence of Low-Cost Electricity and Reductive CO_2 Utilization." *Energy and Environmental Science* 13(2):472–494.

Grubert, E. 2023. "Results from a Survey of Life Cycle Assessment-Aligned Socioenvironmental Priorities in US and Australian Communities Hosting Oil, Natural Gas, Coal, and Solar Thermal Energy Production." *Environmental Research: Infrastructure and Sustainability* 3(1):015007.

Harman, N.J., and H. Wang. 2022. "Electrochemical CO_2 Reduction in the Presence of Impurities: Influences and Mitigation Strategies." *Angewandte Chemie* 134(52):e202213782. https://doi.org/10.1002/anie.202213782.

Hasala, D., S. Supak, and L. Rivers. 2020. "Green Infrastructure Site Selection in the Walnut Creek Wetland Community: A Case Study from Southeast Raleigh, North Carolina." *Landscape and Urban Planning* 196(April):103743. https://doi.org/10.1016/j.landurbplan.2020.103743.

Hauschild, M.Z., and M.A.J. Huijbregts. 2015. "Life Cycle Impact Assessment. Springer Dordrech." https://doi.org/10.1007/978-94-017-9744-3.

Hauschild, M.Z., R.K. Rosenbaum, and S.I. Olsen. 2018. *Life Cycle Assessment: Theory and Practice*. Cham: Springer International Publishing. https://doi. org/10.1007/978-3-319-56475-3.

Hauschild, M.Z., T. E. McKone, K. Arnbjerg-Nielsen, T. Hald, B.F. Nielsen, S.E. Mabit, and P. Fantke. 2022. "Risk and Sustainability: Trade-offs and Synergies for Robust Decision Making." *Environmental Sciences Europe* 34:11.

Huang, Z., R.G. Grim, J.A. Schaidle, and L. Tao. 2021. "The Economic Outlook for Converting CO_2 and Electrons to Molecules." *Energy and Environmental Science* 14(7):3664–3678.

IEA (International Energy Agency). 2020a. "CCUS Around the World in 2021." Paris: IEA. https://www.iea.org/reports/ccus-around-the-world-in-2021.

IEA. 2020b. "Projected Costs of Generating Electricity 2020." Paris: IEA. https://www.iea.org/reports/projected-costs-of-generating-electricity-2020.

IEAGHG. 2023. "Techno-Economic Assessment of Electro-Chemical CO_2 Conversion Technologies." https://ieaghg.org/publications/technical-reports/reports-list/9-technical-reports/1096-2023-03-techno-economic-assessment-of-electro-chemical-co2-conversion-technologies.

Intratec Solutions. 2023. "Construction Cost Location Factors." *Medium*. https://medium.com/intratec-knowledge-base/construction-cost-location-factors-eeaf46d7acbb.

Iribarren, D., R. Calvo-Serrano, M. Martín-Gamboa, Á. Galán-Martín, and G. Guillén-Gosálbez. 2022. "Social Life Cycle Assessment of Green Methanol and Benchmarking Against Conventional Fossil Methanol." *Science of The Total Environment* 824(June):153840. https://doi.org/10.1016/j.scitotenv.2022.153840.

ISO (International Organization for Standardization). 2022. "Net Zero Guidelines." IWA 42:2022. Section 9.2.2. https://www.iso.org/obp/ui/en/#iso:std:iso:iwa:42:ed-1:v1:en.

ISO. 2024. "Circular Economy—Vocabulary, Principles and Guidance for Implementation." https://www.iso.org/standard/80648.html.

Jiang, K., P. Ashworth, S. Zhang, X. Liang, Y. Sun, and D. Angus. 2020. "China's Carbon Capture, Utilization and Storage (CCUS) Policy: A Critical Review." Renewable and Sustainable Energy Reviews 119:109601. https://doi.org/10.1016/j.rser.2019.109601.

Jordaan, S.M., and C. Wang, 2021. "Electrocatalytic Conversion of Carbon Dioxide for the Paris Goals." *Nature Catalysis* 4(11):915–920.

Kähler, F., O. Porc, and M. Carus. 2023. "RCI Carbon Flows Report: Compilation of Supply and Demand of Fossil and Renewable Carbon on a Global and European Level." *Renewable Carbon Initiative*. https://renewable-carbon.eu/publications/product/the-renewable-carbon-initiatives-carbon-flows-report-pdf.

Kowalczyk, Z., S. Twardowski, M. Malinowski, and M. Kuboń, 2023. "Life Cycle Assessment (LCA) and Energy Assessment of the Production and Use of Windows in Residential Buildings." *Scientific Reports* 13(1):19752. Vancouver.

Langhorst, T., S. McCord, A. Zimmermann, L. Müller, L. Cremonese, T. Strunge, Y. Wang, et al. 2022. "Techno-Economic Assessment and Life Cycle Assessment Guidelines for CO_2 Utilization (Version 2.0)." *Global CO_2 Initiative, University of Michigan*. https://doi.org/10.7302/4190.

Lenzen, M. 2008. "Double-Counting in Life Cycle Calculations." *Journal of Industrial Ecology* 12(4):583–599. https://doi.org/10.1111/j.1530-9290.2008.00067.x.

Lettner, M., and F. Hesser. 2019. "Asking Instead of Telling—Recommendations for Developing Life Cycle Assessment Within Technical R&D Projects." Pp. 173–188 in *Progress in Life Cycle Assessment 2019*, S. Albrecht, M. Fischer, P. Leistner, and L. Schebek, eds. Sustainable Production, Life Cycle Engineering and Management. Cham: Springer. https://doi.org/10.1007/978-3-030-50519-6_13.

Levasseur, A., P. Lesage, M. Margni, L. Deschênes, and R. Samson. 2010. "Considering Time in LCA: Dynamic LCA and Its Application to Global Warming Impact Assessments." *Environmental Science and Technology* 44(8):3169–3174. https://doi.org/10.1021/es9030003.

Levasseur, A., P. Lesage, M. Margni, M. Brandão and R. Brandão, and R. Samson. 2012. "Assessing Temporary Carbon Sequestration and Storage Projects Through Land Use, Land-Use Change and Forestry: Comparison of Dynamic Life Cycle Assessment with Ton-Year Approaches." *Climatic Change* 115:759–776. https://doi.org/10.1007/s10584-012-0473-x.

Linder, M., S. Sarasini, and P. Van Loon. 2017. "A Metric for Quantifying Product-Level Circularity." *Journal of Industrial Ecology* 21(3):545–558. https://doi.org/10.1111/jiec.12552.

Lueddeckens, S., P. Saling, and E. Guenther. 2020. "Temporal Issues in Life Cycle Assessment—A Systematic Review." *The International Journal of Life Cycle Assessment* 25:1385–1401. https://link.springer.com/article/10.1007/s11367-020-01757-1.

Macombe, C., and D. Loeillet. 2014. "Social LCA in progress." *4th SocSem*.

Mahmud R., S.M. Moni, K. High, and M. Carbajales-Dale. 2021. "Integration of Techno-Economic Analysis and Life Cycle Assessment for Sustainable Process Design—A Review." *Journal of Cleaner Production* 317:128247. https://www.sciencedirect.com/science/article/abs/pii/S0959652621024641.

Malins, C., L. Pereira, Z. Popstoyanova, and M. Riemer. 2023. "Support to the Development of Methodologies for the Certification of Industrial Carbon Removals with Permanent Storage: Review of Certification Methodologies and Relevant EU Legislation." 330301431. ICF. https://climate.ec.europa.eu/document/download/28698b02-7624-4709-9aec-379b26273bc0_en?filename=policy_carbon_expert_carbon_removals_with_permanent_storage_en.pdf.

Manalal, T., M. Perez-Fortes, and A. Ramirez. n.d. "The Relevance of Hydrogen Origin for the Screening of Alternative Carbon-Based Technologies for Ethylene Production." Article in preparation.

Marler, R.T., and J.S. Arora. 2004. "Survey of Multi-Objective Optimization Methods for Engineering." *Structural and Multidisciplinary Optimization* 26(6):369–395. https://doi.org/10.1007/s00158-003-0368-6.

Mason, F., G.M. Stokes, S. Fancy, and V. Sick. 2023. "Implications of the Downstream Handling of Captured CO_2." *Frontiers in Climate* 5:1286588. https://doi.org/10.3389/fclim.2023.1286588.

Matthews, H., C. Hendrickson, and D. Matthews. 2014. "Life Cycle Assessment: Quantitative Approaches for Decisions That Matter."

Mattioda, R.A., P.T. Fernandes, J.L. Casela, and O. Canciglieri, Jr. 2017. "Social Life Cycle Assessment of Hydrogen Energy Technologies." Pp. 171–188 in *Hydrogen Economy*. Academic Press.

McCord, S., A.V. Zaragoza, P. Styring, L. Cremonese, Y. Wang, T. Langhorst, V. Sick. 2021. "Multi-Attributional Decision Making in LCA and TEA for CCU: An Introduction to Approaches and a Worked Example." Global CO_2 Initiative. University of Michigan. https://dx.doi.org/10.7302/805.

McCord, S., A. Ahmed, G. Cooney, R. Dominguez-Faus, R. Galindo, M. Krynock, M. Leitch, T. Strunge, E. Tan, V. Sick. 2023. "CCU TEA and LCA Guidance 2023—A Harmonized Approach." Workshop. May 16–18. Ann Arbor, MI. https://dx.doi.org/10.7302/8081.

Messagie, M., F. Boureima, N. Sergeant, J.-M. Timmermans, C. Macharis, and J. Van Mierlo. 2012. "Environmental Breakeven Point: An Introduction into Environmental Optimization for Passenger Car Replacement Schemes." Pp. 39–49 in WIT Transactions on The Built Environment. Volume 128. Spain: WIT Press. https://doi.org/10.2495/UT120041.

Miller, S.A., and G.A. Keoleian. 2015. "Framework for Analyzing Transformative Technologies in Life Cycle Assessment." *Environmental Science and Technology* 49(5):3067–3075. https://doi.org/10.1021/es505217a.

Moni, S.M., R. Mahmud, K. High, and M. Carbajales-Dale. 2019. "Life Cycle Assessment of Emerging Technologies: A Review." *Journal of Industrial Ecology* 24(1):52–63. https://doi.org/10.1111/jiec.12965.

Müller, S., P. Lesage, A. Ciroth, C. Mutel, B.P. Weidema, and R. Samson. 2016. "The Application of the Pedigree Approach to the Distributions Foreseen in Ecoinvent v3." *The International Journal of Life Cycle Assessment* 21:1327–1337.

Müller, L.J., A. Kätelhön, M. Bachmann, A. Zimmermann, A. Sternberg, and A. Bardow. 2020. "A Guideline for Life Cycle Assessment of Carbon Capture and Utilization." *Frontiers in Energy Research* 8:15. https://doi.org/10.3389/fenrg.2020.00015.

NASEM (National Academies of Sciences, Engineering, and Medicine). 2019. *Gaseous Carbon Waste Streams Utilization: Status and Research Needs*. Washington, DC: The National Academies Press. https://doi.org/10.17226/25232.

NASEM. 2023a. *Carbon Dioxide Utilization Markets and Infrastructure: Status and Opportunities: A First Report*. Washington, DC: The National Academies Press. https://doi.org/10.17226/26703.

NASEM. 2023b. "Energy Justic and Equity." Chapter 2 in *Accelerating Decarbonization in the United States: Technology, Policy, and Societal Dimensions*. Washington, DC: The National Academies Press. https://doi.org/10.17226/25931.

Newman, A.J.K., G.R.M. Dowson, E.G. Platt, H.J. Handford-Styring, and P. Styring. 2023. "Custodians of Carbon: Creating a Circular Carbon Economy." *Frontiers in Energy Research* 11:1124072. https://doi.org/10.3389/fenrg.2023.1124072.

Norris, G.A. 2001. "Integrating Economic Analysis into LCA. Environmental Quality Management." 10(3):59–64.

Offermann-van Heek, J., K. Arning, A. Linzenich, and M. Ziefle. 2018. "Trust and Distrust in Carbon Capture and Utilization Industry as Relevant Factors for the Acceptance of Carbon-Based Products." *Frontiers in Energy Research* 6(August):73. https://doi.org/10.3389/fenrg.2018.00073.

Offermann-van Heek, J., K. Arning, A. Sternberg, A. Bardow, and M. Ziefle. 2020. "Assessing Public Acceptance of the Life Cycle of CO2-Based Fuels: Does Information Make the Difference?" *Energy Policy* 143(August):111586. https://doi.org/10.1016/j.enpol.2020.111586.

Palstra, S.W.L., and H.A.J. Meijer. 2010. "Carbon-14 Based Determination of the Biogenic Fraction of Industrial CO2 Emissions–Application and Validation." *Bioresource Technology* 101(10):3702–3710.

Parsons, R.E., and L.K. Mottee. 2024. "Exploring Equity in Social Impact Assessment." *Current Sociology* 72(4):732–752. https://doi.org/10.1177/00113921231203170.

Pattnaik, F., B.R. Patra, J.A. Okolie, S. Nanda, A.K. Dalai, and S. Naik. 2022. "A Review of Thermocatalytic Conversion of Biogenic Wastes into Crude Biofuels and Biochemical Precursors." *Fuel* 320:123857. https://doi.org/10.1016/j.fuel.2022.123857.

Pinto, A.S.S., L.J. McDonald, J.L.H. Galvan, and M. McManus. 2024. "Improving Life Cycle Assessment for Carbon Capture and Circular Product Systems." *The International Journal of Life Cycle Assessment* 29(3):394–415. https://doi.org/10.1007/s11367-023-02272-9.

PSILCA. n.d. "PSILCA Understanding Social Impacts." https://psilca.net.

Rajan, K., J. Kennedy, and L. King. 2013. "Is Wealthier Always Healthier in Poor Countries? The Health Implications of Income, Inequality, Poverty, and Literacy in India." *Social Science and Medicine* 88:98–107.

Ramírez, A.R. 2022. "Accounting Negative Emissions. How Difficult Could It Be?" Pp. 57–79 in *Greenhouse Gas Removal Technologies*, M. Bui and N. Mac Dowell, eds. The Royal Society of Chemistry. https://doi.org/10.1039/9781839165245-00057.

Ravikumar, D., G. Keoleian, and S. Miller. 2020. "The Environmental Opportunity Cost of Using Renewable Energy for Carbon Capture and Utilization for Methanol Production." *Applied Energy* 279:115770.

Reijnders, L. 2022. "Life Cycle Sustainability Assessment of Biofuels." In *Reference Module in Earth Systems and Environmental Sciences*. Elsevier. https://doi.org/10.1016/B978-0-323-90386-8.00016-4.

Roussanaly, S., E. Rubin, M. Der Spek, G. Booras, N. Berghout, T. Fout, M. Garcia, et al. 2021. "Towards Improved Guidelines for Cost Evaluation of Carbon Capture and Storage." 1779820. https://doi.org/10.2172/1779820.

Royal Society of Chemistry. n.d. "Life Cycle Assessment." https://www.rsc.org/globalassets/22-new-perspectives/sustainability/progressive-plastics/explainers/progressive-plastics-explainer-8---life-cycle-assessment.pdf.

Rubin, E.S. 2014. "Seven Simple Steps to Improve Cost Estimates for Advanced Carbon Capture Technologies." Presented at the Department of Energy Transformational Carbon Capture Technology Workshop. Pittsburgh, PA. Department of Energy National Energy Technology Laboratory.

Rubin, E.S. 2016. "Evaluating the Cost of Emerging Technologies." Presentation to the CLIMIT Workshop on Emerging CO_2 Capture Technologies. Pittsburgh, Pennsylvania.

Rubin, E.S. 2019. "Improving Cost Estimates for Advanced Low-Carbon Power Plants." *International Journal of Greenhouse Gas Control* 88:1–9.

Ruttinger, A.W., S. Tavakkoli, H. Shen, C. Wang, and S.M. Jordaan. 2022. "Designing an Innovation System to Support Profitable Electro- and Bio-Catalytic Carbon Upgrade." *Energy & Environmental Science* 15(3):1222–1233. https://doi.org/10.1039/D1EE03753F.

Sadasivam, N. 2023. "Why the White House's Environmental Justice Tool Is Still Disappointing Advocates." *Grist*, February 27. https://grist.org/equity/white-house-environmental-justice-tool-cejst-update-race.

Sala, S., A. Vasta, L. Mancini, J. Dewulf, and E. Rosenbaum. 2015. "Social Life Cycle Assessment: State of the Art and Challenges for Supporting Product Policies." Joint Research Centre, European Commission. EUR 27624. Luxembourg: Publications Office of the European Union. JRC99101.

Saltelli, A., M. Ratto, T. Andres, F. Campolongo, J. Cariboni, D. Gatelli, M. Saisana, and S. Tarantola. 2008. "Global Sensitivity Analysis." In *The Primer*. Chichester: John Wiley and Sons Ltd.

Sarswat, A., D.S. Sholl, and R.P. Lively. 2022. "Achieving Order of Magnitude Increases in CO2 Reduction Reaction Efficiency by Product Separations and Recycling. *Sustainable Energy & Fuels* 6(20):4598–4604.

Shou, M., and T. Domenech. 2022. "Integrating LCA and Blockchain Technology to Promote Circular Fashion—A Case Study of Leather Handbags." *Journal of Cleaner Production* 373(November):133557. https://doi.org/10.1016/j.jclepro.2022.133557.

Sick, V., K. Armstrong, G. Cooney, L. Cremonese, A. Eggleston, G. Faber, G. Hackett, et al. 2020. "The Need for and Path to Harmonized Life Cycle Assessment and Techno-Economic Assessment for Carbon Dioxide Capture and Utilization." *Energy Technology* 8(11):1901034. https://doi.org/10.1002/ente.201901034.

Sick, V., G. Stokes, F. Mason, Y.-S. Yu, A. Van Berkel, R. Daliah, O. Gamez, C. Gee, and M. Kaushik. 2022. "Implementing CO2 Capture and Utilization at Scale and Speed." https://doi.org/10.7302/5825.

Sick, V., K. Armstrong, and S. Moni. 2023. "Harmonizing Life Cycle Analysis (LCA) and Techno-Economic Analysis (TEA) Guidelines: A Common Framework for Consistent Conduct and Transparent Reporting of Carbon Dioxide Removal and CCU Technology Appraisal." *Frontiers in Climate* 5:1204840. https://doi.org/10.3389/fclim.2023.1204840.

Singh, U., S. Banerjee, and T.R. Hawkins. 2023. "Implications of CO_2 Sourcing on the Life-Cycle Greenhouse Gas Emissions and Costs of Algae Biofuels." *ACS Sustainable Chemistry and Engineering* 11(39):14435–14444. https://doi.org/10.1021/acssuschemeng.3c02082.

Skone, T. 2019. "NETL CO2U LCA Guidance Document." DOE/NETL-2019/2069. Department of Energy National Energy Technology Laboratory. https://www.osti.gov/biblio/1566144.

Skone, T.J., M. Mutchek, M. Krynock, S. Moni, S. Rai, J. Chou, D.R. Carlson, et al. 2022. "Carbon Dioxide Utilization Life Cycle Analysis Guidance for the U.S. DOE Office of Fossil Energy and Carbon Management (Version 2.0)." https://www.osti.gov/servlets/purl/1845020.

Stegmann, P., V. Daioglou, M. Londo, D.P. Van Vuuren, and M. Junginger. 2022. "Plastic Futures and Their CO2 Emissions." *Nature* 612(7939):272–276. https://doi.org/10.1038/s41586-022-05422-5.

Stewart, D.-L. 2013. "Racially Minoritized Students at U.S. Four-Year Institutions." *Journal of Negro Education* 82(2): 184-197. https://www.jstor.org/stable/10.7709/jnegroeducation.82.2.0184.

Subramanian, V., and J.G. Guinée. 2021. "Implementing Safe by Design in Product Development Through Combining Risk Assessment and Life Cycle Assessment." The Netherlands: University Leiden. An IenW Report.

Tan, L., S.J. Mandley, and W. Peijnenburg, S.L. Waaijers-van Der Loop, D. Giesen, J.B. Legradi, and L. Shen. 2018. "Combining Ex-Ante LCA and EHS Screening to Assist Green Design: A Case Study of Cellulose Nanocrystal Foam." *Journal of Cleaner Production* 178(March):494–506. https://doi.org/10.1016/j.jclepro.2017.12.243.

Tanzer, S.E., K. Blok, and A. Ramírez Ramírez. 2023. "Scoping Costs and Abatement Metrics for Biomass with Carbon Capture and Storage—The Example of BioCCS in Cement." *International Journal of Greenhouse Gas Control* 125:103864. https://doi.org/10.1016/j.ijggc.2023.103864.

Thielges, S., B. Olfe-Kräutlein, A. Rees, J. Jahn, V. Sick, and R. Quitzow. 2022. "Committed to Implementing CCU? A Comparison of the Policy Mix in the US and the EU." *Frontiers in Climate* 4(2022):943387. https://www.frontiersin.org/articles/10.3389/fclim.2022.943387.

Thonemann, N. 2020. "Environmental Impacts of CO_2-Based Chemical Production: A Systematic Literature Review and Meta-Analysis." *Applied Energy* 263:114599.

Thonemann, N., and M. Pizzol. 2019. "Consequential Life Cycle Assessment of Carbon Capture and Utilization Technologies Within the Chemical Industry." *Energy and Environmental Science*. RSC Publishing. 12:2253–2263. https://doi.org/10.1039/C9EE00914K.

Toniolo, S., R.C. Tosato, F. Gambaro, and J. Ren. 2020. "Chapter 3—Life Cycle Thinking Tools: Life Cycle Assessment, Life Cycle Costing and Social Life Cycle Assessment." Pp. 39–56 in *Life Cycle Sustainability Assessment for Decision-Making: Methodologies and Case Studies*. Amsterdam, Netherlands: Elsevier. https://doi.org/10.1016/B978-0-12-818355-7.00003-8.

Trading Economics. 2024. "Steel: Price - Chart - Historical Data - News." https://tradingeconomics.com/commodity/steel.

Tsoy, N., B. Steubing, C. van der Giesen, and J. Guinée. 2020. "Upscaling Methods Used in Ex Ante Life Cycle Assessment of Emerging Technologies: A Review." *International Journal of Life Cycle Assessment* 25:1680–1692. https://link.springer.com/article/10.1007/s11367-020-01796-8.

UNCTAD (United Nations Conference on Trade and Development). 2022. "Leveraging New Technologies' Impact Through Technology Assessment: Note by the UNCTAD Secretariat." https://unctad.org/system/files/official-document/ciid48_en.pdf.

UNEP (United Nations Environment Programme). 2009. "Guidelines for Social Life Cycle Assessment of Products." C. Benoît Norris and B. Mazjin, eds. United National Environment Programme. https://www.lifecycleinitiative.org/wp-content/uploads/2012/12/2009%20-%20Guidelines%20for%20sLCA%20-%20EN.pdf.

UNEP. 2020. "Guidelines for Social Life Cycle Assessment of Products and Organizations 2020." C. Benoît Norris, M. Traverzo, S. Neugebauer, E. Ekener, T. Schaubroeck, and S. Russo Garrido, eds. United Nations Environment Programme. https://wedocs.unep.org/20.500.11822/34554.

van der Spek, M., A. Ramírez Ramírez, and A. Faaij. 2017. "Challenges and Uncertainties of Ex Ante Techno-Economic Analysis of Low TRL CO_2 Capture Technology: Lessons from a Case Study of an NGCC with Exhaust Gas Recycle and Electric Swing Adsorption." *Applied Energy* 208:920–934.

Van Dijk, E., and H. Wilke. 1993. "Differential Interests, Equity, and Public Good Provision." *Journal of Experimental Social Psychology* 29(1):1–16.

Van Harmelen, T., E.K. Zondervan-Van Den Beuken, D.H. Brouwer, E. Kuijpers, W. Fransman, H.B. Buist, T.N. Ligthart, et al. 2016. "LICARA nanoSCAN—A Tool for the Self-Assessment of Benefits and Risks of Nanoproducts." *Environment International* 91:150–160. https://doi.org/10.1016/j.envint.2016.02.021.

Vidal, F., E.R. Van Der Marel, R.W.F. Kerr, C. McElroy, N. Schroeder, C. Mitchell, G. Rosetto, et al. 2024. "Designing a Circular Carbon and Plastics Economy for a Sustainable Future." *Nature* 626(7997):45–57. https://doi.org/10.1038/s41586-023-06939-z.

von der Assen, N., J. Jung, and A. Bardow. 2013. "Life-Cycle Assessment of Carbon Dioxide Capture and Utilization: Avoiding the Pitfalls." *Energy and Environmental Science* 6:2721–2734. RSC Publishing. https://doi.org/10.1039/C3EE41151F.

von der Assen, N., P. Voll, M. Peters, and A. Bardow. 2014. "Life Cycle Assessment of CO_2 Capture and Utilization: A Tutorial Review." *Chemical Society Reviews* 43:7982–7994. RSC Publishing. https://doi.org/10.1039/C3CS60373C.

von der Assen, N., L.J. Müller, A. Steingrube, P. Voll, and A. Bardow. 2016. "Selecting CO_2 Sources for CO_2 Utilization by Environmental-Merit-Order Curves." *Environmental Science and Technology* 50(3):1093–1101. https://doi.org/10.1021/acs.est.5b03474.

Vos, J., A. Ramírez Ramírez, and M. Pérez-Fortes. 2023. "Conceptual Design of Pretreatment Units for Co-Electrolysis of CO_2 and Water." *Computer Aided Chemical Engineering* 52:3153–3158. Elsevier. https://doi.org/10.1016/B978-0-443-15274-0.50503-5.

Vos, J., A. Ramírez Ramírez, and M. Perez-Fortes. 2024. "Is it Time to Move On? Limitations of Techno-Economic Assessments for Low-Temperature CO_2 Electrolysis."

WHEJAC (White House Environmental Justice Advisory Council). 2022. "Recommendations for the Climate and Economic Justice Screening Tool." Letter to Brenda Mallory, Council on Environmental Quality. August 16. https://www.epa.gov/system/files/documents/2022-08/CEJST%20Recommendations%20Letter%208_4_2022%20Final.pdf.

Wright-Mair, R. 2022. "The Costs of Staying: Experiences of Racially Minoritized LGBTQ + Faculty in the Field of Higher Education." *Innovative Higher Education* 48:329–350. https://doi.org/10.1007/s10755-022-09620-x.

Yang, S., K. Ma, Z. Liu, J. Ren, and Y. Man. 2020. "Chapter 5—Development and Applicability of Life Cycle Impact Assessment Methodologies." In *Life Cycle Sustainability Assessment for Decision-Making: Methodologies and Case Studies*. https://doi.org/10.1016/B978-0-12-818355-7.00005-1.

Zeug, W., A. Bezama, and D. Thrän. 2023. "Life Cycle Sustainability Assessment for Sustainable Bioeconomy, Societal-Ecological Transformation and Beyond." Pp. 131–159 in *Progress in Life Cycle Assessment 2021*, F. Hesser, I. Kral, G. Obersteiner, S. Hörtenhuber, M. Kühmaier, V. Zeller, and L. Schebek, eds. Sustainable Production, Life Cycle Engineering and Management. Cham: Springer International Publishing. https://doi.org/10.1007/978-3-031-29294-1_8.

Zhang, A., R.Y. Zhong, M. Farooque, K. Kang, and V.G. Venkatesh. 2020. "Blockchain-Based Life Cycle Assessment: An Implementation Framework and System Architecture." *Resources, Conservation and Recycling* 152:104512.

Zimmermann, A.W., and R. Schomäcker. 2017. "Assessing Early-Stage CO_2 Utilization Technologies—Comparing Apples and Oranges?" *Energy Technology* 5(6):850–860. https://doi.org/10.1002/ente.201600805.

Zuniga-Teran, A.A., A.K. Gerlak, A.D. Elder, and A. Tam. 2021. "The Unjust Distribution of Urban Green Infrastructure Is Just the Tip of the Iceberg: A Systematic Review of Place-Based Studies." *Environmental Science and Policy* 126(December):234–245. https://doi.org/10.1016/j.envsci.2021.10.001.

4

Policy and Regulatory Frameworks Needed for Economically Viable and Sustainable CO_2 Utilization

4.1 INTRODUCTION

Most forms of carbon dioxide (CO_2) utilization will not be competitive without a price on carbon or a subsidy-based model to support a market for CO_2-derived products. Recent federal subsidies signal the beginning of this growing market, but more policy is required to encourage demand for products and to support businesses entering this emerging sector. Developing goal-oriented, adaptable policy that encourages innovative technology could strengthen the impact of a carbon price. The CO_2 utilization sector can become an exemplar for policy that supports a quickly changing industry. Additionally, significant opportunity exists to prioritize justice goals and drive the build-out and implementation of the CO_2 utilization sector while ensuring that its outcomes are multifaceted and equitable.

This chapter addresses policy and regulatory frameworks needed to support the increased development and use of CO_2-derived products, including the societal considerations that policy can incorporate into project development, siting, and selecting processes—with the assumption that there will be an implicit price on carbon for the policy recommendations made. A variety of considerations can be categorized as economic and noneconomic drivers, which can be broken down further into demand- and supply-side considerations, and sector and societal impacts, respectively (see Figure 4-1). The combined impact of the economic and noneconomic drivers can create a sector with economically viable products, a sustainable market, adaptable policy and regulations, and equitable access to sector benefits.

This chapter reviews the existing policy landscape for CO_2 utilization and identifies the gaps and opportunities for policy to shape a market for CO_2-derived products. It then highlights opportunities for the federal government to support business development, particularly for small businesses, to diversify the market. Next, the chapter identifies key equity and justice considerations and best practices for public discourse and community engagement to help ensure that injustices are not created or exacerbated by the emerging sector. The chapter concludes with findings and recommendations related to the policy and regulatory frameworks needed for an economically viable and sustainable CO_2 utilization sector.

4.2 POLICY AND REGULATORY FRAMEWORKS

The committee's first report outlined the regulation and policy that would be needed to support CO_2 capture, utilization, storage, and transportation (NASEM 2023d). The committee identified key barriers and recommended solutions that policy and regulation could address, including internalizing carbon externalities (e.g., with a carbon tax) and subsidizing knowledge creation with grants for fundamental research and tax credits for pilot plants and

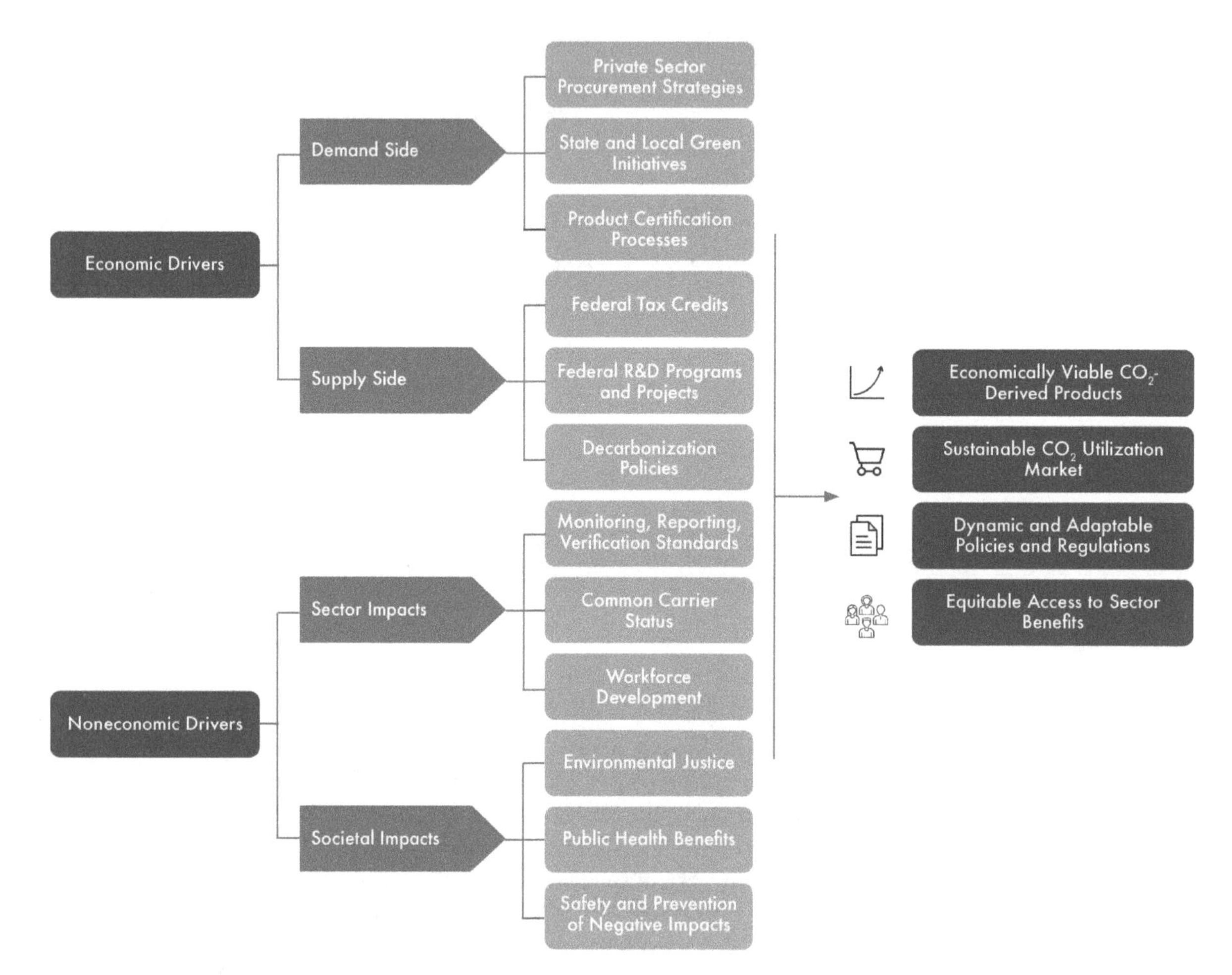

FIGURE 4-1 Summary of the policy and regulatory considerations and impacts for the emerging CO_2 utilization sector.
SOURCE: Icons from the Noun Project, https://thenounproject.com. CC BY 3.0.

demonstration units (NASEM 2023d, Finding 5.1); signaling a commitment to create a market for low-carbon technologies (NASEM 2023d, Finding 5.3); and accounting for distributional impacts of CO_2 utilization projects through processes that include community engagement (NASEM 2023d, Finding 5.9 and Recommendation 5.6). The committee continues to elevate Findings 5.1, 5.3, and 5.9 and Recommendation 5.6 from its first report as critical policy considerations for the CO_2 utilization sector.

This section discusses the existing policy frameworks for CO_2 utilization that aim to make CO_2-derived products economically viable. It reviews the economic and noneconomic drivers that exist and can be utilized as the sector builds out. It then identifies gaps in policy and makes recommendations that will support the production and ongoing market of CO_2-derived products, focusing on policies that deal with both environmental externalities and economic incentives.

4.2.1 Existing Incentives for CO_2 Utilization

4.2.1.1 Economic Drivers

The current cost of CO_2-derived products is greater than their incumbent equivalents in all cases considered by the committee (see Chapter 2). Most of these products are identical commodities and traded on world markets. However, a key difference between CO_2-derived products and their incumbent equivalents is carbon intensity (CI)—the measurement of a product's life cycle CO_2 emissions per unit. The economic rationale for consumers becomes more complex when CO_2-derived products display characteristics superior to incumbents, which provides an additional dimension of value to drive purchasing decisions beyond cost and CI (e.g., cured concrete has demonstrated enhanced structural performance compared to conventional concrete). In the absence of carbon border adjustment mechanisms (CBAMs)[1]—or other public or private policy that ascribes an economic value or promulgates a standard for CI—incentives that consider all dimensions of purchasing decisions are needed to prompt consumer demand.

Both identical substitutes and superior incumbent products are currently in the earliest stages of commercialization. Sustained demand signals and efficiency gains in production will be needed to drive down costs to approach current market prices for incumbents. To support the formation of commercial-scale markets for CO_2-derived products, this section discusses two broad categories of economic drivers: demand-side tools and supply-side tools. Both need to be applied simultaneously to scale up CO_2-derived products in a timely fashion and achieve meaningful market share.

4.2.1.1.1 Demand-Side Tools

Demand-side tools largely focus on CI-based thresholds for products and/or economic offsets for the purchase of CO_2-derived products. The consumers targeted by demand-side tools are mostly government agencies or private sector businesses. However, individual households can benefit from tools that decrease the cost of some CO_2-derived products, such as cleaning supplies.

A tool used to support demand in the private sector is procurement strategy, the purchase of upstream commodities used within a firm's value chain based on CI. This approach has been observed in cases like "green steel," where the European automotive industry finds it economically advantageous to pay a premium for lower-CI steel to meet customer preferences and corporate carbon climate ambitions (e.g., see Boston 2021 and Muslemani et al. 2022). However, it has not been observed for CO_2-derived products, given the abatement cost associated with these products compared to other strategies to meet corporate climate commitments (Comello et al. 2023; Fan and Friedmann 2021). For example, in maritime shipping, it may be less costly to first take energy efficiency measures to reduce emissions than to consider e-methanol or other CO_2-derived fuels (IRENA 2021).

Under current conditions, it is more cost-effective to pursue carbon abatement strategies other than CO_2-derived products to decarbonize scope emissions within a value chain, although this is industry- and brand-specific. For example, an industry standard that goes into effect in 2027 will drive demand for lower-CI aviation fuels (a scope 1 emission for the industry), especially for sustainable aviation fuels (SAFs), which can be derived from CO_2 (ICAO Environment 2023). In contrast, the availability of modular concrete blocks (a scope 3 emission for the housing industry) may not increase demand in the short term if alternative emissions reduction strategies (e.g., more efficient heating and cooling, upgraded insulation, and fuel switching from natural gas to electric) remain more cost-effective (Malinowski 2023). Moreover, even in the case of low-embodied carbon structures, there are lower-cost approaches to meeting design targets than using CO_2-derived products, such as material reuse (Malinowski 2023). Therefore, purchasing CO_2-derived products typically is not a preferred method in the private sector, given that other strategies to abate or transfer emissions are more cost-effective.

Within the public sector, various local, state, and federal programs are creating a demand signal for CO_2-derived products. Across existing "green initiatives," a few policies explicitly mention life cycle factors for product procurement (e.g., the U.S. Environmental Protection Agency [EPA] Environmentally Preferable Purchasing Program [EPA 2024, n.d.(g)]; the Federal Sustainability Plan [CEQ n.d.(a)]; and Orange County, California's Environmentally Preferable Purchasing Policy [Orange County Procurement Office 2022]). Only two federal initiatives explicitly mention CI considerations: the Federal Buy Clean Initiative—which partners with states to consolidate data sources and material standards for a more consistent market for lower-carbon materials[2]—and

[1] CBAM is an emerging policy tool that aims to cut global and national industry emission (e.g., see EU n.d.). However, currently, CO_2 utilization is not the lowest-cost approach to decarbonizing products in many cases and is therefore unlikely to be deployed as a first option for CBAM compliance.

[2] Beyond existing initiatives like Buy Clean, NASEM (2023a) recommended that DOE, EPA, and the National Institute of Standards and Technology develop standardized approaches for determining the CI of industrial products, with associated labeling program for consumer awareness (Recommendation 10-6, NASEM 2023a). Additionally, EPA should establish a tradeable performance standard for domestic and imported industrial products based on declining CI benchmarks for major product families, to be determined by DOE and the Department of Commerce (Recommendation 10-9, NASEM 2023a).

the Department of Energy's (DOE's) Utilization Procurement Grants (UPGrants) program—an economic-based incentive mechanism that provides grants to states, local governments, and public utilities to support the commercialization of technologies that reduce carbon emissions while also procuring and using commercial or industrial products derived from captured carbon emissions (NETL n.d.(b); White House 2023). The UPGrants are unique because they focus on creating a durable demand signal for CO_2-derived products by lowering the relative cost of those products and offering flexibility in how grant money can be used (e.g., a contract-for-difference, auction, reverse auction, or other structure can be employed). As the UPGrants are awarded, actual costs data will be revealed and collected, which will help to inform the potential of various products derived from captured carbon emissions and shape or expand the program to induce a further demand signal.

4.2.1.1.2 Supply-Side Tools

Supply-side tools are largely focused on reducing the cost to produce CO_2-derived products. The 45Q tax credit offered through the Inflation Reduction Act (IRA), which provides \$60/tonne CO_2 captured and utilized, is the most well-known supply-side incentive (H.R. 5376 2022). However, the value of the 45Q tax credit for utilization is less than that for CO_2 captured and permanently sequestered in geologic storage, which has a value of \$85/tonne. The disparity between the two credit values is not directly ascribed to permanence of CO_2 captured that would otherwise have been emitted to the atmosphere. For example, a project converting CO_2 to a long-lived product would still receive a lower tax credit than a project that geologically sequesters CO_2, despite the outcome of both being durable storage of CO_2.

While 45Q is useful in reducing the unit economics of CO_2 utilization, the Infrastructure Investment and Jobs Act (IIJA) offers various cost-share grants to offset the cost of plant, property, and equipment to demonstrate and/or deploy CO_2 utilization technologies at scale (H.R. 3684 2021). See Table 4-1 for a list of IIJA funding for carbon management programs and projects, totaling to about \$20 billion in new funding. These grants largely fall within the carbon management funding opportunities managed by DOE's Office of Fossil Energy and Carbon Management (FECM), of which up to \$46 million is available to develop technologies to remove, capture, and convert or store CO_2 from utility and industrial sources or the atmosphere (DOE 2022a; DOE-FECM n.d.(a)).[3]

Outside of these CO_2 utilization-specific supply-side tools, DOE's Loan Programs Office provides access to low-cost debt, which can significantly reduce the overall unit economics of CO_2-derived products (DOE-LPO n.d.). However, a project cannot receive both a grant and a loan from DOE. To prevent a "double benefit" from occurring, project development requires careful structuring and sequencing. There is no conflict in using a federal grant or a loan in combination with the 45Q tax credit (or any tax credit for that matter) to reduce the supply cost of CO_2-derived products.

4.2.1.2 Noneconomic Drivers

4.2.1.2.1 Existing Workforce and Translational Skill Sets

At present, the CO_2 sourced for utilization relies largely on point-source carbon capture technologies retrofitted onto existing polluting facilities such as industrial or power plants. An analysis of carbon capture retrofits found that more than 70 percent of coal plant retrofits will occur in the near term (by 2035), while about 70 percent of gas plant retrofits will occur in the long term (by 2050) (Larsen et al. 2021). Furthermore, Larsen et al. (2021) project that retrofit operations across the industrial and power sectors in the next 15 years will create up to 43,000 on- and off-site jobs, including installation, maintenance, labor, and chemical and water treatment. See Section 4.3.3 below for more about the upstream labor needs for the CO_2 utilization sector. This workforce will need to be maintained and, in some cases, grown as the facilities expand their capabilities.

Aspects of the CO_2 utilization value chain parallel those in the oil and gas sector, including the siting and development of facilities to capture, maintain, and prepare a resource for subsequent phases of production, transport, and transformation of the resource to a final product or end use. These similar value chain mechanisms mean

[3] DOE has a living list of funding and award announcements related to the IIJA and the IRA here: https://www.energy.gov/infrastructure/clean-energy-infrastructure-program-and-funding-announcements.

TABLE 4-1 Carbon Management Investments from the Infrastructure Investment and Jobs Act

Description	Amount
§ 40302—Carbon Utilization Program	$310 million over a 5-year period
§ 40303—Carbon Capture Technology Program	$100 million over a 5-year period
§ 40304—CO_2 Transportation Finance and Innovation Program	$2.1 billion over a 5-year period
§ 40305—Carbon Storage Validation and Testing	$2.5 billion over a 5-year period
§ 40308—Regional Direct Air Capture Hubs	$3.5 billion over a 5-year period
§ 40314—Regional Clean Hydrogen Hubs	$8 billion over a 5-year period
§ 41004—Carbon Capture Large-Scale Pilot Projects	$937 million over a 4-year period
§ 41004—Carbon Capture Demonstration Projects	$2.5 billion over a 4-year period
§ 41005—Direct Air Capture Technologies Prize Competitions	Precommercial: $15 million for FY 2022 Commercial: $100 million for FY 2022

SOURCES: Adapted from Clean Air Task Force (2021) and DOE-FECM (2022).

there is high transferability across existing professional, technical, and labor sector jobs. For example, Okoroafor et al. (2022) found that a variety of "noncore" technical skill sets—for example, project management, health and safety, and business development—in the oil and gas sector are transferrable to the carbon capture and storage, hydrogen storage, and geothermal energy sectors. Additionally, skills needed to perform extraction activities such as mining, electricity generation, pipeline construction, and manufacturing are prevalent in the fossil fuel sector (Tomer et al. 2021). If coordinated with the build-out of the CO_2 removal industry, the CO_2 utilization sector could develop in a more streamlined and accelerated manner through a reliance on similar workforces. (See Finding 4-2.)

The geography-specific nature of fossil infrastructure and jobs is also an existing incentive for the budding CO_2 utilization industry. Because point-source CO_2 capture relies on heavy-emitting industries, most of the jobs requiring workers with transferable skills from oil and gas will likely exist in similar locations. A survey of oil and gas workers found that Texas, Louisiana, and California have the most workers and residents in the United States in addition to 131 petroleum refineries (as of January 2022) (Biven and Lindner 2023). Furthermore, in states considered for carbon management infrastructure (e.g., North Dakota, Oklahoma, Texas, West Virginia, and Wyoming), fossil-based jobs represent a significant portion of the labor force within smaller counties (30 to 50 percent of all workers are employed in the fossil fuel industry) (Tomer et al. 2021). Carbon management investments can be made in counties where transferable skills and expertise from fossil fuel jobs exist to scale up projects with the speed needed for the energy transition to net zero (Greenspon and Raimi 2022; Pett-Ridge et al. 2023; see Figure 4-2).

Workers in the oil and gas community may be eager to find work that builds on existing skill sets in locales where they have been historically successful, which could bode well for the carbon management industry, and therefore CO_2 utilization. Biven and Liner (2023) found that survey respondents would transition to jobs in well plugging and abandonment (34 percent), pipeline removal (30 percent), or carbon capture and storage (CCS) (15 percent) if skills training and education were free. Increasing the workforce's awareness of declining opportunities in oil and gas, offering more training focused on developing translational skills, and ensuring that these opportunities are accessible to all would support CO_2 utilization workforce pathways. (See Finding 4-2.)

4.2.1.2.2 Environmental Justice Considerations

The federal government's response to environmental justice (EJ) began in 1994 with the Executive Order on Federal Actions to Address Environmental Justice in Minority Populations and Low-Income Populations (E.O. 12898 1994). Over the next decade, EPA (e.g., the National Environmental Policy Act [NEPA] and Considering Environmental Justice During the Development of a Regulatory Action) and the establishment of EPA's Office of Environmental Justice advanced federal EJ considerations (CEQ 1997; EPA 2010). Simultaneously, several states established task forces, commissions, advisory boards, and state offices to address the environmental injustice experienced by minority, low-income, and Indigenous populations. The early adopters include California, Colorado,

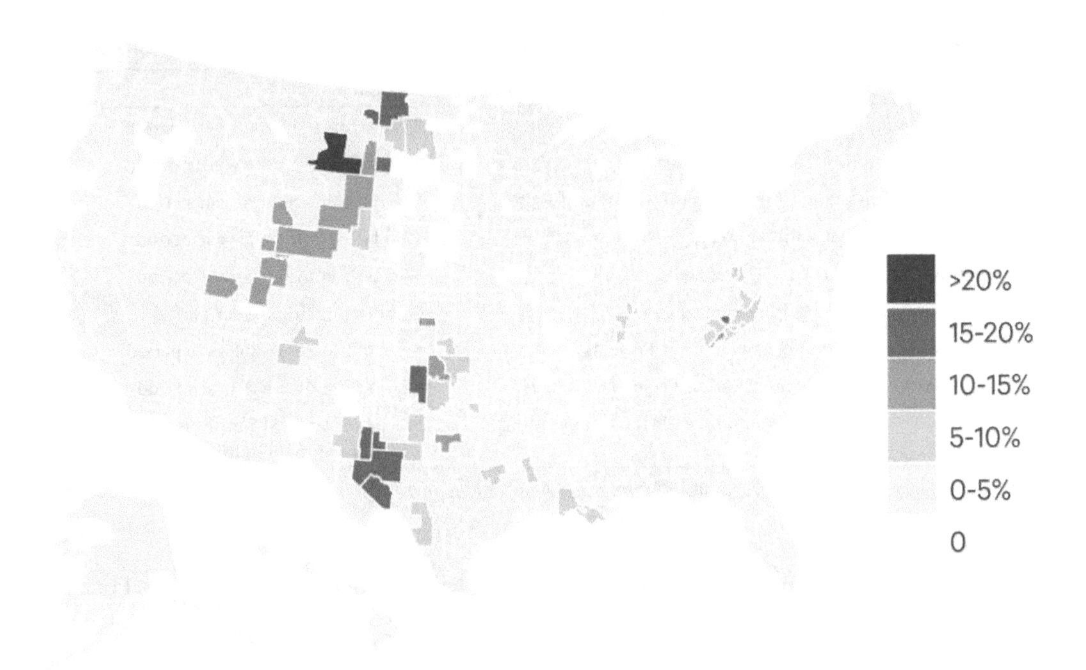

FIGURE 4-2 Fossil fuel employment by commuting zone at risk by 2050 under a net-zero scenario, indicating where people with transferable skills may need new job opportunities and where carbon management investments can be made.
NOTE: A commuting zone uses a hierarchical cluster analysis and U.S. Census Bureau data to reflect where people live and work, combining the nation's counties into 658 groupings.
SOURCE: Greenspon and Raimi (2022).

Illinois, Maryland, Massachusetts, Michigan, New Jersey, New York, Oregon, Pennsylvania, South Carolina, Vermont, Virginia, and Washington (National Conference of State Legislatures 2023).[4] More recently, the robust response of DOE to meet the goals outlined in the Justice40 Initiative creates specific areas of interest through which action eventually can incentivize the continued development of CO_2 utilization (E.O. 14008 2021; White House n.d.(b)). DOE identified eight policy priorities to guide the implementation of EJ across deployment of their programming (DOE n.d.(d)). Of those eight, the carbon management value chain and CO_2 utilization infrastructure have the potential to impact and be impacted by the following:

- **Decrease environmental exposure and burdens for disadvantaged communities.**[5] Low-income communities or communities of color are disproportionately impacted by air pollution and thus are more likely to experience adverse health effects (EPA n.d.(c)). Therefore, the potential for carbon capture equipment to directly address co-pollutants has drawn attention in the larger carbon management and EJ conversation. DOE's Justice40-covered programs, if implemented as intended, may result in a larger build-out of carbon management infrastructure that can support CO_2 utilization and decrease air pollution in affected communities (See Section 4.2.1.2.3.)

[4] See NASEM (2023a, 2023e) for a discussion about current state and federal initiatives to advance the energy transition through holistic programs that seek multifaceted outcomes, including advancing EJ through risk-management planning for stormwater management (e.g., see LA SAFE 2019); applying prices to industrial pollution (e.g., see Cap-and-Invest n.d.); creating working groups to advise policy related to EJ communities (e.g., see IWG 2021 and New York State 2022); and evaluating policy impacts on disadvantaged communities (e.g., see DOE 2023).

[5] Under Justice40, a "disadvantaged community" is a community that is marginalized, underserved, and overburdened by pollution (White House n.d.(b); see also Box 4-3 below).

- **Increase clean energy enterprise creation and contracting in disadvantaged communities.** Owing to the nature of the carbon management value chain and its economics, there are limited opportunities for smaller players, including minority-owned businesses, to partake in the build-out of this sector. The development of small minority and disadvantaged business enterprises has the potential to bring economic development to a disadvantaged community and diversify the carbon management value chain. In particular, the development of these enterprises can create diverse CO_2 utilization opportunities.
- **Increase clean energy jobs, job pipeline, and job training for individuals from disadvantaged communities.** Both racial minorities and women are underrepresented in the U.S. clean energy sector, and racial minorities are less likely to hold executive or leadership roles (DOE-OEJ 2023; E2 et al. 2021; Lehmann et al. 2021). To support the carbon management sector, between 390,000 and 1.8 million jobs are projected to be created in raw materials, engineering and design, construction, and operation and maintenance (Suter et al. 2022). The expansion of the carbon management sector provides an opportunity to diversify the workforce with a focus on offering workforce development opportunities to underserved communities.

It has yet to be determined if Justice40 is a durable policy or if 40 percent of benefits is an achievable target for federal investment, but significant opportunity exists to prioritize these EJ goals to drive the build-out and implementation of the CO_2 utilization sector, while ensuring its outcomes are multifaceted and more equitable. For example, carbon management falls largely under the climate change and clean energy topics of Justice40, but the administrative motivation to create a robust workforce and diversify supply chains also provides incentive to build up CO_2 utilization opportunities in regions transitioning from oil and gas production. Policies that seek to enshrine EJ considerations into CO_2 utilization investments can support the emerging sector in collecting and reporting the outcomes of investments for adaptive management (see Section 4.2.2.1) and by providing direct benefits to overburdened communities (see Section 4.4.3.3).

4.2.1.2.3 Health Co-Benefits from Carbon Management

Once emitted, greenhouse gases (GHGs) last up to thousands of years in the atmosphere and, in addition to climate change, contribute to adverse environmental effects, including air pollution, which leads to an estimated 53,200–355,000 premature deaths annually in the United States (EPA 2022; Mailloux et al. 2022; Vohra et al. 2021). Impacts of air pollution are disproportionately experienced across the nation. For example, *near source* pollution[6] has been found to lead to higher exposures to air contaminants, negatively impacting public health in these areas (EPA n.d.(e)). Emissions mitigation approaches—including carbon capture, focus on reducing emissions, and removing GHGs from the atmosphere—are expected to benefit human and public health. (See Finding 4-3.) Furthermore, replacing processes and products that emit GHGs with low-carbon alternatives will prevent the continued release of GHGs into the atmosphere. Box 4-1 summarizes a study conducted to analyze the potential health benefits from deploying carbon capture technologies on certain facilities.

Providing quantitative data to affected communities and policy makers about how carbon capture technologies can reduce criteria air pollutant emissions, and consequently benefit human health, will further inform the dialogue that is necessary to deploy carbon management technologies (see Section 4.4.1). Furthermore, the results of future-looking reports describing expected health benefits from carbon management can support deployment of CO_2 utilization technologies at the scale necessary to meet climate objectives. (See Recommendation 4-2.)

4.2.2 CO_2 Utilization Policy Gaps

Climate change policies need to be stable and durable such that investments and incentives are maintained while also evolving with new information and changing conditions (NRC 2010). Policy uncertainty hinders investment and adoption of technologies and limits otherwise profitable investments (NASEM 2023d, Finding 5.3). However, because developing a net-zero or circular economy at the scale required to address climate change is a

[6] Living near sources of air pollution, including major roadways, ports, rail yards, and industrial facilities.

BOX 4-1
Potential Health Benefits from Deploying Carbon Capture

Bennett et al. (2023) reviewed 54 facilities in seven industries—cement, coal power plants, ethanol, fertilizer and ammonia, iron and steel, natural-gas power plants, and petroleum refineries—to estimate regional air quality and health benefits that would result from carbon capture deployment. The study used the U.S. Environmental Protection Agency's Co-Benefits Risk Assessment Health Impacts Screening and Mapping Tool (COBRA)—which predicts health outcomes on adult and infant mortality; nonfatal heart attacks; respiratory and cardiovascular-related hospital admissions; acute bronchitis; upper and lower respiratory symptoms; asthma exacerbations and emergency room visits; minor restricted activity days;[a] and work loss days—to identify the air quality and health benefits through the combined removal of CO_2, NO_x, SO_2, and $PM_{2.5}$ via carbon capture. Looking at different regions across the United States, carbon capture on mid-Atlantic facilities is projected to provide the highest reduction in asthma exacerbation and mortality (Bennett et al. 2023). COBRA also was used to find the economic value associated with the changes in health impacts—that is, the monetary value of health benefits from carbon capture. As shown in Figure 4-1-1, the largest monetary value of health benefits is estimated to come from deploying carbon capture on cement, coal, and petroleum refineries. This outcome can be used to inform priority carbon capture investments when the investment goal is to reduce adverse health impacts from emissions of CO_2 and other pollutants. However, the study did not consider additional climate benefits from CO_2 removal or additional economic benefits from installing and maintaining carbon capture technology, both of which could be additional drivers for carbon management.

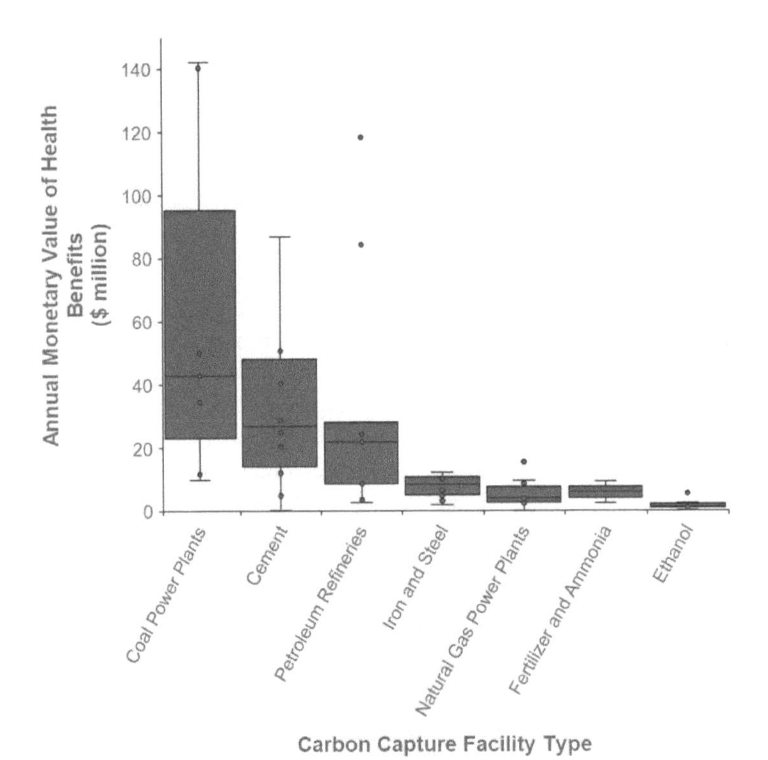

FIGURE 4-1-1 Estimated annual monetary value of health benefits from carbon capture industries.
NOTE: The circles represent the total average value of health benefits calculated for each of the 54 facilities studied.
SOURCE: Based on data from Bennett et al. (2023).

[a] Defined as days on which usual daily activities are reduced, but without falling into work absenteeism.

novel task, there is limited ability to anticipate the ways in which this process can fail, which adds difficulty to policy design (NASEM 2023c). The lack of adaptable policy serves as barrier to the adoption of carbon management infrastructure and CO_2 utilization by creating roadblocks to economic development. For example, there are legislative barriers to updating the 45Q tax credit to include a variety of eligible technology pathways, development, and deployment.

As it stands, the main policy mechanisms for the CO_2 utilization sector are tax credits, permitting and regulatory frameworks, and large omnibus legislation—all of which under the current system are slow moving and difficult to modify, especially when bipartisan consensus is required. (See Finding 4-4.) Adaptable policies can serve the CO_2 utilization sector by matching the pace of market and infrastructure development, and the science as it evolves. Adaptive management, an iterative learning process that produces improved understanding and management over time, is critical to the development of flexible policies (NASEM 2023c). Adaptive management can help identify and avoid unintended consequences like disincentivizing certain capture and removal pathways, while also leading to broader societal acceptance. For example, as more CO_2 utilization technologies and products become commercially available, there will be more data about the direct and indirect impacts experienced by the general public and communities hosting infrastructure. Analysis of these data can be used to modify CO_2 utilization policy to avoid unjust consequences to communities and the environment. The following section outlines how policy for CO_2-derived products can be designed and implemented to support and adapt to an emerging market and identifies potential economic and noneconomic tools to address gaps and barriers. (See Recommendation 4-3.)

4.2.2.1 Economic Tools

As discussed above, the current policy portfolio incentivizes CO_2 capture and production of CO_2-derived products through tax credits such as 45Q and 45V that lower the cost of supply.[7] However, it lacks demand incentives for CO_2-derived product uses or markets, especially relative to other carbon abatement approaches. The lack of a sufficient cost benefit for use of CO_2-derived products prevents uptake. For example, under current policies and prices, the use of SAF from captured CO_2 is more expensive than continuing to use aviation fuel derived directly from fossil sources (Bose 2023). Without financial justification or specific policies incentivizing the use of CO_2-derived products, there will be no economic rationale to drive market adoption. (See Finding 4-1 and Recommendation 4-1.)

Policy needs to create the conditions to both lower the costs and market frictions to produce CO_2-derived products and decrease barriers to demand, at least initially so that a minimal industry can be established. The latter has an analogy in the rise in demand for carbon-free electricity, driven by state-level policies to achieve increasingly less carbon-intensive generation. The rise in demand is accelerated by increasing electrification of economic sectors, like transportation through incentives for and adoption of electric vehicles. The direct policy target for clean grids coupled with increased demand for electricity is creating an enormous demand for the build-out of generation sources like solar and wind energy (e.g., see Motyka et al. n.d. and Wilson and Zimmerman 2023). Economic tools for the consumption of products include policies that support cost parity for consumers between CO_2-derived products and their carbon-intensive alternatives. This section outlines how economic tools can support CO_2 utilization development, including tools that encourage product uptake.

4.2.2.1.1 Energy Mix Uncertainties

There is still significant need and opportunity to grow the zero-carbon electricity share; in 2023, 60 percent of electricity generated by utility-scale facilities in the United States was from fossil fuels (e.g., coal, natural gas, petroleum), while 21 percent was from renewable energy, and 17 percent was from nuclear energy (EIA n.d.). Without first decarbonizing the energy system, an emerging CO_2 utilization sector using CO_2 sourced from carbon capture or removal strategies could be reliant on fossil energy, preventing the desired impact on the nation's climate goals from a life cycle assessment (LCA) perspective. The U.S. grid must continue to diversify while research and

[7] The 45V tax credit for clean hydrogen production has a base rate of $0.60/kg of qualified clean hydrogen produced (H.R. 5376 Sec. 13204).

development (R&D) on low-carbon technologies seeks ways to reduce energy requirements. Both diversification of the U.S. energy portfolio and R&D will support a CO_2 utilization sector that does not rely on fossil fuel combustion.

Recent U.S. legislative vehicles (e.g., the IIJA and the IRA) contain opportunities that encourage the build-out of renewable energy, including investment and production tax credits for installing solar and wind technologies, geothermal, tidal, and hydroelectric energy, and technology agnostic tax credits for clean energy production and investment (EPA n.d.(f)). Global projections show that renewable energy is becoming cost-competitive with fossil fuels, with around 187 gigawatts of all newly commissioned renewable capacity in 2022 having lower costs than fossil fuel–fired electricity (IRENA 2023). Nonetheless, a continual push to decarbonize the electricity mix is needed to develop an ethical and climate-impactful CO_2 utilization market that is less reliant on fossil energy production, providing opportunities for lower-CI pathways.

Regardless of how the carbon capture, removal, and utilization value chain acquires energy and whether the energy is low-carbon, the cost of electricity may be high if facilities do not have access to a wholesale utility-regulated market. The committee's first report discussed how uncertainty around the cost of electricity will influence CO_2 utilization market growth by directly impacting the potential to develop a CO_2 value chain (NASEM 2023d). It also identified that clustering energy supplies (i.e., hubs) could be more cost-effective for carbon capture, utilization, and storage (CCUS) processes and less likely to negatively impact other resources (NASEM 2023b).

4.2.2.1.2 Product Certification Processes and Reporting

Catalyzing CO_2 utilization markets via federal procurement necessitates clear standards and regulation of use for these products. The programs and policies encouraging procurement of CO_2-derived products do not yet have the transparency needed to advance procurement, including the creation of pilot programs or standardization guidelines. EPA has taken steps toward transparency with the Reducing Embodied Greenhouse Gas Emissions for Construction Materials and Products grant program, which helps businesses develop and verify Environmental Product Declarations (EPDs),[8] and create user-friendly standardized labels for products (GSA n.d.(b)). Grants are awarded to projects that fall under five categories, including projects that develop robust, standardized product category standards and projects that support EPD reporting, availability, and verification; standardization of EPD systems; and EPD integration into construction design and procurement systems (GSA n.d.(b)). Grant awardees can help standardize the CO_2-derived product industry by providing transparency on standardized data collection and analysis processes and developing tools and resources for EPD disclosures. However, further policy to support widespread adoption and standardization of CO_2-derived products is necessary.

4.2.2.2 Noneconomic Tools

This section outlines the noneconomic policy tools that can support CO_2 utilization development—namely, common carrier status, clarity regarding LCA standards, building materials standards, and workforce development.

4.2.2.2.1 Absence of Common Carrier Status Rules for CO_2 Transportation

Robust siting frameworks will be needed as demand for CO_2 transport infrastructure increases to support the sector. Historical trends for natural gas and electricity have shown that increased demand led to the development of regulatory frameworks for approving and evaluating infrastructure projects (Brown et al. 2023). State and federal agencies have been granted clear jurisdiction over siting gas pipelines and electricity transmission and have developed processes that are well defined, but not always streamlined. However, CO_2 pipelines may pose greater permitting challenges than gas pipelines or electricity transmission (Brown et al. 2023). For CO_2 midstream, there is uncertainty regarding common carrier rules and status[9] for interstate transportation because common carrier

[8] An EPD is an environmental report that provides that quantified environmental data using predetermined parameters and environmental information is consistent with ISO 14025:2006 (EPA n.d.(d)).

[9] Common carrier status means that conveyance of CO_2 for a fee is made open to the public by the operator, as opposed to private operation, where only specific actors may access such infrastructure.

status varies by state. It is unclear whether the entire pipeline is required to act as a common carrier when it passes through a state with common carrier requirement and a state without the requirement.

The lack of clear rules surrounding pipeline transportation of CO_2 may not be an immediate constraint on market growth, but it will limit the unit economics and ongoing market maturity if not resolved as soon as possible. A significant challenge associated with common carrier status is that pipeline owners have concerns about the chemical composition and potential reactivity of what others may inject for transport in their infrastructure—especially given the many potential sources of CO_2. Chemical impurities can lead to mechanical and metallurgical failures, which would be the responsibility of the pipeline owner. There is a space here for some type of policy or regulatory mechanism to certify CO_2 streams in a common carrier system. However, currently no agreed-upon approach exists to common carrier status that allows certification to happen, and more intentional work is needed to address this.

4.2.2.2.2 Clarity Regarding Assessment Standards

Owing to the myriad pathways CO_2 utilization can take, there is an ongoing discussion around monitoring, reporting, and verification (MRV) and how to standardize "best" practices. Because these practices could change with the development of new techniques and technologies, creating adaptable MRV frameworks is increasingly important.

As key aspects of MRV, LCAs for CO_2 utilization processes need to be better defined and standardized. LCA requirements often lack widespread adoption or clarification outside of these frameworks for federal tax credits or funding opportunities. For example, after an open comment period, the Internal Revenue Service (IRS 2021) determined that an LCA of GHGs—consistent with ISO 14044:2006—has to be submitted in writing and "either performed or verified by a professionally-licensed independent third party," along with the third party's documented qualifications for 45Q tax credit applicants.[10] Ultimately, the LCA needs to quantify the metric tonnes of qualified carbon oxide captured and permanently isolated from the atmosphere or displaced from being emitted into the atmosphere through use of eligible processes. Another example is DOE's Carbon Utilization Program, which requires eligible entities to show significant reductions in life cycle GHG emissions for CO_2-derived products compared to incumbent products using the National Energy Technology Laboratory's (NETL's) LCA Guidance Toolkit as a baseline (DOE n.d.(b); NETL n.d.(a); Skone et al. 2022). Creating standardized processes around LCA requirements and expectations is difficult, and nearly impossible if the purpose is not for a federal credit or funding opportunity.

There are also gaps in the use of social life cycle assessments (s-LCAs), which consider social impacts from a more quantitative perspective, as a part of federal frameworks or other standardized processes.[11] s-LCAs, along with other ways to integrate equity and justice concerns, are not comprehensive or a replacement for community engagement. However, the results of the assessment may enable clear communication of social benefits or the pathway's role in climate mitigation strategies in a way that a more traditional LCA may not. Therefore, these frameworks could play a role in addressing public acceptance issues while integrating social considerations into the traditionally high-level quantitative MRV discussion. Clarity around these systems and consistent regulatory and permitting processes will allow for more transparency among research entities, industry, government, and the public, while creating easier pathways for integrating CO_2 utilization processes and products in our economic system. See Chapter 3 for more information about LCAs and s-LCAs.

4.2.2.2.3 Flexible Policy for CO_2-Derived Building Materials

Strong demand signals exist to produce CO_2-derived building materials—concrete, carbon black additives, and drywall—owing to incentives and requirements for low-embodied carbon in new buildings. These new materials

[10] For these requirements, the IRS defined *life cycle GHG emissions* using the cradle-to-grave boundary, considering the entire product life cycle from raw material extraction until end of life (IRS 2021).

[11] See Ashley et al. (2022) for a proposed equity assessment framework that provides sufficient quantitative information about the effects of federal legislation to inform federal processes.

seek to reduce carbon emissions and minimize adverse environmental impacts from the construction industry. The increasing number of patent applications for CO_2 utilization technologies (a roughly 60 percent increase internationally between 2007 and 2017) reflects the interest from researchers and industries, with investment facilitating technologies to be developed at scale (Norhasyima and Mahlia 2018).[12] The committee's first report identified that CO_2-derived construction materials would motivate the "testing and validation of the new materials, creation of new environmental product declarations, and adaptation of building codes and standards" to support the consumption of these products (NASEM 2023d, p. 64).

Innovative solutions are emerging to address these challenges from different perspectives, such as using renewable energy to produce clinkers for cement, applying alternative materials with lower carbon footprint, capturing CO_2 produced from cement plants, and upcycling construction and demolition materials. (See Box 4-4 below for information about concerns expressed by construction professionals about CO_2-derived materials.) In addition to supporting R&D for construction materials, future policy could incentivize the development of building codes and regulations that are flexible and adaptable as new CO_2- and coal waste–derived materials are validated for use in buildings. For example, Bowles et at. (2022) provides sample language for building codes that could decrease the carbon impacts from the construction industry and support low-CI business models.

4.2.2.2.4 Workforce Development Considerations for Policy

As discussed above, there is an abundant workforce opportunity for carbon management infrastructure as the sector builds out and new prospects for career pathways develop. CO_2 policy design could incorporate the following workforce development considerations:

- *Facilitate localized development of workforce opportunities*. Jobs are frequently cited as an economic benefit to bring communities on board with new projects and investment, position the United States as a leader in manufacturing, and reduce GHG emissions (e.g., see Larsen et al. 2021 and White House 2021). Developing workforce standards and requirements can create measurable benchmarks that can be strategized around. Furthermore, a dedicated commitment at a larger level of workforce development needs to occur simultaneously with a skill building framework implemented through local actors (i.e., providing classes and certifications, and facilitating job placements through a national public university, community college, or trade school) (Coleman 2023).
- *Balance training the new workforce with upgrading the existing workforce*. Skills development is integral to bringing new laborers into the workforce and supporting incumbent workers through the transition. Strategies that will reach both groups of workers include targeted outreach for occupation types and cross-industry partnerships and apprenticeships (Zabin 2020). Opportunities to grow the sector include targeting young people interested in the green economy, which can benefit youth in underserved communities to provide them with a variety of career trajectories. For example, future iterations of the American Climate Corps could include opportunities around carbon management (White House n.d.(a)). Reaching both groups of workers will take intentional and effective outreach to meet the specific needs of the present and future workforce.

4.3 MECHANISMS FOR BUSINESS DEVELOPMENT

Discovering, developing, and commercializing CO_2 utilization processes and products are necessary as the nation transitions to a net-zero economy. This section highlights various considerations for business development, including market fit and access, available federal resources and programs, and potential workforce uncertainties.

[12] See Chapter 2 for more about the market for cement and construction aggregates, Chapter 5 for more about CO_2-derived building materials and the environmental impact of the processes to produce them, and Chapter 9 for more about coal waste–derived building materials.

4.3.1 Market Fit and Access for CO_2-Derived Products

CO_2-derived products have to be considered on a continuum with respect to market fit—the alignment between the specifications of the CO_2-derived product and the needs and preferences of the purchasing consumers and market access—the ability of a product to enter and operate in a particular market successfully (i.e., in an economically sustainable manner) (Aaker and Moorman 2017). This section considers elements of market fit and access, including upstream and downstream partners and commodity gatekeepers.

4.3.1.1 Market Fit

Market fit in the context of CO_2-derived products largely relies on the ability of products to satisfy the claims that they have lower CI than otherwise functionally identical products. For example, SAFs will not be chemically identical to current jet fuels but will have the same functionality and lower CI. Market fit becomes more complex in situations where product performance beyond CI is altered (e.g., concrete blocks that have been cured with CO_2 and thus exhibit greater load-bearing characteristics). Such cases create a new submarket in which customers appropriately pay for functional performance that is greater—or less—than the baseline.

In strict replacement cases, the market fit of CO_2-derived products mostly has been established already by the incumbent. Projections for the evolution of existing markets have to be considered to determine the long-term viability of the CO_2-derived product. In cases where new products cannot be strictly considered replacements, product–market fit analysis has to be continually conducted to determine if—and at what point in time—a sufficiently sized demand signal will emerge to support the economic case of a CO_2-derived product. It may take time for a unique and durable demand signal for new products to appear, as prospective customers need to accumulate knowledge and experience the product. For example, for CO_2-cured concrete blocks, customers must determine if the price premium justifies a one-for-one replacement with existing markets; if new applications can be found that push out incumbent solutions; and if new products perform in the field as expected given standards tests. These considerations require time and experience on the parts of both customers and producers to make informed judgments about the product.

Co-piloting and partnerships are crucial for products to move up the adoption-readiness level ladder. Key upstream partners for CO_2-derived product development include CO_2 supply, specialized capital equipment providers, and specific co-input providers (e.g., providers of emissions-free electricity and clean hydrogen). While production volumes are currently small and uncertain for most CO_2-derived products, key downstream partners are the direct customers that will help prove the commercialization and business case of the company producing the CO_2-derived product. Such agreements are especially beneficial for commodity products where consistent, intentional effort will be required to make the CO_2-derived product relatively cost-competitive with the incumbent that has decades of accumulated knowledge, resources, and market access (DOE n.d.(a)). At-volume, predictable demand over a long timeframe supported by a creditworthy off-taker creates the conditions for cost reductions and market adoption of CO_2-derived products (e.g., see Saiyid 2023).

For the CO_2 utilization company, a partnership agreement provides predictable demand over a long period, which could substantially support capital and operational planning, and a meaningful production/volume target that through accumulated learning effects, know-how, and value-engineering could reduce unit costs. In a sense, a downstream buyer's contract could pave the way for cost reductions in a product, not only for the company with the contract but also for other customers. This, in turn, could create greater demand for products like SAFs (bolstered in part by the International Civil Aviation Organization's emission targets), leading to wider market adoption. Continued engagement between upstream and downstream partners will support both the supply and demand for CO_2-derived products, thus setting the foundation for a CO_2 utilization market.

4.3.1.2 Market Access

Market access relates more to external factors and conditions beyond demand for product features, such as regulatory, legal, competitive, and economic factors that affect product entry into a market. Depending on the product and sector, businesses introducing CO_2-derived materials will need to identify and address the relevant gatekeepers to different commodity markets to gain commercial traction (Ahn 2019). For example, adherence to management requirements such as international quality management standards (e.g., see ASTM International n.d. and ISO 2015) and national chemical purity grading (e.g., see P.L. No. 94-469 1976 and Schieving 2018) assure customers of product reliability. Other gatekeepers for all CO_2-derived products include:

- **Price:** In the absence of sufficient demand driven by incentives or strategic differentiation strategies, price will be the most salient factor affecting market access. Included in this is the concept of margin (price minus cost), which must be sufficient to keep suppliers economically motivated and allow for reinvestment in product improvements. CO_2-derived products will need to have competitive prices, while maintaining adequate margins, to access commodity markets, which are differentiated (e.g., the price of fuels in California versus Texas may create an opportunity for CO_2-derived fuels in the former).
- **Volume:** In commodity markets, volume and production volume certainty are other gating factors to market access. Downstream users of commodities generally optimize their processes around a guaranteed supply through multiple vendors to increase cost efficiency and capacity utilization. Producers of CO_2-derived products will either need the ability to supply sufficient volumes at the onset of entering a market to satisfy procurement needs of customers or have a clear pathway to achieving such through a partnership with an offtaker.
- **Quality:** For a commodity product to be considered marketable, certain standards have to be achieved—typically, a combination of presence or absence criteria and/or tolerance bands with which the product has to comply. For example, the ASTM has several standards specifying, testing, and assessing the physical, mechanical, and chemical properties of plastics that CO_2-derived products would be required to meet.
- **Distribution channels:** Depending on the product, distribution channels can take the form of commodity exchanges, wholesalers and distributors, brokers and agents, or government agencies. Each type is comprised of multiple actors with their own requirements for production volumes, price, insurance, hedging mechanisms, and quality audits. The factors affecting market access for new products are more complex, given the likely need for new submarket formation.

4.3.2 Resources for Emerging CO_2 Utilization Businesses

4.3.2.1 Opportunities Through Federal Funding and Programs

CO_2 utilization may use only a fraction of the total capturable CO_2 otherwise destined for geologic sequestration, as discussed elsewhere in this report. Several funding opportunity announcements (FOAs) support carbon management interfacing with CO_2 utilization, including those for the Regional Direct Air Capture Hubs (DOE-FECM n.d.(b)) and Carbon Capture Demonstration Projects Program (DOE-OCED n.d.). Despite being large in sum, these FOAs may not be the best source of funding for commercializing CO_2 utilization because of focus and time lag to produce usable CO_2. For example, the FOA for Carbon Management is positioned to help demonstrate conversion technologies, but the funding is more similar to R&D than commercial demonstration because it is spread across numerous pathways (DOE-FECM n.d.(c)). Moreover, this funding is unlikely to be sufficient for CO_2 purchase for demonstration purposes.

The need for a demonstration project to claim the available tax credits incentivizes the development of CO_2 utilization demonstration partners. The IIJA contains many funding opportunities to support the build-out of CO_2 utilization infrastructure and R&D on CO_2-derived products (see Table 4-1). For funding to be appropriately used, businesses will need to know the application and reporting criteria to secure funding. Additionally, to take advantage of multiple funding opportunities at once, multiple businesses develop partnerships and site facilities within the same region (i.e., hub design infrastructure). For example, program funding can be used to set up demonstration hubs centered close to ethanol production facilities, which would allow CO_2 utilization to be demonstrated using carbon capture technology that is already commercially proven and available. The CO_2 captured at these hubs would be eligible for the \$60/tonne (or \$130/tonne CO_2 captured using direct air capture [DAC] technologies) tax credit through 45Q, and there is no requirement to geologically sequester. Given that the cost of CO_2 capture from an ethanol facility is \$0–\$55/tonne (Bennett et al. 2023; GAO 2022; Hughes et al. 2022; Moniz et al. 2023;

National Petroleum Council 2019),[13] a CO_2 utilization demonstration hub centered close to ethanol production could offer the CO_2 needed at zero cost. Developing demonstration partners and using hub designs when possible can stretch grant funds by eliminating operational costs from CO_2 utilization unit economics. See Chapter 10 for more on CCUS infrastructure development opportunities.

4.3.2.2 Federal Programs for Building Knowledge About and Skills for CO_2 Utilization

Federal agencies provide business leaders and stakeholders the opportunity to answer questions about proposed programs through various mechanisms like Requests for Information (RFIs) in order to alert the agency to gaps and opportunities in the sector. Frequently, RFIs are used to identify typically underrepresented stakeholders for collaboration, with the purpose of requesting feedback from a variety of stakeholders "all while considering environmental justice, energy transition, tribal, and other impacted communities" (DOE-FECM 2021, p. 3). Responses from small and disadvantaged CO_2 utilization businesses, declaring the need for attention and collaboration, would likely result in additional opportunities in future funding processes.

Similarly, if businesses diversify their collaborations or aim to meet commitments, they improve their odds of success in the CO_2 utilization market. Initiatives such as these create opportunities for businesses to access broader knowledge, perspectives, and skill sets, which can lead to further collaborative possibilities. (See Finding 4-5.) For example, the submission of Promoting Inclusive and Equitable Research Plans, which outline diversification tools such as engagement and collaboration with underserved populations, organizations, and institutions, and provision of professional and learning opportunities for underrepresented populations, such as Black, Indigenous, and other people of color (BIPOC) professionals with science and engineering expertise, are now required for some DOE FOAs (DOE n.d.(e)). These plans seek to advance the federal Small Business Innovation Research/Small Business Technology Transfer (SBIR/STTR) program goal to foster and encourage participation by socially or economically disadvantaged groups in innovation and entrepreneurship.

4.3.2.3 DOE Resources for Small Business Partners

In 2021, Sick et al. (2022) estimated that, of 160 developers active in CO_2 capture and utilization, 39 were new start-ups that had emerged since 2016. Small businesses and start-ups can enter and thrive in the nascent field of CO_2 utilization, and their participation is critical to the development and diversification of CO_2-derived product markets. For example, within the design of hubs, small businesses would have to build their own niche based on what is needed in the system, which provides both a challenge and an opportunity. By securing a specialized role in a hub, small businesses can expect to develop an expanded role as the sector grows and more capacity is required.

The federal government has initiatives that target small businesses and encourage opportunities that will grow the CO_2 utilization sector. For example, through cross-agency coordination, the General Services Administration (GSA) administers awards on behalf of clients in participating agencies and provides information on its website about how to undergo certification processes (GSA n.d.(a), n.d.(c)). The SBIR and STTR programs are another opportunity, and can help small businesses or start-ups initiate relationships with DOE. Projects are awarded in three distinct phases: Phases I and II provide R&D funding, and Phase III—during which federal agencies may award follow-on grants or contracts for products or processes that meet the mission needs of those agencies, or for further R&D—provides nonfederal capital to pursue commercial applications of that earlier R&D (SBIR n.d.). Small businesses experience various challenges to accessing these opportunities or being successful in this nascent sector, including limited awareness of relevant funding calls; limited ability to access facilities that could help their business development and/or result in meaningful partnerships to close gaps in their processes; and barriers in navigating available federal funding and required reporting. More support is needed for small businesses to overcome issues with accessing a broader market in addition to federal funding and resources.

DOE national laboratories provide a unique entry point for business leaders to engage with federal initiatives and programs while developing their business to better meet the needs of the sector. For example, Argonne National Laboratory's Small Business Program provides business owners with technical assistance related to procurement and

[13] Range includes first-of-a-kind and *n*th-of-a-kind facilities.

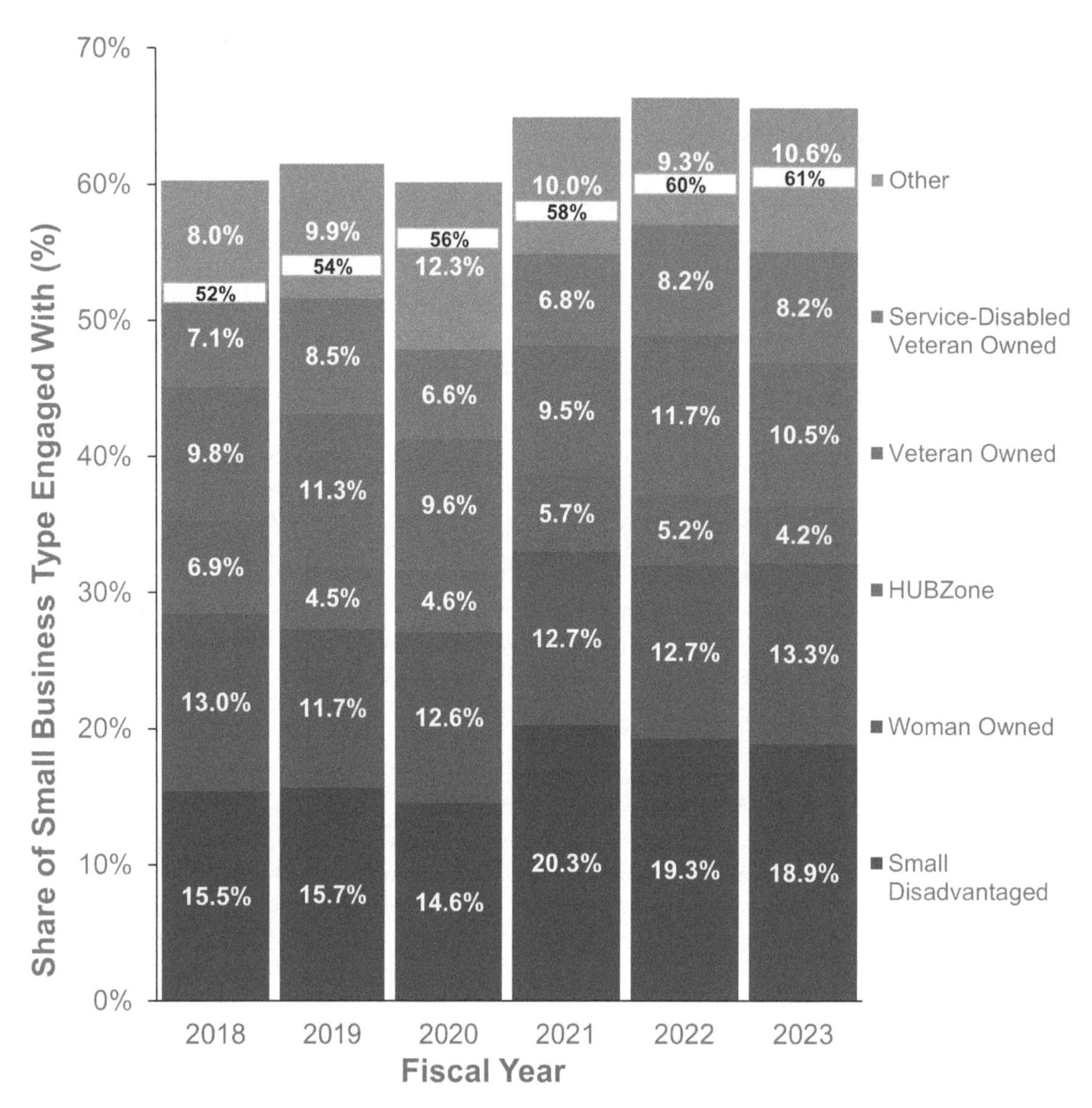

FIGURE 4-3 Sandia National Laboratories' small business engagement goals (white bar) and accomplishments (purple bars), including breakdown by type of small business, FY 2018–2023.
SOURCE: Based on data from SNL (n.d.).

development, and access to a streamlined registration and certification system (ANL n.d.). While no overarching organization provides cross-laboratory information for businesses, most national laboratories provide internal programming with collaboration opportunities. Sandia National Laboratories (SNL), for example, consistently exceeds its small business collaboration goals, reporting $1.1 billion in subcontracts to small and diverse businesses in FY 22, including small disadvantaged, woman-owned, veteran-owned, and service-disabled-veteran-owned businesses and businesses located in Historically Underutilized Business Zones (HUBZones)[14] (Peery 2023; see Figure 4-3). These data are particularly encouraging representations of the opportunity that currently exists, and the potential trajectory for collaboration between national laboratories and small businesses, indicating that the Sandia model could perhaps be mapped successfully to other national laboratories across the country.

[14] HUBZone businesses are part of the U.S. Small Business Administration's program for small companies that operate and employ those in "Historically Under-Utilized Business Zones" (SBA n.d.).

DOE also works with third-party organizations to support the accelerated deployment of technologies. For example, ENERGYWERX, DOE's first intermediary partner, works to increase joint activities between the agency and small business, higher-education institutions, and nontraditional partners to expand the deployment of clean energy solutions (DOE-OTT n.d.(b)). In growing the carbon management and CO_2 utilization sectors, DOE can capitalize on the important role of national laboratories and third-party organizations in developing and commercializing new technologies and MRV methods. Box 4-2 highlights existing opportunities for small businesses to partner with national laboratories and third-party organizations for technology development and deployment support: the Gateway for Accelerated Innovation in Nuclear (GAIN) program and the Voucher Program. The best practices, beneficial components, and lessons learned from both of these programs can be applied to a program developed to aid businesses entering the CO_2 utilization sector. (See Recommendation 4-4.)

BOX 4-2
Existing Opportunities for Commercialization Partnerships Through the Department of Energy

GAIN Program

The GAIN program, administered and led by Idaho National Laboratory in collaboration with Oak Ridge National Laboratory and Argonne National Laboratory, is a public–private partnership framework dedicated to rapid and cost-effective development of innovative nuclear energy technologies and market readiness. Its mission is to provide the nuclear energy industry with access to the technical, regulatory, and financial support needed to commercialize innovative nuclear energy technologies at an accelerated and cost-effective pace (GAIN n.d.(b)). Aside from communication and education programming, GAIN offers a host of valuable resources, including (1) physical access to unique experimental and testing capabilities housed within the national laboratory system; (2) computational and simulation tools; (3) data, information, and sample materials from previous research at national laboratories to inform future experiments; (4) use and site information for demonstration facilities; and (5) experts in nuclear science, engineering, materials science, licensing, and financing (DOE-NE n.d.). Access to these resources generally comes through the GAIN Nuclear Energy Voucher Program, which are not grants, but rather competitively awarded tokens that send funds directly to the national laboratory partner for laboratory time, materials, and equipment for the awardee. Since 2016, the GAIN program has awarded $34.2 million in vouchers to 57 different companies (GAIN n.d.(a)). While there are no size restrictions on applicant companies, special consideration is given to small companies.

DOE's Voucher Program for Energy Technology Innovation

The DOE Voucher Program, overseen by the Office of Technology Transitions (OTT), Office of Clean Energy Demonstrations (OCED), FECM, and Office of Energy Efficiency and Renewable Energy (EERE), is funded by IIJA's Technology Commercialization Fund (DOE-OTT 2023). The program will provide more than $32 million in commercialization support to businesses, including small businesses (DOE-OTT n.d.(a)). The support offered by the program includes (1) manufacturing or supply chain assessments, community benefits assessments, and other technoeconomic analyses; (2) third-party evaluation of technology performance under operating conditions that are certification-relevant; (3) considerations for technology benefits and challenges and siting and permitting best practices, and the development of streamlined processes for permitting and community engagement; (4) business plan, market research, and other commercialization strategy assistance; and (5) independent MRV practices and performance validation support (DOE-OTT 2023, n.d.(a)).

As part of the Voucher Program, businesses work directly with ENERGYWERX to connect with relevant third-party organizations, subject matter experts, and testing facilities. Lessons learned from this initial round of vouchers can guide future iterations of the programs and serve as a model for other commercialization programs.

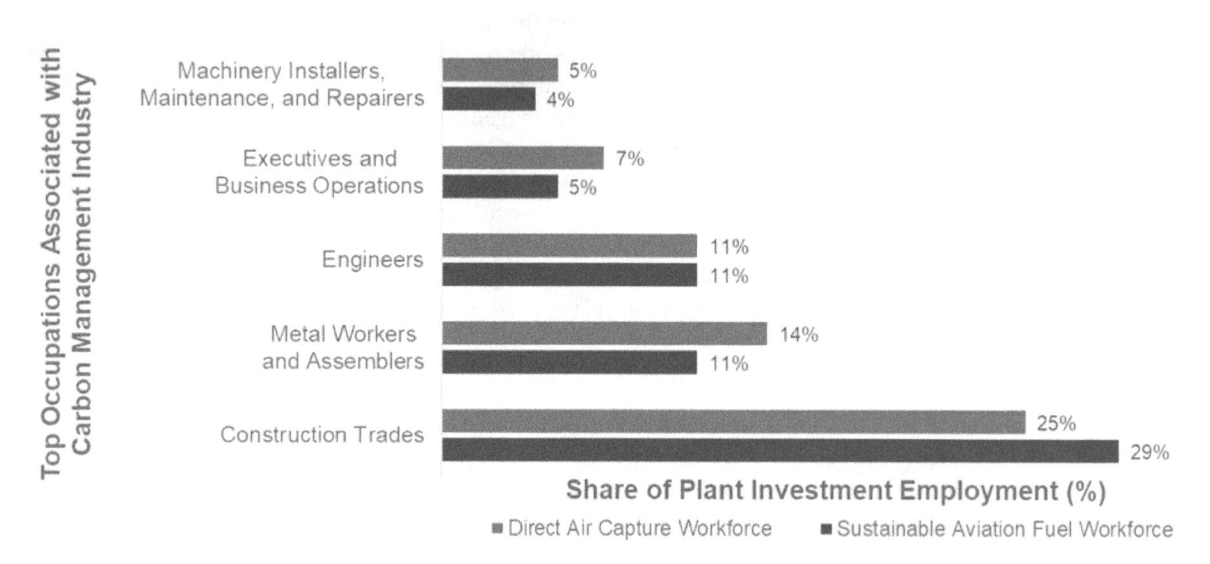

FIGURE 4-4 Upstream labor predicted for the direct air capture and sustainable aviation fuel workforces for top occupations associated with industry.
SOURCES: Based on data from Jones et al. (2023) and O'Rear et al. (2023).

For emerging CO_2 utilization businesses, especially small ones, the available DOE resources and funding need to be appropriately communicated so that diverse types of businesses can access them. Additionally, using programs for other energy-related technologies, like the GAIN and Voucher programs, as a model for CO_2 utilization programs can support the entrance of small businesses into this space by connecting them with market and technology experts.

4.3.3 Workforce Uncertainties for CO_2 Utilization Businesses

CCUS at scale has been estimated to support 177,000 to 295,000 jobs, while, for comparison, 3.1 million clean energy jobs are aligned with DOE's net-zero definition[15] (DOE-OEJ 2023; MacNair and Callihan 2019). However, regardless of the industry, there are not enough employees in the upstream labor force to support the infrastructure build-out that investors seek to fund. This is not owing to a lack of available jobs. The Bureau of Labor Statistics reported an average of 438,000 open construction jobs per month for November 2023 through January 2024 (BLS 2024). Despite the high number of job openings, contractors have reported difficulty finding willing and skilled workers in recent years (NASEM 2023g). For example, the Associated General Contractors of America and Autodesk Cloud Construction (2023) workforce survey found that 68 percent of firms surveyed had trouble filling openings because candidates lacked the skills to work in the industry. This employment trend will persist even without a transition to cleaner energy and products.

Figure 4-4 compares the upstream labor needs predicted for the DAC and SAF workforces. The Rhodium Group estimates that once a facility is built, DAC facilities will need 340 ongoing jobs, and SAF facilities will need 1440 ongoing jobs to support operations (Jones et al. 2023; O'Rear et al. 2023). Beyond the ongoing jobs related to maintenance, executive and business operations will comprise 11 percent of the ongoing employment for DAC facilities, and agricultural workers and managers will comprise 25 percent of the ongoing employment for SAF production (Jones et al. 2023; O'Rear et al. 2023). Small businesses aiming to participate in either field will need to match the skilled labor required to maintain facilities if they hope to compete with larger, more developed businesses.

---

[15] DOE defines clean energy jobs aligned with a net-zero future as relating to "renewable energy; grid technologies and storage; traditional electricity transmission and distribution for electricity; nuclear energy; a subset of energy efficiency that does not involve fossil fuel burning equipment; biofuels; and plug-in hybrid, battery electric, and hydrogen fuel cell vehicles and components" (DOE-OEJ 2023, p. viii).

Very little research has been done to predict the workforce needs for CO_2 utilization-specific businesses. However, the skill sets required for CO_2 utilization projects are expected to translate from existing processes and skill sets for fossil fuel refining and chemical industries, as discussed in Section 4.2.1.2.1. Specialized training—which may be required for R&D-related workforces—will play a key role in workforce development for the growing sector, especially depending on its accessibility or lack thereof. Individuals with specialized skills tend to make more money while being a lower percentage of a sector's workforce. For example, in oil and gas extraction in 2022, 1700 geoscientists (a specialized occupation in the field) were employed with an average annual salary of $145,660, compared to the 9340 wellhead pumpers employed with an average annual salary of $69,770 (BLS n.d.). This presents a challenge for small businesses because they likely will have to hire highly skilled employees at a high cost. Incentives are needed to encourage the development of a sustainable labor force for CO_2 utilization that additionally support the access of small businesses to these skill sets. Special attention will need to be paid at federal and state levels to address the challenges and barriers and allow for the diversification of the CO_2 utilization sector as it builds out.

4.4 SOCIETAL CONSIDERATIONS FOR THE EMERGING CO_2 UTILIZATION SECTOR

Societal dimensions need to be considered and appropriately addressed in CO_2 utilization policy, project design, and workforce development as the sector continues to build out. These considerations include the meaningful engagement of publics and communities, intentional focus on remediating and avoiding environmental harms, and equitable access to economic and workforce benefits of the emerging sector. Without these societal considerations, the CO_2 utilization sector runs the risk of perpetuating past and current environmental and social injustices. This section defines relevant equity and justice terms, summarizes best practices for public and community engagement, elevates select principles of EJ, and identifies key economic considerations. Box 4-3, modified from the first report's Box 5-1, includes definitions of key concepts of justice and equity discussed throughout this section.

BOX 4-3
Concepts of Justice and Equity

- **Justice**—Social arrangements that permit all (adult) members of society to interact with one another as peers (Nancy Fraser, quoted in Cochran and Denholm 2021). The principles of justice discussed in this section are: *recognitional justice*—understanding the historical and present bias regarding societal inequalities, especially about the treatment of communities (Cochran and Denholm 2021); *procedural justice*—the ability of people to be involved in fair decision-making processes (Cochran and Denholm 2021; Kosar and Suarez 2021); *restorative justice*—the act of repairing the impact of past injustices to restore communities and the environment to their original position (Hazrati and Heffron 2021); and *distributive justice*—equitable allocation of resources, risks, impacts, and benefits and burdens across society (Cochran and Denholm 2021; Kosar and Suarez 2021).
- **Environmental justice**—The just treatment and meaningful involvement of all people—regardless of income, race, color, national origin, Tribal affiliation, or disability—in decision-making activities related to climate change, the cumulative impacts of environmental and other burdens, and the legacy of racism or other structural or systemic barriers that disproportionately and adversely affect human health and the environment (EPA n.d.(b)).
- **Social justice**—A situation in which (1) benefits and burdens in society are dispersed with the allocation of a set of principles; (2) procedures and rules that govern decision making preserve the basic rights and entitlements of individuals and groups; and (3) individuals are treated with dignity and respect by authorities and other individuals (Jost and Kay 2010).

continued

BOX 4-3 Continued

- **Equity**—Achieved results where advantage and disadvantage are not distributed based on social identities (Initiative for Energy Justice 2019). In this section, emphasis is placed on equitable access to a healthy, sustainable, and resilient environment in which to live, play, work, learn, grow, worship, and engage in cultural and subsistence practices.
- **Disadvantaged community**—A community that is marginalized by underinvestment and suffers the most from a combination of health, economic, and environmental burdens, including high unemployment, air and water pollution, and poverty (Kosar and Suarez 2021; White House n.d.(b)). Relatedly, *marginalized/underrepresented communities* are groups that experience societal barriers such as social, political, or economic exclusion or discrimination (Machado et al. 2021; Nakintu 2021). In this section, these terms are used interchangeably.

4.4.1 Public and Community Engagement Considerations

CO_2 utilization is part of a suite of carbon management practices that are being designed to support the nation's net-zero goals. In general, decarbonization pathways face a spectrum of responses from the public—from acceptance to opposition—which is common for emerging technologies (Boudet 2019; NASEM 2023f). Opposition to technologies can be broken into two dimensions: (1) concerns inherent to a technology (e.g., how will a project impact everyday life?); and (2) concerns related to the institutions that govern the technology (e.g., are the regulatory systems effective and competent and is a community being meaningfully consulted in deployment?) (NASEM 2023f, Table 8-1). Figure 4-5 illustrates the factors that affect public perceptions of and related responses

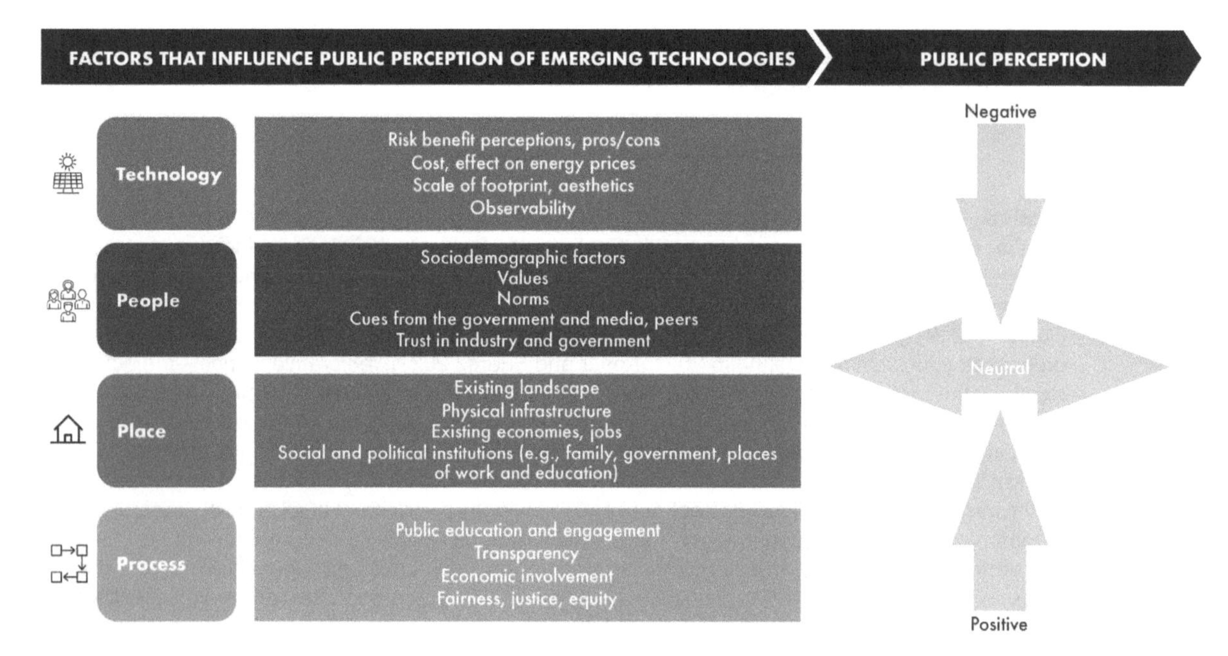

FIGURE 4-5 Factors that shape public perception of and responses to new energy technologies
SOURCES: Modified from H.S. Boudet, 2019, "Public Perceptions of and Responses to New Energy Technologies," *Nature Energy* 4:446–455, Springer Nature. Icons from the Noun Project, https://thenounproject.com. CC BY 3.0.

to new technologies.[16] Studies show that providing the public with more information can lead to a shift in public support of new technologies (Stedman et al. 2016; Stoutenborough and Vedlitz 2016). However, the views of the media, peers, and trusted messengers (i.e., academics or social movement activists) also shape public responses to energy technologies (Boudet 2019).

Carbon management technologies and processes as a whole have been described as "false solutions"[17] by EJ advocates (e.g., see Chemnick 2023a; Earthjustice and Clean Energy Program 2023; Just Transition Alliance 2020; New Energy Economy n.d.). Additionally, there is a public perception that investment in carbon management is outsized compared to the limited contribution that such technologies are expected to make to climate change mitigation (Jones et al. 2017). Skepticism that investment in CO_2 utilization and its value chain is disproportionately large relative to its climate mitigation potential motivates negative public discourse. However, Seltzer (2021) found that 80 percent of the U.S. public either does not know of CCUS technology or cannot definitively recognize it. Transparency about how products are made, how widely used CO_2-derived products are, and how R&D investments compare to the products' GHG impacts can support the public's understanding of CO_2 utilization in relation to carbon management efforts. This section examines opportunities to use public and community engagement to (1) expand public understanding of CO_2 utilization; (2) confront justice and equity questions that shape perceptions of the sector; and (3) communicate CO_2-derived product pathways that align with public and community needs. (See Finding 4-7.)

4.4.1.1 Current Public Discourse Around Utilization

Most societal acceptance research about carbon management uses CCS to gauge an individual's understanding before following up with questions on CO_2 utilization, either because carbon capture is a source for CO_2-derived products or because CCS is more widely discussed. For example, Offermann-van Heek et al. (2018) used semistandardized interviews to identify the most important factors concerning trust and acceptance of CO_2 utilization. The results of the qualitative study were incorporated in a quantitative online survey of 127 participants, which found that a lack of knowledge or awareness of CCS was owing to misconceptions, misleading information, or pseudo-opinions. When questions focused on CO_2 utilization, Offermann-van Heek et al. (2018) found individual differences in preferences for end products (e.g., long-lasting cement versus fuels), skepticism about whether investment in CCS and CO_2 utilization is worthwhile (e.g., preventing actual societal change, maintaining *business as usual*), and the acceptable amount of risk for health, sustainability, product quality, and the environment.

Offermann-van Heek et al. (2018) also found that customers have concerns about manufacturing considerations (e.g., the sustainability of production) and company considerations (e.g., company environmental management) for CO_2-derived products. These concerns align with public considerations for conventional products, suggesting that customers do not view CO_2-derived products differently. However, there are some challenges within the construction industry for using CO_2-derived building materials (see Box 4-4).[18] The concerns of both consumers and the construction industry can be addressed through low-stress testing projects, such as sidewalks and driveways (Derouin 2023). The transparency of testing products in public spaces can also support public communication of results.

Studies specific to CO_2 utilization are sparse, and findings vary greatly. According to a meta-narrative review of 53 peer-reviewed publications by Nielsen et al. (2022), this variance of findings results from a lack of cohesive definitions—and therefore, a lack of cohesive metrics—for acceptance, community, and impacts. For example, a study that conceptualized acceptance as a lack of public resistance would frame their questions differently from

---

[16] For examples of EJ concerns about decarbonization technologies, see Appendix E in NASEM (2023b).

[17] The term *false solutions* is used to connote pathways that are viewed as continually extractive, leading to concentration of political and economic power, likely to continue poisoning or displacing communities, and reductive of the climate crisis to a solely carbon-based focus (Climate Justice Alliance 2019).

[18] See Chapter 9, Section 9.3.5 for information about the safety of coal waste applications. See also Bhide and Sengupta (2024) for an overview of insurance sector considerations for the net-zero transition, including considerations about CO_2 utilization technologies.

BOX 4-4
Discourse About CO_2 Utilization Within the Construction Sector

A strong demand signals exist to produce CO_2-derived building materials—concrete, carbon black additives, and drywall—owing to incentives and requirements for low-embodied carbon in new buildings. However, the construction industry is often risk-averse with regard to using new materials because "structural engineering or building infrastructure that impacts human health and safety is scrutinized" (Derouin 2023). Therefore, there exist challenges to adoption of these new CO_2-derived building materials. For example, a mixed-method study used a survey and a series of interviews to identify economic, technical, practical, and cultural barriers to adopting building materials with lower embodied carbon from the perspective of 47 construction professionals (Giesekam et al. 2015). While respondents viewed the architect, client, or contractor as having the greatest influence over construction material selection, most felt that they had some influence on material selection (Giesekam et al. 2015). The perception held by construction professionals can shape the perception of the general public about these new materials.

Across different studies, key barriers to the adoption of CO_2-derived building materials include

- Extensive training needed for engineers, builders, and contractors to develop skills needed to work with novel materials (Althoey et al. 2023).
- High costs of materials—such as fuels (i.e., cost to prepare fuels and cost of transporting fuels to facilities) and fly ash—needed to support utilization processes (Althoey et al. 2023; van Oss and Padovani 2003).
- Impact of local building codes, material regulations, and project specifications on types of materials used (Althoey et al. 2023; van Oss and Padovani 2003).
- Lack of communication with the public about safety considerations addressed by the sector (Althoey et al. 2023; Derouin 2023).
- Lengthy and expensive permitting processes (van Oss and Padovani 2003).
- Logistics regarding the physical movement of material (van Oss and Padovani 2003).
- Regional differences in resource availability (Althoey et al. 2023).
- Reliance on financial investment to establish and maintain novel technologies and processes, including during research and development efforts (Althoey et al. 2023).
- Risk averse stakeholders, including the construction sector itself (Althoey et al. 2023).

a study that conceptualized acceptance as consensus within a group (Nielsen et al. 2022). Figure 4-6 shows the major underlying dynamics of acceptance, community, and impacts, and the related conceptualization for each that influence public perception of CO_2 utilization projects. Future studies on acceptance of CO_2 utilization will have to consider not only the origins of the CO_2 but also the impacts of the CO_2 utilization sector itself, especially as they intersect with communities, to gain a better understanding of what characteristics are integral to defining public acceptance. (See Recommendation 4-6.)

4.4.1.2 Strengthening Public Understanding of CO_2 Utilization Through Engagement

As the CO_2 utilization sector continues to build out, meaningful public engagement can help to address public concerns about and strengthen public perception of CO_2 utilization. NASEM 2023f found that generative dialogue—conversations that expand understanding through inclusive engagement—have the potential to facilitate early understanding of concerns and values of technologies and projects being proposed, make information more accessible and digestible, and allow the public to form their own views about a technology through interactions with experts. For example, a conjoint experiment-based study by Offermann-van Heek et al. (2020) found that carefully designed information allowed respondents without any technical knowledge about CO_2 utilization to

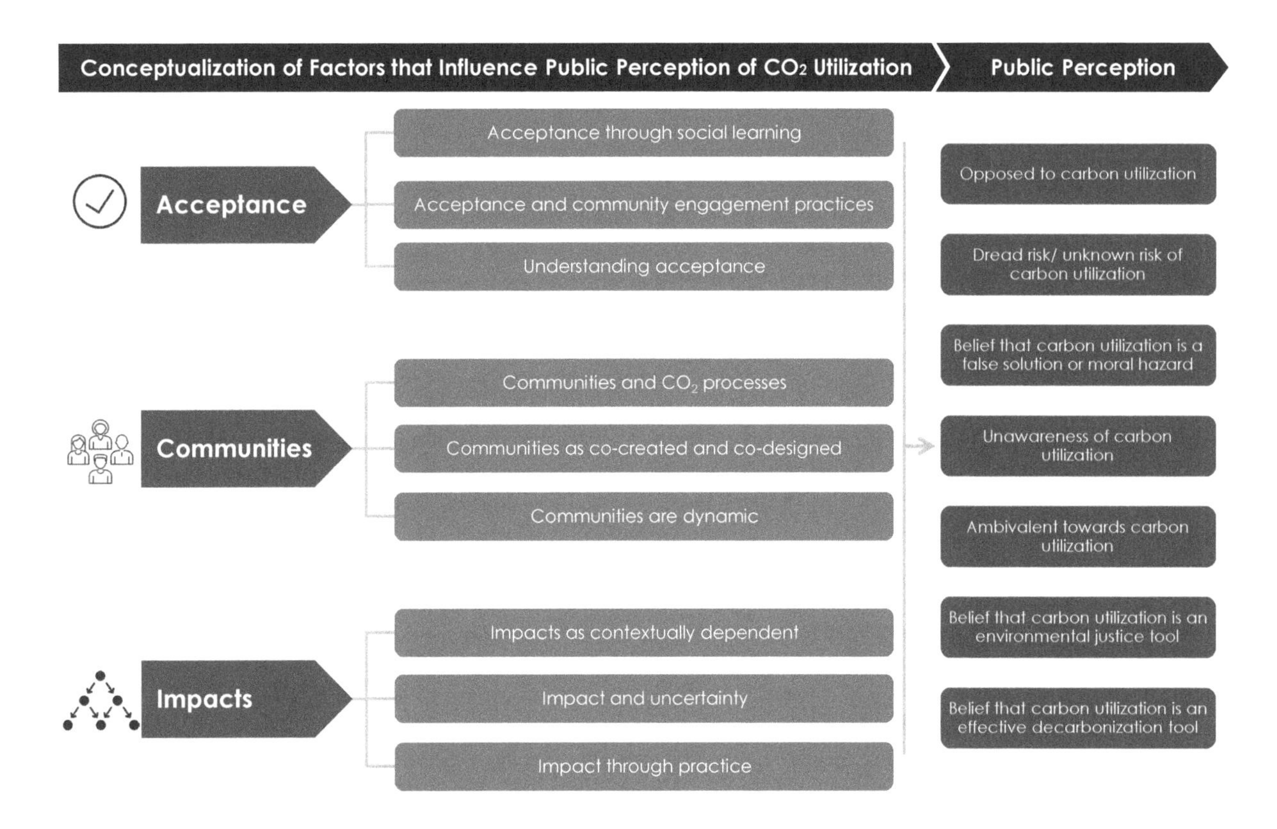

FIGURE 4-6 How the conceptualization of the underlying dynamics of acceptance, community, and impacts influences the public discourse around CO_2 utilization.
SOURCES: Modified from Nielsen et al. (2022), https://doi.org/10.1371/journal.pone.0272409. CC BY 4.0. Icons from the Noun Project, https://thenounproject.com. CC BY 3.0.

formulate informed decisions and update their preferences. However, the authors underscored that while people may update their preferences when presented with new, technically correct, comprehensible, and timely information, public perception is typically formed based on heuristics derived from deeply ingrained social preferences (Offermann-van Heek et al. 2020). Hence, perceptions and public acceptance need to be assessed early in the development cycle of new technologies to learn which decisions can be modified by providing new information.

In contrast, Buttorff et al. (2020) assessed public opinion and attitudes toward carbon management, support for carbon mitigation policies and R&D for decarbonization, and the willingness to pay for energy and products derived from carbon management. The online survey–based study did not provide CCS-based context and found support for the adoption of carbon management and its incentivization through governmental action and policy changes (Buttorff et al. 2020). The study also found that respondents were willing to pay higher prices—to the extent they are considered affordable and without internalizing the full cost of low-carbon alternatives—when presented with information on how carbon management technologies would impact the price they pay for electricity and other products like low-carbon fuels. Another study surveyed likely voters in Wyoming, Texas, Louisiana, and Colorado and found that most respondents supported turning captured CO_2 into long-lived materials and were more skeptical of permanent underground storage (National Wildlife Federation 2024).

Effective and meaningful public engagement can ensure that relevant information is shared with the public, beyond the communities impacted by projects, including "How are the benefits of CO_2 utilization market demand distributed?" and "What are the potential negative impacts and burdens of CO_2 utilization?" However, engagement practices and strategies have to be location-specific and consider the history of and identities that exist in a region.

Such engagement, especially through public awareness and education efforts, can help empower people with the right technical information and help them form heuristics for decision making on new technologies incentivized by public money (Chailleux 2019; Offermann-van Heek et al. 2020). Furthermore, building public support through larger community education will develop trust, accountability, and transparency between project designers and developers and the general public and potential host communities (Meckling et al 2022). Holistic public engagement approaches that recognize a larger social movement to advance equity and justice across sectors—such as housing, education, and health care—facilitate a truly just cross-society transition and are foundational to any strategy. (See Finding 4-8 and Recommendation 4-6.)

4.4.1.3 Engagement of Affected Communities and Project Consent

A lack of engagement with affected communities, in particular, may result in reluctance to adopt or rejection and opposition to new low-carbon technologies, especially if there are perceived high costs and negative social impacts (NASEM 2023b). Meaningful community engagement, especially in the project development period, can provide on-the-ground knowledge that can shape projects to meet community needs, respect the rights and boundaries of underserved populations within those communities, and retool procedures that might pose barriers to overburdened communities (WHEJAC 2022a). Recommendations around community engagement emphasize accountability of federal agencies and project developers to the community—promises need to be delivered on if they are made, and the better a developer or government entity can track the dissemination of benefits, the more enfranchised community members can be (BW Research Partnership and Climate Equity Initiative 2023; WHEJAC 2022a). These recommended practices will not ensure that a project is selected by a community; instead, they can support community and public engagement during the design and build-out of projects, as appropriate.[19] (See Recommendation 4-7.)

Within meaningful community engagement, consent—a collective decision made by those being engaged with—has to be sought from and granted or withheld by a community based on the unique relationship between elected officials and representatives within each community (FAO et al. 2016). Collective consent can be subject to change upon receiving new information about the project and can be given or withheld in phases of the project while also giving the community rights to govern what occurs after the decision is made (FAO et al. 2016; IHRB 2022). In support of consent, successful community engagement processes are designed to collect and address feedback. The outcome of the feedback process, which needs to continue into the implementation phase of a project, can be defined by how well project managers incorporate community concerns and solutions into the agreement (FAO et al. 2016). Both consent-seeking actions and feedback processes need to be place-based and flexible to match the needs of a community.

While no comprehensive list of "best" practices for community engagement exists, below are high-level themes and principles for meaningful public engagement. These practices are not formal guidance but promising engagement strategies. Elements can be incorporated into existing efforts by developers with federal funding for CO_2 utilization projects and can provide the foundation for future projects seeking to expand the sector.

- **Avoid past mistakes.** Historically, disenfranchised communities have been forced to live with the impacts of decisions made without their input. To move forward and create space for a dialogue around CO_2 utilization infrastructure, mistakes of past energy and technology transitions need to be avoided, especially mistakes specific to a potential host community. NASEM (2023f) identified two categories of failed communication strategies: (1) the "engineer's myth"—the idea that technical modifications can alter the risk calculation and change public attitudes to a technology's deployment—which has not been shown to be true; and (2) a lack of understanding or information rather than a lack of public trust. Early and frequent

[19] Note that engagement, with the public or communities, is not appropriate for early technology readiness level (TRL) due to the uncertainty about what the impact of a process or project will be, what and where a project will be, and who will be impacted. For lower-TRL projects (e.g., those with small amounts of funding or compressed timelines), community engagement opportunities may not be meaningful and may act as a barrier to the project. For higher-TRL projects (e.g., those with extensive timelines), community engagement opportunities can be additive to the process. See Chapter 3, Section 3.4.2 for more about uses of s-LCA at different TRLs.

engagement facilitates communication between developers and communities through which project intent and community benefits can be discussed and negotiated (see Section 4.4.2.3.2). Awareness of factors that lead to poor public engagement and focused efforts to improve engagement center principles of recognitional and procedural justice, allowing new projects to avoid failures of the past.

- **Build trust and maintain relationships.** Trust is fundamental to any type of community-based project and is especially necessary for the deployment of emerging technologies. To build trust, developers can (1) avoid using scientists, engineers, regulators, and policy makers who aggressively support the technology when engaging with potential host communities; (2) compare new technology to what communities are familiar with to support individual judgments about whether scientific knowledge is trustworthy; and (3) avoid the *parachute method*, in which developers drop into communities for the express purpose of developing a project and then disappear once the project is complete (NASEM 2023f). Outsiders, whether developers or scholars, need to avoid approaching the community with predetermined solutions or methodologies, and instead co-create strategies with the community (Deaton 2022). Often, trust-building results from developing long-standing relationships with reliable actors within the community.
- **Center enfranchising frameworks.** Communities can be reticent to engage during development processes because they are often met with an approach focused on what they lack—whether information, infrastructure, or democratic power. Two frameworks that instead seek to enfranchise communities are
  - *Asset-Based Community Development (ABCD).* The ABCD framework builds on a community's assets through mobilized action from community members and local institutions (CNT n.d.). Fundamentally, ABCD highlights that communities can drive development processes by responding to and creating economic opportunities at a local level (CNT n.d.).
  - *Free, Prior, and Informed Consent (FPIC).* The FPIC framework is centered on the fact that all people have the right to self-determination (FAO et al. 2016). FPIC, often discussed in the context of development on Tribal lands, allows a group to give consent to a project (see Section 4.4.2.3.2). While the process does not guarantee consent, it allows a group to conduct its own independent, collective decision making with project information that is continually provided and discussed (FOA et al. 2016).
- **Provide resources for engagement.** Communities, especially underserved populations, often do not have the bandwidth to engage on every single development issue they may face, while also confronting barriers to participation such as insufficient housing, employment and familial responsibilities, clothing, and food (Donnelly et al. 2015). As such, providing honorariums, transportation to and from events, food or vouchers at events, and wrap-around services like childcare for participating in engagement, acknowledges that engagement strategies do not exist in a vacuum and encourages them to reflect the complexity of the larger cultural ecosystem in which the project is operating (Barnes and Schmitz 2016; Donnelly et al. 2015; Langness et al. 2023). Additionally, providing access to mapping products, data, or community science frameworks may increase their ability to engage (Kimmell et al. 2021).
- **Employ various ownership models.** While the entire carbon management sector may not lend itself to ownership models similar to community solar or neighborhood cooperatives, it is worth considering how to transform traditionally corporate capitalist systems into more community-centered models with direct benefits. For example, a model in which local stakeholders own most of the project and define expected collective benefits can be adapted to CO_2 utilization projects.
- **Transparent reporting and information sharing for meeting metrics.** Working alongside the community to develop metrics and goals to measure progress can provide clear benchmarks for project accountability. This could involve convening project-specific community advisory committees to identify potential impacts; preparing documents to communicate findings with community members; and using clear language to support understanding at varied levels of knowledge and reading proficiency (EJ IWG and NEPA Committee 2016). Goals and metrics have to be consistently revisited to ensure that they are being met, with the ability for community input to update and inform those outcomes.

Figure 4-7 illustrates how the outlined themes and principles can be integrated into a standard project cycle: project identification, project formulation, project selection, project implementation, and project closure.

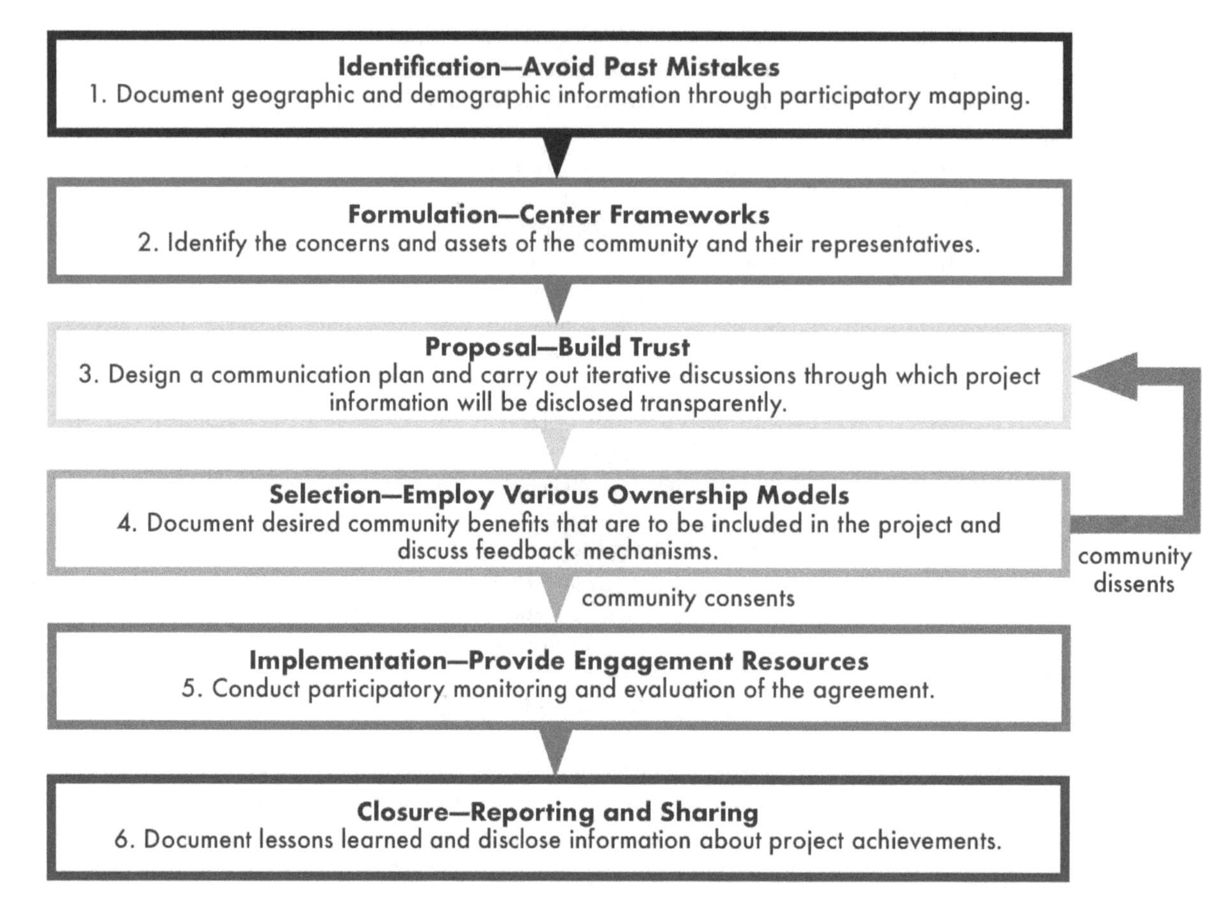

FIGURE 4-7 Project cycle incorporating best practices for community engagement.

4.4.2 Environmental Justice Considerations

The mainstream environmental movement has historically neglected social justice and equity issues (DeLuca 2007). For example, critics assert that early work in the environmental movement had "little regard to underlying social inequalities that drive differential exposures to pollution and did not incorporate voices of people of color and the working classes in solving them" (Mohai et al. 2009, p. 3). Even as it attempts to prioritize discussion on and address racial inequality, the scholarly EJ movement has often neglected to highlight well-established social scientific theories of race and racism or focused on race and class as dividing lines without acknowledging the context from which they emerged (Park and Pellow 2004). There is still scholarly and practical debate about how both environmental and social justice movements can and should work together to achieve common goals. As the CO_2 utilization sector expands and considers EJ in the creation of environmental and industrial policy, it is imperative to recognize that EJ is inextricably linked to racism, in addition to other socioeconomic factors that lead to the marginalization of communities and groups.

EJ does not just apply to the creation of *new* infrastructure; part of the consideration of the movement is repairing historic harm, which includes existing and historic infrastructure. One example is mine remediation efforts, which thus far largely have focused on the technical aspects of clean-up but have important political, social, and cultural implications in addition to environmental ones (Beckett and Keeling 2019). Similar parallels may exist in the context of coal waste utilization as a method to commercialize carbon-based goods while also cleaning up existing mines. The value chain that supports CO_2 utilization processes will also undoubtedly include sites that represent historical burdens in certain communities. Principles of justice, both historical and forward-looking, need to be continually applied as the CO_2 utilization sector builds out to strive toward EJ as an outcome. This section discusses how to operationalize foundational principles of EJ, how to measure EJ in projects, and EJ considerations for project selection and siting.

4.4.2.1 Operationalizing Foundational Principles of EJ

A defining moment in the EJ movement with significant relevance to infrastructure development was the creation of the 17 Principles of Environmental Justice at the 1991 National People of Color Environmental Leadership Summit. The goals for the principles include to build a national and international movement to fight the destruction of land and communities; to respect and celebrate cultures and beliefs about the natural world; and to promote economic alternatives that contribute to the development of environmentally safe livelihoods (Madison et al. 1992). These principles are still considered foundational in the climate movement and can be used as guidelines to evaluate specific projects (e.g., see O'Laughlin 2021). Given the substantial opportunity for developing CO_2 utilization and the accompanying capture and removal infrastructure, the sector has to develop in the most just way possible, not only to bolster public acceptance but also to do good for the public. Table 4-2 describes how Principles 3, 5, 6, 7, 8, 11, and 16 can be incorporated into considerations for the emerging CO_2 utilization sector. (See Finding 4-9 and Recommendation 4-8.)

While the principles serve as a rubric through which communities can evaluate specific projects, the EJ movement has also served communities by enlarging the constituency through the incorporation of more disadvantaged communities; building community capacity through educational campaigns that draw direct connections between the EJ movement and the surrounding environment of many disadvantaged communities; and facilitating community empowerment through grassroots efforts (Faber and McCarthy 2001). These structures enfranchise communities as decision makers and reflect the nature of EJ as a dynamic movement, constantly pushing against antiquated and often exclusionary processes to ensure equity across all aspects of the project value chain.

4.4.2.2 Holistically Measuring EJ

Federal and state agencies designed and implemented policies and initiatives to advance EJ as early as 1994 (see Section 4.2.1.2.2). However, there is no standard way to measure elements of EJ or to locate communities impacted by environmental injustices. For example, CEQ's new Climate and Economic Justice Screening Tool (CEJST; Version 1.0) is one of more than 30 EJ screening tools that exist across federal, state, and local agencies (Dean and Esling 2023).[20] Assessment tools—such as CEJST and EPA's EJScreen—can be beneficial for the process of deciding where infrastructure and retrofits are sited and can be used to track EJ outcomes (CEQ n.d.(b); DOE n.d.(d); EPA n.d.(a)). However, there are criticisms about how sufficiently these tools address certain factors, such as weighing race as a key demographic that indicates disproportionate impact or "disadvantaged" categorization (Sadasivam 2023; WHEJAC 2022b). Additionally, while environmental impact assessments are useful and integral to the siting process, they are better suited to address "potential harm and cumulative impacts of a proposed project, and supporting a decision to relocate, mitigate, or even stop a project" (Wang et al. 2023, p. 73). This results in the lack of ability to identify and address actual adverse outcomes of project and demonstrates that while these assessment tools have a place in the design and siting process, they alone are not comprehensive enough to be used without additional community considerations that often involve the history, racial makeup, and concentration of power in a community.[21]

While a lack of standardization makes quantitative measurement of EJ challenging, there exist foundational framings of holistic benefits that might flow into communities in response to direct investments in EJ-related projects. For example, Dr. Bunyan Bryant, a former University of Michigan School for Environment and Sustainability professor and noted EJ leader, stated that EJ is "supported by decent paying and safe jobs; quality schools and recreation; decent housing and adequate health care; democratic decision-making and personal empowerment; and communities free of violence, drugs, and poverty" (Bryant 1995, p. 6). Because communities are not monolithic, the specific benefits a community wants and experiences will differ from project to project. Federal agencies have recommended that programs work with stakeholders to define program and project benefits, specifically in the context of Justice40-covered programs (e.g., see WHEJAC 2021 and Young et al. 2021). These recommendations seek to avoid harm and maximize federal investments, including directing investments in geography and

[20] For more information about how these screening tools intersect, see the Environmental Policy Innovation Center's EJ Tools Map at https://epic-tech.shinyapps.io/ej-tools-beta, accessed August 5, 2024.

[21] See Chapter 3 for more information about assessment tools.

TABLE 4-2 How to Operationalize the Principles of Environmental Justice in the CO_2 Utilization Sector

Principle	Description	Operationalization
3. "Environmental Justice mandates the right to ethical, balanced, and responsible uses of land and renewable resources in the interest of a sustainable planet for humans and other living things."	The most sustainable, climate-benefiting strategies can be applied across the carbon management value chain. This is especially crucial if infrastructure is located near overburdened communities.	Consider the long-term impact on the development of a circular economy, especially through meaningful consideration of the most appropriate uses of resources.
5. "Environmental Justice affirms the fundamental right to political, economic, cultural, and environmental self-determination of all peoples."	Community engagement, especially that of underserved populations, is an imperative factor and thus processes should be inclusive. And, as CO_2 utilization remains novel, the education of stakeholders and decision makers is necessary.	Consider and prioritize self-determination of the community to create just precedents for infrastructure build out and informed understanding of CO_2 utilization.
6. "Environmental Justice demands the cessation of the production of all toxins hazardous wastes, and radioactive materials, and that all past and current producers be held strictly accountable to the people for detoxification and the containment at the point of production."	Entities in the CO_2 utilization sector that partner with fossil industry or infrastructure need to be aware that many EJ groups feel that reparative and recognitional justice are core to confronting the history of toxic industries to transform communities positively.	Strategize around waste products and pollutants that might arise from the creation of CO_2-derived products or feedstocks, in addition to understanding the historical impact of certain industries on communities.
7. "Environmental Justice demands the right to participate as equal partners at every level of decision making, including needs assessment, planning, implementation, enforcement, and evaluation."	Communities feel their participation in all processes of development is necessary for a process to be considered just. Furthermore, engagement strategies will not be effective if only utilized at the beginning of a development process.	Consider ways to actively integrate community input and reflect that input through adaptations of a project plan throughout its lifetime.
8. "Environmental Justice affirms the right of all workers to a safe and healthy work environment without being forced to choose between an unsafe livelihood and unemployment. It also affirms the right of those who work at home to be free from environmental hazards."	The histories of industries have informed community desires to protect the health of the working class in these sectors, as well as prevent other environmental hazards from impacting the surrounding community. It is critical that working conditions in new infrastructure or businesses are conducive to healthy and safe work environments.	Ensure that workers and communities are provided with access to new job opportunities and related benefits of the sector, while prioritizing the health and well-being of the community and their employees.
11. "Environmental Justice must recognize a special legal and natural relationship of Indigenous Nations to the U.S. government through treaties, agreements, compacts, and covenants affirming sovereignty and self-determination."	The U.S. Indigenous community has its own relationship to the development of the energy industry, as well as sovereignty over its territories. Decision making and implementation processes need to acknowledge these relationships and authorities over respective territories.	Consider community approaches to decision making and implementation when engaging with Indigenous nations and communities that respect existing relationships with and authorities over respective territories.
16. "Environmental Justice calls for the education of present and future generations which emphasizes social and environmental issues based on our experience and an appreciation of our diverse cultural perspectives."	For communities to continually engage in project decision making and development it is imperative that education around technologies and their uses is accessible and further incorporates the values and perspectives of diverse groups and reflects a holistic historical perspective.	Consider education around CO_2 utilization pathways and technologies and their deployment whether through academic institutions, community benefits plans, and/or apprenticeship opportunities that incorporates the lived experience of decision makers and their values.

SOURCE: Based on data from Madison et al. (1992).

people, making indirect and direct investments in a community, and providing essential services to a community by external direct investments (WHEJAC 2021). Efforts to advance EJ can benefit from a core set of overarching benefit types from which projects can outline specific benefits to potential host communities.[22] (See Finding 4-6 and Recommendation 4-5.)

[22] For example, NASEM (2023b) recommended that federal legislation require the collection and reporting of standardized metrics for direct impacts on jobs, public health, and access to technologies and programs (see NASEM 2023a, Recommendation 2-1). For more about direct benefits from initiatives focused on equity and justice, see NASEM (2023a).

4.4.2.3 Considerations for Selecting Projects

When considering the role of EJ practices in societal and community acceptance of proposed projects, the mechanisms and processes producing inequities across various institutions, industries, and frameworks for deployment need to be evaluated, as do the interaction of rules, attitudes, and politics (Foster 1998). This section outlines factors that impact acceptance of projects, EJ considerations for resource consumption, and procedural justice in selection processes.

4.4.2.3.1 EJ Considerations for CO_2 Utilization Resource Consumption

Avoiding adverse impacts of CO_2 utilization infrastructure on communities requires consideration of the facilities' resource consumption, especially for geographies where shared resources might be particularly scarce. The committee's first report included discussion of the electricity, hydrogen, water, and energy storage needed to support CO_2 utilization infrastructure (see NASEM 2023b, p. 96). While LCAs are adept at predicting resource consumption for a particular deployment scenario, the specific location where infrastructure might be sited also has to be studied during project design—for example, via an environmental impact assessment.[23]

Energy is one of the driving factors of GHG emissions and is therefore integral to the boundary considerations of the system (Terlouw et al. 2021). While the United States has advanced its commitments to deploy renewables, many underserved communities across the country still face substantial energy burdens and barriers to accessing zero-carbon energy infrastructure (DOE n.d.(c)). In these communities, energy justice—the goal of achieving equitable participation in the energy system while remediating the disproportionate social, economic, and health burdens of the current energy system—is integral for siting carbon management infrastructure, especially in cases where substantial energy requirements are necessary (DOE 2022b).[24] In addition to the land used for electricity and fuel production itself, the environmental impacts from transporting and storing electricity and chemical feedstocks have to be considered when developing CO_2 utilization projects. Similar to other renewable energy options (e.g., solar or wind), CO_2 utilization is estimated to require a lot of land for facilities and transport infrastructure. For example, one study estimates that 68,000 miles of CO_2 pipelines will be needed across the United States to meet the demand of a CO_2 utilization sector (Larson et al. 2020; Thomley 2023b).[25] While relying on a clean electricity grid can reduce the direct land transformation of a particular project, the externalities of this choice deserve acknowledgment, and co-locating electricity generation and storage facilities to minimize land use needs to be considered.

Water consumption for carbon management will vary across pathways, with biomass- and biochar-associated techniques typically resulting in intensive land use and water consumption (Rosa et al. 2021; Terlouw et al. 2021). Water as a resource has a long history in EJ communities, receiving more attention in recent years owing to events like the Flint, Michigan, water crisis, which led to contaminated drinking water and eroded trust of the local government, and the Jackson, Mississippi, water treatment facility failure, which left 150,00 residents without drinkable water (Denchak 2018; O'Neill 2023).[26] Water is a highly scrutinized resource in many resource-scarce communities, with nearly 2.2 million Americans living in homes without running water or basic plumbing (O'Neill 2023).

For the carbon management sector, there are growing concerns about how the technologies will impact communities with scarce water resources. For example, the Central Valley—which has seen substantial oil, gas, and agricultural booms—is being considered for several DOE-funded DAC projects announced in fall 2023 (California Resources Corporation 2023). However, droughts in 2021 put "a massive strain on many households and farms in the area," leading to losses of $1.7 billion and more than 14,000 jobs (Alonso and Ferrell 2023; DeLonge 2022) and creating competition for necessary resources like water and nonpolluted air (Cox 2020). Because recent analyses demonstrate that high water use hinders certain DAC approaches, trying to site facilities in drier regions is common (Küng et al. 2023). However, with existing burdens and competition of resources as a result of climate change in the Central Valley, any project development that could potentially increase resource insecurity in the region needs to be meaningfully considered (e.g., see Chemnick 2023b and Fernandez-Bou et al. 2023).

[23] See Chapter 3 for more on LCAs and environmental impact assessments.

[24] Although zero-carbon energy sources are favorable for carbon management infrastructure nationwide, these pathways come with their own resource costs on communities, such as the mining of critical minerals for lithium-ion batteries, or the potential air quality impacts resulting from the production of cement and steel needed for wind energy deployment (IEA 2021).

[25] See Chapter 10 for more on pipeline development.

[26] See also Tabuchi and Migliozzi (2023) for more about the impact fracking has made on water availability.

The committee's first report identified that, while water requirements for CO_2 utilization processes will not significantly increase water demand at a national level, local water impacts will vary based on geographical region (NASEM 2023b, Finding 4.14). Thus, DOE should work with its national laboratories to analyze the effect of CO_2 utilization on local water demands and identify the regions where there are opportunities for water infrastructure to serve multiple projects while considering local and EJ impacts (NASEM 2023b, Recommendation 4.6).[27] The committee still recommends that these analyses be conducted for CO_2 utilization processes that require water, including waste mineralization, vacuum production, product rinsing, dilution, and distillation, and that the results are appropriately considered during the planning stage for CO_2 utilization infrastructure.

4.4.2.3.2 Incorporating Procedural Justice in Project Selection

To address societal acceptance, several organizations outline engagement frameworks rooted in themes of justice and equity. For example, the Jemez Principles for Democratic Organizing feature agreements for organizing across diverse cultures, organizations, and politics (Solís 1997). These principles can help lay foundations for positive partnerships and engagement with communities that build trust and might, therefore, result in higher rates of acceptance of the CO_2 utilization sector as it builds out. (See also the Just Transition Principles outlined by the Climate Justice Alliance 2019 and the Principles of Working Together adopted at the Second People of Color Environmental Leadership Summit [see Energy Justice Network n.d.].) Engagement processes with justice practices prioritized early can result in iterative learning processes and may prevent community rejection. However, if consent is not reached, the aspects of the project that the community rejects need to be identified and, where possible, updated and modified to address objections. Box 4-5 describes how a feedback process can be developed to approach a situation in which a community says "no."

BOX 4-5
What Happens When a Community Says "No"?

The practice of EJ involves achieving environmental and socially just outcomes where there is development. It may be argued that neither of these outcomes can be achieved if the community is denied the right to refuse a "project deemed incompatible with their needs, even if a project comes with significant economic or labor benefits," something with which preliminary polls of voters agree (Fraser 2023, p. 18). A community saying "no" to a particular project design has significant implications for developers to consider related to why exactly the community has refused the project. If a developer continues forward with a project despite community opposition—especially if the community is vulnerable owing to historic disenfranchisement—this action has the potential to set a negative tone for the carbon management and CO_2 utilization sectors as they develop. Pushing a project forward despite community rejection may result in other unintended roadblocks, such as public outcry and even legal cases brought by the community, as demonstrated in this example:

> When giant wind turbines were being planned on indigenous Saami reindeer herding lands in northern Sweden, the impacted communities argued that the project was in breach of Saami rights. In response, the Swedish government argued that renewable energy development had to be prioritised over the rights of the indigenous Saami. It is reported that the financier of the project, KfW IPEX-Bank, used the Swedish government's statement to absolve itself of responsibility towards the Indigenous communities. The bank considered that Swedish law was sufficient to protect Saami rights. However, the Norwegian Supreme Court ruled that the wind project was illegal and it was to be discontinued. (IHRB 2022)

[27] As noted in the committee's first report, these analyses are being conducted for algal cultivation systems by Argonne National Laboratory, with support from DOE's Bioenergy Technology Office and Office of Fossil Energy, and for different mineralization processes that take place under aqueous conditions. For more information about these analyses, see DOE-BETO (2021), Naraharisetti et al. (2017), and Xu et al. (2019).

BOX 4-5 Continued

Treating Tribal and underrepresented communities as "stakeholders" rather than "decision makers" or "rights-holders" who "[have] freedom and autonomy over their lives and their territories" can breed distrust between parties (IHRB 2022). When the government is also a primary decision maker, there may be times when a site is selected contrary to community desires. In this case, communities still need to be engaged throughout the process to ensure that their right to self-determination is preserved.

4.4.2.4 Just Practices for Permitting of CO_2 Utilization Infrastructure

While widespread permitting gaps still exist and need to be prioritized for the development of the CO_2 utilization sector, permitting may not result in the accountability communities seek from what is ultimately supposed to be a protective process. Permitting processes can be analyzed at both the federal and state level; some permitting will need federal guidance when infrastructure for carbon management, hydrogen, and utilization intersect with other sectors, with the understanding that states have significant knowledge of their own localities. Upon reviewing permitting processes for CO_2 utilization in its first report, the committee recommended that these processes be coordinated by a single agency or entity that would also guide developers through the process of engaging with states and localities (NASEM 2023d, Recommendation 5.4).[28] For technical information about CO_2 utilization infrastructure, see Chapter 10.

To advance community voice in a permitting process, developers can share examples of the expected, place-based impact the CO_2 infrastructure might have and allow time for community members to ask questions and express their concerns related to the information provided. The incorporation of just practices is a critical part of permitting, as these permits can ultimately determine impactful outcomes (Guana 2015). For example, employing the pillars of procedural justice—neutrality, respect, voice, and trustworthiness—can ensure that all necessary parties are respected and heard throughout the permitting process (Yale Law School n.d.). Procedural justice strategies in infrastructure planning, however, do not guarantee community endorsement or acceptance. NASEM (2023e) identified key siting features that could reduce conflict and delays in the permitting process and highlighted the necessary inclusion of stakeholders that otherwise would not be included in the process, the need to elevate Indigenous knowledge, and requirements for following federal guidance memos.[29] Ultimately, for communities to build further trust with permitting entities, there has to be evidence of fair and equitably distributed outcomes alongside the integration of community perspective throughout the permitting process and its ongoing reassessment.

4.4.2.4.1 New and Existing Industrial Facilities

The procurement of CO_2 is a good starting point when considering how to center justice practices in the CO_2 utilization value chain. There are two categories in which this infrastructure can be placed: (1) existing infrastructure, such as industrial facilities that produce CO_2 as a by-product, and (2) new infrastructure, such as the DAC and hydrogen (H_2) hubs that, respectively, will remove atmospheric CO_2 or capture CO_2 from natural gas-based H_2 production. While existing infrastructure still has opportunities to address long-standing EJ issues in the surrounding community and employ reparative and restorative justice principles, new infrastructure should be considered carefully throughout the design and project development stages to address community concerns and needs.

Existing facilities. Studies have linked historical racist policies and practices to the fact that sources of air pollution are disproportionately located in and adversely impact communities of color (Bravo et al. 2016; Cushing et al. 2023; Lane et al. 2022; Liu et al. 2021; Ringquist 2005; Rothstein 2017; Tessum et al. 2019; Woodruff et al. 2023). In particular, racial zoning, redlining, and segregation have isolated racial and ethnic minorities within built environments with inadequate physical infrastructure (e.g., stormwater drainage, green space, and energy systems)

---

[28] For more information, see Section 5.2.1 through Section 5.2.3 in NASEM (2023d).

[29] Appendix G describes key features of effective siting and permitting processes from scholars and practitioners as outlined in NASEM (2023e).

(Bullard 2020; Hendricks and Van Zandt 2021) or near hazardous industrial facilities (Agyeman et al. 2002; GAO 1983; James et al. 2012; Linder et al. 2008; Mohai et al. 2009). These patterns of environmental injustice were "shaped by power and privilege" and created "areas of both prosperity and disadvantage" (Hendricks and Van Zandt 2021, p. 1). These historic inequities have laid the foundation for the industrial sector that persists to this day.

While an existing industrial facility's location will not necessarily be revisited, retrofitting the facility with new technology can strive to incorporate EJ and democratic community organizing principles. An important first step is for the developers and advocates to familiarize themselves with the historical experience of the most disenfranchised in the immediately impacted community. Then, restorative justice can be considered; before new work can begin at an existing site where harm has taken place, especially if that harm is generational, the actors that have perpetuated harm accept responsibility and act to repair the harm done while reducing the likelihood of creating new injustices (Hazrati and Heffron 2021). Beyond just capturing CO_2 from the facility, additional infrastructure in these locations can seek to alleviate existing burdens in the community where possible. For example, new infrastructure can pursue pollution reduction and include provisions for people's science to conduct data MRV for the outcomes of retrofits.

New facilities. While new developments featuring emerging technologies benefit from widespread site selection, they also face similar public acceptance challenges to other climate-mitigation technologies.[30] In the case of DOE's hub programs, developers are often encouraged to look at or partner with Opportunity Zones—economically distressed areas of the United States—with the goal of "[spurring] economic growth and job creation in low-income communities while providing tax benefits to investors" (IRS n.d.). While siting of new CO_2-based infrastructure in Opportunity Zones may bring positive economic impacts, the fact that these communities are already economically depressed or overburdened indicates the presence of disenfranchised groups and potential structural inequities. Advocates and developers will need to center community history and EJ practices in site selection for hubs. The process of designing hubs is intricate and complex owing to the variety of potential industries, actors, and build-out involved, and could benefit from ongoing studies around social understanding and acceptance (Gough and Mander 2022; Upham et al. 2022). General pros and cons of clustered siting that could apply to DAC, H_2, or CO_2 facilities are listed in Table 4-3. See Chapter 10 for more discussion of infrastructure co-location.

TABLE 4-3 Pros and Cons of New Clustered Siting

Pros	Cons
1. Multiyear hub planning strategies may prepare neighboring communities for the construction and development of a cluster, outlining phases of build-out that gives the community insight into ongoing development and goals that can impact efficiency.	1. Hubs that cross various state or county lines might hinder ways to track community engagement, especially engagement that prioritizes overburdened populations most impacted by the infrastructure development.
2. With larger swaths of federal funding to build out hubs, there may be an ability to designate more resources to communities through legal agreements. This can help build stakeholder networks and establish trust with communities.	2. Hubs that cross various state or county lines might make it difficult to measure and ensure that communities impacted by the hubs are receiving direct benefits.
3. Hubs can provide localized economic benefits—like construction jobs—and develop lasting relationships with the community, and signal the willingness to work with local businesses, universities, and organizations.	3. Large areas of land might be necessary for some hub siting, which has implications for the surrounding environment and wildlife.
4. Hubs can support diverse participation of industry owing to the shared resources from clustered infrastructure (e.g., local pipelines), allowing smaller entities to participate.	4. Hub siting that incorporates new and existing technologies may not adequately consider the local history of its geography with respect to previous infrastructure and could therefore exacerbate issues the community may already be facing.

SOURCES: Based on data from Gough and Mander (2022), Sovacool et al. (2023), and Upham et al. (2022).

[30] For example, see NASEM (2022) for a summary of the societal challenges facing advanced nuclear.

New infrastructure does not necessarily connote *good* infrastructure. Practices that do not address the social and spatial impacts that remain within communities that have experienced adverse and extractive industry or infrastructure risk reproducing harms (Heck 2021). For example, in 2011, a St. Louis utility negotiated with EPA to reduce the cost of a redevelopment project for the St. Louis wastewater infrastructure. This negotiation resulted in a $1.3 million decrease in the project's cost by eliminating the need to improve existing infrastructure, which resulted in a lack of benefits to marginalized communities directly impacted by failing sewer and stormwater infrastructure (Heck 2021). The siting of new infrastructure, especially on the scale of hubs, which will cross state lines and impact a variety of geographies, has to be considered in the context of "a progressive lens that views physical infrastructure as an extension of social circumstances," thus reflecting historically racialized frameworks and policies that target disenfranchised groups (Hendricks and Van Zandt 2021, p. 1).

An example of a progressive lens is the *social construction of technology* framework, which states that technology and society are mutually constructed together (Bijker and Law 1994). This framework identifies the complex nature of projects in which the relevant social groups have different ideas about what a technology does; therefore, a project's negotiation will have to acknowledge different technological frames—the goals, theories, and solutions that capture the interactions between social groups and the unique ideas about a proposed technology or project—before coming to consensus about a project (Sovacool et al. 2023). The complex relationship between social groups, technological frames, technology, solutions, and risks within a social construction of technology framework (see Figure 4-8) highlights that the design and implementation of net-zero megaprojects, such as CCUS hubs, have economic, political, and socioenvironmental dimensions that need to be identified and addressed throughout the project's life cycle.

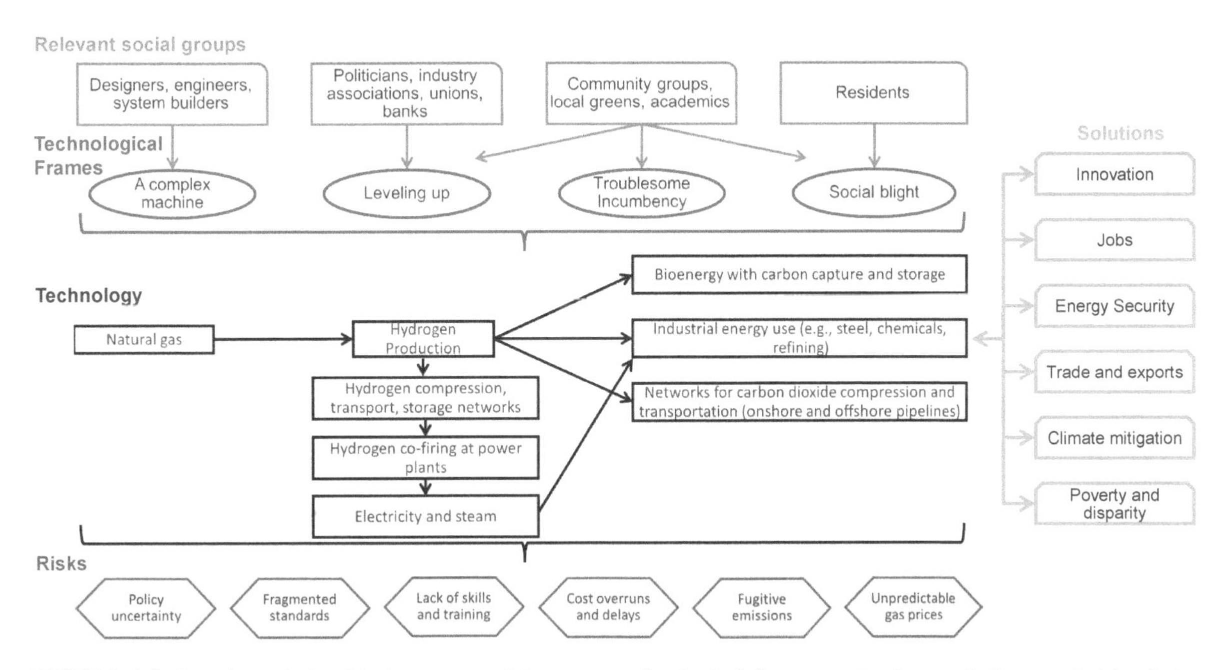

FIGURE 4-8 Complex relationship between social groups, technological frames, technology, solutions, and risks during project design and proposal.
NOTES: The arrows depict how the social groups (shown in blue) interact with frames (shown in red) which also shape both technology (shown in black) and relevant solutions provided by technology (shown in green). The technologies in turn respond to or are impeded by risks (shown in purple).
SOURCE: Based on Sovacool et al. (2023), https://doi.org/10.1016/j.techfore.2023.122332. CC BY 4.0.

4.4.2.4.2 Pipelines

Pipelines are a critical part of energy infrastructure development and have controversial perceptions among the public, especially related to oil and gas development in the United States owing to the potential localized environmental and social impacts (e.g., land transformation and safety risks) (Jensen 2017). Responses to pipelines are often a result of factors such as technology (e.g., risks and benefits), process, place (e.g., physical infrastructure), and people (Boudet 2019; Janzwood 2023; see also Figure 4-5). Given that overburdened and disenfranchised communities are often impacted by harmful infrastructure, the intersection between pipeline infrastructure and environmental injustice has to be analyzed. The connection between pipelines and autonomy is especially pertinent in the case of Indigenous nations and communities who are vulnerable to and have limited capacity to mitigate pipeline-related incidents (Datta and Hurlbert 2020).

Concerns around pipelines—whether transporting oil, natural gas, or CO_2—remain the same: impact on community resources (e.g., water and air); protection of culturally valuable sites (e.g., prayer and grave sites); impact on public and environmental safety and health; and adequate approval from local groups (Strube et al. 2021). The cancellation of the Mackenzie Valley Pipeline after the Berger Inquiry in 1974 is a notable example that voiced the concerns of Indigenous tribes. The proposed pipeline would have brought Canadian natural gas to U.S. markets, but a comprehensive social, environmental, and economic assessment recommended that no pipelines should be built in the Northern Yukon region inhabited by Tribes. More recently, Indigenous Tribes in the United States protested the Keystone XL and Dakota Access pipelines (Suls 2017). In the case of Dakota Access, protesters stated that the U.S. Army Corps of Engineers had failed to adequately consult Tribe members before approving the pipeline and had violated the National Historic Preservation Act (Herscher 2017).[31] The general public reticence around CO_2 pipelines is markedly similar. The cancellation of a 1300-mile CO_2 pipeline across five Midwestern states in October 2023 was the outcome of public opposition and citations of a lack of certainty in the regulatory and permitting processes (Phillips 2023). Another Iowa-based CO_2 pipeline project faces delays from comparable opposition and an unclear and difficult regulatory process (Tomich et al. 2023).[32] There is also the overarching sentiment that pipelines are a part of the infrastructure that furthers the lifetime of fossil-based industries (e.g., see Earthjustice 2021).

Public health and safety are also important considerations when it comes to pipeline siting and practices. Public attention was drawn to safety and regulation of CO_2 pipelines in February 2020 when a pipeline carrying CO_2 and other chemicals ruptured in Satartia, Mississippi. While the failure was a consequence of natural force damage and did not result in any fatalities, the inadequate response by the pipeline operator led to an evacuation of 200 people and 45 hospitalizations. This event eroded public trust in CO_2 pipelines and raised concerns about the safety and regulatory capabilities of the Pipeline and Hazardous Materials Safety Association (PHMSA). PHMSA's official failure investigation report pointed to technical shortcomings—such as the failure of pipeline owner Denbury's atmospheric models for emergencies to consider the locality of incident—and critical problems with the response to the incident (DOT-PHMSA 2022). Specifically, "[l]ocal emergency responders were not informed by Denbury of the rupture and the nature of the unique safety risks of the CO_2 pipeline" (DOT-PHMSA 2022, p. 2). Because emergency responders were not trained to handle CO_2 leaks, mitigation measures were not sufficient to prevent hospitalizations. This incident prompted PHMSA to improve its pipeline and emergency regulations.

While pipeline safety is left up to federal regulators, there are opportunities to learn from the Satartia incident to encourage community engagement and education on pipeline safety protocols, including what to do in the event of a rupture. Resources and funding for emergency training are possible preventative measures that might create "awareness of nearby CO_2 pipeline and pipeline facilities and what to do if a CO_2 release occurs" (DOT-PHMSA 2022, p. 2). Box 4-6 outlines lessons learned in Indigenous–Canadian communities from pipeline spills. These strategies, which have been recommended for Canadian government and Indigenous community interaction, could

[31] Following the May 2021 announcement that the pipeline would remain in operation, another environmental impact statement was court-ordered. The Army Corps of Engineers received 200,000 comments during the open comment period of the trial, which ended in March 2024 without the presentation of closing arguments (Dalrymple 2024; Streurer and Dalrymple 2024). A final decision is expected in late 2024.

[32] See Chapter 10 for more information about these proposed CO_2 pipelines and recommended solutions to overcome the barriers public sentiment creates.

BOX 4-6
Government Interaction with Underrepresented Communities in Pipeline Siting

Pipeline siting is an opportunity to prioritize culturally relevant knowledge and practices to develop community-based solutions—a strategy that is often not employed. Effective processes for developing community-based solutions center a community's worldview and consider their preferred research methodologies and frameworks (Datta and Hurlbert 2020). When addressing community concerns around resources and culturally significant sites, tools that seek to model a variety of scenarios while weighing both social and environmental impacts can be useful, especially when combined with a robust and transparent community procedural process (Shih et al. 2022). The development of community-based solutions requires robust community engagement. Recommendations for government interaction with Indigenous communities as developed by Datta and Hurlbert (2020) are listed below and can be applied to developer interactions with U.S. communities:

1. **Support community-based participatory action research.** Community-led research initiatives can provide understanding of community perspectives from within and allow communities to invest in frameworks that support interventions relevant to the community while staying up to date with new science and protocols as they emerge. The outcome of community-led research can bridge gaps between outside advocates like scholars or nongovernmental organizations and the communities with whom they are working.
2. **Develop a database.** Access to information is paramount for communities to advocate for themselves and understand the impacts of infrastructure on their health and resources, while further informing government agencies and other stakeholders.
3. **Consider community perspectives.** To build projects in a community, the perspective of community groups needs to be centered and developed around the issues that impact them.
4. **Provide funding.** It is critical to provide communities with resources to act in times of crisis, perform their own analyses and science, create action-based solutions, and educate the community before an emergency occurs.
5. **Develop community-led programs.** Internal programs can work to maintain databases, perform outreach and education, report issues, and implement strategies to find and deploy effective community-centered solutions.

be a valuable addition to U.S. siting processes to ensure the incorporation of community perspectives in remedying pipeline hazards and developing pipeline safety plans. CO_2 pipeline safety is discussed further in Chapter 10.

Federal review procedures conducted under the National Historic Preservation Act (NHPA) and the NEPA have been criticized for failing to adequately advance EJ or result in environmentally just outcomes, despite having EJ considerations built in. For example, there are concerns that the outcomes of the NHPA are based on the level of public engagement and thus will not ensure the preservation of culturally relevant or significant locations and that the NEPA has been implemented through guidance documents, not legally binding processes (Lockman 2023; Sassman 2021). Furthermore, these federal processes do not capture state-level nuances that need to be identified to achieve granular critiques of the siting and its impacts (e.g., the use of "Natural Resource for Public Purpose" or "Enhanced Oil Recovery for Public Purpose" in Idaho, Wyoming, and Colorado, or "Eminent Domain for Public Purpose" in North Dakota, Montana, or "Common Carriers" in Texas [Righetti 2017]).

Public transparency around pipeline protocols is imperative to ensure the health and safety of surrounding communities and increase their preparedness in the case of an incident. Models that evaluate technical, environmental, and social impacts can be used to expand accurate and transparent information about protocols. Studies have shown that "the optimal path is very sensitive to environmental and social impact considerations at even low weights, in that a small increase in pipeline length (and cost) significantly avoids large environmental and social impacts" (Shih et al. 2022, p. 1). With a procedurally just approach focusing on restorative principles, consensus may be reached for the siting of a pipeline, especially if existing rights of way might be repurposed for CO_2. However, consensus is not guaranteed just because certain practices are employed. Ultimately, when it comes to pipeline siting, developers need to consider the relationship of the pipeline to potential resources and culturally relevant sites. (See Finding 4-10 and Recommendation 4-9.)

The committee's first report identified that special consideration is needed when selecting pipeline materials to ensure that appropriate mechanical properties are used to resist ductile and brittle propagating fractures (NASEM 2023b). The committee recommended, and still recommends, that DOE collaborate with national laboratories, university researchers, PHMSA (which is responsible for regulating liquid pipelines, including their design, construction, operation, corrosion control, and testing, maintenance, and reporting requirements), and industry to develop and test rigorous fluid-structure models (NASEM 2023b, Recommendation 4.3).

4.4.3 Just Economic Considerations

The CO_2 utilization sector will have to consider how to equitably distribute economics in its planning. Otherwise, the potential capabilities for the nation's workforce can become a missed opportunity, resulting in the loss of growth and productivity for the nation (Jacobs 2013). This section lays out how CO_2-derived products contribute to a circular economy, what workforce considerations the sector needs to acknowledge, and how the economic benefits of the sector should be equitable distributed.

4.4.3.1 Product End Use and Contributing to a Circular Economy

Historically, underserved communities have hosted facilities that contribute to the linear economy, which is fundamentally based in extraction and waste, and then have been burdened with the climate impacts derived from these decisions (Berry et al. 2021; Schröder 2020). Aspects of these injustices can be addressed with a well-designed and inclusive circular economy (EPA n.d.(h)). Integral to creating a more just circular economy is considering the end-use impact of that CO_2, in addition to the selection of which facilities go in what communities. For example, enhanced oil recovery, the primary current use of CO_2 in industrial processes, is an economically useful and viable practice but does not create materials and products with a long lifetime, thus offering minimal contribution to a circular economy. See Chapter 2 for more about how CO_2 utilization contributes to a circular economy.

Acceptance of CO_2 utilization projects is likely to be determined by the level of trust that facilities will be run safely and the extent to which a project and its processes contribute to a more circular economy (Jones et al. 2017; Schröder 2020). Additionally, public acceptance will depend on policies and practices identifying the long-term consequences of the sector and how public investment aligns with the perceived rationale for CO_2 utilization as it relates to climate mitigation strategies (Jones et al. 2017). As the transition to a circular economy is a response to climate change, further alignment of the CO_2 utilization sector with goals the public understands is critical. These goals include sustainable practices that restore natural systems, design waste out, and choose biological or renewable materials over nonrenewable ones (Schröder 2020). Furthermore, a circular economy needs to include frameworks that incorporate principles of procedural and distributive justice for it to be just (Berry et al. 2021).

4.4.3.2 Workforce Considerations

Communities have shown particular concern that extractive industries and carbon management infrastructure will inflict ongoing harm on disenfranchised communities. Specifically, "concrete examples of environmental inequity leading directly to unequal health status can be found in occupational health literature and among . . . clinics which serve populations that include low wage workers and workers of color" (Friedman-Jiménez 1994, p. 605). To reform perceptions and experiences with the carbon management sector and set a more equitable tone for CO_2 utilization, a breakdown of the *jobs versus environment* paradigm, which consistently pits the health and well-being of workers in potentially dangerous industrial jobs against their need for economic stability, needs to occur. This dichotomy has hindered the development of "efforts to build solidarity amongst local environmental justice goals on the one hand, and workers and union aspirations for secure, quality jobs on the other" (Evans and

Phelan 2016, p. 529). Providing workers with safe and healthy work environments needs to be prioritized in the build-out and siting of CO_2 utilization projects.

An emerging CO_2 utilization sector is also an opportunity to diversify the workforce, in terms of race and class, across the entire CO_2 utilization value chain (Taylor 2011). Whether it is the skilled-labor force or positions that require higher education, there is room to further reflect the diversity of the U.S. population. The DOE Research Experience in Carbon Sequestration program, for example, promotes diversity and inclusion strategies like mentorship programs between program alumni and BIPOC graduates and early-career professionals, targeted recruitment strategies, intensive short courses, and internships (Cao and Tomski 2023). The intentional targeting of historically marginalized communities aims to overcome barriers to access to career pathways for unemployed adults, workers with low-wage jobs, and high-school-to-career transitions with significant data showing that "workforce development strategies can build pathways out of poverty" (Zabin 2020, p. 116). The inclusion of underserved and disenfranchised populations also can shape the future development of the CO_2 utilization sector, encouraging an emphasis on a just economy moving forward.

Principles of developing a robust workforce that supports an inclusive economy include the following (Coleman 2023):

- **Ground the workforce system.** This requires both a more precise and a more expansive definition of individuals' skills and needs based on the diverse needs of the population.
- **Provide access to services and resources.** To support adult learners who may face daily challenges that affect program participation and completion, workforce systems need to mitigate the personal and financial constraints that make training and education difficult to sustain.
- **Coordinate cross-sector action.** An effective workforce system—one that advances inclusive economic growth—engages with multiple institutions and sectors to meet the skill and transparency requirements necessary to produce economic mobility at scale.

While the CO_2 utilization sector will vary by location, the preceding principles are important considerations for increasing access and cementing impactful workforce philosophies.

4.4.3.3 Working to Ensure Just Economic Outcomes

The distribution of infrastructure, benefits, and resources that come from building out the sector is another important consideration when selecting projects, especially given the U.S. history of toxic infrastructure disproportionately impacting underrepresented communities. Frameworks of distributive justice, a key principle of EJ, acting in cohesion with recognitional and procedural justice, have to be incorporated into processes that identify potential costs and benefits of proposed projects.

Increasing opportunities exist for businesses to incorporate social justice principles into their dealings, including through community benefits from the development and build-out of the CO_2 utilization sector, which can include jobs and educational opportunities, preservation of natural resources, contributions to local trust funds, guarantees to pay workers local living wages, and reduction of toxic pollutants. Two types of community benefit frameworks that can be part of project negotiations are community benefits agreements (CBAs), which provide a transparent way for community members to work with developers to determine clear benefits the project will bring the community, and project labor agreements (PLAs),[33] which are agreements between a developer and relevant labor unions to set terms for labor requirements on a project (Fraser 2023). There is evidence to suggest that voters across party lines, and especially those of color, support the use of these agreements in project design in their communities. In particular, Fraser (2023) found that support for CBAs is highest for Latino (86 percent) and

[33] For example, the Memorandum of Understanding between the United Association of Union Plumbers and Pipefitters and Texas-based carbon capture company CapturePoint Solutions LLC created direct-entry career and technical education pathways for high schoolers to train in pipeline, plumbing, and steam-fitting industries, including providing updated facilities and workshops, student transportation, and instructors (Contractor 2023).

Black (77 percent) voters.[34] These plans, when designed through a collaborative framework that is legally binding and adaptable, can serve as a pulse check on a project and can support distributive justice and equitable outcomes.

Community benefits frameworks also offer up obligations meant to diversify the value chain of a project, which can create and solidify inroads for underrepresented groups. For example, a small business developing partnerships with underserved groups might find it particularly beneficial to negotiate a PLA to ensure equitable access to employment opportunities. Additionally, as a start-up begins to develop its initial relationships across the supply chain, choices regarding company principles and workforce decisions have the potential to diversify the CO_2 utilization supply chain and increase small business participation. However, the rights of people and communities need to be recognized and protected during a project's selection process. Important considerations around these rights and just cross-sectoral distribution include "How are marginalized circular economy views and narratives, knowledge and values recognized and integrated into dominant narratives? How can competing development interests be resolved through participatory processes? And which institutions can guarantee recognition and protection of rights during the transition processes?" (Schröder 2020, p. 14).

The distributive implications of the CO_2 utilization sector go beyond the U.S. domestic context, as the sector is globally connected across extensive supply chains. Sectors like textiles, chemicals, and waste management and recycling—in which CO_2 utilization could play a key role as it develops—are intertwined globally. Holistic approaches and multistakeholder collaboration are required to create "decent, high value work throughout the value chain and with a focus on vulnerable communities and regions" affected by the transition to a more circular economy (Smith 2017, p. 18). Therefore, consideration of impacts from CO_2 utilization will have a global element, of which the rights of countries and cultures will need to be protected.

4.5 CONCLUSIONS

The right combination of policy, regulatory, and engagement frameworks can supplement existing opportunities for the CO_2 utilization sector to be economically viable and sustainable and to contribute to a circular economy. This chapter reviewed existing policy and regulatory mechanisms for the CO_2 utilization sector. Passage of congressional legislation such as the IIJA and the IRA provided economic and noneconomic incentives to build out this sector. The IIJA alone appropriates about $20 billion to support R&D and commercialization for carbon management technologies, including CO_2 utilization. Additionally, federal- and state-level efforts to incorporate EJ goals into policy can ensure that communities receive the public health (e.g., Justice40) and workforce benefits that may stem from investment in carbon management. In addition to developing more robust and adaptable policy mechanisms through which to incorporate EJ, opportunities also exist to develop more demand-side policy for CO_2-derived products.

Because discovering, developing, and commercializing CO_2 utilization processes and CO_2-derived products will be necessary as the nation transitions to a net-zero economy, integrating business development opportunities into the expansion of the CO_2 utilization sector is critical. The chapter identified key areas where policy can continue to support businesses—including small businesses—pursuing CO_2 utilization. Additionally, programs that support partnerships among businesses, customers, and national laboratories will advance the market for CO_2-derived products during R&D and commercialization phases. At the federal level, existing R&D programs, such as the GAIN Program and the Voucher Program, can serve as models for CO_2-utilization-specific programs. Such a CO_2-utilization-specific program could be especially beneficial for small businesses trying to find their niche in the emerging CO_2 utilization sector.

The chapter identified an opportunity for project developers to address the lack of public awareness of CO_2 utilization through public engagement and education practices that include meaningful engagement with impacted communities, which can provide substantial guidance on how to invest in communities holistically, regardless of

[34] Columbia University's Sabin Center for Climate Change Law compiled a database of publicly available CBAs and released a report with best practices for developing and implementing CBAs (Columbia University Sabin Center n.d.). Additionally, there is a wealth of scholarly literature around the efficacy of CBAs as a positive tool when implemented thoughtfully and as a drawback when actual community views are not reflected (e.g., see Been 2010; De Barbieri 2016; Fraser 2022, 2023; Marantz 2015; Wolf-Powers 2010).

infrastructure type (see Figure 4-7). The nascency of the CO_2 utilization sector means that EJ principles can be meaningfully considered and appropriately incorporated into sector project and workforce development, as well as policy. Specifically, consideration of the 3rd, 5th, 6th, 7th, 8th, 11th, and 16th Principles of Environmental Justice can help ensure that the CO_2 sector advances ethical and responsible uses of land and renewable resources; ensures all communities the right to self-determination; ameliorates waste products and potential pollutants; actively integrates community input into project design; provides workers and communities with access to new job opportunities and related benefits; incorporates community approaches into decision making and implementation; and provides education opportunities emphasizing social perspectives and histories on environmental issues. Furthermore, through the consideration of just economic outcomes, the emerging CO_2 utilization sector can advance equitable access to the circular economy, new jobs associated with the emerging sector, and direct benefits to communities from projects.

This section enumerates the findings and recommendations related to the policy and regulatory frameworks needed for economically viable and sustainable CO_2 utilization and concludes with the research agenda item associated with this chapter.

4.5.1 Findings and Recommendations

Finding 4-1: Demand incentives—Various supply-side economic incentives support carbon utilization, in contrast to minimal demand-side policy incentives. Multiple approaches to lower the carbon intensity (CI) of purchased commodities have been shown to be more cost effective than manufacture of CO_2-derived products, given the current levels of decarbonization required by corporate goals in the private and public sectors. Adopting low-CI products produced from captured CO_2 can support private and public sector decarbonization goals and also support the development of a CO_2 market, especially as decarbonization targets become more stringent. Without the simultaneous incentivization of the demand for CO_2-derived products or the imposition of low CI requirements for new products, CO_2-derived products will remain costly compared to incumbents.

Recommendation 4-1: Targeted procurement standards—The General Services Administration (GSA) should implement and provide guidance to state and local procurement offices to ensure that programs are designed to motivate CO_2-derived product demand. The GSA guidance should outline the benefits of procurement programs that either set product quotas or set a scoring system based on the prevalence of products procured through CO_2 utilization processes, taking into consideration locally relevant conditions as appropriate, for state and local offices to consider. At least in the short term, these products may have higher costs and potentially higher carbon intensity (CI) than non-CO_2-derived products, but demonstrations are critical to obtain the necessary learnings to drive down costs through scaling. If the products continue to have higher CI, then a progressive CI target per carbon utilized that rachets down over time should be considered.

Finding 4-2: Workforce development—The workforce needed to support carbon utilization will be small in comparison with that for other clean energy jobs, but there are not enough employees in the upstream labor force to support the infrastructure build-out that investors seek to fund. To support the expansion of a market for CO_2-derived products, a sustainable workforce is needed for upstream labor for the deployment and maintenance of CO_2 utilization facilities and technologies. Creating this workforce will require innovative approaches to make the labor force appealing and competitive to prospective employees. The development of a CO_2 utilization workforce can use the existing oil and gas workforce that has transferable professional, technical, and labor skill sets, including project management, business development, pipeline construction, and manufacturing. The CO_2 utilization sector can develop a diverse workforce by building on the awareness that opportunities in oil and gas are declining, offering more training focused on developing and using translatable skills that evolve with the industry, and ensuring emerging workforce opportunities are accessible to all.

Finding 4-3: Co-benefits—Preventing emissions of greenhouse gases and short-lived climate pollutants, including carbon dioxide, black carbon, and methane, is expected to have positive impacts on human and public health, especially in communities and regions most adversely impacted by harmful emissions and air pollutants. Assessing and communicating the expected health impacts of carbon management technologies and developments will help inform project developers and potential host communities about what can be achieved and, in turn, will better inform decisions made during project selection and program implementation.

Recommendation 4-2: Disclosure of health impacts—The Department of Energy (DOE) should require the use of the U.S. Environmental Protection Agency's Co-Benefits Risk Assessment Health Impacts Screening and Mapping Tool (COBRA) to analyze and communicate the projected health impacts of carbon capture, utilization, and sequestration technologies for DOE projects. The specificity of COBRA's estimates will ensure that the maximum health benefits are achieved when siting and selecting carbon management projects.

Finding 4-4: Policy drivers—The main policy mechanisms for the CO_2 utilization sector are tax credits, permitting and regulatory frameworks, and large omnibus legislation. These mechanisms largely incentivize the supply of CO_2-utilization-derived materials and products and are difficult to modify as market conditions change. Both supply- and demand-side mechanisms need to be applied simultaneously to scale up CO_2-derived products in a timely fashion and achieve meaningful market share. The absence of durable and adaptable policy serves as a barrier to successful market and infrastructure development for the CO_2 utilization sector, because there is no clear and consistent direction for carbon management, which hinders the investment in CO_2 utilization infrastructure or markets and the adoption of CO_2 utilization technologies. The novel task of developing a circular economy at the scale required to meet the nation's net-zero targets creates the opportunity to implement innovative policy that is flexible to meet targets and adaptable to course correct when relevant.

Recommendation 4-3: Policy flexibility—Congress and federal agencies that work on carbon management (e.g., the Department of Energy and the U.S. Environmental Protection Agency) should include a variety of directionally consistent policy mechanisms to support ongoing research, development, demonstration, and deployment; allow flexibility for updating and changing information used to shape incentives based on evolving market conditions; and develop financial mechanism alternatives to corporate tax credits to support adaptive policy development. These federal-level mechanisms should be further flexible to state- and local-level conditions for carbon management infrastructure to support decarbonization and/or economic development goals and enable a stable utilization market.

Finding 4-5: Emerging business support—There is an opportunity to diversify supply chains and types of small businesses that participate across the carbon management value chain as the market for CO_2-derived products is created. Support through policy mechanisms or incentives for local, and veteran-, woman-, or minority-owned companies and start-ups to participate will allow new CO_2 utilization opportunities to be more accessible than incumbent industries. Co-piloting and demonstration partnerships between key upstream and downstream partners of CO_2 utilization will help prove the commercialization of emerging businesses of any kind. However, there is a lack of programs that connect emerging CO_2 utilization businesses with the facilities and technologies needed to upscale.

Recommendation 4-4: Small business support—The General Services Administration, in collaboration with the Department of Energy (DOE), should develop opportunities for small businesses to upscale, modeled on the Gateway for Accelerated Innovation in Nuclear (GAIN) program run by the DOE national laboratories and DOE's Voucher Program for energy technologies, to better support small businesses entering this field. The key components that should be adopted include:

a. From the GAIN program: Access to testing capabilities; use of demonstration facilities; and connection to experts in material science, licensing, and finance. Like the GAIN program, a voucher program should be established to fund time and effort for the national laboratories to work with businesses. Vouchers for laboratory time and effort should be awarded on a competitive basis with meaningful consideration of how applicants can diversify the supply chain.

b. **From the Voucher Program: Predemonstration support to address key adoption risk areas; performance validation and certification support; commercialization support to access market research and business plan strategy assistance or to access independent monitoring, reporting, and verification technologies and practices. This support will enable small businesses to access information from subject-matter experts and opportunities through testing facilities that drive their businesses to the next level.**

Finding 4-6: Host community impacts—The distribution of costs, benefits, and resources that stem from infrastructure need to be communicated to and considered by potential host communities of CO_2 utilization infrastructure projects. Meaningful engagement and communication between a project's developers and potential host community members is required to determine what aspects are considered costs or benefits. To supplement community engagement, assessment frameworks that quantify societal impacts—such as social life cycle assessments—can be used to communicate the societal benefits and risks of a project's role in climate mitigation to the public. The combination of engagement and assessment can address barriers to public understanding of the emerging CO_2 utilization sector.

Recommendation 4-5: Community impact tracking—The Department of Energy's (DOE's) Office of Energy Justice and Equity should work in partnership with community-centered councils and agencies like the White House Environmental Justice Advisory Council to develop robust pathways for defining, tracking, and measuring impacts of CO_2 utilization projects and infrastructure to determine whether and how those impacts are being equitably distributed, while communicating expected outcomes to communities. These pathways should be incorporated into community benefit and project labor agreements and plans as relevant. If the impacts are determined to be benefits by the community, DOE should work on developing a framework to track how these project benefits also meet environmental justice metrics (e.g., taking into account the guidance issued by Office of Management of Budget on Justice40 implementation).

Finding 4-7: Community education—There is limited public knowledge about CO_2 utilization compared with other technologies that will support a net-zero economy. Most of what the public understands is centered around limited familiarity with carbon management technologies, including opposition following similar concerns about other emerging technologies like hydrogen (now) and nuclear (decades ago). These concerns include how a project will impact everyday life in a community and if the required regulatory systems are effective and competent. Communities and consumers want to see connections between the contributions of CO_2 utilization infrastructure and its products and climate mitigation, proving that the technology investment is yielding worthwhile results while prioritizing processes that value public engagement. Additionally, because the narrative around CO_2 utilization is nascent, there are opportunities to confront justice-related challenges and shape a positive narrative by employing justice-centered practices.

Recommendation 4-6: Educational material development—Nongovernmental organizations and research-conducting entities—such as national laboratories, think tanks, and universities—should identify gaps in knowledge, sharing their data and findings about societal acceptance of or opposition to the CO_2 utilization sector through the following actions:

a. **Nongovernmental organizations (including universities) and national laboratories should conduct targeted research to develop transparent resources to communicate findings and support improved education related to the direct impact of the CO_2 utilization sector on climate change mitigation.**
b. **Research-conducting entities should continue to conduct social analyses to determine what consumers and communities think about the CO_2 utilization sector, both in relation to and separate from the carbon management value chain, filling the gaps in knowledge about societal acceptance and potential opposition, to better understand and address the concerns of the public. These data and conclusions of this research should be shared through a centralized and broadly accessible framework.**

c. **Research-conducting entities should use analyses that combine techno-economic and life cycle assessment objectives to determine a levelized cost of CO_2 abatement to be used to assess the desirability of projects as a function of CO_2 source (fossil or biogenic point source, direct air capture, or direct ocean capture) and product durability. The results of these analyses and assessments should be communicated to the Department of Energy and other entities funding carbon utilization projects to inform them of factors that can influence community acceptance of projects and expected outcomes.**

Finding 4-8: Public acceptance—To best support and fit into a circular economy, the CO_2 utilization sector faces and must overcome barriers to public acceptance. Holistic approaches to public and community engagement that acknowledge the intersection of identities and priorities within a community and recognize environmental justice as a facet of a larger social movement facilitate a just, cross-society transition to a circular economy. In particular, meaningful engagement across the value chain can include opportunities for public education about the benefits and risks of the sector, diversifying the sector's workforce and business types (e.g., small versus start-up versus private business), incorporating labor and community benefits agreements into project design, and consent-based siting and selecting practices. The application of these practices will be diverse, based on differences in project scale and scope, impacted and engaged community, and context within the project cycle.

Recommendation 4-7: Meaningful community engagement—The Department of Energy (DOE) should prioritize projects that can prove that meaningful community engagement frameworks were incorporated into their decision-making processes, as appropriate for the technology readiness level of the project. Additionally, to support DOE's efforts:

a. **Private project designers that conduct life cycle assessments should communicate their findings transparently with community advocates as a part of planning and design processes. This protocol will ensure that a community is made aware of risks to shared resources such as energy, land, and water, and that any barriers are assessed before the project is selected so that place-based solutions can be developed.**
b. **Project designers and funders, in both the public and private sector, should maintain transparency when co-developing projects and reporting emergency protocols to a community. This will ensure that public health and safety are prioritized during siting and emergency response processes. If the site includes infrastructure that might have potential safety impacts, developers should fund community engagement specifically targeting first responders to equip them with the necessary understanding of safety protocols in the event of an emergency.**
c. **Project designers and funders, in both the public and private sector, should use diverse methods to incorporate community engagement into assessments to determine whether a project should move forward in a particular community. These methods include community benefits and project labor agreements (which are more robust if legally binding) and plans, and frameworks to employ elements of Free, Prior, and Informed Consent that incorporate vital community input into the decision-making process.**

Finding 4-9: Environmental justice—It is critical for environmental justice considerations to be incorporated throughout the project process to ensure that the emerging CO_2 utilization sector develops justly and does good for the public. Although project siting and selecting processes need to be place-based and adaptable to each community and project, a just utilization sector can rely on incorporating the Principles of Environmental Justice, adopted at the National People of Color Environmental Leadership Summit in 1991, into project design processes. Of the 17 principles, seven principles can be particularly useful as guidelines through which communities evaluate proposed carbon utilization projects: Principles 3, 5, 6, 7, 8, 11, and 16.

Recommendation 4-8: Principles of environmental justice—New CO_2 utilization infrastructure development should apply justice principles and learn from other emerging technology pathways, while

seeking to deploy tangible benefits to surrounding communities beyond remediation of the climate, during the planning and design process. To support this, the Department of Energy should incorporate the Principles of Environmental Justice into the requirement for community benefit plans and the review of funding applications for CO_2 utilization projects that require infrastructure development or changes. Specifically, the following principles should be incorporated to prioritize the just prior and ongoing development of the carbon utilization sector:

a. **Principle 3: Environmental Justice calls for the right to ethical, balanced, and responsible uses of land and renewable resources. Carbon management infrastructure, including for transport infrastructure to utilization facilities, should consider the most appropriate uses of energy, land, and related impacts through using this framework.**
b. **Principle 5: Environmental Justice affirms the right to cultural, political, economic, and environmental self-determination of all. The self-determination of the community should be considered and prioritized by project planners and designers to create just precedents for infrastructure build-out.**
c. **Principle 6: Environmental Justice calls for the end of hazardous wastes and radioactive materials production, while demanding past and current producers be held accountable to the public. The utilization sector should identify and plan to ameliorate waste products and potential pollutants from their facilities that might arise from the creation of CO_2-derived products or feedstocks.**
d. **Principle 7: Environmental Justice calls for the right to participate at every level of decision making as equal partners. The utilization sector should consider ways to actively integrate community input and reflect that input through adaptations of their project plans.**
e. **Principle 8: Environmental Justice affirms the right of all workers to a safe and healthy work environment. The utilization sector should ensure that workers and communities are provided with access to new job opportunities, healthy working conditions, and related benefits of the sector.**
f. **Principle 11: Environmental Justice must recognize a special legal and natural relationship of Indigenous nations to the U.S. government. The U.S. government should consider community approaches to decision making and implementation when engaging with Indigenous groups, as well as their authority over respective territories.**
g. **Principle 16: Environmental Justice calls for the holistic education of present and future generations, including relevant social perspectives and histories on environmental issues. Prioritizing education through appropriate social lenses allows for the participation of underrepresented communities and supports the development of project frameworks that call for just outcomes.**

Finding 4-10: Host community engagement as a part of project development—Project designers need to consider the benefits of available engagement models and select one that ensures just outcomes, including project co-benefits, for the potential host community. When addressing community concerns around resources and culturally significant sites, tools that seek to model a variety of scenarios while weighing both social and environmental impacts are critical for just planning and siting. However, processes dictated by the National Historic Preservation Act or the National Environmental Policy Act may not address appropriately state-level and hyper-local nuances, especially for projects that cross state boundaries. CO_2 utilization developers can use lessons learned from infrastructure accidents (e.g., the 2020 pipeline rupture in Satartia, Mississippi) or failed siting processes (e.g., the terminated wind project in Sweden) to avoid past mistakes and apply successful models to CO_2 utilization projects. Furthermore, legal frameworks, such as community benefits agreements or project labor agreements, have the potential to support transparent project development and provide direct workforce benefits to a community.

Recommendation 4-9: Using past experience to develop new projects—The Department of Energy should develop standards and guidelines for other federal projects to encourage the co-siting of infrastructure based on lessons learned from their direct air capture and hydrogen hubs. Furthermore, the lessons learned from these hubs should be made public. These exemplars can be used by the private

sector looking to develop clustered CO_2 utilization infrastructure to ensure that specific considerations are made during siting processes and to minimize adverse impacts on communities. These considerations include the sharing of resources such as energy, land, and water; the development of pipelines; and synchronized construction periods.

4.5.2 Research Agenda for Policy and Regulatory Frameworks

The committee was tasked with (1) identifying and assessing the progress of technologies and approaches for carbon utilization that may play an important role in a circular carbon economy; (2) assessing and identifying gaps in research efforts to address barriers to commercialization of carbon utilization technologies; and (3) updating the 2019 National Academies' report *Gaseous Carbon Waste Streams Utilization: Status and Research Needs* (NASEM 2019). Policy and regulatory considerations are considered *enabling* research items because they contribute to a net-zero future but are not research on CO_2 utilization technologies or processes themselves; therefore, the research agenda item from this chapter contributes to task (2).

While about 80 percent of the U.S. public either does not know what CCUS technology is or cannot definitively recognize it, there are polarizing opinions about the carbon management sector (see Section 4.4.1.1). As the CO_2 utilization sector emerges, there is an opportunity to invest in public engagement and education that contributes to an informed understanding of the technologies and the expected societal and environmental impacts of the processes. For example, there are opportunities to encourage public and community engagement and education around pipeline safety protocols through existing state-level initiatives. Table 4-4 describes the policy and regulatory

TABLE 4-4 Policy and Regulatory Frameworks Research Agenda

Research, Development, and Demonstration Need	Funding Agencies or Other Actors	Basic, Applied, Demonstration, or Enabling	Research Area	Product Class	Long- or Short-Lived	Research Barrier Addressed	Source
4-A. Knowledge gaps in public perception of carbon utilization technologies, and factors that influence community acceptance.	Nongovernmental organizations Universities National laboratories Other research-conducting entities Department of Energy	Enabling	Societal Impacts	All	Long-lived Short-lived	Environmental and societal considerations for CO_2 and coal waste utilization technologies	Fin. 4-7 Rec. 4-6

Recommendation 4-6: Educational material development—Nongovernmental organizations and research-conducting entities—such as national laboratories, think tanks, and universities—should identify gaps in knowledge, sharing their data and findings about societal acceptance of or opposition to the CO_2 utilization sector through the following actions:

a. **Nongovernmental organizations (including universities) and national laboratories should conduct targeted research to develop transparent resources to communicate findings and support improved education related to the direct impact of the CO_2 utilization sector on climate change mitigation.**
b. **Research-conducting entities should continue to conduct social analyses to determine what consumers and communities think about the CO_2 utilization sector, both in relation to and separate from the carbon management value chain, filling the gaps in knowledge about societal acceptance and potential opposition, to better understand and address the concerns of the public. These data and conclusions of this research should be shared through a centralized and broadly accessible framework.**
c. **Research-conducting entities should use analyses that combine techno-economic and life cycle assessment objectives to determine a levelized cost of CO_2 abatement to be used to assess the desirability of projects as a function of CO_2 source (fossil or biogenic point source, direct air capture, or direct ocean capture) and product durability. The results of these analyses and assessments should be communicated to the Department of Energy and other entities funding carbon utilization projects to inform them of factors that can influence community acceptance of projects and expected outcomes.**

frameworks research agenda, including related research agenda recommendations. The table includes the relevant funding agencies or other actors; whether the need is basic research, applied research, technology demonstration, or enabling technologies and processes for CO_2 utilization, the research theme that the research need falls into; the relevant research area and product class covered by the research need; whether the relevant products are long- or short-lived; and the source of the research need. The committee's full research agenda can be found in Chapter 11.

4.6 REFERENCES

Aaker, D.A., and C. Moorman. 2017. *Strategic Market Management*. 11th Edition. John Wiley & Sons.

Agyeman, J., R.D. Bullard, and B. Evans. 2002. "Exploring the Nexus: Bringing Together Sustainability, Environmental Justice, and Equity." *Space and Polity* 6(1):77–90. https://doi.org/10.1080/13562570220137907.

Ahn, D.P. 2019. *Principles of Commodity Economics and Finance*. The MIT Press.

Alonso, A., and J. Ferrell. 2023. "Carbon Removal in California: Striving Toward Environmental Justice in the Central Valley." *National Wildlife Federation: Environmental Justice* (blog), November 2. https://blog.nwf.org/2023/11/carbon-removal-in-california.

Althoey, F., W.S. Ansari, M. Sufian, and A.F. Deifalla. 2023. "Advancements in Low-Carbon Concrete as a Construction Material for the Sustainable Built Environment." *Developments in the Built Environment* 16(December):100284. https://doi.org/10.1016/j.dibe.2023.100284.

ANL (Argonne National Laboratory). n.d. "Small Business Resources." https://www.anl.gov/partnerships/small-business-resources.

Ashley, S., G. Acs, S. Brown, M. Deich, G. MacDonald, D. Marron, R. Balu, et al. 2022. *Scoring Federal Legislation for Equity: Definition, Framework, and Potential Application*. Washington, DC: Urban Institute and PolicyLink. https://www.urban.org/sites/default/files/2022-06/Scoring%20Federal%20Legislation%20for%20Equity.pdf.

Associated General Contractors of America and Autodesk Cloud Construction. 2023. *2023 Workforce Survey Results: National Results*. Associated General Contractors of America. https://www.agc.org/sites/default/files/Files/Communications/2023_Workforce_Survey_National_Final.pdf.

ASTM International. n.d. "Construction Standards." https://www.astm.org/products-services/standards-and-publications/standards/construction-standards.html.

Barnes, M., and P. Schmitz. 2016. "Community Engagement Matters (Now More Than Ever)." *Stanford Social Innovation Review* 14(2):32–39. https://doi.org/10.48558/J83Z-0440.

Beckett, C., and A. Keeling. 2019. "Rethinking Remediation: Mine Reclamation, Environmental Justice, and Relations of Care." *Local Environment* 24(3):216–230. https://doi.org/10.1080/13549839.2018.1557127.

Been, V. 2010. "Community Benefits Agreements: A New Local Government Tool or Another Variation on the Exactions Theme?" *The University of Chicago Law Review* 77(1):5–35. https://www.jstor.org/stable/40663024.

Bennett, J., R. Kammer, K. Eidno, M. Ford, S. Henao, N. Holwerda, E. Middleton, et al. 2023. "Carbon Capture Co-Benefits: Carbon Capture's Role in Removing Pollutants and Reducing Health Impacts." *Great Plains Institute*. https://carboncaptureready.betterenergy.org/carbon-capture-co-benefits.

Berry, B., B. Farber, F.C. Rios, M.A. Haedicke, S. Chakraborty, S.S. Lowden, M.M. Bilec, and C. Isenhour. 2021. "Just by Design: Exploring Justice as a Multidimensional Concept in US Circular Economy Discourse." *The International Journal of Justice and Sustainability* 27(10–11):1225–1241. https://doi.org/10.1080/13549839.2021.1994535.

Bhide, P., and S. Sengupta. 2024. "Insuring the Net-Zero Transition: A Risk Management Strategy to Support Zurich North America's Efforts to Accelerate the Adoption of Net-Zero Technologies." *University of Michigan School for Environment and Sustainability*. https://dx.doi.org/10.7302/22618.

Bijker, W.E., and J. Law. 1992. "Shaping Technology/Building Society." *Bulletin of Science, Technology & Society* 14(4):240–241. https://doi.org/10.1177/027046769401400468.

Biven, M.M., and L. Lindner. 2023. "The Future of Energy and Work in the United States: The American Oil and Gas Worker Survey." *True Transition*. https://www.truetransition.org/_files/ugd/0ad80c_069ea867b3f044afba4dae2a1da8d737.pdf?index=true.

BLS (Bureau of Labor Statistics). 2024. "Table 1. Job Openings Levels and Rates by Industry and Region, Seasonally Adjusted." In *Economic News Release*. https://www.bls.gov/news.release/jolts.t01.htm.

BLS. n.d. "Industries at a Glance: Oil and Gas Extraction: NAICS 211." https://www.bls.gov/iag/tgs/iag211.htm.

Bose, S. 2023. "U.S. Sustainable Aviation Fuel Production Target Faces Cost, Margin Challenges." *Reuters*, November 1. https://www.reuters.com/sustainability/us-sustainable-aviation-fuel-production-target-faces-cost-margin-challenges-2023-11-01.

Boston, W. 2021. "Green Steel Becomes a Hot Commodity for Big Auto Makers." *The Wall Street Journal*, September 13. https://www.wsj.com/articles/green-steel-becomes-a-hot-commodity-for-big-auto-makers-11631525401.

Boudet, H.S. 2019. "Public Perceptions of and Responses to New Energy Technologies." *Nature Energy* 4(2019):446–455. https://doi.org/10.1038/s41560-019-0399-x.

Bowles, W., K. Cheslak, and J. Edelson. 2022. "Lifecycle GHG Impacts in Building Codes." *New Buildings Institute*. https://newbuildings.org/wp-content/uploads/2022/04/LifecycleGHGImpactsinBuildingCodes.pdf.

Bravo, M.A., R. Anthopolos, M.L. Bell, and M.L. Miranda. 2016. "Racial Isolation and Exposure to Airborne Particulate Matter and Ozone in Understudied US Populations: Environmental Justice Applications of Downscaled Numerical Model Output." *Environment International* 92–93:247–255. https://doi.org/10.1016/j.envint.2016.04.008.

Brown, J.D., S.D. Comello, M. Jeong, M. Downey, and M.I. Cohen. 2023. "Turning CCS Projects in Heavy Industry and Power into Blue Chip Financial Investments." *Energy Futures Initiative*. https://efifoundation.org/wp-content/uploads/sites/3/2023/02/20230212-CCS-Final_Full-copy.pdf.

Bryant, B., ed. 1995. "Environmental Justice: Issues, Policies, and Solutions." Covelo, CA: Island Press.

Bullard, R.D. 2000. *Dumping in Dixie: Race, Class, and Environmental Quality*. Third edition. Routledge. https://doi.org/10.4324/9780429495274.

Buttorff, G., F. Cantu, R. Krishnamoorti, P. Pinti, A. Datta, and Y. Olapade. 2020. "Carbon Management: Changing Attitudes and an Opportunity for Action." University of Houston. https://uh.edu/uh-energy/research/research-reports/carbon-management-report-2020.

BW Research Partnership and Climate Equity Initiative. 2023. "Perspectives from Environmental Justice Communities: A National Survey." *Clean Air Task Force*. https://cdn.catf.us/wp-content/uploads/2023/07/11093912/perspectives-environmental-justice-communities-national-survey.pdf.

California Resources Corporation. 2023. "Carbon TerraVault's California DAC Hub Consortium Selected for U.S. DOE Funding to Bring Direct Air Capture and Storage to the Golde State." https://s202.q4cdn.com/682408967/files/doc_news/2023/Aug/california-dac-hub-award-press-release-final-8-11-2023-1015am.pdf.

Cao, X.E., and P. Tomski. 2023. "The Sustainability Workforce Shift: Building a Talent Pipeline and Career Network." *Matter* 6:2471–2475. https://doi.org/10.1016/j.matt.2023.06.016.

Cap-and-Invest. n.d. "Reducing Pollution, Investing in Communities, Creating Jobs, and Preserving Competitiveness." New York State. https://capandinvest.ny.gov.

CEQ (Council on Environmental Quality). 1997. "Environmental Justice: Guidance Under the National Environmental Policy Act." https://www.energy.gov/nepa/articles/environmental-justice-guidance-under-nepa-ceq-1997.

CEQ. n.d.(a). "Net-Zero Emissions Procurement by 2050." https://www.sustainability.gov/federalsustainabilityplan/procurement.html.

CEQ. n.d.(b). "Climate and Economic Justice Screening Tool: Frequently Asked Questions." https://screeningtool.geoplatform.gov/en/frequently-asked-questions.

Chailleux, S. 2019. "Making the Subsurface Political: How Enhanced Oil Recovery Techniques Reshaped the Energy Transition." *Environment and Planning C: Politics and Space* 38(4):773–750. https://doi.org/10.1177/2399654419884077.

Chemnick, J. 2023a. "'False promise': DOE's Carbon Removal Plans Rankle Community Advocates." *E&E News by Politico: ClimateWire*. https://www.eenews.net/articles/false-promise-does-carbon-removal-plans-rankle-community-advocates.

Chemnick, J. 2023b. "The Carbon Removal Project That Puts Communities in the Driver's Seat." E&E News. *ClimateWire*, October 26. https://www.eenews.net/articles/the-carbon-removal-project-that-puts-communities-in-the-drivers-seat.

Clean Air Task Force. 2021. "Carbon Management Provisions in the Infrastructure Investments and Jobs Act." https://cdn.catf.us/wp-content/uploads/2021/12/13104556/carbon-management-provisions-iija-1.pdf.

Climate Justice Alliance. 2019. "Climate Justice Alliance: Just Transition Principles." https://climatejusticealliance.org/wp-content/uploads/2019/11/CJA_JustTransition_highres.pdf.

CNT (Collaborative for Neighborhood Transformation). n.d. "What Is Asset Based Community Development (ABCD)." *ABCD Toolkit*. https://resources.depaul.edu/abcd-institute/resources/Documents/WhatisAssetBasedCommunityDevelopment.pdf.

Cochran, J., and P. Denholm, eds. 2021. *LA100: The Los Angeles 100% Renewable Energy Study*. NREL/TP-6A20-79444. Golden, CO: National Renewable Energy Laboratory. https://maps.nrel.gov/la100.

Coleman, K.M. 2023. "Modernizing Workforce Development for a Healthy and Inclusive Economy." *Harvard Advanced Leadership Initiative Social Impact Review*. March 21. https://www.sir.advancedleadership.harvard.edu/articles/modernizing-workforce-development-for-healthy-inclusive-economy.

Comello, S.D., J. Reichelstein, and S. Reichelstein. 2023. "Corporate Carbon Reporting: Improving Transparency and Accountability." *One Earth* 6(7):803–810. https://doi.org/10.1016/j.oneear.2023.06.002.

Contractor. 2023. "UA Signs Memorandum of Understanding to Establish the 'Capturing Better Futures' Initiative." *Contractor Training*, June 22. https://www.contractormag.com/training/article/21268303/ua-signs-memorandum-of-understanding-to-establish-the-capturing-better-futures-initiative.

Cox, J. 2020. "Climate Change Report Forecasts Hard Time for Kern ag." *Bakersfield.com News*, August 14. https://www.bakersfield.com/news/climate-change-report-forecasts-hard-times-for-kern-ag/article_a9b0f9e2-ddb3-11ea-b024-bbc9636fdb74.html.

Cushing, L.J., S. Li, B.B. Steiger, and J.A Casey. 2023. "Historical Red-Lining Is Associated with Fossil Fuel Power Plant Siting and Present-Day Inequalities in Air Pollutant Emissions." *Nature Energy* 8:52–61. https://doi.org/10.1038/s41560-022-01162-y.

Dalrymple, A. 2024. "200,000 Comments Submitted on Dakota Access Pipeline Environmental Review." *South Dakota Searchlight*, February 28. https://southdakotasearchlight.com/briefs/200000-comments-submitted-on-dakota-access-pipeline-environmental-review.

Datta, R., and M.A. Hurlbert. 2020. "Pipeline Spills and Indigenous Energy Justice." *Sustainability* 12(47). http://dx.doi.org/10.3390/su12010047.

De Barbieri, E.W. 2016. *Do Community Benefits Agreements Benefit Communities?* Brookly Law School. Legal Studies Paper No. 462. http://ssrn.com/abstract=2802409.

Dean, B., and P. Esling. 2023. "CEJST Is a Simple Map, with Big Implications—and Attention to Cumulative Burdens Matters." *Intersections* (blog). https://cnt.org/blog/cejst-is-a-simple-map-with-big-implications-and-attention-to-cumulative-burdens-matters.

Deaton, D. 2022. "Environmental Justice Practices and Resources for Rural Communities." *Aspen Institute Energy and Environment* (blog), June 16. https://www.aspeninstitute.org/blog-posts/rural-environmental-justice.

DeLonge, M. 2022. "In California's Central Valley, Drought Is a Growing Threat to Farms, Food, and People." *Union of Concerned Scientists* (blog), March 15. https://blog.ucsusa.org/marcia-delonge/in-californias-central-valley-drought-is-a-growing-threat-to-farms-food-and-people.

DeLuca, K.M. 2007. "A Wilderness Environmentalism Manifesto: Contesting the Infinite Self-Absorption of Humans." Pp. 27–56 in *Environmental Justice and Environmentalism: The Social Justice Challenge to the Environmental Movement*, Sandler, R., and P.C. Pezzullo, eds. Cambridge, MA: The MIT Press. https://doi.org/10.7551/mitpress/2781.003.0005.

Denchak, M. 2018. "Flint Water Crisis: Everything You Need to Know." *Natural Resources Defense Council*, November 8. https://www.nrdc.org/stories/flint-water-crisis-everything-you-need-know#summary.

Derouin, S. 2023. "Is Low- to No-Carbon Cement the Future of Construction?" *American Society of Civil Engineers*, October 2. https://www.asce.org/publications-and-news/civil-engineering-source/civil-engineering-magazine/article/2023/10/is-low--to-no-carbon-cement-the-future-of-construction.

DOE (Department of Energy). 2022a. "DOE Announces $46 Million to Explore New Technologies That Convert Carbon and Waste into Clean Energy." August 31. https://www.energy.gov/articles/doe-announces-46-million-explore-new-technologies-convert-carbon-and-waste-clean-energy.

DOE. 2022b. "How Energy Justice, Presidential Initiatives, and Executive Orders Shape Equity at DOE." https://www.energy.gov/justice/articles/how-energy-justice-presidential-initiatives-and-executive-orders-shape-equity-doe.

DOE. 2023. "Office of Energy Justice Policy and Analysis." https://www.energy.gov/diversity/office-energy-justice-policy-and-analysis.

DOE. n.d.(a). "Adoption Readiness Levels (ARL): A Complement to TRL." https://www.energy.gov/technologytransitions/adoption-readiness-levels-arl-complement-trl.

DOE. n.d.(b). "Carbon Utilization Program." https://www.energy.gov/fecm/carbon-utilization-program.

DOE. n.d.(c). "Energy Justice Dashboard (BETA)." https://www.energy.gov/diversity/energy-justice-dashboard-beta.

DOE. n.d.(d). "Justice40 Initiative." https://www.energy.gov/diversity/justice40-initiative.

DOE. n.d.(e). "Promoting Inclusive and Equitable Research (PIER) Plan Guidance." https://science.osti.gov/sbir/Applicant-Resources/PIER-Plan.

DOE-BETO (Bioenergy Technologies Office). 2021. "2021 Project Peer Review." Washington, DC: Department of Energy. https://www.energy.gov/sites/default/files/2022-06/beto-00-2021-peer-review-report.pdf.

DOE-FECM (Office of Fossil Energy and Carbon Management). 2021. "Deployment and Demonstration Opportunities for Carbon Reduction and Removal Technologies." DE-FOA-0002660. https://www.fedconnect.net/FedConnect/PublicPages/PublicSearch/Public_OpportunitySummary.aspx?ReturnUrl=%2ffedconnect%2f%3fdoc%3dDE-FOA-0002660%26agency%3dDOE&doc=DE-FOA-0002660&agency=DOE.

DOE-FECM. 2022. "The Infrastructure Investment and Jobs Act: Opportunities to Accelerate Deployment in Fossil Energy and Carbon Management Activities." https://www.energy.gov/sites/default/files/2022-09/FECM%20IIJA%20BIL%20Factsheet_revised%20September%202022.pdf.

DOE-FECM. n.d.(a). "Carbon Conversion." https://www.energy.gov/fecm/carbon-conversion.

DOE-FECM. n.d.(b). "Funding Notice: Bipartisan Infrastructure Law: Regional Direct Air Capture Hubs." https://www.energy.gov/fecm/funding-notice-bipartisan-infrastructure-law-regional-direct-air-capture-hubs.

DOE-FECM. n.d.(c). "Funding Notice: Carbon Management." https://www.energy.gov/fecm/funding-notice-carbon-management.

DOE-LPO (Loan Programs Office). n.d. "About LPO." https://www.energy.gov/lpo/loan-programs-office.

DOE-NE (Office of Nuclear Energy). n.d. "Gateway for Accelerated Innovation in Nuclear (GAIN)." https://www.energy.gov/ne/gateway-accelerated-innovation-nuclear-gain.

DOE-OCED (Office of Clean Energy Demonstrations). n.d. "Funding Notice: Carbon Capture Demonstration Projects Program." https://www.energy.gov/oced/funding-notice-carbon-capture-demonstration-projects-program.

DOE-OEJ (Office of Energy Jobs). 2023. "United States Energy and Employment Report 2023." DOE/OP-0020. https://www.energy.gov/sites/default/files/2023-06/2023%20USEER%20REPORT-v2.pdf.

DOE-OTT (Office of Technology Transitions). 2023. "DOE Announces New $27.5M Voucher Program to Bring Innovative Energy Technologies to Market." https://www.energy.gov/technologytransitions/articles/doe-announces-new-275m-voucher-program-bring-innovative-energy.

DOE-OTT. n.d.(a). "Bipartisan Infrastructure Law Technology Commercialization Fund." https://www.energy.gov/technologytransitions/bipartisan-infrastructure-law-technology-commercialization-fund.

DOE-OTT. n.d.(b). "DOE Partnership Intermediary Agreement." https://www.energy.gov/technologytransitions/doe-partnership-intermediary-agreement.

Donnelly, K., S. Doerge, J. Vandergrift, M. Jubinville-Stafford, G. Blinick, G. Keefe, D. Paris-Mackay, N. Drouin, B. Gilligan, and T. Corner. 2015. "Resident Honorarium—Guiding Principles and Promising Practices." *Community Development Framework*. https://cdfcdc.ca/wp-content/uploads/2015/02/Honorarium-Guiding-Principles-and-Considerations-Final-July-2015.pdf.

DOT-PHMSA (Department of Transportation Pipeline and Hazardous Materials Safety Administration). 2022. "Failure Investigation Report—Denbury Gulf Coast Pipelines, LLC—Pipeline Rupture/Natural Force Damage." Pipeline and Hazardous Materials Safety Administration—Office of Pipeline Safety. https://www.phmsa.dot.gov/sites/phmsa.dot.gov/files/2022-05/Failure%20Investigation%20Report%20-%20Denbury%20Gulf%20Coast%20Pipeline.pdf.

E2, Alliance to Save Energy, American Association of Blacks in Energy, Energy Efficiency for All, Black Owners of Solar Services, and BW Research Partnership. 2021. *Help Wanted: Diversity in Clean Energy*. E2R:21-07-F. https://e2.org/wp-content/uploads/2021/09/E2-ASE-AABE-EEFA-BOSS-Diversity-Report-2021.pdf.

Earthjustice. 2021. "Why Are Fossil Fuel Pipelines Bad for Our Climate and Communities?" July 14. https://earthjustice.org/feature/fighting-pipelines-fossil-fuels-oil-and-gas

Earthjustice and Clean Energy Program. 2023. "Carbon Capture: The Fossil Fuel Industry's False Climate Solution." *EarthJustice*. https://earthjustice.org/article/carbon-capture-the-fossil-fuel-industrys-false-climate-solution.

EIA (U.S. Energy Information Administration). n.d. "Frequently Asked Questions (FAQs): What Is U.S. Electricity Generation by Energy Source?" https://www.eia.gov/tools/faqs/faq.php?id=427&t=3.

EJ IWG (Federal Interagency Working Group on Environmental Justice) and NEPA (National Environmental Policy Act) Committee. 2016. "Promising Practices for EJ Methodologies in NEPA Reviews." EPA 300B16001. https://www.epa.gov/sites/default/files/2016-08/documents/nepa_promising_practices_document_2016.pdf.

Energy Justice Network. n.d. "Principles of Working Together." Presented at Second People of Color Leadership Summit. October 26. Washington, DC. https://www.ejnet.org/ej/workingtogether.pdf.

E.O. (Executive Order) 12898. 1994. "Federal Actions to Address Environmental Justice in Minority Populations and Low-Income Populations."

E.O. 14008. 2021. "Tackling the Climate Crisis at Home and Abroad."

EPA (U.S. Environmental Protection Agency). 2010. "EPA's Action Development Process Interim Guidance on Considering Environmental Justice During the Development of an Action." https://www.epa.gov/sites/default/files/2015-03/documents/considering-ej-in-rulemaking-guide-07-2010.pdf.

EPA. 2022. "Climate Change Indicators: Greenhouse Gases." https://www.epa.gov/climate-indicators/greenhouse-gases https://www.epa.gov/climate-indicators/greenhouse-gases.

EPA. 2024. "Recommendations of Specifications, Standards, and Ecolabels for Federal Purchasing." https://www.epa.gov/greenerproducts/recommendations-specifications-standards-and-ecolabels-federal-purchasing.

EPA. n.d.(a). "EJScreen: Environmental Justice Screening and Mapping Tool." https://www.epa.gov/ejscreen.

EPA. n.d.(b). "Environmental Justice." https://www.epa.gov/environmentaljustice.

EPA. n.d.(c). "EPA Research: Environmental Justice and Air Pollution." https://www.epa.gov/ej-research/epa-research-environmental-justice-and-air-pollution.

EPA. n.d.(d). "Grant Program: Reducing Embodied Greenhouse Gas Emissions for Construction Materials and Products." https://www.epa.gov/greenerproducts/grant-program-reducing-embodied-greenhouse-gas-emissions-construction-materials-and.
EPA. n.d.(e). "Research on Near Roadway and Other Near Source Air Pollution." https://www.epa.gov/air-research/research-near-roadway-and-other-near-source-air-pollution.
EPA. n.d.(f). "Summary of Inflation Reduction Act Provisions Related to Renewable Energy." https://www.epa.gov/green-power-markets/summary-inflation-reduction-act-provisions-related-renewable-energy.
EPA. n.d.(g). "Sustainable Marketplace: Greener Products and Services: About the Environmentally Preferable Purchasing Program." https://www.epa.gov/greenerproducts/about-environmentally-preferable-purchasing-program.
EPA. n.d.(h). "What Is a Circular Economy? Why Is It Important?" https://www.epa.gov/circulareconomy/what-circular-economy.
EU (European Union). n.d. "Carbon Border Adjustment Mechanism." https://taxation-customs.ec.europa.eu/carbon-border-adjustment-mechanism_en#cbam.
Evans, G., and L. Phelan. 2016. "Transition to a Post-Carbon Society: Linking Environmental Justice and Just Transition Discourses." *Energy Policy* 99:329–339. https://doi.org/10.1016/j.enpol.2016.05.003.
Faber, D., and D. McCarthy. 2001. The Evolving Structure of the Environmental Justice Movement in the United States: New Models for Democratic Decision-Making. *Social Justice Research* 14:405–421. https://doi.org/10.1023/A:1014602729040.
Fan, Z., and J. Friedmann. 2021. "Low-Carbon Production of Iron and Steel: Technology Options, Economic Assessment, and Policy." *Joule* 5(4):829–862. https://doi.org/10.1016/j.joule.2021.02.018.
FAO (Food and Agriculture Organization of the United Nations), Action Against Hunger, Action Aid, Deusche Gesellschaft für Internationale Zusammenarbeit, International Federation of Red Cross and Red Crescent Societies, Agencia Española de Cooperación Internacional para el Desarrollo, and World Vision International. 2016. "Free Prior and Informed Consent: An Indigenous Peoples' Right and a Good Practice for Local Communities." *Food and Agriculture Organization of the United Nations*. https://www.fao.org/3/i6190e/i6190e.pdf.
Fernandez-Bou, A.S., J.M. Rodriguez-Flores, A. Guzman, J.P. Ortiz-Partida, L.M. Classen-Rodriguez, P.A. Sandchez-Perez, J. Valero-Fandino, et al. 2023. "Water, Environment, and Socioeconomic Justice in California: A Multi-Benefit Cropland Repurposing Framework." *Science of the Total Environment* 858(2023):159963. http://dx.doi.org/10.1016/j.scitotenv.2022.159963.
Foster, S. 1998. "Justice from the Ground Up: Distributive Inequities, Grassroots Resistance, and the Transformative Politics of the Environmental Justice Movement." *California Law Review* 86(4):775–842. https://ir.lawnet.fordham.edu/faculty_scholarship/295.
Fraser, C. 2022. "Community Benefits Agreements Offer Meaningful Opportunities to Include Voter's Voices in Development." *Data for Progress*, July 6. https://www.dataforprogress.org/blog/2022/7/5/community-benefits-agreements-offer-meaningful-opportunities-to-include-voters-voices-in-development.
Fraser, C. 2023. "Community and Labor Benefits in Climate Infrastructure: Lessons for Equitable, Community-Centered Direct Air Capture Hub Development." *Data for Progress*. https://www.filesforprogress.org/memos/community-and-labor-benefits-in-climate-infrastructure.pdf.
Friedman-Jiménez, G. 1994. "Achieving Environmental Justice: The Role of Occupational Health." *Fordham Urban Law Journal* 21(3):605–632. https://ir.lawnet.fordham.edu/ulj/vol21/iss3/8.
GAIN (Gateway for Accelerated Innovation in Nuclear). n.d.(a). "NA Vouchers." https://gain.inl.gov/SitePages/Nuclear%20Energy%20Vouchers.aspx.
GAIN. n.d.(b). "What Is GAIN?" https://gain.inl.gov/about.
GAO (U.S. Government Accountability Office). 1983. "Siting of Hazardous Waste Landfills and Their Correlation with Racial and Economic Status of Surrounding Communities." GAO/RCED-83-168. Gaithersburg, MD: General Accounting Office.
GAO. 2022. *Decarbonization: Status, Challenges, and Policy Options for Carbon Capture, Utilization, and Storage*. Technology Assessment. GAO-22-105274. https://www.gao.gov/assets/730/723198.pdf.
Giesekam, J., J.R. Barrett, and P. Taylor. 2015. "Construction Sector Views on Low Carbon Building Materials." *Building Research and Information* 44(4):423–444. https://doi.org/10.1080/09613218.2016.1086872.
Gough, C., and S. Mander. 2022. "CCS Industrial Clusters: Building a Social License to Operate." *International Journal of Greenhouse Gas Control* 119(September):103713. https://doi.org/10.1016/j.ijggc.2022.103713.
Greenspon, J., and D. Raimi. 2022. *Matching Geographies and Job Skills in the Energy Transition*. Resources for the Future. WP 22-25. https://media.rff.org/documents/WP_22-25_PnkcURf.pdf.
GSA (General Services Administration). n.d.(a). "Certify as a Small Business." https://www.gsa.gov/sell-to-government/step-2-compete-for-a-contract/certify-as-a-small-business?topnav=sell-to-government.

GSA. n.d.(b). "Environmental Product Declarations (EPDs)." https://sftool.gov/plan/402/environmental-product-declarations-epds.

GSA. n.d.(c). "Small Business Innovation Research (SBIR) and Small Business Technology Transfer (STTR) Programs." https://www.gsa.gov/small-business/small-business-resources/small-business-innovation-research-sbir-and-small-business-technology-transfer-sttr-programs.

Guana, E. 2015. "Federal Environmental Justice Policy in Permitting." Chapter 3 in *Failed Promises: Evaluating the Federal Government's Response to Environmental Justice*, D.M. Konisky, ed. Cambridge, MA: The MIT Press. http://www.jstor.org/stable/j.ctt17kk8mr.

Hazrati, M., and R.J. Heffron. 2021. "Conceptualising Restorative Justice in the Energy Transition: Changing the Perspectives of Fossil Fuels." *Energy Research and Social Sciences* 78:102115. https://doi.org/10.1016/j.erss.2021.102115.

Heck, S. 2021. "Greening the Color Line: Historicizing Water Infrastructure Redevelopment and Environmental Justice in the St. Louis Metropolitan Region." *Journal of Environmental Policy and Planning* 23(5):565–580.

Hendricks, M.D., and S. Van Zandt. 2021. "Unequal Protection Revisited: Planning for Environmental Justice, Hazard Vulnerability, and Critical Infrastructure in Communities of Color." *Environmental Justice* 14(2):87–97. https://doi.org/10.1089/env.2020.0054.

Herscher, R. 2017. "Key Moments in the Dakota Access Pipeline Fight." *The Two-Way*, February 22. https://www.npr.org/sections/thetwo-way/2017/02/22/514988040/key-moments-in-the-dakota-access-pipeline-fight.

H.R. 3684. 2021. *Infrastructure Investment and Jobs Act*. 117th Congress (2021–2022). November 15. https://www.congress.gov/bill/117th-congress/house-bill/3684.

H.R. 5376. 2022. *Inflation Reduction Act of 2022*. 117th Congress (2021–2022). August 16. https://www.congress.gov/bill/117th-congress/house-bill/5376.

Hughes, S., A. Zoelle, M. Woods, S. Henry, S. Homsy, S. Pidaparti, N. Kuehn, et al. 2022. *Cost of Capturing CO_2 from Industrial Sources*. Technical Report. DOE/NETL-2022/3319. National Energy Technology Laboratory. https://doi.org/10.2172/1887586.

ICAO (International Civil Aviation Organization) Environment. 2023. "CORSIA: Carbon Offsetting and Reduction Scheme for International Aviation Implementation Plan Brochure." Montreal, QC, Canada: ICAO. https://www.icao.int/environmental-protection/CORSIA/Documents/CORSIA%20Brochure/2023%20Edition/CORSIA-Brochure2023-EN-WEB.pdf.

IEA (International Energy Agency). 2021. "Executive Summary." In *The Role of Critical Minerals in Clean Energy Transitions*. World Energy Outlook Special Report. Paris: IEA. https://www.iea.org/reports/the-role-of-critical-minerals-in-clean-energy-transitions/executive-summary.

IHRB (Institute for Human Rights and Business). 2022. "What Is Free, Prior and Informed Consent (FPIC)?" *Institute for Human Rights and Business Explainers*, December 13. https://www.ihrb.org/explainers/what-is-free-prior-and-informed-consent-fpic.

Initiative for Energy Justice. 2019. *The Energy Justice Workbook*. Boston, MA. https://iejusa.org/wp-content/uploads/2019/12/The-Energy-Justice-Workbook-2019-web.pdf.

IRENA (International Renewable Energy Agency). 2021. "A Pathway to Decarbonise the Shipping Sector by 2050." International Renewable Energy Agency. Abu Dhabi, United Arab Emirates. https://www.irena.org/-/media/Files/IRENA/Agency/Publication/2021/Oct/IRENA_Decarbonising_Shipping_2021.pdf?rev=b5dfda5f69e741a4970680a5ced1ac1e.

IRENA. 2023. "Renewables Competitiveness Accelerates, Despite Cost Inflation." Press Release. https://www.irena.org/News/pressreleases/2023/Aug/Renewables-Competitiveness-Accelerates-Despite-Cost-Inflation.

IRS (Internal Revenue Service). 2021. "Credit for Carbon Oxide Sequestration." *Federal Register* 86:4728. https://www.federalregister.gov/documents/2021/01/15/2021-00302/credit-for-carbon-oxide-sequestration.

IRS. n.d. "Business Credits and Deductions: Opportunity Zones." https://www.irs.gov/credits-deductions/businesses/opportunity-zones.

ISO (International Organization for Standardization). 2015. *Quality Management Systems—Requirements* (ISO Standard No. 9001). https://www.iso.org/standard/62085.html.

IWG (Interagency Working Group on Coal and Power Plant Communities and Economic Revitalization). 2021. "Initial Report to the President on Empowering Workers Through Revitalizing Energy Communities." https://netl.doe.gov/sites/default/files/2021-04/Initial%20Report%20on%20Energy%20Communities_Apr2021.pdf.

Jacobs, E. 2013. *Principles for Reforming Workforce Development and Human Capital Policies in the United States*. Washington, DC: The Brookings Institution Governance Studies. https://www.brookings.edu/wp-content/uploads/2016/06/FedRole-WorkforceDev.pdf.

James, W., C. Jia, and S. Kedia. 2012. "Uneven Magnitude of Disparities in Cancer Risks from Air Toxics." *International Journal of Environmental Research and Public Health* 9(12):4365–4385.

Janzwood, A. 2023. "Pipeline Politics and the Future of Environmental Justice Struggles in North America." *Global Environmental Politics* 23(3):120–126. https://doi.org/10.1162/glep_r_00731.

Jensen, N.M. 2017. "Eminent Domain and Oil Pipelines: A Slippery Path for Federal Regulation." *Fordham Environmental Law Review* 29(2):320–348. https://ir.lawnet.fordham.edu/elr/vol29/iss2/6.

Jones, C.R., B. Olfe-Krautlein, H. Naims, and K. Armstrong. 2017. "The Social Acceptance of Carbon Dioxide Utilisation: A Review and Research Agenda." *Frontiers Energy Research* 5(11). https://doi.org/10.3389/fenrg.2017.00011.

Jones, W., G. Hiltbrand, E.G. O'Rear, B. King, and N. Pastorek. 2023. "Direct Air Capture Workforce Development: Opportunities by Occupation." *Rhodium Group*, October 12. https://rhg.com/research/direct-air-capture-workforce-development.

Jost, J.T., and A.C. Kay. 2010. "Social Justice: History, Theory, and Research." Pp. 1122–1165 in *Handbook of Social Psychology*, S.T. Fiske, D.T. Gilbert, and G. Lindzey, eds. John Wiley & Sons, Inc. https://psycnet.apa.org/doi/10.1002/9780470561119.socpsy002030.

Just Transition Alliance. 2020. "False Solutions to Address Climate Change." https://jtalliance.org/wp-content/uploads/2020/02/False-Solutions.pdf.

Kimmell, K., A. Boyle, Y. Si, and M. Sotolongo. 2021. "A User's Guide to Environmental Justice: Theory, Policy, and Practice." Northeastern University School of Public Policy and Urban Affairs. https://cssh.northeastern.edu/policyschool/wp-content/uploads/sites/2/2021/07/Users-Guide-to-Environmental-Justice-6.22.21-clean.pdf.

Kosar, U., and V. Suarez. 2021. "Removing Forward: Centering Equity and Justice in a Carbon-Removing Future." *Carbon180*. https://static1.squarespace.com/static/5b9362d89d5abb8c51d474f8/t/6115485ae47e7f00829083e1/1628784739915/Carbon180+RemovingForward.pdf.

Küng, L., S. Aeschlimann, C. Charalambous, F. McIlwaine, J. Young, N. Shannon, K. Strassel, et al. 2023. "A Roadmap for Achieving Scalable, Safe, and Low-Cost Direct Air Carbon Capture and Storage." *Energy and Environmental Science* 16:4280–4304. https://doi.org/10.1039/d3ee01008b.

LA SAFE (Louisiana's Strategic Adaptations for Future Environments). 2019. "Regional and Parish Adaptation Strategies." https://lasafe.la.gov.

Lane, H.M., R. Morello-Frosch, J.D. Marshall, and J.S. Apte. 2022. "Historical Redlining Is Associated with Present-Day Air Pollution Disparities in U.S. Cities." *Environmental Science and Technology Letters* 9(4):345–350. https://doi.org/10.1021/acs.estlett.1c01012.

Langness, M., J.W. Morgan, S. Cedano, E. Falkenburger. 2023. *Equitable Compensation for Community Engagement Guidebook*. Urban Institute: Washington, DC. https://www.urban.org/sites/default/files/2023-08/Equitable%20Compensation%20for%20Community%20Engagement%20Guidebook.pdf.

Larsen, J., W. Herndon, G. Hiltbrand, and B. King. 2021. "The Economic Benefits of Carbon Capture: Investment and Employment Opportunities for the Contiguous United States. Phase III." *Rhodium Group*. https://rhg.com/wp-content/uploads/2021/04/The-Economic-Benefits-of-Carbon-Capture-Investment-and-Employment-Opportunities_Phase-III.pdf.

Larson, E., C. Greig, J. Jenkins, E. Mayfield, A. Pascale, C. Zhang, J. Drossman, et al. 2020. *Net-Zero America: Potential Pathways, Infrastructure, and Impacts*. Interim Report. Princeton, NJ: Princeton University. https://netzeroamerica.princeton.edu/img/Princeton_NZA_Interim_Report_15_Dec_2020_FINAL.pdf.

Lehmann, S., N. Hunt, C. Frongillo, and P. Jordan. 2021. "Diversity in the U.S. Energy Workforce: Data Findings to Inform State Energy, Climate, and Workforce Development Policies and Programs." *BW Research Partnership*. https://www.naseo.org/data/sites/1/documents/publications/Workforce%20Diversity%20Data%20Findings%20MASTER%20Final42.pdf.

Linder, S.H., D. Marko, and K. Sexton. 2008. "Cumulative Cancer Risk from Air Pollution in Houston: Disparities in Risk Burden and Social Disadvantage." *Environmental Science and Technology* 42(12):4312–4322. https://doi.org/10.1021/es072042u.

Liu, J., L.P. Clark, M.J. Bechle, A. Hajat, S.-Y. Kim, A.L. Robinson, L. Sheppard, A.A. Szpiro, and J.D. Marshall. 2021. "Disparities in Air Pollution Exposure in the United States by Race/Ethnicity and Income, 1990–2010." *Environmental Health Perspectives* 129(12):127005. https://ehp.niehs.nih.gov/doi/abs/10.1289/EHP8584.

Lockman, M. 2023. "*Permitting CO_2 Pipelines: Assessing the Landscape of Federal and State Regulation*." New York: Sabin Center for Climate Change Law. https://scholarship.law.columbia.edu/sabin_climate_change/207.

Machado, A.A., S.A. Edwards, M. Mueller, and V. Saini. 2021. "Effective Interventions to Increase Routine Childhood Immunization Coverage in Low Socioeconomic Status Communities in Developed Countries: A Systematic Review and Critical Appraisal of Peer-Reviewed Literature." *Vaccine* 39(22):2938–2964.

MacNair, D., and R. Callihan. 2019. "Appendix D—ERM Memo: Economic Impacts of CCUS Deployment." In *Meeting the Dual Challenge: A Roadmap to At-Scale Deployment of Carbon Capture, Use, and Storage*. November 20. https://www.energy.gov/sites/default/files/2022-10/CCUS-Appendix_D_Final.pdf.

Madison, I., V. Miller, and C. Lee. 1992. "The Principles of Environmental Justice: Formation and Meaning." In *The First National People of Color Environmental Leadership Summitt Proceedings*. United Church of Christ Commission for Racial Justice. Washington, DC. http://rescarta.ucc.org/jsp/RcWebImageViewer.jsp?doc_id=32092eb9-294e-4f6e-a880-17b8bbe02d88/OhClUCC0/00000001/00000070&pg_seq=1&search_doc=.

Mailloux, N.A., D.W. Abel, T. Holloway, and J.A. Patz. 2022. "Nationwide and Regional $PM_{2.5}$-Related Air Quality Health Benefits from the Removal of Energy-Related Emissions in the United States." *GeoHealth* 6(5):e2022GH000603. https://doi.org/10.1029/2022gh000603.

Malinowski, M. 2023. "CALGreen Mandatory Measures for Embodied Carbon Reduction: Frequently Asked Questions." Sacramento, CA: American Institute of Architects California. https://aiacalifornia.org/news/calgreen-mandatory-measures-for-embodied-carbon-reduction.

Marantz, N.J. 2015. "What Do Community Benefits Agreements Deliver? Evidence from Los Angeles." *Journal of the American Planning Association* 81(4):251–267. https://doi.org/10.1080/01944363.2015.1092093.

Meckling, J., J.E. Aldy, M.J. Kotchen, S. Carley, D.C. Esty, P.A. Raymond, B. Tonkonogy, et al. 2022. "Busting the Myths Around Public Investment in Clean Energy." *Nature Energy* 7(7):563–565.

Mohai, P., D. Pellow, and J.T. Roberts. 2009. "Environmental Justice." *Annual Review of Environment and Resources* 34:405–430. https://doi.org/10.1146/annurev-environ-082508-094348.

Moniz, E.J., J.D. Brown, S.D. Comello, M. Jeong, M. Downey, and M.I Cohen. 2023. "Turning CCS Projects in Heavy Industry and Power into Blue Chip Financial Investments." Energy Future Initiative. https://efifoundation.org/wp-content/uploads/sites/3/2023/02/20230212-CCS-Final_Full-copy.pdf.

Motyka, M., J. Thomson, K. Hardin, and C. Amon. n.d. "2024 Renewable Energy Industry Outlook." *Deloitte*. https://www2.deloitte.com/us/en/insights/industry/renewable-energy/renewable-energy-industry-outlook.html.

Muslemani, H., F. Ascui K. Kaesehage, X. Liang, and J. Wilson. 2022. *Steeling the Race: "Green Steel" as the New Clean Material in the Automotive Sector*. The Oxford Institute for Energy Studies. https://www.oxfordenergy.org/wpcms/wp-content/uploads/2022/03/Green-steel-as-the-new-clean-material-in-the-automotive-sector-ET09.pdf.

Nakintu, S. 2021. "Diversity, Equity and Inclusion: Key Terms and Definitions." *National Association of Counties*, November 29. https://www.naco.org/resources/featured/key-terms-definitions-diversity-equity-inclusion.

Naraharisetti, P.K., T.Y. Yeo, and J. Bu. 2017. "Factors Influencing CO_2 and Energy Penalties of CO_2 Mineralization Processes." *ChemPhysChem* 18(22):3189–3202. https://doi.org/10.1002/cphc.201700565.

NASEM (National Academies of Sciences, Engineering, and Medicine). 2019. *Gaseous Carbon Waste Streams Utilization: Status and Research Needs*. Washington, DC: The National Academies Press. https://doi.org/10.17226/25232.

NASEM. 2023a. "Energy Justic and Equity." Chapter 2 in *Accelerating Decarbonization in the United States: Technology, Policy, and Societal Dimensions*. Washington, DC: The National Academies Press. https://doi.org/10.17226/25931.

NASEM. 2023b. "Infrastructure Considerations for CO_2 Utilization." Chapter 4 in *Carbon Dioxide Utilization Markets and Infrastructure: Status and Opportunities: A First Report*. Washington, DC: The National Academies Press. https://doi.org/10.17226/26703.

NASEM. 2023c. "Introduction." Chapter 1 in *Accelerating Decarbonization in the United States: Technology, Policy, and Societal Dimensions*. Washington, DC: The National Academies Press. https://doi.org/10.17226/25931.

NASEM. 2023d. "Policy, Regulatory, and Societal Considerations for CO_2 Utilization Systems." Chapter 5 in *Carbon Dioxide Utilization Markets and Infrastructure: Status and Opportunities: A First Report*. Washington, DC: The National Academies Press. https://doi.org/10.17226/26703.

NASEM. 2023e. "Public Engagement to Build a Strong Social Contract for Deep Decarbonization." Chapter 5 in *Accelerating Decarbonization in the United States: Technology, Policy, and Societal Dimensions*. Washington, DC: The National Academies Press. https://doi.org/10.17226/25931.

NASEM. 2023f. "The Social Acceptance Challenge." Chapter 8 in *Laying the Foundation for New and Advanced Nuclear Reactors in the United States*. Washington, DC: The National Academies Press. https://doi.org/10.17226/26630.

NASEM. 2023g. "Workforce Needs, Opportunities, and Support." Chapter 4 in *Accelerating Decarbonization in the United States: Technology, Policy, and Societal Dimensions*. Washington, DC: The National Academies Press. https://doi.org/10.17226/25931.

National Conference of State Legislatures. 2023. "State and Federal Environmental Justice Efforts." https://www.ncsl.org/environment-and-natural-resources/state-and-federal-environmental-justice-efforts.

National Petroleum Council. 2019. "Chapter Two—CCUS Supply Chains and Economics." In *Meeting the Dual Challenge: A Roadmap to At-Scale Deployment of Carbon Capture, Use, and Storage*. https://dualchallenge.npc.org/files/CCUS-Chap_2-030521.pdf.

National Wildlife Federation. 2024. "Public Perceptions of Carbon Dioxide Removal in Wyoming, Texas, Louisiana, and Colorado." https://www.nwf.org/-/media/Documents/PDFs/NWF-Reports/2024/Public-Perceptions-of-Carbon-Dioxide-Removal-WY-TX-LA-CO.pdf.

NETL (National Energy Technology Laboratory). n.d.(a). "NETL CO2U LCA Guidance Toolkit." https://netl.doe.gov/LCA/co2u.

NETL. n.d.(b). "Utilization Procurement Grants (UPGrants)." https://www.netl.doe.gov/upgrants.

New Energy Economy. n.d. "Opposing False Solutions." https://www.newenergyeconomy.org/opposing-false-solutions.

New York State. 2022. "Climate Justice Working Group." *New York's Climate Leadership and Community Protection Act*. https://climate.ny.gov/resources/climate-justice-working-group.

Nielsen, J.A.E., K. Stavrianakis, and Z. Morrison. 2022. "Community Acceptance and Social Impacts of Carbon Capture, Utilization and Storage Projects: A Systemic Meta-Narrative Literature Review." *PLoS ONE* 17(8):e0272409. https://doi.org/10.1371/journal.pone.0272409.

Norhasyima, R.S., and T.M.I. Mahlia. 2018. "Advances in CO_2 Utilization Technology: A Patent Landscape Review." *Journal of CO_2 Utilization* 26(July):323–335. https://doi.org/10.1016/j.jcou.2018.05.022.

NRC (National Research Council). 2010. "Policy Durability and Adaptability." Chapter 8 in *Limiting the Magnitude of Future Climate Change*. Washington, DC: The National Academies Press. https://doi.org/10.17226/12785.

Offermann-van Heek, J., K. Arning, A. Linzechich, and M. Ziefle. 2018. "Trust and Distrust in Carbon Capture and Utilization Industry as Relevant Factors for the Acceptance of Carbon-Based Products." *Frontiers Energy Research* 6(73). https://doi.org/10.3389/fenrg.2018.00073.

Offermann-van Heek, J., K. Arning, A. Sternberg, A. Bardow, and M. Ziefle. 2020. "Assessing Public Acceptance of the Life Cycle of CO_2-Based Fuels: Does Information Make the Difference?" *Energy Policy* 143:111586.

Okoroafor, E.R., C.P. Offor, and E.I. Prince. 2022. "Mapping Relevant Petroleum Engineering Skillsets for the Transition to Renewable Energy and Sustainable Energy." Presented at the SPE Nigeria Annual International Conference and Exhibition. Lagos, Nigeria. https://doi.org/10.2118/212040-MS.

O'Laughlin, T.T. 2021. "30 Years Ago, Leaders Declared the Principles of Environmental Justice. They're Still Fighting to Make Them Heard." *Grist*, November 5. https://grist.org/fix/justice/1991-national-people-of-color-environmental-leaderships-sum-declared-principles-environmental-justice.

O'Neill, R. 2023. "Addressing a Growing Water Crisis in the U.S." *CDC Foundation* (blog), March 22. https://www.cdcfoundation.org/blog/addressing-growing-water-crisis-us.

Orange County Procurement Office. 2022. "Environmentally Preferable Purchasing Policy." https://cpo.ocgov.com/sites/cpo/files/2022-05/EPP%20Policy%20Document%20upgrade%20040122.pdf.

O'Rear, E.G., W. Jones, G. Hiltbrand, M. Adeyemo, B. King, and N. Pastorek. 2023. "Sustainable Aviation Fuel Workforce Development: Opportunities by Occupation." *Rhodium Group*, October 25. https://rhg.com/research/sustainable-aviation-fuel-workforce-development.

Park, L., and D.N. Pellow. 2004. "Racial Formation, Environmental Racism, and the Emergence of Silicon Valley." *Ethnicities* 4(3):403–424. https://doi.org/10.1177/1468796804045241.

Peery, J.S. 2023. "Letter to Sandia National Laboratory." https://www.sandia.gov/app/uploads/sites/113/2023/10/Commitment-to-Small-Business-Letter-2023.08.29.pdf.

Pett-Ridge, J., H.Z. Ammar, A. Aui, M. Ashton, S.E. Baker, B. Basso, M. Bradford, et al. 2023. "Chapter 9: Energy, Equity, and Environmental Justice Impacts." In *Roads to Removal: Options for Carbon Dioxide Removal in the United States*. LLNL-TR-852901. Lawrence Livermore National Laboratory.

Phillips, A. 2023. "Major Carbon Pipeline Is Cancelled as Opposition Grows and Regulations Remain Elusive." *Environmental Integrity Project Oil and Gas Watch*. October 25. https://news.oilandgaswatch.org/post/major-carbon-pipeline-is-cancelled-as-opposition-grows-and-regulations-remain-elusive.

P.L. No. (Public Law Number) 94-469. 1976. *Toxic Substances Control Act*. Amended December 27, 2022. https://www.govinfo.gov/content/pkg/COMPS-895/pdf/COMPS-895.pdf.

Righetti, T. K. 2017. "Siting Carbon Dioxide Pipelines." *Oil and Gas, Natural Resources and Energy Journal* 3(4):907.

Ringquist, E.J. 2005. "Assessing Evidence of Environmental Inequities: A Meta-Analysis." *Journal of Policy Analysis and Management* 24(2):223–247. https://doi.org/10.1002/pam.20088.

Rosa, L.D., L. Sanchez, and M. Mazzott. 2021. "Assessment of Carbon Dioxide Removal Potential via BECCS in a Carbon-Neutral Europe." *Energy and Environmental Science* 14:3086–3097.

Rothstein, R. 2017. "The Color of Law: A Forgotten History of How Our Government Segregated America." First edition. New York: Liveright Publishing Corporation, W.W. Norton.

Sabin Center (Columbia University School Sabin Center for Climate Change Law). n.d. "Community Benefits Agreements Database." https://climate.law.columbia.edu/content/community-benefits-agreements-database.

Sadasivam, N. 2023. "Why the White House's Environmental Justice Tool Is Still Disappointing Advocates." *Grist*, February 27. https://grist.org/equity/white-house-environmental-justice-tool-cejst-update-race.

Saiyid, A.H. 2023. "Exclusive: United Inks Largest Deal for Clean Jet Fuel with Novel Method." *Cipher*, September 13. https://ciphernews.com/articles/exclusive-united-inks-largest-deal-for-clean-jet-fuel-with-novel-method.

Sassman, W.G. 2021. "Community Empowerment in Decarbonization: NEPA's Role." *Washington Law Review* 96(4):1551–1566. https://digitalcommons.law.uw.edu/cgi/viewcontent.cgi?article=5196&context=wlr.

SBA (Small Business Administration). n.d. "HUBZone Program." https://www.sba.gov/federal-contracting/contracting-assistance-programs/hubzone-program.

SBIR (Small Business Innovation Research). n.d. "About: The SBIR and STTR Programs." https://www.sbir.gov/about.

Schieving, A. 2018. "The Most Common Grades of Reagents and Chemicals. https://www.labmanager.com/the-seven-most-common-grades-for-chemicals-and-reagents-2655.

Schröder, P. 2020. *Promoting a Just Transition to an Inclusive Circular Economy*. London: Chatham House. https://www.chathamhouse.org/sites/default/files/2020-04-01-inclusive-circular-economy-schroder.pdf.

Seltzer, M. 2021. "Americans Are Unaware of Carbon Capture and Sequestration Technology, According to a New Study." *Princeton University Andlinger Center for Policy Research on Energy and the Environment*. June 15. https://cpree.princeton.edu/news/2021/americans-are-unaware-carbon-capture-and-sequestration-technology-according-new-study.

Shih, J.-S., B. Chen, A. Krupnick, A. Thompson, D. Livingston, R. Pratt, and R. Pawar. 2022. *Modeling Ecological Constraints on a CO2 Pipeline Network*. Proceedings of the 16th Greenhouse Gas Control Technologies Conference (GHGT-16). https://dx.doi.org/10.2139/ssrn.4282306.

Sick, V., G. Stokes, F. Mason, Y.-S. Yu, A. Van Berkel, R. Daliah, O. Gamez, C. Gee, and M. Kaushik. 2022. "Implementing CO_2 Capture and Utilization at Scale and Speed." *University of Michigan Global CO_2 Initiative*. https://dx.doi.org/10.7302/5825.

Skone, T.J., M. Mutchek, M. Krynock, G. Cooney, S. Moni, S. Rai, J. Chou, et al. 2022. *Carbon Dioxide Utilization Life Cycle Analysis Guidance for the U.S. DOE Office of Fossil Energy and Carbon Management Version 2.0*. Pittsburgh, PA: National Energy Technology Laboratory. https://www.osti.gov/biblio/1845020.

Smith, S. 2017. *Just Transition: A Report for the OECD*. Just Transition Centre. https://www.oecd.org/environment/cc/g20-climate/collapsecontents/Just-Transition-Centre-report-just-transition.pdf.

SNL (Sandia National Laboratories). n.d. "Sandia's Small Business Goals." https://www.sandia.gov/working-with-sandia/prospective-suppliers/small-business/sandias-small-business-goals-2.

Solís, R. 1997. "Jemez Principles for Democratic Organization." *SouthWest Organizing Project*. https://epa.illinois.gov/content/dam/soi/en/web/epa/documents/environmental-justice/jemez-principles.pdf.

Sovacool, B.K., M. Iskandarova, and F.W. Geels. 2023. "Bigger Than Government: Exploring the Social Construction and Contestation of Net-Zero Industrial Megaprojects in England." *Technological Forecasting and Social Change* 188(March):122332. https://doi.org/10.1016/j.techfore.2023.122332.

Stedman, R.C., D. Evensen, S. O'Hara, and M. Humphrey. 2016. "Comparing the Relationship Between Knowledge and Support for Hydraulic Fracturing Between Residents of the United States and the United Kingdom." *Energy Research and Social Science* 20(October):142–148. https://doi.org/10.1016/j.erss.2016.06.017.

Stoutenborough, J.W., and A. Vedlitz. 2016. "The Role of Scientific Knowledge in the Public's Perceptions of Energy Technology Risks." *Energy Policy* 96(September):206–216. https://doi.org/10.1016/j.enpol.2016.05.031.

Streurer, M., and A. Dalrymple. 2024. "Dakota Access Pipeline Protest Trial Ends, Ruling Still Months Out." *Source New Mexico*, March 15. https://sourcenm.com/2024/03/15/dakota-access-pipeline-protest-trial-ends-ruling-still-months-out.

Strube, J., B. Thiede, and W. Auch. 2021. "Proposed Pipelines and Environmental Justice: Exploring the Association Between Race, Socioeconomic Status, and Pipeline Proposals in the United States." *Rural Sociology* 86(4):647–672. https://doi.org/10.1111%2Fruso.12367.

Suls, R. 2017. "Public Divided Over Keystone XL, Dakota Pipelines; Democrats Turn Decisively Against Keystone." *Pew Research Center*. February 21. https://www.pewresearch.org/short-reads/2017/02/21/public-divided-over-keystone-xl-dakota-pipelines-democrats-turn-decisively-against-keystone.

Suter, J., B. Ramsey, T. Warner, R. Vactor, and C. Noack. 2022. *Carbon Capture, Transport, and Storage: Supply Chain Deep Dive Assessment*. Department of Energy. DOE/OP-0001. https://www.energy.gov/sites/default/files/2022-02/Carbon%20Capture%20Supply%20Chain%20Report%20-%20Final.pdf.

Tabuchi, H., and B. Migliozzi. 2023. "Uncharted Waters: 'Monster Fracks' Are Getting Far Bigger. And Far Thirstier." *The New York Times*, September 25. https://www.nytimes.com/interactive/2023/09/25/climate/fracking-oil-gas-wells-water.html.

Taylor, D.E. 2011. "Green Jobs and the Potential to Diversify the Environmental Workforce." *Utah Environmental Law Review* 31(1):47–78. https://heinonline.org/HOL/Page?handle=hein.journals/lrel31&div=6&id=&page=&collection=journals.

Terlouw, T., K. Treyer, C. Bauer, and M. Mazzotti. 2021. "Life Cycle Assessment of Direct Air Carbon Capture and Storage with Low-Carbon Energy Sources." *Environmental Science and Technology* 55(16):11397–11411. https://doi.org/10.1021/acs.est.1c03263.

Tessum, C.W., J.S. Apte, A.L. Goodkind, N.Z. Muller, K.A. Mullins, D.A. Paolella, S. Polasky, et al. 2019. "Inequity in Consumption of Goods and Services Adds to Racial–Ethnic Disparities in Air Pollution Exposure." *Proceedings of the National Academy of Sciences* 116(13):6001–6006. https://doi.org/10.1073/pnas.1818859116.

Thomley, E. 2023b. "Carbon Dioxide Transport 101." https://betterenergy.org/blog/carbon-dioxide-transport-101.

Tomer, A., J.W. Kane, and C. George. 2021. "How Renewable Energy Jobs Can Uplift Fossil Fuel Communities and Remake Climate Politics." *Brookings*. February 23. https://www.brookings.edu/articles/how-renewable-energy-jobs-can-uplift-fossil-fuel-communities-and-remake-climate-politics.

Tomich, J., J. Plautz, and N.H. Farah. 2023. "Scuttled CO_2 Pipeline Renews Debate About State Hurdles." *E&E News by Politico*, October 23. https://www.eenews.net/articles/scuttled-co2-pipeline-renews-debate-about-state-hurdles.

Upham, P., B. Sovacool, and B. Ghosh. 2022. "Just Transitions for Industrial Decarbonisation: A Framework for Innovation, Participation, and Justice." *Renewable and Sustainable Energy Reviews* 167:112699.

van Oss, H.G., and A.C. Padovani. 2003. "Cement Manufacture and the Environment: Part II: Environmental Challenges and Opportunities." *Journal of Industrial Ecology* 7(1):93–126. https://onlinelibrary.wiley.com/doi/pdf/10.1162/108819803766729212.

Vohra, K., A. Vodonos, J. Schwartz, E.A. Marais, M.P. Sulprizio, and L.J. Mickley. 2021. "Global Mortality from Outdoor Fine Particle Pollution Generated by Fossil Fuel Combustion: Results from GEOS-Chem." *Environmental Research* 195:110754. https://doi.org/10.1016/j.envres.2021.110754.

Wang, J., N. Ulibarri, T.A. Scott, and S.J. Davis. 2023. "Environmental Justice, Infrastructure Provisioning, and Environmental Impact Assessment: Evidence from the California Environmental Quality Act." *Environmental Science and Policy* 146:66–75. https://doi.org/10.1016/j.envsci.2023.05.003.

WHEJAC (White House Environmental Justice Advisory Council). 2021. *Final Recommendations: Justice40 Climate and Economic Justice Screening Tool and Executive Order 12898 Revisions*. May 21. https://www.epa.gov/sites/default/files/2021-05/documents/whiteh2.pdf.

WHEJAC. 2022a. *Justice40 Initiative Implementation: Phase 1 Recommendations*. White House Environmental Justice Advisory Council. https://www.epa.gov/system/files/documents/2022-08/WHEJAC%20J40%20Implementation%20Recommendations%20Final%20Aug2022b.pdf.

WHEJAC. 2022b. "Recommendations for the Climate and Economic Justice Screening Tool." Letter to Brenda Mallory, Council on Environmental Quality. August 16. https://www.epa.gov/system/files/documents/2022-08/CEJST%20Recommendations%20Letter%208_4_2022%20Final.pdf.

White House. 2021. "FACT SHEET: The American Jobs Plan." Statements and Releases. March 31. https://www.whitehouse.gov/briefing-room/statements-releases/2021/03/31/fact-sheet-the-american-jobs-plan.

White House. 2023. *Federal-state Buy Clean Partnership Principles*. https://www.sustainability.gov/pdfs/federal-state-partnership-principles.pdf.

White House. n.d.(a). "American Climate Corps." https://www.whitehouse.gov/climatecorps.

White House. n.d.(b). "Justice40." https://www.whitehouse.gov/environmentaljustice/justice40.

Wilson, J.D., and Z. Zimmerman. 2023. "The Era of Flat Power Demand Is Over." *GridStrategies*. https://gridstrategiesllc.com/wp-content/uploads/2023/12/National-Load-Growth-Report-2023.pdf.

Wolf-Powers, L. 2010. "Community Benefits Agreements and Local Government." *Journal of the American Planning Association* 76(2):141–159. https://doi.org/10.1080/01944360903490923.

Woodruff, T.J., J.D. Parker, A.D. Kyle, and K.C. Schoendorf. 2003. "Disparities in Exposure to Air Pollution During Pregnancy." *Environmental Health Perspectives* 111(7):942–946. https://doi.org/10.1289/ehp.5317.

Xu, H., U. Lee, A.M. Coleman, M.S. Wigmosta, and M. Wang. 2019. "Assessment of Algal Biofuel Resource Potential in the United States with Consideration of Regional Water Stress." *Algal Research* 37(January):30–39. https://doi.org/10.1016/j.algal.2018.11.002.

Yale Law School. n.d. "Procedural Justice." https://law.yale.edu/justice-collaboratory/procedural-justice.

Young, S.D., B. Mallory, and G. McCarthy. 2021. "Interim Implementation Guidance for the Justice40 Initiative." Memorandum for the Heads of Departments and Agencies. M-21-28. Washington, DC: Office of Management and Budget, Executive Office of the President. https://www.whitehouse.gov/wp-content/uploads/2021/07/M-21-28.pdf.

Zabin, C. 2020. "Supply-Side Workforce Development Strategies: Preparing Workers for the Low-Carbon Transition." Chapter 3 in *Putting California on the High Road: A Jobs and Climate Action Plan for 2030*. Berkeley, CA: California Workforce Development Board. https://laborcenter.berkeley.edu/wp-content/uploads/2020/08/Chapter-3-Supply-Side-Workforce-Development-Strategies-Putting-California-on-the-High-Road.pdf.

5

Mineralization of CO_2 to Inorganic Carbonates

Mineralization of CO_2 is a key approach to carbon management, both to geologically store carbon and to produce inorganic carbonate solids that can be used in various ways, including to support the defossilization of the built environment. The mineralization process converts thermodynamically stable, gaseous CO_2 into an insoluble solid with similar thermodynamic stability and a lifetime on geologic timescales. In nature, mineralization occurs through the weathering of minerals and rocks containing alkaline metals such as calcium (Ca) and magnesium (Mg). These minerals react with moisture and CO_2, dissolve, and precipitate as solid carbonates. However, this natural weathering occurs too slowly to contribute to carbon utilization solutions; timescales for natural mineralization can be hundreds or thousands of years, depending on the concentration of CO_2 and rock type. Hence, research and development (R&D) have increased in recent years to produce carbon mineralization technologies that chemically and physically accelerate mineral dissolution and carbonate precipitation processes to useful rates and scales. This report's scope includes CO_2 conversion to tradable inorganic carbonate commodities that are used in applications such as building materials, pigments and fillers, and excludes in situ mineralization processes where the product is not extracted from the environment, such as enhanced rock weathering and ocean alkalinity adjustment.

As mentioned in Chapter 2, construction materials like cement are derived from carbon-bearing minerals and rocks (e.g., limestone). In 2021, cement production emitted about 2.69 gigatonne (Gt) CO_2 globally (Andrew 2023), in part from emissions from fossil fuel combustion used to heat the reaction, and in part from the CO_2 released from minerals as they are transformed. An opportunity exists to significantly reduce global CO_2 emissions by utilizing inorganic carbonates produced from captured CO_2 in physical infrastructure to displace higher-carbon-emitting products. In recent years, new feedstocks and unconventional resources, including alkaline industrial wastes and alkaline materials derived from seawater and brine, have been identified for carbon mineralization technologies. The integration of renewable energy and carbon mineralization with mining and mineral processing (called carbon-negative mining[1]) also has opened up new applications and strategies for carbon management in the form of inorganic carbonates while creating additional economic benefits (e.g., recovery of energy-relevant critical minerals as by-products).

This chapter focuses on the status, challenges, and R&D needs for the mineralization of CO_2 into inorganic carbonates for use in durable building materials at a scale relevant to climate change mitigation. Carbon mineralization technologies are at a more advanced stage as compared to CO_2 utilization technologies discussed in other

[1] Carbon-negative mining is a mining process in which CO_2 emissions produced during mining as well as CO_2 from other industrial emissions are stored in mined rocks (e.g., as solid carbonates) and geologic formations.

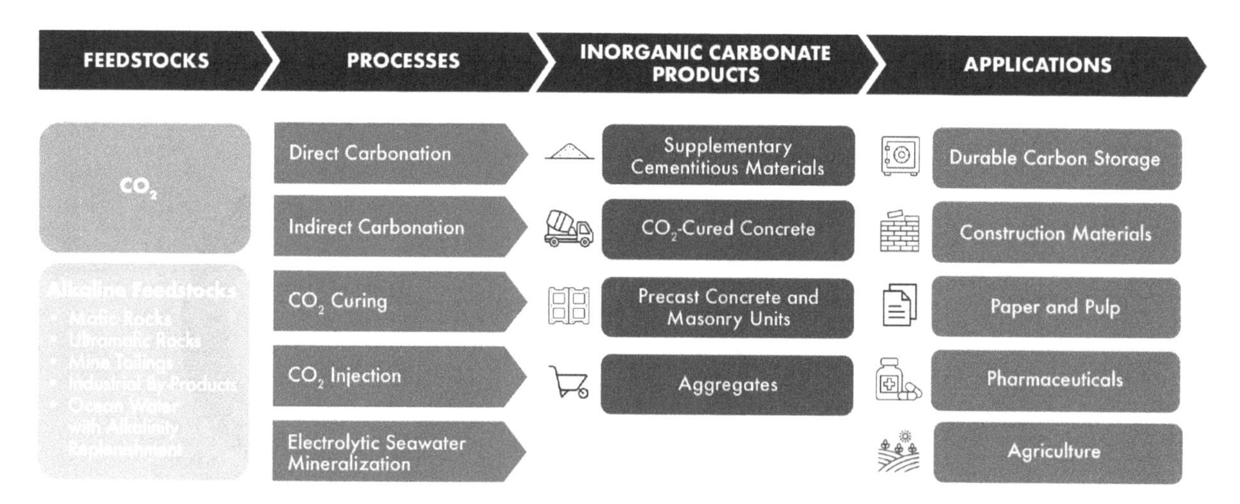

FIGURE 5-1 Summary of the feedstock inputs, processes, products, and applications for mineral carbon utilization processes to form inorganic carbonates.
SOURCE: Icons from the Noun Project, https://thenounproject.com. CC BY 3.0.

chapters, ranging from fundamental research to industrial deployment. Thus, this chapter discusses how U.S. R&D investment from both government and industrial funding sources (e.g., large corporations, start-ups, venture capital, investment funds, and banks) could accelerate critical discoveries and advance carbon mineralization technologies to make transformational impact at scale. Carbonation of fly ash and carbon-negative mining using mine tailings and alkaline industrial waste are discussed in the present chapter, while coal waste as an unconventional feedstock for carbon utilization is covered in Chapter 9. Figure 5-1 shows the major features of mineral carbon utilization, including feedstock inputs, processes, products, and applications. The remainder of the chapter describes the current status and research, development, and demonstration (RD&D) needs for processes and resulting products, noting relevant applications where appropriate.

5.1 OVERVIEW OF CO_2 CONVERSION TO INORGANIC CARBONATES

5.1.1 Inorganic Carbonate Products That Can Be Derived from CO_2

The current construction materials industry faces a dual challenge: increasing expectations for regulation mandating reduced carbon emissions and simultaneous rapid global demand growth for physical infrastructure. The market for concrete and construction aggregates continues to expand, while conventional construction material manufacturing processes are associated with significant environmental impacts (e.g., carbon emissions, mining-related water contamination, and air pollution). Carbonate minerals and rocks, such as limestone ($CaCO_3$) and dolomite ($CaMg(CO_3)_2$), have been widely used in large quantities to produce cement and refractory materials, respectively (Haldar 2020). While the most common use of calcium carbonate is as a feedstock to produce construction materials, it is also used in nonconstruction applications. For example, high-purity (90–99 percent, depending on the application) calcium carbonate ($CaCO_3$) is used as a filler in paper, paint, rubber, and plastic, among other applications (Ropp 2013; Tanaka et al. 2022).

The mining, quarrying, transportation, and processing of carbonate minerals and rocks for construction materials have resulted in significant carbon emissions contributing to climate change despite serving as key processes for physical infrastructure build-out. Primary mineral construction materials include cementitious materials, concrete (a mixture of sand, cement, water, and aggregates), mortar (a mixture of cement and sand), and masonry materials (bricks). Figure 5-2 shows the estimated global CO_2 emissions associated with major construction materials—cement and aggregates—currently used in the built environment. The continued use of natural carbonate minerals and rocks to create cement and concrete is not compatible with achieving net-zero emissions.

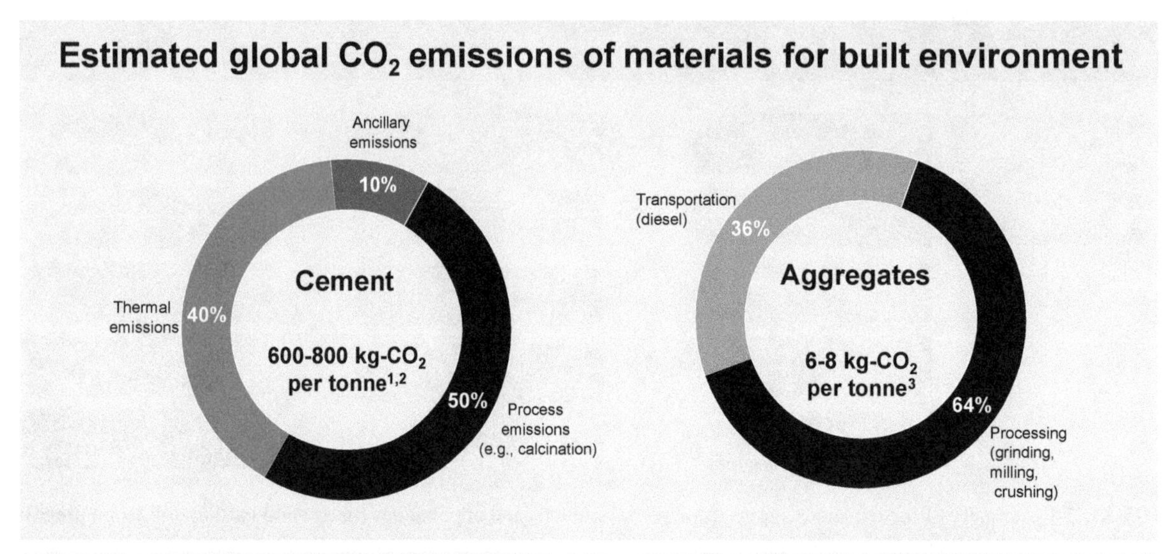

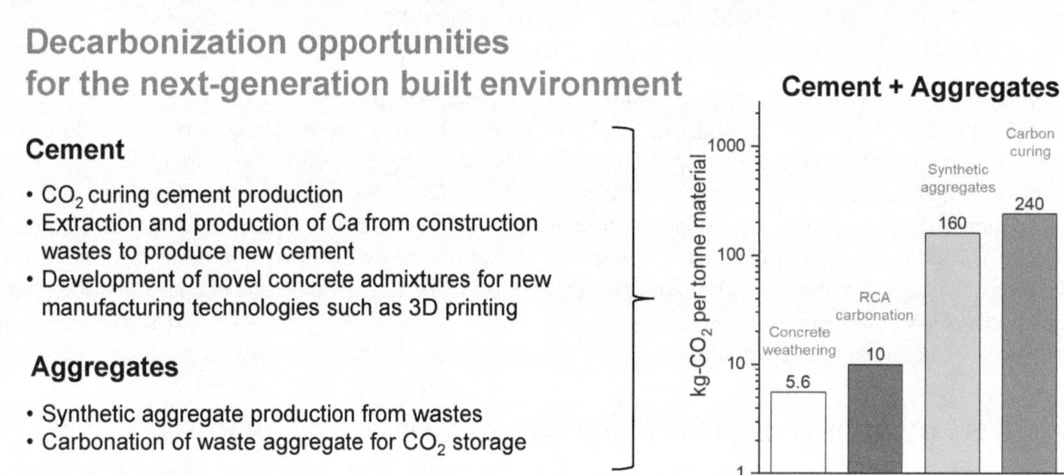

FIGURE 5-2 Global CO_2 emissions associated with cement and aggregates in the built environment and multifaceted opportunities for their decarbonization.
NOTE: RCA = recycled concrete aggregate.
SOURCES: Adapted from Park et al. (2024), https://doi.org/10.3389/fenrg.2024.1388516. CC BY 4.0. Data from Gerres et al. (2021); Ho et al. (2021); Holappa (2020); IEA (2022); Lehne and Preston (2018); Mayes et al. (2018); Rosa et al. (2022); Seddik Meddah (2017); Zhang et al. (2020a).

Concrete is produced conventionally by combining Ordinary Portland Cement (OPC) powder (clinker + gypsum) with water, sand, and gravel. More than 4 billion tons of cement were produced in 2021; the demand for cement continues to rise and is expected to reach 6.2 billion tons by 2050 (GCCA 2021; IEA 2022). The CO_2 emitted from manufacturing cement is responsible for about 7–8 percent of global carbon emissions (Andrew 2023; IEA 2022). As shown in Figure 5-2, these emissions are primarily industrial process emissions from the chemical reaction of limestone decarbonation to produce clinker[2] and from the carbon-intensive fuels required to reach the high temperatures needed in cement kilns (up to 1400°C), together accounting for 90 percent of the carbon emissions

[2] Desired clinker phases are alite (Ca_2SiO_4, sometimes formulated as $3CaO{\cdot}SiO_2$ [C_3S in cement chemist notation]), belite (Ca_2SiO_4, sometimes formulated as $2CaO{\cdot}SiO_2$ [C_2S in cement chemist notation]), aluminate (Al_2O_3), and ferrite (Fe_2O_3).

from cement production (Fennell et al. 2021; Park et al. 2024). The current global production rate of aggregate is approximately 40 billion metric tons per year (Global Aggregates Information Network 2023). Producing 1 metric ton of cement and 1 metric ton of aggregate can lead to emissions of approximately 0.6–1 metric tons CO_2 and 6–20 kg CO_2, respectively (Czigler et al. 2020; Fennell et al. 2021; IPCC 2023; Monteiro et al. 2017). From an economic perspective, cement production accounts for the most emissions per revenue dollar at about 6.9 kg of CO_2 per dollar of revenue generated (Czigler et al. 2020).

Figure 5-2 (bottom) lists innovative and transformative technological options currently being developed and employed to reduce CO_2 emissions associated with cement and aggregate materials. Strategies to mitigate carbon emissions in the construction materials industry cover different domains: mining (e.g., quarrying and transportation with improved efficiencies and the use of sustainable energy); integrated process design (e.g., using renewable energy to produce cement clinkers, capturing or purifying CO_2 from cement plants, CO_2 curing for concrete); new and alternative materials production with lower carbon footprints (e.g., supplementary cementitious materials, admixtures); and sustainable demolition and upcycling processes (e.g., upcycling demolished materials) (Miller et al. 2021; Ostovari et al. 2021; Tiefenthaler et al. 2021).

It is not easy to reduce the amount of cement in concrete because cement is the binder required to provide the mechanical properties of concrete (e.g., compressive strength, durability). The required amount of binder varies by application (e.g., structural versus nonstructural), and standards vary by country. An alternative material that can replace cement binder in concrete is called a supplementary cementitious material (SCM). SCMs react with water (hydraulic reaction) and/or calcium hydroxide (pozzolanic reaction), enhancing material strength and durability while reducing the overall life cycle CO_2 emissions of concrete. Limestone ($CaCO_3$) is calcined to produce $Ca(OH)_2$ leading to CO_2 emissions. The use of end-of-life carbonate wastes such as $CaCO_3$ in demolition wastes could provide a net-zero pathway to produce $Ca(OH)_2$. If silicate minerals (e.g., wollastonite, $CaSiO_3$) are used to produce $Ca(OH)_2$, the net carbon intensity of the produced concrete would be lowered further. Some minerals and alkaline industrial wastes react with CO_2 to form SCMs. Each U.S. state has at least one type of prescriptive specification that either requires certain proportions of traditional cement and concrete materials, limits substitution rates of SCMs and other alternative materials, or restricts which materials are acceptable for certain applications (Kelly et al. 2024). Because developers and construction firms carefully follow state and local codes, prescriptive specifications encourage traditional materials over deployment of innovative materials. Several states are pursuing performance specifications based on desired engineering performance (durability, strength, flexibility, temperature-tolerance) rather than mandating particular material mixes (e.g., ASTM International 2023a). Performance standards require significant investments in training and technical capacity at the state and local levels, where code enforcement occurs, as well as in benchmarking and performance testing equipment and protocols.

Other carbon mineralization approaches the cement industry is taking to reduce their CO_2 emissions include the reincorporation of CO_2 back into the concrete product, either up front during mixing and curing of concrete or via treatment of concrete demolition waste at the conclusion of its service life (Winnefeld et al. 2022). For example, concrete masonry units (CMUs)—tiles, bricks, or blocks with a mixture of powdered Ca-rich steel slag, water, and aggregate—have been produced without cement by incorporating SCMs (e.g., steel slag). These CMUs are cured in a chamber with CO_2 captured from industrial sources, allowing steel slag to react with CO_2 to produce $CaCO_3$. This incorporates additional carbon into solid carbonate and reduces the energy requirement for concrete curing. The newly formed carbonates act as binders, which eliminates the need for much of the carbon-intensive cement paste. Researchers have tested the performance of products with up to 75 percent cement replacement and are pursuing blends with up to 100 percent cement replacement (George 2023; Jin et al. 2024; Nukah et al. 2023; Phuyal et al. 2023; Shah et al. 2022; Srubar et al. 2023). The current products are used in precast concrete road pavement blocks, river embankment blocks, or ceilings (Li et al. 2022a).

CMUs produced using steel slag as SCMs have been reported to have higher compressive strength than cement-based CMUs, but CMUs using steel slag SCMs can also be more brittle or porous, affecting potential applications (Newtson et al. 2022; Nguyen et al. 2020; Parron-Rubio et al. 2019; Taha et al. 2014). Some of these alternative cements have been certified as meeting ASTM C90 or C150 standards as construction materials. Their performance and quality can be maintained using a mixture of OPC and SCMs, which cuts greenhouse gas emissions in proportion to the fraction of SCM used. Other approaches to produce concrete with carbon storage include the addition of γ-C2S (made from calcium hydroxide [$Ca(OH)_2$] and silica) into the concrete (forming a water, cement, aggregate, γ-C2S, industrial wastes mixture). The strength and durability of the γ-C2S concrete improves upon reaction with CO_2 during the curing process.

The construction materials industry is developing technologies to utilize captured CO_2 for curing concrete, carbonating natural minerals or industrial wastes to produce SCMs, and to produce synthetic aggregate to store more CO_2 in construction materials (e.g., fillers) (Norhasyima and Mahlia 2018). Carbon mineralization technologies are amenable to being optimized to utilize dilute concentrations of CO_2, such as flue gas (~15 percent CO_2), directly to form carbonate products, eliminating the need for energy-intensive CO_2 capture and compression. The co-location of CO_2 sources with alkaline wastes is desirable, in order to minimize transportation and cost. Using CO_2 to produce synthetic aggregate could reduce mining of carbonated minerals and rocks, which naturally store CO_2, and avoid greenhouse gas emissions, ecological degradation, and human health risk.

5.1.2 Conversion Routes

Carbon mineralization (also known as mineral carbonation or CO_2 mineralization) is a chemical phenomenon in which divalent alkaline metal ions such as Ca^{2+} and Mg^{2+} react with CO_2 to produce solid carbonates. In nature, minerals containing Mg, Ca, or iron (Fe) such as serpentine ($Mg_3Si_2O_5(OH)_4$), olivine (Mg_2SiO_4), and wollastonite ($CaSiO_3$), can react with CO_2 in the air to form a stable and inert carbonate rock, a process called weathering (Blondes et al. 2019; Gadikota et al. 2014; Kashim et al. 2020; Park and Fan 2004). While these minerals contain varied concentrations of Fe and other mineral phases, here the chemical formulas are written only using Mg and Ca for simplicity. There are three key reaction steps in natural or engineered carbon mineralization, illustrated in the following reactions: (1) CO_2 hydration (Reactions 5.1–5.3); (2) mineral dissolution (Reactions 5.4–5.6); and (3) formation of solid inorganic carbonates (Reactions 5.7–5.8). While there are numerous fundamental studies of the kinetics of CO_2 hydration, mineral dissolution, and formation of carbonates, the coupled effects of pH, temperature, and partial pressure of CO_2 on coupled mineral dissolution and carbonation behavior can vary widely depending on the complexity of the starting minerals, rocks, or industrial waste. The carbon mineralization reactions for three major Mg- and Ca-bearing silicate minerals are as follows:

CO_2 hydration:

$$CO_{2(g)} \rightarrow CO_{2(aq)} \quad (R5.1)$$

$$CO_{2(aq)} + H_2O \rightarrow H_2CO_{3(aq)} \leftrightarrow H^+_{(aq)} + HCO_3^-{}_{(aq)} \quad (R5.2)$$

$$HCO_3^-{}_{(aq)} \rightarrow CO_3^{2-}{}_{(aq)} + H^+_{(aq)} \quad (R5.3)$$

Forsterite dissolution:

$$Mg_2SiO_{4(s)} + 4H^+_{(aq)} \rightarrow 2Mg^{2+}_{(aq)} + SiO_{2(s)} + 2H_2O \quad (R5.4)$$

Wollastonite dissolution:

$$CaSiO_{3(s)} + 2H^+_{(aq)} \rightarrow Ca^{2+}_{(aq)} + SiO_{2(s)} + H_2O \quad (R5.5)$$

Serpentine dissolution:

$$Mg_3Si_2O_5(OH)_{4(s)} + 6H^+_{(aq)} \rightarrow 3Mg^{2+}_{(aq)} + 2SiO_{2(s)} + 5H_2O \quad (R5.6)$$

Carbonate formation:

$$Mg^{2+}_{(aq)} + CO_3^{2-}{}_{(aq)} \rightarrow MgCO_{3(s)} \quad (R5.7)$$

$$Ca^{2+}_{(aq)} + CO_3^{2-}{}_{(aq)} \; CaCO_{3(s)} \quad (R5.8)$$

Carbon mineralization can occur via utilization or non-utilization modes, as illustrated in Figure 5-3. Panel 4 of Figure 5-3 shows in situ carbon mineralization, which occurs when CO_2 is injected into reactive geologic formations with high Ca and Mg content. Panel 3 shows enhanced rock weathering, where alkaline feedstock is spread in the environment to react with CO_2 and be stored in dispersed, solid carbonates in the environment. While both in situ carbon mineralization and enhanced rock weathering could increase the CO_2 sequestration potential of the carbon storage reservoir or natural environments, respectively, they would not result in inorganic carbonate commercial products, and so are out of scope of this report. The focus of this chapter is the direct and indirect utilization approaches to carbon mineralization shown in panels 1 and 2 of Figure 5-3, which could produce high-value products like concrete, aggregates, fillers, and pigments, as discussed in Section 5.1.1.

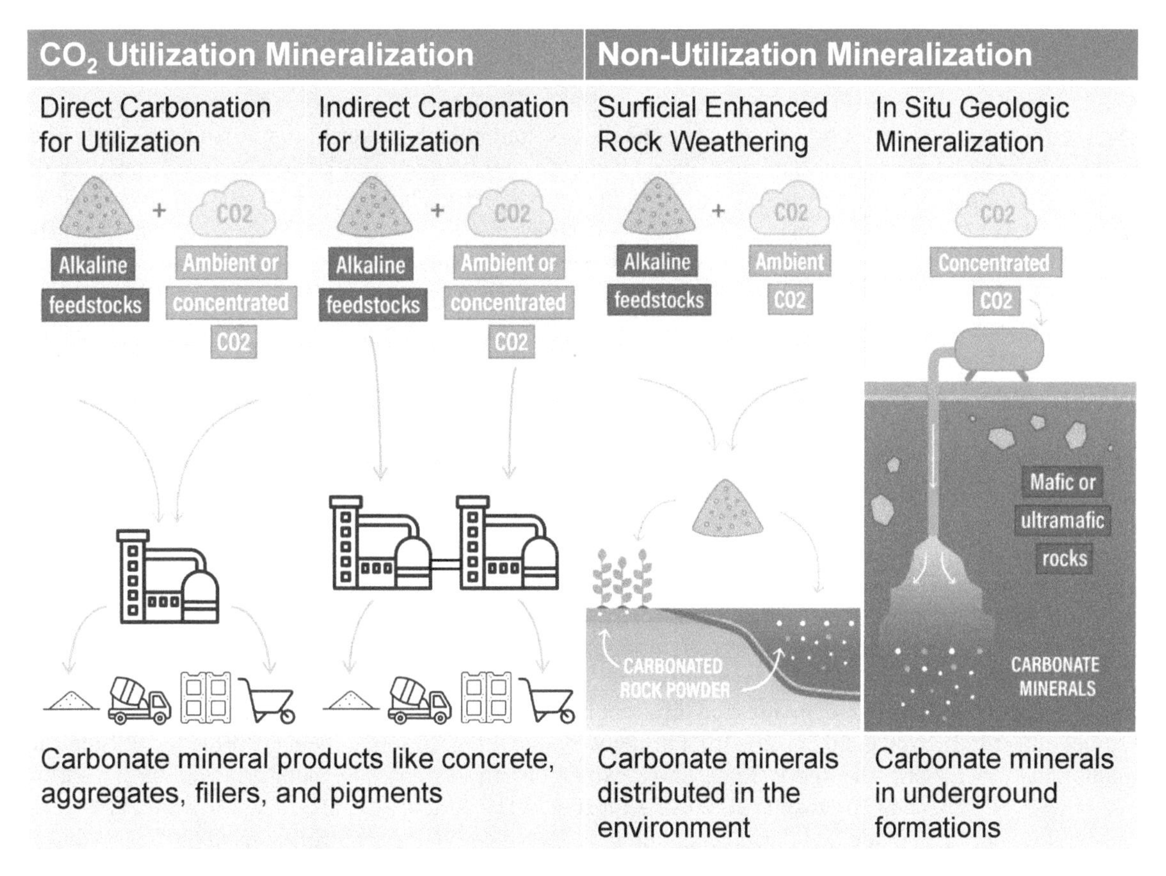

FIGURE 5-3 Types of carbon mineralization approaches, including utilization, and non-utilization approaches.
NOTES: Panels 1 and 2 of the figure show direct and indirect approaches, respectively, that create mineral carbon products and are explored further in this report. Panels 3 and 4 show surficial enhanced rock weathering and in situ geologic mineralization, which do not create products and are out of scope of this report. Alkaline feedstock = alkaline mine tailings, some industrial by-products and certain types of mined rock (e.g., silicate minerals such as serpentine, olivine and wollastonite).
SOURCES: Adapted from Riedl et al. (2023). Icons from the Noun Project, https://thenounproject.com. CC BY 3.0.

Carbon mineralization technologies generally fall into two operating modes: direct and indirect carbonation. Direct carbonation involves a single-step reaction of CO_2 and materials (e.g., minerals, rocks, alkaline industrial wastes), whereas the indirect process comprises multistep reactions and separations (i.e., inorganic solid dissolution at lower pH [<4] facilitated by leaching agents such as acids and chelating agents targeting Ca and Mg, followed by carbonation at higher pH [>8]). Various factors can impact mineralization and subsequently influence the properties of produced construction materials. For example, indirect carbonation allows reaction and separation steps to be optimized individually, therefore enabling production of higher purity products and making it easier to produce inorganic carbonates and by-products (e.g., high surface area SiO_2 that can replace silica fume) with tailored chemical and physical properties. However, the energy (e.g., heat and electricity) and chemical (e.g., acids and ligands) inputs required for the overall indirect carbonation (and attendant, potentially hazardous, liquid waste disposal) would be greater compared to direct carbonation where CO_2 is the main input. The net carbon benefits and environmental impacts for each technology need to be carefully evaluated via life cycle assessment (LCA).

Direct carbonation is a process in which CO_2 is introduced into solid materials or aqueous slurry/solutions rich in Ca/Mg to form solid carbonates (Gadikota et al. 2015). It is simple and capable of handling materials in a single process to generate metal carbonates (Swanson et al. 2014). Many of the current commercialized construction materials emissions mitigation technologies in the construction materials industry react alkaline industrial wastes with CO_2 to form Ca or Mg carbonates via direct carbonation. These multiphase reactions typically involve gas-solid and gas-liquid-solid processes, similar to natural weathering reactions, but with faster rates than those of natural processes (Campbell et al. 2022; Pan et al. 2018). The interaction between the minerals and CO_2 still involves a sequence of processes, including hydration, dissolution, and carbonation, but within a single reactor. In addition to reaction temperature, the hydration level (or water amount) can influence significantly the direct carbonation rates and polymorphs of carbonate products (e.g., nesquehonite [$MgCO_3 \cdot 3H_2O$], hydromagnesite [$(MgCO_3)_4 \cdot Mg(OH)_2 \cdot 4H_2O$] and magnesite [$MgCO_3$]) (Fricker et al. 2013, 2014). Generally, natural silicate minerals—including serpentine—cannot be converted to carbonate minerals via gas-solid reactions owing to their very slow reaction kinetics without water. However, more reactive materials such as fly ash and cement kiln dust can be directly converted to carbonate products via gas-solid reactions. Most direct carbonation processes employ a slurry reactor that operates at high temperature (up to 185°C) and CO_2 pressure (up to 150 atm) to achieve significant carbonation within a few hours.

Indirect carbonation refers to the multistep process of leaching out active metals from minerals (i.e., Ca and Mg) using solvents containing acids (weak acids or CO_2 bubbling can be used as a sustainable acid) and chelating agents targeting Ca and Mg (e.g., citrate, acetate, and oxalate) (Gadikota et al. 2014). The pH of the Ca- and Mg-rich solution then is increased to a pH >9 to promote the formation of solid carbonates while injecting/bubbling CO_2 into the reactor. The dissolution and carbonation processes can be controlled and optimized by varying pH and temperature. The carbonate products obtained from indirect carbonation are high purity (>99 percent if the solid residue from the mineral dissolution reactor is removed before the carbonation step), and their polymorphs can be tailored for different applications in various industries, such as construction, paper, and rubber (Zhang and Moment 2023; Zhao et al. 2023a, 2023b). For example, the polymorphs of $CaCO_3$ include vaterite, aragonite and calcite, and recent studies have demonstrated different reactivities of these polymorphs in cement pastes (Zhang and Moment 2023; Zhao et al. 2023a, 2023b). Furthermore, the use of different ligands during a multistep carbon mineralization process also allows for the selective extraction of other valuable metals, including rare earth elements (REEs), Ni, Co, and Cu (Hong et al. 2020; Kim et al. 2021; Sim et al. 2022). The extraction and recovery of these critical metals provide additional economic benefits for carbon mineralization technologies.

Carbon mineralization also can be used to durably store CO_2 within the carbonates in SCMs, recycled aggregates, or CO_2-cured concrete (Supriya et al. 2023; Zajac et al. 2022). As described above, SCMs can reduce significantly the life cycle CO_2 emissions of construction materials, and the feedstock can be natural minerals (Mg/Ca silicates) or industrial wastes (e.g., fly ash, steel slags, mine tailings, brines, red mud), which are emerging as promising alternative resources owing to their widespread availability (see Chapter 9 for more on the use of fly ash as a coal waste stream). In some cases, these alternative feedstocks not only can lower environmental impact but also can improve the performance of concrete products. Recent techno-economic assessment (TEA) has revealed that using SCMs produced via carbon mineralization reaction can be profitable, with approximately $35 more revenue per metric ton of cement produced and with CO_2 emission reductions of 8–33 percent compared to conventional cement (Strunge et al. 2022). The economic value results from the higher quality of produced SCMs and the value of the carbon storage associated with carbonates incorporated into cement.

This chapter describes a number of innovative pathways to produce sustainable construction materials via carbon mineralization. Current bottlenecks for viable mineral carbonation processes on an industrial scale include large energy requirements for mining and mineral processing, and the need to further accelerate both mineral dissolution and carbonation rates. Additionally, new formulations of materials, such as concrete derived from carbon mineralization, will require testing and property validation before being accepted by users and construction materials market regulators. Existing and emerging carbon mineralization technologies and their specific challenges and opportunities are discussed in the next section.

5.2 EXISTING AND EMERGING PRODUCTS, PROCESSES, CHALLENGES, AND RESEARCH AND DEVELOPMENT OPPORTUNITIES

The current status of carbon mineralization technologies spans across technology readiness levels (TRLs) from fundamental research to commercialization. The carbonation of alkaline industrial wastes, such as fly ash and iron and steel slag, has been deployed at an industrial scale, in part because of beneficial economic and regulatory incentives. This technology not only can capture and store CO_2 in carbonate products but also helps manage solid waste. Innovative technologies based on carbon mineralization also are emerging that produce new products and provide CO_2 utilization options. Sections 5.2.1–5.2.6 discuss these existing and emerging carbon mineralization approaches—carbonation of natural minerals and rocks, alkaline industrial wastes and demolition wastes, enhanced carbon uptake by construction materials, electrolytic seawater mineralization, alternative cementitious materials with increased CO_2 utilization potential, and integrated carbon mineralization technologies—providing analysis of their challenges and R&D opportunities.

5.2.1 Carbonation of Natural Minerals and Rocks

As discussed in Section 5.1.2, a wide range of earth-abundant Mg- and Ca-rich natural silicate minerals are available for carbon mineralization. These silicate minerals and rocks are not as reactive as alkaline industrial wastes (e.g., fly ash, iron and steel slags, and cement kiln dust), which are generally amorphous with high surface area that increases reactivity. An exception is asbestos, which does have high surface area. Thus, carbon mineralization processes for natural rocks need to be engineered to accelerate the rate of mineral dissolution, which is often rate-limiting. This section describes pathways for carbon mineralization using natural minerals and rocks.

5.2.1.1 Current Technology

A common approach to accelerate the carbonation of natural silicate minerals and rocks is feedstock activation, which is achieved through thermal pretreatment (thermal activation) or mechanical pretreatment (mechanical activation) (Rim et al. 2020a, 2021). Thermal activation is an effective strategy to enhance the reactivity of minerals for dissolution, but it also consumes significant energy, reducing the net CO_2 utilization potential (Rim et al. 2020b, 2021). Recently, researchers have started to use renewable energy to thermally treat minerals (e.g., the calcination of solid carbonates using solar thermal energy [Kelemen et al. 2020]); those technologies may be able to lower the carbon intensity of mineral activation.

Recent studies have shown that the reactivity of silicate minerals can be predicted based on their structures. SiO_4 tetrahedra are the building blocks of most silicate minerals, and their connectivity or lack thereof (Figure 5-4) determines the overall structure of minerals (Ashbrook and Dawson 2016). The degree of polymerization of SiO_4 provides a simple metric of connectivity and is denoted by the symbol Q^n ($n = 0, 1, 2, 3, 4$), where n is the number of shared oxygens that "bridge" silicon in other SiO_4 tetrahedra (Rim et al. 2020b). Natural hydrous magnesium silicate mineral (serpentine) consists of predominantly Q^3. When it is heated beyond 600°C, new silicate structures (Q^0, Q^1, Q^2, Q^4, and altered Q^3) are formed as the chemically bonded hydroxyl group (OH) is released from the mineral (Balucan and Dlugogorski 2013; Balucan et al. 2011; Chizmeshya et al. 2006; Dlugogorski and Balucan 2014; Liu and Gadikota 2018; McKelvy et al. 2004; Rim et al. 2020b). The newly formed Q structures impact the dissolution behavior of silicate minerals. Q structures can be examined using solid-state ^{29}Si magic angle spinning nuclear magnetic resonance (NMR) and X-ray photoelectron spectroscopy (XPS) techniques, as shown in Figure 5-4.

NMR (as illustrated in Figure 5-4) and XPS techniques can be employed to examine the chemical shifts within silicate structures in both unreacted and reacted minerals and rocks, facilitating the investigation of dissolution mechanisms. Rim et al. (2020b) showed that heat-treated serpentine is a mixture of amorphous phases of Q^1 (dehydroxylate I), Q^2 (enstatite), and Q^4 (silica), as well as crystalline phases of Q^0 (forsterite) and Q^3 (dehydroxylate II and serpentine). The dissolution of amorphous silicate structures is significantly easier than those in crystalline phases (Rim et al. 2020b). Thus, heat activation of silicate minerals can promote the formation of Q^1 (dehydroxylate I) and Q^2 (enstatite) structures while minimizing Q^3 (serpentine) structures to accelerate mineral dissolution for carbon mineralization.

Mineral dissolution is hindered also by mass transfer limitations caused by the formation of an Si-rich passivation layer on the surface of mineral particles (Rim et al. 2021). The Si-rich passivation layer can be removed or reduced by in situ grinding, where grinding medium is added to the slurry reactor to refresh the surface of mineral particles during their dissolution. The in situ grinding requires extra energy input. Rim et al. (2020a) found that the grinding media stress intensity, which can be used to estimate the energy requirement, needs to be optimized for the target extent of mineral dissolution enhancement (Rim et al. 2020a). Two mechanisms of physical grinding activation exist: fragmentation (which requires more energy but creates a large reactive surface area)

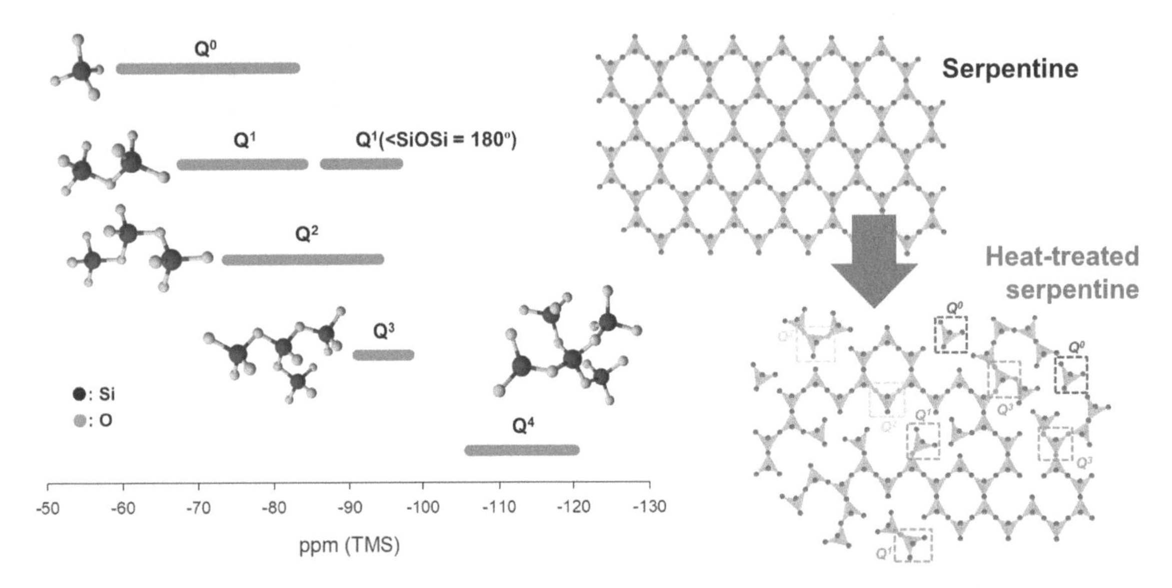

FIGURE 5-4 Classification of silicate minerals according to the degree of polymerization (Q^n) of SiO_4, showing their chemical shifts in ^{29}Si NMR (left) and the structural changes of serpentine during heat treatment (right). Chemical shift is reported in parts per million (ppm) versus a tetramethylsiloxane (TMS) standard.
SOURCE: Reprinted from G. Rim, A.K. Marchese, P. Stallworth, S.G. Greenbaum, and A.-H.A. Park. 2020b, "^{29}Si Solid State MAS NMR Study on Leaching Behaviors and Chemical Stability of Different Mg-silicate Structures for CO_2 Sequestration," *Chemical Engineering Journal* 396:125204. Copyright (2020), with permission from Elsevier.

and abrasion (which requires less energy but only refreshes the existing mineral surface). The in situ grinding in fragmentation mode is more effective in improving the Mg leaching rate from serpentine compared to abrasion mode (Rim et al. 2020a). The operational mode of in situ grinding should be determined based on the mineral dissolution rate (which varies for different minerals and rocks) and the energy requirement per mole of Mg or Ca extracted from the minerals.

The dissolution of minerals also can be accelerated by using Mg- and Ca-targeting chelating agents, although strong metal-ligand bonds might prevent subsequent carbonation (Gadikota et al. 2014; Park and Fan 2004). Thus, ligands with moderate binding energy (e.g., acetate, citrate, oxalate) are used to accelerate the dissolution of silicate minerals for carbon mineralization. All these methods (e.g., heat treatment, in situ grinding, Mg- and Ca-targeting ligands) can be used together to enhance mineral dissolution.

Once Mg and Ca are leached out into the solution phase, one increases the pH of the solution a pH >9 and introduces CO_2 to form solid carbonates. Although mineral dissolution generally is considered to be the rate-limiting step in carbon mineralization, the hydration of CO_2 (Reactions 5.1–5.3 in Section 5.1.2) may need to be accelerated as well. One strategy utilizes an enzymatic catalyst, carbonic anhydrase, which increases the rate of CO_2 hydration by improving proton transfer between H_2O and CO_2 at a zinc ion (Zn^{2+}) active site (Patel et al. 2013, 2014a, 2014b).

A relatively high metal cation extraction efficiency for Mg^{2+} and Ca^{2+} (>60 percent) can be achieved in 30 minutes, when the abovementioned chemical and physical enhancement strategies are used on ground mineral and rock feedstocks for mineral dissolution. The final products from the carbonation of natural silicate minerals and rocks include solid inorganic carbonates (e.g., $MgCO_3$ and $CaCO_3$), silica by-product, and unreacted mineral/rock residues. These materials can be used as cement replacement and clinker substitutions as discussed in Section 5.1. Besides carbonates, the solid residues after leaching can be utilized as SCMs to decrease life cycle CO_2 emissions further (Hargis et al. 2021). This reactive $CaCO_3$ is in the vaterite phase, and it achieves high compressive strength after a polymorphic transformation to stable aragonite (Hargis et al. 2021). Different polymorphs of calcium carbonates (vaterite, aragonite, and calcite) can be produced by tuning the reaction temperature and time as well as by introducing seed material to promote rapid formation of metastable carbonate phases (Zhang and Moment 2023; Zhao et al. 2023a, 2023b). The use of carbonates and by-products from the carbonation of natural silicate minerals and rocks as construction materials can significantly reduce the life cycle CO_2 emissions associated with the built environment.

5.2.1.2 Challenges

Carbon mineralization of Mg- and Ca-bearing silicate minerals has been studied and developed over the past three decades as an important carbon storage method with long-term stability because it produces chemically stable inorganic carbonates. The challenge has been the slow reaction kinetics associated with mineral dissolution and carbonation. While various methods (e.g., heat treatment, physical activation via in situ grinding, and the use of Ca- and Mg-targeting ligands) have been well studied and developed, additional challenges remain. Heat treatment and in situ grinding increase the overall energy requirement, and the use of ligands and acids adds significant costs and life cycle impacts associated with those chemicals. Because most of the carbon mineralization technologies involve slurry reactions, the water requirement for mineral dissolution and carbonation reactions is also a substantial challenge when these technologies are deployed at industrial scale.

If freshly mined silicate minerals are used for carbon mineralization, the mining, quarrying, processing, and transport costs also would be very high (Mazzotti 2005). Furthermore, the environmental impacts associated with large-scale mining and transportation of feedstocks and products would need to be addressed. Thus, it will be important to work with the mining industry to determine any potential impacts before carbon mineralization technologies can be deployed at scale.

5.2.1.3 R&D Opportunities

Mg- and Ca-bearing silicate minerals are earth-abundant and thus offer tremendous potential to help mitigate climate change by converting them to inorganic carbonates and thus durably storing carbon via carbonation. Because these processes for natural minerals and rocks will start from mining (carbonation of mine tailings is discussed in Section 5.2.2), R&D is needed to develop efficient mining processes and technologies. The use of renewable energy in mining processes also will improve the net CO_2 benefit of carbon mineralization.

Mineral dissolution and carbonate formation reactions typically require acid and base, as well as chemical additives such as ligands. Thus, to the extent a particular process makes use of them, the sustainable production of these chemicals also will play a key role in the process's CO_2 utilization potential. A number of technologies are being developed to produce "renewable" acids and bases. For example, an electrochemical bipolar membrane system can produce acids and bases (e.g., HCl and NaOH) via salt splitting using renewable electricity (Talabi et al. 2017). These "renewable" acids and bases could play an important role in decarbonizing the mining industry, but significant R&D is required before these electrochemical processes can be scaled up economically. The fermentation of biogenic wastes can also produce organic acids including acetic acid and citric acid. The effect of these acids on the mineral dissolution is relevant to properly evaluate the net CO_2 utilization potential of such processes.

While acids and bases are consumed during carbon mineralization, ligands may be recycled. Thus, developing efficient methods to recycle ligands and other chemical additives throughout the carbon mineralization processes will be important. Because carbon mineralization technologies consist of multiple reaction and separation steps, systems integration research to optimize the process is also of interest. A summary of RD&D needs for carbon mineralization is compiled in Section 5.3 and integrated with the RD&D needs for other carbon utilization pathways in Chapter 11.

5.2.2 Carbonation of Alkaline Industrial Wastes and Demolition Wastes

As discussed above, there is strong industrial interest in developing carbon mineralization technologies for alkaline industrial wastes such as fly ash and slag from iron and steel owing to multifaceted environmental and economic benefits, including solid waste management and decarbonization. Because these industries (e.g., power plants and chemical, cement, iron, and steel manufacturing plants) are also large CO_2 emitters, carbon mineralization processes can benefit from the co-location of CO_2 and alkaline industrial wastes, minimizing CO_2 compression and transportation costs. While alkaline industrial wastes are often more reactive than the natural Ca- and Mg-rich silicate minerals discussed in Section 5.2.1, they can be challenged by impurities in waste streams. Thus, different chemistries and reactor/separation systems are being developed for specific alkaline wastes. One of the emerging feedstocks for carbon mineralization is demolition wastes, which will start to play a more important role as aging infrastructure is replaced, while aiming to create a materials circularity in the built environment.

5.2.2.1 Current Technology

In addition to natural minerals, alkaline industrial by-products and wastes—including slags, fly ash, mine tailings, recycled aggregates, and reactive demolition wastes—have the capacity to form $CaCO_3$ and $MgCO_3$ when exposed to CO_2 (Hanifa et al. 2023; Supriya et al. 2023; Zajac et al. 2022). A significant amount of CO_2 (on the order of Gt per year) can be removed by combining it with alkaline industrial wastes (Pan et al. 2020; Rim et al. 2021). Moreover, this process is economically favored by converting two waste streams (i.e., CO_2 from point sources and alkaline industrial waste) to generate value-added products. Therefore, a number of start-ups and corporations are actively developing technologies using this approach, aiming to produce low-carbon construction materials and foster a circular economy.

5.2.2.1.1 Slags

Slags (e.g., steel, blast furnace, and basic oxygen furnace [BOF] slags) are by-products from steelmaking industries and are rich in alkaline metal oxides (CaO, Al_2O_3, SiO_2, MgO, Fe_2O_3), which can react with CO_2 to produce SCMs (Juenger et al. 2019; Pan et al. 2017). The reactivity of ironmaking and steelmaking slags (as well as ashes from combustion/incineration processes) permits a wide range of possible process routes and applications, including the generation of higher-value products and greater uptake of CO_2 as a carbon sink.

Ironmaking slags have been explored for their potential use in the construction industry. These tailings typically contain fine granulometry, high silica content, iron oxides, alumina, and other minerals, which make them suitable for various construction applications. Bodor et al. (2016) investigated the use of carbonated ironmaking slag (specifically BOF slag) as a partial replacement of natural aggregate in cement mortars, with key objectives to (1) stabilize free lime (CaO) in the slag, which causes detrimental swelling of the construction material; and (2) limit the mobility of heavy metals contained in the slag. To ensure the suitability of its intended use, BOF slag was crushed to suitable particle size (<0.5 mm), carbonated as a slurry in an aqueous solution of carbonic acid (to 10–16 wt% CO_2 uptake), and utilized to replace 50 percent of natural sand aggregate in cement mortars. The results showed satisfactory performance for all considered aspects (paste consistency, soundness, compressive strength, and leaching tendency) of the mortar sample containing 37.5 wt% carbonated BOF slag of <0.5 mm particle size (Bodor et al. 2016).

Salman et al. (2014) produced construction materials using exclusively steelmaking slag (specifically argon oxygen decarburization slag), which was carbonated after being mixed with water only. CO_2 uptake reached 4–8 wt% depending on the carbonation conditions and the compressive strength of the produced concrete containing slag surpassed 30 MPa. Leaching of heavy metals was within prescribed limits but the study highlighted the risk of metalloid leaching, as these elements are not captured by carbonate phases and rely on physical entrapment or another form of chemical sequestration to limit mobility (Salman et al. 2014).

5.2.2.1.2 Fly Ash

Coal fly ash is a by-product derived from coal-fired power plants and is the most common SCM used to react with CO_2. (See Chapter 9 for a discussion of fly ash availability and its direct use in pavement and concrete applications.) It contains CaO and SiO_2, which react with CO_2 to form $CaCO_3$ and silicate. Its fine particle size and high surface area result in superior reactivity compared to other untreated industrial wastes. The ultra-small particles of fly ash expedite carbonation, making them an efficient SCM. According to ASTM C618 standards, coal fly ash can be classified as two types based on its chemical and physical properties (particularly Ca content): Type C (>10 percent Ca) and Type F (<10 percent Ca) (ASTM International 2023b).

Replacing a portion of the OPC in concrete with fly ash reduces the life cycle CO_2 emissions of that concrete by sequestering the reacted CO_2 as carbonate, and the pozzolanic properties of fly ash enhance the concrete's strength and durability. Pozzolanic materials are siliceous, or siliceous and aluminous, materials that are not cementitious inherently, but fine particulates of pozzolans in water react with $Ca(OH)_2$ at ambient temperatures to form cementitious materials. More cementitious compounds form in concrete made using fly ash cement than OPC, ultimately making it harder and more durable (Nayak et al. 2022). Concrete and mortars incorporating fly ash exhibit comparable compressive strength to conventional composites after carbonation (Bui Viet et al. 2020). The replacement proportion of SCMs has to be controlled carefully to maintain strength because too much calcium-silicate-hydrate (C-S-H) can form with extraneous addition of SCMs, leading to a deleterious porous structure in the cement (Wu and Ye 2017). Other ashes (e.g., waste-to-energy plant ashes) also can be used for carbon mineralization technologies. Furthermore, air pollution control residues (e.g., the solid reaction products and residues from the SO_x scrubbers at power plants, which often contain unreacted $Ca(OH)_2$) can be carbonated with CO_2 to produce recycled aggregates; this technology has been commercialized (GEA n.d.; Hills et al. 2020).

5.2.2.1.3 Mine Tailings

Mine tailing waste continues to grow around the world, causing significant environmental impacts including water contamination. Thus, a technology that can utilize mine tailings would address multiple environmental problems. Araujo et al. (2022) view the production of construction materials as one of the main applications for recycling mine waste and a significant area of R&D in the field of mine waste management. Construction materials derived from mine waste offer several advantages, such as reducing the demand for natural resources by utilizing mine wastes generated for other industrial uses (e.g., metal recovery), minimizing environmental impacts, and providing a sustainable solution for waste utilization. In the construction industry, mine waste materials are utilized as additives in cement for manufacturing various products. One common application is the incorporation of mine waste, such as copper mine tailings, into concrete block manufacturing. The use of copper mine tailings in road- and highway-pavement concrete and brick production also has been explored. Most mine tailings can be used as filler materials for nonstructural concrete, an application for which their reactivity or Ca and Mg content is not critical. However, for SCM production, selecting mine tailings that contain significant amounts of Ca or Mg (>10 percent) is important. The processes of Ca and Mg extraction and carbonate formation would be similar to those developed for natural silicate minerals (discussed in Section 5.2.1).

Chakravarthy et al. (2020) explored the potential use of carbonated kimberlite[3] tailings, a waste product from diamond mining, as a partial substitute for cement in the production of concrete bricks. The utilized kimberlite was sourced from the De Beers Gahcho Kué mine in the Northwest Territories, Canada. The carbonated kimberlite tailings were produced through a thin-film carbonation process and then were used to cast bricks. The study investigated different carbonation conditions, including varying levels of CO_2 concentration, moisture content, and temperature. The results showed that carbonated kimberlite can be used as a partial replacement for cement in concrete bricks, with improvements in compressive strength observed. The study highlights the potential of utilizing mine tailings to sequester CO_2 and produce sustainable building materials with lower life cycle CO_2 emissions but recommends further research to optimize the carbonation process and to investigate the long-term durability of the carbonated kimberlite bricks.

The production of construction materials from mine waste offers a sustainable and resource-efficient approach to waste management in the mining industry. However, the adoption of these materials still faces challenges, such as transportation costs, considering that many mine tailings are stored or generated in remote locations. Additionally, the environmental and health implications of using mine waste in construction materials must be assessed to ensure that the materials meet regulatory standards for safety and performance.

5.2.2.1.4 Recycled Aggregates

The substitution of unconventional or recycled aggregates improves sustainability through CO_2 curing, achieving compressive strengths comparable to those attained through conventional curing methods (Yi et al. 2020). Aggregates are granular materials that are mixed with cement, water, and often other additives to produce concrete, providing strength and durability. They account for 60 percent to 80 percent of the volume and 70 percent to

[3] Kimberlites are high-pressure igneous rocks with a complex mixture of minerals, low in silica and high in magnesium. They are derived from the Earth's upper mantle, in which, under the right conditions, carbon may occur as diamond, a high-pressure form. Not all kimberlites are diamond-bearing.

85 percent of the weight of concrete. Typically, aggregates are made of sand, gravel, or crushed stones, mixing fine and coarse aggregates in different proportions.

Aggregates from demolished materials can be recycled to produce fresh materials. However, recycled aggregates have a more porous structure than fresh aggregates, which results in lower compressive strength (Tam et al. 2020). Carbonation reactions offer a solution for this weakness by generating $CaCO_3$ to fill these pores, achieved through either carbonating the recycled aggregates prior to concrete production or employing CO_2 curing during the concrete-making process (Tam et al. 2020). As discussed in Section 5.1.2, there are also technologies to produce recycled aggregates via carbon mineralization, reducing the quantity of mined aggregates needed for concrete and utilizing waste CO_2 from industrial processes.

5.2.2.1.5 Reactive Demolition Wastes

Demolished materials include reactive feedstocks that can capture and sequester CO_2 to produce fresh concrete (Li et al. 2022a, 2022b; Zajac et al. 2021). A process is required to separate the aggregates (inert portion) and cement to use these materials efficiently (Li et al. 2022a, 2022b). The current recycling practice of using demolished concrete typically is limited to recovering the steel rebar and coarse aggregates. However, the hydrated paste phase (cement part of the demolition waste) contains desirable chemical components for fillers and SCMs (i.e., Ca, Si, Al) and is the most expensive and carbon-intensive portion of concrete. The reactive demolition waste is currently an untapped alkaline waste source but can be upcycled via emerging CO_2 mineralization schemes such as direct and indirect wet carbonation. Direct carbonation would be more straightforward, but the final product is a mixture of carbonate and alumina-silica gel (Zajac et al. 2020a, 2020b), limiting its purity.

On the other hand, in indirect carbonation, where leaching and carbonation occur in two steps via a pH swing, pure products can be formed—$CaCO_3$ (to serve as filler) and silica-rich residue (to serve as an SCM) (Rim et al. 2021). In a study that applied a two-step leaching and carbonation method to hydrated cement paste (to simulate waste concrete), the derived carbonates demonstrated comparable performance to conventional limestone filler in terms of hydration kinetics and compressive strength development (Rim et al. 2021). Furthermore, this leaching and carbonation process allows for the formation of different $CaCO_3$ polymorphs in high purity—calcite, aragonite, and vaterite (Zhang and Moment 2023).

Owing to its needle-like morphology, aragonite was found to be an effective rheological modifier, specifically in enhancing the structural build-up behavior of cement pastes, which points to potential applications in 3D concrete printing (Zhao et al. 2023a, 2023b). The metastable polymorphs (i.e., aragonite and vaterite) stabilize in the cement-based system, thereby showing promise as functional fillers that may have benefits for other key concrete properties such as shrinkage and durability. Additionally, many of these derived carbonates likely would fall under the category of "limestone filler" and thereby adhere to current code, potentially accelerating the deployment of such carbonates. In addition to the carbonates, a silica-rich residue remains after the Ca is leached out, which can perform comparably to silica fume, a high-value SCM.

5.2.2.2 Challenges

Although the carbon mineralization potential of alkaline industrial wastes and demolition wastes is significant, the composition of these wastes is not consistent over the time, location, or processes in which they are collected. It is very difficult to develop carbon mineralization technology that can dynamically adjust processing based on the composition of incoming waste feedstock. Chemical additives (e.g., acids and ligands) need to be changed, or their concentration and type adjusted, depending on the compositions and mineralogy of waste incoming to the carbon mineralization process. For example, the presence of Fe in wastes can significantly reduce the purity of inorganic carbonate products that impact brightness, and thus, an additional separation step is needed prior to the carbonation reactor to produce highly pure $CaCO_3$ or $MgCO_3$ for paper filler applications.

5.2.2.3 R&D Opportunities

Of the CO_2 reincorporation approaches mentioned above, the one with the largest potential to reduce CO_2 emissions associated with the cement industry is the carbonation of concrete demolition waste. If waste building materials can be used directly to produce new construction materials onsite, its circular economy could be achieved without transporting large amounts of heavy materials. While there is great potential to significantly improve the overall sustainability of the construction industry via CO_2 utilization, there exists a wide range of R&D needs to develop new materials to replace carbon-intensive construction materials.

Carbonated industrial wastes can be used for clinker substitution, which involves replacing OPC clinker in concrete with SCMs. One of the most promising combinations of SCMs currently available on a global scale is calcined kaolin clay and ground limestone, where clinker substitution levels of 50 to 60 percent are being pursued (Scrivener et al. 2018). The main advantage of the clinker substitution approach is the avoidance of CO_2 emissions in the first place owing to the reduced amount of OPC clinker being used. The production of ground limestone and calcined clay has associated CO_2 emissions, but much reduced compared with OPC powder. As such, on a binder basis (OPC powder + water + SCMs), the limestone calcined clay cement has approximately 40 percent lower CO_2 emissions compared with neat OPC binder when using 60 percent clinker substitution. With these substitution levels, the amount of CaO available for reaction with injected/incorporated CO_2 will be minimal. Thus, the carbonated alkaline industrial wastes discussed in the present section should be investigated as a new class of SCMs and the overall process should be developed while maximizing net carbon storage and minimizing waste generation and mining of fresh mineral and rocks.

The ability of a material to react with CO_2 and form stable carbonates depends on the amount of alkaline earths (CaO and MgO) available for formation of $CaCO_3$ and $MgCO_3$-type phases. As such, in addition to carbonate cement based on carbonation of pseudo-wollastonite/rankinite, an emerging R&D area is the formation of carbonates from highly alkaline industrial by-products such as steel slags (Beerling et al. 2020) (rich in nonhydraulic calcium silicate phases) and underutilized coal ashes. The availability of such industrial by-products tends to be (1) already fully utilized in concrete production as SCMs, as is the case of blast furnace slag and good quality coal fly ashes (Habert et al. 2020), or (2) somewhat limited in availability, as is the case of steel slags (190 to 280 Mt/yr globally[4] [Tuck 2022]) and coal ashes in the future as the amount of legacy coal ashes decreases. However, the economic growth of different countries, particularly developing countries, and the replacement of aging infrastructures in the United States could significantly increase the production of iron and steel and lead to a continuous supply of alkaline industrial wastes. One area that requires more research for some of these industrial by-products (e.g., steel slag) is understanding the effects of carbonation on the leachability of trace elements, including heavy metals from these materials to ensure the safety of the produced construction materials.

As discussed above, one of the largest challenges of waste carbon mineralization is feedstock variability and complexity. Thus, a technology that can rapidly identify compositions and mineralogy of the feedstock would be extremely valuable. With the rapid advancement in artificial intelligence and machine learning, as well as in-line (e.g., infrared) sensor systems, waste sorting and characterization techniques can be developed and integrated into the carbon mineralization process to provide operational stability at scale.

Carbon-intensive construction materials will have to be increasingly manufactured, used, and upcycled via a circular carbon economy to minimize the use of natural resources. As illustrated in Figure 5-5, CO_2 can be reincorporated back into construction material at the end of its service life and the overall process can be electrified using renewable energy (e.g., a sustainable electrodialysis strategy to create renewable acids and bases). In concrete demolition waste, reactive CaO is available in the form of C-S-H gel and portlandite ($Ca(OH)_2$) and can react readily with dissolved CO_2 in the pore solution[5] to form $CaCO_3$. The opportunity for this pathway is significant, as 7 billion metric tons of concrete demolition waste are produced each year (Krausmann et al. 2017), with the majority being directly disposed of with minimal uptake of CO_2. The challenge of this upcycling process includes crushing and grinding the demolition wastes to expose the available CaO. Thus, research is needed to develop integrated physical and chemical separation technologies to minimize the energy requirements for carbon mineralization and waste upcycling. Focusing on low-energy separation pathways, such as membranes, would reduce the life cycle CO_2 emissions of such processes. Design of bipolar membranes for electrodialysis that are

---

[4] As a point of comparison, worldwide production of cement in 2022 was 3.7 billion metric tons, with 96 million metric tons produced in the United States alone.

[5] Pore solution refers to the alkaline solution present in the pores of hardened concrete. The composition of the pore solution changes over the course of the concrete's useful life and plays an important role in concrete durability. Pore solution composition influences the potential for steel corrosion, concrete spalling, and other degradations of reinforced concrete products (Diamond 2007).

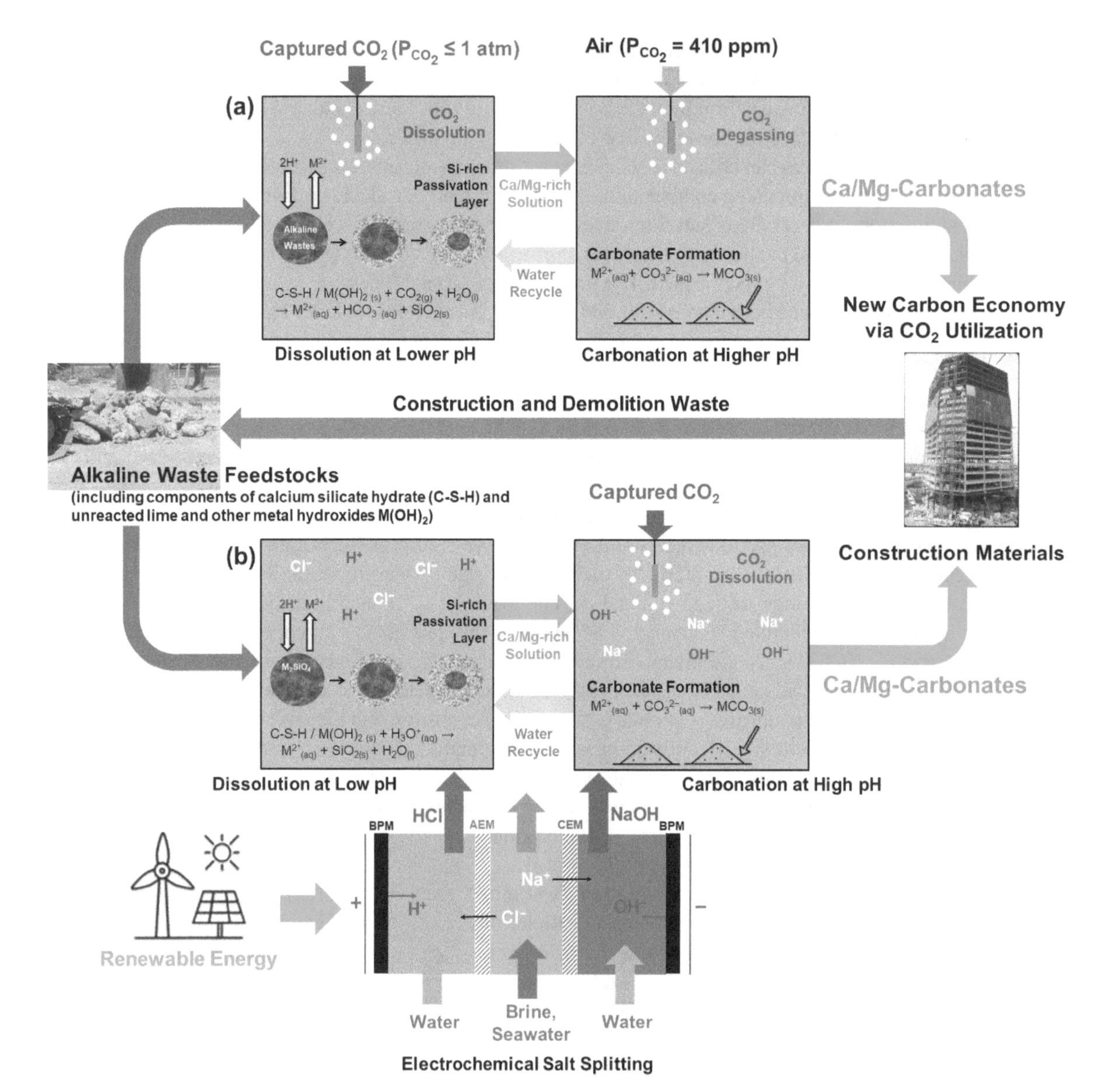

FIGURE 5-5 Carbon mineralization of alkaline industrial waste feedstocks (e.g., construction and demolition waste, mine tailings, fly ash, iron and steel slag, and red mud) and silicate minerals (a) via P_{CO_2} swing and (b) via pH swing using acid and base generated from electrolysis with a bipolar membrane system. Equations are generalized and not balanced.
NOTES: AEM = anion exchange membrane; BPM = bipolar membrane; CEM = cation exchange membrane; M = Ca or Mg. All take place in an aqueous phase between 30 and 90°C.
SOURCE: Icons from the Noun Project, https://thenounproject.com. CC BY 3.0.

stable under a wide variety of operating conditions and have high conductivity, fast water dissociation kinetics, low ion crossover, and long lifetime should be a priority.

Furthermore, to use produced carbonates and other solid products as construction materials for a wide range of applications including structural concrete, they have to be carefully tested and certified to ensure their performance. Also, new syntheses and formulation methods have to be developed for different concrete manufacturing processes (e.g., CO_2 curing, three-dimensional [3D] printing). These new processes may provide applications for waste streams that previously did not have a use case; for example, work on 3D printable concrete materials has found that the fine, powdery condition of mineralized wastes is advantageous, in contrast to traditional concrete production requiring large aggregates.

Last, a better database (based on industrial data) for accurate LCA of produced carbonate products and their different uses would be beneficial. The definition of permanence of CO_2 storage in buildings and infrastructure is still being debated, so further discussions are needed to estimate the carbon storage potential of carbonates used in infrastructure to provide appropriate carbon credits. A summary of RD&D needs for carbon mineralization is compiled in Section 5.3 and integrated with the RD&D needs for other carbon utilization pathways in Chapter 11.

5.2.3 Enhanced Carbon Uptake by Construction Materials

The previous two sections (5.2.1 and 5.2.2) described how different feedstocks (natural minerals, alkaline industrial wastes, and demolition wastes) can be converted to solid carbonates and by-products via carbon mineralization and how they can be used as value-added products. This section describes other technologies that can directly enhance carbon uptake by construction materials—CO_2 injection and CO_2 curing. It discusses how these processes work and the fate of injected CO_2.

5.2.3.1 Current Technology

5.2.3.1.1 CO_2 Injection

Injection of a small amount of CO_2 as gas or solid during mixing of OPC concrete has been shown to increase short- and long-term strength (Cannon et al. 2021). This is thought to be owing to the immediate formation of $CaCO_3$ (amorphous or nanocrystalline) that then provides additional nucleation sites to accelerate precipitation of the main strength-giving phase in OPC concrete, C-S-H gel (Monkman et al. 2018). However, there is an upper limit to the amount of CO_2 that can be added during mixing, beyond which added CO_2 is found to be reduce compressive strength (Monkman and McDonald 2017; Ravikumar et al. 2021; Shaqraa 2024).

5.2.3.1.2 CO_2 Curing

CO_2 curing refers to the process used in the production of concrete and cementitious materials to accelerate the process of hardening materials in the presence of gaseous CO_2. The conventional curing process uses water to hydrate and solidify the materials. During the CO_2 curing process, a carbonation reaction occurs between $Ca(OH)_2$ or calcium silicates and CO_2, forming $CaCO_3$ in the concrete matrix as shown in Reactions 5.9, 5.10, and 5.11.

$$Ca(OH)_2 + CO_2 \rightarrow CaCO_3 + H_2O \quad (R5.9)$$
$$3CaO \cdot SiO_2 + 3CO_2 + yH_2O \rightarrow SiO_2 \cdot yH_2O + 3CaCO_3 \quad (R5.10)$$
$$2CaO \cdot SiO_2 + 2CO_2 + yH_2O \rightarrow SiO_2 \cdot yH_2O + 2CaCO_3 \quad (R5.11)$$

This process can reduce water consumption significantly in concrete production and sequester CO_2 within the concrete (Monkman and MacDonald 2017; Ravikumar et al. 2021). It also increases the compressive strength of concrete owing to an optimal microstructure formation (Wang et al. 2022a).

The reaction mechanisms of CO_2 curing include diffusion-controlled reactions between the hydrated reactants (e.g., hydrated $Ca(OH)_2$ and CO_2 not in slurry or solution form) in capillary channels, and a wet route involving CO_2 dissolution in the aqueous phase (Wang et al. 2022b; Yi et al. 2020). CO_2 curing is influenced by multiple factors, such as CO_2 concentration, relative humidity, temperature, and intrinsic composition of materials (von Greve-Dierfeld et al. 2020). For example, the concentration of CO_2 can affect the carbonation rates and polymorph of produced $CaCO_3$ (e.g., calcite, aragonite, and vaterite), which may be different from natural limestone (von Greve-Dierfeld et al. 2020). Elevated CO_2 partial pressure can accelerate the hardening process and lead to enhanced mechanical properties of concrete cured within a given time (Yi et al. 2020; Zhan et al. 2016).

Carbonation curing is another avenue being pursued by the cement industry as a means of reducing CO_2 emissions. Carbonation curing involves exposure of OPC concrete to a CO_2-rich environment shortly after it has been poured for a duration of a couple of hours to a few days (Ravikumar et al. 2021). A number of factors influence the amount of CO_2 that can be incorporated in concrete using this approach, primarily its porosity, permeability, and degree of water saturation (Winnefeld et al. 2022). The uptake of CO_2 is associated with available CaO, which in general is attributed to the calcium originating from OPC powder and thus the decomposition of limestone ($CaCO_3$). For OPC concrete reinforced with steel, there will be concerns regarding the impact of CO_2 curing in reducing the internal pore solution alkalinity in the vicinity of the embedded steel. The pore solution pH of OPC concrete is found to be between approximately 13 and 14, which protects the steel and prevents its corrosion. However, CO_2 acidifies this pore solution, and steel will begin to corrode below a pH of ~11.

5.2.3.2 Challenges

Unlike the carbon mineralization processes described in Sections 5.2.1 and 5.2.2, CO_2 injection and curing technologies require a relatively high concentration of CO_2 (e.g., >90 percent) to achieve a sufficient carbonation rate. Because CO_2 is introduced to already prepared and mixed construction materials, including cement and SCMs, it has to be free of impurities like SO_2 and NO_x. The reliance on a stable, high-purity CO_2 supplier might interrupt manufacturing if CO_2 availability is insufficient. Although CO_2 injection and curing technologies are relatively easy to implement because they do not require complex systems to scale up, the total amount of CO_2 utilized in concrete is less than that of other carbon mineralization processes (e.g., the incorporation of solid carbonates as fillers or aggregates). Thus, technologies to further increase CO_2 uptake by these ready-mix construction materials need to be developed. Also, the pH profile created by the even mixture of carbonates and unreacted cement and SCMs in concrete may lead to faster corrosion of steel inside concrete.

5.2.3.3 R&D Opportunities

With an easier scale-up process, CO_2 injection and curing technologies have already been demonstrated and commercialized for a few conventional concrete industries. Because pure CO_2 (i.e., dry ice or gaseous CO_2) is used for these technologies, systems integration and process intensification with carbon capture processes from various sources are desired. It has been reported that CO_2 curing shortens the time required to harden concrete and significantly reduces the overall energy requirement. Further development of these processes should be carried out via pilot-scale demonstrations under a wide range of reaction conditions (e.g., temperature, CO_2 concentration and pressure, curing time, and concrete mix compositions) to determine the optimized process parameters and to confirm that the performance of produced concrete meets code.

As discussed earlier, the pH near steel bars inside the concrete needs to remain high to prevent their corrosion. There are concerns about depassivation of steel rebar in reinforced concrete owing to natural carbonation, where atmospheric CO_2 reacts with cement hydration products ($Ca(OH)_2$ and C-S-H) and reduces the overall pH during its service life (Stefanoni et al. 2018). Although CO_2 curing is different from weathering carbonation, as it occurs at very early ages and within a short amount of time, durability concerns still remain for reinforced concrete produced via CO_2 injection and curing. Studies on this aspect are few (Zhang and Shao 2016), as CO_2 curing is still mostly limited to nonstructural, unreinforced elements, so more fundamental studies and pilot-scale demonstrations are needed to fill this knowledge gap regarding durability.

Fast diffusion of CO_2 into the bulk concrete material is needed to maximize CO_2 intake and solid carbonate formation. However, the depth of carbonation can be limited under accelerated CO_2 curing conditions, where progressive carbonation from the exposed surface will lead to a denser microstructure and decrease subsequent CO_2 transport into the concrete. Curing methods and conditions, as well as mix design, can impact CO_2 diffusion behavior (Zhang et al. 2017), and thus, more systematic investigations are needed to scale up this technology.

While the altered pH profile within cured concrete may pose a problem with steel corrosion, this problem may be addressed by replacing steel with alternative reinforcement materials (e.g., glass fiber reinforced plastic rebar, engineered bamboo, and plastic fiber). Recently, carbon fiber and carbon nanotubes are also being tested as alternative reinforcement materials. Because these solid carbon materials can be produced via CO_2 conversions described in Chapter 6, their incorporation into the construction materials could further increase the CO_2 utilization and storage potential of the built environment.

R&D for new manufacturing technologies, including 3D printing that does not employ steel reinforcement, needs to be conducted for producing new construction materials via CO_2 injection and curing methods. Emerging 3D printing technologies for cement-based materials eliminate the need for formwork, which can reduce labor and improve construction efficiency (Paul et al. 2018). However, to overcome the absence of formwork, the material must exhibit precise flow and solidification behavior so that it can flow during pumping and deposition but rapidly gain structure immediately after deposition to achieve shape stability (Marchon et al. 2018). Furthermore, the printed material must continue to harden rapidly to support subsequent layers and avoid collective buckling of the assembled system, as there is not formwork to protect the materials during early curing. Because concrete mixes for 3D printing have higher proportions of cement than traditional poured concrete and CMUs, products can be vulnerable to evaporation and subsequent shrinkage-induced cracking (Moelich et al. 2022). All of these challenges must be addressed to develop 3D printing of cement-based systems. A summary of RD&D needs for carbon mineralization is compiled in Section 5.3 and integrated with the RD&D needs for other carbon utilization pathways in Chapter 11.

5.2.4 Electrolytic Brine and Seawater Mineralization and Biological Enhancement

As discussed above, natural minerals, rocks, and wastes derived from those mineral resources are good sources of Ca and Mg. Other unconventional resources also have been considered for carbon mineralization, one of which is an ocean-based approach. As shown in Figure 5-6, the ocean is one of the largest sinks for CO_2, as it contains

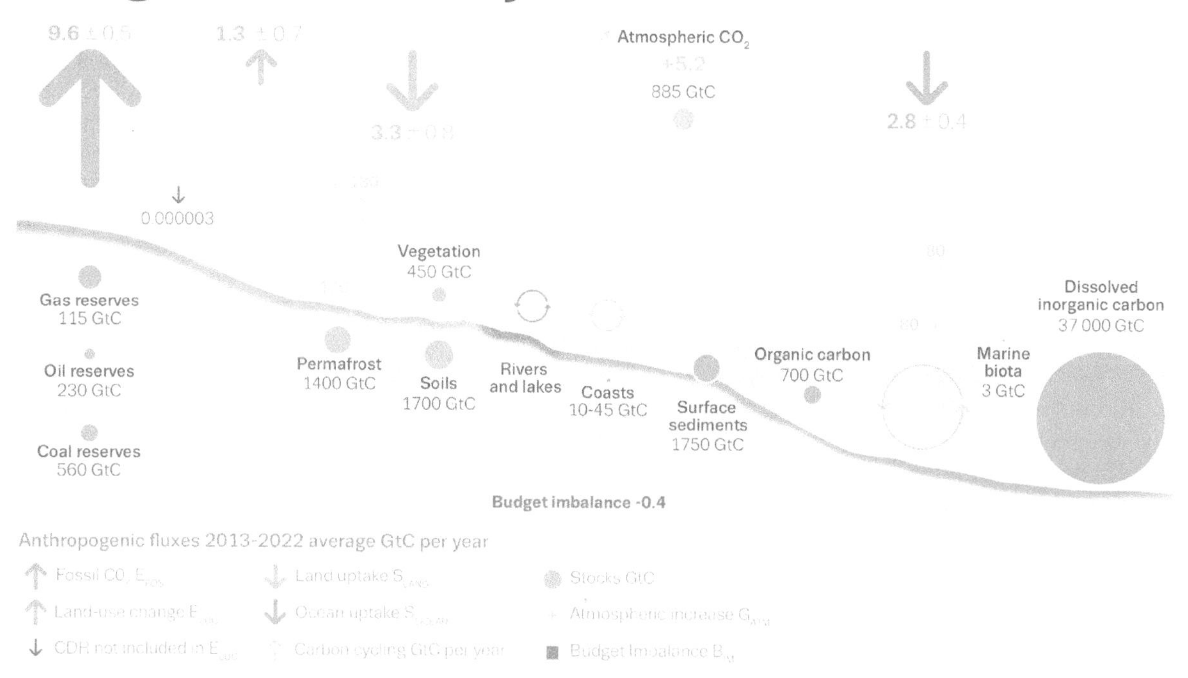

FIGURE 5-6 Distribution of carbon on earth, showing that the majority of earth's carbon is in the form of dissolved inorganic carbon. This resource is in equilibrium with CO_2 in the atmosphere, and so consumption of ocean CO_2, such as through mineralization, can form inorganic carbonate products that durably store carbon (data used from the Global Carbon Budget 2020).
NOTE: B_{IM} = an estimated imbalance between the estimated emissions and the estimated changes in the atmosphere, land, and ocean; E_{FOS} = emissions from fossil fuel combustion and oxidation from all energy and industrial processes, including cement production and carbonation; E_{LUC} = emissions resulting from deliberate human activities on land, including those leading to land-use change; G_{ATM} = growth rate of atmospheric CO_2 concentration; GtC = gigatonnes of carbon; S_{LAND} and S_{OCEAN} = the uptake of CO_2 on land and by the ocean, respectively.
SOURCE: Friedlingstein et al. (2022), https://doi.org/10.5194/essd-14-4811-2022. CC BY 4.0.

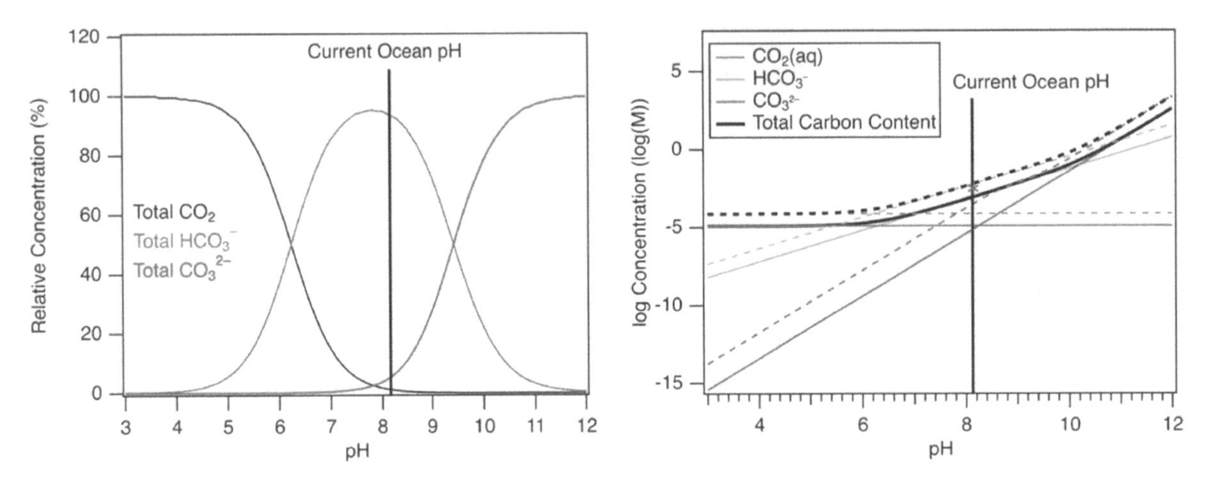

FIGURE 5-7 (Left) Bjerrum plot showing the speciation of dissolved inorganic carbon in a background of seawater at fixed total carbon concentration at 25°C; (Right) CO_2 speciation versus pH for pure water (solid lines) and seawater (dotted lines) at fixed partial pressure in an open CO_2 system at 25°C. Calculations were performed with ChemEQL—A Software for the Calculation of Chemical Equilibria. The asterisk denotes the total carbon content of water as calculated in Riebesell et al. 2010. SOURCE: Vibbert and Park (2022), https://doi.org/10.3389/fenrg.2022.999307. CC BY 4.0.

enormous amounts of alkaline metals (concentrations of International Association for the Physical Sciences of the Oceans standard seawater: 1300 ppm Mg and 400 ppm Ca). As CO_2 is dissolved into seawater, it is stored mostly in the form of bicarbonate (HCO_3^-) and carbonate (CO_3^{2-}), the speciation of which is a strong function of pH (Figure 5-7). In recent years, researchers have started to engineer ocean chemistry to capture CO_2 (direct ocean capture [DOC]) or even produce inorganic carbonates as products. In this section, electrolytic seawater mineralization is discussed as a new route to form products via carbon mineralization. A 2022 National Academies' study, *A Research Strategy for Ocean-Based Carbon Dioxide Removal and Sequestration* (NASEM 2022), explores a broad range of potential ocean-based carbon dioxide removal strategies, including and beyond those described here.

5.2.4.1 Current Technology

5.2.4.1.1 Direct Ocean Capture Mineralization

Electrolysis of seawater can produce locally high concentrations of NaOH, along with H_2 and Cl_2 (the latter is produced preferentially instead of oxygen at most anodes). The locally high concentration of electrolytically produced hydroxide shifts the bicarbonate-carbonate equilibrium near the cathode to favor carbonate as in Figure 5-8, thereby enabling $CaCO_3$ to form; $MgCO_3$ is kinetically hindered from forming and requires further processing (La Plante et al. 2021, 2023). An alkaline mineral hydroxide (e.g., $Mg(OH)_2$) formed in the process can further equilibrate with CO_2 to form additional carbonates. This "continuous electrolytic pH pump" can precipitate $CaCO_3$, $Mg(OH)_2$, and hydrated magnesium (bi)carbonates depending on the pH conditions (La Plante et al. 2023). An electrolytic flow reactor and integrated rotary drum filter process was developed (La Plante et al. 2021).

The process is an example of integrated carbon capture and conversion, as the process uses DOC and does not require separation or purification of CO_2 prior to forming the final product. A large deployment of the electrochemical DOC would require the use of the by-products (e.g., H_2 and Cl_2/HCl) in order to maximize the energy and atomic efficiencies of renewable energy utilization. The production of large amounts of multiple products allows the development of novel technologies, for example, the co-production of H_2, O_2, NaOH, and HCl from brine via electrolysis. There is a large market for green H_2 and H_2 can also be used along with captured CO_2 to produce hydrogenated products. NaOH and HCl can be used for pH swing carbon mineralization technologies to co-recover energy-relevant critical metals, silica, and calcium and magnesium hydroxide from alkaline residues to react with CO_2 to produce the respective carbonates. Advances in electrochemical processes including direct electrolysis of brine with or without the use of bipolar membranes (Kumar et al. 2019; Tian et al. 2022) are needed

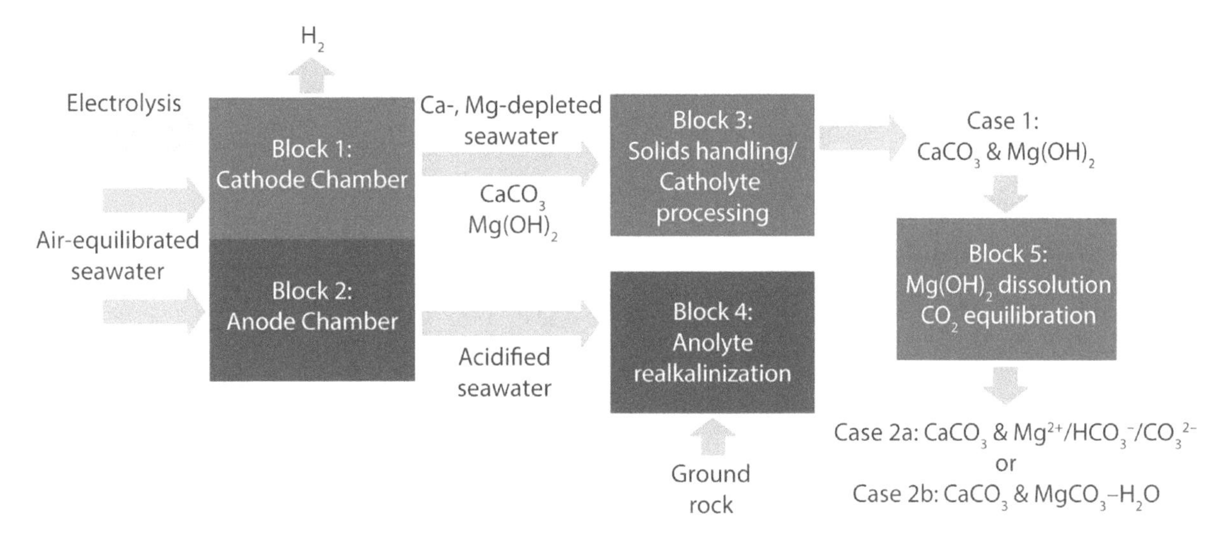

FIGURE 5-8 A schematic of the process showing major inlet and outlet feeds of the primary steps for CO_2 removal associated with the formation of: carbonate solids and (aqueous) dissolved CO_2 (Cases 1, 2a) and carbonate solids only (Case 2b). The major energy inputs include electricity for electrolysis, water processing and pumping, and rock grinding.
SOURCE: La Plante et al. (2023), https://doi.org/10.1021/acsestengg.3c00004. CC BY 4.0.

to enable scalable deployment of renewable acid and base technologies. The produced acid can also be used to dissolve silicate minerals, neutralizing the acid and producing metal ions (e.g., Ca^{2+} and Mg^{2+}) that can be returned to the ocean to replace those used in the mineralization process. In most cases, DOC mineralization processes could be applied to either CO_2 utilization or CDR, the former if produced solids are collected for use as aggregates or other products, and the latter if they are returned to the ocean or spread on land.

Glasser et al. (2016) showed that lightweight nequehonite-based ($MgCO_3 \cdot 3H_2O$) cement can be produced with brines containing 30 wt% CO_2. $Mg(OH)_2$ generated from Mg^{2+} in seawater with membraneless electrolyzers can be used as a precursor in Mg-based cement, with a comparable compressive strength (i.e., 20 MPa) for a 2-day CO_2 curing (Badjatya et al. 2022). Figure 5-9 shows how a membraneless electrolyzer works to split seawater into acidic and alkaline streams, or renewable acids and bases. The desalinated brines (concentrated saltwater rich in

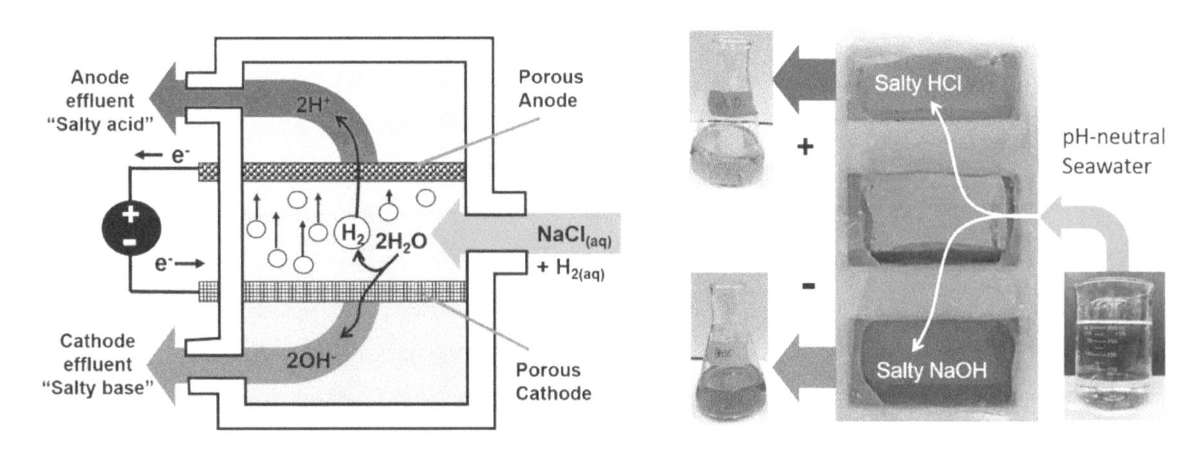

FIGURE 5-9 (Left) Schematic of a membraneless electrolyzer used to split seawater into acidic and alkaline streams and (Right) photograph of a membraneless electrolyzer during steady-state electrolysis of natural seawater in the presence of a pH indicator dye, which turns purple or red in alkaline or acidic environments, respectively.
SOURCE: Badjatya et al. (2022), https://doi.org/10.1073/pnas.2114680119. CC BY-NC-ND 4.0.

Mg/Ca) also have the potential to be used for metal carbonate production through aqueous mineralization (Glasser et al. 2016; Zhang et al. 2019, 2020b). The use of a membraneless process to produce acid and base from brines using angled mesh flow-through electrodes (Talabi et al. 2017) may reduce the precipitation of undesired solid phases (e.g., $Mg(OH)_2$) on the electrode surface.

5.2.4.1.2 Biologically Inspired Technologies Using Carbonic Anhydrase

Carbonic anhydrases are a metalloenzyme family found in all mammals, plants, algae, fungi, and bacteria, that catalyzes CO_2 hydration and dehydration (Elleuche and Poggeler 2009). Carbonic anhydrases play a key role in the ocean's carbon balance by accelerating rate-limiting steps of CO_2 uptake by the ocean. They also are involved in CO_2 homeostasis, biosynthetic reactions, lipogenesis, ureagenesis, and calcification, among other processes relevant to life in the ocean (Supuran 2016). Carbonic anhydrases are thought to mediate the hydration of CO_2 through the mechanism proposed in Figure 5-10, shown with a Zn^{2+} metal center. By adding carbonic anhydrase into a carbonation reactor, the formation rate of solid carbonates can be accelerated (Patel et al. 2014a, 2014b). The use of pure enzymatic catalyst would be not economical owing to its costly purification steps, and thus, whole cell biocatalyst (e.g., surface display of small peptides on E-coli) has been developed to deploy carbonic anhydrase for carbon mineralization (Patel et al. 2014a, 2014b). This technology has been demonstrated at laboratory scale (Fu et al. 2018). There are few start-ups and industrial demonstrations that utilize carbonic anhydrase for carbon capture processes. Similar technologies can be used to accelerate carbon mineralization processes by improving CO_2 hydration rates.

5.2.4.2 Challenges

Because ocean-based carbon mineralization technologies would intake seawater and discharge seawater after the carbonation reaction, it is critical to investigate its potential environmental and ecological impacts. Seawater

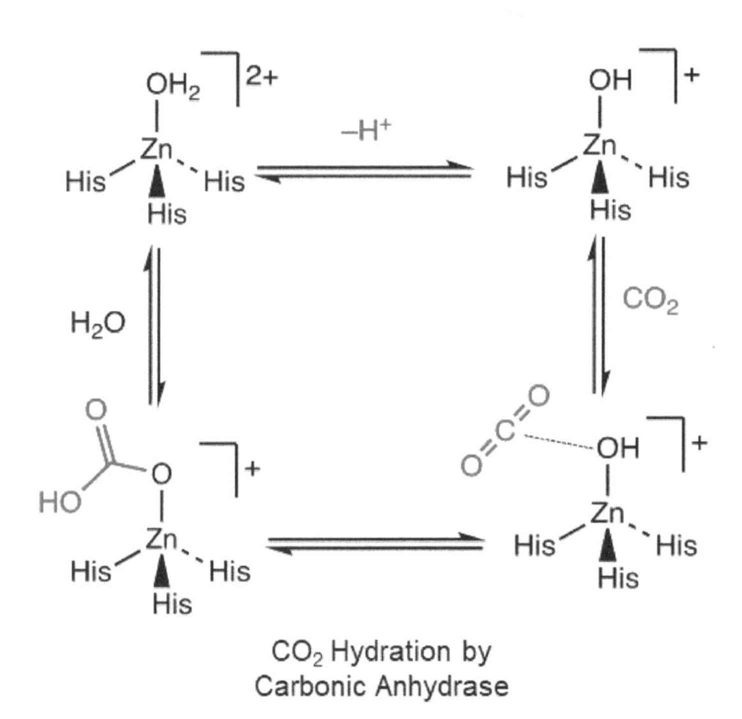

FIGURE 5-10 Illustration of the Zn-based carbonic anhydrase active site, showing how it can be used to accelerate the production of solid carbonates and the proposed mechanism for CO_2 hydration.
NOTE: His = histidine.
SOURCE: Modified from Vibbert and Park (2022), https://doi.org/10.3389/fenrg.2022.999307. CC BY 4.0.

contains a wide range of ionic species, and they may precipitate out via undesired side reactions and foul the membrane and reactor systems. Monitoring, reporting, and verification (MRV) will play a crucial role in ocean-based carbon capture and conversion, because the amount of CO_2 utilized and durably stored is more difficult to measure and monitor. MRV provides a transparent and accurate assessment of the amount of carbon being captured and stored; ensures that the technology meets the requirements of international climate agreements; and contributes to the R&D of more effective technologies (Ho et al. 2023). MRV requirements may increase the cost of ocean-based carbon mineralization technologies and their product costs.

Another challenge with ocean-based carbon capture and conversion is the local depletion of alkalinity in the ocean. This effect could impact the local ecosystem, and further, would impact the ocean's ability to store CO_2 unless alkalinity is replenished. This ocean-based technology can be integrated with the dissolution of natural silicate minerals or alkaline industrial wastes to continue the supply of alkalinity into the ocean while producing solid carbonate products.

Further improvements are needed to electrolytically supplied, local alkalinity driven metal carbonate formation—for example, by incorporating not only biocatalysts such as illustrated in Figure 5-10 but other co-catalysts that can overcome the kinetic inhibition to $MgCO_3$ formation. Improved rates of $MgCO_3$ formation could make better use of the more abundant alkaline earth cation (Mg) in the ocean, and simplify the DOC process by removing additional processing units dedicated to Mg (e.g., block 5 in Figure 5-8). Computational modeling has provided numerous insights into carbon dioxide mineralization, largely based on classical simulations (see recent reviews by Sun et al. 2023 and Abdolhosseini Qomi et al. 2022). Recently, high-level quantum-based dynamics simulations investigating carbon dioxide dissolution and reaction (Martirez and Carter 2023), as well as fundamental differences in free energetics and pH dependencies for Ca and Mg dehydration and carbonate formation (Boyn and Carter 2023a, 2023b), have begun to appear. Such modeling that reveals mechanisms and key influences on reactions, along with machine learning approaches to get to longer timescales and length scales, could help improve processes and catalyst design.

The availability of affordable renewable energy is also a critical requirement for electrolytic brine and seawater mineralization processes. Offshore wind energy could be a great option to integrate and other renewable energy systems (e.g., wave energy) should be considered, depending on the scale and deployment schemes of the developed ocean-based carbon mineralization processes. Offshore applications would require significant automation to minimize the maintenance issues, and the intermittency of renewable energy should be addressed during the process design and optimization.

5.2.4.3 R&D Opportunities

Electrolytic brine and seawater mineralization is one emerging research area for CO_2 capture and utilization with great potential. This technology could be improved with the development of more efficient and robust electrochemical systems, as well as a better understanding of the precipitation of undesired solid phases at the membrane surface. Discovery of new catalysts or processes that help overcome the kinetic inhibition observed for Mg carbonate formation are needed. Computational modeling to understand the fundamental processes of dissolution, speciation, hydration-dehydration, ion-pairing and nucleation and growth, with and without catalysts, could help catalyst or process design, in tandem with experiments. The solubilities of ionic species are a strong function of pH. Thus, it is important to accurately measure the pH of the reaction environment. While the measurement of the bulk pH is relatively easy, localized pH measurements near electrode surfaces are challenging. New measurement techniques should be developed to probe the local reaction conditions (e.g., at the electrode-electrolyte interface or at the membrane-electrolyte interface). These dynamic local measurements would provide insights into any undesired reactions and potential fouling issues at the membrane.

The electrolytic seawater mineralization technologies can generate a large amount of acid as a by-product. To manage the co-produced acid stream and provide a beneficial use, it can be reacted with silicate minerals. Silicate minerals are basic materials and thus neutralize the acid and form alkali metal ions (e.g., Mg^{2+}). The Mg and Ca ions can be either used to replenish those removed from the ocean, or carbonated to produce additional carbonate products as described earlier in Section 5.2.1.

Greater systems integration and intensification should be developed to improve the overall sustainability of electrolytic seawater mineralization technologies while utilizing available renewable energy. Process design and optimization should aim to align throughput rates across the separate reaction steps for maximum energy efficiency. MRV technologies and LCA/TEA frameworks should be developed considering unique environmental and ecological challenges in the ocean system. A summary of RD&D needs for carbon mineralization is compiled in section 5.3, and synthesized with the RD&D needs for other carbon utilization pathways in Chapter 11.

5.2.5 Alternative Cementitious Materials and Alternative Mineralization Pathways

As discussed in Section 5.1.1, conventional cementitious materials are calcium-bearing because Ca-based cementitious materials are known to provide great mechanical strength. Even SCMs are produced by extracting Ca from different feedstocks such as fly ash. As demand for cement and concrete rapidly increases, alternative cementitious materials not derived from carbonate minerals such as limestone will be needed to reduce CO_2 emissions. In particular, Mg-based cementitious materials are emerging as alternatives, and efforts are ongoing to improve their performance as construction materials. Mg is abundant in both earth (e.g., silicate minerals such as olivine and serpentine) and ocean systems (i.e., 1300 ppm Mg^{2+}). This section introduces several examples of innovative cement construction materials.

5.2.5.1 Current Technology

Beyond OPC-based concrete, a range of alternative cement technologies are being pursued with varying degrees of CO_2 emission reduction. These include alkali-activated cement, carbonate cement, super-sulfated cement, and sulfoaluminate cement, covering a broad range of chemistries and reaction mechanisms, some of which actively reincorporate CO_2 during production. For example, carbonate cement involves the carbonation of nonhydraulic calcium silicate minerals (e.g., pseudo-wollastonite and rankinite) where exposure of the calcium silicate to a humid CO_2 environment leads to dissolution of the calcium silicate mineral and precipitation of calcium carbonate as the binder along with silica gel. Carbonate cement is incompatible with steel reinforcement owing to the CO_2 environment necessary for the formation of calcium carbonate, specifically lower pH conditions compared with OPC concrete, leading to potential corrosion issues as discussed earlier. The nonhydraulic calcium silicates are typically manufactured using existing cement kiln infrastructure. Thus, as is the case with OPC clinker, the nonhydraulic calcium silicates are formed via the decomposition of limestone and calcination at higher temperatures to obtain the desired phases (~1200°C). Other alternative cements, such as alkali-activated cements, have inherently lower CO_2 emissions compared with OPC binder owing to utilization of alternative chemistries with reduced CO_2 emissions associated with the final product (Alventosa et al. 2021).

Recent studies have shown that Mg-based cements have the potential to exhibit comparable or superior properties to OPC, with reduced carbon emissions (Bernard et al. 2023; Walling and Provis 2016). For CO_2 mineralization, reactive MgO is of particular interest as it hardens through a carbonation reaction to form a solid carbonate binder. Alternatively, ferrous oxalate cement containing iron oxalate hydrate ($FeC2O_4 \cdot 2H_2O$) can be produced via reactions between iron-rich copper slag and oxalic acid (Luo et al. 2021). This new type of acid-base cement is interesting in terms of CO_2 utilization and storage potential in construction materials because the carbon:metal ratio can be doubled (e.g., limestone $CaCO_3$ has a C:Ca ratio of 1 whereas $FeC2O_4$ has a C:Fe ratio of 2). Thus, with ferrous oxalate cement, there is a potential to store more CO_2 in the built environment. Furthermore, the reaction between oxalic acid and dissolved alkaline metals is known to be fast (Luo et al. 2021).

5.2.5.2 Challenges

While Mg-based cement is promising, it is typically sourced from magnesite ($MgCO_3$), which introduces challenges such as availability (it is limited to certain geographic regions) and emissions associated with processing (calcination of magnesite for MgO is analogous to that of limestone for OPC, releasing CO_2 in its formation). Furthermore, MgO cements need to be carbonation cured, and CO_2 diffusion can be limited in these dense microstructures.

The performance of Mg-based cements also should be carefully evaluated for various applications (e.g., structural concrete versus road covering) because their physical and chemical property data are mostly unavailable. Sufficient data would need to be collected to certify them for commercial applications.

The new oxalate-based cements potentially could utilize more CO_2 in building materials than conventional cements. However, scaling the amount of oxalic acid required could be difficult. Moreover, this oxalic acid should be derived from captured CO_2. Both technology readiness and the availability of renewable energy inputs would limit the deployment of Fe oxalate and Mg cements.

5.2.5.3 R&D Opportunities

Potentially scalable, low-carbon, and economical alternative sources of Mg include emerging chemical/electrochemical schemes that can source Mg from non-carbon-containing feedstocks, like Mg-bearing silicates and seawater/brine (Badjatya et al. 2022; Bernard et al. 2023). As discussed in previous sections, Mg is abundant in both silicate minerals and oceans. Thus, MgO cement technology should be evaluated for Mg feedstocks derived from those carbon-free feedstocks.

CO_2 diffusion, which correlates with mechanical strength, remains a challenge for MgO carbonated cements. Additive manufacturing schemes can help enhance CO_2 diffusion through control of the material architecture, where carbonated MgO cement pastes have been demonstrated to achieve higher compressive strength via 3D printing compared to conventional mold casting (Khalil et al. 2020). More research on alternative cement systems and supporting manufacturing schemes will be warranted as new feedstocks become available.

The carbonation behavior of $MgO/Mg(OH)_2$ is different from that of $CaO/Ca(OH)_2$. Without elevated temperature / CO_2 pressure conditions, only hydrated magnesium carbonate phases form, which are metastable and thus lead to durability concerns (Bernard et al. 2023). R&D efforts therefore should focus on understanding phase and polymorph changes of MgO and hydrated magnesium carbonates under various carbonation and hydration conditions. The carbonation extent of MgO and $Mg(OH)_2$ needs to be improved under carbon curing.

Other alternative mineralization pathways, including those that may be enabled by CO_2-derived organic acids (e.g., H_2C2O_4 and $NaHC2O_4$) and have significantly higher carbon mineralization efficiency, also should be explored. Exemplar reactions of such alternative mineralization pathways are:

$$MSiO_3 + H_2C_xO_y + zH_2O \rightarrow M + C_xO_y \cdot (z + 1)H_2O + SiO_2(aq) \quad \text{(R5.12)}$$

$$MO + H_2C_xO_y + zH_2O \rightarrow M + C_xO_y \cdot (z + 1)H_2O \quad \text{(R5.13)}$$

where M denotes divalent metals (e.g., Ca, Mg, Fe, Cu, Pb, Ni); $H_2C_xO_y$ denotes an organic acid derived from CO_2, where x ≥2; z equals zero or 1; and $SiO_2(aq)$ can be precipitated silica, aqueous silicic acid, or any of its acid dissociation products. The key requirements for the organic acid are that it can (1) be produced from CO_2 using renewable energy and (2) readily react with the M^{2+}-containing minerals to form thermodynamically stable and insoluble products (e.g., the magnesium oxalate cement mentioned earlier (Fricker and Park 2013; Zhang et al. 2020b). A preliminary study showed that oxalic acid can react with olivine at a rate of 2×10^{-6} mol/m^2/s at 25°C/atm, while the fastest carbonation reactions of natural minerals occur at 10^{-12}–10^{-7} mol/m^2/s (Zhang et al. 2019).

The major challenge of this alternative mineralization pathway is the sustainable production of organic acids (e.g., oxalic acid). If CO_2 conversion technologies from Chapter 7 can be integrated with this oxalate cement technology to create first oxalic acid from waste CO_2 and then use it to create oxalate cement, it could create a paradigm shift in how carbon is utilized and stored in the built environment. Last, long-term durability and testing data of produced cement should be carefully collected and analyzed to develop a protocol and certification scheme for these newly developed construction materials, for their performance as well as CO_2 utilization potential. A summary of RD&D needs for carbon mineralization is compiled in Section 5.3 and integrated with the RD&D needs for other carbon utilization pathways in Chapter 11.

5.2.6 Integrated Carbon Mineralization Technologies

The overall sustainability of CO_2 utilization can be improved substantially if carbon capture and conversion are integrated. This is a particularly important aspect of carbon mineralization technologies, because CO_2 conversion to inorganic carbonates often does not require high CO_2 concentration in the feedstock gas stream if the reaction conditions (e.g., pH and temperature) are optimized to favor the precipitation of carbonates. A number of innovative pathways of reactive carbon capture have been developed to integrate carbon capture and mineralization reactions (e.g., amine regeneration after carbonation reaction). Furthermore, as discussed in Section 5.1.2, multistep carbon mineralization allows the extraction and separation of other valuable metals (e.g., REEs, Ni, Co, and Cu) during CO_2 utilization. Thus, there are active efforts to develop carbon-negative mining of energy-relevant critical elements by integrating carbon mineralization technologies. In this section, examples of such integrated carbon mineralization technologies are introduced.

5.2.6.1 Current Technology

5.2.6.1.1 Integrated Carbon Capture and Conversion

While carbon mineralization has been developed to harness high-purity CO_2 or CO_2 in a concentrated flue gas stream, a less explored but potentially transformative strategy involves coupling CO_2 capture from ultra-dilute sources with mineralization. This scheme involves harnessing solvents or solid sorbents to capture CO_2 and directly reacting CO_2-loaded solvents or sorbents with Ca- and Mg-bearing resources to produce Ca- and Mg-bearing carbonates with inherent regeneration of the solvent/sorbent, as shown in Figure 5-11. This approach

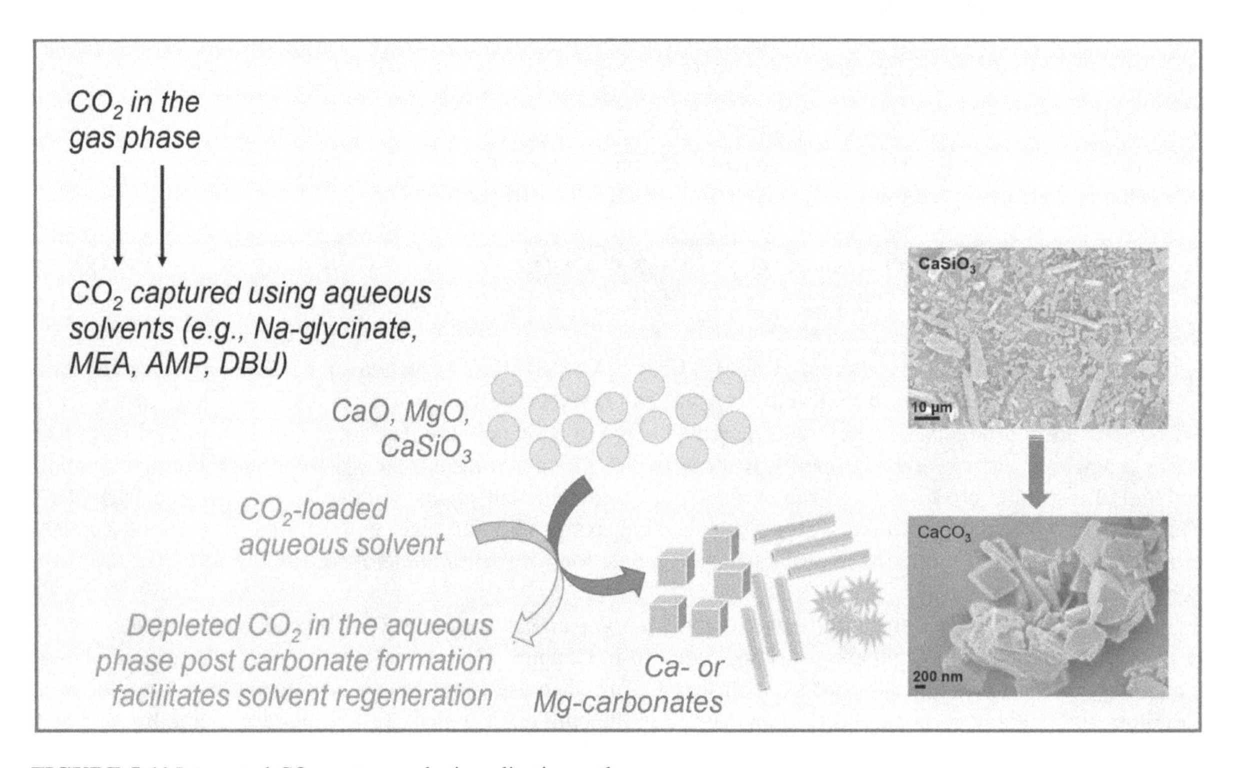

FIGURE 5-11 Integrated CO_2 capture and mineralization pathways.
SOURCE: Reprinted with permission from M. Liu, A. Hohenshil, and G. Gadikota, 2021, "Integrated CO_2 Capture and Removal via Carbon Mineralization with Inherent Regeneration of Aqueous Solvents," *Energy & Fuels* 35(9):8051–8068, https://doi.org/10.1021/acs.energyfuels.0c04346. Copyright 2021 American Chemical Society.

is energy-efficient because it couples an endothermic CO_2 capture reaction with exothermic mineralization, thus potentially lowering the overall energy needs if the heat from the latter can be recuperated for use by the former (Gadikota 2020, 2021; Liu et al. 2021). The coupling of solvent regeneration with carbon mineralization eliminates additional unit operations for producing high-purity CO_2 and then solubilizing this CO_2 for carbon mineralization. Comprehensive investigations of the influence of solvents on CO_2 capture and mineralization have shown that amino acid salts such as Na-glycinate and potassium sarcosinate are as effective at capturing CO_2 as amines such as monoethanolamine with the advantage of being more environmentally benign and far less corrosive (Dashti et al. 2021; Kasturi et al. 2023; Liu and Gadikota 2018, 2020; Ramezani et al. 2022; Yin et al. 2022). The energy- and material-efficiency of this molecularly integrated and intensified approach makes it uniquely suited for integration with the capture of ultra-dilute CO_2 sources. Note also that the DOC mineralization scheme of La Plante et al. (2021, 2023) described in Section 5.2.4.1 is another type of integrated carbon capture and mineralization technology.

5.2.6.1.2 Carbon-Negative Mining

As discussed earlier, natural silicate minerals and alkaline industrial wastes often contain other valuable elements besides Mg and Ca. In fact, serpentine is an ore containing 0.2–2 percent Ni (Grant 1968; Hseu et al. 2018; Morrison et al. 2015). Because critical minerals and REEs play a key role in the clean energy transition, it is economically and environmentallly synergistic to combine carbon mineralization with metal recovery. Such an approach could help secure supply chains of critical metals and maintain environmental advantages. As described in Section 5.2.1, mineral dissolution targeting Mg and Ca leaching can also target other valuable metals (which may be in much smaller quantities) and downstream processing can be designed to recover and separate those metals before the carbonation step, as illustrated in Figure 5-12. Leaching agents can vary from supercritical CO_2,

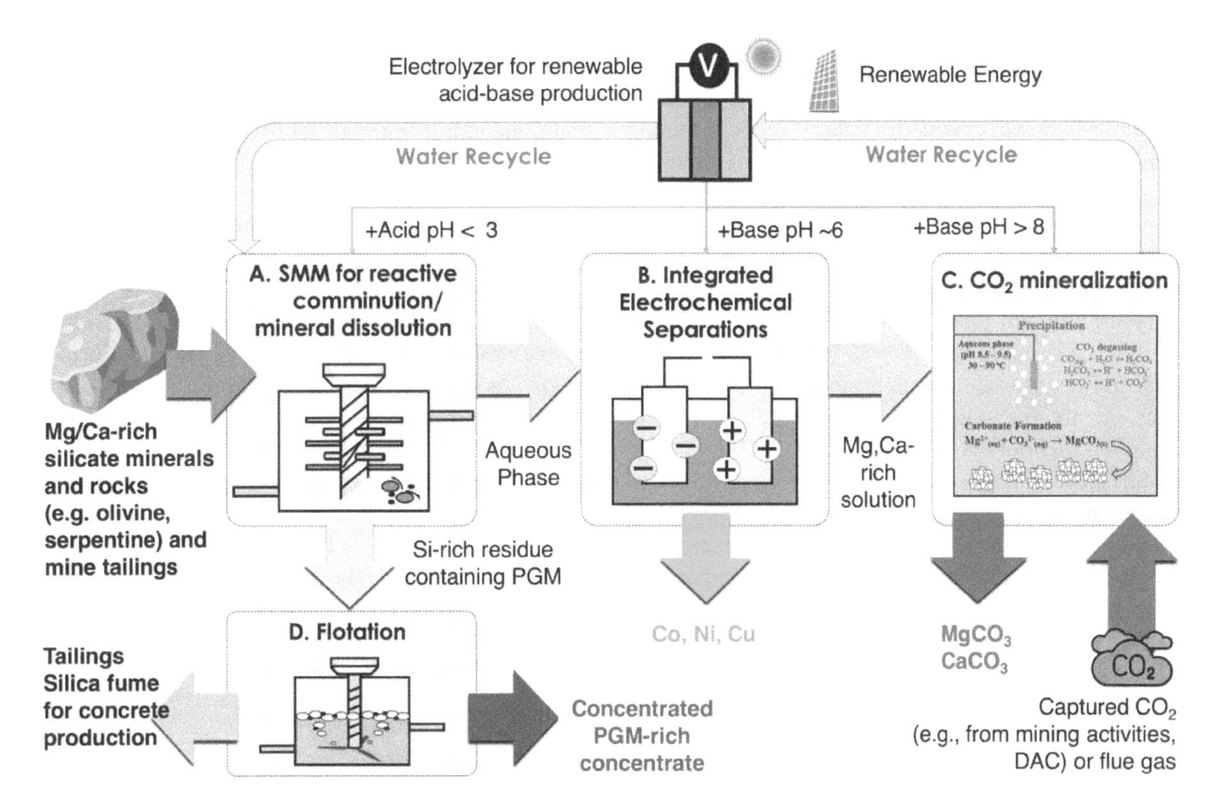

FIGURE 5-12 Scheme of carbon negative mining integrated with carbon mineralization.
NOTE: DAC = direct air capture; PGM = platinum group metal; SMM = stirred media mill.

inorganic/organic acids, to chelating agents targeting specific metals (Hong et al. 2020; Sim et al. 2022). Subsequent separation and purification processes, such as electrodeposition, electrowinning, and precipitation, can aid the production of readily deployable metal products while keeping Mg and Ca in the solution phase for the carbonation step (Kim et al. 2021). The extent of CO_2 utilization via carbon mineralization as well as the overall energy requirements would determine the net carbon intensity of recovered metals.

5.2.6.2 Challenges

There are many technologies that can be integrated to produce solid carbonate products while either utilizing energy or materials for another application. While integrated carbon capture and conversion and carbon negative mining have unique potential to improve further the energy and material efficiencies of carbon mineralization process, they are challenged by the complex integration of multistep reactions and separations. The specific challenges would include—but are not limited to—the accumulation of impurities throughout the process (e.g., process water recycling issues), the difficulty of heat management between reaction and separation units (e.g., rapid heating and cooling are not easy in slurry/liquid systems), and potential undesired reactions (e.g., co-precipitation of Mg and Ca with REEs and other metals). Notably, comprehensive LCAs/TEAs may be difficult to perform.

5.2.6.3 R&D Opportunities

While it could be challenging, there is a clear advantage to developing integrated carbon capture and conversion schemes. Instead of releasing captured CO_2 as gaseous CO_2 during the regeneration step, it should be released as bicarbonate or carbonate ions in solution to integrate with carbon mineralization processes. Thus, similar to the process shown in Figure 5-12, water-soluble carbon capture materials could be investigated for a pH swing to release captured CO_2 into an aqueous phase for a subsequent carbonation reaction.

Carbon-negative mining has substantial economic potential because REEs and energy-relevant metals are high in value. However, they are often present in very small quantities: >300 ppm of REEs are considered economical; however it is difficult to recover metals at that concentration range. Thus, new separation technologies should be integrated to improve the overall metal recovery efficiencies. For example, liquid-liquid extraction of REEs can be performed while continuously removing REEs electrochemically from the organic phase. These reactive separation approaches can provide innovative ways to recover metals without losing Mg and Ca from the solution phase. Ion-selective membranes would provide another low-energy route to recover REEs. The transport of such components through membranes, particularly from complex mixtures, is rarely studied and therefore is an area ripe for basic research. Hybrid processes, consisting of, for example, membranes coupled with liquid-liquid extraction or other technologies also should be explored to provide opportunities to use each separation method under conditions where it performs best. The LCA/TEA of produced metals and solid carbonates should be performed carefully to assess the overall sustainability of the developed technology.

Structured ligand systems also can be designed to create multiple binding sites and cavity-matches the size of target metals to improve the efficiency of metal recovery in the presence of high concentrations of Mg and Ca (Vibbert et al. 2024). There are a wide range of ligands and scaffolds available for such structural ligands but it is also important to design and synthesize ligands that can be robust under relevant reaction conditions (e.g., higher than mineral dissolution $pH\ 4 < pH < 8$ (lower than the carbonation pH). A summary of RD&D needs for carbon mineralization is compiled in Section 5.3, and integrated with the RD&D needs for other carbon utilization pathways in Chapter 11.

5.3 CONCLUSIONS

As shown in Figure 5-13, there are many possibilities to utilize CO_2 through carbon mineralization, enabling a circular economy in the built environment and beyond. Based on the discussions presented in Section 5.2, the committee makes the following findings and recommendations about R&D needs and opportunities for mineralization of CO_2 to inorganic carbonates.

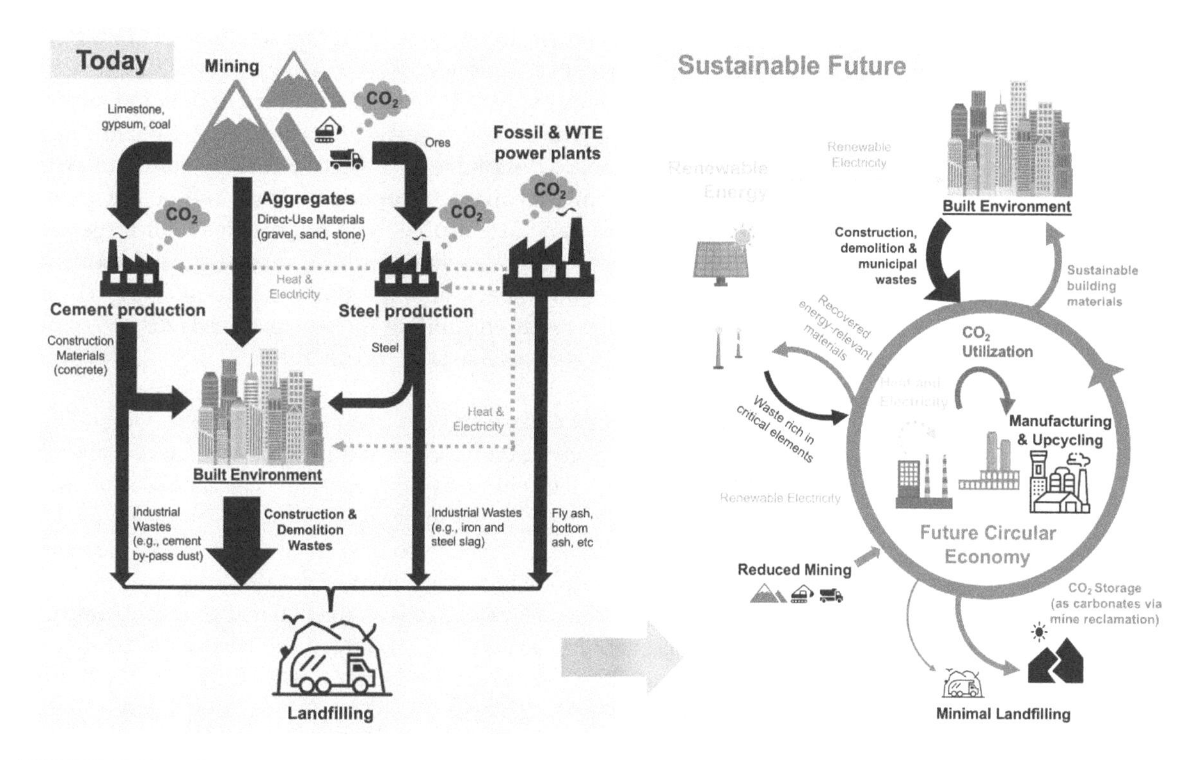

FIGURE 5-13 An illustration of a potential future built environment within the circular economy, showing how CO_2 utilization, along with highly integrated resource recovery and advanced manufacturing technologies, address needs for manufacturing and upcycling of materials for the built environment.
SOURCE: Park et al. (2024), https://doi.org/10.3389/fenrg.2024.1388516. CC BY 4.0.

5.3.1 Findings and Recommendations

Finding 5-1: Challenges and opportunities in carbon mineralization—Carbon mineralization is an important CO_2 utilization approach that produces thermodynamically stable solid products from Ca- and Mg-bearing minerals and rocks with long-term carbon storage potential. This engineered weathering reaction also can convert asbestos, mine tailings, and alkaline industrial wastes to low-risk carbonates while producing valuable products. Although carbon mineralization has been demonstrated at multiple scales, it still faces challenges in terms of energy requirements, process efficiency, product selectivity, and ability to scale to the gigatonne level. It also may result in negative environmental impacts owing to the mining of alkaline minerals or generation of various hazardous by-products that will need to be minimized. The use of legacy and newly produced industrial wastes can minimize mining and maximize environmental benefits by reducing landfill of those waste streams. The co-location of CO_2 with alkaline sources will maximize the carbon utilization potential while helping to minimize energy requirements. Connecting CO_2 sources to sinks/utilization sites is particularly important for carbon mineralization technologies because both reactants (e.g., minerals, rocks and industrial wastes) and products (e.g., solid inorganic carbonates) are energy intensive to transport.

Recommendation 5-1: Support research and development to link alkaline resources to carbon mineralization sites—The availability of alkaline resources, including minerals and industrial wastes, and their chemical and physical properties should be carefully evaluated and mapped for carbon mineralization integrated with different CO_2 sources to create a carbon mineralization atlas. This effort should be funded by the Department of Energy (DOE) and the U.S. Geological Survey. DOE and the Department of Transportation should support research and development efforts focused on how to link CO_2 and mineral sources to carbon mineralization sites and product markets by designing multimodal solutions based on new and existing infrastructure to process and transport efficiently large amounts of solids (both feedstock and carbonate products).

Recommendation 5-2: Support research and development to scale carbon mineralization technologies—The National Science Foundation (NSF), Department of Energy's (DOE's) Office of Basic Energy Sciences and Advanced Research Projects Agency–Energy, and U.S. Geological Survey should support fundamental, experimental and theoretical, and translational research into emerging carbon mineralization approaches (e.g., electrochemically driven carbon mineralization). Funding should be available for university–industry–national laboratory collaborations to rapidly scale up and deploy carbon mineralization technologies and address challenges associated with energy requirements for large-scale mining and mineral processing (e.g., grinding), process integration, chemical recycling, environmental challenges (e.g., handling asbestos), and water requirements, among others. Additionally, DOE and NSF should provide more fundamental and applied research funding for new materials discovery and characterization that would enable new processing such as 3D-printed concrete.

Finding 5-2: Early-stage research and development needs in ocean-based carbon mineralization—Ocean-based carbon mineralization has great potential to integrate the capture of carbon and production of solid carbonate products. Electrolytic seawater mineralization also can produce alternative cementitious materials such as MgO-based cement that could provide increased carbon uptake while maintaining mechanical properties required for construction. Such reactive CO_2 capture and utilization technologies can be developed using renewable energy and have a nearly unlimited scale. However, ocean-based carbon mineralization technologies, depending on their implementation, could significantly alter ocean alkalinity, with attendant environmental and ecological impacts. Thus, ocean-based carbon mineralization technologies need to be carefully monitored and environmental impacts addressed as these technologies scale up (e.g., co-produced acids can be used to dissolve Ca and Mg-bearing minerals and wastes to replenish or add additional alkalinity to the ocean).

Recommendation 5-3: Support research and development for ocean-based carbon utilization technologies—The Department of Energy (DOE), along with other relevant agencies (Department of Defense [DoD], U.S. Environmental Protection Agency [EPA], and National Oceanic and Atmospheric Administration [NOAA]) should support research and development to understand the local environmental and ecological impacts of ocean-based CO_2 utilization solutions, which are still largely at early stages of development. A testing facility should be developed and installed (similar to the National Carbon Capture Center) to provide a platform where various ocean-based carbon mineralization concepts and technologies can be evaluated in actual ocean conditions but with minimal environmental impacts. DOE, DoD, EPA, and NOAA should also develop an environmental protocol to assess and mitigate unexpected environmental and ecological impacts because an acute local spike or gradual change in alkalinity could significantly impact the ocean environment.

Finding 5-3: Research needs for electrochemically driven CO_2 mineralization—The use of electrochemistry for carbon mineralization is an emerging technology area with great potential. "Renewable" acids and bases produced via electrodialysis using renewable energy (e.g., solar, wind, and geothermal energy) could not only decarbonize the mining industry but also provide renewable energy stored in the acid and base to drive carbon mineralization processes. The recyclability of process water with spent acids and bases as well as dissolved ions may lead to additional challenges, including the precipitation of undesired solid phases (e.g., hydroxides and carbonates) in an electrochemical reactor.

Recommendation 5-4: Support research, development, and demonstration for electrochemically driven CO_2 mineralization under a wide range of electrolyte conditions—The Department of Energy's (DOE's) Office of Basic Energy Sciences, Office of Energy Efficiency and Renewable Energy, and Office of Fossil Energy and Carbon Management should support a full range of research, development, and deployment (RD&D) activities for electrochemically driven CO_2 mineralization under a wide range of electrolyte conditions, including seawater and brine. Specifically, DOE should increase support for fundamental experimental and theoretical research into catalyst development, electrochemical cell design, membrane materials, and overall systems engineering and integration with carbon mineralization. As part of this

effort, DOE should also fund RD&D on monitoring side reactions (e.g., seawater oxidation at the anode producing chlorine gas) and membrane fouling issues (e.g., precipitation of undesired solid phases) that could impact the robustness of electrochemically driven carbon mineralization; developing tools to monitor local reaction environments at electrode/electrolyte interfaces; and evaluating the possibility of recycling process water with spent acids, bases, and dissolved ions.

Finding 5-4: Potential for CO_2 mineralization integrated with metal recovery and separation—Minerals and rocks as well as alkaline industrial wastes (e.g., fly ash, iron, and steel slag) that are feedstocks for carbon mineralization processes often contain minor/trace amounts of energy-relevant critical minerals (e.g., low-grade Ni ore such as serpentine contains less than 1% Ni). The grades of these metal ores have been considered too low to be economical for refining. But the recent development of carbon mineralization technologies integrated with metal recovery and separation allows the production of metals as valuable by-products (e.g., rare earth elements, Ni, Co, and Cu). Carbon-negative mining would reduce the overall mining needs for the clean energy transition while providing economic benefits to carbon mineralization by simultaneously producing metals and carbonate products.

Recommendation 5-5: Increase support for research into CO_2 mineralization integrated with metal recovery and separation—The Department of Energy's Offices of Basic Energy Sciences, Energy Efficiency and Renewable Energy, Fossil Energy and Carbon Management, and Advanced Research Projects Agency–Energy (ARPA-E) should increase support for basic and applied research into carbon mineralization integrated with metal recovery (e.g., the MINER program at ARPA-E). The research should focus on, but not be limited to, the development of energy-efficient grinding/comminution of minerals and rocks, selective separation of metals in the presence of large amounts of competing ions (e.g., Ca and Mg ions), improved recycling of chemicals (e.g., ligands), reduced emissions (e.g., mine tailings), and systems integration and optimization. Funding should also be available for university–industry–national laboratory collaborations to rapidly scale up and deploy carbon-negative mining technologies with large CO_2 utilization potential for producing solid carbonates.

Finding 5-5: Challenges and opportunities for widespread adoption of carbon mineralization products—New construction materials (e.g., supplementary cementitious materials, Ca- and Mg-bearing carbonates and oxalates) for new manufacturing processes such as 3D printing and CO_2 curing can be developed using natural and waste alkaline materials, as well as unconventional Ca and Mg resources from the ocean and brines. The resulting carbon mineralization products are promising as additives to increase the strength of concrete while storing CO_2 in building materials. These technologies are not fully developed and therefore are still underutilized, so there is an opportunity for new entrants, as well as optimization and scale up. Adequate testing, standardization, and certification of new products will need to be developed to drive the adoption of these new construction materials. The overall energy savings and the net carbon benefit need to be estimated carefully using life cycle assessment based on analysis of the permanency of carbon utilized via carbon mineralization.

Recommendation 5-6: Develop performance-based standards for construction materials to enable and encourage use of innovative materials—The National Institute of Standards and Technology should collaborate with industrial associations (e.g., the Portland Cement Association, National Concrete Masonry Association) to develop testing, standardization, and certification systems for replacement building materials such as low-carbon or carbon-negative cement and aggregates from CO_2 mineralization in terms of the product performance (e.g., compressive strength) and carbon content. This work should result in an industry standard and certification process to provide to carbon mineralization companies.

5.3.2 Research Agenda for CO_2 Mineralization to Inorganic Carbonates

Table 5-1 presents the committee's research agenda for CO_2 mineralization inorganic carbonates, including research needs (numbered by chapter), and related research agenda recommendations (a subset of research-related recommendations from the chapter). The table includes the relevant funding agencies or other actors; whether the

TABLE 5-1 Research Agenda for CO_2 Mineralization to Inorganic Carbonates

Research, Development, and Demonstration Need	Funding Agencies or Other Actors	Basic, Applied, Demonstration, Enabling	Research Area	Product Class	Long- or Short-Lived	Research Themes	Source
5-A. Evaluation and expansion of mapping of alkaline resources, including minerals and industrial wastes, as well as their chemical and physical properties.	DOE-FECM USGS EPA	Enabling	Mineralization	Construction materials	Long-lived	Resource mapping	Fin. 5-1 Rec. 5-1
5-B. Multimodal optimization of existing and new infrastructure to link feedstocks, including CO_2 and reactant minerals, to sites of carbon mineralization and product markets.	DOE-FECM DOT	Enabling	Mineralization	Construction materials	Long-lived	Enabling technology and infrastructure needs Market opportunities	Fin. 5-1 Rec. 5-1
5-C. Fundamental and translational research on emerging approaches to carbon mineralization to improve energy efficiency, process efficiency, product selectivity, and ability to scale to the gigatonne level.	NSF DOE-BES DOE-ARPA-E DoD USGS	Basic Applied Demonstration	Mineralization	Construction materials	Long-lived	Reactor design and reaction engineering Energy efficiency, electrification, and alternative heating	Fin. 5-1 Rec. 5-2
Recommendation 5-1: Support research and development to link alkaline resources to carbon mineralization sites—The availability of alkaline resources, including minerals and industrial wastes, and their chemical and physical properties should be carefully evaluated and mapped for carbon mineralization integrated with different CO_2 sources to create a carbon mineralization atlas. This effort should be funded by the Department of Energy (DOE) and the U.S. Geological Survey. DOE and the Department of Transportation should support research and development efforts focused on how to link CO_2 and mineral sources to carbon mineralization sites and product markets by designing multimodal solutions based on new and existing infrastructure to process and transport efficiently large amounts of solids (both feedstock and carbonate products).							
Recommendation 5-2: Support research and development to scale carbon mineralization technologies—The National Science Foundation (NSF), Department of Energy's (DOE's) Office of Basic Energy Sciences and the Advanced Research Projects Agency–Energy, and U.S. Geological Survey should support fundamental, experimental and theoretical, and translational research into emerging carbon mineralization approaches (e.g., electrochemically driven carbon mineralization). Funding should be available for university–industry–national laboratory collaborations to rapidly scale up and deploy carbon mineralization technologies and address challenges associated with energy requirements for large-scale mining and mineral processing (e.g., grinding), process integration, chemical recycling, environmental challenges (e.g., handling asbestos), and water requirements, among others. Additionally, DOE and NSF should provide more fundamental and applied research funding for new materials discovery and characterization that would enable new processing such as 3D-printed concrete.							
5-D. Understanding of local environmental and ecological impacts of ocean-based CO_2 utilization and development of an environmental protocol to assess and mitigate unexpected environmental and ecological impacts from pH changes.	DOE-ARPA-E DOE-EERE DOE-FECM DoD-ONR NOAA EPA	Applied	Mineralization—ocean-based CO_2 utilization	Construction materials	Long-lived	Environmental and societal considerations for CO_2 and coal waste utilization technologies	Fin. 5-2 Rec. 5-3
5-E. Development of a testing facility platform, similar to the National Carbon Capture Center, where various ocean-based carbon mineralization concepts and technologies can be evaluated in real ocean conditions with minimal environmental impacts.	NOAA DOE-FECM	Demonstration	Mineralization—ocean-based CO_2 utilization	Construction materials	Long-lived	Research centers and facilities	Fin. 5-2 Rec. 5-3

Recommendation 5-3: Support research and development for ocean-based carbon utilization technologies—The Department of Energy (DOE), along with other relevant agencies (Department of Defense [DoD], U.S. Environmental Protection Agency [EPA], and National Oceanic and Atmospheric Administration [NOAA]) should support research and development to understand the local environmental and ecological impacts of ocean-based CO_2 utilization solutions, which are still largely at early stages of development. A testing facility should be developed and installed (similar to the National Carbon Capture Center) to provide a platform where various ocean-based carbon mineralization concepts and technologies can be evaluated in actual ocean conditions but with minimal environmental impacts. DOE, DoD, EPA, and NOAA should also develop an environmental protocol to assess and mitigate unexpected environmental and ecological impacts because a local pH spike or alkalinity change could significantly impact the ocean environment.

5-F. Full spectrum of research, development, and demonstration (RD&D) activities for electrochemically driven CO_2 mineralization under a wide range of electrolyte conditions, including seawater and brine. RD&D needs to include catalyst development, electrochemical cell design, membrane materials, and overall systems engineering and integration with carbon mineralization.	DOE-BES DOE-EERE DOE-FECM	Basic, Applied Demonstration	Mineralization—Electrochemical	Construction materials	Long-lived	Catalyst innovation and optimization Reactor design and reaction engineering Integrated systems Computational modeling and machine learning	Fin. 5-3 Rec. 5-4
5-G. Monitoring of side reactions (e.g., seawater oxidation at the anode producing chlorine gas) and membrane fouling issues (e.g., precipitation of undesired solid phases). Development of tools to monitor local reaction environments at electrode/electrolyte interfaces and evaluation of recyclability of process water with spent acids and bases as well as dissolved ions.	DOE-EERE DOE-FECM DOE-BES	Basic Applied	Mineralization—Electrochemical	Construction materials	Long-lived	Fundamental knowledge Reactor design and reaction engineering Environmental and societal considerations for CO_2 and coal waste utilization technologies	Fin. 5-3 Rec. 5-4

Recommendation 5-4: Support research, development, and demonstration for electrochemically driven CO_2 mineralization under a wide range of electrolyte conditions—The Department of Energy's (DOE's) Office of Basic Energy Sciences, Office of Energy Efficiency and Renewable Energy, and Office of Fossil Energy and Carbon Management should support a full range of research, development, and deployment (RD&D) activities for electrochemically driven CO_2 mineralization under a wide range of electrolyte conditions, including seawater and brine. Specifically, DOE should increase support for fundamental experimental and theoretical research into catalyst development, electrochemical cell design, membrane materials, and overall systems engineering and integration with carbon mineralization. As part of this effort, DOE should also fund RD&D on monitoring side reactions (e.g., seawater oxidation at the anode producing chlorine gas) and membrane fouling issues (e.g., precipitation of undesired solid phases) that could impact the robustness of electrochemically driven carbon mineralization; developing tools to monitor local reaction environments at electrode/electrolyte interfaces; and evaluating the possibility of recycling process water with spent acids, bases, and dissolved ions.

5-H. Fundamental to translational research of carbon mineralization integrated with metal recovery, focused on energy-efficient grinding/comminution, selective separation, improved recycling, reduced emissions, and systems integration and optimization.	DOE-BES DOE-EERE DOE-FECM DOE-ARPA-E	Basic Applied	Mineralization	Construction materials	Long-lived	Reactor design and reaction engineering Integrated systems Energy efficiency, electrification, and alternative heating Separations	Rec. 5-5

continued

TABLE 5-1 Continued

5-I. University–industry–national laboratory collaborations to rapidly scale up and deploy carbon-negative mining technologies with large CO_2 utilization potential.	DOE-EERE DOE-FECM DOE-ARPA-E	Applied Demonstration	Mineralization	Construction materials	Long-lived	Research centers and facilities	Rec. 5-5
Recommendation 5-5: Increase support for research into CO_2 mineralization integrated with metal recovery and separation—The Department of Energy's Offices of Basic Energy Sciences, Energy Efficiency and Renewable Energy, Fossil Energy and Carbon Management, and Advanced Research Projects Agency–Energy (ARPA-E) should increase support for basic and applied research into carbon mineralization integrated with metal recovery (e.g., the MINER program at ARPA-E). The research should focus on, but not be limited to, the development of energy-efficient grinding/comminution of minerals and rocks, selective separation of metals in the presence of large amounts of competing ions (e.g., Ca and Mg ions), improved recycling of chemicals (e.g., ligands), reduced emissions (e.g., mine tailings), and systems integration and optimization. Funding should also be available for university–industry–national laboratory collaborations to rapidly scale up and deploy carbon-negative mining technologies with large CO_2 utilization potential, producing solid carbonates.							
5-J. Testing, standardization, and certification systems for replacement construction materials produced from CO_2.	NIST Industrial associations	Enabling	Mineralization	Construction materials	Long-lived	Certification and standards	Rec. 5-6
5-K. Materials discovery and characterization of new forms of mineral carbonates to enable new processing like 3D-printed concrete.	DOE-BES DOE-EERE-AMMTO NSF	Basic Applied	Mineralization	Construction materials	Long-lived	Fundamental knowledge	Rec. 5-6
Recommendation 5-6: Develop performance-based standards for construction materials to enable and encourage use of innovative materials—The National Institute of Standards and Technology should collaborate with industrial associations (e.g., the National Concrete Association) to develop testing, standardization, and certification systems for replacement building materials such as low-carbon or carbon-negative cement and aggregates from CO_2 mineralization in terms of the product performance (e.g., compressive strength) and carbon content. This work should result in an industry standard and certification process to provide to carbon mineralization companies.							

NOTE: AMMTO = Advanced Materials and Manufacturing Technologies Office; ARPA-E = Advanced Research Projects Agency–Energy; BES = Basic Energy Sciences; DoD = Department of Defense; DOE = Department of Energy; DOT = Department of Transportation; EERE = Office of Energy Efficiency and Renewable Energy; EPA = U.S. Environmental Protection Agency; FECM = Office of Fossil Energy Carbon Management; NIST = National Institute of Standards and Technology; NSF = National Science Foundation; ONR = Office of Naval Research; USGS = U.S. Geological Survey.

need is for basic research, applied research, technology demonstration, or enabling technologies and processes for CO_2 utilization; the research theme(s) that the research need falls into; the relevant research area and product class covered by the research need; whether the relevant product(s) are long- or short-lived; and the source of the research need (chapter section, finding, or recommendation). The committee's full research agenda can be found in Chapter 11.

5.4 REFERENCES

Abdolhosseini Qomi, M.J., Q.R.S. Miller, S. Zare, H.T. Schaef, J.P. Kaszuba, and K.M. Rosso. 2022. "Molecular-Scale Mechanisms of CO_2 Mineralization in Nanoscale Interfacial Water Films." *Nature Reviews Chemistry* 6(9):598–613. https://doi.org/10.1038/s41570-022-00418-1.

Alventosa, K.M.L., and C.E. White. 2021. "The Effects of Calcium Hydroxide and Activator Chemistry on Alkali-Activated Metakaolin Pastes." *Cement and Concrete Research* 145(July). https://doi.org/10.1016/j.cemconres.2021.106453.

Andrew, R. 2023. "Global CO_2 Emissions from Cement Production." Zenodo Database. Version 231222. https://doi.org/10.5281/zenodo.10423498.

Araujo, F.S.M., I. Taborda-Llano, E.B. Nunes, and R.M. Santos. 2022. "Recycling and Reuse of Mine Tailings: A Review of Advancements and Their Implications." *Geosciences* 12(9). https://doi.org/10.3390/geosciences12090319.

Ashbrook, S.E., and D.M. Dawson. 2016. "NMR Spectroscopy of Minerals and Allied Materials." Pp. 1–52 in *Nuclear Magnetic Resonance* 45, V. Ramesh, ed. Cambridge: Royal Society of Chemistry. https://doi.org/10.1039/9781782624103-00001.

ASTM International. 2023a. "ASTM, C1157/C1157M-23, Standard Performance Specification for Hydraulic Cement." In *Annual Book of ASTM Standards*. Volume 04.01.

ASTM International. 2023b. "ASTM, C618-23e1, Standard Specification for Coal Fly Ash and Raw or Calcined Natural Pozzolan for Use as a Mineral Admixture in Concrete." In *Annual Book of ASTM Standards*. Volume 04.02.

Badjatya, P., A.H. Akca, D.V. Fraga Alvarez, B. Chang, S. Ma, X. Pang, E. Wang, Q. van Hinsberg, D.V. Esposito, and S. Kawashima. 2022. "Carbon-Negative Cement Manufacturing from Seawater-Derived Magnesium Feedstocks." *Proceedings of the National Academy of Sciences* 119(34):e2114680119. https://doi.org/10.1073/pnas.2114680119.

Balucan, R.D., and B.Z. Dlugogorski. 2013. "Thermal Activation of Antigorite for Mineralization of CO_2." *Environmental Science and Technology* 47:182–190.

Balucan, R.D., E.M. Kennedy, J.F. Mackie, and B.Z. Dlugogorski. 2011. "Optimization of Antigorite Heat Pre-Treatment Via Kinetic Modeling of the Dehydroxylation Reaction for CO_2 Mineralization." *Greenhouse Gases* 1:294–304.

Beerling, D.J., E.P. Kantzas, M.R. Lomas, P. Wade, R.M. Eufrasio, P. Renforth, B. Sarkar, et al. 2020. "Potential for Large-Scale CO_2 Removal via Enhanced Rock Weathering with Croplands." *Nature* 583(7815):242–248. https://doi.org/10.1038/s41586-020-2448-9.

Bernard, E., H. Nguyen, S. Kawashima, B. Lothenbach, H. Manzano, J. Provis, A. Scott, C. Unluer, F. Winnefeld, and P. Kinnunen. 2023. "MgO-Based Cements—Current Status and Opportunities." *RILEM Technical Letters* 8(November):65–78. https://doi.org/10.21809/rilemtechlett.2023.177.

Blondes, M.S., M.D. Merrill, S.T. Anderson, and C.A. DeVera. 2019. "Carbon Dioxide Mineralization Feasibility in the United States: U.S. Geological Survey Scientific Investigations Report 2018–5079." https://doi.org/10.3133/sir20185079.

Bodor, M., R.M. Santos, G. Cristea, M. Salman, Ö. Cizer, R.I. Iacobescu, Y.W. Chiang, K. van Balen, M. Vlad, and T. van Gerven. 2016. "Laboratory Investigation of Carbonated BOF Slag Used as Partial Replacement of Natural Aggregate in Cement Mortars." *Cement and Concrete Composites* 65:55–66. https://doi.org/10.1016/j.cemconcomp.2015.10.002.

Boyn, J.-N. and E.A. Carter. 2023a. "Characterizing the Mechanisms of Ca and Mg Carbonate Ion-Pair Formation with Multi-Level Molecular Dynamics/Quantum Mechanics Simulations." *The Journal of Physical Chemistry B* 127(50):10824–10832. https://doi.org/10.1021/acs.jpcb.3c05369.

Boyn, J.-N., and E.A. Carter. 2023b. "Probing PH-Dependent Dehydration Dynamics of Mg and Ca Cations in Aqueous Solutions with Multi-Level Quantum Mechanics/Molecular Dynamics Simulations." *Journal of the American Chemical Society* 145(37):20462–20472. https://doi.org/10.1021/jacs.3c06182.

Bui Viet, D., W.P. Chan, Z.H. Phua, A. Ebrahimi, A. Abbas, and G. Lisak. 2020. "The Use of Fly Ashes from Waste-to-Energy Processes as Mineral CO_2 Sequesters and Supplementary Cementitious Materials." *Journal of Hazardous Materials* 398:122906. https://doi.org/10.1016/j.jhazmat.2020.122906.

Campbell, J.S., S. Foteinis, V. Furey, O. Hawrot, D. Pike, S. Aeschlimann, C.N. Maesano, et al. 2022. "Geochemical Negative Emissions Technologies: Part I. Review." *Frontiers in Climate* 4. https://doi.org/10.3389/fclim.2022.879133.

Cannon, C., V. Guido, and L. Wright. 2021. "Concrete Solutions Guide: Six Actions to Lower the Embodied Carbon of Concrete." *RMI*. http://www.rmi.org/concretesolutions-guide.

Chakravarthy, C., S. Chalouati, Y.E. Chai, H. Fantucci, and R.M. Santos. 2020. "Valorization of Kimberlite Tailings by Carbon Capture and Utilization (CCU) Method." *Minerals* 10(7). https://doi.org/10.3390/min10070611.

Chizmeshya, A.V.G., M.J. McKelvy, G.H. Wolf, R.W. Carpenter, D.A. Gormley, J.R. Diefenbacher, and R. Marzke. 2006. "Enhancing the Atomic-Level Understanding of CO_2 Mineral Sequestration Mechanisms via Advanced Computational Modeling." Tempe: Arizona State University.

Czigler, T., S. Reiter, P. Schulze, and K. Somers. 2020. "Laying the Foundation for Zero-Carbon Cement." McKinsey & Company.

Dashti, A., F. Amirkhani, A.-S. Hamedi, and A.H. Mohammadi. 2021. "Evaluation of CO_2 Absorption by Amino Acid Aqueous Solution Using Hybrid Soft Computing Methods." *ACS Omega* 6(19):12459–12469. https://doi.org/10.1021/acsomega.0c06158.

Diamond, S. 2007. "2–Physical and Chemical Characteristics of Cement Composites." Pp. 10–44 in *Durability of Concrete and Cement Composites*, C.L. Page and M.M. Page, eds. Woodhead Publishing Series in Civil and Structural Engineering. Woodhead Publishing. https://doi.org/10.1533/9781845693398.10.

Dlugogorski, B.Z., and R.D. Balucan. 2014. "Dehydroxylation of Serpentine Minerals: Implications for Mineral Carbonation." *Renewable Sustainable Energy Review* 31:353-367.

Elleuche, S., and S. Poggeler. 2009. "Evolution of Carbonic Anhydrases in Fungi." *Current Genetics* 55(2):211–222.

Fennell, P.S., S.J. Davis, and A. Mohammed. 2021. "Decarbonizing Cement Production." *Joule* 5(6):1305–1311. https://doi.org/10.1016/j.joule.2021.04.011.

Fricker, K.J., and A.-H.A. Park. 2013. "Effect of H_2O on $Mg(OH)_2$ Carbonation Pathways for Combined CO_2 Capture and Storage." *Chemical Engineering Science* 100:332–341. https://doi.org/10.1016/j.ces.2012.12.027.

Fricker, K.J., and A.-H.A. Park. 2014. "Investigation of the Different Carbonate Phases and Their Formation Kinetics During $Mg(OH)_2$ Slurry Carbonation." *Industrial and Engineering Chemistry Research* 53(47):18170–18179. https://doi.org/10.1021/ie503131s.

Friedlingstein, P., M. O'Sullivan, M.W. Jones, R.M. Andrew, J. Hauck, A. Olsen, G.P. Peters, et al. 2020. "Global Carbon Budget 2020." *Earth System Science Data* 12(4):3269–3340.

Fu, Y., Y.-B. Jiang, D. Dunphy, H. Xiong, E. Coker, S.S. Chou, H. Zhang, et al. 2018. "Ultra-Thin Enzymatic Liquid Membrane for CO_2 Separation and Capture." *Nature Communications* 9(1):990. https://doi.org/10.1038/s41467-018-03285-x.

Gadikota, G. 2020. "Multiphase Carbon Mineralization for the Reactive Separation of CO_2 and Directed Synthesis of H_2." *National Reviews Chemistry* 4(2):78–89. https://doi.org/10.1038/s41570-019-0158-3.

Gadikota, G. 2021. "Carbon Mineralization Pathways for Carbon Capture, Storage and Utilization." *Communications Chemistry* 4(1):23.

Gadikota, G., E.J. Swanson, H. Zhao, and A.-H.A. Park. 2014. "Experimental Design and Data Analysis for Accurate Estimation of Reaction Kinetics and Conversion for Carbon Mineralization." *Industrial and Engineering Chemistry Research* 53(16):6664–6676. https://doi.org/10.1021/ie500393h.

Gadikota, G., K. Fricker, S.-H. Jang, and A.-H.A. Park. 2015. "Carbonation of Silicate Minerals and Industrial Wastes and Their Potential Use as Sustainable Construction Materials." *Advances in CO_2 Capture, Sequestration, and Conversion* 295–322.

GCCA (Global Cement and Concrete Association). 2021. "2050 Cement and Concrete Industry Roadmap for Net Zero Concrete." https://gccassociation.org/concretefuture/wp-content/uploads/2021/10/GCCA-Concrete-Future-Roadmap-Document-AW.pdf.

GEA. n.d. "Removal of SO_x by Wet Scrubbing." https://www.gea.com/en/products/emission-control/sorption/sox-removal-by-wet-scrubbing.

George, V. 2023. "CarbonBuilt to Make Commercial Concrete with No Embodied Carbon." *Carbon Herald* (blog), September 22. https://carbonherald.com/carbonbuilt-to-make-commercial-concrete-with-no-embodied-carbon.

Gerres, T., J. Lehne, G. Mete, S. Chenk, and C. Swalec. 2021. "Green Steel Production: How G7 Countries Can Help Change the Global Landscape." *LeadIt* 6.

Glasser, F.P., G. Jauffret, J. Morrison, J.-L. Galvez-Martos, N. Patterson, and M.S.-E. Imbabi. 2016. "Sequestering CO_2 by Mineralization into Useful Nesquehonite-Based Products." *Frontiers in Energy Research* 4. https://doi.org/10.3389/fenrg.2016.00003.

Global Aggregates Information Network. 2023. "Gain World Map." gain.ie.

Grant, R.W. 1968. "Mineral Collecting in Vermont." Vermont Geological Survey, Special Publication 2:25.

Habert, G., S.A. Miller, V.M. John, J.L. Provis, A. Favier, A. Horvath, and K.L. Scrivener. 2020. "Environmental Impacts and Decarbonization Strategies in the Cement and Concrete Industries." *Nature Reviews Earth & Environment* 1(11):559–573. https://doi.org/10.1038/s43017-020-0093-3.

Haldar, S.K. 2020. "Chapter 6—Sedimentary Rocks." Pp. 187–268 in *Introduction to Mineralogy and Petrology* (Second Edition), S.K. Haldar, ed. Elsevier. https://doi.org/10.1016/B978-0-12-820585-3.00006-5.

Hanifa, M., R. Agarwal, U. Sharma, P.C. Thapliyal, and L.P. Singh. 2023. "A Review on CO_2 Capture and Sequestration in the Construction Industry: Emerging Approaches and Commercialised Technologies." *Journal of CO_2 Utilization* 67. https://doi.org/10.1016/j.jcou.2022.102292.

Hargis, C.W., I.A. Chen, M. Devenney, M.J. Fernandez, R.J. Gilliam, and R.P. Thatcher. 2021. "Calcium Carbonate Cement: A Carbon Capture, Utilization, and Storage (CCUS) Technique." *Materials* 14(11). https://doi.org/10.3390/ma14112709.

Hills, C.D., N. Tripathi, and P.J. Carey. 2020. "Mineralization Technology for Carbon Capture, Utilization, and Storage." *Frontiers in Energy Research* 8. https://doi.org/10.3389/fenrg.2020.00142.

Ho, D.T., L. Bopp, J.B. Palter, M.C. Long, P.W. Boyd, G. Neukermans, and L.T. Bach. 2023. "Monitoring, Reporting, and Verification for Ocean Alkalinity Enhancement." *Guide to Best Practices in Ocean Alkalinity Enhancement Research*. Copernicus Publications. https://doi.org/10.5194/sp-2-oae2023.

Ho, H.-J., A. Iizuka, E. Shibata, H. Tomita, K. Takano, and T. Endo. 2021. "Utilization of CO_2 in Direct Aqueous Carbonation of Concrete Fines Generated from Aggregate Recycling: Influences of the Solid–Liquid Ratio and CO_2 Concentration." *Journal of Cleaner Production* 312(August):127832. https://doi.org/10.1016/j.jclepro.2021.127832.

Holappa, L. 2020. "A General Vision for Reduction of Energy Consumption and CO2 Emissions from the Steel Industry." *Metals* 10(9):1117. https://doi.org/10.3390/MET10091117.

Hong, S., H.D. Huang, G. Rim, Y. Park, and A.-H.A. Park. 2020. "Integration of Two Waste Streams for Carbon Storage and Utilization: Enhanced Metal Extraction from Steel Slag Using Biogenic Volatile Organic Acids." *ACS Sustainable Chemistry and Engineering* 8(50):18519–18527. https://doi.org/10.1021/acssuschemeng.0c06355.

Hseu, Z.-Y., F. Zehetner, K. Fujii, T. Watanabe, and A. Nakao. 2018. "Geochemical Fractionation of Chromium and Nickel in Serpentine Soil Profiles Along a Temperate to Tropical Climate Gradient." *Geoderma* 327(October):97–106. https://doi.org/10.1016/j.geoderma.2018.04.030.

IEA (International Energy Agency). 2020. "Cement, Energy Technology Perspectives 2020." Paris: IEA.

IEA. 2022. "Cement." Paris: IEA. https://www.iea.org/reports/cement.

IPCC (International Panel on Climate Change). 2023. "Climate Change 2023: Synthesis Report." A Report of the Intergovernmental Panel on Climate Change. Contribution of Working Groups I, II and III to the Sixth Assessment Report of the Intergovernmental Panel on Climate Change.

Jin, F., M. Zhao, M. Xu, and L. Mo. 2024. "Maximising the Benefits of Calcium Carbonate in Sustainable Cements: Opportunities and Challenges Associated with Alkaline Waste Carbonation." *npj Materials Sustainability* 2(1):1–7. https://doi.org/10.1038/s44296-024-00005-z.

Juenger, M.C.G., R. Snellings, and S.A. Bernal. 2019. "Supplementary Cementitious Materials: New Sources, Characterization, and Performance Insights." *Cement and Concrete Research* 122:257–273. https://doi.org/10.1016/j.cemconres.2019.05.008.

Kashim, M.Z., H. Tsegab, O. Rahmani, Z.A.A. Bakar, and S.M. Aminpour. 2020. "Reaction Mechanism of Wollastonite In Situ Mineral Carbonation for CO_2 Sequestration: Effects of Saline Conditions, Temperature, and Pressure." *ACS Omega* 5(45):28942–28954. https://doi.org/10.1021/acsomega.0c02358.

Kasturi, A., G.G. Jang, A.D.-T. Akin, A. Jackson, J. Jun, D. Stamberga, R. Custelcean, D.S. Sholl, S. Yiacoumi, and C. Tsouris. 2023. "An Effective Air–Liquid Contactor for CO_2 Direct Air Capture Using Aqueous Solvents." *Separation and Purification Technology* 324(November):124398. https://doi.org/10.1016/j.seppur.2023.124398.

Kelemen, P.B., N. McQueen, J. Wilcox, P. Renforth, G. Dipple, and A.P. Vankeuren. 2020. "Engineered Carbon Mineralization in Ultramafic Rocks for CO_2 Removal from Air: Review and New Insights." *Chemical Geology* 550. https://doi.org/10.1016/j.chemgeo.2020.119628.

Kelly, C., R. Ghani, C. Kardish, T. Moran, E. Tucker, E. Sheff, and A. Masinter. 2024. "Paving the Way to Innovation: Moving from Prescriptive to Performance Specifications to Unlock Low-Carbon Cement, Concrete and Asphalt Innovations. Clearpath, Center for Climate and Energy Solutions (C2ES), Clean Air Task Force (CATF)." https://clearpath.org/wp-content/uploads/sites/44/2024/02/202402_PBS-Research-Report_Final.pdf.

Khalil, A., X. Wang, and K. Celik. 2020. "3D Printable Magnesium Oxide Concrete: Towards Sustainable Modern Architecture." *Additive Manufacturing* 33. https://doi.org/10.1016/j.addma.2020.101145.

Kim, K., R. Candeago, G. Rim, D. Raymond, A.A. Park, and X. Su. 2021. "Electrochemical Approaches for Selective Recovery of Critical Elements in Hydrometallurgical Processes of Complex Feedstocks." *iScience* 24(5):102374. https://doi.org/10.1016/j.isci.2021.102374.

Krausmann, F., D. Wiedenhofer, C. Lauk, W. Haas, H. Tanikawa, T. Fishman, A. Miatto, H. Schandl, and H. Haberl. 2017. "Global Socioeconomic Material Stocks Rise 23-Fold Over the 20th Century and Require Half of Annual Resource Use." *Proceedings of the National Academy of Sciences* 114(8):1880–1885. https://doi.org/10.1073/pnas.1613773114.

Kumar, A., K.R. Phillips, G.P. Thiel, U. Schröder, and J.H. Lienhard. 2019. "Direct Electrosynthesis of Sodium Hydroxide and Hydrochloric Acid from Brine Streams." *Nature Catalysis* 2(2):106–113.

La Plante, E.C., D.A. Simonetti, J. Wang, A. Al-Turki, X. Chen, D. Jassby, and G.N. Sant. 2021. "Saline Water-Based Mineralization Pathway for Gigatonne-Scale CO_2 Management." *ACS Sustainable Chemistry and Engineering* 9(3):1073–1089. https://doi.org/10.1021/acssuschemeng.0c08561.

La Plante, E.C., X. Chen, S. Bustillos, A. Bouissonnie, T. Traynor, D. Jassby, L. Corsini, D.A. Simonetti, and G.N. Sant. 2023. "Electrolytic Seawater Mineralization and the Mass Balances That Demonstrate Carbon Dioxide Removal." *ACS ES&T Engineering* 3(7):955–968. https://doi.org/10.1021/acsestengg.3c00004.

Lehne, J., and F. Preston. 2018. "Making Concrete Change: Innovation in Low-Carbon Cement and Concrete." Chatham House Report. The Royal Institute of International Affairs. https://www.chathamhouse.org/2018/06/making-concrete-change-innovation-low-carbon-cement-and-concrete-0/about-authors.

Li, L., Q. Liu, T. Huang, and W. Peng. 2022b. "Mineralization and Utilization of CO_2 in Construction and Demolition Wastes Recycling for Building Materials: A Systematic Review of Recycled Concrete Aggregate and Recycled Hardened Cement Powder." *Separation and Purification Technology* 298. https://doi.org/10.1016/j.seppur.2022.121512.

Li, N., L. Mo, and C. Unluer. 2022a. "Emerging CO_2 Utilization Technologies for Construction Materials: A Review." *Journal of CO_2 Utilization* 65(November):102237. https://doi.org/10.1016/j.jcou.2022.102237.

Liu, M., and G. Gadikota. 2018. "Integrated CO_2 Capture, Conversion, and Storage to Produce Calcium Carbonate Using an Amine Looping Strategy." *Energy and Fuels* 33(3):1722–1733.

Liu, M., and G. Gadikota. 2020. "Single-step, Low Temperature and Integrated CO_2 Capture and Conversion Using Sodium Glycinate to Produce Calcium Carbonate." *Fuel* 275:117887.

Liu, M., A. Hohenshil, and G. Gadikota. 2021. "Integrated CO_2 Capture and Removal via Carbon Mineralization with Inherent Regeneration of Aqueous Solvents." *Energy and Fuels* 35(9):8051–8068. https://doi.org/10.1021/acs.energyfuels.0c04346.

Luo, Z., Y. Ma, H. He, W. Mu, X. Zhou, W. Liao, and H. Ma. 2021. "Preparation and Characterization of Ferrous Oxalate Cement—A Novel Acid-Base Cement." *Journal of the American Ceramic Society* 104(2):1120–1131. https://doi.org/10.1111/jace.17511.

Marchon, D., S. Kawashima, H. Bessaies-Bey, S. Mantellato, and S. Ng. 2018. "Hydration and Rheology Control of Concrete for Digital Fabrication: Potential Admixtures and Cement Chemistry." *Cement and Concrete Research* 112(October):96–110. https://doi.org/10.1016/j.cemconres.2018.05.014.

Martirez, J.M.P., and E.A. Carter. 2023. "Solvent Dynamics Are Critical to Understanding Carbon Dioxide Dissolution and Hydration in Water." *Journal of the American Chemical Society* 145(23):12561–12575. https://doi.org/10.1021/jacs.3c01283.

Mayes, W.M., A.L. Riley, H.I. Gomes, P. Brabham, J. Hamlyn, H. Pullin, et al. 2018. "Atmospheric CO2 Sequestration in Iron and Steel Slag: Consett, County Durham, United Kingdom." *Environmental Science and Technology* 52(14):7892–7900. https://doi.org/10.1021/acs.est.8b01883.

Mazzotti, M., J.C. Abanades, R. Allam, K.S. Lackner, F. Meunier, E. Rubin, J.C. Sanchez, K. Yogo, and R. Zevenhoven. 2005. "7 | Mineral Carbonation and Industrial Uses of Carbon Dioxide." Pp. 319–338 in *Carbon Dioxide Capture and Storage*. Cambridge, UK: Cambridge University Press. https://www.ipcc.ch/report/carbon-dioxide-capture-and-storage.

McKelvy, M.J., A.V.G. Chizmeshya, J. Diefenbacher, H. Bearat, and G. Wolf. 2004. "Exploration of the Role of Heat Activation in Enhancing Serpentine Carbon Sequestration Reactions." *Environmental Science and Technology* 38:6897–6903.

Miller, S.A., G. Habert, R.J. Myers, and J.T. Harvey. 2021. "Achieving Net Zero Greenhouse Gas Emissions in the Cement Industry via Value Chain Mitigation Strategies." *One Earth* 4(10):1398–1411. https://doi.org/10.1016/j.oneear.2021.09.011.

Moelich, G.M., P.J. Kruger, and R. Combrinck. 2022. "A Plastic Shrinkage Cracking Risk Model for 3D Printed Concrete Exposed to Different Environments." *Cement and Concrete Composites* 130:104516. https://doi.org/10.1016/j.cemconcomp.2022.104516.

Monkman, S., and M. MacDonald. 2017. "On Carbon Dioxide Utilization as a Means to Improve the Sustainability of Ready-Mixed Concrete." *Journal of Cleaner Production* 167:365–375. https://doi.org/10.1016/j.jclepro.2017.08.194.

Monkman, S., P.A. Kenward, G. Dipple, M. MacDonald, and M. Raudsepp. 2018. "Activation of Cement Hydration with Carbon Dioxide." *Journal of Sustainable Cement-Based Materials* 7(3):160–181. https://doi.org/10.1080/21650373.2018.1443854.

Monteiro, P.J.M., S.A. Miller, and A. Horvath. 2017. "Towards Sustainable Concrete." *Nature Materials* 16(7):698–699. https://doi.org/10.1038/nmat4930.

Morrison, J.M., M.B. Goldhaber, C.T. Mills, G.N. Breit, R.L. Hooper, J.M. Holloway, S.F. Diehl, and J.F. Ranville. 2015. "Weathering and Transport of Chromium and Nickel from Serpentinite in the Coast Range Ophiolite to the Sacramento Valley, California, USA." *Applied Geochemistry* 61:72–86. https://doi.org/10.1016/j.apgeochem.2015.05.018.

NASEM (National Academies of Sciences, Engineering, and Medicine). 2019. *Gaseous Carbon Waste Streams Utilization: Status and Research Needs*. Washington, DC: The National Academies Press. https://doi.org/10.17226/25232.

NASEM. 2022. *A Research Strategy for Ocean-Based Carbon Dioxide Removal and Sequestration*. Washington, DC: The National Academies Press. https://doi.org/10.17226/26278.

Nayak, D.K., P.P. Abhilash, R. Singh, R. Kumar, and V. Kumar. 2022. "Fly Ash for Sustainable Construction: A Review of Fly Ash Concrete and Its Beneficial Use Case Studies." *Cleaner Materials* 6(December):100143. https://doi.org/10.1016/j.clema.2022.100143.

Newtson, C.R., S. Mousavinezhad, G.J. Gonzales, W.K. Toledo, and J.M. Garcia. "Alternative Supplementary Cementitious Materials in Ultra-High Performance Concrete." Transportation Consortium of South-Central States (Tran-SET). https://repository.lsu.edu/cgi/viewcontent.cgi?article=1141&context=transet_pubs.

Nguyen, T.-T.-H., D.-H. Phan, H.-H. Mai, and D.-L. Nguyen. 2020. "Investigation on Compressive Characteristics of Steel-Slag Concrete." *Materials* 13(8):1928. https://doi.org/10.3390/ma13081928.

Norhasyima, R.S., and T.M.I. Mahlia. 2018. "Advances in CO_2 Utilization Technology: A Patent Landscape Review." *Journal of CO_2 Utilization* 26:323–335. https://doi.org/10.1016/j.jcou.2018.05.022.

Nukah, P.D., S.J. Abbey, C.A. Booth, and G. Nounu. 2023. "Mapping and Synthesizing the Viability of Cement Replacement Materials via a Systematic Review and Meta-Analysis." *Construction and Building Materials* 405:133290. https://doi.org/10.1016/j.conbuildmat.2023.133290.

Ostovari, H., L. Muller, J. Skocek, and A. Bardow. 2021. "From Unavoidable CO_2 Source to CO_2 Sink? A Cement Industry Based on CO_2 Mineralization." *Environmental Science & Technology* 55(8):5212–5223. https://doi.org/10.1021/acs.est.0c07599.

Pan, S.-Y., T.C. Chung, C.C. Ho, C.J. Hou, Y.H. Chen, and P.C. Chiang. 2017. "CO_2 Mineralization and Utilization Using Steel Slag for Establishing a Waste-to-Resource Supply Chain." *Scientific Reports* 7(1):17227. https://doi.org/10.1038/s41598-017-17648-9.

Pan, S.-Y., T.-C. Ling, A.-H.A. Park, and P.-C. Chiang. 2018. "An Overview: Reaction Mechanisms and Modelling of CO_2 Utilization via Mineralization." *Aerosol and Air Quality Research* 18(4):829–848. https://doi.org/10.4209/aaqr.2018.03.0093.

Pan, S.-Y., Y.-H. Chen, L.-S. Fan, H. Kim, X. Gao, T.-C. Ling, P.-C. Chiang, S.-L. Pei, and G. Gu. 2020. "CO_2 Mineralization and Utilization by Alkaline Solid Wastes for Potential Carbon Reduction." *Nature Sustainability* 3(5):399–405. https://doi.org/10.1038/s41893-020-0486-9.

Park, A.-H.A., and L.-S. Fan. 2004. "CO_2 Mineral Sequestration: Physically Activated Dissolution of Serpentine and pH Swing Process." *Chemical Engineering Science* 59(22–23):5241–5247. https://doi.org/10.1016/j.ces.2004.09.008.

Park, A.-H.A., J.M. Williams, J. Friedmann, D. Hanson, S. Kawashima, V. Sick, M.R. Taha, and J. Wilcox. 2024. "Challenges and Opportunities for the Built Environment in a Carbon-Constrained World for the Next 100 Years and Beyond." *Frontiers in Energy Research* 12(March 18):1388516. https://doi.org/10.3389/fenrg.2024.1388516.

Parron-Rubio, M.E., F. Perez-Garcia, A. Gonzalez-Herrera, M. José Oliveira, and M.D. Rubio-Cintas. 2019. "Slag Substitution as a Cementing Material in Concrete: Mechanical, Physical and Environmental Properties." *Materials* 12(18):2845. https://doi.org/10.3390/ma12182845.

Patel, T.N., A.-H.A. Park, and S. Banta. 2013. "Periplasmic Expression of Carbonic Anhydrase in Escherichia coli: A New Biocatalyst for CO_2 Hydration." *Biotechnology and Bioengineering* 110(7):1865–1873. https://doi.org/10.1002/bit.24863.

Patel, T.N., A.-H.A. Park, and S. Banta. 2014a. "Genetic Manipulation of Outer Membrane Permeability: Generating Porous Heterogeneous Catalyst Analogs in Escherichia coli." *ACS Synthetic Biology* 3(12):848–854. https://doi.org/10.1021/sb400202s.

Patel, T.N., A.-H.A. Park, and S. Banta. 2014b. "Surface Display of Small Peptides on *Escherichia coli* for Enhanced Calcite Precipitation Rates." *Biopolymers (PeptideScience)* 102(2):191–196. https://doi.org/10.1002/bip.22466.

Paul, S.C., G.P.A.G. Van Zijl, M.J. Tan, and I. Gibson. 2018. "A Review of 3D Concrete Printing Systems and Materials Properties: Current Status and Future Research Prospects." *Rapid Prototyping Journal* 24(4):784–798. https://doi.org/10.1108/RPJ-09-2016-0154.

Phuyal, K., U. Sharma, J. Mahar, K. Mondal, and M. Mashal. 2023. "A Sustainable and Environmentally Friendly Concrete for Structural Applications." *Sustainability* 15(20):14694. https://doi.org/10.3390/su152014694.

Ramezani, R., S. Mazinani, and R. Di Felice. 2022. "State-of-the-Art of CO_2 Capture with Amino Acid Salt Solutions." *Reviews in Chemical Engineering* 38(3):273–299. https://doi.org/10.1515/revce-2020-0012.

Ravikumar, D., D. Zhang, G. Keoleian, S. Miller, V. Sick, and V. Li. 2021. "Carbon Dioxide Utilization in Concrete Curing or Mixing Might Not Produce a Net Climate Benefit." *Nature Communication* 12(1):855. https://doi.org/10.1038/s41467-021-21148-w.

Riebesell, U., V.J. Fabry, L. Hansson, and J.-P. Gattuso, eds. 2010. "Part 1: Seawater Carbonate Chemistry." Luxembourg: Publications Office of the European Union.

Riedl, D., Z. Byrum, S. Li, H. Pilorgé, P. Psarras, and K. Lebling. 2023. "5 Things to Know About Carbon Mineralization as a Carbon Removal Strategy." https://www.wri.org/insights/carbon-mineralization-carbon-removal.

Rim, G., D. Wang, M. Rayson, G. Brent, and A.-H.A. Park. 2020a. "Investigation on Abrasion Versus Fragmentation of the Si-Rich Passivation Layer for Enhanced Carbon Mineralization via CO_2 Partial Pressure Swing." *Industrial and Engineering Chemistry Research* 59(14):6517–6531. https://doi.org/10.1021/acs.iecr.9b07050.

Rim, G., A.K. Marchese, P. Stallworth, S.G. Greenbaum, and A.-H.A. Park. 2020b. "^{29}Si Solid State MAS NMR Study on Leaching Behaviors and Chemical Stability of Different Mg-Silicate Structures for CO_2 Sequestration." *Chemical Engineering Journal* 396:125204. https://doi.org/10.1016/j.cej.2020.125204.

Rim, G., N. Roy, D. Zhao, S. Kawashima, P. Stallworth, S.G. Greenbaum, and A.A. Park. 2021. "CO_2 Utilization in Built Environment via the P_{CO2} Swing Carbonation of Alkaline Solid Wastes with Different Mineralogy." *Faraday Discuss* 230(0):187–212. https://doi.org/10.1039/d1fd00022e.

Ropp, R.C. 2013. "Group 14 (C, Si, Ge, Sn, and Pb) Alkaline Earth Compounds." Pp. 351–480 in *Encyclopedia of the Alkaline Earth Compounds*. Elsevier. https://doi.org/10.1016/B978-0-444-59550-8.00005-3.

Rosa, L., V. Becattini, P. Gabrielli, A. Andreotti, and M. Mazzotti. 2022. "Carbon Dioxide Mineralization in Recycled Concrete Aggregates Can Contribute Immediately to Carbon-Neutrality." *Resources, Conservation and Recycling* 184(September):106436. https://doi.org/10.1016/j.resconrec.2022.106436.

Salman, M., Ö. Cizer, Y. Pontikes, R.M. Santos, R. Snellings, L. Vandewalle, B. Blanpain, and K. Van Balen. 2014. "Effect of Accelerated Carbonation on AOD Stainless Steel Slag for Its Valorisation as a CO_2-Sequestering Construction Material." *Chemical Engineering Journal* 246:39–52. https://doi.org/10.1016/j.cej.2014.02.051.

Scrivener, K., F. Martirena, S. Bishnoi, and S. Maity. 2018. "Calcined Clay Limestone Cements (LC3)." *Cement and Concrete Research* 114(December):49–56. https://doi.org/10.1016/j.cemconres.2017.08.017.

Seddik Meddah, M. 2017. "Recycled Aggregates in Concrete Production: Engineering Properties and Environmental Impact." *MATEC Web Conference* 101:1–8. https://doi.org/10.1051/matecconf/201710105021.

Shah, I.H., S.A. Miller, D. Jiang, and R.J. Myers. 2022. "Cement Substitution with Secondary Materials Can Reduce Annual Global CO_2 Emissions by Up to 1.3 Gigatons." *Nature Communications* 13(1):5758. https://doi.org/10.1038/s41467-022-33289-7.

Shaqraa, A.A. 2024. "Measuring the Performance of CO_2 Injection into Field-Cured Concrete in the Arabian Gulf Climate: An Experimental Study." *HBRC Journal* 20(1):465–482. https://doi.org/10.1080/16874048.2024.2333681.

Sim, G., S. Hong, S. Moon, S. Noh, J. Cho, P.T. Triwigati, A.-H.A. Park, and Y. Park. 2022. "Simultaneous CO_2 Utilization and Rare Earth Elements Recovery by Novel Aqueous Carbon Mineralization of Blast Furnace Slag." *Journal of Environmental Chemical Engineering* 10(2). https://doi.org/10.1016/j.jece.2022.107327.

Srubar, M., N. Fischetti, W.V. Bockelman. 2023. "Solving Cement's Massive Carbon Problem." *Scientific American*. https://www.scientificamerican.com/article/solving-cements-massive-carbon-problem.

Stefanoni, M., U. Angst, and B. Elsener. 2018. "Corrosion Rate of Carbon Steel in Carbonated Concrete—A Critical Review." *Cement and Concrete Research* 103:35–48. https://doi.org/10.1016/j.cemconres.2017.10.007.

Strunge, T., P. Renforth, and M. Van der Spek. 2022. "Towards a Business Case for CO_2 Mineralisation in the Cement Industry." *Communications Earth and Environment* 3(1). https://doi.org/10.1038/s43247-022-00390-0.

Sun, L., Y. Liu, Z. Cheng, L. Jiang, P. Lv, and Y. Song. 2023. "Review on Multiscale CO_2 Mineralization and Geological Storage: Mechanisms, Characterization, Modeling, Applications and Perspectives." *Energy and Fuels* 37(19):14512–14537. https://doi.org/10.1021/acs.energyfuels.3c01830.

Supriya, R., Chaudhury, U. Sharma, P.C. Thapliyal, and L.P. Singh. 2023. "Low-CO_2 Emission Strategies to Achieve Net Zero Target in Cement Sector." *Journal of Cleaner Production* 417. https://doi.org/10.1016/j.jclepro.2023.137466.

Supuran, C.T. 2016. "Structure and Function of Carbonic Anhydrases." *Biochem Journal* 473(14):2023–2032.

Swanson, E.J., K.J. Fricker, M. Sun, and A.H. Park. 2014. "Directed Precipitation of Hydrated and Anhydrous Magnesium Carbonates for Carbon Storage." *Physical Chemistry Chemical Physics* 16(42):23440–23450. https://doi.org/10.1039/c4cp03491k.

Taha, R., N. Alnuaimi, A. Kilayli, and A. Salem. 2014. "Use of Local Discarded Materials in Concrete." *International Journal of Sustainable Built Environment* 3. https://doi.org/10.1016/j.ijsbe.2014.04.005.

Talabi, O.O., A.E. Dorfi, G.D. O'Neil, and D.V. Esposito. 2017. "Membraneless Electrolyzers for the Simultaneous Production of Acid and Base." *Chemical Communications* 53(57):8006–8009. https://doi.org/10.1039/C7CC02361H.

Tam, V.W.Y., A. Butera, K.N. Le, and W. Li. 2020. "Utilising CO_2 Technologies for Recycled Aggregate Concrete: A Critical Review." *Construction and Building Materials* 250. https://doi.org/10.1016/j.conbuildmat.2020.118903.

Tanaka, S., K. Takahashi, M. Abe, M. Noguchi, and A. Yamasaki. 2022. "Preparation of High-Purity Calcium Carbonate by Mineral Carbonation Using Concrete Sludge." *ACS Omega* 7(23):19600–19605. https://doi.org/10.1021/acsomega.2c01297.

Tian, H., X. Yan, F. Zhou, C. Xu, C. Li, X. Chen, and X. He. 2022. "Effect of Process Conditions on Generation of Hydrochloric Acid and Lithium Hydroxide from Simulated Lithium Chloride Solution Using Bipolar Membrane Electrodialysis." *SN Applied Sciences* 4(2):47.

Tiefenthaler, J., L. Braune, C. Bauer, R. Sacchi, and M. Mazzotti. 2021. "Technological Demonstration and Life Cycle Assessment of a Negative Emission Value Chain in the Swiss Concrete Sector." *Frontiers in Climate* 3.

Tuck, C. 2022. "Iron and Steel Slag." *Mineral Commodity Summary, U.S. Geological Survey*. https://pubs.usgs.gov/periodicals/mcs2022/mcs2022-iron-steel-slag.pdf.

Vibbert, H.B., and A.H.A. Park. 2022. "Harvesting, Storing, and Converting Carbon from the Ocean to Create a New Carbon Economy: Challenges and Opportunities." *Frontiers in Energy Research—Carbon Capture, Utilization and Storage* 10(21 Sept). https://doi.org/10.3389/fenrg.2022.999307.

Vibbert, H.B., A.W.S. Ooi, and A.-H.A. Park. 2024. "Selective Recovery of Cerium as High-Purity Oxides via Reactive Separation Using CO_2 -Responsive Structured Ligands." *ACS Sustainable Chemistry and Engineering* 12(13):5186–5196. https://doi.org/10.1021/acssuschemeng.3c08103.

von Greve-Dierfeld, S., B. Lothenbach, A. Vollpracht, B. Wu, B. Huet, C. Andrade, C. Medina, et al. 2020. "Understanding the Carbonation of Concrete with Supplementary Cementitious Materials: A Critical Review by RILEM TC 281-CCC." *Materials and Structures* 53(6):136. https://doi.org/10.1617/s11527-020-01558-w.

Walling, S.A., and J.L. Provis. 2016. "Magnesia-Based Cements: A Journey of 150 Years, and Cements for the Future?" *Chemical Reviews* 116(7):4170–4204.

Wang, T., Z. Yi, J. Song, C. Zhao, R. Guo, and X. Gao. 2022a. "An Industrial Demonstration Study on CO_2 Mineralization Curing for Concrete." *iScience* 25(5):104261. https://doi.org/10.1016/j.isci.2022.104261.

Wang, X., M.-Z. Guo, and T.-C. Ling. 2022b. "Review on CO_2 Curing of Non-Hydraulic Calcium Silicates Cements: Mechanism, Carbonation and Performance." *Cement and Concrete Composites* 133. https://doi.org/10.1016/j.cemconcomp.2022.104641.

Winnefeld, F., A. Leemann, A. German, and B. Lothenbach. 2022. "CO_2 Storage in Cement and Concrete by Mineral Carbonation." *Current Opinion in Green and Sustainable Chemistry* 38(December):100672. https://doi.org/10.1016/j.cogsc.2022.100672.

Wu, B., and G. Ye. 2017. "Development of Porosity of Cement Paste Blended with Supplementary Cementitious Materials After Carbonation." *Construction and Building Materials* 145:52–61. https://doi.org/10.1016/j.conbuildmat.2017.03.176.

Yi, Z., T. Wang, and R. Guo. 2020. "Sustainable Building Material from CO_2 Mineralization Slag: Aggregate for Concretes and Effect of CO_2 Curing." *Journal of CO_2 Utilization* 40. https://doi.org/10.1016/j.jcou.2020.101196.

Yin, T., S. Yin, A. Srivastava, and G. Gadikota. 2022. "Regenerable Solvents Mediate Accelerated Low Temperature CO_2 Capture and Carbon Mineralization of Ash and Nano-Scale Calcium Carbonate Formation." *Resources, Conservation and Recycling* 180. https://doi.org/10.1016/j.resconrec.2022.106209.

Zajac, M., J. Skibsted, P. Durdzinski, F. Bullerjahn, J. Skocek, and M. Ben Haha. 2020a. "Kinetics of Enforced Carbonation of Cement Paste." *Cement and Concrete Research* 131. https://doi.org/10.1016/j.cemconres.2020.106013.

Zajac, M., J. Skibsted, J. Skocek, P. Durdzinski, F. Bullerjahn, and M. Ben Haha. 2020b. "Phase Assemblage and Microstructure of Cement Paste Subjected to Enforced, Wet Carbonation." *Cement and Concrete Research* 130. https://doi.org/10.1016/j.cemconres.2020.105990.

Zajac, M., J. Skocek, J. Skibsted, and M. Ben Haha. 2021. "CO_2 Mineralization of Demolished Concrete Wastes into a Supplementary Cementitious Material—A New CCU Approach for the Cement Industry." *RILEM Technical Letters* 6:53–60. https://doi.org/10.21809/rilemtechlett.2021.141.

Zajac, M., J. Skocek, M. Ben Haha, and J. Deja. 2022. "CO_2 Mineralization Methods in Cement and Concrete Industry." *Energies* 15(10). https://doi.org/10.3390/en15103597.

Zhan, B.J., D.X. Xuan, C.S. Poon, and C.J. Shi. 2016. "Effect of Curing Parameters on CO_2 Curing of Concrete Blocks Containing Recycled Aggregates." *Cement and Concrete Composites* 71:122–130. https://doi.org/10.1016/j.cemconcomp.2016.05.002.

Zhang, N., Y.E. Chai, R.M. Santos, and L. Šiller. 2020b. "Advances in Process Development of Aqueous CO_2 Mineralisation Towards Scalability." *Journal of Environmental Chemical Engineering* 8(6). https://doi.org/10.1016/j.jece.2020.104453.

Zhang, N., H. Duan, T.R. Miller, V.W.Y. Tam, G. Liu, and J. Zuo. 2020a. "Mitigation of Carbon Dioxide by Accelerated Sequestration in Concrete Debris." *Renewable and Sustainable Energy Reviews* 117(January):109495. https://doi.org/10.1016/j.rser.2019.109495.

Zhang, D., Z. Ghouleh, and Y. Shao. 2017. "Review on Carbonation Curing of Cement-Based Materials." *Journal of CO_2 Utilization* 21:119–131. https://doi.org/10.1016/j.jcou.2017.07.003.

Zhang, N., and A. Moment. 2023. "Upcycling Construction and Demolition Waste into Calcium Carbonates: Characterization of Leaching Kinetics and Carbon Mineralization Conditions." *ACS Sustainable Chemistry and Engineering* 11(3):866–879. https://doi.org/10.1021/acssuschemeng.2c04241.

Zhang, N., R.M. Santos, S.M. Smith, and L. Šiller. 2019. "Acceleration of CO_2 Mineralisation of Alkaline Brines with Nickel Nanoparticles Catalysts in Continuous Tubular Reactor." *Chemical Engineering Journal* 377. https://doi.org/10.1016/j.cej.2018.11.177.

Zhang, D., and Y. Shao. 2016. "Effect of Early Carbonation Curing on Chloride Penetration and Weathering Carbonation in Concrete." *Construction and Building Materials* 123:516–526. https://doi.org/10.1016/j.conbuildmat.2016.07.041.

Zhao, D., J.M. Williams, A.-H.A. Park, and S. Kawashima. 2023a. "Rheology of Cement Pastes with Calcium Carbonate Polymorphs." *Cement and Concrete Research* 172. https://doi.org/10.1016/j.cemconres.2023.107214.

Zhao, D., J.M. Williams, Z. Li, A.-H.A. Park, A. Radlińska, P. Hou, and S. Kawashima. 2023b. "Hydration of Cement Pastes with Calcium Carbonate Polymorphs." *Cement and Concrete Research* 173. https://doi.org/10.1016/j.cemconres.2023.107270.

6

Chemical CO_2 Conversion to Elemental Carbon Materials

6.1 INTRODUCTION TO ELEMENTAL CARBON PRODUCTS

Carbon accounts for ~27.3 wt% of the total mass of CO_2. Captured CO_2 could supply the elemental carbon needed for products in a net-zero future. These elemental carbon products can be divided into two categories according to the need for additional reactants and subsequent processes: directly and indirectly derived elemental carbon materials. The products in each category are described in Section 6.2. Elemental carbon materials can be directly derived from CO_2 noncatalytically or catalytically by CO_2 decomposition or by reacting with a reducing agent. Other elemental carbon materials are produced indirectly with CO_2 as their carbon source via multiple steps (e.g., reduction or decomposition of CO_2-derived chemicals). There are several CO_2 reduction reaction pathways by which captured CO_2 can result in useful elemental carbon materials, as discussed in Section 6.3.

As introduced in Chapter 2, the market for elemental carbon materials could increase by 400 percent from 2020 to 2050 (see Table 2-1), owing to increased demand for materials with novel structural and electronic properties. Increased understanding of structures and characteristics of such materials has expanded potential applications to include energy conversion (e.g., supercapacitors [Luo et al. 2023]), novel chemical and material syntheses (e.g., nonprecious-metal electrocatalysts [Collins et al. 2023]), construction materials, health care (e.g., for bioimaging), and environmental protection (e.g., photocatalytic degradation of pollutants [Yao et al. 2023]), among others (Dabees et al. 2023; Malode et al. 2023; Sasikumar et al. 2023; Son et al. 2023; Zhang et al. 2023a). As discussed in Section 2.2.5.5 of this report, some elemental carbon products of CO_2 utilization can provide durable storage of carbon, can replace highly carbon-emitting processes via material substitution, or can produce high value products, although most products in this class are likely to be small-volume and thus will not utilize large amounts of CO_2. Status and research needs for CO_2 conversion to carbon nanotubes (CNTs) were discussed in Chapter 4, Chemical Utilization of CO_2 into Chemicals and Fuels, in the 2019 National Academies report *Gaseous Carbon Waste Streams Utilization: Status and Research Needs* (NASEM 2019), although other classes of elemental carbon materials were not addressed. They were briefly discussed as long-lived CO_2-derived products, or Track 1 products, in Section 3.3.4 of the first report of this committee, *Carbon Dioxide Utilization Markets and Infrastructure: Status and Opportunities: A First Report* (NASEM 2023). Chapter 2 of the present report provides market information on elemental carbon materials, and this chapter provides an up-to-date review of research and development (R&D) on CO_2 conversion to elemental carbon materials.

CO_2-derived carbon materials can adopt multidimensional (zero- to three-dimensional [0D–3D]) structures from among more than 1500 hypothetical 3D-periodic allotropes of carbon found so far (Hoffman et al. 2016). Zero-dimensional (0D) nanocarbon materials are those having three dimensions only at the nanoscale, with no dimension larger than ~100 nm—for example, carbon quantum dots (CQDs) in sphere form and with crystal lattice or cross-linked network structures consisting of sp^2 and sp^3 hybridized[1] carbon and heteroatoms, such as O and N; graphene quantum dots (GQDs) in the form of a single truncated atomic layer of graphite (Bacon et al. 2014); and fullerene, representing a class of carbon allotropes in the form of spherical, cage molecules with carbon atoms located at the surface and vertices of a polyhedral structure consisting of pentagons and hexagons (Dhall 2023). One-dimensional (1D) nanocarbon materials are those with only one dimension beyond the nanoscale—that is, larger than ~100 nm; for example, carbon nanorods (CNRs), CNTs, carbon nanowires (CNWs), and carbon tubular clusters (CTCs). Despite their common 1D structures, CNRs, CNTs, and CNWs each possess distinguishing features. For example, CNRs have aspect ratios (length/width) of 3–5 with lengths typically of 10–120 nm (Abraham et al. 2021), whereas CNW aspect ratios can be higher than 10^3, while CNT aspect ratios can be $>10^8$. Thus, their aspect ratios are significantly different. Also, CNWs have very high specific surface areas. Moreover, CNTs are chiral materials, which governs their metallic or semiconducting character. Among CNTs are novel CTCs, which are very stable. Interestingly, the electronic properties of metallic CTCs are not affected by their diameters, number of walls, or chirality. Two-dimensional (2D) nanocarbon materials have only two dimensions beyond the nanoscale—that is, larger than ~100 nm. These include carbon nanofilms with thin layers of material spanning from a fraction of a nanometer to several micrometers in thickness (Ranzoni and Cooper 2017); carbon nanolayers with monolayer, bilayers, and multilayers (Schaefer 2010); graphene; and nanocoatings. Three-dimensional (3D) carbon nanomaterials (CNMs) are those with three dimensions beyond nanoscale—that is, all larger than ~100 nm; for example, bulk carbon powders, graphite, carbon fibers (CFs), carbon foams, carbon-carbon composites (CCCs), bundles of CNWs and CNTs, as well as multinanolayers, and dispersions of nanoparticles (Chung 2002; Park 2015; Spradling and Guth 2003; Terrones et al. 1998; Windhorst and Blount 1997; Wissler 2006; Zhao et al. 2023).

Each type of 0D–3D carbon material consists of atoms with unique electronic orbital hybridization state(s), and thus specific characteristics and applications. As shown in Figure 6-2, the carbon materials with sp, sp^2, and sp^3 hybridizations, created intrinsically and extrinsically via different defect engineering approaches, typically have different applications in energy conversion and storage areas—for example, Li-ion, Na-ion, and K-ion batteries (Rajagopalan et al. 2020), supercapacitors, the hydrogen evolution reaction, and the oxygen reduction reaction (Luo et al. 2023). CQDs typically have sp^2 carbon cores and sp^3 carbon shells terminated with –OH, –COOH, $–NH_2$, and other functional groups resulting from their preparation processes, which determine the properties and applications of CQDs. Also, the amount of sp^2 carbon and the ratio of sp^2 to sp^3 carbon can significantly affect the photoluminescence properties, and thus the bioimaging ability, of CQDs (Jana and Dev 2022; Jhonsi 2018). Furthermore, the amount of $sp^2 + sp^3$ bonds in the carbon shell or on the surface of CQDs has a substantial effect on the solubilities of CQDs in water and other solvents (e.g., ethanol). The water solubilities of CQDs determine their various applications in many fields, including solar cells (Kim et al. 2021), drug delivery (Jana and Dev 2022; Zoghi et al. 2023), and gene delivery (Rezaei and Hashemi 2021). The carbon atom hybridization structures of 0D–3D carbon materials can change significantly upon chemical modification—for example, ozone and high-temperature oxidation and irradiation. For example, all carbon atoms in pristine or pure graphite have sp^2 hybridization. Pristine graphene with its sp^2 hybridization, although extremely interesting owing to its large specific surface area, unusual physicochemical properties and extraordinary anisotropic mechanical strength, and exceptional thermal and electronic conductivity, is a zero-band-gap semiconductor (a semi-metal), which makes it uninteresting for a number of device applications (Gui et al. 2008; Mbayachi et al. 2021; Rani and Jindal 2013). To make graphene widely useful in the electronics space, its band gap needs to be opened and tunable according to application requirements. Pristine graphene therefore needs to be modified in various ways (e.g., UV-light-assisted oxidation [Güneş et al. 2011]) to enable its carbon atom hybridization structures to change from 100 percent sp^2 to mixtures of sp^2 and sp^3 with different sp^2/sp^3 ratios to achieve improved properties and more applications, including in electronic, electromagnetic, and optical devices and for catalysis. Representative 0D–3D carbon materials, their typical carbon hybridization characteristics, and their major properties are summarized in Table 6-1.

[1] Orbital hybridization of carbon atoms in different types of elemental carbon materials are described as sp, sp^2, and sp^3, describing the amount of s- and p-type orbital character of the carbon atoms. The amount of s or p character impacts the bonding between the carbon atoms, with sp having linear, sp^2 having trigonal planar, and sp^3 tetrahedral character. See Figure 6-2 for images showing the orbital hybridization and illustration of the impact on bonding in elemental carbon materials.

TABLE 6-1 0D–3D Carbon Materials and Their Structural Characteristics and Major Properties

	0D	1D	2D	3D
Representative carbon materials	- Carbon quantum dots - Graphene quantum dots - Fullerenes	- CNTs - CNWs - CNRs - CTCs - Carbon nanofibers	- Graphene - Carbon nanofilms - Carbon nanolayers	- Bulk carbon powders - Graphite - CFs - Carbon foams - CCCs
Typical orbital hybridization characteristics of their carbon atoms	- sp^2 (e.g., fullerene and the aromatic domain or core of CQDs) - sp^2 + sp^3 (e.g., CQDs and GQD)	- sp + sp^2 (e.g., CNWs) - sp^2 (e.g., CNFs, CNTs and CTCs) - sp^2 +sp^3 (e.g., CNTs)	- sp^2 (e.g., pristine graphene) - sp^2 + sp^3 (e.g., carbon nanolayers, carbon nanofilms)	- sp^2 (e.g., graphite) - sp^3 (e.g., diamond) - sp^2 +sp^3 (e.g., CFs)
Major properties	- Low toxicity - Biocompatibility - Outstanding photostability - Tunable - Multicolored emission - High-water solubility - Good dispersibility - Easy surface grafting for different applications	- Good conductivity - Abundant reactive hydroxyl groups distributed on surface of major 1D product—CNTs - More active sites, resulting from surface modification - Hydrogen storage capacity - Thermodynamically favorable for mass transfer and diffusion of reaction substances - Diathermancy - High mechanical strength	- High surface area - Superior mechanical flexibility - High electronic mobility - Abundant photo-electrochemically reactive sites - Low recombination of photogenerated charges - Good mechanical properties	- High surface area - Designability - Well-developed porous channels for ion transfer - Large specific surface area - Good reaction micro-environment - More active sites - Potentially very high mechanical strength

NOTES: Orbital hybridization patterns of carbon atoms in different types of materials are described as sp, sp^2, and sp^3, where the s and p describes the amount of s- or p-type orbital character around the carbon atoms. The amount of s or p character impacts the bonding around the carbon atoms, with sp having linear, sp^2 having trigonal planar, and sp^3 having tetrahedral character which in turn affects the material properties. See Figure 6-2 for images showing the orbital hybridization and illustration of the impact on bonding.
SOURCES: Based on data from Budyka et al. (2017); Cartwright et al. (2014); He et al. (2021); Khan and Alamry (2022); Lee et al. (2020); Lesiak et al. (2018); Liu et al. (2020a, 2020b); Tsang et al. (2006); Van Tran et al. (2022); Zhang et al. (2023b). Zhao et al. (2003); Zhou and Zhang (2021).

Carbon materials with different hybridization structures sometimes have been (at laboratory scale) utilized similarly, as evidenced in Figure 6-2. For example, both sp- and sp^2-hybridized carbon materials could be used for synthesizing supercapacitors owing to their similar structures and properties, such as high conductivity and stability, and thus can provide the same function (Luo et al. 2023). Both CQDs and CNRs can be used as sensing materials because both contain sp^2 hybridized structures; their sensing capabilities increase with the amount of sp^2 hybridized carbon present. CQDs are being used as components of light-emitting diodes, luminescent solar concentrators, and photovoltaic cells; they also can be functionalized to act as catalysts (Zhou et al. 2024).

As indicated in Figure 6-1, a variety of pathways exist for producing 0D–3D carbon materials from CO_2. The major challenge of converting CO_2 into elemental carbon materials is to reduce the formal oxidation state of C from +4 to 0 via various reduction reactions, which may be realized with thermochemical, electrochemical, photochemical, plasmachemical, or hybrid/tandem processes. All of these technologies are still in the R&D phase at the present time, and each has its advantages and disadvantages from energy consumption and environmental impact perspectives, as discussed in the following sections of this chapter.

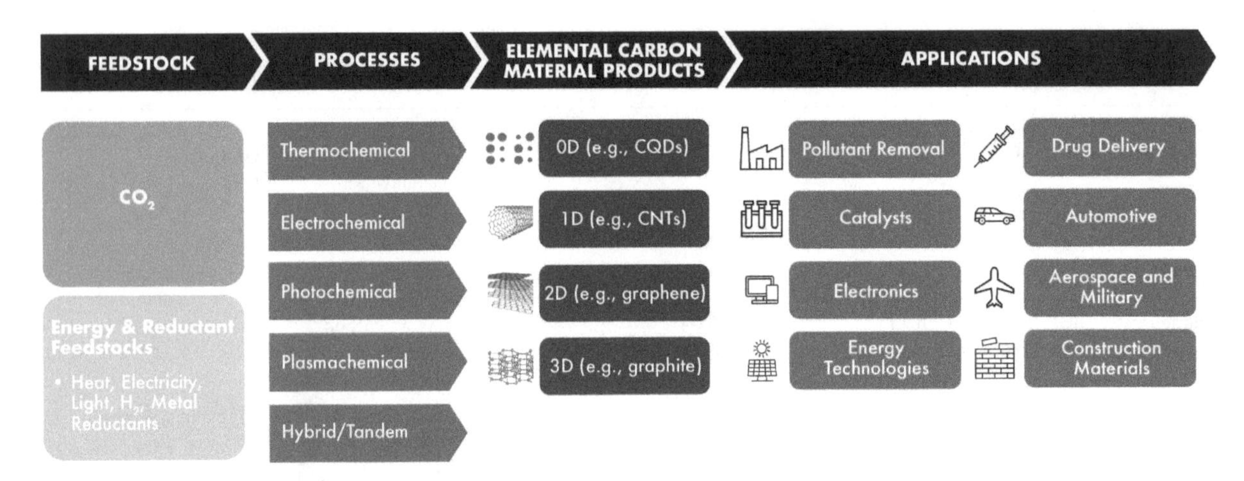

FIGURE 6-1 Summary of the feedstock inputs, processes, products, and applications for 0D–3D carbon materials.
SOURCE: Icons from the Noun Project, https://thenounproject.com. CC BY 3.0.

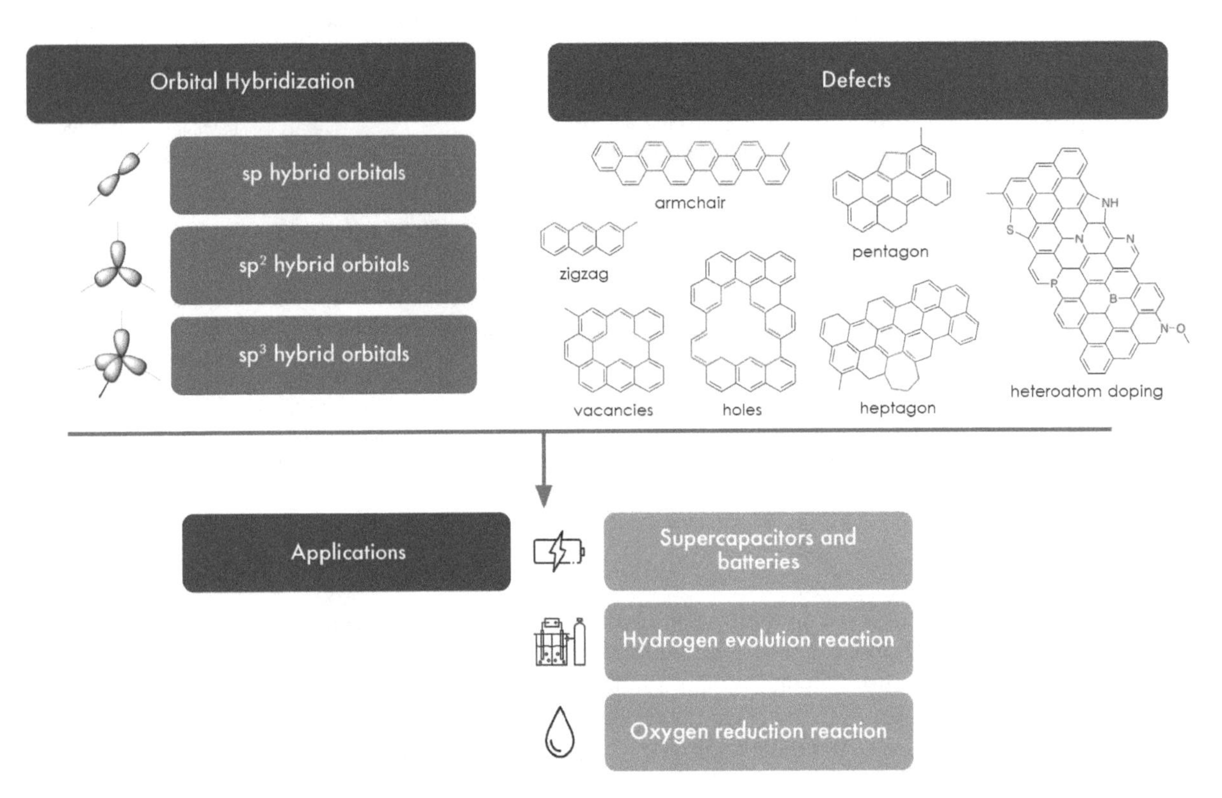

FIGURE 6-2 Applications of differently hybridized and defective carbon materials, as shown, in energy conversion and storage.
SOURCES: Adapted from X. Luo, H. Zheng, W. Lai, P. Yuan, S. Li, D. Li, and Y. Chen, 2023, "Defect Engineering of Carbons for Energy Conversion and Storage Applications," *Energy and Environmental Materials*, Wiley. Orbitals sourced from University of Saskatchewan – Hybrid Orbitals, https://openpress.usask.ca/intro-organic-chemistry/chapter/1-5. CC BY NC-SA 4.0. Icons from the Noun Project, https://thenounproject.com. CC BY 3.0.

6.2 CO_2-DERIVED ELEMENTAL CARBON MATERIALS

6.2.1 Directly Derived Elemental Carbon Materials

6.2.1.1 Fullerenes

Fullerenes consist of sp^2 hybridized carbon atoms linked by single and double bonds, forming a spherical- or cylindrical-shaped closed structure (Ramazani et al. 2021). The most famous of the closed spherical fullerenes is C_{60}, for the 60 carbon atoms in the molecule, also known as buckminsterfullerene or a buckyball, for the molecule's resemblance to both the geodesic domes of Buckminster Fuller and a standard soccer ball. Fullerene C_{60} has a van der Waals diameter of about 1.1 nm and a nucleus-to-nucleus diameter of about 0.71 nm. Open-ended cylindrical fullerenes are known as CNTs; single-walled CNTs typically have 0.5–2 nm diameters. Owing to their unique structure and electronic characteristics, the applications of fullerenes are extensive, including photodynamic therapy, drug and gene delivery, nano-sensors, battery electrodes, and organic solar cells, covering many industries (Anctil et al. 2011; Ramazani et al. 2021). Hence, there is significant interest among chemical and material scientists in developing fullerene synthesis methods. General strategies for fullerene synthesis involve generation of fullerene-rich soot through arc discharges, combustion, laser ablation, or microwaves, and then purification of the soot by toluene or benzene washing, Soxhlet extraction, or active carbon filtering (Keypour et al. 2013; Komatsu et al. 2004; Parker et al. 1992; Ramazani et al. 2021; Taylor et al. 1993).

The conventional carbon sources for fullerenes are graphite, aliphatic and olefinic hydrocarbons, chloroform, and aromatics (e.g., naphthalene) (Ramazani et al. 2021); however, there is increasing interest in using captured CO_2 as a feedstock for fullerene synthesis (Chen and Lou 2009; Motiei et al. 2001). Chen and Lou (2009) reported that CO_2 can be successfully reduced to C_{60} by metallic lithium at 700°C and 100 MPa, as confirmed by ultraviolet (UV)-visible absorption spectroscopy, matrix-assisted laser desorption/ionization (MALDI) time-of-flight mass (TOF) mass spectrometry (MS), and high-performance liquid chromatography. Higher fullerenes (C_{70}, C_{78}, etc.) were not detected using this synthetic approach. The authors postulated that CO_2 radical anion ($CO_2^{\bullet-}$) or other single carbon radicals, resulting from the transfer of an electron from elemental Li to supercritical CO_2 (whose polarity increases with its density, thus facilitating electron transfer [Tucker 1999]), are possible key intermediates during the reduction of CO_2 to C_{60}. The generation of the $CO_2/CO_2^{\bullet-}$ couple in the reaction system is the key step for the formation of C_{60} owing to the fact that the standard potential of the redox $CO_2/CO_2^{\bullet-}$ couple in an aprotic solvent such as N,N-dimethylformamide (DMF) containing a counter-cation (tetraethylammonium, NEt_4^+) can be as negative as −2.2 V versus saturated calomel electrode (Bhugun et al. 1996) and the standard potential of the redox couple Li^+/Li, 3.04 V, although the potentials of the redox $CO_2/CO_2^{\bullet-}$ and Li^+/Li couples in this reaction system were not measured (Chen and Lou 2009; Motiei et al. 2001). The finding by Motiei et al. (2001) was used by Chen and Lou to explain the feasibility of C_{60} formation.

6.2.1.2 Hollow Carbon Spheres

Unlike 0D fullerene, hollow carbon spheres (HCSs) can have either 2D or 3D structure (Deshmukh et al. 2010; Liu et al. 2019) and both sp^2 and sp^3 hybridized carbon atoms on their surface (Deshmukh et al. 2010). HCSs have diameters between 2 nm and several microns. HCSs have unique properties, including encapsulation ability, controllable permeability, surface functionality, high surface-to-volume ratios, and excellent chemical and thermal stabilities (Li et al. 2016). HCSs can be synthesized with hard-templating, soft-templating, and template-free processes. The properties of HCSs vary, depending on the raw carbon materials and synthesis conditions used. For example, the surface areas of HCS can change from a few m^2/g to more than 1000 m^2/g, which determines their potential applications, especially those of functionalized HCSs.

Synthesis of HCSs from CO_2 has been demonstrated using the microbubble-effect-assisted electrolytic method in a $CaCO_3$-containing LiCl–KCl melt electrolyte at 450°C (Deng et al. 2017). The authors employed precise control of the electrode potential to tune the electrochemical reduction rate of carbonate ions and the CO microbubble effect to shape the hollow spheres within the resultant carbon sheets. The produced HCSs exhibited good plasticity and capacitance, which are desirable properties of HCSs used for battery, capacitor, and fuel cell applications, and composite materials. Li et al. (2018) also synthesized HCSs from CO_2 using molten carbonate electrolyzers and proposed the following mechanism for the conversion:

Cathode reaction: $CO_3^{2-} + 4e^- \rightarrow C + 3O^{2-}$ (R6.1)

Anode reaction: $2O^{2-} - 4e^- \rightarrow O_2$ (R6.2)

Electrolyte reproduction: $O^{2-} + CO_2 \rightarrow CO_3^{2-}$ (R6.3)

6.2.1.3 Carbon Nanofiber

1D carbon nanofibers (CNFs) with sp^2 hybridized carbon atoms (Deng et al. 2013; Zhou et al. 2020) can be used for energy storage (Zhang et al. 2016), electrochemical catalysis (Shakoorioskooie et al. 2018), sensor manufacturing (Sengupta et al. 2020), and high-strength building material development (Ren et al. 2015). Traditional CNF synthesis methods include electrospinning/carbonization, arc/plasma, and chemical vapor deposition (CVD), with electrospinning/carbonization being the principal approach. Conventional raw materials for CNF preparation include polyacrylonitriles, pitch, acetone, and hydrocarbon gases (Ren et al. 2015) but recently, renewable feedstocks (e.g., lignin and cellulose) for CNF manufacturing have received considerable attention (Lallave et al. 2007; Wu et al. 2013).

Several research groups have demonstrated that CO_2 could be a good candidate for CNF synthesis (Lau et al. 2016; Novoselova et al. 2007; Ren et al. 2015; Xie et al. 2024). Ren et al. (2015) reported a one-pot synthesis method for synthesizing CNFs via electrolytic conversion of CO_2. The technology, based on Li_2O looping for CO_2 reduction, employs low-cost, scalable nickel and steel electrodes to decompose CO_2 into CNFs and O_2, producing CNFs with diameters of 200–300 nm and lengths of 20–200 μm. The Coulombic efficiency is higher than 80 percent and can be close to 100 percent (i.e., complete decomposition) if all the products can be collected. Lau et al. (2016) developed a system integrating Li_2CO_3 electrolysis with a combined cycle natural gas power plant to produce CNFs and pure O_2. The system consumes 219 kJ to convert 1 mol of CO_2 to 1 mol of carbon, while the pure O_2 generated from CO_2 decomposition is sent back to the gas turbine, improving electricity generation efficacy. Xie et al. (2024) reported an electrochemical–thermochemical tandem catalysis system to convert CO_2 to CNF using renewable hydrogen, which achieved an average yield of 2.5 $g_{carbon}\ g_{metals}^{-1}h^{-1}$ ("metals" here refers to the catalyst used for conversion; in Xie et al. [2024] this was an FeCo alloy and extra metallic Co). In summary, CO_2 conversion to CNFs shows promise as a method to reduce CO_2 emissions while generating high-value CNFs with market potential in nanoelectronics, energy storage, and construction materials.

6.2.1.4 Carbon Nanotubes

Carbon nanotubes (CNTs) are rolled graphene sheets. CNTs are 1D carbon materials with sp^2 or $sp^2 + sp^3$ hybridized carbon atoms and are classified as single-walled nanotubes (SWNTs) or multiwalled nanotubes (MWNTs). SWNTs are open-ended cylinders with only one wrapped graphene sheet, while MWNTs are an assembly of homocentric SWNTs. The dimensions of SWNTs and MWNTs differ in both length and diameter, leading to considerably different properties (Kozinsky and Mazari 2006; Rathinavel et al. 2021). For example, unlike SWNTs, the mechanical properties (e.g., Young's modulus) of MWNTs depend not only on diameter and chirality but also on the number of sidewalls (Rathinavel et al. 2021). Various methods have been developed for preparing CNTs, including arc discharge, laser ablation, CVD, and injection of carbon atoms into metal particles (Rodríguez-Manzo et al. 2007). Factors affecting CNT syntheses include the carbon source (e.g., hydrocarbons, alcohols), catalyst (e.g., Al_2O_3), temperature, pressure, and flow of gases. CNTs have been explored for use in many applications, including sensing (Wang et al. 2023), cancer therapy (Mishra et al. 2023), preparation of biological fuel cells (ul Haque et al. 2023), light-weight reinforced high-strength materials (Hong et al. 2023), Li/Na-ion batteries (Qu et al. 2024), and others (Kordek-Khalil et al. 2024).

CO_2 can be used as a carbon source for CNT synthesis. Research to-date primarily has explored electrochemical conversion processes (Douglas et al. 2018; Li et al. 2018; Licht et al. 2016; Moyer et al. 2020). For example,

Licht et al. (2016) reported that ambient CO_2 could be successfully captured and converted to CNTs and CNFs in molten lithium carbonates at high yield via electrolysis using inexpensive steel electrodes, with the resultant carbon materials exhibiting good, stable capacities for energy storage. Douglas et al. (2018) synthesized CNTs with small diameters (~10 nm) using ambient CO_2 and Fe catalysts. They concluded that (1) the energy input costs for the conversion of CO_2 into CNTs are $\$50/kg_{CNT}$ and $\$5/kg_{CNT}$ using Al_2O_3 and ZrO_2 as thermal insulation materials, respectively, and (2) the CO_2 to small-diameter CNTs technology is superior to other CO_2 conversion technologies with lower-value materials as their products. Li et al. (2018) found that the electrolyte composition in a molten carbonate electrolyzer that captures CO_2 from air and converts it to CNTs and HCSs plays a key role in the selectivity toward CNTs, as well as in determining the diameter of the synthesized CNTs.

6.2.1.5 Graphene

2D graphene, with its sp^2 hybridized carbon atoms and very stable structure, is the thinnest (sheet thickness of 0.34 nm) and strongest nanomaterial known (Yu et al. 2020). Graphene itself has limited applications owing to its easy agglomeration, and difficult processing (Yu et al. 2020), and it requires modification and functionalization to increase its potential applications. Graphene synthesis techniques include exfoliation of graphite, reduction of graphene oxide, thermal and plasma CVD of hydrocarbons, thermal decomposition of silicon carbide, and unzipping of CNTs. Graphene functionalization processes are based on (1) the formation of covalent bonds between graphene and introduced functional groups (e.g., –OH and –COOH); (2) the formation of non-covalent bonds (e.g., π–π interactions, hydrogen, ionic, and dative bonding); and (3) element doping (Yu et al. 2020). The primary challenges facing production and use of graphene and functionalized graphene are high costs and carbon/environmental footprint, determined by the characteristics of typical synthesis methods. For example, graphene oxide (GO) reduction requires the use of highly corrosive agents and thus a long washing process after reduction, energy-intensive CVD, and low quality of large-scale production (Liu et al. 2020c; Urade et al. 2023). Nonetheless, graphene and functionalized graphene could have a wide range of applications—including in supercapacitors, solar cells, electrodes, and e-textiles—owing to their many desirable properties (Su et al. 2020; Urade et al. 2023; Yu et al. 2024), including rich functional group variability and density, environmental stability and compactness (Su et al. 2020). For example, the deleterious ion migration that reduces operational stability of iodide perovskite solar cells synthesized with organic–inorganic halide perovskite materials, resulting from the weak Coulomb interactions in the perovskite lattice, can be suppressed by using graphene, which has a lattice parameter (0.246 nm) smaller than the radius of I^- (0.412 nm) (Su et al. 2020).

CO_2 also has been explored as a feedstock for graphene synthesis (Chakrabarti et al. 2011; Hu et al. 2016; Liu et al. 2020c; Strudwick et al. 2015; Wei et al. 2016). For example, Liu et al. (2020c) used molten carbonate electrolysis to synthesize graphene, where the conversion occurs via carbonate formation in Li_2O, electrolysis of Li_2CO_3, and then exfoliation of the resultant carbon nanoplatelets:

$$\text{Chemical dissolution and carbonate formation: } CO_2 + Li_2O \rightarrow Li_2CO_3 \quad \text{(R6.4)}$$

$$\text{Electrolysis: } Li_2CO_3 \rightarrow C_{platelets} + O_2 + Li_2O \quad \text{(R6.5)}$$

$$\text{Exfoliation (DC voltage): } C_{platelets} \rightarrow C_{graphene} \quad \text{(R6.6)}$$

Addition of zinc and increased electrolysis current led to the selective (more than 95 percent yield) formation of high-purity carbon nanoplatelets rather than CNTs, and exfoliation of the carbon nanoplatelets produced graphene in 83 percent yield by mass of the original carbon nanoplatelets (Liu et al. 2020c).

6.2.1.6 Graphite

Graphite is a 3D material with a stacked planar sp^2-hybridized C6 fused ring structure—that is, stacked layers of graphene with AB stacking (Jara et al. 2019). It is known for its high specific surface area, thermal conductivity, fracture strength, and special charge transport phenomena. It can be obtained as naturally occurring graphite or produced by synthesis (Surovtseva et al. 2022). Natural occurring graphite is mined but requires energy- and chemical-intensive beneficiation and purification thereafter, especially for its use in batteries (Surovtseva et al. 2022). Synthetic graphite is tunable in microstructure and morphology. The synthesis typically includes two sequential steps: formation of amorphous carbon via carbonization of a carbon precursor and subsequent graphitization of the amorphous carbon. Both steps occur at high temperature and are therefore energy-intensive and generally CO_2-emitting. Synthetic graphite can be prepared by various methods, including graphitizing nongraphitic carbons (e.g., cokes [Gharpure and Vander Wal 2023]), processing hydrocarbons (e.g., agricultural wastes or biomass materials [Yap et al. 2023]), CVD at >2500°C, and decomposing unstable carbides. Graphite is used in batteries (Kim et al. 2024), refractories (Chandra and Sarkar 2023), metallurgical processing (Li et al. 2023a), and other fields.

Some researchers have synthesized graphite from CO_2 (Chen et al. 2017a; Hu et al. 2015, 2019; Hut et al. 1986; Liang et al. 2021; Ognibene et al. 2003; Yu et al. 2021). For example, Liang et al. (2021) prepared synthetic graphite submicroflakes by heating CO_2 in the presence of lithium aluminum hydride ($LiAlH_4$) at 126°C. This synthetic graphite was compared to commercial graphite to test its potential application as an anode for lithium storage materials, and both showed stable reversible capacities around 320 mAh g^{-1} from the 1st to 100th cycles at a current density of 0.1 A g^{-1}. After 100 cycles, the synthetic graphite and commercial graphite achieved 99 percent and 95.4 percent retention efficiencies, respectively, suggesting that the synthetic graphite prepared with CO_2 is superior to its commercial counterpart, especially considering that it was generated without a separate graphitization step. Electrochemical methods also can be used to produce graphite from CO_2, with Chen et al. (2017a) demonstrating that CO_2 captured from synthetic flue gas by a molten salt ($Li_2CO_3 - Na_2CO_3 - K_2CO_3 - Li_2SO_4$) at as low as 775°C (the electrolysis temperature) without the use of any catalyst can produce nano-structured graphite.

6.2.2 Indirectly Derived Carbon-Rich Carbon Materials

6.2.2.1 Carbon Fiber

Carbon fibers (CFs), with a $7.1 billion market in 2023 and annual growth rate of 12.6 percent (marketsandmarkets 2024), have extremely useful properties, including high elastic moduli, compressive and tensile strengths, and thermal and electrical conductivities, as well as low coefficients of thermal expansion (Hiremath et al. 2017). CFs are reinforcing materials widely used in airplanes, cars, and wind turbine blade manufacturing (Liu and Kumar 2012). The most common carbon source for CF synthesis is polyacrylonitrile (Le and Yoon 2019). Polyacrylonitrile-based CF manufacturing includes fiber spinning, stabilization, carbonization, and graphitization steps (Kaur et al. 2016). Synthesis of polyacrylonitrile is based on the free-radical polymerization of acrylonitrile (Pillai et al. 1992), which is produced via catalytic ammoxidation of propylene. Propylene, in turn, is generally produced from the reaction of ethylene with 2-butene, where the latter is synthesized via ethylene dimerization (Pillai et al. 1992). Thus, ethylene is a critical intermediate in CF production.

As discussed in Chapter 7 of this report, ethylene can be produced via thermochemical, electrochemical, or potentially photochemical conversion of CO_2, and thus, CO_2 can play an indirect role in CF syntheses (Li et al. 2020; Pappijn et al. 2020). The development of highly active, selective, and stable CO_2 to ethylene catalysts would facilitate the CO_2-to-CF pathway.

6.2.2.2 Carbon-Carbon Composites

As noted in Section 6.2.2.1, CO_2 can be a raw material for producing the critical precursor to CFs, which then could be used subsequently to produce CCCs. CCCs are lightweight, high-strength materials with good electrical properties, making them attractive for a wide spectrum of applications. The CCC manufacturing process involves saturation (impregnation) of other materials into carbon matrices, followed by graphitization or carbonization (a pyrolysis process) to form a graphitic structure (Windhorst and Blount 1997). During pyrolysis, voids form because of volatilization, which is deleterious for the mechanical properties of the CCCs. Repeated impregnation and carbonization can address this problem, but repetition increases manufacturing time and thus cost. Additionally, as-made CCCs may not possess adequate microstructure, porosity, interlaminar shear strength, flexural, ultrasonic and vibration damping behavior for certain applications (Bansal et al. 2013). Further modifications may be necessary to achieve the desired qualities, including physical treatments (e.g., plasma-based surface changes) and chemical treatments or use of additives.

Among the possible additives are carbon nanomodifiers (CNMOs), or the carbon nanostructure materials introduced in Table 6-1. Some CNMOs, such as nanographene, have been synthesized from CO_2, as discussed in Section 6.2.1. CNMOs might be able to overcome the abovementioned challenges faced during CCC manufacturing, combining with other matrices, such as pitch, to enhance the properties of CCCs. CNMOs can mitigate shrinkage and tailor the properties of CCCs. For example, Bansal et al. (2013) introduced nanographene platelets (NGPs) to fill defects such as pores and cracks during manufacturing of CCCs. When the NGP-to-CCC ratio was 1.5 wt%, the interlaminar shear strength, flexural strength, and Young's modulus of the CCCs increased by 22 percent, 27 percent, and 15 percent, respectively, compared to CCCs that did not contain NGPs. Meanwhile, the porosity of the modified CCCs was reduced by 17.5 percent. Eslami et al. (2015) filled carbon-fiber/phenolic composites with MWNTs. When the sample was modified by 1 wt% MWNTs, the thermal stability of the CCCs increased, according to thermogravimetric analysis, and the linear and mass ablation rates decreased by about 80 percent and 52 percent, respectively. Scanning electron microscopy showed the formation of a strong carbon network in CCCs resulting from the addition of MWNTs. These examples suggest that CO_2-derived CNMs can play an important role in the development of high-quality CCCs.

6.3 EMERGING TECHNOLOGIES TO REDUCE CO_2 TO ELEMENTAL CARBON

6.3.1 Introduction

As introduced above, the reduction of CO_2 to elemental carbon (CTEC) can be performed by four major chemical reaction processes: thermochemical, photochemical, electrochemical, and plasmachemical reduction (see Figure 6-1). Table 6-2 summarizes the strengths and shortcoming of these four chemical conversion pathways for CTEC. The four pathways could be coordinated to develop potentially more efficient and less expensive CTEC processes by combining their strengths and overcoming their shortcomings, as listed in Table 6-2. For example, the temperature required for thermochemical CO_2 conversion to CNTs can be as high as 700°C (Lou et al. 2006), however, a combined photo-thermochemical CO_2-to-CNT process can proceed at as low as 80°C (Duan et al. 2013).

The majority of CTEC conversion technologies are still in the lab-scale study phase, and thus their development status can be described as "emerging," although the specific technology readiness levels (TRLs) of different conversion technologies vary. Initial research on CTEC processes examined thermochemical pathways, and as early as 1978, cation-excess magnetite was used to convert CO_2 to carbon via CTEC with an efficiency of nearly 100 percent at 290°C, although the structure of the generated carbon wasn't reported by the researchers (Tamaura and Tahata 1990). Thermochemical CTEC processes are at higher TRLs than other pathways, with bench-scale[2] and pilot-scale[3] conversions being successfully demonstrated. As of April 2024, no CTEC technology has been commercialized.

Some CTEC processes are effective at forming specific high-value carbon materials but are very energy-intensive. For example, high-quality CNMs, such as CNFs and CNTs, can be produced by CVD, but this method requires very high temperatures and low pressures over long periods of time and is not easily scalable. As a result, this method is estimated to have an unusually large carbon footprint of up to 600 tonnes of CO_2 emitted per tonne of CNM produced (Wang et al. 2020). Alkali and alkaline earth metals, such as lithium, sodium, magnesium, and calcium, can be used as reductants to reduce CO_2 to various carbon products, including carbon spheres, graphene, and CNTs. However,

[2] For example, 100 percent direct conversion of CO_2 to C has been demonstrated at the bench scale with $Ni_{0.39}Fe_{2.6}O_4$ (reaction conditions: flow rate of simulated flue gas: 9 dm^3/h; composition of the simulated flue gas flue gas: 20% CO_2 and 80% N_2) (Taylor et al. 1993).

[3] The bench-scale result described in Taylor et al. (1993) was tested at pilot scale by designing a system with the capacity to treat 1000 Nm^3 h^{-1} flue gas (composition: 9.4% CO_2, 74.8% N_2, 0.8% O_2, 15% H_2O, 50 ppm NO_x) from a liquefied natural gas combustion boiler (Yoshida et al. 1997).

TABLE 6-2 Comparison of the Strengths and Weaknesses of the Four Major State-of-the-Art (SOA) Chemical Reduction Processes Being Explored for CTEC Material Conversion

	Thermochemical Reduction	Photochemical Reduction	Electrochemical Reduction	Plasmachemical Reduction
Strengths	- High conversion rate - Easy scale up - Relatively mature in terms of material preparation and regeneration, as well as equipment manufacturing and operation - When combined with photochemical reduction, relatively easy to synergize, disperse, and activate catalysts	- Moderate reaction conditions (e.g., temperature as low as 80°C) - Relatively low energy requirements for reaction process - Easy to avoid side reactions and thus by-products with photocatalyst defect engineering - Relatively easy to recycle spent catalysts	- Relatively easy to realize high selectivity for desired product(s) - Environmentally benign reaction process - Short starting time - Relatively less expensive	- Quick to reach reaction conditions - Easy to increase internal energy of reactants
Weaknesses	- Relatively high temperature requirements for reaction, which could deactivate materials via coking, and thus decrease CO_2 conversion rates - Relatively low overall energy utilization efficiency	- Suboptimal conversion of electric energy into radiation energy of desired wavelengths of light-emitting diodes (LEDs) - Relatively high loss of the heat generated from some light sources - Difficult to achieve high conversion and scale up owing to limited direct light access (surface area)	- Slow mass transfer of reactants to the active surface area of electrode - Not cost-effective owing to the use of precious metal electrocatalysts - Frequently requires expensive product separation methods (e.g., membranes)	- Generally low product selectivity - Energy generation rate of plasma system is much higher than the total energy consumption rate of reactions - Relatively high heat loss owing to conduction leads to low energy efficiency - Low catalyst target selectivity

SOURCES: Based on data from the following: Thermochemical: Álvarez et al. (2017); Kondratenko et al. (2013); Kosari et al. (2022); Ye et al. (2019); Zuraiqi et al. (2022). Photochemical: Duan et al. (2013); Han et al. (2023a); Li et al. (2014); Yaashikaa et al. (2019). Electrochemical: Lu and Jiao (2016); Overa et al. (2022); Pérez-Gallent et al. (2020); Sajna et al. (2023); Spinner et al. (2012); Tackett et al. (2019). Plasmachemical: George et al. (2021); Lerouge et al. (2001); Martirez et al. (2021); Snoeckx and Bogaerts (2017).

regeneration of these reductants is also very energy intensive. For CTEC processes to be competitive against other alternative carbon sources or processes, energy efficiency must be maximized and low-carbon energy sources used.

The distribution percentages of the 161 journal papers found with CO_2 being the sole carbon source in the four CTEC research areas is shown in Figure 6-3. Clearly, research work in the thermochemical area dominates all those four areas, accounting for 70 percent of the total work reported in published papers. The thermochemical research largely involves CTEC studies of decomposition-based reactions between CO_2 and cation-excess materials and reactions between CO_2 and strong reducing agents. Note that only published journal papers as of January 2024 were collected in the analysis given in Figure 6-3.

The history of the annual publications resulting from global CTEC R&D efforts is presented in Figure 6-4, where it is evident that this is an understudied phenomenon, yet to take off. The first CTEC paper was published in 1978, followed by a drought of CTEC research for 12 years. CTEC research productivity was stable from 1990 to 2001, during which about three papers were published annually. The average quantity of annual CTEC publications has tripled since 2015, a significant increase. However, the pace of CTEC R&D activities has been much slower than other CO_2 utilization technologies. Figure 6-4 reveals that thermochemical methods have dominated CTEC technology history, while photochemical and plasmachemical approaches have only been studied occasionally. Note that researchers are increasingly interested in developing electrochemical CTEC technologies owing to their advantages compared to thermochemical ones, as given in Appendix K.

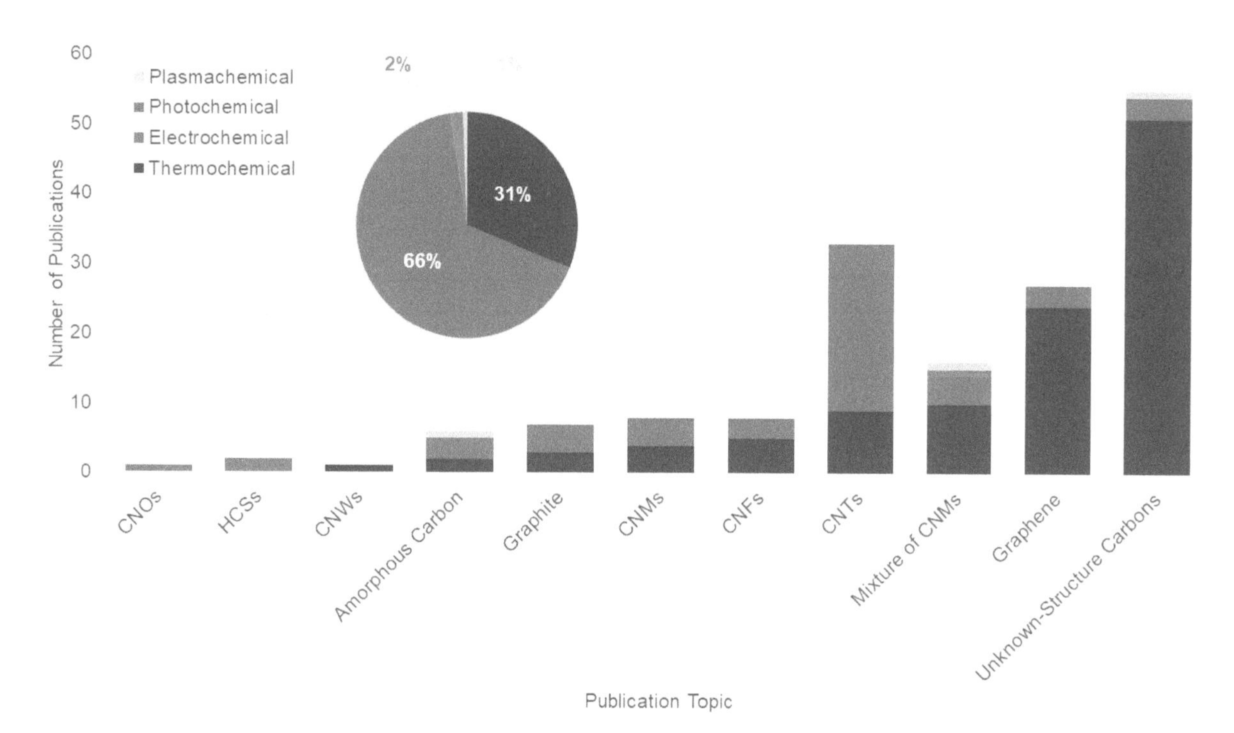

FIGURE 6-3 The distributions of published papers versus the carbon products generated in the four CTEC areas.
NOTES: CNF = carbon nanofiber; CNM = carbon nanomaterial; CNO = carbon nano-onion; CNT = carbon nanotube; CNW = carbon nanowire; HCS = hollow carbon sphere. See Appendix K for the full literature review informing this figure.

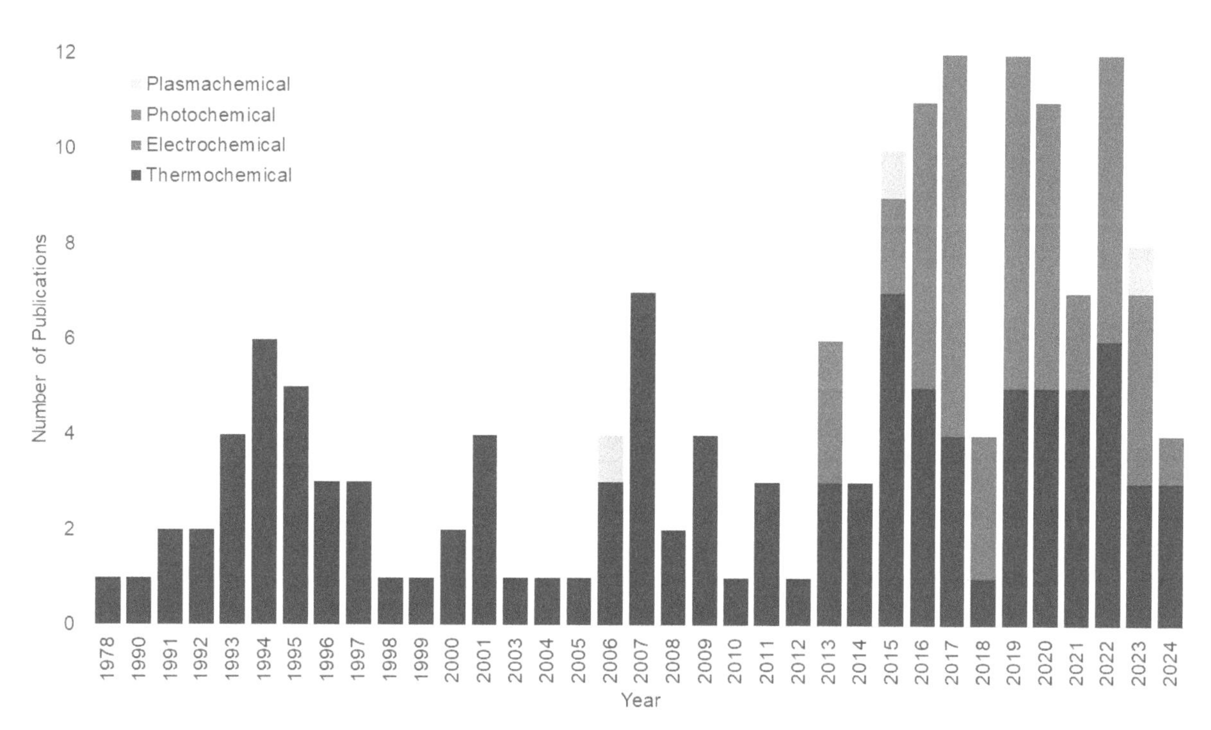

FIGURE 6-4 CTEC publications within four technological areas in different years.
NOTE: See Appendix K for the full literature review informing this figure.

6.3.2 Reaction Processes

6.3.2.1 Thermochemical Conversion Pathways

Thermochemical CTEC can involve CO_2 decomposition or CO_2 reaction with strong reducing agents—for example, H_2 or alkali and alkaline earth metals, with and without the use of other reactants or catalysts. The strengths and weaknesses of these two types of CTEC technologies are described below.

6.3.2.1.1 Decomposition-Based CTEC

The CO_2 decomposition into C can be via the R6.7 pathway:

$$CO_2 \leftrightarrow C + O_2 \quad \Delta G^0_{R6.7} = 394.4 \text{ kJ/mol} \tag{R6.7}$$

Synthesizing elemental carbon from CO_2 via R6.7 is difficult because of its large, positive $\Delta G°$ value. Research efforts have aimed to identify materials that can enable more favorable reduction pathways and thus allow lower operating temperatures to be used (Kim et al. 2001, 2019; Kormarneni et al. 1997; Lin et al. 2011; Sim et al. 2020; Tamaura and Tahata 1990; Tsuji et al. 1996a; Yoshida et al. 1997). The strategy of exposing CO_2 to a cation-excess metal oxide works because the material is oxygen-deficient and therefore at much lower temperatures than in the gas phase it is possible to strip oxygen from CO_2 by absorbing that oxygen in the metal oxide lattice, filling oxygen vacancy sites. For example, Tamaura and Tahata (1990) found that reacting CO_2 with cation-excess magnetite ($Fe_{3+\delta}O_4$, $\delta = 0.127$) can reduce CO_2 to carbon with an efficiency of nearly 100 percent at 290°C. During CO_2 reduction, all of the oxygen in CO_2 transfers as described above, in the form of O^{2-} to the cation-excess magnetite, because only carbon and no CO was detected (Tamaura and Tahata 1990).

Tsuji et al. (1996a) investigated the reactivity toward CO_2 decomposition of metallic iron formed on oxygen-deficient Ni(II)-bearing ferrite. 86 percent CO_2 conversion was observed, yielding 97 percent elemental carbon and 3 percent CO. The same group also achieved excellent elemental carbon selectivity with an ultrafine Ni(II) ferrite prepared with coprecipitation of Ni^{2+}, Fe^{2+}, and Fe^{3+} at 60°C with 36 percent Ni^{2+} substitution for Fe^{2+} in magnetite at 300°C (Tsuji et al. 1996b). The associated CO_2 decomposition mechanism can be written as:

$$M_xFe_{3-x}O_{4-\delta} + (\delta - \delta')/2\ CO_2 \rightarrow M_xFe_{3-x}O_{4-\delta'} + (\delta - \delta')/2\ C(\delta > \delta') \tag{R6.8}$$

where M represents divalent metals, and δ and δ' are the initial oxygen deficiency and the oxygen deficiency at any reaction time (t), respectively (Tsuji et al. 1996b). The change in oxygen deficiency ($\delta - \delta'$) directly reflects the degree of CO_2 conversion to elemental carbon.

Kim et al. (2019) examined $SrFeCo_{0.5}O_x$ for its ability to reduce CO_2 to elemental carbon. The highest CO_2 decomposition efficiency achieved with $SrFeCo_{0.5}O_x$ reached ~90 percent, while decomposition efficiencies of ≥80 percent at 550°C to 750°C lasted for more than 60 minutes. The reaction mechanism of the interaction between $SrFeCo_{0.5}O_x$ and CO_2, proposed by Kim et al. (2019), is shown in Figure 6-5.

The same research group examined another reactant, $SrFeO_{3-\delta}$, for CO_2 reduction under the same test conditions, achieving a CO_2 decomposition efficiency of ≥90 percent, with decomposition ≥80 percent lasting for ~170 min, 70 percent longer than that realized with $SrFeCo_{0.5}O_x$ (Sim et al. 2020). However, the elemental carbon production selectivity achieved with $SrFeCo_{0.5}O_x$ and $SrFeO_{3-\delta}$, despite their activity, stability, and reproducibility, are not as good as that obtained with other solid oxide reactants—for example, Ni(II) ferrites (Tsuji et al. 1996b). In other words, unlike Ni(II) ferrites, both $SrFeCo_{0.5}O_x$ and $SrFeO_{3-\delta}$ based CTEC processes have a common shortcoming—that is, generation of CO as a by-product, which needs to be overcome when elemental carbons are the targeted products. One possible method is to add another reactor to consequently split the CO generated in Step I in Figure 6-5 subsequently into elemental carbon. Another difference between Kim et al. (2019) and previously reported work in this area is that a continuous CO_2 decomposition system was used, which is beneficial to eventual commercialization of the cation-excess materials based catalytic CTEC technology.

Note that use of these solid-oxide reactants to decompose CO_2 is closely related to research and development being pursued for solar thermochemical syngas production, in which oxygen-deficient metal oxides at high

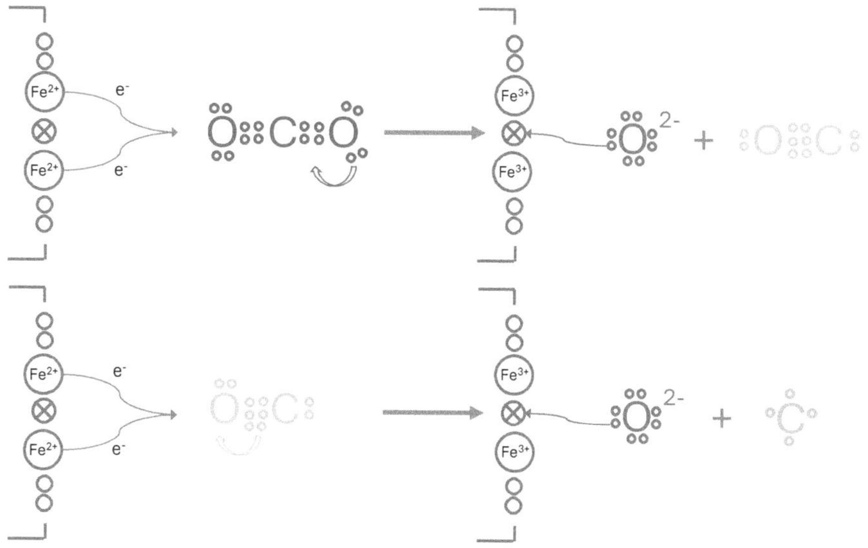

FIGURE 6-5 The suggested mechanism of CO_2 decomposition with $SrFeCo_{0.5}O_x$.
SOURCE: Reprinted from S.-H. Kim, J.T. Jang, J. Sim, J.-H. Lee, S.-C. Nam, and C.Y. Park, 2019, "Carbon Dioxide Decomposition Using SrFeCo0.5Ox, a Nonperovskite-Type Metal Oxide," *Journal of CO_2 Utilization* 34:709–715, Copyright (2019), with permission from Elsevier.

temperature (heated by concentrated sunlight) strip oxygen from water and/or carbon dioxide to produce hydrogen and carbon monoxide (Wexler et al. 2023a; Zhai et al. 2022). Some of the same materials as mentioned above have been explored for this purpose (Gautam et al. 2020; Wexler et al. 2021, 2023b). As reduction to elemental carbon by these materials therefore likely involves stepwise stripping of oxygen from carbon dioxide to form carbon monoxide (the end goal in solar thermochemical syn gas production) first, it is unsurprising that it is even more difficult to form elemental carbon, given that the second C-O chemical bond in carbon monoxide is a stronger triple bond whereas the first C-O bond to be stripped is a double bond (see Figure 6-5).

6.3.2.1.2 CTEC with Strong Reducing Agents

In this type of CO_2 to elemental carbon conversion, H_2, alkali metals, and alkaline earth metals are used as reductants—for example,

$$CO_2 + 2H_2 \leftrightarrow C + 2H_2O \qquad \Delta G^0 = -79.8 \text{ kJ} \qquad \text{(R6.9)}$$

$$CO_2 + 2Mg \rightarrow C + 2MgO \qquad \Delta G^0 = -743.6 \text{ kJ} \qquad \text{(R6.10)}$$

All CTEC reactions of this type are thermodynamically favorable under standard conditions. The temperatures required for R6.9 and R6.10 are lower or much lower than those for the above-mentioned decomposition-based reduction reactions, depending on the reducing agent used. Magnesium (Mg) is the primary metal used as a strong reducing agent to react with CO_2 and produce CNMs in this type of CTEC process (R6.10). The major elemental carbon product of such processes is graphene, which can combine with Mg to form Mg matrix composites (Li et al. 2022c, 2023b; Samiee and Goharshadi 2014; Wei et al. 2022; Zhang et al. 2014). R6.10 was first reported in 1978 (Driscoll 1978). In recent years, multiple research groups have demonstrated its use for simultaneous reduction of CO_2 and generation of elemental carbon materials (Li et al. 2022c, 2023b; Samiee and Goharshadi 2014; Wei et al. 2022; Zhang et al. 2014).

Ideally, the products of R6.9 and R6.10 are elemental carbon and H_2O or metal oxides. However, completely avoiding the generation of by-products is difficult. For example, R6.11 could occur simultaneously with R6.9 (Pease and Chesebro 1928)

$$CO_2 + 4H_2 \leftrightarrow CH_4 + 2H_2O \qquad \Delta G^0 = -113.3 \text{ kJ}, \qquad \text{(R6.11)}$$

and form an undesirable product, CH_4. Also, MgO generated in R6.10 can further react with CO_2 to form $MgCO_3$ via

$$3/2\ CO_2 + Mg \rightarrow 3/2\ C + MgCO_3 \qquad \Delta G^0 = -743.6 \text{ kJ}. \qquad \text{(R6.12)}$$

The formation of undesired products (e.g., CH_4, CO, carbonates) in R6.11 and R6.12 decreases elemental carbon selectivity and yield and complicates the separation of elemental carbon during decomposition-based strong reducing agents based CTEC processes. Carbon materials—for example, graphene, produced via the metal reduction CTEC process can exhibit excellent practical properties. For example, Wei et al. (2022) reported successful preparation of few-layer graphene through the reaction between molten Mg and CO_2 gas at 750°C, 100°C above the Mg melting temperature. The produced graphene, exhibiting a high degree of graphitization and nanoscale thickness, served as a lithium storage material and achieved excellent rate capability and cycling performance with a reversible capacity of 130 m Ah g^{-1} after 1,000 cycles at a current density of 1.0 A g^{-1}.

Compared to decomposition-based CTEC technologies, CTEC using strong reducing agents has several notable advantages. First, the reactions of CO_2 with strong reducing agents typically have negative Gibbs free energy (ΔG^0) and enthalpy (ΔH^0) changes, indicating their thermodynamic feasibility under standard conditions. Second, the very negative ΔG^0 and ΔH^0 values for reactions of CO_2 with strong reducing agents to form elemental carbon materials imply that 100 percent CO_2 conversion efficiency potentially can be achieved, a precondition for reaching a high yield of elemental carbon. Also, this class of CTEC processes does not require catalysts because moderate temperature elevation alone can significantly increase the rates of the reactions between CO_2 and strong reducing agents (e.g., H_2, alkali or alkaline earth metals). Moreover, the reaction set-ups and operation are relatively simple, which will reduce costs of the CTEC processes.

Despite having advantages of thermodynamic feasibility, high conversion efficiency, and less dependence on catalysts, CTEC processes using strong reducing agents are not without shortcomings. Selecting an appropriate reactive metal or reducing agent for the CTEC process is crucial, as the choice will affect CO_2 conversion efficiency and elemental carbon selectivity. When a metal is used instead of H_2 as the reducing agent, the reaction has to proceed at temperatures above the melting point of that metal. These temperatures are typically higher than those needed for the thermal-decomposition-based CTEC (~300°C). Moreover, achieving precise control over the morphology and properties of the elemental carbon produced is not easy, especially when specific structures of elemental carbon materials are desired—for example, 2D graphene and graphite with sp^2 carbon hybridization, and 0D CQDs with both sp^2 core and sp^3 shell carbon hybridizations. However, as noted above, the metal reduction technology generates metal oxides and sometimes by-products like carbonates that are not only long-lived materials to store carbon but that also could be sold for industrial applications owing to their high purities after they are separated from the elemental carbon materials and the metal oxides. Separation of carbon and noncarbon materials is relatively easy. The resulting metal oxides can be reduced to elemental metals (e.g., Mg) for cyclic CTEC conversion but this requires the use of energy-intensive processes. Also, as with any technology, transitioning from the current laboratory-scale experiments to large-scale production for practical applications could present additional challenges.

Last, although using CO_2 as a feedstock to produce long-lived elemental carbon materials could provide environmental benefits (Cossutta and McKechnie 2021; Goerzen et al. 2024; NASEM 2023), the overall environmental impact of thermochemical CTEC conversion needs to be assessed holistically, considering requirements for temperature, pressure, product separation, and reactant material recycling, if applicable.

6.3.2.2 Electrochemical and Electrically Driven Thermal Conversion Pathways

6.3.2.2.1 Electrochemical

The use of high-temperature molten carbonate electrolysis to produce elemental carbon products from CO_2 has advanced considerably over the past decade. Derived from earlier solar thermal electrochemical processes (Ren et al. 2019), these systems use metal carbonate electrolytes, which have high melting points, so the reactions are typically run at 450°C–700°C. High-temperature molten carbonate electrolysis can form a large spectrum of carbon products, including carbon powder, nanoplatelets, graphene, nano-onions, high-quality nanotubes, and other nanocarbon allotropes. Additives such as calcium or lithium chloride or fluorides can improve the performance of these systems by increasing the solubility of CO_2 and the calcium oxide (Tomkute et al. 2013). The mechanism of this reaction is currently unclear (Han et al. 2023b).

High-temperature molten carbonate electrolysis is scalable because all the elemental components are abundant and economical, although high temperatures are required to prevent the molten carbonate electrolyte from solidifying. The molten carbonate precursor, often an alkali metal oxide, reacts with CO_2 selectively at low concentrations, including from air, to generate the electrolyte. Thus, this method can be used to integrate CO_2 capture from atmospheric or flue gas streams for conversion. Such integration of molten salt CO_2 capture and electrochemical transformation eliminates the need for distinct separations processes and the accompanying energy and capital requirements. These systems might be able to operate on waste heat from high-temperature flue gas exhaust in a fossil fuel plant, and the oxygen generated at the anode could be used to facilitate combustion. It has been suggested that this method might be able to achieve a 100× price reduction for the production of CNTs compared to conventional CVD methods (Ren and Licht 2016).

A more recent development is room-temperature CO_2 electrolysis into carbon materials, which uses metals that are liquid at room temperature, such as gallium, alloyed with redox-active metals, such as cerium. In one such system, analysis of the liquid revealed a thin film of cerium oxide on the surface, and it is believed redox reactions occur by cycling cerium between the zero and tetravalent oxidation state (Esrafilzadeh et al. 2019). The proposed reactions are shown below, where the R6.13–R6.16 reduction reactions occur at the liquid metal working electrode (a cathode made of a gallium-indium-tin alloy, galinstan), and R6.17 is the oxidation reaction at the anode. The carbonaceous nanoflakes have been applied as carbon-based supercapacitor materials (Esrafilzadeh et al. 2019).

(1) $2Ce(galinstan) + 1.5\ O_2(air) \rightarrow Ce_2O_3$ (R6.13)

(2) $Ce_2O_3 + 3H_2O + 6e^- \rightarrow 2\ Ce(0) + 6OH^-$ (R6.14)

(3) $Ce(0) + CO_2 \rightarrow CeO_2 + C$ (R6.15)

(4) $CeO_2 + 2H_2O + 4e^- \rightarrow Ce + 4OH^-$ (R6.16)

(5) $4OH^- \rightarrow O_2 + 2\ H_2O + 4e^-$ (R6.17)

This approach has also been reported with other liquid metal electrode formulations (Irfan et al. 2023; Ye et al. 2023). Other examples of CO_2 conversion to carbon materials that use liquid metals apply mechanical energy (Tang et al. 2022) or heat (Zuraiqi et al. 2022) as the energy source. The use of liquid metals provides several advantages over CVD and high-temperature molten carbonate electrolysis. The components are abundant and economical and have low toxicity and vapor pressure. Unlike the other electrochemical methods discussed, the process operates at low or room temperature. The high surface tension of the liquid metal also acts as an intrinsic coking-resistant surface. The carbonaceous products spontaneously flake off the surface of the electrode, which facilitates separations and prevents catalyst poisoning or inhibition.

6.3.2.2.2 Electrically Driven Thermal Reduction

Solid oxide electrolyzer cells (SOECs) can be used to combine thermal and electrochemical energy to reduce CO_2 to elemental carbon materials. CO_2 electrolysis in the SOEC forms carbon monoxide (CO) and solid carbon at the cathode, while oxygen is evolved at the anode. The electrodes again are typically made of oxide perovskites similar to the materials discussed earlier. This method has been applied to produce CNTs, albeit at high temperatures (>800°C) (Tao et al. 2014).

Cathode: $CO_2 + 2e^- \rightarrow CO + O^{2-}$ (R6.18)

$CO + 2e^- \rightarrow C + O^{2-}$ (R6.19)

Anode: $2O^{2-} \rightarrow O_2 + 4e^-$ (R6.20)

A related technology is the use of tandem electro-thermochemical looping to access carbon materials from CO_2. In these systems, CO_2 is initially reduced to CO in a low-temperature electrolytic cell or a high-temperature SOEC (Luc et al. 2018; Mori et al. 2016). The CO is then heated to high temperatures (500°C–700°C), where it undergoes disproportionation in a Boudouard reaction, forming solid carbon and CO_2. This method has been used to produce densely packed and aligned CNTs. Challenges to this approach include integration of the CO stream to the reactor, as well as the purity of the former. Minimizing concomitant evolution of H_2 with CO is critical to prevent hydrogen gas accumulation in the system. Additionally, recovered CO_2 must be free of carbon materials before being looped back into the electrolysis cell to avoid poisoning the electrodes in the low-temperature electrolytic cell (Luc et al. 2018).

Boudouard Reaction: $2CO \leftrightarrow CO_2 + C$ (solid) (R6.21)

When CO_2 decomposes to CO and ½ O_2, CO can disproportionate via the Boudouard reaction to produce ½ CO_2 and ½ C(solid), which, with recycle of CO_2, can result in an overall conversion of $CO_2 \rightarrow$ C(solid) + O_2 via this two-step pathway. Solid carbon is more readily generated from CO than CO_2 below 971K (Chery et al. 2015; Han et al. 2023b). Hence, pathways through CO can provide indirect conversion of CO_2 to solid carbon products.

6.3.2.3 Photochemical Conversion Pathways

Photocatalytic CO_2 reduction has been widely researched to produce C1 and C2 gas and liquid products (Fu et al. 2019). The direct production of solid carbon materials from CO_2 by photocatalysis is challenging due the limited reaction pathways in photochemistry (Li et al. 2019a). However, many photocatalytic products like methane and CO are valuable substrates for downstream carbon material production through chemical synthesis (Duan et al. 2013). A viable approach for carbon material production leveraging photocatalysis is a tandem process, where photocatalytic CO_2 reduction first converts CO_2 into gaseous products like methane and CO, followed by conversion of methane and CO into carbon materials via (e.g., pyrolytic) decomposition (Anisimov et al. 2010; Shen and Lua 2015).

6.3.2.4 Plasmachemical Conversion Pathways

Plasmachemical processes can facilitate thermodynamically unfavorable reactions, such as CO_2 activation, at relatively low temperatures. Several nonthermal plasma sources have been applied to transform CO_2, including dielectric barrier discharge, microwave discharge, and gliding arc (Mei et al. 2014). Plasma processing parameters such as discharge power, gas composition, feed flow rate, and dielectric material affect the conversion of reactants. Catalysts can be used to assist plasma reactions, especially when packed-bed reactors are employed, which can integrate catalysts to enhance the synergies between catalysts and plasmas.

Plasma catalysis can be performed via two main types of reactor configurations, in-plasma catalysis and post-plasma catalysis (George et al. 2021; Wang et al. 2018). For the in-plasma catalysis configuration, catalysts are placed in the plasma discharge region, which facilitates a direct reaction between the plasma species and the catalyst surface, potentially enhancing the conversion of reactants to desired products (e.g., nanocarbon materials). In post-plasma catalysis, the reaction occurs in two steps. First, CO_2 could participate in in-plasma reactions if plasma is used to split CO_2 into CO and O_2, while in the second stage, CO and/or CO_2 could undergo catalytic surface reactions. The chemical and physical interactions between the nonthermal plasma and catalyst strongly affect the percentage of CO_2 transformed and the product selectivity (George et al. 2021). Plasma can impact the catalyst by generating excited species and radicals, lowering the activation barrier and enhancing pathways for surface reactions. At the same time, the catalyst can affect the electric field of the plasma, discharge type, generation of micro discharge in pores, and impurity concentration in the plasma. Consequently, plasma catalysis-sis may be able to increase energy efficiency, reaction rate, product yield, and catalyst durability; enhance the concentration of active species; and improve selectivity (Neyts et al. 2015). However, Bogaerts et al. (2020) note that good thermal CO_2 activation catalysts might not be good plasma catalysts for CTEC conversion because the plasma could change the properties of the catalyst surface and introduce new species (e.g., excited species and reactive species) for surface reactions, thereby changing surface interactions with the catalyst and thus the reaction pathways.

As of April 2024, research on splitting CO_2 into carbon materials and O_2 with nonthermal plasma remains at the concept stage. Only one paper is known to have been published in this area, which discusses the feasibility of the concept in detail (Centi et al. 2021). However, no experiments have been performed using plasma with or without the help of catalysts to confirm the reaction.

6.3.3 Potential Alternative or Competitor Routes to Elemental Carbon Materials

In a net-zero future, alternative routes to produce elemental carbon materials include methane pyrolysis and lignocellulosic biomass processing. While a detailed discussion of these approaches is out of scope for this study, brief descriptions are provided below, along with their advantages and disadvantages relative to the production of elemental carbon materials from CO_2. Life cycle and techno-economic assessments (see Chapter 3) will be needed to help evaluate and compare the emissions impacts and economic viability of these different approaches for producing elemental carbon materials.

6.3.3.1 Methane Pyrolysis

Methane pyrolysis entails the decomposition of methane to form solid carbon and hydrogen (Amin et al. 2011) and is a potential sustainable pathway to produce both elemental carbon products and clean hydrogen, provided the solid carbon products remain sequestered. The process is endothermic (Lewis et al. 2001) with its enthalpy change being 74.5 kJ/mol-CH_4 or 37.3 kJ/mol-H_2 produced (Catalan et al. 2023):

$$CH_4 \rightarrow C(s) + 2\ H_2 \qquad \text{(R6.22)}$$

Methane pyrolysis is attractive as a source of hydrogen because, in principle, solid carbon can be separated from the gas-phase hydrogen product without CO_2 capture and sequestration, as would be required for sustainable conventional steam methane reforming (Korányi et al. 2022; Sánchez-Bastardo et al. 2021; Timmerberg et al. 2020). Formation of carbon fibers and nanotubes from methane (fully reduced carbon) may be less difficult than a similar pathway from CO_2 (fully oxidized carbon), given that a number of the CNT pathways from CO_2 are postulated to proceed through methane or other reduced hydrocarbon intermediates (Kim et al. 2020a, 2020b).

6.3.3.2 Lignocellulosic Biomass Processing

Elemental carbon materials derived from plant lignocellulosic biomass represents a competitive platform to CO_2-derived carbon materials. Of course, biomass-based carbon materials also are derived from CO_2 ultimately, as plants fix CO_2 through photosynthesis and convert it into biopolymers such as cellulose and lignin in the plant cell wall. These biopolymers subsequently can be converted into carbon materials like carbon fiber, graphene, CNTs, carbon foam, and others (Li et al. 2022b). Among different biopolymers, lignin has the greatest potential for carbon material manufacturing owing to its high carbon content and aromatic ring structure (Zhang et al. 2022). Recent studies have advanced fundamental understanding of structure–function relationships of how lignin composition (molecular weight, uniformity, linkage profile, and functional group) can impact its structure and performance. This understanding guides the design of new lignin structures to improve the quality and performance of the resulting carbon materials (Li et al. 2022a, 2022b).

6.3.4 Challenges for CO_2 Conversion to Elemental Carbon Materials

As noted in Section 6.3.2, except for plasmachemical processes, research into CTEC processes has been ongoing for quite some time, especially thermochemical CTEC research that started as early as 1990 (Tamaura and Tahata 1990). However, CTEC research activities and processes are still limited, despite the large market potential for elemental carbon materials discussed in Chapter 2. All four major types of CTEC technologies (thermochemical, electrochemical, photochemical, and plasmachemical) are still at low TRLs, in concept development stages. The sections below identify challenges common to all four CTEC approaches and those unique to each CTEC approach.

6.3.4.1 Common Challenges for the Four CTEC Approaches

1. **Research efforts have been limited:** Although research on CTEC conversions, especially in thermochemical approaches, has been ongoing for several decades, research intensity in the four areas have not been high. The first thermochemical Mg-CO_2–to–MgO-C conversion reaction was reported in 1978, and the first thermochemical CO_2 splitting with cation-excess magnetite was published in 1990. Nonetheless, only 161 papers have been published on thermochemical, electrochemical, photochemical, and plasmachemical CTEC conversion as of April 2024. Therefore, theoretical and experimental work on CTEC conversions are still limited, especially in the areas of photochemical and plasmachemical CTEC, as illustrated in Figure 6-3.
2. **Difficult to compare CTEC approaches:** Research activities on the four CTEC approaches are not balanced; thus, it is difficult to assess the advantages and disadvantages of the different reaction pathways.
3. **Substantial energy requirements:** Breaking carbon-oxygen bonds in CO_2 is energy intensive; thus, all four CTEC approaches require substantial external energy. For these products and processes to have net-zero emissions, this energy will have to be provided by zero-carbon-emission sources of electricity or heat.
4. **Limited understanding of catalyst/reactant material stability and carbon product selectivity:** The stabilities of the catalysts/reactant materials used and carbon materials generated during CTEC processes have not been well characterized.
5. **A grand challenge for all of these CTEC technologies is how to convert CO_2 selectively at high yield to a particular morphological form of solid carbon.**

6.3.4.2 Specific Challenges for Each CTEC Approach

6.3.4.2.1 Thermochemical CTEC

Both the decomposition-based and strong reducing agents–based CTEC have low overall energy utilization efficiencies owing to the low mass and heat transfer efficiencies of fixed bed reactors in SOA thermochemical processes, the need to heat the reaction systems to moderately high temperatures to initiate the reaction, and loss of the heat released during the reaction. The two thermochemical CTEC technologies also have their own specific challenges.

- Decomposition-based CTEC:
  - Fast reactant material deactivation and the resultant slow reaction kinetics owing to coking via direct active site poisoning and pore plugging from the nanoscale carbon particles generated from CO_2 reduction, which lowers the reaction rate.
  - Stabilities of cation-excess reactants during their reactions with CO_2 at high temperatures are not well studied.
  - The purity and structure of the carbon materials generated have not been systematically evaluated.
- CTEC using strong reducing agents:
  - Initiation of the elemental-metals-based CTEC processes are slow owing to the need to reach high temperatures (higher than metal melting points) prior to the initiation of the reactions.

6.3.4.2.2 Electrochemical CTEC

Most electrochemical CTEC processes require high temperatures, and therefore greater energy input. Room-temperature electrolysis to produce elemental carbon materials is comparatively nascent and does not generate as high-value materials as the high-temperature systems. Deposition of carbon on electrode surfaces can lead to the significant reduction of active sites and a decrease in electrode activity. It is difficult to successfully convert CO_2 from gas to solid carbon owing to the complexity in overcoming kinetic barriers and achieving efficient nucleation and solid carbon structure growth.

6.3.4.2.3 Photochemical CTEC

Only one paper has been published using this CTEC technology (Duan et al. 2013). The research was performed with the help of temperature management. Thus, knowledge of reaction mechanisms for photocatalytic CTEC area is entirely lacking.

6.3.4.2.4 Plasmachemical CTEC

This concept has been proposed but not yet confirmed experimentally.

6.3.5 R&D Opportunities for CO_2 Conversion to Elemental Carbon Materials

There are many R&D activities that can advance CTEC processes. Research is needed into how the reaction conditions (e.g., temperature, pressure, composition of CO_2-containing feedstock), use of catalyst, and other factors affect the types and qualities of carbon materials produced. Work is also needed to fully characterize the carbon materials generated via CTEC technologies and identify their corresponding markets, including the preparation of organic solar cells and light-emitting diodes, supercapacitors, batteries, sensors, and catalysts. The following lays out specific R&D opportunities across thermochemical, electrochemical, photochemical, and plasmachemical CTEC.

Thermochemical CTEC:

- Decomposition-based CTEC:
  - There is an opportunity to build on knowledge of solid-oxide reactants, including ferrites and $SrFeO_3$-σ, to develop new types of cation-excess materials with high CO_2 conversion activity, ~100 percent carbon formation selectivity, and good stability and regeneration ability for decomposition-based CTEC. Utilizing knowledge already gleaned from the solar thermochemical hydrogen solid-oxide materials research could prove fruitful (Wexler et al. 2021, 2023b). Similarly, exploiting knowledge already gathered from solid oxide fuel cell materials characterization and optimization also could be helpful (Muñoz-García et al. 2014; Ritzmann et al. 2016).
  - The carbon materials obtained with decomposition-based CTEC, mainly CNT and graphene, can differ from those generated by reacting CO_2 with strong reducing agents because their reaction temperatures are different. Thus, more R&D is needed to discover how to generate different high-value carbon materials with decomposition-based CTEC processes.
- Strong-reducing-agents-based CTEC:
  - R&D is needed to diversify the elemental carbon products that can be made from the reaction of Mg and CO_2. Current systems primarily yield CNTs and graphene. By changing the reaction conditions, other products such as CNFs and CQDs might be accessible.
  - The primary metal used as a strong reducing agent for CTEC is Mg. R&D is needed to explore the pros and cons of using non-Mg metals for metal reduction CTEC processes.

Electrochemical CTEC:

- Lowering the operating temperature for molten electrolysis or solid oxide electrolyzer cells or coupling these processes with exothermic reactions could lower the overall energy requirements.

- Low-temperature electrolysis with liquid metals is a promising approach for generating elemental carbon materials, but more research is required to access higher-value carbon products. In the reported electrochemical processes, the mechanism of reduction is not well understood.

Photochemical CTEC:

- According to Duan et al. (2013), the photochemical method does not work on its own; thermocatalysis needs to be coupled to photocatalysis for the reduction of CO_2 to C to occur. Thus, dual-function catalysts that can simultaneously accelerate both thermal- and photo-splitting reactions may play a key role and could be a promising future research direction in this area.

Plasmachemical CTEC:

- A plasmachemical CTEC scheme (Mei et al. 2014) was proposed about 10 years ago. However, no realizable experimental data have been published to confirm the concept. Accordingly, multiple R&D opportunities in this area exist in this area, including:
  - Theory, modeling, and simulation—While such research has begun for some proposed methods (e.g., methane pyrolysis), none is available yet for understanding plasmachemical CO_2 reduction mechanisms.
  - Experiments—Experiments need to be conducted to confirm the feasibility of plasmachemical CTEC, especially regarding the use of catalysts both in situ and post-plasma.

6.4 CONCLUSIONS

6.4.1 Findings and Recommendations

Finding 6-1: CO_2 to elemental carbon technologies are far from commercialization—Thermochemical, electrochemical, photochemical, and plasmachemical conversion of CO_2 to elemental carbon technologies are generally at technology readiness levels of 3, 2 (for room-temperature electrolysis), 1, and 1, respectively, indicating that all are far from commercialization.

> **Recommendation 6-1: Support basic research to advance CO_2 to elemental carbon technologies—Basic Energy Sciences within the Department of Energy's Office of Science and the National Science Foundation should invest in building the knowledge foundation and accelerating the maturities of the four CO_2 to elemental carbon technology areas: thermochemical, electrochemical, photochemical, and plasmachemical.**

Finding 6-2: Demanding materials and energy requirements for CO_2 to elemental carbon technologies—All CO_2 to elemental carbon (CTEC) technologies need strong reducing agents (e.g., Mg or H_2) or very negative electrochemical potentials, oxygen-deficient reactant materials (e.g., cation-excess magnetite), and other materials (e.g., molten salts) as part of their conversion process. To develop CTEC technologies with net-zero or net-negative CO_2 footprints, the materials used in the CTEC technologies need to be generated from low-carbon-emission sources. In addition, all CTEC technologies need external energy to initiate and/or maintain the reactions.

Finding 6-3: Challenges with activity, selectivity, and stability of redox-active materials key to CO_2 to elemental carbon conversion—Redox-active materials are key to the success of CO_2 to elemental carbon technologies (CTEC). However, current catalysts for these technologies lack sufficient activity, selectivity, and stability to achieve high performance. Carbon nanotubes and graphene are the primary carbon materials generated from CO_2 via CTEC, with CO_2-derived graphene performing better than its commercial counterpart. Several other carbon materials, such as carbon nanofibers, are less frequently produced from CO_2.

Recommendation 6-2: Fund research into catalysts and materials for CO_2 to elemental carbon conversion—Basic Energy Sciences within the Department of Energy's Office of Science (DOE-BES) and the National Science Foundation should fund basic research into the discovery of high-performance catalysts that are active, morphologically selective, and robust for low-cost CO_2 to elemental carbon (CTEC) conversion. DOE-BES, DOE's Office of Energy Efficiency and Renewable Energy, and DOE's Office of Fossil Energy and Carbon Management should, jointly or independently, fund research on materials (e.g., catalysts, reducing agents) used in CO_2 to elemental carbon processes. These investigations should aim to discover new, optimal materials for catalysis and separation; understand how to control CTEC reactions to increase the diversity of products and selectively generate desired morphologies; and increase energy efficiency of CTEC reaction processes that can be powered by clean energy.

Finding 6-4: Tandem systems have potential to optimize CO_2 to elemental carbon conversion—Combining multiple CO_2 to elemental carbon technologies—for example, photo/thermal or electro/thermal combinations, either in one-pot or sequential systems, could be more efficient than any single process alone. These superior efficiencies could include increased carbon yield, optimized systems, minimized energy input, and control of desired carbon material morphology.

Recommendation 6-3: Fund the development of tandem CO_2 to elemental carbon technologies to maximize economic and environmental benefits—Basic Energy Sciences within the Department of Energy's (DOE's) Office of Science, DOE's Office of Energy Efficiency and Renewable Energy, and the Advanced Research Projects Agency–Energy should fund independently and/or collectively the development of tandem CO_2 to elemental carbon (CTEC) technologies that can combine the advantages of different types of CTEC processes to maximize the economic and environmental benefits of the converted carbon materials.

Finding 6-5: Combined capture and conversion of CO_2 to elemental carbon can lead to savings—Combining CO_2 capture with CO_2 to elemental carbon (CTEC) conversion can lead to reductions in capital and operation costs, and carbon footprints, of CTEC technologies, in contrast to discrete operations of CO_2 capture and subsequent CO_2 reduction to carbon materials.

Recommendation 6-4: Fund research on integrated CO_2 capture and conversion to elemental carbon materials to maximize economic and environmental benefits—The Department of Energy's Office of Fossil Energy and Carbon Management, Office of Energy Efficiency and Renewable Energy, and the Advanced Research Projects Agency–Energy should fund research on integrated CO_2 capture and conversion to elemental carbon materials, with particular consideration of technology integration and economic and environmental benefit enhancement.

6.4.2 Research Agenda for Chemical CO_2 Conversion to Elemental Carbon Materials

Table 6-3 presents the committee's research agenda for chemical CO_2 conversion to elemental carbon materials, including research needs (numbered by chapter), and related research agenda recommendations (a subset of research-related recommendations from the chapter). The table includes the relevant funding agencies or other actors; whether the need is for basic research, applied research, technology demonstration, or enabling technologies and processes for CO_2 utilization; the research theme(s) that the research need falls into; the relevant research area and product class covered by the research need; whether the relevant product(s) are long- or short-lived; and the source of the research need (chapter section, finding, or recommendation). The committee's full research agenda can be found in Chapter 11.

TABLE 6-3 Research Agenda for CO_2 Conversion to Elemental Carbon Materials

Research, Development, and Demonstration Need	Funding Agencies or Other Actors	Basic, Applied, Demonstration, Enabling	Research Area	Product Class	Long- or Short-Lived	Research Themes	Source
6-A. Foundational knowledge of thermochemical, electrochemical, photochemical, and plasma processes to make elemental carbon products from CO_2.	DOE-BES NSF	Basic	Chemical	Elemental Carbon Materials	Long-lived Short-lived	Fundamental knowledge	Rec. 6-1
Recommendation 6-1: Support basic research to advance CO_2 to elemental carbon technologies—Basic Energy Sciences within the Department of Energy's Office of Science and the National Science Foundation should invest in building the knowledge foundation and accelerating the maturities of the four CO_2 to elemental carbon technology areas: thermochemical, electrochemical, photochemical, and plasmachemical.							
6-B. Novel and improved catalysts and low-energy reaction processes to produce elemental carbon products from CO_2.	DOE-BES DOE-EERE DOE-FECM	Basic Applied	Chemical	Elemental Carbon Materials	Long-lived Short-lived	Catalyst innovation and optimization Reactor design and reaction engineering. Energy efficiency, electrification, and alternative heating	Fin. 6-2 Rec. 6-2
6-C. Catalysts and processes that are selective for particular material morphologies.	DOE-BES DOE-EERE DOE-FECM	Basic Applied	Chemical	Elemental Carbon Materials	Long-lived Short-lived	Catalyst innovation and optimization	Fin. 6-2 Rec. 6-2
6-D. Enhanced activity, selectivity, and stability of catalysts to achieve high performance of reactions transforming CO_2 to elemental carbon products.	DOE-BES NSF	Basic	Chemical	Elemental Carbon Materials	Long-lived Short-lived	Catalyst innovation and optimization	Fin. 6-3 Rec. 6-2
6-E. Understanding and control of processes that produce CO_2-derived elemental carbon products.	DOE-BES NSF	Basic	Chemical	Elemental Carbon Materials	Long-lived Short-lived	Fundamental knowledge	Fin. 6-3 Rec. 6-2
6-F. Reaction electrification and heat integration including plasma processes (thermochemical, plasmachemical, etc.).	DOE-BES DOE-EERE DOE-FECM	Basic Applied	Chemical	Elemental Carbon Materials	Long-lived Short-lived	Reactor design and reaction engineering Energy efficiency, electrification, and alternative heating	Fin. 6-3 Rec. 6-2

6-G. Separation of catalyst from solid carbon products, and different elemental carbon materials from each other.	DOE-BES DOE-EERE DOE-FECM	Basic Applied	Chemical	Elemental Carbon Materials	Long-lived Short-lived	Separations	Fin. 6-3 Rec. 6-2

Recommendation 6-2: Fund research into catalysts and materials for CO_2 to elemental carbon conversion—Basic Energy Sciences within the Department of Energy's Office of Science (DOE-BES) and the National Science Foundation should fund basic research into the discovery of high-performance catalysts that are active, morphologically selective, and robust for low-cost CO_2 to elemental carbon (CTEC) conversion. DOE-BES, DOE's Office of Energy Efficiency and Renewable Energy, and DOE's Office of Fossil Energy and Carbon Management should, jointly or independently, fund research on materials (e.g., catalysts, reducing agents) used in CO_2 to elemental carbon processes. These investigations should aim to discover new, optimal materials for catalysis and separation; understand how to control CTEC reactions to increase the diversity of products and selectively generate desired morphologies; and increase energy efficiency of CTEC reaction processes that can be powered by clean energy.

6-H. Development of tandem processes to produce elemental carbon products from CO_2.	DOE-BES DOE-EERE DOE-ARPA-E	Basic Applied	Chemical	Elemental Carbon Materials	Long-lived Short-lived	Integrated systems Reactor design and reaction engineering	Fin. 6-4 Rec. 6-3

Recommendation 6-3: Fund the development of tandem CO_2 to elemental carbon technologies to maximize economic and environmental benefits—Basic Energy Sciences within the Department of Energy's (DOE's) Office of Science, DOE's Office of Energy Efficiency and Renewable Energy, and the Advanced Research Projects Agency–Energy should fund independently and/or collectively the development of tandem CO_2 to elemental carbon (CTEC) technologies that can combine the advantages of different types of CTEC processes to maximize the economic and environmental benefits of the converted carbon materials.

6-I. Integrated CO_2 capture and conversion to elemental carbon materials including improved technology integration and enhanced economic and/or environmental benefits.	DOE-FECM DOE-EERE DOE-ARPA-E	Applied	Chemical	Elemental Carbon Materials	Long-lived Short-lived	Integrated systems	Fin. 6-5 Rec. 6-4

Recommendation 6-4: Fund research on integrated CO_2 capture and conversion to elemental carbon materials to maximize economic and environmental benefits—The Department of Energy's Office of Fossil Energy and Carbon Management, Office of Energy Efficiency and Renewable Energy, and the Advanced Research Projects Agency–Energy should fund research on integrated CO_2 capture and conversion to elemental carbon materials, with particular consideration of technology integration and economic and environmental benefit enhancement.

NOTE: ARPA-E = Advanced Research Projects Agency–Energy; BES = Basic Energy Sciences; DOE = Department of Energy; EERE = Office of Energy Efficiency and Renewable Energy; FECM = Office of Fossil Energy Carbon Management; NSF = National Science Foundation.

6.5 REFERENCES

Abraham, J., R. Arunima, K.C. Nimitha, S.C. George, and S. Thomas. 2021. "Chapter 3—One-Dimensional (1D) Nanomaterials: Nanorods and Nanowires; Nanoscale Processing." Pp. 71–101 in *Nanoscale Processing*, S. Thomas and P. Balakrishnan, eds. Micro and Nano Technologies. Elsevier. https://doi.org/10.1016/B978-0-12-820569-3.00003-7.

Álvarez, A., A. Bansode, A. Urakawa, A.V. Bavykina, T.A. Wezendonk, M. Makkee, J. Gascon, and F. Kapteijn. 2017. "Challenges in the Greener Production of Formates/Formic Acid, Methanol, and DME by Heterogeneously Catalyzed CO_2 Hydrogenation Processes." *Chemical Reviews* 117(14):9804–9838.

Amin, A.M., E. Croiset, and W. Epling. 2011. "Review of Methane Catalytic Cracking for Hydrogen Production." *International Journal of Hydrogen Energy* 36(4):2904–2935. https://doi.org/10.1016/j.ijhydene.2010.11.035.

Anctil, A., C.W. Babbitt, R.P. Raffaelle, and B.J. Landi. 2011. "Material and Energy Intensity of Fullerene Production." *Environmental Science and Technology* 45(6):2353–2359.

Anisimov, A.S., A.G. Nasibulin, H. Jiang, P. Launois, J. Cambedouzou, S.D. Shandakov, and E.I. Kauppinen. 2010. "Mechanistic Investigations of Single-Walled Carbon Nanotube Synthesis by Ferrocene Vapor Decomposition in Carbon Monoxide." *Carbon* 48(2):380–388. https://doi.org/10.1016/j.carbon.2009.09.040.

Bacon, M., S.J. Bradley, and T. Nann. 2014. "Graphene Quantum Dots." *Particle and Particle Systems Characterization* 31(4):415–428. https://doi.org/10.1002/ppsc.201300252.

Bansal, D., S. Pillay, and U. Vaidya. 2013. "Nanographene-Reinforced Carbon/Carbon Composites." *Carbon* 55:233–244.

Bhugun, I., D. Lexa, and J.-M. Savéant. 1996. "Catalysis of the Electrochemical Reduction of Carbon Dioxide by Iron (0) Porphyrins. Synergistic Effect of Lewis Acid Cations." *The Journal of Physical Chemistry* 100(51):19981–19985.

Bogaerts, A., X. Tu, J.C. Whitehead, G. Centi, L. Lefferts, O. Guaitella, F. Azzolina-Jury, et al. 2020. "The 2020 Plasma Catalysis Roadmap." *Journal of Physics D: Applied Physics* 53(44):443001.

Budyka, M.F., E.F. Sheka, and N.A. Popova. 2017. "Graphene Quantum Dots: Theory and Experiment." *Reviews on Advanced Materials Science* 51(1).

Cartwright, R.J., S. Esconjauregui, R.S. Weatherup, D. Hardeman, Y. Guo, E. Wright, D. Oakes, S. Hofmann, and J. Robertson. 2014. "The Role of the sp2: sp3 Substrate Content in Carbon Supported Nanotube Growth." *Carbon* 75:327–334.

Catalan, L.J.J., B. Roberts, and E. Rezaei. 2023. "A Low Carbon Methanol Process Using Natural Gas Pyrolysis in a Catalytic Molten Metal Bubble Reactor." *Chemical Engineering Journal* 462(April 1):142230. https://doi.org/10.1016/j.cej.2023.142230.

Centi, G., S. Perathoner, and G. Papanikolaou. 2021. "Plasma Assisted CO_2 Splitting to Carbon and Oxygen: A Concept Review Analysis." *Journal of CO_2 Utilization* 54:101775.

Chakrabarti, A., J. Lu, J.C. Skrabutenas, T. Xu, Z. Xiao, J.A. Maguire, and N.S. Hosmane. 2011. "Conversion of Carbon Dioxide to Few-Layer Graphene." *Journal of Materials Chemistry* 21(26):9491–9493.

Chandra, K.S., and D. Sarkar. 2023. "Nanoscale Reinforcement Efficiency Analysis in Al2O3–MgO–C Refractory Composites." *Materials Science and Engineering: A* 865:144613.

Chen, C., and Z. Lou. 2009. "Formation of C_{60} by Reduction of CO_2." *The Journal of Supercritical Fluids* 50(1):42–45. https://doi.org/10.1016/j.supflu.2009.04.008.

Chen, Z., Y. Gu, L. Hu, W. Xiao, X. Mao, H. Zhu, and D. Wang. 2017a. "Synthesis of Nanostructured Graphite via Molten Salt Reduction of CO_2 and SO_2 at a Relatively Low Temperature." *Journal of Materials Chemistry A* 5(39):20603–20607.

Chery, D., V. Lair, and M. Cassir. 2015. "CO2 Electrochemical Reduction into CO or C in Molten Carbonates: A Thermodynamic Point of View." *Electrochimica Acta* 160:74–81. https://doi.org/10.1016/j.electacta.2015.01.216.

Chung, D.D.L. 2002. "Review Graphite." *Journal of Materials Science* 37(8):1475–1489. https://doi.org/10.1023/A:1014915307738.

Collins, G., P.R. Kasturi, R. Karthik, J.-J. Shim, R. Sukanya, and C.B. Breslin. 2023. "Mesoporous Carbon-Based Materials and Their Applications as Non-Precious Metal Electrocatalysts in the Oxygen Reduction Reaction." *Electrochimica Acta* 439(January):141678. https://doi.org/10.1016/j.electacta.2022.141678.

Cossutta, M., and J. McKechnie. 2021. "Environmental Impacts and Safety Concerns of Carbon Nanomaterials," Pp. 249–278 in *Carbon Related Materials*, S. Kaneko, M. Aono, A. Pruna, M. Can, P. Mele, M. Ertugrul, and T. Endo, eds. Singapore: Springer Singapore. https://doi.org/10.1007/978-981-15-7610-2_11.

Dabees, S., T. Osman, and B.M. Kamel. 2023. "Mechanical, Thermal, and Flammability Properties of Polyamide-6 Reinforced with a Combination of Carbon Nanotubes and Titanium Dioxide for Under-the-Hood Applications." *Journal of Thermoplastic Composite Materials* 36(4):1545–1575.

Deng, B., X. Mao, W. Xiao, and D. Wang. 2017. "Microbubble Effect-Assisted Electrolytic Synthesis of Hollow Carbon Spheres from CO_2." *Journal of Materials Chemistry A* 5(25):12822–12827.

Deng, L., R.J. Young, I.A. Kinloch, Y. Zhu, and S.J. Eichhorn. 2013. "Carbon Nanofibres Produced from Electrospun Cellulose Nanofibres." *Carbon* 58:66–75.

Deshmukh, A.A., S.D. Mhlanga, and N.J. Coville. 2010. "Carbon Spheres." *Materials Science and Engineering: R: Reports* 70(1–2):1–28.

Dhall, S., ed. 2023. *Carbon Nanomaterials and Their Nanocomposite-Based Chemiresistive Gas Sensors: Applications, Fabrication and Commercialization*. 1st edition. Amsterdam, Netherlands: Elsevier.

Douglas, A., R. Carter, M. Li, and C.L. Pint. 2018. "Toward Small-Diameter Carbon Nanotubes Synthesized from Captured Carbon Dioxide: Critical Role of Catalyst Coarsening." *ACS Applied Materials and Interfaces* 10(22):19010–19018.

Driscoll, J.A. 1978. "A Demonstration of Burning Magnesium and Dry Ice." *Journal of Chemical Education* 55(7):450.

Duan, Y.Q., T. Du, X.W. Wang, F.S. Cai, Z.H. Yuan. 2013. "Photoassisted CO_2 Conversion to Carbon by Reduced $NiFe_2O_4$." *Advanced Materials Research* 726–731:420–424. https://doi.org/10.4028/www.scientific.net/AMR.726-731.420.

Eslami, Z., F. Yazdani, and M.A. Mirzapour. 2015. "Thermal and Mechanical Properties of Phenolic-Based Composites Reinforced by Carbon Fibers and Multiwall Carbon Nanotubes." *Composites Part A: Applied Science and Manufacturing* 72:22–31.

Esrafilzadeh, D., A. Zavabeti, R. Jalili, P. Atkin, J. Choi, B.J. Carey, R. Brkljača, et al. 2019. "Room Temperature CO_2 Reduction to Solid Carbon Species on Liquid Metals Featuring Atomically Thin Ceria Interfaces." *Nature Communications* 10(1):865. https://doi.org/10.1038/s41467-019-08824-8.

Fu, Z., Q. Yang, Z. Liu, F. Chen, F. Yao, T. Xie, Y. Zhong, et al. 2019. "Photocatalytic Conversion of Carbon Dioxide: From Products to Design the Catalysts." *Journal of CO_2 Utilization* 34(December):63–73. https://doi.org/10.1016/j.jcou.2019.05.032.

Gautam, G.S., E.B. Stechel, and E.A. Carter. 2020. "Exploring Ca-Ce-M-O (M = 3d Transition Metal) Oxide Perovskites for Solar Thermochemical Applications," *Chemistry Mater* 32(9964). https://doi.org/10.1021/acs.chemmater.0c02912.

George, A., B. Shen, M. Craven, Y. Wang, D. Kang, C. Wu, and X. Tu. 2021. "A Review of Non-Thermal Plasma Technology: A Novel Solution for CO_2 Conversion and Utilization." *Renewable and Sustainable Energy Reviews* 135:109702.

Gharpure, A., and R. Vander Wal. 2023. "Improving Graphenic Quality by Oxidative Liberation of Crosslinks in Non-Graphitizable Carbons." *Carbon* 209:118010.

Goerzen, D., D.A. Heller, and R. Meidl. 2024. "Balancing Safety and Innovation: Shaping Responsible Carbon Nanotube Policy." Baker Institute. https://www.bakerinstitute.org/research/balancing-safety-and-innovation-shaping-responsible-carbon-nanotube-policy.

Gui, G., J. Li, and J. Zhong. 2008. "Band Structure Engineering of Graphene by Strain: First-Principles Calculations." *Physical Review B* 78(7):075435.

Güneş, F., G.H. Han, H.-J. Shin, S.Y. Lee, M. Jin, D.L. Duong, S.J. Chae, et al. 2011. "UV-Light-Assisted Oxidative sp3 Hybridization of Graphene." *Nano* 6(5):409–418.

Han, G.H., J. Bang, G. Park, S. Choe, Y.J. Jang, H.W. Jang, S.Y. Kim, and S.H. Ahn. 2023a. "Recent Advances in Electrochemical, Photochemical, and Photoelectrochemical Reduction of CO_2 to C_{2+} Products." *Small* 19(16):2205765. https://doi.org/10.1002/smll.202205765.

Han, X., K.K. Ostrikov, J. Chen, Y. Zheng, and X. Xu. 2023b. "Electrochemical Reduction of Carbon Dioxide to Solid Carbon: Development, Challenges, and Perspectives." *Energy and Fuels* 37(17):12665–12684. https://doi.org/10.1021/acs.energyfuels.3c02204.

He, B., M. Feng, X. Chen, and J. Sun. 2021. "Multidimensional (0D–3D) Functional Nanocarbon: Promising Material to Strengthen the Photocatalytic Activity of Graphitic Carbon Nitride." *Green Energy and Environment* 6(6):823–845.

Hiremath, N., J. Mays, and G. Bhat. 2017. "Recent Developments in Carbon Fibers and Carbon Nanotube-Based Fibers: A Review." *Polymer Reviews* 57(2):339–368.

Hoffmann, R., A.A. Kabanov, A.A. Golov, and D.M. Proserpio. 2016. "Homo Citans and Carbon Allotropes: For an Ethics of Citation." *Angewandte Chemie International Edition* 55(37):10962–10976. Database updated February 15, 2024. https://www.sacada.info/sacada_3D.php.

Hong, S.-H., J.-S. Choi, S.-J. Yoo, and Y.-S. Yoon. 2023. "Structural Benefits of Using Carbon Nanotube Reinforced High-Strength Lightweight Concrete Beams." *Developments in the Built Environment* 16:100234.

Hu, L., Y. Song, J. Ge, J. Zhu, and S. Jiao. 2015. "Capture and Electrochemical Conversion of CO_2 to Ultrathin Graphite Sheets in $CaCl_2$-based Melts." *Journal of Materials Chemistry A* 3(42):21211–21218.

Hu, L., Y. Song, S. Jiao, Y. Liu, J. Ge, H. Jiao, J. Zhu, J. Wang, H. Zhu, and D.J. Fray. 2016. "Direct Conversion of Greenhouse Gas CO_2 into Graphene via Molten Salts Electrolysis." *ChemSusChem* 9(6):588–594.

Hu, L., W. Yang, Z. Yang, and J. Xu. 2019. "Fabrication of Graphite via Electrochemical Conversion of CO_2 in a $CaCl_2$ Based Molten Salt at a Relatively Low Temperature." *RSC Advances* 9(15):8585–8593.

Hut, G., H.G. Östlund, and K. van der Borg. 1986. "Fast and Complete CO_2-to-Graphite Conversion for 14C Accelerator Mass Spectrometry." *Radiocarbon* 28(2A):186–190.

Irfan, M., K. Zuraiqi, C.K. Nguyen, T.C. Le, F. Jabbar, M. Ameen, C.J. Parker, et al. 2023. "Liquid Metal-Based Catalysts for the Electroreduction of Carbon Dioxide into Solid Carbon." *Journal of Materials Chemistry A* 11(27):14990–14996. https://doi.org/10.1039/D3TA01379K.

Jana, P., and A. Dev. 2022. "Carbon Quantum Dots: A Promising Nanocarrier for Bioimaging and Drug Delivery in Cancer." *Materials Today Communications* 32(August):104068. https://doi.org/10.1016/j.mtcomm.2022.104068.

Jara, A.D., A. Betemariam, G. Woldetinsae, and J.Y. Kim. 2019. "Purification, Application and Current Market Trend of Natural Graphite: A Review." *International Journal of Mining Science and Technology* 29(5):671–689.

Jhonsi, M.A. 2018. "Carbon Quantum Dots for Bioimaging." In *State of the Art in Nano-Bioimaging*. IntechOpen. https://doi.org/10.5772/intechopen.72723.

Kaur, J., K. Millington, and S. Smith. 2016. "Producing High-Quality Precursor Polymer and Fibers to Achieve Theoretical Strength in Carbon Fibers: A Review." *Journal of Applied Polymer Science* 133(38).

Keypour, H., M. Noroozi, and A. Rashidi. 2013. "An Improved Method for the Purification of Fullerene from Fullerene Soot with Activated Carbon, Celite, and Silica Gel Stationary Phases." *Journal of Nanostructure in Chemistry* 3:1–9.

Khan, A., and K.A. Alamry. 2022. "Surface Modified Carbon Nanotubes: An Introduction." Pp. 1–25 in *Surface Modified Carbon Nanotubes Volume 1: Fundamentals, Synthesis and Recent Trends*. American Chemical Society.

Kim, A., J.K. Dash, P. Kumar, and R. Patel. 2021. "Carbon-Based Quantum Dots for Photovoltaic Devices: A Review." *ACS Applied Electronic Materials* 4(1):27–58.

Kim, G.M., W.Y. Choi, J.H. Park, S.J. Jeong, J.-E. Hong, W. Jung, and J.W. Lee. 2020a. "Electrically Conductive Oxidation-Resistant Boron-Coated Carbon Nanotubes Derived from Atmospheric CO_2 for Use at High Temperature." *ACS Applied Nano Materials* 3(9):8592–8597. https://doi.org/10.1021/acsanm.0c01909.

Kim, G.M., W.-G. Lim, D. Kang, J.H. Park, H. Lee, J. Lee, and J.W. Lee. 2020b. "Transformation of Carbon Dioxide into Carbon Nanotubes for Enhanced Ion Transport and Energy Storage." *Nanoscale* 12(14):7822–7833. https://doi.org/10.1039/C9NR10552B.

Kim, J., M. Shin, S.H. So, S. Hong, D.Y. Park, C. Kim, and C.R. Park. 2024. "Correlation Between Structural Characteristics of Edge Selectively-Oxidized Graphite and Electrochemical Performances Under Fast Charging Condition." *Carbon* (218):118664.

Kim, J.-S., J.-R. Ahn, C.W. Lee, Y. Murakami, and D. Shindo. 2001. "Morphological Properties of Ultra-fine (Ni, Zn)-Ferrites and Their Ability to Decompose CO_2." *Journal of Materials Chemistry* 11(12):3373–3376.

Kim, S.-H., J.T. Jang, J. Sim, J.-H. Lee, S.-C. Nam, and C.Y. Park. 2019. "Carbon Dioxide Decomposition Using SrFeCo0.5Ox, a Nonperovskite-Type Metal Oxide." *Journal of CO_2 Utilization* 34:709–715.

Komarneni, S., M. Tsuji, Y. Wada, and Y. Tamaura. 1997. "Nanophase Ferrites for CO2 Greenhouse Gas Decomposition." *Journal of Materials Chemistry* 7(12):2339–2340. https://doi.org/10.1039/a705849g.

Komatsu, N., T. Ohe, and K. Matsushige. 2004. "A Highly Improved Method for Purification of Fullerenes Applicable to Large-Scale Production." *Carbon* 42(1):163–167.

Kondratenko, E.V., G. Mul, J. Baltrusaitis, G.O. Larrazábal, and J. Pérez-Ramírez. 2013. "Status and Perspectives of CO_2 Conversion into Fuels and Chemicals by Catalytic, Photocatalytic and Electrocatalytic Processes." *Energy and Environmental Science* 6(11):3112–3135.

Korányi, T.I., M. Németh, A. Beck, and A. Horváth. 2022. "Recent Advances in Methane Pyrolysis: Turquoise Hydrogen with Solid Carbon Production." *Energies* 15(17):6342. https://doi.org/10.3390/en15176342.

Kordek-Khalil, K., A. Moyseowicz, P. Rutkowski, and G. Gryglewicz. 2024. "Excellent Performance of Nitrogen-Doped Carbon Nanofibers as Electrocatalysts for Water Splitting Reactions." *International Journal of Hydrogen Energy* 52:494–506.

Kosari, M., A.M.H. Lim, Y. Shao, B. Li, K.M. Kwok, A.M. Seayad, A. Borgna, and H.C. Zeng. 2022. "Thermocatalytic CO_2 Conversion by Siliceous Matter: A Review." *Journal of Materials Chemistry A*.

Kozinsky, B., and N. Marzari. 2006. "Static Dielectric Properties of Carbon Nanotubes from First Principles." *Physical Review Letters* 96(16):166801.

Lallave, M., J. Bedia, R. Ruiz-Rosas, J. Rodríguez-Mirasol, T. Cordero, J.C. Otero, M. Marquez, A. Barrero, and I.G. Loscertales. 2007. "Filled and Hollow Carbon Nanofibers by Coaxial Electrospinning of Alcell Lignin Without Binder Polymers." *Advanced Materials* 19(23):4292–4296.

Lau, J., G. Dey, and S. Licht. 2016. "Thermodynamic Assessment of CO_2 to Carbon Nanofiber Transformation for Carbon Sequestration in a Combined Cycle Gas or a Coal Power Plant." *Energy Conversion and Management* 122:400–410.

Le, T.-H., and H. Yoon. 2019. "Strategies for Fabricating Versatile Carbon Nanomaterials from Polymer Precursors." *Carbon* 152:796–817.

Lee, K.H., S.H. Lee, and R.S. Ruoff. 2020. "Synthesis of Diamond-Like Carbon Nanofiber Films." *ACS nano* 14(10):13663–13672.

Lerouge, S., M.R. Wertheimer, and L.H. Yahia. 2001. "Plasma Sterilization: A Review of Parameters, Mechanisms, and Limitations." *Plasmas and Polymers* 6:175–188.

Lesiak, B., L. Kövér, J. Tóth, J. Zemek, P. Jiricek, A. Kromka, and N.J.A.S.S. Rangam. 2018. "C sp2/sp3 Hybridisations in Carbon Nanomaterials–XPS and (X) AES study." *Applied Surface Science* 452:223–231.

Lewis, M.A., M. Serban, C.L. Marshall, and D. Lewis. 2001. "Direct Contact Pyrolysis of Methane Using Nuclear Reactor Heat." Lemont, IL: Argonne National Laboratory. https://publications.anl.gov/anlpubs/2001/11/41245.pdf.

Li, A., Q. Cao, G. Zhou, B.V.K.J. Schmidt, W. Zhu, X. Yuan, H. Huo, J. Gong, and M. Antonietti. 2019a. "Three-Phase Photocatalysis for the Enhanced Selectivity and Activity of CO_2 Reduction on a Hydrophobic Surface." *Angewandte Chemie International Edition* 58(41):14549–14555. https://doi.org/10.1002/anie.201908058.

Li, F., A. Thevenon, A. Rosas-Hernández, Z. Wang, Y. Li, C.M. Gabardo, A. Ozden, et al. 2020. "Molecular Tuning of CO_2-to-Ethylene Conversion." *Nature* 577(7791):509–513. https://doi.org/10.1038/s41586-019-1782-2.

Li, J., C. Hu, Y.-Y. Wang, X. Meng, S. Xiang, C. Bakker, K. Plaza, A.J. Ragauskas, S.Y. Dai, and J.S. Yuan. 2022a. "Lignin Molecular Design to Transform Green Manufacturing." *Matter* 5(10):3513–3529. https://doi.org/10.1016/j.matt.2022.07.011.

Li, J., C. Hu, J. Arreola-Vargas, K. Chen, and J.S. Yuan. 2022b. "Feedstock Design for Quality Biomaterials." *Trends in Biotechnology* 40(12):1535–1549. https://doi.org/10.1016/j.tibtech.2022.09.017.

Li, J., Z. Xu, R. Yang, L. Ren, L. Ma, G. Huang, S. Zhang, and Z. Huang. 2023a. "Carbon Cycle in a Steelmaking Mill: Recycling of Kish Graphite and Its Subsequent Application for Steelmaking Carburant."

Li, K., X. An, K.H. Park, M. Khraisheh, and J. Tang. 2014. "A Critical Review of CO_2 Photoconversion: Catalysts and Reactors." *Catalysis Today* 224:3–12.

Li, S., A. Pasc, V. Fierro, and A. Celzard. 2016. "Hollow Carbon Spheres, Synthesis and Applications—A Review." *Journal of Materials Chemistry A* 4(33):12686–12713.

Li, X., H. Shi, X. Wang, X. Hu, C. Xu, and W. Shao. 2022c. "Direct Synthesis and Modification of Graphene in Mg Melt by Converting CO_2: A Novel Route to Achieve High Strength and Stiffness in Graphene/Mg Composites." *Carbon* 186:632–643.

Li, X., X. Wang, X. Hu, C. Xu, W. Shao, and K. Wu. 2023b. "Direct Conversion of CO_2 to Graphene via Vapor–Liquid Reaction for Magnesium Matrix Composites with Structural and Functional Properties." *Journal of Magnesium and Alloys* 11(4):1206–1212.

Li, Z., D. Yuan, H. Wu, W. Li, and D. Gu. 2018. "A Novel Route to Synthesize Carbon Spheres and Carbon Nanotubes from Carbon Dioxide in a Molten Carbonate Electrolyzer." *Inorganic Chemistry Frontiers* 5(1):208–216.

Liang, C., Y. Chen, M. Wu, K. Wang, W. Zhang, Y. Gan, H. Huang, et al. 2021. "Green Synthesis of Graphite from CO_2 Without Graphitization Process of Amorphous Carbon." *Nature Communications* 12(1):119.

Licht, S., A. Douglas, J. Ren, R. Carter, M. Lefler, and C.L. Pint. 2016. "Carbon Nanotubes Produced from Ambient Carbon Dioxide for Environmentally Sustainable Lithium-ion and Sodium-ion Battery Anodes." *ACS Central Science* 2(3):162–168.

Lin, K.-S., A.K. Adhikari, Z.-Y. Tsai, Y.-P. Chen, T.-T. Chien, and H.-B. Tsai. 2011. "Synthesis and Characterization of Nickel Ferrite Nanocatalysts for CO_2 Decomposition." *Catalysis Today* 174(1):88–96.

Liu, H., S. Wu, N. Tian, F. Yan, C. You, and Y. Yang. 2020a. "Carbon Foams: 3D Porous Carbon Materials Holding Immense Potential." *Journal of Materials Chemistry A* 8(45):23699–23723.

Liu, J., R. Li, and B. Yang. 2020b. "Carbon Dots: A New Type of Carbon-Based Nanomaterial with Wide Applications." *ACS Central Science* 6(12):2179–2195.

Liu, T., L. Zhang, B. Cheng, and J. Yu. 2019. "Hollow Carbon Spheres and Their Hybrid Nanomaterials in Electrochemical Energy Storage." *Advanced Energy Materials* 9(17):1803900.

Liu, X., X. Wang, G. Licht, and S. Licht. 2020c. "Transformation of the Greenhouse Gas Carbon Dioxide to Graphene." *Journal of CO_2 Utilization* 36:288–294.

Liu, Y., and S. Kumar. 2012. "Recent Progress in Fabrication, Structure, and Properties of Carbon Fibers." *Polymer Reviews* 52(3):234–258.

Lou, Z., C. Chen, H. Huang, and D. Zhao. 2006. "Fabrication of Y–Junction Carbon Nanotubes by Reduction of Carbon Dioxide with Sodium Borohydride." *Diamond and Related Materials* 15(10):1540–1543. https://doi.org/10.1016/j.diamond.2005.12.044.

Lu, Q., and F. Jiao. 2016. "Electrochemical CO_2 reduction: Electrocatalyst, Reaction Mechanism, and Process Engineering." *Nano Energy* 29:439–456.

Luc, W., M. Jouny, J. Rosen, and F. Jiao. 2018. "Carbon Dioxide Splitting Using an Electro-Thermochemical Hybrid Looping Strategy." *Energy and Environmental Science* 11(10):2928–2934. https://doi.org/10.1039/C8EE00532J.

Luo, X., H. Zheng, W. Lai, P. Yuan, S. Li, D. Li, and Y. Chen. 2023. "Defect Engineering of Carbons for Energy Conversion and Storage Applications." *Energy and Environmental Materials* 6(3):e12402. https://doi.org/10.1002/eem2.12402.

Malode, S.J., M.M. Shanbhag, R. Kumari, D.S. Dkhar, P. Chandra, and N.P. Shetti. 2023. "Biomass-Derived Carbon Nanomaterials for Sensor Applications." *Journal of Pharmaceutical and Biomedical Analysis* 222:115102.

marketsandmarkets. 2022. "Carbon Fiber Market, Industry Share Forecast Trends Report." https://www.marketsandmarkets.com/Market-Reports/carbon-fiber-396.html.

Martirez, J.M.P., J.L. Bao, and E.A. Carter. 2021. "First-Principles Insights into Plasmon-Induced Catalysis." *Annual Review of Physical Chemistry* 72:99–119.

Mbayachi, V.B., E. Ndayiragije, T. Sammani, S. Taj, and E.R. Mbuta. 2021. "Graphene Synthesis, Characterization and Its Applications: A Review." *Results in Chemistry* 3:100163.

Mei, D., X. Zhu, Y.-L. He, J.D. Yan, and X. Tu. 2014. "Plasma-Assisted Conversion of CO_2 in a Dielectric Barrier Discharge Reactor: Understanding the Effect of Packing Materials." *Plasma Sources Science and Technology* 24(1):015011.

Mishra, S., S. Kumari, A.C. Mishra, R. Chaubey, and S. Ojha. 2023. "Carbon Nanotube–Synthesis, Purification and Biomedical Applications." *Current Nanomaterials* 8(4):328–335.

Mori, S., N. Matsuura, L.L. Tun, and M. Suzuki. 2016. "Direct Synthesis of Carbon Nanotubes from Only CO_2 by a Hybrid Reactor of Dielectric Barrier Discharge and Solid Oxide Electrolyser Cell." *Plasma Chemistry and Plasma Processing* 36(1):231–239. https://doi.org/10.1007/s11090-015-9681-2.

Motiei, M., Y. Rosenfeld Hacohen, J. Calderon-Moreno, and A. Gedanken. 2001. "Preparing Carbon Nanotubes and Nested Fullerenes from Supercritical CO_2 by a Chemical Reaction." *Journal of the American Chemical Society* 123(35): 8624–8625. https://doi.org/10.1021/ja015859a.

Moyer, K., M. Zohair, J. Eaves-Rathert, A. Douglas, and C.L. Pint. 2020. "Oxygen Evolution Activity Limits the Nucleation and Catalytic Growth of Carbon Nanotubes from Carbon Dioxide Electrolysis via Molten Carbonates." *Carbon* 165:90–99.

Muñoz-García, A.B., A.M. Ritzmann, M. Pavone, J.A. Keith, and E.A. Carter. 2014. "Oxygen Transport in Perovskite-Type Solid Oxide Fuel Cell Materials: Insights from Quantum Mechanics," *Accounts of Chemical Research* 47(3340). https://doi.org/10.1021/ar4003174.

NASEM (National Academies of Sciences, Engineering, and Medicine). 2019. *Gaseous Carbon Waste Streams Utilization: Status and Research Needs*. Washington, DC: The National Academies Press. https://doi.org/10.17226/25232.

NASEM. 2023. *Carbon Dioxide Utilization Markets and Infrastructure: Status and Opportunities: A First Report*. Washington, DC: The National Academies Press.

Neyts, E.C., K. Ostrikov, M.K. Sunkara, and A. Bogaerts. 2015. "Plasma Catalysis: Synergistic Effects at the Nanoscale." *Chemical Reviews* 115(24):13408–13446.

Novoselova, I.A., N.F. Oliynyk, and S.V. Volkov. 2007. "Electrolytic Production of Carbon Nanotubes in Chloride-Oxide Melts Under Carbon Dioxide Pressure." Pp. 459–465 in *Hydrogen Materials Science and Chemistry of Carbon Nanomaterials*. D.V. Schur, B. Baranowski, A.P. Shpak, V.V. Skorokhod, A. Kale, T.N. Veziroglu, S.Y. Zaginaichenko, eds. Berlin: Springer.

Ognibene, T.J., G. Bench, J.S. Vogel, G.F. Peaslee, and S. Murov. 2003. "A High-Throughput Method for the Conversion of CO_2 Obtained from Biochemical Samples to Graphite in Septa-Sealed Vials for Quantification of 14C via Accelerator Mass Spectrometry." *Analytical Chemistry* 75(9):2192–2196.

Overa, S., B.H. Ko, Y. Zhao, and F. Jiao. 2022. "Electrochemical Approaches for CO_2 Conversion to Chemicals: A Journey Toward Practical Applications." *Accounts of Chemical Research* 55(5):638–648.

Pappijn, C.A.R., M. Ruitenbeek, M.-F. Reyniers, and K.M. Van Geem. 2020. "Challenges and Opportunities of Carbon Capture and Utilization: Electrochemical Conversion of CO_2 to Ethylene." *Frontiers in Energy Research* 8:557466.

Park, S.-J. 2015. *Carbon Fibers*. Vol. 210. Springer Series in Materials Science. Dordrecht: Springer Netherlands. https://doi.org/10.1007/978-94-017-9478-7.

Parker, D.H., K. Chatterjee, P. Wurz, K.R. Lykke, M.J. Pellin, L.M. Stock, and J.C. Hemminger. 1992. "Fullerenes and Giant Fullerenes: Synthesis, Separation, and Mass Spectrometric Characterization." *Carbon* 30(8):1167–1182.

Pease, R.N., and P.R. Chesebro. 1928. "Equilibrium in the Reaction CH4 + 2H2O ⇄ CO2 + 4H21." *Journal of the American Chemical Society* 50(5):1464–1469. https://doi.org/10.1021/ja01392a035.

Pérez-Gallent, E., C. Sánchez-Martínez, L.F.G. Geers, S. Turk, R. Latsuzbaia, and E.L.V. Goetheer. 2020. "Overcoming Mass Transport Limitations in Electrochemical Reactors with a Pulsating Flow Electrolyzer." *Industrial and Engineering Chemistry Research* 59(13):5648–5656. https://doi.org/10.1021/acs.iecr.9b06925.

Pillai, S.M., G.L. Tembe, and M. Ravindranathan. 1992. "Dimerization of Ethene to 2-Butene and Metathesis with 1-Butene by Sequential Use of Homogeneous Catalyst Systems." *Applied Catalysis A: General* 81(2):273–278.

Qu, Q., J. Guo, H. Wang, K. Zhang, and J. Li. 2024. "Carbon Nanofibers Implanted Porous Catalytic Metal Oxide Design as Efficient Bifunctional Electrode Host Material for Lithium-Sulfur Battery." *Electrochimica Acta* 473:143454.

Rajagopalan, R., Y. Tang, X. Ji, C. Jia, and H. Wang. 2020. "Advancements and Challenges in Potassium Ion Batteries: A Comprehensive Review." *Advanced Functional Materials* 30(12):1909486.

Ramazani, A., M.A. Moghaddasi, A.M. Malekzadeh, S. Rezayati, Y. Hanifehpour, and S.W. Joo. 2021. "Industrial Oriented Approach on Fullerene Preparation Methods." *Inorganic Chemistry Communications* 125:108442.

Rani, P., and V.K. Jindal. 2013. "Designing Band Gap of Graphene by B and N Dopant Atoms." *RSC Advances* 3(3):802–812.

Ranzoni, A., and M.A. Cooper. 2017. "The Growing Influence of Nanotechnology in Our Lives." Pp. 1–20 in *Micro- and Nanotechnology in Vaccine Development*, M. Skwarczynski and I. Tóth, eds. Micro and Nano Technologies Series. Amsterdam and Boston: Elsevier/William Andrew.

Rathinavel, S., K. Priyadharshini, and D. Panda. 2021. "A Review on Carbon Nanotube: An Overview of Synthesis, Properties, Functionalization, Characterization, and the Application." *Materials Science and Engineering: B* 268:115095.

Ren, J., and S. Licht. 2016. "Tracking Airborne CO_2 Mitigation and Low Cost Transformation into Valuable Carbon Nanotubes." *Scientific Reports* 6(1):27760. https://doi.org/10.1038/srep27760.

Ren, J., F.-F. Li, J. Lau, L. González-Urbina, and S. Licht. 2015. "One-Pot Synthesis of Carbon Nanofibers from CO_2." *Nano Letters* 15(9):6142–6148.

Ren, J., A. Yu, P. Peng, M. Lefler, F.-F. Li, and S. Licht. 2019. "Recent Advances in Solar Thermal Electrochemical Process (STEP) for Carbon Neutral Products and High Value Nanocarbons." *Accounts of Chemical Research* 52(11):3177–3187. https://doi.org/10.1021/acs.accounts.9b00405.

Rezaei, A., and E. Hashemi. 2021. "A Pseudohomogeneous Nanocarrier Based on Carbon Quantum Dots Decorated with Arginine as an Efficient Gene Delivery Vehicle." *Scientific Reports* 11(1):13790.

Ritzmann, A.M., J.M. Dieterich, and E.A. Carter. 2016. "Density Functional Theory + U Analysis of the Electronic Structure and Defect Chemistry of LSCF (La0.5Sr0.5Co0.25Fe0.75O3-δ)." *Physical Chemistry Chemical Physics* 18:12260–12269. https://doi.org/10.1039/C6CP01720G.

Rodríguez-Manzo, J.A., M. Terrones, H. Terrones, H.W. Kroto, L. Sun, and F. Banhart. 2007. "In Situ Nucleation of Carbon Nanotubes by the Injection of Carbon Atoms into Metal Particles." *Nature Nanotechnology* 2(5):307–311.

Sajna, M.S., S. Zavahir, A. Popelka, P. Kasak, A. Al-Sharshani, U. Onwusogh, M. Wang, H. Park, and D. S. Han. 2023. "Electrochemical System Design for CO_2 Conversion: A Comprehensive Review." *Journal of Environmental Chemical Engineering* 11(5):110467.

Samiee, S., and E.K. Goharshadi. 2014. "Graphene Nanosheets as Efficient Adsorbent for an Azo Dye Removal: Kinetic and Thermodynamic Studies." *Journal of Nanoparticle Research* 16:1–16.

Sánchez-Bastardo, N., R. Schlögl, and H. Ruland. 2021. "Methane Pyrolysis for Zero-Emission Hydrogen Production: A Potential Bridge Technology from Fossil Fuels to a Renewable and Sustainable Hydrogen Economy." *Industrial and Engineering Chemistry Research* 60(32):11855–11881. https://doi.org/10.1021/acs.iecr.1c01679.

Sasikumar, B., S.A.G. Krishnan, M. Afnas, G. Arthanareeswaran, P.S. Goh, and A.F. Ismail. 2023. "A Comprehensive Performance Comparison on the Impact of MOF-71, HNT, SiO2, and Activated Carbon Nanomaterials in Polyetherimide Membranes for Treating Oil-in-Water Contaminants." *Journal of Environmental Chemical Engineering* 11(1):109010.

Schaefer, H.-E. 2010. *Nanoscience: The Science of the Small in Physics, Engineering, Chemistry, Biology and Medicine*. Berlin, Heidelberg: Springer. https://doi.org/10.1007/978-3-642-10559-3.

Sengupta, D., S. Chen, A. Michael, C.Y. Kwok, S. Lim, Y. Pei, and A.G. Kottapalli Prakash. 2020. "Single and Bundled Carbon Nanofibers as Ultralightweight and Flexible Piezoresistive Sensors." *NPJ Flex Electron* 4(1):9.

Shakoorioskooie, M., Y.Z. Menceloglu, S. Unal, and S.H. Soytas. 2018. "Rapid Microwave-Assisted Synthesis of Platinum Nanoparticles Immobilized in Electrospun Carbon Nanofibers for Electrochemical Catalysis." *ACS Applied Nano Materials* 1(11):6236–6246.

Shen, Y., and A.C. Lua. 2015. "Synthesis of Ni and Ni–Cu Supported on Carbon Nanotubes for Hydrogen and Carbon Production by Catalytic Decomposition of Methane." *Applied Catalysis B: Environmental* 164(March):61–69. https://doi.org/10.1016/j.apcatb.2014.08.038.

Sim, J., S.-H. Kim, J.-Y. Kim, K. Bong Lee, S.-C. Nam, and C.Y. Park. 2020. "Enhanced Carbon Dioxide Decomposition Using Activated SrFeO3–δ." *Catalysts* 10(11):1278.

Snoeckx, R., and A. Bogaerts. 2017. "Plasma Technology—A Novel Solution for CO 2 Conversion?" *Chemical Society Reviews* 46(19):5805–5863.

Son, D.-H., D. Hwangbo, H. Suh, B.-I. Bae, S. Bae, and C.-S. Choi. 2023. "Mechanical Properties of Mortar and Concrete Incorporated with Concentrated Graphene Oxide, Functionalized Carbon Nanotube, Nano Silica Hybrid Aqueous Solution." *Case Studies in Construction Materials* 18:e01603.

Spinner, N.S., J.A. Vega, and W.E. Mustain. 2012. "Recent Progress in the Electrochemical Conversion and Utilization of CO_2." *Catalysis Science Technology* 2(1):19–28. https://doi.org/10.1039/C1CY00314C.

Spradling, D.M., and R.A. Guth. 2003. "Carbon Foams." *Advanced Materials and Processes* 161(11):29–31.

Strudwick, A.J., N.E. Weber, M.G. Schwab, M. Kettner, R.T. Weitz, J.R. Wünsch, K. Müllen, and H. Sachdev. 2015. "Chemical Vapor Deposition of High Quality Graphene Films from Carbon Dioxide Atmospheres." *ACS Nano* 9(1):31–42.

Su, H., T. Wu, D. Cui, X. Lin, X. Luo, Y. Wang, and L. Han. 2020. "The Application of Graphene Derivatives in Perovskite Solar Cells." *Small Methods* 4(10):2000507.

Surovtseva, D., E. Crossin, R. Pell, and L. Stamford. 2022. "Toward a Life Cycle Inventory for Graphite Production." *Journal of Industrial Ecology* 26(3):964–979.

Tackett, B.M., E. Gomez, and J.G. Chen. 2019. "Net Reduction of CO_2 via Its Thermocatalytic and Electrocatalytic Transformation Reactions in Standard and Hybrid Processes." *Nature Catalysis* 2(5):381–386. https://doi.org/10.1038/s41929-019-0266-y.

Tamaura, Y., and M. Tahata. 1990. "Complete Reduction of Carbon Dioxide to Carbon Using Cation-Excess Magnetite." *Nature* 346(6281):255–256.

Tang, J., J. Tang, M. Mayyas, M.B. Ghasemian, J. Sun, M.A. Rahim, J. Yang, et al. 2022. "Liquid-Metal-Enabled Mechanical-Energy-Induced CO_2 Conversion." *Advanced Materials* 34(1):2105789. https://doi.org/10.1002/adma.202105789.

Tao, Y., S.D. Ebbesen, W. Zhang, and M.B. Mogensen. 2014. "Carbon Nanotube Growth on Nanozirconia Under Strong Cathodic Polarization in Steam and Carbon Dioxide." *ChemCatChem* 6(5):1220–1224. https://doi.org/10.1002/cctc.201300941.

Taylor, R., G.J. Langley, H.W. Kroto, and D.R.M. Walton. 1993. "Formation of C60 by Pyrolysis of Naphthalene." *Nature* 366(6457):728–731.

Terrones, M., W.K. Hsu, A. Schilder, H. Terrones, N. Grobert, J.P. Hare, Y.Q. Zhu, et al. 1998. "Novel Nanotubes and Encapsulated Nanowires." *Applied Physics A: Materials Science and Processing* 66(3).

Timmerberg, S., M. Kaltschmitt, and M. Finkbeiner. 2020. "Hydrogen and Hydrogen-Derived Fuels Through Methane Decomposition of Natural Gas—GHG Emissions and Costs." *Energy Conversion and Management: X* 7:100043. https://doi.org/10.1016/j.ecmx.2020.100043.

Tomkute, V., A. Solheim, and E. Olsen. 2013. "Investigation of High-Temperature CO_2 Capture by CaO in $CaCl_2$ Molten Salt." *Energy and Fuels* 27(9):5373–5379. https://doi.org/10.1021/ef4009899.

Tsang, W.M., S.J. Henley, V. Stolojan, and S.R.P. Silva. 2006. "Negative Differential Conductance Observed in Electron Field Emission from Band Gap Modulated Amorphous-Carbon Nanolayers." *Applied Physics Letters* 89(19).

Tsuji, M., Y. Wada, T. Yamamoto, T. Sano, and Y. Tamaura. 1996a. "CO_2 Decomposition by Metallic Phase on Oxygen-Deficient Ni (II)-Bearing Ferrite." *Journal of Materials Science Letters* 15:156–158.

Tsuji, M., T. Kodama, T. Yoshida, Y. Kitayama, and Y. Tamaura. 1996b. "Preparation and CO_2 Methanation Activity of an Ultrafine Ni (II) Ferrite Catalyst." *Journal of Catalysis* 164(2):315–321.

Tucker, S.C. 1999. "Solvent Density Inhomogeneities in Supercritical Fluids." *Chemical Reviews* 99(2):391–418.

ul Haque, S., A. Nasar, N. Duteanu, and S. Pandey. 2023. "Carbon Based-Nanomaterials Used in Biofuel Cells—A Review." *Fuel* 331:125634.

Urade, A.R., I. Lahiri, and K.S. Suresh. 2023. "Graphene Properties, Synthesis and Applications: A Review." *JOM* 75(3):614–630.

Van Tran, V., E. Wi, S.Y. Shin, D. Lee, Y.A. Kim, B.C. Ma, and M. Chang. 2022. "Microgels Based on 0D–3D Carbon Materials: Synthetic Techniques, Properties, Applications, and Challenges." *Chemosphere* 307(Pt 3):135981.

Wang, R., L. Sun, X. Zhu, W. Ge, H. Li, Z. Li, H. Zhang et al. 2023. "Carbon Nanotube-Based Strain Sensors: Structures, Fabrication, and Applications." *Advanced Materials Technologies* 8(1):2200855.

Wang, X., X. Liu, G. Licht, and S. Licht. 2020. "Calcium Metaborate Induced Thin Walled Carbon Nanotube Syntheses from CO_2 by Molten Carbonate Electrolysis." *Scientific Reports* 10(1):15146.

Wang, Z., Y. Zhang, E.C. Neyts, X. Cao, X. Zhang, B.W.-L. Jang, and C. Liu. 2018. "Catalyst Preparation with Plasmas: How Does It Work?" *ACS Catalysis* 8(3):2093–2110.

Wei, W., K. Sun, and Y.H. Hu. 2016. "Direct Conversion of CO2 to 3D Graphene and Its Excellent Performance for Dye-Sensitized Solar Cells with 10% Efficiency." *Journal of Materials Chemistry A* 4(31):12054–12057.

Wei, S., H. Shi, X. Li, X. Hu, C. Xu, and X. Wang. 2022. "A Green and Efficient Method for Preparing Graphene Using CO_2@Mg In-Situ Reaction and Its Application in High-Performance Lithium-Ion Batteries." *Journal of Alloys and Compounds* 902:163700.

Wexler, R.B., G.S. Gautam, E.B. Stechel, and E.A. Carter. 2021. "Factors Governing Oxygen Vacancy Formation in Oxide Perovskites," *Journal of the American Chemical Society* 143(33):13212–13227. https://doi.org/10.1021/jacs.1c05570.

Wexler, R.B., G.S. Gautam, R. Bell, S. Shulda, N.A. Strange, J.A. Trindell, J.D. Sugar, E. Nygren, S. Sainio, A.H. McDaniel, D. Ginley, E.A. Carter, and E.B. Stechel. 2023a. "Multiple and Nonlocal Cation Redox in Ca–Ce–Ti–Mn Oxide Perovskites for Solar Thermochemical Applications." *Energy & Environmental Science* 16:2550–2560. https://doi.org/10.1039/D3EE00234A.

Wexler, R.B., E.B. Stechel, and E.A. Carter. 2023b. "Materials Design Directions for Solar Thermochemical Water Splitting," Pp. 3–64 in *Solar Fuels*, N.D. Sankir and M. Sankir, eds. Vol. 3. Wiley-Scrivener. https://doi.org/10.1002/9781119752097.ch1.

Windhorst, T., and G. Blount. 1997. "Carbon-Carbon Composites: A Summary of Recent Developments and Applications." *Materials and Design* 18(1):11–15.

Wissler, M. 2006. "Graphite and Carbon Powders for Electrochemical Applications." *Journal of Power Sources* 156(2):142–150. https://doi.org/10.1016/j.jpowsour.2006.02.064.

Wu, Z.-Y., C. Li, H.-W. Liang, J.-F. Chen, and S.-H. Yu. 2013. "Ultralight, Flexible, and Fire-Resistant Carbon Nanofiber Aerogels from Bacterial Cellulose." *Angewandte Chemie* 125(10):2997–3001.

Xie, Z., E. Huang, S. Garg, S. Hwang, P. Liu, and J.G. Chen. 2024. "CO_2 Fixation into Carbon Nanofibres Using Electrochemical–Thermochemical Tandem Catalysis." *Nature Catalysis* 7(1):98–109. https://doi.org/10.1038/s41929-023-01085-1.

Yaashikaa, P.R., P.S. Kumar, S.J. Varjani, and A. Saravanan. 2019. "A Review on Photochemical, Biochemical and Electrochemical Transformation of CO_2 into Value-Added Products." *Journal of CO_2 Utilization* 33:131–147.

Yao, L., C. Sun, H. Lin, G. Li, Z. Lian, R. Song, S. Zhuang, and D. Zhang. 2023. "Electrospun Bi-Decorated BixTiyOz/TiO2 Flexible Carbon Nanofibers and Their Applications on Degradating of Organic Pollutants Under Solar Radiation." *Journal of Materials Science and Technology* 150(July):114–123. https://doi.org/10.1016/j.jmst.2022.07.066.

Yap, Y.W., N. Mahmed, M.N. Norizan, S.Z.A. Rahim, M.N.A. Salimi, K.A. Razak, I.S. Mohamad, M.M. Al-Bakri Abdullah, and M.Y.M. Yunus. 2023. "Recent Advances in Synthesis of Graphite from Agricultural Bio-Waste Material: A Review." *Materials* 16(9):3601.

Ye, L., N. Syed, D. Wang, J. Guo, J. Yang, J. Buston, R. Singh, M.S. Alivand, G.K. Li, and A. Zavabeti. 2023. "Low-Temperature CO_2 Reduction Using Mg–Ga Liquid Metal Interface." *Advanced Materials Interfaces* 10(3):2201625. https://doi.org/10.1002/admi.202201625.

Ye, R.-P., J. Ding, W. Gong, M.D. Argyle, Q. Zhong, Y. Wang, C.K. Russell, et al. 2019. "CO_2 Hydrogenation to High-Value Products via Heterogeneous Catalysis." *Nature Communications* 10(1):5698.

Yoshida, T., M. Tsuji, Y. Tamaura, T. Hurue, T. Hayashida, and K. Ogawa. 1997. "Carbon Recycling System Through Methanation of CO_2 in Flue Gas in LNG Power Plant." *Energy Conversion and Management* 38:S443–S448.

Yu, R., B. Deng, K. Du, D. Chen, M. Gao, and D. Wang. 2021. "Modulating Carbon Growth Kinetics Enables Electrosynthesis of Graphite Derived from CO_2 via a Liquid–Solid–Solid Process." *Carbon* 184:426–436.

Yu, W., L. Sisi, Y. Haiyan, and L. Jie. 2020. "Progress in the Functional Modification of Graphene/Graphene Oxide: A Review." *RSC Advances* 10(26):15328–15345.

Yu, Z., S. Wang, Z. Xiao, F. Xu, C. Xiang, L. Sun, and Y. Zou. 2024. "Graphene-Supported Cu_2O-CoO Heterojunctions for High Performance Supercapacitors." *Journal of Energy Storage* 77:110009.

Zhai, S., J. Nam, G.S. Gautam, K. Lim, J. Rojas, M.F. Toney, E.A. Carter, I.-H. Jung, W.C. Chueh, and A. Majumdar. 2022. "Thermodynamic Guiding Principles of High-Capacity Phase Transformation Materials for Splitting H_2O and CO_2 by Thermochemical Looping," *Journal of Materials Chemistry A* 10:3552–3561. https://doi.org/10.1039/D1TA10391A.

Zhang, B., F. Kang, J. Tarascon, and J. Kim. 2016. "Recent Advances in Electrospun Carbon Nanofibers and Their Application in Electrochemical Energy Storage." *Progress in Materials Science* 76:319–380.

Zhang, J., T. Tian, Y. Chen, Y. Niu, J. Tang, and L.-C. Qin. 2014. "Synthesis of Graphene from Dry Ice in Flames and Its Application in Supercapacitors." *Chemical Physics Letters* 591:78–81.

Zhang, P., J. Su, J. Guo, and S. Hu. 2023a. "Influence of Carbon Nanotube on Properties of Concrete: A Review." *Construction and Building Materials* 369:130388.

Zhang, W., X. Qiu, C. Wang, L. Zhong, F. Fu, J. Zhu, Z. Zhang, Y. Qin, D. Yang, and C.C. Xu. 2022. "Lignin Derived Carbon Materials: Current Status and Future Trends." *Carbon Research* 1(1):14. https://doi.org/10.1007/s44246-022-00009-1.

Zhang, X., R. Han, Y. Liu, H. Li, W. Shi, X. Yan, X. Zhao, Y. Li, and B. Liu. 2023b. "Porous and Graphitic Structure Optimization of Biomass-Based Carbon Materials from 0D to 3D for Supercapacitors: A Review." *Chemical Engineering Journal* 460:141607.

Zhao, L., Y. Chang, S. Qiu, H. Liu, J. Zhao, and J. Gao. 2023. "High Mechanical Energy Storage Capacity of Ultranarrow Carbon Nanowires Bundles by Machine Learning Driving Predictions." *Advanced Energy and Sustainability Research* 4(11):2300112. https://doi.org/10.1002/aesr.202300112.

Zhao, X., Y. Ando, Y. Liu, M. Jinno, and T. Suzuki. 2003. "Carbon Nanowire Made of a Long Linear Carbon Chain Inserted Inside a Multiwalled Carbon Nanotube." *Physical Review Letters* 90(18):187401.

Zhou, D., C. Zheng, Y. Niu, D. Feng, H. Ren, Y. Zhang, and H. Yu. 2024. "Hydrogen Storage Property Improvement of Ball-Milled Mg2. 3Y0. 1Ni Alloy with Graphene." *International Journal of Hydrogen Energy* 50(2024):123–135.

Zhou, T., and T. Zhang. 2021. "Recent Progress of Nanostructured Sensing Materials from 0D to 3D: Overview of Structure–Property–Application Relationship for Gas Sensors." *Small Methods* 5(9):2100515.

Zhou, X., B. Liu, Y. Chen, L. Guo, and G. Wei. 2020. "Carbon Nanofiber-Based Three-Dimensional Nanomaterials for Energy and Environmental Applications." *Materials Advances* 1(7):2163–2181.

Zoghi, M., M. Pourmadadi, F. Yazdian, M.N. Nigjeh, H. Rashedi, and R. Sahraeian. 2023. "Synthesis and Characterization of Chitosan/Carbon Quantum Dots/Fe2O3 Nanocomposite Comprising Curcumin for Targeted Drug Delivery in Breast Cancer Therapy." *International Journal of Biological Macromolecules* 249:125788.

Zuraiqi, K., A. Zavabeti, J. Clarke-Hannaford, B.J. Murdoch, K. Shah, M.J.S. Spencer, C.F. McConville, T. Daeneke, and K. Chiang. 2022. "Direct Conversion of CO2 to Solid Carbon by Ga-Based Liquid Metals." *Energy and Environmental Science* 15(2):595–600.

7

Chemical CO_2 Conversion to Fuels, Chemicals, and Polymers

Carbon dioxide (CO_2) is a potential feedstock for sustainable synthesis of carbon-based materials in a net-zero greenhouse gas (GHG) emissions future. As noted in previous chapters, most of the carbon in products that are manufactured and used today is derived from fossil feedstocks like natural gas and petroleum. Chapter 2 described markets for future products and intermediates derived from CO_2, as well as the competitive alternatives of electrification and clean hydrogen to replace carbon-based fuels for energy and energy storage, biomass and recycled plastic or material waste as feedstocks for carbon-based products, and extensive cradle-to-grave carbon capture and storage with continued fossil production of chemical and material products. This chapter focuses on chemical transformations of CO_2 into organic products where CO_2 utilization has some competitive advantages in a net-zero future, at a scale and impact that warrants national U.S. research and development (R&D) investment (see Sections 2.2.5.2, 2.2.5.3, and 2.2.5.4). As shown in Figure 7-1, these products include fuels, chemical intermediates, commodity chemicals, and polymers and their precursors, and they can be produced by a variety of chemical processes. The remainder of the chapter describes the current status and R&D needs for chemical CO_2 conversion processes and the resulting products, noting relevant applications where appropriate.

7.1 OVERVIEW OF CHEMICAL CONVERSION ROUTES FROM CO_2 TO ORGANIC PRODUCTS

7.1.1 Organic Chemical Products That Can Be Derived from CO_2

In principle, any carbon-based product can be formed chemically from CO_2. This report focuses on conversion of CO_2 to priority products in a net-zero future, including single carbon (C1) products, such as carbon monoxide, methanol, formic acid, urea, and methane, and multicarbon products, such as polycarbon oxygenates (alcohols, aldehydes, carboxylic acids, organic carbonates), olefins, aromatics, and hydrocarbons, including fuels (Figure 7-2). (See Chapter 2 for a detailed discussion of potential future market needs.) This chapter builds on Chapter 4 of the 2019 National Academies' report *Gaseous Carbon Waste Streams Utilization: Status and Research Needs* (NASEM 2019).

Carbon-based organic chemicals include compounds of carbon and hydrogen, with or without additional elements such as oxygen and nitrogen. The modern chemical industry developed to use petroleum as a source of both carbon and energy that is inexpensive, easy to ship, and contains advantageous carbon-carbon bonds. Large-volume

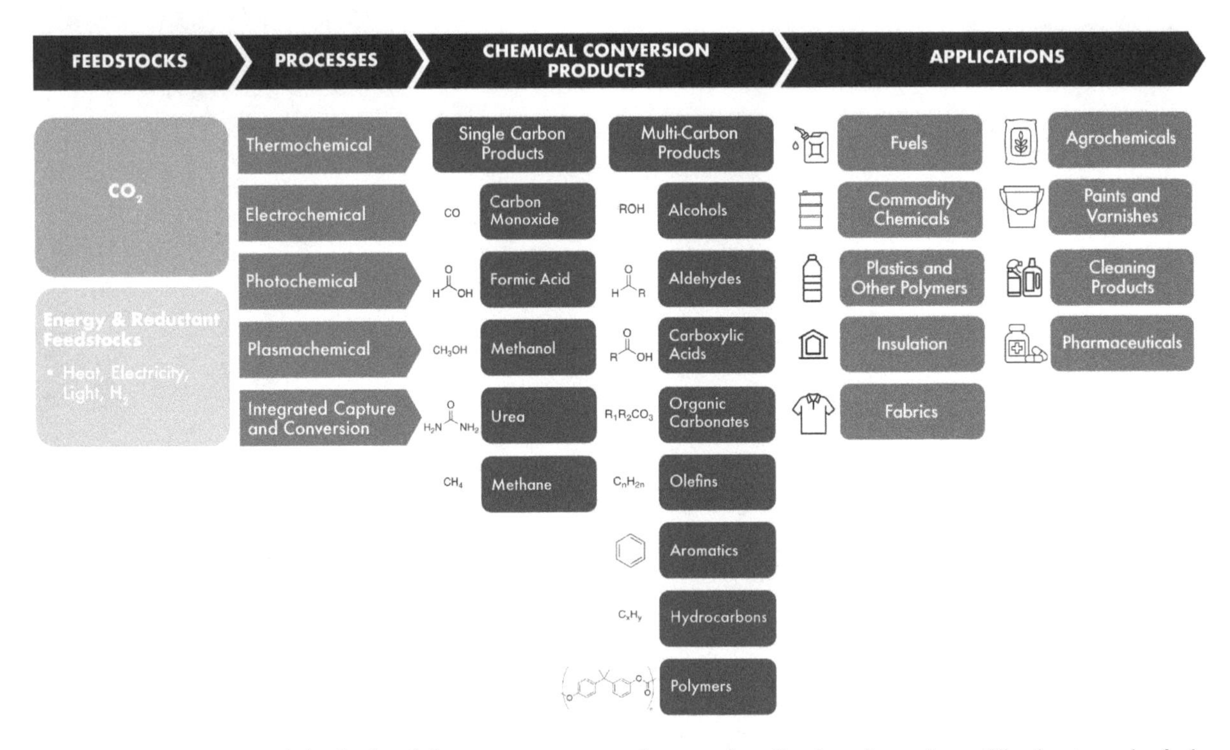

FIGURE 7-1 Summary of the feedstock inputs, processes, products, and applications for carbon utilization to make fuels, chemicals, and polymers.
SOURCE: Icons from the Noun Project, https://thenounproject.com. CC BY 3.0.

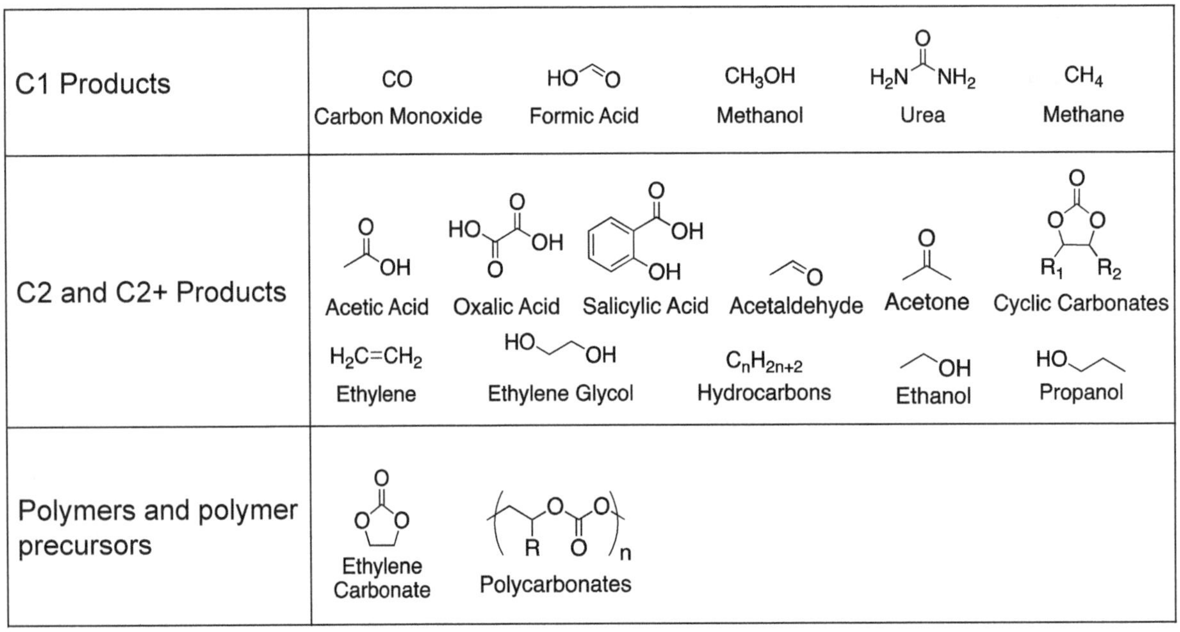

C1 Products	CO Carbon Monoxide; HO⁀O Formic Acid; CH_3OH Methanol; H_2N–C(=O)–NH_2 Urea; CH_4 Methane
C2 and C2+ Products	Acetic Acid; Oxalic Acid; Salicylic Acid; Acetaldehyde; Acetone; Cyclic Carbonates; $H_2C{=}CH_2$ Ethylene; HO⁀OH Ethylene Glycol; C_nH_{2n+2} Hydrocarbons; ⁀OH Ethanol; HO⁀ Propanol
Polymers and polymer precursors	Ethylene Carbonate; Polycarbonates

FIGURE 7-2 Example organic chemical products that can be derived from CO_2, illustrating common features such as carboxyl and alcohol groups, and highlighting priority products, including carbon monoxide, methanol, ethanol, ethylene, and urea.

chemical intermediates such as methanol, ethylene, propylene, benzene, toluene, and xylenes (Ellis et al. 2023) underlie the production of most other final products and so are particularly important parts of the chemical market. Manufactured organic chemical products pervade modern life. Their applications include fuels; plastics and other polymers for pipes, insulation, and fabrics; agrochemicals including fertilizers; paints and varnishes; cleaning products; pharmaceuticals; and more. As the global economy transitions to one with net-zero emissions, the need for these products will remain, but they will have to be produced from a non-fossil-carbon feedstock.

Alternatives to chemical production from petroleum have been explored and developed when access to petroleum was constrained (such as during wars or trade embargoes), when other resources were abundant and inexpensive relative to petroleum, or when there was an interest to diversify potential carbon sources away from only petroleum, such as domestic biofuel (EPA 2018; Lamprecht 2007; NRC 2006). Technologies and processes to use alternative carbon feedstocks of coal, natural gas, and biomass and its derivatives were developed, including the production of hydrocarbon chemicals and fuels via "syngas" (carbon monoxide and hydrogen). In a net-zero future, petroleum use as a chemical feedstock likely will be highly constrained owing to costs or limits on the resulting CO_2 emissions from the product life cycle. In this scenario, CO_2 is one option of sustainable carbon feedstock to replace petroleum. In addition to a change in feedstocks for chemicals production, the routes to produce chemicals could proceed via different priority intermediates than those currently used in the chemical industry, owing to the different properties of CO_2 as a feedstock as compared to petroleum. This chapter describes routes to potential future priority intermediates as well as final products.

Although not discussed in detail in this chapter, simply using CO_2 as a feedstock does not eliminate net GHG emissions from the life cycle of organic chemical production. CO_2 and other GHG emissions associated with the production of CO_2 and other feedstocks, transformation of the feedstocks into the product, delivery of the product to the user, use of the product, and its eventual disposal or recycling also have to be eliminated. See Chapter 3 for more discussion of life cycle assessments for CO_2 utilization.

7.1.2 Conversion Routes

There are several approaches to chemical conversion of CO_2 to organic products, all of which are geared toward overcoming the main challenges of using CO_2 as a chemical feedstock: its stability/nonreactivity, lack of carbon-carbon bonds, and presence as a dilute gas under ambient conditions. The ability of CO_2 to serve as a sustainable carbon feedstock is tied to these properties, as it is the primary waste product of combustion and other organic-molecule decomposition processes. Formation of most organic products from CO_2 requires energy input to overcome reaction barriers, some portion of which becomes energy stored in the product. Catalysts are often required to facilitate faster, more selective reactions. Thermochemical, electrochemical, photochemical, and plasmachemical reactions, as well as integrated CO_2 capture and conversion, can incorporate both energy input and catalysis into CO_2 conversion processes. Comparisons of "practical" energy requirements for different conversion pathways is challenging, as researchers often report different metrics for efficiency. Calculating the free energy of CO_2 conversion to a given product is possible, but unproductive, as the actual amount of energy required will exceed the theoretical limit and vary by process. The committee's first report quantified energy requirements for various carbon capture and hydrogen production processes, as well as the stoichiometric hydrogen requirements for several carbon-based products (see Figures 3-6, 3-7, and 3-8 in NASEM 2023). The following sections highlight status, challenges, and R&D opportunities for chemical CO_2 conversion processes in a net-zero future.

7.2 EXISTING AND EMERGING PROCESSES, CHALLENGES, AND R&D OPPORTUNITIES

Chemical conversions of CO_2 into organic chemicals span across technology readiness levels (TRLs). Figure 7-3 illustrates different pathways—thermochemical, electrochemical, photochemical, and plasmachemical—to produce organic chemicals from CO_2 and describes the technical maturity of the most advanced design of each process type. The following sections discuss current technologies, challenges, and R&D opportunities for each of these conversion pathways, as well as for integrated capture and conversion and the production of polymers from CO_2.

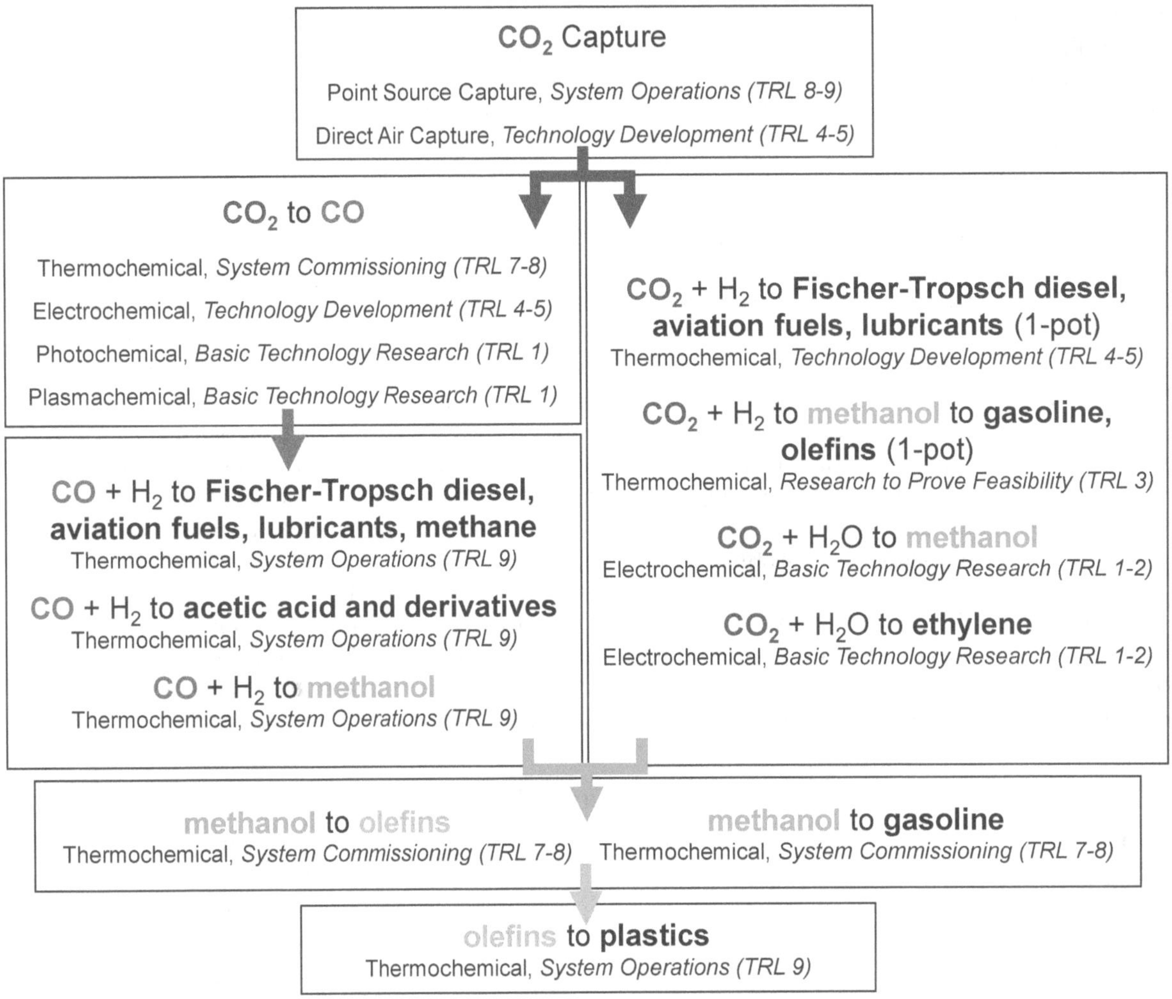

CO_2 Capture
Point Source Capture, *System Operations (TRL 8-9)*
Direct Air Capture, *Technology Development (TRL 4-5)*

CO_2 to CO
Thermochemical, *System Commissioning (TRL 7-8)*
Electrochemical, *Technology Development (TRL 4-5)*
Photochemical, *Basic Technology Research (TRL 1)*
Plasmachemical, *Basic Technology Research (TRL 1)*

CO + H_2 to **Fischer-Tropsch diesel, aviation fuels, lubricants, methane**
Thermochemical, *System Operations (TRL 9)*

CO + H_2 to **acetic acid and derivatives**
Thermochemical, *System Operations (TRL 9)*

CO + H_2 to methanol
Thermochemical, *System Operations (TRL 9)*

CO_2 + H_2 to **Fischer-Tropsch diesel, aviation fuels, lubricants** (1-pot)
Thermochemical, *Technology Development (TRL 4-5)*

CO_2 + H_2 to methanol to **gasoline, olefins** (1-pot)
Thermochemical, *Research to Prove Feasibility (TRL 3)*

CO_2 + H_2O to methanol
Electrochemical, *Basic Technology Research (TRL 1-2)*

CO_2 + H_2O to **ethylene**
Electrochemical, *Basic Technology Research (TRL 1-2)*

methanol to olefins
Thermochemical, *System Commissioning (TRL 7-8)*

methanol to **gasoline**
Thermochemical, *System Commissioning (TRL 7-8)*

olefins to **plastics**
Thermochemical, *System Operations (TRL 9)*

FIGURE 7-3 Schematic of CO_2 utilization processes to produce priority chemicals, fuels, and intermediates in a net-zero future, including maturity (approximate technology readiness level [TRL]) of the most advanced design of each process type: thermochemical, electrochemical, photochemical, or plasmachemical.
NOTES: All paths begin with CO_2 capture (point source capture or direct air capture), then conversion of the CO_2 (blue) into intermediates of CO (pink) or methanol (green), or directly to products (black bold). Conversion of CO or methanol to products, or to olefin intermediates (yellow) is also shown, along with olefin conversion to plastics.

7.2.1 Thermochemical Conversion Pathways

This section describes pathways for thermochemical conversion of CO_2 into the following products: carbon monoxide (CO) and synthesis gas ("syngas": a mixture of CO and H_2); methanol and its derivatives; formate/formic acid; C2+ hydrocarbons, oxygenates, and intermediates; C2 and C2+ carboxylic acids; fuels from Fischer-Tropsch synthesis using CO produced from CO_2; and polymer precursors. It begins by describing current technologies and processes for these conversions, followed by discussions of the challenges with and R&D opportunities for thermochemical CO_2 conversion. More emphasis is placed on the importance of CO_2-derived intermediates than final products, as the steps to produce final products are well known once key intermediates are produced. Thermochemical conversion pathways and associated products are shown in Figure 7-4 and Table 7-1.

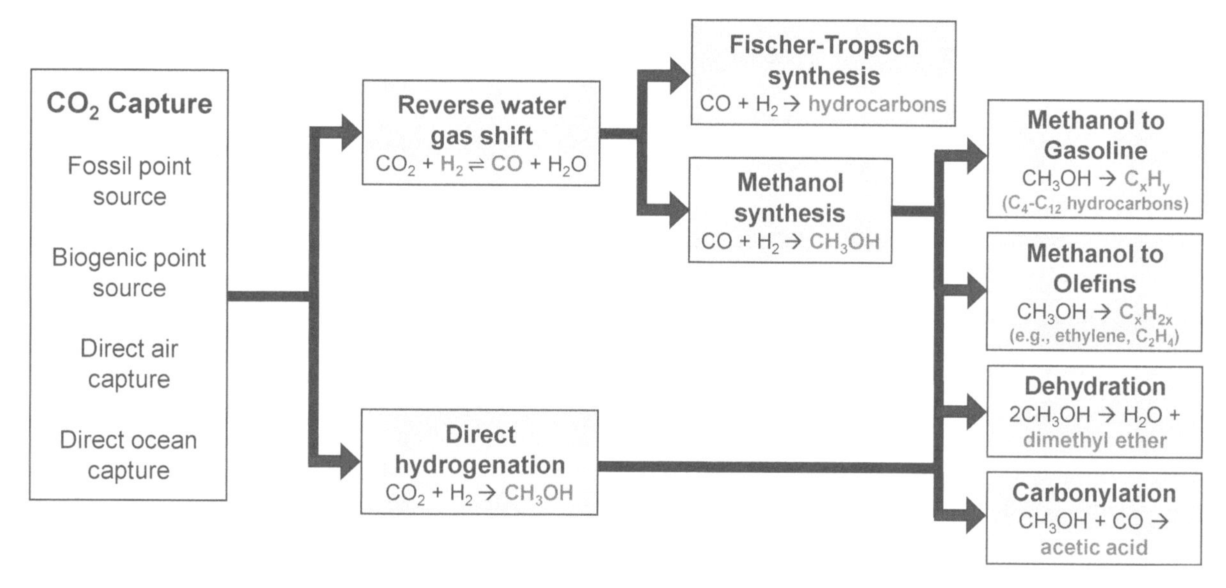

FIGURE 7-4 Pathways for thermochemical conversion of CO_2 to various organic products.
NOTE: CO_2 source in gray, processes in purple, products in pink.

7.2.1.1 Current Technology

Carbon Monoxide and Syngas

CO_2 conversion to CO is a key initial reaction for thermocatalytic pathways to hydrocarbon products. Processes for converting CO and its mixture with H_2 (syngas) into various hydrocarbons have been subject to ongoing research for continuous improvement for more than a century. Once syngas is formed, a full set of proven commercial pathways are known for comprehensive chemical synthesis across all molecules comprising the current hydrocarbon chemicals economy (Cho et al. 2017; Xie and Olsbye 2023).

TABLE 7-1 Key Products from Thermochemical CO_2 Conversion and Processes for Their Formation

Product	Processes for Formation from CO_2
Carbon monoxide and syngas	• Hydrogenation of CO_2 via reverse water gas shift • High-temperature solar thermochemical CO_2/H_2O splitting • Catalytic dry reforming of methane (CO_2 incorporated as one-half of the carbon in the syngas mixture)
Methanol and derivatives	• Direct hydrogenation • Production from syngas
Formate and formic acid	• Molecular catalysis • Heterogeneous catalysis • Direct thermocatalytic hydrogenation
C2+ hydrocarbons, oxygenates, intermediates	• Coupling with epoxides (half of carbon comes from epoxide) • Fischer-Tropsch (especially fuels) • From methanol (especially gasoline, olefins, acetic acid, aromatics) • Carboxylation of organometallic reagents, organic (pseudo)halides, unsaturated hydrocarbons, and sp, sp^2, and sp^3 hybridized C–H bonds
Polymer precursors	• From ethylene oxide and CO_2 to ethylene carbonate

CO forms readily by hydrogenating CO_2 in the reverse water-gas shift (RWGS) reaction (Reaction 7.1).[1] Albeit typically in the presence of a catalyst. The RWGS (and forward water-gas shift [WGS]) reaction is equilibrium constrained, with conversion affected by temperature, which impacts the equilibrium and kinetics, and to a moderate degree by pressure (influencing reaction rates); RWGS reaction kinetics are well documented (Bustamante et al. 2004; Chen et al. 2020).

RWGS reaction: $CO_2 + H_2 \rightleftharpoons CO + H_2O$ (R7.1)

The RWGS reaction typically uses copper or platinum, palladium, or rhodium catalysts supported on redox catalysts or supports such as ceria (CeO_2) (Ye et al. 2019; Zhou et al. 2017), rather than the alumina-supported catalysts used for the forward WGS reaction, to avoid acidity that can lead to coking at the higher temperatures used for RWGS. While well-known catalysts for the forward reaction can also perform the reverse reaction, new formulations could offer improved performance at the temperatures and pressures required and with impurities present. For example, platinum-doped cerium oxide catalysts offer higher RWGS reaction rates and yields, at the expense of requiring a noble metal (Ampelli et al. 2015). For syngas production via RWGS to be economically viable, improvements in CO yield, CO productivity, and catalyst durability are also needed. As of yet, the RWGS reaction has not been fully developed because there has been no economic incentive to do so in the absence of a price on fossil carbon. However, several studies have examined potential catalysts for the reaction: Dimitriou et al. (2015) review the approach for liquid fuels production; Chen et al. (2020) describe formulation of metal versus metal-oxide catalyst to improve tolerance to poisons; Zhang et al. (2022a) present a molybdenum phosphide–based catalyst to avoid use of noble metals; and Daza and Kuhn (2016) review catalyst options for producing liquid fuels by CO_2 hydrogenation. See Section 7.2.1.3 for more on catalyst development opportunities.

Another pathway for generating syngas is high-temperature solar thermochemical splitting of CO_2 and water (Al-Shankiti et al. 2017; Pullar et al. 2019; Wenzel et al. 2016). Use of solar thermochemical technologies for hydrogen generation[2] has lagged photovoltaic (PV)-electrolysis as a promising pathway for renewable hydrogen production, given the lower costs of PV electricity generation. However, solar thermochemical technologies for CO_2 splitting may be cost-effective for hydrocarbon product synthesis because of the ability to integrate with energy/heat storage and recuperation to enable 24/7 industrial operations. Thermal cycling of the working redox materials and differential thermal expansion are issues for system design and durability.

Syngas also can be produced by catalytic dry reforming of methane, which uses CO_2 as a soft oxidant and additional source of carbon: $CO_2 + CH_4 \rightleftharpoons 2CO + 2H_2$, where half of the carbon and all of the oxygen come from CO_2, and half of the carbon and all of the hydrogen come from methane (see, e.g., Shi et al. 2013). Dry reforming produces a syngas composition with CO/H_2 ratio of 1, which is too rich in CO for methanol or other chemical synthesis, other than addition to olefins or epoxides via hydroformylation, which has limited market size and hence limited ability to uptake CO_2 into products. Water-gas shift to remove some of the CO yields more H_2 but results in additional CO_2 formation, which via the subsequent reaction network of C1 chemistry (including the large endothermic heat of reaction for CO_2 conversion) results in a net increase rather than consumption of CO_2 (Sandoval-Diaz et al. 2022). CO separation via chemical looping or carbon rejection via solid nanofibers, or injection of additional clean H_2 is needed to render dry reforming a viable pathway for chemical production (Challiwala et al. 2021), except for limited market volume products (e.g., dimethyl ether) where a 1:1 syngas ratio is directly consumed. Some process flue gas compositions also may benefit from a 1:1 syngas composition for some retrofit applications. Dry reforming can be used in conjunction with renewable methane[3] to expand sequestration of carbon into products and synergistically produce solid carbon (see, e.g., Azara et al. 2019 and Chapter 6 of this report). However, the high temperatures required for CO_2 conversion

[1] RWGS is a stoichiometric reaction that converts CO_2 into CO by consumption of H_2. Typically, one adds an excess of H_2 so that once a given amount of CO_2 is "shifted" to CO, one has the desired ratio of H_2/CO for subsequent reactions. Alternatively, one can add excess H_2 after the RWGS reaction to obtain a desired ratio.

[2] For more information on solar thermochemical technologies for hydrogen production, see Wexler et al. (2023).

[3] Renewable methane is methane sourced from nonfossil feedstocks, like biomass, municipal solid waste, and other waste carbon-containing materials, like plastics.

by thermochemical dry reforming leads to substantial catalyst coking which, together with the need for carbon rejection to address the problem of incorrect (low) syngas ratio for most large-scale products, has limited commercial applications (Sandoval-Diaz et al. 2022).

Methanol and Derivatives

Pathways to convert CO_2 to methanol include direct hydrogenation and a two-step process of RWGS followed by methanol production from syngas (Elsernagawy et al. 2020). The status of commercial pathways for methanol production from syngas has been reviewed by the National Energy Technology Laboratory (NETL n.d.(a)). In general, conversions and yields for direct CO_2 hydrogenation to methanol are lower than for the two-step process via syngas under standard conditions owing to poorer activity and formation of additional water as a coproduct.

Research efforts have targeted improvements in yield and selectivity of direct CO_2 conversion to methanol (Jiang et al. 2020; Ye et al. 2019). For example, increasing H_2/CO_2 ratios to 10 and operating at higher pressure (35 MPa) allows direct conversion to methanol above 95 percent yield with 98 percent selectivity for a conventional copper/zinc oxide/alumina ($Cu/ZnO/Al_2O_3$) catalyst (Bansode and Urakawa 2014). Use of dispersed copper nanoparticles encapsulated in metal organic frameworks via strong support interactions shows enhanced activity, near 100 percent selectivity to methanol, and reduced catalyst sintering while preventing agglomeration of the copper nanoparticles (Rungtaweevoranit et al. 2016). Zirconium dioxide acts as a promoter and support in copper-based catalysts for CO_2 conversion to methanol (Lam et al. 2018). Catalysts incorporating indium (III) oxide on nickel or nickel-indium-aluminum/silica ($Ni\text{-}In\text{-}Al/SiO_2$) enhance rates for low-pressure methanol synthesis (Richard and Fan 2017). Bifunctional catalysts are being developed that couple CO_2 hydrogenation to methanol via copper, indium, or zinc-based catalysts with methanol dehydration or coupling using zeolites (Ye et al. 2019).

The CAMERE process provided an early pilot of two-step methanol production via RWGS and methanol synthesis (Joo et al. 1999). Samimi et al. (2018) examined addition of an in situ membrane for water removal during methanol synthesis using the CAMERE process, which showed improvements in methanol yields, as removal of water is expected to improve catalyst life. Subsequent analysis identified membrane options for enhancing RWGS in packed-bed membrane reactors (Dzuryk and Rezaei 2022).

From methanol, the subsequent steps to produce gasoline or olefins are fully developed and have initial commercial units in China (Gogate 2019). Methanol-to-olefins results in a ratio of $C2^{=}/C3^{=}$ product from 0.7 to 1.1, whereas current technology from ethane (cracking) or propane (dehydrogenation) can give better than 90 percent yields of a specific olefin (Tian et al. 2015). Selective conversion of methanol to light olefins (ethylene, propylene) is essential for providing key intermediates for the chemical economy and can be achieved via catalyst and reactor optimization (Jiao et al. 2016; Tian et al. 2015). A process for converting methanol to gasoline has been demonstrated, and commercial operations are planned (NETL n.d.(b)). Methanol conversion to aromatics is also known and would allow coverage of a full spectrum of CO_2 to polymer and chemical intermediates (Sibi et al. 2022). Methanol carbonylation is fully commercial at industrial scale. However, use of earth abundant metals in place of rhodium and iridium and avoidance of corrosive halogen promoters remain goals for practice of more sustainable, green chemistry (Kalck et al. 2020).

Formate and Formic Acid

Thermocatalytic routes to synthesize formate or formic acid from CO_2 have been reported using molecular (homogeneous) catalysts, which also have relevance for direct methanol synthesis from CO_2 (Wang et al. 2015a). Behr and Nowakowski (2014) reviewed both homogeneous and heterogeneous catalysts as well as attempts to develop commercial systems. Homogeneous catalysts based on ligand-modified platinum group metals are active for formate/formic acid production, with ruthenium, rhodium, and iridium showing highest activity, but are challenged by performance, cost, and low element abundance at commercial scale, which impedes industrial consideration. Significant activity is only achieved in the presence of base to produce formate salts instead of formic acid, which drives the endergonic reaction but inhibits product separation and adds cost. A wide variety of mono- or bidentate phosphine or amine ligands impart changes in steric effects and electron density that modify activity and selectivity, giving rise to a rich domain for experimentation. Recent developments include improved rates for iron- or cobalt-based homogeneous catalysts promoted with phosphine ligands, but these systems again require expensive promoters for activity, either a base as in the precious metals catalysts, or a Lewis acid (Bernskoetter and Hazari 2017; Filonenko et al. 2018). Heterogeneous catalysts were known as early as 1932 (Raney nickel; Covert and Adkins 1932) but give poorer yields. Overall, direct thermocatalytic CO_2 hydrogenation to formic acid or formate has progressed significantly because extensive exploration began in the 1970s, but the relatively low turnover frequencies, difficult separations, and expensive components have limited commercial deployment. Electrochemical approaches could be highly competitive in this space (see Section 7.2.2). Currently, formic acid is made on a 0.8 kiloton per annum global scale via thermocatalytic carbonylation of methanol to methyl formate, followed by base-catalyzed hydrolysis to formic acid and methanol (Hietala et al. 2016); it can potentially be made for small-scale markets via bioprocessing. The use of formic acid or formate at a larger scale—for instance, as a transport medium for syngas—would require process technology optimization and scale up, if this were found to be a competitive pathway versus methanol production.

One-Step C2+ Hydrocarbons, Oxygenates, and Intermediates

C-C bond coupling to form C2+ hydrocarbons and oxygenates is a challenge for thermochemical CO_2 activation, although it is an area of active research (Fors and Malapit 2023; Pescarmona 2021; Zhang and Hou 2013). Coupling with epoxides is one means of activating CO_2 (Kothandaraman and Heldebrant 2020).[4] Multistep pathways to C2+ products via syngas formation followed by Fischer-Tropsch are described below, and multistep pathways to C2+ products via syngas formation followed by methanol synthesis and subsequent reactions to olefins or gasoline were described above.

"One-pot" synthesis of C–C bonded products can be attempted via either the methanol or Fischer-Tropsch synthesis routes, to save capital expenditure and simplify the number of process steps (Ye et al. 2019). The initial reaction of CO_2 with H_2 (RWGS) must overcome the high reaction activation energy of CO_2 and equilibrium, and hence requires high temperature (950°C). To attempt a one-pot synthesis to yield methanol as an intermediate for coupling to olefins or dimethyl ether, the subsequent reaction(s) also must be able to take place selectively at high temperature. While multistep pathways from CO_2 to methanol and derivatives or to Fischer-Tropsch products can exhibit high conversion at C–C yields of 80 percent or better, one-pot synthesis yields are restricted to 50 percent or lower because the subsequent conversion steps have to be conducted at the same high temperature as the RWGS reaction (Ye et al. 2019). Nonetheless, considerable research efforts continue for one-pot synthesis routes, given their potential lower costs.

C2 and C2+ Carboxylic Acids

CO_2 has been widely explored as a carboxylation agent in the production of C2 and C2+ carboxylic acids, enabling more sustainable syntheses compared to current industrial methods, although catalytic approaches remain largely at the basic research stage (Cauwenbergh et al. 2022; Davies et al. 2021; Tortajada et al. 2018; Wang et al. 2017; Zhang et al. 2024). Unlike other CO_2 conversions discussed in this chapter, these carboxylation reactions do not use CO_2 as the source of all carbon atoms in the target compound, but rather as the source of a carboxyl group. Both heterogeneous (Zhang et al. 2024) and homogeneous (Cauwenbergh et al. 2022; Tortajada et al. 2018) catalytic systems have been studied, with palladium, rhodium, nickel, copper, and cobalt being among the most common metals for catalysis. A wide range of products are accessible through the various catalytic reaction pathways for carboxylation with CO_2, which include nucleophilic addition of organometallic reagents, reductive coupling with organic (pseudo)halides, reaction with unsaturated hydrocarbons, and functionalization of sp, sp^2, and sp^3 hybridized C–H bonds (Davies et al. 2021; Tortajada et al. 2018). Synthesis of acrylic acid from CO_2 and ethylene is of particular industrial interest given the widespread applications of these compounds in manufacturing and consumer products (Davies et al. 2021; Tortajada et al. 2018; Wang et al. 2017).

[4] Epoxides are unstable molecules that require high energy for synthesis. For sustainable processing, the C2 epoxide co-reactant would have to be made from CO_2 as well, via formation of syngas, synthetic methanol-to-olefins, and epoxidation of ethylene, for example. The other feedstocks and energy inputs for CO_2-to-epoxide conversion would also need to have net-zero emissions on a life cycle basis.

Hydrocarbons from Fischer-Tropsch Synthesis

Fischer-Tropsch synthesis is a surface chain-growth polymerization reaction (Anderson-Schulz-Flory distribution) that converts syngas into a range of hydrocarbon products. Conversion of syngas to diesel and chemical products (waxes and lubricants) is fully commercial at the industrial refinery scale (see NETL n.d.(c)). Thus, generation of syngas from CO_2, as described above, can enable production of fuels and commodity chemicals using already existing commercial methods, although this is not currently viable at scale (see Section 7.2.1.2).

The synthesis reactions between CO and H_2 can be written as (Martín and Grossman 2011):

$$nCO + (n+m/2)H_2 \rightarrow C_nH_m + nH_2O, \qquad \text{(R7.2)}$$

which occurs by a chain growth mechanism to add $-CH_2-$ units:

$$CO + 2H_2 \rightarrow -CH_2- + H_2O \quad \Delta H_r = -165 \text{ kJ/mol} \qquad \text{(R7.3)}$$

Preferred industrial catalysts are cobalt and iron operating at temperatures between 200 and 350°C and pressures from 10–40 bar (Martín and Grossman 2011). Iron has higher WGS activity, leading to higher consumption of CO and of the produced water, which increases the H_2/CO ratio, decreasing the probability for chain growth but reducing catalyst deactivation caused by water (Bukur et al. 2016). Iron catalysts also exhibit greater selectivity to unsaturated olefins because of additional surface intermediates formed. Product distributions can be modeled via a probability for chain growth, which depends on total pressure, H_2:CO ratio (typically 1:1 to 2:1), the extent of WGS activity along the reactor, temperature, catalyst design, and pore structure of the support (Bukur et al. 2016).

Polymer Precursors

CO_2 is used in the production of some monomers for polymerization reactions (Grignard et al. 2019). (Section 7.2.6 discusses direct polymerization of CO_2.) For example, in 2012, Asahi Kasei Corporation industrialized a process to make bisphenol-A polycarbonate (BisA-PC) starting from ethylene oxide and CO_2 (see Figure 7-5; Asahi Kasei n.d.;

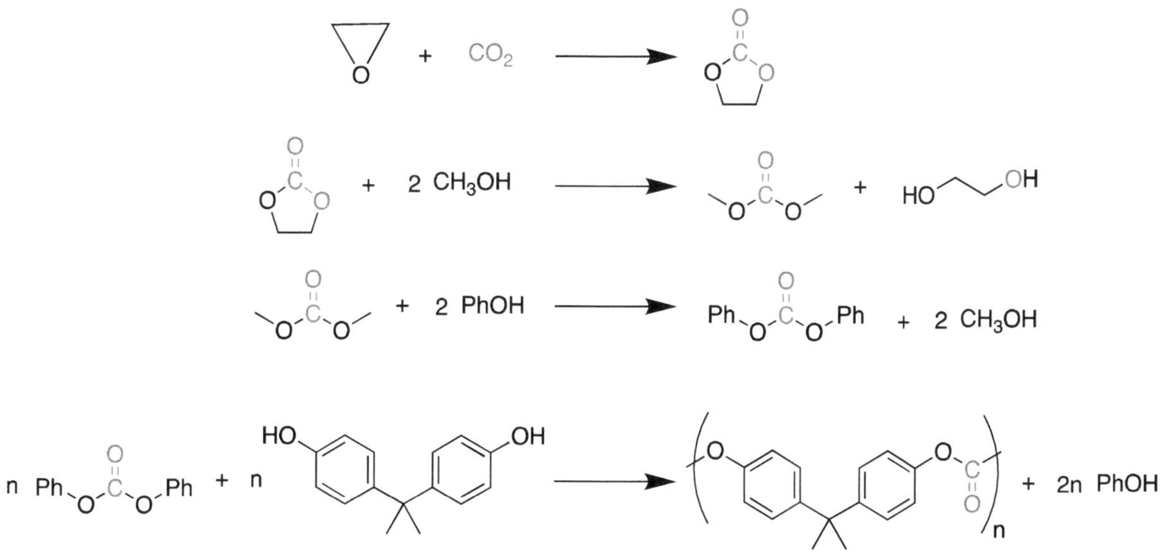

FIGURE 7-5 Asahi-Kasei process to produce BisA-PC from ethylene oxide and CO_2. Atoms from CO_2 are shown in blue.
SOURCE: Reprinted with permission from S. Fukuoka, I. Fukawa, T. Adachi, H. Fujita, N. Sugiyama, and T. Sawa, 2019, "Industrialization and Expansion of Green Sustainable Chemical Process: A Review of Non-Phosgene Polycarbonate from CO_2," *Organic Process Research & Development* 23(2):145–169, https://doi.org/10.1021/acs.oprd.8b00391. Copyright (2019). American Chemical Society.

Fukuoka et al. 2019). The process first reacts ethylene oxide and CO_2 to make ethylene carbonate, which then reacts with methanol to produce dimethyl carbonate. Dimethyl carbonate is converted to diphenyl carbonate via reactive distillation, and last, diphenyl carbonate and bisphenol-A are reacted to produce BisA-PC. In addition to being an opportunity for CO_2 utilization, the route to BisA-PC via CO_2 and ethylene oxide avoids the use of phosgene and the associated safety concerns of the traditional synthesis route.

7.2.1.2 Challenges

Thermochemical CO_2 conversions, like all large-scale industrial catalytic processes, are subject to continuous improvement in product yield, catalyst durability, and reactor performance (conversion per unit mass of catalyst). The near-term industry focus for thermochemical CO_2 conversion is improving the RWGS reaction because all subsequent process routes utilizing syngas are proven at scale, albeit not with sustainable energy inputs and circularity constraints. The RWGS reaction is thermodynamically favorable at high temperature, but under these conditions, catalyst deactivation owing to sintering, coke formation, reduction of active species, and/or CO poisoning can be a challenge (Chen et al. 2020; Goguet et al. 2004; Tang et al. 2021; Zhou et al. 2023). Industry is exploring the noncatalytic RWGS reaction as a potentially more cost-effective option, but this approach requires higher temperature, pressure, and metallurgy for high per-pass yields to minimize the need for CO_2 recycling.

The difficulty of activating CO_2 makes catalyst development for RWGS—and, indeed, for all thermochemical conversions of CO_2—particularly challenging. Typical copper-based RWGS catalysts are not stable at the high temperatures required for reaction, and it is difficult to achieve high CO selectivity because of undesired methanation (Chen et al. 2017). Perovskite oxides can act as oxygen donor-acceptors to minimize methanation side reactions (Chen et al. 2020). Strong metal-support interactions, structure sensitivity to dispersed metal particle size, introduction of a second metal or metal oxide, and alkali promoters all provide opportunities for commercial improvement. Supported noble metals (e.g., platinum, rhodium, palladium, gold) and first-row transition metals (e.g., copper, iron) have been examined as catalysts, but the high cost for noble metals renders them impractical. For carboxylation reactions, challenges with catalyst (and catalytic system) development include poor stereo-, regio-, and enantio-selectivity; limited mechanistic understanding; and the requirement for stoichiometric reductant, alkylation agent, strong base, and/or toxic solvent (Cauwenbergh et al. 2022; Davies et al. 2021; Tortajada et al. 2018; Zhang et al. 2024).

Further catalyst development has to consider availability and sustainability of the elements chosen; thus iridium -and ruthenium-based catalysts are industrially or economically challenged. Future metal catalyst functionality from abundant materials is a goal, including use of transition-metal carbides. To this end, the use of transition-metal catalysts supported on metal oxides shows promise for RWGS, provided methane formation can be suppressed via techniques such as metal-support interactions for structure-sensitive hydrogen activation and CO_2 hydrogenation reactions (Chen et al. 2020). An additional challenge will be developing catalysts that can tolerate specific CO_2 feed stream compositions, including impurities. See Tables 4.3 and 4.4 in the committee's first report (NASEM 2023) for an overview of impurities in CO_2 streams from different sources (reproduced in Appendix H as Tables H-1 and H-2).

Other challenges involve reactor design and scaling of processes. The RWGS reaction exhibits a relatively low heat of reaction, and reactor design and scale up do not present significant challenges using fixed beds of catalysts with interstage cooling (Saw and Nandong 2016). Design of new gas-solid catalytic reactors with low volumetric heat transfer rates can be done from design principles without requiring demonstration. Producing fuels and other chemicals by coupling the RWGS reaction with Fischer-Tropsch or methanol synthesis requires that the high-temperature RWGS reactor outlet is cooled before sending the syngas to a Fischer-Tropsch or methanol synthesis reactor because catalysts for those reactions are not selective at high temperature. Because CO_2 is difficult to activate, per pass conversions at the lower temperatures where "CO_2 hydrogenation" (i.e., the forward direction of RWGS) can be coupled with Fischer-Tropsch or methanol synthesis are low, currently less than about 30 percent (Dang et al. 2019; Saeidi et al. 2021; Zhang et al. 2021). In such cases when CO_2 conversion is below 95 percent, it has to be separated, recycled, and reheated, which reduces energy efficiency. Nonetheless, deployment to date is limited not by technology scale up, but by the lack of economic competitiveness of products derived from CO via RWGS. Demonstrations have produced small amounts of product as a showcase (e.g., Dineen 2023). RWGS also can be performed at higher temperature via a noncatalytic thermal conversion reactor, where temperature is used to compensate for lack of catalyst. In this case, heat transfer can require larger-scale demonstration for reliable scale up. Overall heat transfer rates are not large, however, so this would be an optimization exercise and not a showstopper for industry.

Owing to the efficiency of large-scale chemical synthesis, the CO_2-to-products industry of the future likely will entail large-scale plants in locations favorable to their deployment, and liquid or solid products will be shipped to market. Facilities that convert CO_2 to CO and CO to products likely would need to be co-located, as it is impractical to build pipelines or other commercial transportation of CO, a toxic and reactive gas, beyond short commercial-unit trunklines. Where conditions do not favor large plants owing to water restrictions or land use or other limitations for CO_2 capture and renewable power generation, distributed modular plants for integrated conversion of CO_2 to CO and subsequent CO to liquid or solid products can be considered. The challenge for distributed modular processing, in competition with global mega-scale plants with low-cost shipping of products, is that capital costs and process scale do not increase at the same rate. Termed the "0.6 power rule," capital costs for thermochemical reactions conducted in bulk equipment typically increase only at the 0.6 to 0.7 power of process scale, reflecting the fact that essential tasks for engineering design and fabrication must be performed regardless of scale, such that costs per unit of production decrease as production rates or annual capacities are increased (Timmerhaus and Peters 1991). Mini- and micro-channel reactors with improved heat transfer and membrane reactors are rare but scale closer to 1.0 power (i.e., capital costs increase proportionally to production volume), such that scale up requires "numbering up" smaller units rather than increasing the scale of a given process unit. To remain favorable despite their smaller production scale, distributed production plants have to be highly integrated, volumetrically efficient, and employ low capital expenditure approaches. Small-scale, stranded natural gas conversion facilities face similar constraints, and have generally chosen physical transport of the stranded gas as liquefied natural gas rather than reactive conversion to products on a distributed basis. Related considerations for CO_2 transport versus small-scale conversion may come to the same conclusion.

Fischer-Tropsch synthesis, the demonstrated pathway for creating a chemical economy from CO, has never been economically competitive at small or intermediate scales despite numerous showcase demonstration projects (De Klerk 2014; Dieterich et al. 2020). Fischer-Tropsch is a C1 oligomerization process, which produces an Anderson-Schulz-Flory statistical distribution of hydrocarbons with a broad range of carbon numbers, including diesel through aviation (C9+) to heavy waxes for lubricants (C35+, which can also be cracked back to smaller molecular weight), as depicted in Figure 7-6. The multiple processing steps required result in high capital expenditure and poor ability to scale down. Catalytic studies seek to reduce or eliminate the heavy end wax (lubricants) formation but also to avoid using commercially nonviable metals such as ruthenium. Additionally, olefin yields for the Fischer-Tropsch pathway have been limited to around 50 percent, compared to around 80 percent for the methanol-pathway alternative (Ye et al. 2019)—that is, CO_2 hydrogenation to methanol and subsequent conversion to olefins (He et al. 2019).

Thermochemical conversion of CO_2 to chemicals and fuels has higher costs than current production from fossil hydrocarbons, even for the CO_2 utilization processes that are well known. Under future conditions for sustainable synthesis of circular carbon chemicals and fuels, the large amount of capital and high energy required to capture CO_2 from air and upgrade it from a thermodynamically degraded state into synthetic hydrocarbons will present significant hurdles for CO_2 utilization. Production of sufficient clean hydrogen to meet demand for CO_2 utilization and other applications could be particularly challenging, warranting additional R&D to facilitate scale up (NPC 2024). Delivery of the low-carbon-intensity, high-temperature process heat needed for thermochemical CO_2 conversion could occur via electrification (see Section 7.2.1.3 for more on electrified reactors), redesign of furnaces to support use of clean hydrogen as a fuel, or implementation of carbon capture on existing fossil fuel furnaces (see Section 10.3.2.1 for more on these retrofitting options). Decisions about the optimal approach will require consideration of trade-offs in carbon intensity and reaction efficiency. Additionally, CO_2 conversion pathways, such as RWGS, that require additional heat at high temperatures in the presence of hydrogen can present challenges in reactor design and metallurgy. Given high capital intensity of capture and conversion facilities, energy storage may be required for 24/7 access to renewable H_2 for conversion of CO_2. Two examinations of techno-economic

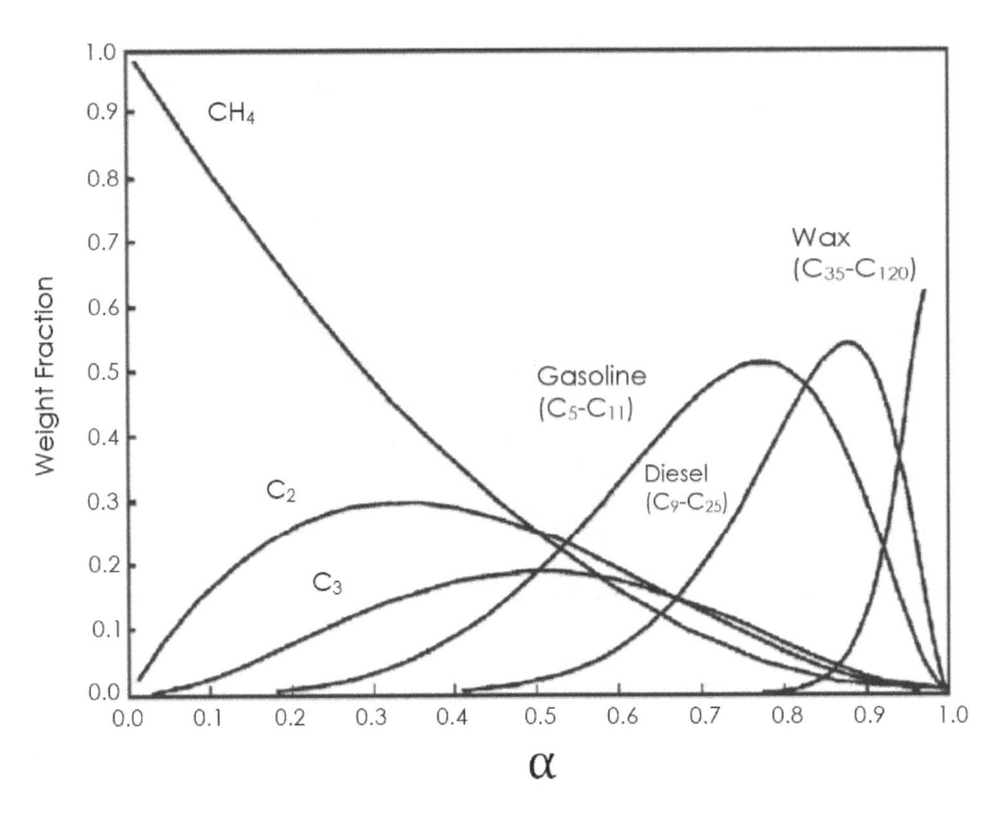

FIGURE 7-6 Anderson-Schulz-Flory product distribution for Fischer-Tropsch synthesis.
SOURCE: Reprinted with permission from M. Martín and I.E. Grossmann, 2011, "Process Optimization of FT-Diesel Production from Lignocellulosic Switchgrass," *Industrial & Engineering Chemistry Research* 50(23):13485–13499, https://doi.org/10.1021/ie201261t. Copyright (2011) American Chemical Society.

potential for manufacture of CO_2-derived aviation fuels found prices were about 2–7 times higher than current fuels, and there was little expectation that technology innovation could overcome that barrier, so strong policy support would be needed (Freire Ordóñez et al. 2022; Soler et al. 2022). Chapter 4 discusses policy options to support CO_2 utilization in a net-zero emissions future.

7.2.1.3 R&D Opportunities

R&D opportunities for thermochemical CO_2 conversion include integrated catalyst development and multiscale reactor optimization, process and systems integration, use of advanced characterization and discovery techniques, and development of electrified reactors. A related R&D opportunity, integrated capture and conversion (i.e., reactive capture), is discussed in Section 7.2.5. General descriptions of each R&D area are outlined below, as many are shared across conversion pathways and product targets. Examples are provided for conversion to specific products where relevant.

Integrated Catalyst Development and Multiscale Reactor Optimization

Catalyst discovery and development for RWGS is an active topic of research for CO_2 utilization, given the bifunctional nature of RWGS catalysts (metal/metal oxide), observed structure sensitivity, and relevance of surface or lattice oxygen storage in controlling the reaction pathways. For CO_2 hydrogenation reactions involving zeolite catalysts, synthesis of zeolites with desired structures is a key research area for integrated bifunctional catalyst performance, including impacts of acidity, silicon/aluminum ratio, and pore structure. Furthermore, catalysts or catalytic systems should tolerate specific CO_2 feed stream compositions, including impurities. These properties also affect the subsequent conversion of methanol to gasoline or dimethyl ether (with HZSM-5 zeolite) or olefins (with SAPO molecular sieves) (Ye et al. 2019). Computational design, including use of quantum mechanical simulations (e.g., density functional theory), can identify desired structures and compositions, but multiscale modeling to consider integrated reaction and transport properties is key for defining reactor-scale performance (Ye et al. 2019). Research on Fischer-Tropsch synthesis to enhance yields of lower molecular weight olefin products has long examined iron-based catalysts (Storch et al. 1961) with promotion by potassium, manganese, or copper as dopants (Dorner et al. 2010).

The switch from traditional Fischer-Tropsch synthesis with syngas as feed to "CO_2 Hydrogenation" with CO_2 and H_2 as feed favors iron rather than cobalt-based catalysts owing to the former's higher WGS potential (and hence RWGS potential). Olefin yields can be increased up to four-fold with reduced methane by-product (Saeidi et al. 2021). Lin et al. (2022) describe strategies for controlling Fischer-Tropsch product selectivity, including metal particle size, strong metal-support interactions, use of alkali and other cationic promoters, use of bifunctional catalysis to couple conventional carbon chain growth via cobalt or iron metal catalysts with carbide (Co_2C or Fe_2C) catalysts that enable nondissociated CO insertion for higher alcohol synthesis. Dual functional cobalt-manganese (Co-Mn), iron-manganese (Fe-Mn), and zinc/chromium-oxide (Zn/Cr-oxide) catalysts can achieve olefin selectivity greater than 40 percent, although CO_2 and methane formation are problematic. Two key challenges to address are (1) developing a dual functional catalyst having comparable rates for CO formation from CO_2 and subsequent hydrogenation of the CO-derived intermediate and (2) mechanistic understanding of possible formate or ketene intermediate species and their impact on observed product distributions that exceed limitations of the Anderson-Schulz-Flory mechanism.

As noted above, one challenge for thermochemical CO_2 activation is C–C coupling to form C2+ products. The development of tandem catalysts and better understanding of C–C coupling mechanisms could improve selectivity for CO_2 conversions to long-chain hydrocarbons (Gao et al. 2020). Tandem catalysis has been demonstrated for integrated ethylene, propylene, and aromatics production from CO_2 (Gao et al. 2017, 2020; Saeidi et al. 2021; Zhang et al. 2019a). For example, Zhang et al. (2019a) prepared a zinc oxide/zirconium oxide (ZnO/ZrO_2)-ZSM-5 tandem catalyst, with CO_2 hydrogenation provided by ZnO/ZrO_2 and C–C bond coupling and aromatization by H-ZSM-5. This system showed aromatic selectivity of 70 percent at 9 percent CO_2 conversion and 613 K (340°C), indicating opportunities for further improvement. Detailed mechanistic modeling and catalyst characterization also can help catalyst development for production of C2+ chemicals. For example, Gao et al. (2017) developed a bifunctional In_2O_3/HZSM-5 catalyst with 4 nm pores that yielded 78.6 percent of C5+ liquid gasoline product from CO_2 hydrogenation using a single integrated reactor, under conditions where Anderson-Schulz-Flory distribution would have limited C5+ production to less than 48 percent. More research efforts are needed to optimize catalyst compositions for tandem reactions in a single reactor and to improve reactor design for tandem processes involving multiple reactors.

Process and Systems Integration

The use of separate reaction steps—RWGS to form syngas, and then subsequent reactions of syngas to generate desired products—allows independent control of reaction conditions and catalyst formulation to achieve high yields for each step. Process intensification or integration of steps via coupling of endothermic and exothermic reactions could decrease capital costs, improve heat integration, and reduce energy use. System integration for CO formation will also be important, as CO is a stranded gas and further conversion is essential for rendering a commercial intermediate or product (González-Castaño et al. 2021). This integration will require new optimization of syngas catalysts for subsequent conversion steps, unlocking yet another era of interest in syngas catalyst optimization. For example, the endothermic RWGS reaction could be combined with exothermic methanol synthesis and further heat integrated into a methanol-to-product step (e.g., methanol to olefins). Traditional copper-zinc oxide (Cu-ZnO) catalysts for methanol synthesis from syngas are also active in WGS and hence could be considered for integrated CO_2 hydrogenation to methanol (Ampelli et al. 2015). Use of small pore zeolites as supports allows a third-step integration of methanol to olefins (e.g., ethylene, propylene), although intermediate dehydration can be a process challenge. Additional catalyst development and optimization are needed to improve kinetics and selectivity starting from CO_2 rather than CO. Research is also needed on integrating catalyst design into the reaction and process system for optimization, as illustrated by González-Castaño et al. (2021) in their examination of the characteristics of copper-, cobalt-, iron-, and platinum-based catalysts for the RWGS reaction, including the impact of poisons.

Syngas generation from CO_2 via RWGS could be integrated with the Fischer-Tropsch reaction pathways to provide an integrated route from CO_2 to olefins or other C–C bonded products. One can use a coupled reaction sequence where the final reaction is not equilibrium constrained (e.g., methanol to olefins or fuels), to pull the reaction equilibrium constraint of RWGS and methanol synthesis to get a single reactor system that may operate at a lower temperature and obtain high conversion to fuels in a single pot. Such process intensification and integration would require integrated catalyst performance, as one cannot independently control reaction parameters for each elementary reaction step. Nonetheless, given the highly competitive nature of commercial industrial chemicals, this opportunity is driving innovative research in catalyst design and architecture for multifunctional syntheses, as well as reactor design and potential integrated separations. Research efforts are devoted to finding catalysts that improve the "CO_2 hydrogenation" step and can be integrated with Fischer-Tropsch synthesis, methanol synthesis, and derivative conversion reactions to olefins or fuels so that the overall reactor can operate inexpensively at lower temperature and pressure, yet still give high per-pass yields (Gao et al. 2017; Saeidi et al. 2021). Research in catalyst and reactor design for the methanol-to-olefins process that targets incumbent distributions of olefin products would allow direct integration with existing petrochemical facilities and downstream processes, avoiding the need for extensive changes in equipment and operations.

Additional synergistic opportunities include integration of exothermic syngas reactions with high-temperature solid-oxide electrolyzer cells (SOECs) for hydrogen generation from water splitting, where thermal energy integration can improve electrical energy efficiency to near 100 percent (Hauch et al. 2020). A second synergy occurs with high-temperature solar thermochemical processes to split both CO_2 and H_2O to make syngas, integrated with thermal energy storage to allow 24/7 operation. More R&D is needed on thermal cycling of the working redox materials and differential thermal expansion to improve system design and durability. Given that production of net-zero fuels and chemicals requires use of atmospheric or biogenic (and not fossil point source) CO_2, integrated capture and conversion of CO_2 represents an essential opportunity to increase adsorption strength for low-concentration (420 ppm) atmospheric CO_2, with synergistic use of chemical reactions to regenerate via conversion to preferred chemical products (see Section 7.2.5 for more detail).

Advanced Characterization and Discovery

Characterization techniques such as atomic force microscopy, transmission electron microscopy, and X-ray absorption spectroscopy provide in-depth analysis of supported nanoparticles. In addition, temporal analysis of reactors with gas chromatography-mass spectrometry analysis to obtain reaction rate data can complement the more established in situ Fourier-transform infrared spectroscopy approaches. Enhanced data acquisition combined with artificial intelligence/machine learning can provide important guidance for catalyst discovery.

Electrified Reactors

Given that low-cost, zero-emissions electricity can be produced directly from wind, solar, and other power sources (Lazard 2024), direct electrical heating of chemical reactors potentially can provide zero-carbon energy (heat) to drive the chemical conversions required for CO_2 utilization. Chemicals and petroleum refining are responsible for approximately 50 percent of U.S. manufacturing CO_2 emissions (EIA 2023), and the Department of Energy's (DOE's) industrial decarbonization roadmap highlights electrification (using zero-carbon electricity) as a key opportunity for decarbonizing these subsectors (DOE 2022). Simply replacing fossil-based electricity with clean electricity for currently electrified processes in the chemical industry could reduce the sector's emissions by 35 percent, and further reductions are possible if additional processes (e.g., CO_2 conversion) are electrified (Eryazici et al. 2021). Life cycle and techno-economic assessments of electrified reactors for syngas generation from CO_2 will be critical for verifying GHG emissions reductions relative to conventional methods (Cao et al. 2022).

Catalyst options for electrically heated reactors have been investigated (Centi and Perathoner 2023). For example, Zheng et al. (2023) showed the efficacy of porous silicon carbide foams for Joule heating of RWGS catalysts at 650°C–700°C. Thor Wismann et al. (2022) examined electrically heated RWGS for methanol synthesis over a nickel-based catalyst, finding that routing the synthesis via methane formation reduced carbon deposition (coking). Dong et al. (2022) demonstrated that periodic pulsed heating can enhance selectivity to C2 hydrocarbons for methane reductive coupling, relative to steady-state operation. The "co-benefits" provided by the ability to rapidly pulse heat relative to conventional approaches with respect to concentration forcing are over and above the simpler replacement of fuel heat with low-carbon electrical energy to drive endothermic reactions. Similarly, microwave heating of catalyst particles, as demonstrated for methane conversion, may provide enhanced selectivity for endothermic CO_2 reactions (i.e., RWGS) and reduce energy losses relative to bulk heating (Hunt et al. 2013). However, as opposed to direct heating via electric energy, microwave heating will suffer energy losses from conversion of electrical energy to electromagnetic radiation.

These examples show the potential for the emerging field of "electrified thermochemical" reactors (as opposed to the traditional "electrochemical" reactors) to provide new performance breakthroughs, especially for endothermic reactions such as CO_2 reduction. The ability to rapidly change and control temporal and spatial heating to selectively heat catalyst surfaces versus bulk fluids, manipulate time constants for multistep reactions to improve selectivity, and reduce catalyst poisoning by operation under rapidly varying dynamic heating conditions (unlike traditional "concentration forcing" conditions), or use microwaves for selective catalyst heating provide new handles for catalyst and reaction control that are yielding promising results.

Solar thermochemical hydrogen production and solar thermochemical CO_2 conversion can be readily integrated with thermal energy storage and improve the economic viability of high capital intensity processes (e.g., methanol and Fischer-Tropsch syntheses) that require 24/7 operation. R&D is needed to examine this synergy and consider its relative cost versus using solar photovoltaic energy plus battery storage to maintain 24/7 operability of Fischer-Tropsch or methanol synthesis.

7.2.2 Electrochemical Conversion Pathways

7.2.2.1 Current Technology

Significant progress has been made in developing electrocatalysts and electrochemical devices for the CO_2 reduction reaction (CO_2RR) to produce value-added chemicals, including C1 (carbon monoxide, methane, methanol, formic acid), C2 (ethylene, ethanol, acetic acid), and some C2+ (acetone, propanol, etc.) products. Low-temperature electrochemical conversion of CO_2 to C1 products is occurring at the pilot scale (Grim et al. 2023; Masel et al. 2021; Xia et al. 2022), and high-temperature conversion of CO_2 to CO is nearing commercialization (Hauch et al. 2020; Küngas 2020). Many studies have identified CO_2RR electrocatalysts that are selective toward specific products. For example, noble metals such as silver and palladium are efficient in producing CO. First-row transition metals such as cobalt and nickel produce both CO and CH_4. Main group metals such as tin, indium, and bismuth, as well as their oxides, are selective for formic acid production. A proton exchange membrane system using a lead/lead sulfate cathode was recently reported to produce formic acid with high selectivity and durability (Fang et al. 2024). At present, copper is the primary element identified that can catalyze CO_2RR to C2 and C2+ products (Nitopi et al. 2019; Yan et al. 2023). Some studies have explored the possibility of enhancing the activity of copper using copper-based bimetallic alloys (Lee et al. 2018), and some have reported the production of long-chain hydrocarbons using non-copper-based catalysts (Zhou et al. 2022).

The status of theoretical simulations of electrochemical CO_2RR was summarized by Xu and Carter (2019a). Modeling efforts are mostly based on density functional theory, which has difficulties describing key intermediates (CO) and electron-transfer reactions owing to errors in its electron exchange-correlation functionals. Recent work has shown that such errors can be corrected by including accurate wavefunction descriptions of exchange-correlation via embedding methods (e.g., Zhao et al. 2021). Multiscale modeling of bipolar membranes for electrochemical systems provides insights into structure–property–performance relationships that can help inform the design of CO_2 electrolyzers (Bui et al. 2024).

TABLE 7-2 Electrochemical CO_2RR Products with Equilibrium Potentials

Reaction	E^0 [V versus RHE]	Products
$CO_2 + 2H^+ + 2e^- \rightarrow HCOOH_{(aq)}$	−0.12	Formic acid
$CO_2 + 2H^+ + 2e^- \rightarrow CO_{(g)} + H_2O$	−0.10	Carbon monoxide
$CO_2 + 6H^+ + 6e^- \rightarrow CH_3OH_{(aq)} + H_2O$	0.03	Methanol
$CO_2 + 8H^+ + 8e^- \rightarrow CH_{4(g)} + 2H_2O$	0.17	Methane
$CO_2 + 4H^+ + 4e^- \rightarrow C_{(s)} + 2H_2O$	0.21	Graphite
$2CO_2 + 2H^+ + 2e^- \rightarrow (COOH)_{2(s)}$	−0.47	Oxalic acid
$2CO_2 + 8H^+ + 8e^- \rightarrow CH_3COOH_{(aq)} + 2H_2O$	0.11	Acetic acid
$2CO_2 + 10H^+ + 10e^- \rightarrow CH_3CHO_{(aq)} + 3H_2O$	0.06	Acetaldehyde
$2CO_2 + 12H^+ + 12e^- \rightarrow C_2H_5OH_{(aq)} + 3H_2O$	0.09	Ethanol
$2CO_2 + 12H^+ + 12e^- \rightarrow C_2H_{4(g)} + 4H_2O$	0.08	Ethylene
$2CO_2 + 14H^+ + 14e^- \rightarrow C_2H_{6(g)} + 4H_2O$	0.14	Ethane
$3CO_2 + 16H^+ + 16e^- \rightarrow C_2H_5CHO_{(aq)} + 5H_2O$	0.09	Propionaldehyde
$3CO_2 + 18H^+ + 18e^- \rightarrow C_3H_7OH_{(aq)} + 5H_2O$	0.10	Propanol

NOTE: RHE = reversible hydrogen electrode.
SOURCE: Adapted from Nitopi et al. (2019).

Molecular electrocatalysts have also been studied extensively for CO_2RR, with more than 100 different catalysts identified. Catalysts have been reported with 13 different transition metals, and a few examples exist of non-metal-containing catalysts (Francke et al. 2018). Although these catalysts operate under a wide variety of conditions, including organic and aqueous solvents, only a few different products have been reported. Under protic conditions, CO is the most common product, followed by formate or formic acid. Under nonprotic conditions, typical products are oxalate, CO, and carbonate (CO_3^{2-}). A handful of systems have been reported that catalyze the six-electron reduction to methanol or the eight-electron reduction to methane, although some of these are not strictly homogeneous, but instead molecular catalysts immobilized onto electrode surfaces (Boutin and Robert 2021).

Among all the potential products from CO_2RR, as shown in Table 7-2, CO and formic acid (HCOOH) are generally considered to be the most commercially viable molecules based on a recent review (Nitopi et al. 2019) and techno-economic assessment (Aresta et al. 2014). Conversions of CO_2 to CO or HCOOH require only two electrons and are kinetically more facile than the multiple-electron and multiple bond formation-scission processes for products containing two or more carbons. Equally important, CO and HCOOH can be produced using catalysts that do not contain copper, therefore allowing the utilization and optimization of a wide range of electrocatalysts. HCOOH is a bulk chemical that can be used as a feedstock for the chemical industry and for energy storage (Aresta et al. 2014). One advantage of converting CO_2 to CO is the higher efficiency of converting CO to value-added products, either through electrochemical (Jouny et al. 2019) or thermochemical (see Section 7.2.1) upgrading reactions. Converting CO_2 to CO is also advantageous in terms of carbon utilization efficiency. The alkaline environment required to achieve high reaction rates for CO_2RR results in large amounts of (bi)carbonate production, which has limited CO_2RR selectivity for multiple carbon products (Nitopi et al. 2019). In contrast, CO_2RR to CO can be carried out in a nonalkaline environment with high CO selectivity without producing (bi)carbonates. Furthermore, because CO is a gas, it is easier to separate from the electrolyte than liquid products (i.e., liquid-liquid separations are not required). The production of several C2 molecules, including ethylene ($C2H_4$), ethanol ($C2H_5OH$), and acetic acid (CH_3COOH), has also been investigated extensively. It is widely accepted that these reactions proceed via the formation of a *CO-containing surface intermediate followed by its dimerization and subsequent reduction to form C2 products (Nitopi et al. 2019).

Although most current research on CO_2RR focuses on low-temperature electrochemical devices, there are efforts in using intermediate-temperature (molten carbonate electrolysis) and high-temperature (solid oxide electrolysis) devices (Küngas 2020). For example, Figure 7-7 compares different electrochemical devices for CO_2

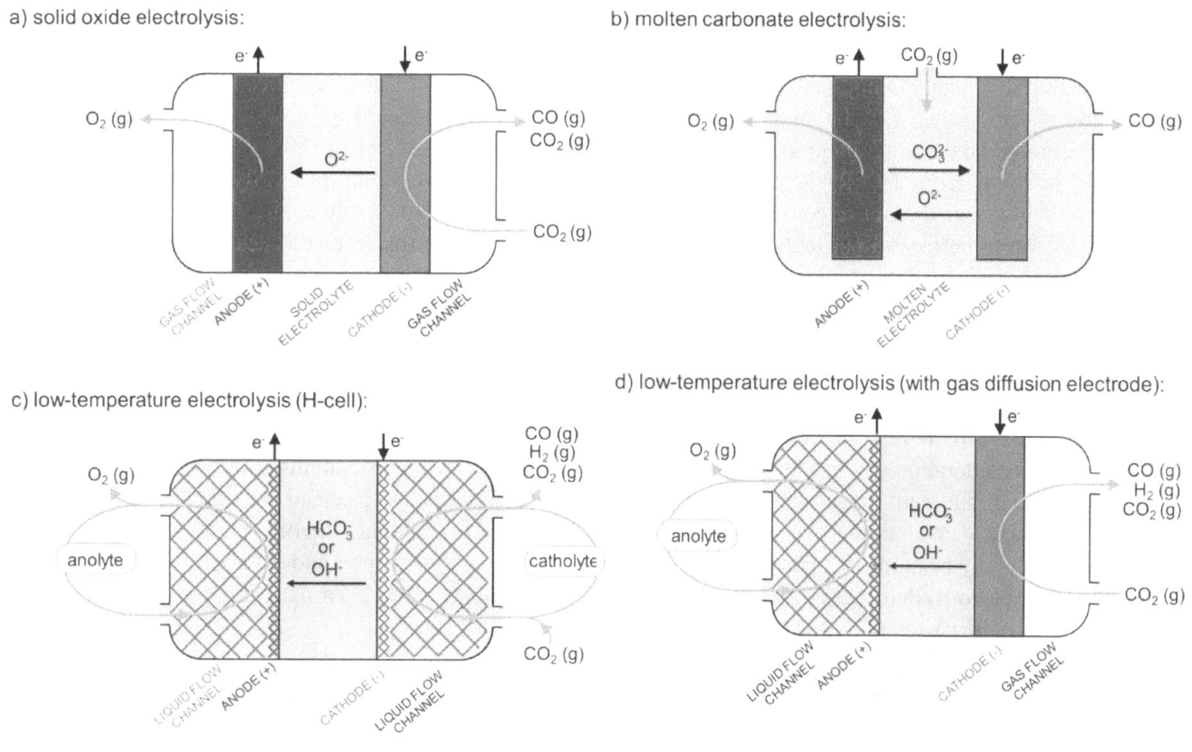

FIGURE 7-7 Comparison of electrochemical devices for CO_2 conversion to CO at various temperatures, including high-temperature solid oxide electrolysis (a), intermediate-temperature molten carbonate electrolysis (b), and low-temperature electrolysis with an H-cell configuration (c) or gas diffusion electrode (d).
SOURCE: Küngas (2020), https://doi.org/10.1149/1945-7111/ab7099. CC BY-NC-ND 4.0.

reduction to CO. Based on an analysis by Hauch et al. (2020) high-temperature solid oxide electrolysis of CO_2 to CO is considered to be approaching commercialization with promising catalytic rates and long-term durability (Hauch et al. 2020; Küngas 2020).

Electrochemical carboxylation reactions using CO_2 are also of interest as a more environmentally friendly method of producing industrially relevant carboxylic acids (Ton et al. 2024; Vanhoof et al. 2024). These reactions can involve a variety of co-substrates, including alkenes, alkynes, benzyl, aryl, and alkyl halides, and aryl aldehydes and ketones, and thus are able to form a diverse range of products. The systems often require a sacrificial anode (commonly magnesium or zinc) to provide stabilizing metal ions for the radical anion species formed during electroreduction of substrate or CO_2, but there are efforts to develop sacrificial-anode-free systems through electrolyte, substrate, and cell design to reduce cost and improve sustainability.

7.2.2.2 Challenges

Although electrocatalytic CO_2RR can produce several C1 products (CO, HCOOH, methanol, and methane), one critical challenge is the selective production of >2e^- reduced products, and C2+ products in particular, in high yield. The energy efficiency is further complicated by the competing hydrogen evolution reaction. As noted above, at present copper is the primary element identified that can catalyze CO_2RR to C2+ products with appreciable Faradaic efficiency.[5]

[5] Faradaic efficiency is a measure of selectivity of an electrochemical reaction, calculated as a ratio of the amount of product formed over the theoretical maximum amount based on the charge passed (Kempler and Nielander 2023).

As with heterogeneous systems, homogeneous CO_2RR is also challenged by product selectivity for a single carbon-based product and the competing hydrogen evolution reaction, although there are examples of catalysts with high selectivity for CO and formate (Francke et al. 2018). However, few examples exist of homogeneous electrocatalysts that reduce CO_2 beyond two electrons, and none that demonstrate C-C bond coupling to form C2+ products except for oxalate (Francke et al. 2018).

Most low-temperature CO_2RR studies are at an early stage of development, primarily owing to issues with long-term stability and product selectivity (Grim et al. 2023; Küngas 2020). While some progress has been made in improving these metrics, in particular through developments in gas diffusion electrodes, challenges remain in reducing overpotential and improving stability at high current densities, as well as decreasing energy losses from carbonate formation (Wakerley et al. 2022). Although high-temperature CO_2RR is considered to be at higher TRL, its feasibility only has been demonstrated for CO_2 conversion to CO.

Electrochemical reactors exhibit a unit scaling factor (with capacity) of near 1.0, such that one must "number up" to achieve a large scale of production. This inability to reduce costs at increasing scale has been an issue for achieving cost-effective production of H_2 via water electrolysis, and likely would be for CO_2 electrolysis as well. Another consideration when scaling up electrochemical CO_2 conversion systems is the energy requirement compared to that of alternative tandem electrocatalysis-thermocatalysis routes, as clean electricity availability may be limited by supply chain constraints. Where electrical efficiency for direct CO_2 conversion is poor (i.e., high overpotential), it may be more efficient to generate clean H_2 via water electrolysis and use that H_2 to form syngas, which can then be converted thermochemically to fuels and chemicals using existing technologies, as described in Section 7.2.1.1 (Eryazici et al. 2021).

7.2.2.3 R&D Opportunities

A key R&D opportunity for electrochemical CO_2 conversion is to expand the number of catalysts that can generate C2+ products with relatively high yields. One approach to enhance C2+ product generation is to modify copper with an element that is efficient for CO_2 to CO conversion. The resulting bimetallic electrocatalysts could effectively convert CO_2 to CO, which subsequently could be converted to C2+ products by copper. The CO-rich reaction environment also should inhibit the competing hydrogen evolution reaction that would otherwise reduce the selectivity for C2+ products on pure copper. Silver and gold are attractive options because of their high CO_2RR selectivity to CO and their immiscibility with copper, which prevents changes in electrocatalytic properties owing to formation of bimetallic alloys. Utilization of multiple catalysts consisting of copper and a CO-producing electrocatalyst, in the form of either physical mixtures or segmented catalyst beds, has been demonstrated for CO_2 conversion to multicarbon products (Yin et al. 2022). One common practice uses one metal as a catalytically active and conductive substrate onto which the second metal is deposited. Segmented electrodes also have been studied recently to control the separation between distinct catalysts. For example, two catalysts can be deposited adjacent to each other to produce a high concentration of CO that then flows over a C2-producing catalyst (Zhang et al. 2022b). As another approach, recent efforts have explored the utilization of electrocatalytic-thermocatalytic tandem processes to produce C2+ oxygenates and hydrocarbons (Biswas et al. 2022b; Lee et al. 2023), although more studies are needed to determine whether such tandem processes can be economically competitive. Computational modeling, specifically advanced quantum mechanics methods that go beyond density functional theory (see, e.g., Martirez et al. 2021) when needed for simulating electron-transfer reactions, combined with ab-initio molecular dynamics for solvent configurational sampling (see, e.g., Martirez and Carter 2023), along with machine-learned force-field molecular dynamics (Poltavsky and Tkatchenko 2021; Unke et al. 2021; Wu et al. 2023) to sample longer time and larger sample sizes, will be the methods of choice in the future. Additional research on multiscale modeling of mass transport effects is needed to improve understanding and optimization of electrochemical device design (Stephens et al. 2022).

In molecular systems, continued work on mechanisms, modifying the electronic properties of active sites, and understanding and modifying secondary coordination sphere interactions have provided some insight into inhibiting competitive hydrogen evolution and/or steering product selectivity (Barlow and Yang 2019). Trade-offs (e.g., scaling relationships) between activity and overpotential have been identified (see, e.g., Bernatis et al. 1994; Nie and McCrory 2022). However, these scaling relations can be broken with appropriate secondary sphere effects such as proton-relays/hydrogen-bonding interactions (Costentin et al. 2012), charge (Azcarate et al. 2016; Margarit et al. 2020), or simultaneous changes in multiple reaction parameters (Klug et al. 2018; Martin et al. 2020). Redox-active ligands also have been used to delocalize charge to access catalytic intermediates at milder potentials (Queyriaux 2021). Some of these strategies are bio-inspired by mimicking either the electronic structure or local environment of enzymes that catalyze these reactions efficiently (Shafaat and Yang 2021). Immobilizing molecular catalysts onto certain types of electrodes also appears to result in different selectivity (Boutin et al. 2019). Ligand modifications can be applied to study local environmental effects that tune selectivity or inhibit hydrogen evolution. These strategies may be translatable to heterogeneous systems (Banerjee et al. 2019).

In addition to optimizing catalysts that are selective, stable, and scalable for CO_2RR, it is important to develop scalable electrochemical devices. For example, stable and cost-effective anode catalysts, which are required to complete electrochemical systems for CO_2RR, need to be identified. In particular, if the oxygen evolution reaction is used as the anodic reaction under acidic conditions, the costs of the iridium oxide (IrO_2) catalysts need to be considered as a potential barrier for large-scale CO_2RR. Use of inexpensive, alkaline-electrolyte-based anodes for the oxygen evolution reaction, enabled by dipolar membranes, may overcome this cost barrier, but high overpotential might still be a limiting issue (Nitopi et al. 2019). Pairing CO_2RR with a different anodic reaction, which may produce a more valuable product than oxygen, could also be explored (Francke et al. 2018; van den Bosch et al. 2022). The reactor components (electrodes, catalysts, supports, membrane, electrolyte) and reaction conditions (pH values of electrolytes, flow rate, temperature, pressure) also need to be optimized (Sarswat et al. 2022; Stephens et al. 2022; Wakerley et al. 2022). Continued development of semi-empirical CO_2 electrolyzers models could help inform scale up beyond lab- and pilot-scale systems (Edwards et al. 2023). Furthermore, because many CO_2 sources contain various potential contaminants, it is also important to evaluate the tolerance of CO_2RR electrocatalysts and membranes (Nitopi et al. 2019).

For electrocarboxylation reactions, primary R&D opportunities include further development of sacrificial-anode-free systems, experimental and theoretical studies to improve mechanistic understanding and facilitate catalyst design, and improvements to enantioselectivity (Ton et al. 2024; Vanhoof et al. 2024). Focusing research efforts on the most common industrial chemicals, developing flow systems, and designing more robust electrocatalysts could facilitate eventual scale up.

7.2.3 Photochemical/Photoelectrochemical Conversion Pathways

7.2.3.1 Current Technology

The use of light to directly drive CO_2 reduction to fuels or other chemicals has been pursued via several different motifs. These include homogeneous systems that use molecular photosensitizers to absorb light (Figure 7-8) and systems that use a heterogeneous light absorber to generate the voltage required for CO_2 reduction (Figure 7-9). In the latter, catalysis can occur directly at the semiconductor interface, with a heterogeneous or molecular catalyst appended to the semiconductor interface, or with molecular catalysts in solution (Kumar et al. 2012). These systems can be completely photo-driven, where no external voltage or energy source is needed, or photo-assisted, where light energy is used to provide a portion of the energy and reduce the applied voltage required to complete the chemical process.

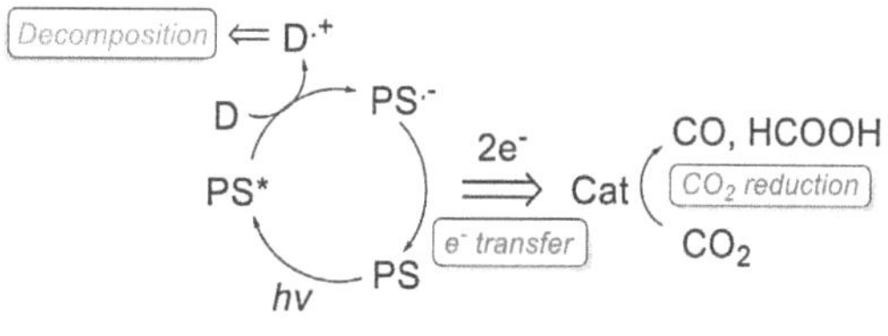

FIGURE 7-8 Schematic example reaction mechanism for photocatalytic multielectron CO_2-reduction reactions.
SOURCE: Reprinted from Y. Yamazaki, H. Takeda, and O. Ishitani, 2015, "Photocatalytic Reduction of CO_2 Using Metal Complexes," *Journal of Photochemistry and Photobiology C: Photochemistry Reviews* 25:109, Copyright (2015), with permission from Elsevier.

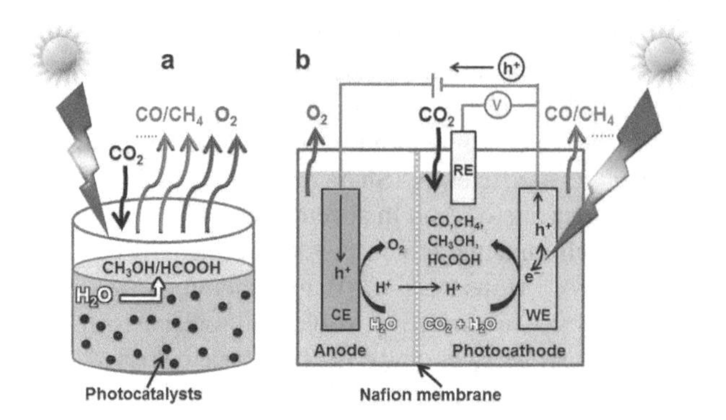

FIGURE 7-9 (a) Photocatalyst particles suspended in a CO_2-containing electrolyte performing both photocatalytic CO_2 reduction and water oxidation reactions. (b) A photoelectrochemical cell with a photocathode as a working electrode (WE) for CO_2 reduction, a counter electrode (CE) for water oxidation and a reference electrode (RE) immersed in a CO_2-containing electrolyte.
SOURCE: Used with permission of the Royal Society of Chemistry from X. Chang, T. Wang, and J. Gong, 2016, "CO_2 Photo-Reduction: Insights into CO_2 Activation and Reaction on Surfaces of Photocatalysts," *Energy & Environmental Science* 9(7):2177–2196, https://doi.org/10.1039/C6EE00383D; permission conveyed through Copyright Clearance Center, Inc.

In addition to the metrics of Faradaic efficiency (product selectivity) and energetic efficiency (overpotential) used to evaluate electrochemical CO_2 conversion, photochemical systems are also described by their photochemical quantum yield (Φ) that evaluates the efficiency in which absorbed photons generate product (Reaction 7.4; Kumar et al. 2012), where

$$\Phi = (\text{moles product/absorbed photons}) \times (\text{electrons needed for conversion}) \quad \text{(R7.4)}$$

Homogeneous photocatalytic systems typically have a photosensitizer, electron donor, and catalyst (Dalle et al. 2019). The photosensitizer absorbs light to generate the electron donor, which reduces the catalyst to initiate CO_2 reduction. Most catalysts with activity toward electrochemical reduction (dark electrocatalysis) also have activity toward photocatalysis with an appropriate photosensitizer and donor. In some cases, the catalyst itself can serve as the photosensitizer (Das et al. 2022; Hawecker et al. 1986). The most common electron donors are aliphatic amines, NAD(P)H model compounds, ascorbate, and imidazole compounds. The choice of electron donor impacts the overall efficiency and stability of photocatalytic systems and can be involved in other reactivity (Sampaio et al. 2020). While the use of these sacrificial electron donors is common, they do not represent a sustainable method for photochemical reduction. Ideally, the electron donor would be water, but water is typically an insufficient reductant to drive CO_2 reduction.

A number of strategies have been applied to improve the performance of heterogeneous photocatalytic (PC) systems for CO_2 reduction. The semiconductor materials must have a suitable band gap (neither too large nor too small) to enable efficient visible light absorption while also being large enough to drive the reaction. The potential of the conductive and valence bands must be sufficient for CO_2 reduction and water oxidation (Kalamaras et al. 2018; Liao and Carter 2013; Mayer 2023). Theoretical approaches to simulating (photo) electrochemical CO_2RR and water splitting at the atomic scale with quantum mechanics modeling have helped elucidate the roles of the structure and composition of the electrochemical interface, absolute band edge positions relative to the redox potentials, charge carrier transport, and proton, electron, and hydride transfers (Govind Rajan et al. 2020; Liao and Carter 2013; Xu and Carter 2019b). Advancements have been made by focusing on materials architecture, which includes quantum dots, nanotubes and nanorods, two-dimensional materials, and more advanced nanostructures (Gui et al. 2021). Additionally, various dopants, sensitizers, and co-catalysts have been introduced to achieve the desired light-absorbing and catalytic properties. To prevent oxidation of the product by photogenerated holes on the photoabsorber, hole scavengers such as hydrogen peroxide (H_2O_2), sodium sulfite (Na_2SO_3), and alcohols are sometimes used (Chang et al. 2016). Several different photoreactors, both batch and continuous types, also have been engineered to improve overall solar-to-product efficiencies. These reactor types are broadly categorized as slurry, fixed bed, and membrane (Khan and Tahir 2019). Key considerations include using geometry to maximize light absorption; using materials (photoabsorber/catalyst, reactor), heat exchange, mixing, and flow characteristics to maintain high contact between the reactants and catalyst; and product separation.

The most common configuration of photoelectrochemical (PEC) cells is composed of a semi-conductor photoelectrode and a counter electrode (White et al. 2015). Compared to PC systems, PEC systems may achieve higher efficiency, because electron-hole recombination is slowed by the external potential. Additionally, a greater variety of materials and configurations can be used. In most cases, CO_2 reduction is accelerated using co-catalysts (Gui et al. 2021), which are often nanoparticles of metals or oxides. Molecular catalysts have also been attached onto surfaces to accelerate CO_2 reduction (White et al. 2015). Other systems use solution-based co-catalysts to promote catalysis. P-type gallium phosphide (GaP) semiconductors have shown the direct photoelectrochemical reduction of CO_2 to methanol with pyridinium additives (Barton Cole et al. 2010; Cohen et al. 2022; Sears and Morrison 1985; Xu and Carter 2019b; Xu et al. 2018). However, different optimal conditions, products, and yields have been reported (Costentin et al. 2018). Nanostructured electrodes have been used to enhance photocatalytic activity by engineering the band structure, increasing the surface area for catalysis, enhancing light absorption, and minimizing electron-hole recombination. Optimization of adsorbed cocatalysts may also help with selectivity and activity (Xu and Carter 2019a).

More recently, researchers have been exploring the use of localized surface plasmon resonance for light-driven CO_2 reduction (Figure 7-10). This phenomenon is a result of the resonant photon-induced collective oscillation of valence electrons and is most commonly observed on nanostructured gold, silver, copper, and aluminum surfaces of nanoparticles (Robatjazi et al. 2021). Plasmonic photocatalysis can contribute to CO_2 reduction by reducing the substrate (or catalyst if used) and providing local thermal heating (Verma et al. 2021; Wang et al. 2023a; Zhang

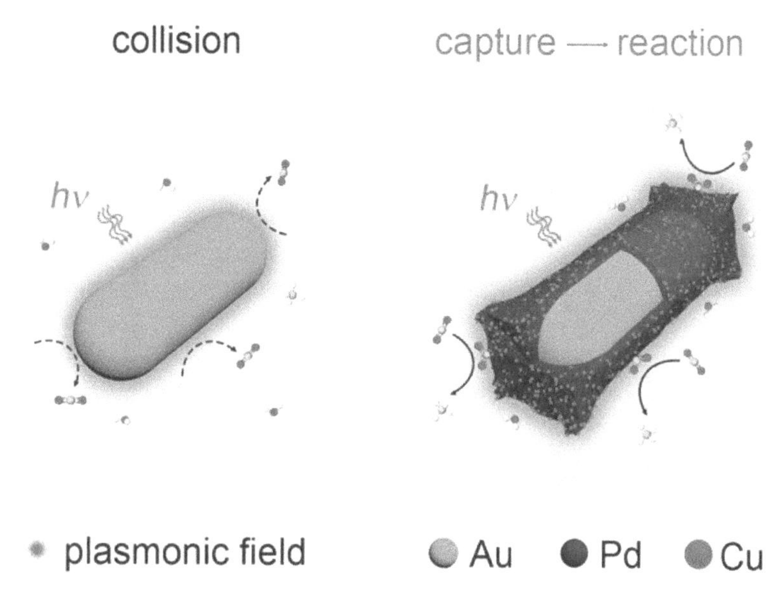

FIGURE 7-10 Plasmon-induced CO_2RR performance. Schematic illustration of the role of a CuPd co-catalyst in capturing CO_2 molecules. Plasmonic catalysis usually takes place very close to the catalyst surface (i.e., within the range of the plasmon-induced local field). For pure Au nanorods, the probability of the pure Au nanorods and CO_2 molecules to contact through collisions is very low (left), resulting in low CO_2 conversion efficiency. For the Au rod@CuPd, the CuPd co-catalyst can capture CO_2 molecules and enhance the CO_2 concentration on the catalyst surface, increasing the opportunity for their further activation and conversion (right).
SOURCE: Hu et al. (2023), https://doi.org/10.1038/s41467-023-35860-2. CC BY 4.0.

et al. 2023). A few promising examples of plasmonic photocatalysis have been reported so far. For example, gold nanoparticles have been used to generate C1 and C2+ products from CO_2 in water (Hu et al. 2023) and an ionic liquid solution (Yu and Jain 2019). Plasmonic photocatalytic systems have also been shown to accelerate the dry reforming of methane (CH_4 + CO_2) into syngas, although the formation of coke limits the lifetime of these catalytic systems (Cai and Hu 2019; Chen et al. 2019; Dieterich et al. 2020; Han et al. 2016; Jang et al. 2019; Zhou et al. 2020).

7.2.3.2 Challenges

The direct use of light to drive CO_2 reduction in photochemical processes is at a basic research stage. Integration of photochemical processes has the potential to reduce the balance of systems costs but comes with other obstacles to be competitive with a photovoltaic and electrolyzer configuration (PV-EC), which already has demonstrated a solar-to-chemical-to-energy conversion efficiency of 21.3 percent for CO production (Liu et al. 2023). Several studies have described the benefits and drawbacks of these two configurations for hydrogen production, and many details from these analyses are also applicable for CO_2 reduction (Ardo et al. 2018; Grimm et al. 2020; Rothschild and Dotan 2017; Shaner et al. 2016). In these analyses, PEC devices need significant improvements to both efficiency and stability. Additional advances in device architectures and operation schemes, such as the power management and light management scheme, are also critical to improve the competitiveness of using PEC versus PV-EC systems. Like electrochemical CO_2 reduction, photochemical CO_2 reduction also contends with product selectivity, particularly with respect to H_2 co-generation and slow kinetics for CO_2 reduction.

In homogeneous systems, most photosensitizers are composed of precious metals, although recent work has focused on the use of abundant components (Ho et al. 2023; Wang et al. 2023b; Xie et al. 2023; Zhang et al. 2019b). While turnover numbers are now reported in the tens of thousands (Dalle et al. 2019), systems with greater long-term stability are needed. Additionally, practical systems will need to demonstrate a catalytic cycle that does not require the use of sacrificial electron donors but instead uses water as the reductant.

Photochemical conversions traditionally have been limited by proximity and surface area contact requirements for photochemical energy. Reactors that combine high surface areas and/or deep penetration zone can overcome these limitations, as can use of high efficiency light-emitting diode arrays (essentially a new form of electrified reactor). These systems likely will have a scaling factor close to 1.0, thus requiring smaller units and numbering up to achieve large-scale production. Such designs tend to favor a modular approach that may make distributed production attractive, but which could limit the overall operating scale.

Additional challenges with photocatalytic systems include charge recombination, the requirement for hole scavengers, and product separations. High rates of charge recombination lead to lower overall quantum efficiency. The use of hole scavengers to prevent reoxidation of product at the photoabsorber adds to the overall cost. Because the cathodic and anodic products are co-generated and are often small molecules of similar sizes, product separation is also a challenge. Low-energy separation strategies, such as membranes, can reduce the CO_2 footprint of the separation/purification process and require considerable future study. Fundamental transport properties of relevant solutes in these solutions are not widely available, which frustrates efforts to design improved membranes for such separations. Other separation methods, such as adsorption and distillation, considered alone or in combination, may play an important role in such separations (Sarswat et al. 2022). Careful consideration of process design and energy requirements would enable informed separation system design. As with membranes, fundamental studies of separation processes are needed using relevant, multicomponent systems.

Research on PEC cells has generated a greater fundamental understanding of semiconductor physics and electronic structure (Xu and Carter 2019b), and the architecture provides a variety of potential materials. However, most photoabsorbers that have been examined either spontaneously corrode (often under aqueous conditions) or experience photocorrosion during operation. Various protective layers, often metal oxides, have been used to improve stability (Lichterman et al. 2016). While plasmonic photocatalysis holds promise, the reported systems currently suffer from high energy input, low product yields, and high costs. In many cases, stability is also an issue. However, new bimetallic alloys have demonstrated resistance to coking for dry methane reforming (Zhou et al. 2020). Additionally, aside from a few reactions (Martirez et al. 2021; Zhou et al. 2020), the mechanisms of catalysis are not well understood, making rational improvements to activity challenging.

7.2.3.3 R&D Opportunities

There are many R&D opportunities for photochemically driven systems. Intrinsically, all motifs require light absorption, generation and separation of electron-hole pairs, and then catalytic reduction of CO_2 paired with another oxidative half reaction. All three of these processes could be improved with further research. There are also specific challenges in the different motifs. While many possible architectures for light-driven CO_2 reduction exist, critical analysis toward scalability, stability, and overall light-to-product efficiency is important, particularly with respect to other methods of carbon-neutral CO_2 utilization.

In homogeneous systems, progress is needed in the use of abundant elements. While photocatalytic systems now operate with turnover frequencies in the thousands per hour, they require operation with longer-term stability for practical application (turnover numbers are typically around 10^4; Dalle et al. 2019). Additionally, systems need to demonstrate operation without the use of sacrificial electron donors, preferably with water and a closed catalytic cycle.

Computational modeling also needs further development, specifically faster, more accurate quantum methods for computing band gaps, absolute band edge positions in the presence of electrolyte, and charge carrier transport and reactions, combined with ab initio molecular dynamics for solvent configuration sampling, along with machine-learned force field molecular dynamics to sample longer time and larger sample sizes.

Photochemical systems continue to benefit from new materials architectures and formulation, which improve overall quantum efficiency and product selectivity. Continued research into reactor design can inform scalable design and performance metrics. Photoelectrochemical systems require improved methods to inhibit corrosion for greater stability.

As in electrochemical systems, catalyst selectivity, rates, and stability are also important. The dominant products in these systems tend to be C1. Obtaining C2+ products will require additional development; an improved understanding of mechanistic pathways for coupling C1 products could benefit this line of inquiry. Photochemical carboxylation to form more complex C2 and C2+ products has been demonstrated, but more research is needed to expand reactivity to unsaturated hydrocarbons and unactivated alkenes, in addition to addressing the other challenges for photocatalytic systems mentioned above (Cauwenbergh et al. 2022; Davies et al. 2021; Tortajada et al. 2018; Zhang et al. 2024). Integration of catalysts into photochemical systems also requires compatibility under operating conditions, as well as robust and stable methods of attachment that do not inhibit both light absorption and catalyst activity.

For plasmonic photocatalysis, the preparation of lower-cost and higher-efficiency noble metal nanoparticles is needed. At this point, the materials space has been minimally explored (Wang et al. 2023a; Zhang et al. 2023). The parameter space that includes the size, shape, and composition of the nanomaterials, as well as the reaction medium and absorption wavelengths, is not well mapped. The mechanisms of plasmonic photocatalysis are not well understood holistically from light absorption through chemistry (despite substantial analysis for individual components of the phenomenon), and advanced characterization may be required to understand the light-matter interaction at an atomic scale. These studies would also be used to inform more sophisticated computational models. Additionally, a better understanding of the molecule-metal interface in hybrid materials may open new routes for more selective or efficient catalysis (Verma et al. 2021; Wang 2023a; Zhang et al. 2023).

7.2.4 Plasmachemical Conversion Pathways

7.2.4.1 Current Technology

Among the possible means to replace fossil fuel–driven thermal conversion with carbon-emission-free electrically driven processes, the potential use of plasma—the phase of matter consisting of gaseous ions and free electrons, formed by passing electricity through a gas—is being explored. Plasmachemical pathways could provide the ability to tune separately reactive ion and electron properties, which may offer unique opportunities for CO_2 conversion. Products of plasmachemical CO_2 activation depend on the types of plasma used, reaction conditions (temperature, pressure, flow rate, and molar ratio of feeds), and the nature of co-reactants. In the absence of other co-reactants, plasma activation of CO_2 generates a nonequilibrium ionized gas that enables the cleavage of the C=O bond in CO_2 to produce CO and O_2 (Snoeckx and Bogaerts 2017), which has been demonstrated using several types of plasma sources, including glow, radiofrequency, and microwave discharges (Ashford and Tu 2017; Xu et al. 2021).

When hydrogen-containing molecules (e.g., H_2, H_2O, CH_4) are included as co-reactants, CO_2 can be converted into a wide range of hydrocarbons and oxygenates (e.g., methanol, formaldehyde, and acetic acid) (Liu et al. 2020).

In principle, plasma activation has potential advantages over conventional processes. For example, high-temperature activation is required to react CO_2 and CH_4 owing to the inert nature of the C=O (E_{diss} = 5.5 eV) and C–H (E_{diss} = 4.5 eV) bonds in CO_2 and CH_4, respectively. This high-temperature condition limits the production of oxygenates thermodynamically. In contrast, CO_2 and CH_4 activation can be achieved at room temperature using a nonthermal plasma. The plasma-induced high-energy electrons in the nonequilibrium ionized gas can activate CO_2 and CH_4 molecules at low bulk gas temperatures to produce oxygenates. Nonthermal plasmachemical reactions of CO_2 and CH_4 have been used to produce both hydrocarbons and oxygenates (Liu et al. 2020). Plasmachemical reactions of CO_2 and ethane also have been explored. A corona plasma was investigated for oxidative dehydrogenation of ethane with CO_2 to produce CO, H_2, and hydrocarbons. A dielectric barrier discharge plasma was used to convert ethane and CO_2 to syngas (CO + H_2) and formaldehyde. Reaction of ethane with CO_2 activated by a dielectric barrier discharge plasma produced C1–C3 alcohols, aldehydes, and acids, in addition to hydrocarbons and CO (Biswas et al. 2022a).

7.2.4.2 Challenges

One of the main challenges in plasmachemical processes is controlling selectivity toward the desired products. Although plasma activation provides a promising route to achieve direct oxidation of light alkanes with CO_2 to produce valuable oxygenated products, the involvement of and interactions between various reactive species results in a wide range of products, as illustrated in Figure 7-11 for plasmachemical reactions of CO_2 and CH_4. On the

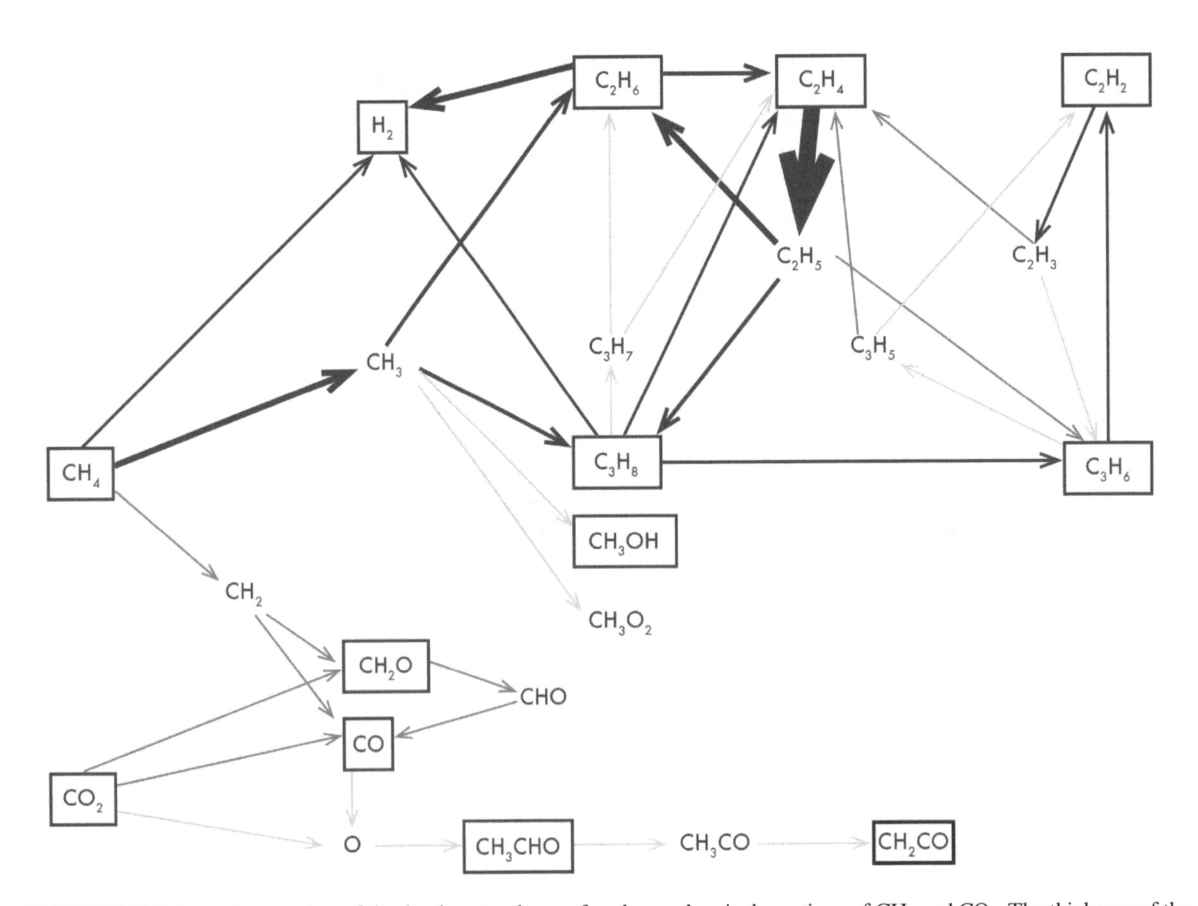

FIGURE 7-11 Schematic overview of the dominant pathways for plasmachemical reactions of CH_4 and CO_2. The thickness of the arrows is correlated to the importance of the reaction pathway. Boxes denote stable products in contrast to transient intermediates. SOURCE: A. Bogaerts, C. De Bie, R. Snoeckx, and T. Kozák, 2017, "Plasma Based CO_2 and CH_4 Conversion: A Modeling Perspective," *Plasma Processes and Polymers* 14(6):1600070. Copyright (2017), with permission from Wiley.

other hand, for certain applications in which nonselective chemical mixtures are desired outcomes, such as for jet fuels, plasmachemical reaction of CO_2 to hydrocarbon mixtures within the desired range of carbon numbers could be an attractive option. Beyond selectivity, challenges for plasmachemical CO_2 conversion include process scale up (scaling factor of close to 1.0) and energy losses resulting from the conversion of electrical energy to plasma energy.

7.2.4.3 R&D Opportunities

At present, research on plasmachemical CO_2 activation is primarily at the stage of lab-scale fundamental studies. Little to no work has been done on computational modeling of plasmachemical CO_2 activation to date. The variety of gaseous and liquid products from plasmachemical CO_2 activation necessitates post-reaction product separation and reduces the energy efficiency. Research focusing on low-energy separation strategies, such as membranes, would be helpful. Membranes today are not designed to separate such complex mixtures, so fundamental research on structure-property-processing of viable membrane candidates is needed. Catalysts may be employed with plasma to provide additional control of reaction selectivity, but significant challenges remain to achieve effective coupling of plasma and catalytic reactions. Understanding plasma-catalyst interactions will require characterization methods to accommodate the complexity of the reaction systems. For example, the chemical properties, surface area, porosity, and dielectric properties of catalyst materials can modify plasma properties. Conversely, the plasma can modify the nature of the catalyst as well. Furthermore, the size and form of the packing material in the catalyst bed can also affect plasma-catalytic activity and selectivity. From the perspective of reactor design, post-plasma-catalysis configurations need to be explored. Moving catalysts outside of the plasma discharge enables the differentiation of interactions with short-lived plasma species from catalytic reactions involving long-lived intermediates and products with the catalyst bed. Further understanding of the complex interactions within plasma-catalyst systems is critical for developing practical plasma-catalytic technologies for selective conversion of CO_2 to desired products.

7.2.5 Integrated Capture and Conversion of CO_2

Most work on CO_2 utilization uses pure and concentrated CO_2 streams as the substrate. However, CO_2 is often found in a dilute stream, with concentrations that range from 0.04 percent in air, to 4–5 percent in natural gas-fired power generation, to greater than 95 percent in some industrial point sources (e.g., ethanol fermentation off-gas) (GAO 2022; NETL n.d.(d)). The composition of the balance of gases also depends on the CO_2 stream, but they commonly contain water, oxygen, and inert gases such as N_2. Industrial streams can also contain lower amounts of gases such as NO_x and SO_x (see Appendix H). Technologies at high readiness levels currently exist for both point source and direct air capture and concentration of CO_2.

CO_2 capture and utilization can be performed independently and in sequence. An example of a sequential commercial process exists in the George Olah Renewable Methanol Plant, which hydrogenates CO_2 isolated from geothermal plant emissions to 4000 tons of methanol per year (Carbon Recycling International n.d.). Integration of capture and utilization provides advantages in process intensification, reducing capital and operational expenses. Integration of these two steps—capture and utilization—is often called reactive capture, defined here as the direct utilization of CO_2 from dilute streams without going through a purified CO_2 intermediate (Freyman et al. 2023). In most CO_2 capture and concentration systems, CO_2 capture is relatively passive except for air-handling, while regeneration of the sorbent to release and compress the CO_2 requires most of the energy input. Direct use of dilute CO_2 or sorbed CO_2 reduces the need for the energy-intensive CO_2 release/concentration step, as well as the need for CO_2 transport or compression, as illustrated in Figure 7-12. The overall energetics of integrated capture and conversion will depend on the sorbent used and product formed (Heldebrant et al. 2022). For example, a techno-economic assessment has shown that integration of CO_2 capture with conversion to methyl formate can save up to 46 percent of the overall energy compared to the sequential process, and up to 8 and 7 percent of the cost and GHG emissions, respectively (Jens et al. 2019). Another analysis indicates that an integrated capture and conversion process can reduce the energy intensity for methanol production from CO_2 by 50 percent compared to sequential capture and conversion approaches, with a 38 percent reduction in capital expenditure (Freyman et al. 2023).

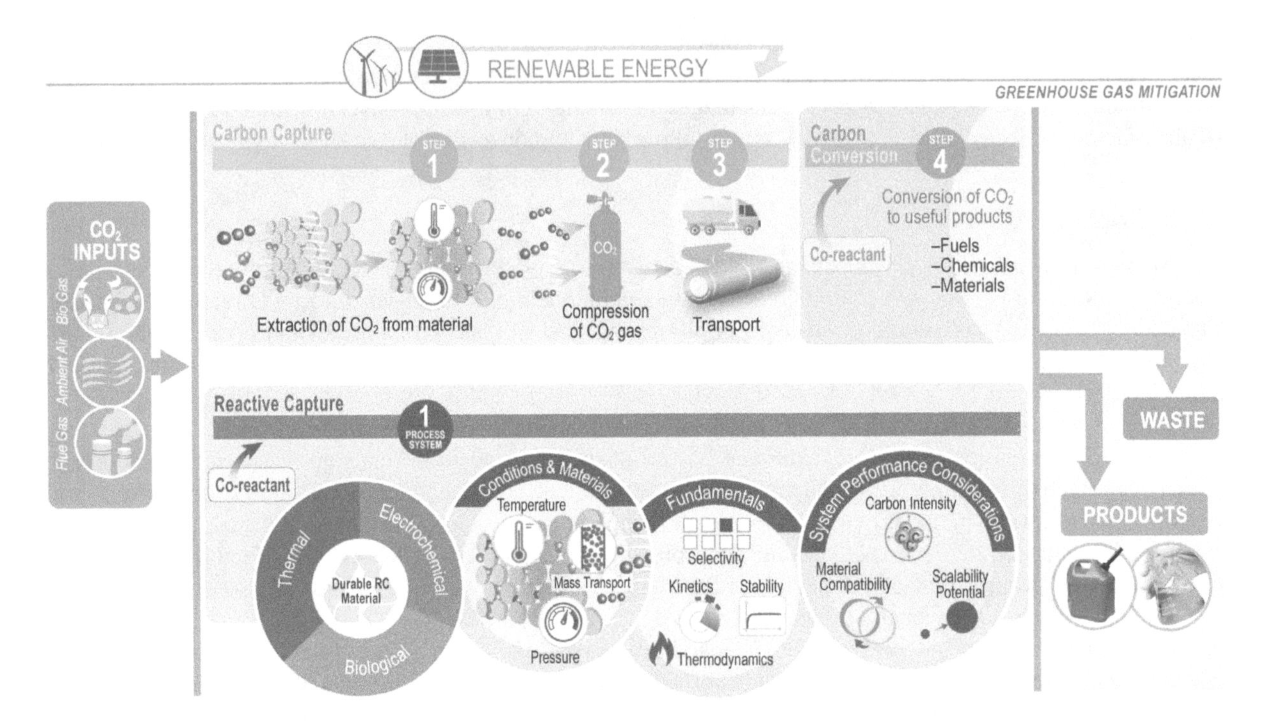

FIGURE 7-12 Schematic of sequential capture and conversion (top) compared to reactive capture (i.e., integrated capture and conversion) (bottom). Reactive capture schematic includes key research and development needs.
SOURCE: Reprinted from M.C. Freyman, Z. Huang, D. Ravikumar, et al., 2023, "Reactive CO_2 Capture: A Path Forward for Process Integration in Carbon Management," *Joule* 7(4):631–651, https://doi.org/10.1016/j.joule.2023.03.013. Copyright (2023), with permission from Elsevier.

Various approaches have been made toward integrated capture and conversion of CO_2, including reductions via electrochemical or thermal routes, synthesis of cyclic carbonates, and biological utilization.

7.2.5.1 Current Technology

An example of electrochemically driven integrated capture and conversion at a high TRL is the use of molten metal oxides to generate solid carbon, as described in Chapter 6. High-temperature (650–900°C) molten salts can capture CO_2 from dilute streams, including flue gases and air, with high selectivity and form carbon materials upon reduction (Zhu et al. 2023). Because the carbon materials have to be removed from the electrode, this system is typically run in a batch mode with reuse of the molten salts (Carbon 2023).

Other examples of electrochemical integrated capture and conversion occur at lower temperatures. Aqueous approaches capitalize on the favorable reaction of hydroxide anions with CO_2 to capture CO_2 and form carbonate or bicarbonate, depending on the pH (Ghobadi et al. 2016). Bicarbonate electrolyzers use a cation exchange membrane or bipolar membrane to generate an acidic environment at the cathode/membrane interface, which reacts with bicarbonate to release CO_2, which is then reduced at the cathode. The hydroxide formed as a product of CO_2 reduction is regenerated as the sorbent. The major product depends on the electrocatalyst, with silver making CO and copper with a cationic surfactant giving CH_4 (Lees et al. 2022; Zhang et al. 2022c). Another approach uses a CO_2-binding organic liquid (ethylene glycol and choline hydroxide) to capture CO_2 from simulated flue gas, which is released after transport through an anion exchange membrane. Electrolysis with a copper mesh cathode produces multiple carbon-based products, including CO, HCOOH, CH_4, C_2H_4, C_2H_5OH, and C_3H_7OH. Under optimized conditions, a high of 64 percent Faradaic efficiency for carbon-based products is achieved (Prajapati et al. 2022).

In both configurations, the overall product selectivity depends on the local environment at the electrode (including water concentration) and optimizing the rate of substrate transport, both as captured CO_2 and released CO_2.

In the heterogeneous integrated capture and conversion systems that use hydroxide as the capture agent, CO_2 is believed to be released at the electrode for reduction. There are, however, examples of homogeneous electrocatalysts that are believed to directly reduce bicarbonate. For example, $[Ru^{III}(edta)(H_2O)]^-$ directly reduces HCO_3^- to formate with Faradaic efficiencies as high as 90 percent electrochemically (Chatterjee et al. 2014) or photochemically (Mondal and Chatterjee 2016). A dinuclear copper complex that cooperatively binds carbonate has also shown photocatalytic activity toward the production of CO (Liu et al. 2012a).

Other CO_2 capture solvents have been used for direct electrochemical reduction. Ionic liquids (ILs) with high solubility and selectivity for CO_2 have been explored. ILs are characterized by their low vapor pressure, which minimizes evaporative losses during CO_2 capture. They generally have large electrochemical windows (i.e., resistance to oxidation and reduction) and sufficient conductivity to not require the addition of external electrolytes. Early studies of ILs for electrosynthesis used CO_2 with co-substrates to form more complex products (Alvarez-Guerra et al. 2015), where the choice of co-reductant guides the product. For example, CO_2 is co-reduced with alcohols or alkyl iodides to form organic carbonates, most commonly dimethyl carbonate. Reduction of olefins with CO_2 is a route to carboxylic acids, while the use of epoxides with CO_2 forms cyclic organic carbonates. In studies with organic co-reductants, the most common heterogeneous electrode catalysts used are copper, platinum, and nickel. In contrast to the complex products produced with co-reductants, direct electrochemical reduction of CO_2 in ILs by heterogeneous catalysts typically forms CO, a combination of CO and H_2 (syngas), or more rarely, formate. The most common ILs for electrochemical reduction with no co-reductant are composed of imidazolium salts with fluorinated anions. Catalyst materials have varied but are most commonly metals supported by a carbon electrode (Alvarez-Guerra et al. 2015).

Homogeneous catalysts have been explored for CO_2 reduction in ILs. Early studies focused on the highly selective catalyst $Re(bpy)(CO)_3Cl$ and an imidazolium cation-based IL with tetracyanoborate as the counteranion, demonstrating continued high selectivity for the product CO with a significantly reduced overpotential compared to operation in organic solvents (Grills et al. 2014). Further studies have demonstrated photocatalytic activity in supramolecular systems (Grills and Fujita 2010). The higher efficiency of these catalysts in ILs is attributed to the interaction between the imidazolium cation with the rhenium complexes through hydrogen-bonding interactions, leading to a milder reduction potential (Matsubara et al. 2015).

Thermochemical systems for integrated capture and conversion of CO_2 use co-reductants to valorize the captured carbon. In heterogeneous systems, these are often called integrated carbon capture and utilization (ICCU). Most examples of heterogeneous integrated capture and conversion use "dual functional materials," which capture CO_2 in metal oxide materials (Omodolor et al. 2020; Shao et al. 2022). The input gas is then switched to a reductant. When hydrogen is used, methane is the predominant product, although some catalysts can promote the RWGS reaction (ICCU-RWGS). Alternatively, in ICCU-DRM (where DRM is dry reforming of methane), light alkanes such as methane can be used as the co-reductant to produce syngas (Kim et al. 2018; le Saché and Reina 2022) or reduce ethane to ethylene (Gambo et al. 2021). Both the capture and conversion steps are typically conducted at high temperatures, with the conversion step often requiring temperatures >500°C and/or high pressures (Sun et al. 2021).

There is precedent for using molecular catalysts for the hydrogenation of solubilized CO_2. Multiple pincer-type catalysts have been tested for the hydrogenation of carbonate solution (hydroxide-captured CO_2) to generate formate (Kar et al. 2018a). These hydrogenation reactions can be carried out at milder temperatures (80°C) and 50 bar H_2, regenerate the hydroxide sorbent, and have turnover frequencies of 10^3 hr^{-1}. Amine solutions, which capture CO_2 to form ammonium carbamates, can also be hydrogenated to both ammonium formate (Kothandaraman et al. 2016a) and methanol (Kar et al. 2018b, 2019b; Kothandaraman et al. 2016b). To improve recyclability in these systems, work has been performed on immobilizing the amine sorbents and sorbent generation (Kar et al. 2019a).

A nonredox method of CO_2 functionalization capitalizes on its reaction with epoxides to form cyclic carbonates. Various methods have been employed to perform this reaction, including the use of specially designed covalent organic frameworks (Talapaneni et al. 2015; Wang et al. 2015b), metal organic frameworks (Ding and Jiang 2018; Liu et al. 2016a; Zhang et al. 2016), and ILs that combine capture functionalities with catalytic active sites (Liu et al. 2016b).

Captured CO_2 solutions have also been useful for other applications that do not involve conversion to chemicals or fuels. Several studies indicate that solvent-based capture can enhance CO_2 mineralization to carbonates and related solids while regenerating the sorbent (Heldebrant et al. 2022). Another approach uses direct seawater electrolysis to promote the formation of mineral carbonates from oceanic dissolved inorganic carbon (La Plante et al. 2021, 2023), which is discussed further in Section 5.2.4 of Chapter 5.

An alternative form of integrated capture and conversion is sacrificial capture and utilization (Marocco Stuardi et al. 2019). In this case, the CO_2–sorbent bond is retained in the final product, so the sorbent is not regenerated. This method has been used with amine-based capture agents to form alkylcarbamates, urethanes, and alkylureas, as well as reduced products such as formamides, formamidines, and methylamines. Sacrificial capture and utilization open the door for a greater variety of fine chemicals synthesized with CO_2 as a C1 precursor.

7.2.5.2 Challenges

The challenges for electrochemical integrated capture and conversion mirror some of those in direct electrochemical reduction of CO_2. Obtaining products with high selectivity while suppressing the hydrogen evolution reaction is an important goal. Several heterogeneous systems, particularly bicarbonate electrolyzers, use protons to release CO_2 to the cathode. Thus, controlling the release of substrate and water to match arrival at the catalyst interface are important design aspects. Direct bicarbonate or carbonate reduction at heterogeneous electrodes may be challenging as the carbon-containing substrate is anionic, which lowers accessibility to a negatively charged electrode. Homogeneous systems have been proposed for direct electrochemical reduction of carbonate or bicarbonate, and may be better suited for such reactions, given that carbonate and bicarbonate are common ligands in transition-metal complexes (Krishnamurty et al. 1970). Unique to integrated systems is the challenge of matching the timescales of capture and conversion, which depends on the system architecture. There may also be a mismatch in the thermodynamics of the capture and conversion steps such that the integrated system has a larger energy requirement than the two processes separately (Appel and Yang 2024). Additionally, low conversion efficiencies can require additional downstream purification. Lastly, compared to pure CO_2 reduction, fewer products are currently available in combined capture and conversion systems.

Thermochemical integrated capture and conversion has been explored using heterogeneous and homogeneous catalysts with promising results. Heterogeneous systems provide a diversity of potential reactions, including methanation, dry reforming of light alkanes, and RWGS, albeit at high temperatures. Homogeneous systems can produce formate and methanol at relatively high rates. The overall carbon footprint of these systems will depend on the operating temperature, as well as the source of the co-reductant. Most hydrogen is currently generated from fossil sources, so a major challenge in minimizing carbon emissions from CO_2 utilization will be economical and abundant sources of clean hydrogen.

7.2.5.3 R&D Opportunities

All electrochemical integrated capture and conversion systems are still in the R&D stage except for the high-temperature molten carbonate systems that produce elemental carbon products. Research directions include aiming for a better understanding of CO_2 speciation, concentration, and transport, including capture and release mechanisms. In homogeneous systems, there is an opportunity to use ligand design to capture CO_2 in the secondary coordination sphere, or otherwise activate it for reduction (Sung et al. 2017). Additionally, some studies indicate that metal-ligand bonds can be used to capture CO_2 for reduction, but systematic studies on this reactivity have not been performed (Sattler and Parkin 2014).

Ionic liquids are promising for electrochemical integrated capture and conversion because of their high solubility for CO_2, large electrochemical windows, and general inertness to common contaminants in industrial flue gases. The exact speciation of CO_2 in solution is still not entirely understood, as CO_2 can exist in a soluble form or chemically interact with the imidazolium cations of the ILs. The mechanism of CO_2 reduction—both directly and with co-reductants, is still not clear, and thus there are few catalyst design rules to guide development. However, there is significant room for improvement. In addition to developing new catalysts, ILs themselves are highly tunable, as their cations and anions can be tailored for specific properties. For electrochemical reduction of CO_2-sorbing ILs, the viscosity of the solvent needs to be considered in electrolyzer design as it affects mass transport of substrate. Many of the electrolyzers have improved kinetics at lower temperatures, where CO_2 is more soluble but ILs become even more viscous. ILs also tend to be hydroscopic, which can result in mixed solvents that have to be controlled under practical conditions. ILs are more expensive than most solvents and electrolyte combinations used in electrolysis, which could add scalability challenges. Very few studies have considered this cost, but in one case it was estimated to increase the total capital cost for solvent from less than 1 percent to 14 percent (Chang et al. 2021). More detailed studies are needed to determine the primary cost contributors more accurately.

Most integrated capture and conversion systems regenerate the sorbent, but, as introduced above, there are opportunities to use CO_2 as a C1 precursor in sacrificial captures, where the sorbent–C bond is retained. A sacrificial capture system also can be used to synthesize valuable heteroatom-carbon bonds, as well as to introduce stereocenters. Further reduction expands the accessible functionalities; for example, syntheses of arylcarbamates, oxazolidine, urethanes, and alkylureas have been described (Bernoud et al. 2017; Feroci et al. 2005; Liu et al. 2012b; McGhee et al. 1995). Commercial amine reagents that capture CO_2 to form carbamates can be reduced to form formamides, formamidines, methylamines, and aminals (Tlili et al. 2015). As the sorbent is a stoichiometric reagent, the reaction must provide significant added value to be cost-effective; thus, synthesis of fine chemicals typically has been targeted.

Because current methods of CO_2 capture and concentration require significant inputs and infrastructure, the case of integrated capture and conversion is compelling. However, integrated systems will have to be evaluated holistically to determine whether they are advantageous over systems that perform sequential capture and concentration followed by utilization. While sorbents can kinetically activate nonpolar CO_2 molecules, they also result in greater thermodynamic stability, requiring more energy for subsequent conversion. Integrated capture and conversion methods that use sorbents will have to consider their initial cost and recyclability (regeneration) in the overall techno-economic assessment of these processes. Incomplete capture and/or conversion of CO_2 may also lead to downstream separations costs. High-performance, low-energy separation technologies, such as membranes, may play a key role in augmenting the electrochemical processes. These factors are important in evaluating the overall value propositions of integrated capture and conversion schemes.

7.2.6 Polymers

7.2.6.1 Current Technology

Direct polymerization of CO_2 to make poly(CO_2) is possible, but owing to the low reactivity of CO_2, the synthesis conditions are exceptionally challenging, such as 1800 K (1527°C) and 40,000 MPa (Huang et al. 2020; Iota et al. 1999). There are no currently known efforts to further develop poly(CO_2).

There is, however, commercial activity to synthesize polycarbonates from CO_2, taking advantage of the specific chemistry of the CO_2 molecule. As noted in Chapter 2, polycarbonate production occurs at a scale of about 1.5 Mt/year globally (Neelis et al. 2007), making it a promising opportunity for CO_2 utilization. Inoue and coworkers first reported the polymerization of CO_2 with oxiranes, such as propylene oxide, to form polycarbonates in 1969 (Inoue et al. 1969a, 1969b; see Figure 7-13). Numerous studies since have developed this and similar polymerization reactions further (Appaturi et al. 2021; Fukuoka 2012; Rehman et al. 2021; Tabanelli et al. 2019; Tan et al. 2021; Wołosz et al. 2022), motivated in part by the safety and toxicity benefits of using CO_2 as a feedstock compared to traditional methods using phosgene.

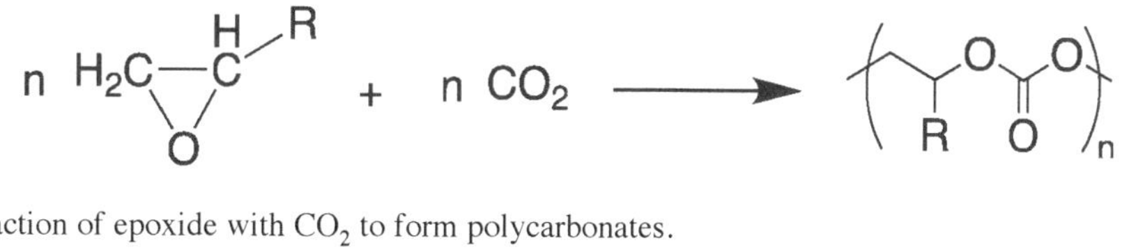

FIGURE 7-13 Reaction of epoxide with CO_2 to form polycarbonates.

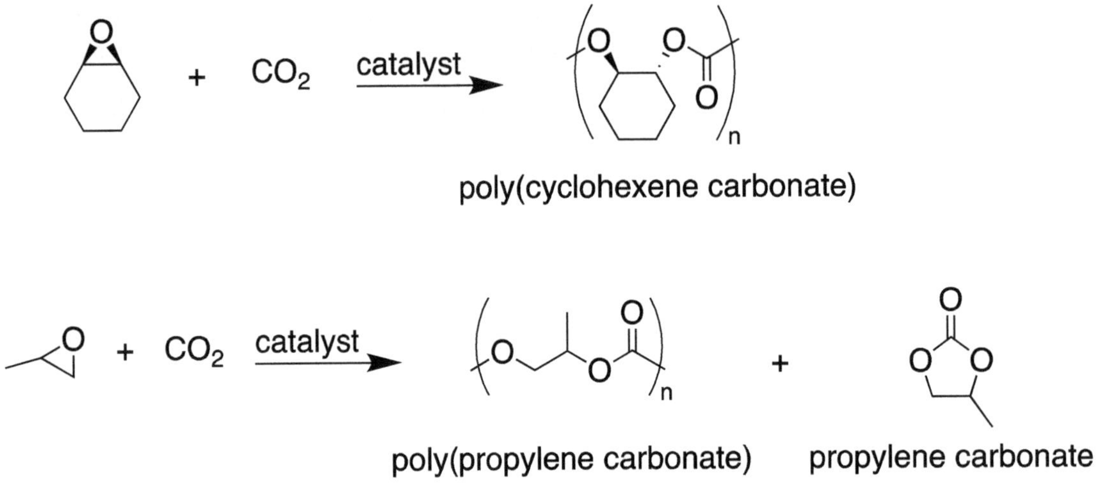

+ CO_2 catalyst → poly(cyclohexene carbonate)

+ CO_2 catalyst → poly(propylene carbonate) + propylene carbonate

FIGURE 7-14 Reaction schemes for formation of aliphatic polycarbonates from CO_2.
SOURCE: Used with permission of Angewandte Chemie, adapted from G.W. Coates and D.R. Moore, 2004, "Discrete Metal-Based Catalysts for the Copolymerization of CO_2 and Epoxides: Discovery, Reactivity, Optimization, and Mechanism," *Angewandte Chemie International Edition* 43(48):6618–6639; permission conveyed through Copyright Clearance Center, Inc.

Benefits to this approach include the industrial-scale availability of oxiranes such as ethylene oxide and propylene oxide, the lack of by-products produced in the reaction, the lack of need for stoichiometric co-reagents, and the fact that products can contain up to 50 percent CO_2/O_2 (Dabral and Schaub 2019). There has been extensive research on forming aliphatic polycarbonates using ring-strained monomers, such as cyclohexene oxide, ethylene oxide, propylene oxide, and others (Coates and Moore 2004; Darensbourg and Holtcamp 1996; Grignard et al. 2019; Huang et al. 2020; Kember et al. 2011; Liu and Lu 2023; Yeung et al. 2023) (Figure 7-14). However, the glass transition (i.e., softening) temperature of polypropylene carbonate is 35°C–40°C,[6] and it decomposes at 250°C (Langanke et al. 2015), which limits its utility in conventional engineering thermoplastics applications (Coates and Moore 2004; von der Assen and Bardow 2014). Rather, aliphatic polycarbonates typically are used as binders in adhesives and ceramics (Langanke et al. 2015). This use case only provides extremely short-duration storage of CO_2, as the binders are sacrificial and designed to decompose in the ceramic formation, rereleasing CO_2. Polyetherol carbonates also can be produced via reaction of ring-strained monomers with CO_2 (Dabral and Schaub 2019). Together, these aliphatic polycarbonates, prepared with hydroxyl end groups (i.e., polyols), are produced commercially (Grignard et al. 2019; Liu and Lu 2023). An important use of these polyols is to react them with isocyanates to produce polyurethanes (Liu and Lu 2023).

A substantial amount of research today is focused on developing catalysts to perform these reactions (Huang et al. 2020; Lidston et al. 2022). For example, the reaction of CO_2 with ethylene oxide and propylene oxide occurs at high pressure[7] using a variety of catalysts, such as organometallic compounds (e.g., $ZnEt_2$), ammonium and phosphonium salts, alkali metal iodides, various aluminum and manganese catalysts, chromium catalysts, cobalt catalysts, lanthanide series catalysts, Lewis acids, and ion exchangers containing ammonium or phosphonium groups (Buysch 2011; Coates and Moore 2004; Darensbourg 2007; Darensbourg and Holtcamp 1996; Huang et al. 2020; Liu and Lu 2023). Bifunctional catalysts (e.g., alkali metal and zinc halide) are used under milder conditions. Most of these catalysts are metal based, including various catalysts involving aluminum and manganese, chromium or cobalt, as well as lanthanide series catalysts (Coates and Moore 2004; Darensbourg 2007; Darensbourg and Holtcamp 2007; Huang et al. 2020; Liu and Lu 2023; Yeung et al. 2023).

[6] For comparison, the glass transition temperature of BisA-PC, a widely used engineering thermoplastic, is ~147°C.

[7] For example, in their review of catalysts for copolymerization of CO_2 and epoxides, Coates and Moore (2004) list pressure requirements of 7–135 atm for various catalytic systems.

Other families of polymers that have used CO_2 directly in polymerization include polyureas synthesized from CO_2 and diamines, polyesters from CO_2 copolymerization with ethylene and other olefins, and poly(urethanes) from reacting CO_2 with aziridines or with amino alcohols (Grignard et al. 2019). However, these polymers have not been commercialized and appear to be at the laboratory or bench scale at this time.

7.2.6.2 Challenges

Current challenges for deriving polymers from CO_2 include the lack of routes to incorporate CO_2 directly into polymerizations that involve aromatic compounds that yield the high glass-transition, tough, ductile engineering thermoplastics that are dominant in the polycarbonate field currently (e.g., BisA-PC). Much of the work to date has focused on polymerizing oxiranes and related compounds with CO_2, and whether the portfolio of accessible monomers can be expanded to produce a wider variety of polymers with property profiles better matched to commercial needs is not well understood. Whether catalysts can be developed that permit rapid, economical polymerization of precision (i.e., stereochemistry-controlled) materials is unclear. More efficient catalysts are needed to expand the commercial opportunities for polymerizations that directly use CO_2 (Huang et al. 2020). Many current catalysts do not exhibit the high productivity needed to drive production of CO_2-based polymers to the same scale as, for example, polyolefins (Liu and Lu 2023). How the cost and property profile of polymers made with CO_2 will compare to those of polymers made by conventional routes is not well-defined. While CO_2-derived polycarbonates and polyols for polyurethane have been commercially available for more than a decade, they have seen limited market penetration; for example, CO_2-derived aromatic polycarbonates are produced at about 0.90 Mt/year, representing only 16 percent of the total global annual production of this polymer (Nova-Institut 2023). The carbon cost of obtaining CO_2 at sufficient purity and quantities to conduct such polymerizations is not well understood. Last, because one cannot rely on poly(CO_2) to address the need for CO_2-containing polymers, the question of where the co-monomers and reactants will come from (e.g., from fossil sources or carbon-neutral renewable resources) is not yet resolved at a commercial scale.

7.2.6.3 R&D Opportunities

The R&D opportunities for polymerizations involving CO_2 stem from the challenges outlined above. Fundamental studies related to advanced catalyst discovery are needed to obtain further control over the stereochemistry of resulting polymers and increase the productivity of the catalysts to enable large-scale polymerization. Exploration of monomers beyond those that have been considered to date, coupled with catalyst development to permit their polymerization with CO_2, could open new classes of materials to production directly from CO_2 (Song et al. 2022). Combining electrochemical and organometallic catalysts may provide routes to additional materials than can be achieved by organometallic catalysts alone (Dodge et al. 2023). Advancing research to derive carbon-neutral, sustainable co-monomers for use in CO_2 polymerizations is critical. Defining carbon-neutral routes to incorporation of aromatics in CO_2-derived polycarbonates will broaden substantially their market opportunities. Most of the research on catalysts has focused on homogeneous catalytic approaches, so further exploration of heterogeneous catalysts may be fruitful (Huang et al. 2020).

7.3 CONCLUSIONS

Based on the discussions presented in Section 7.2, the committee highlights the following R&D needs and opportunities for chemical CO_2 conversion:

- There are potential advantages using tandem catalysis combining two or more of the chemical conversion routes: thermochemical, electrochemical, photochemical, and plasmachemical processes. In some cases, the tandem strategy can lead to products that a single process cannot achieve, such as the conversion of CO_2 to C3 oxygenates using tandem electrochemical-thermochemical reactions (Biswas et al. 2023; Garg et al. 2024).

- Thermochemical CO_2 conversion typically requires high temperature, leading to challenges in controlling catalytic selectivity and catalyst stability. Alternative heating methods, such as (pulsed) electrical heating, could have potential advantages in better catalytic performance, energy savings from not heating the entire reactor, and reduced GHG emissions by using clean electricity (Zheng et al. 2023).
- Low-temperature electrochemical CO_2 conversion faces challenges in terms of long-term catalyst stability and sensitivity to impurities. Copper remains the primary element that catalyzes the production of hydrocarbons and oxygenates containing two or more carbon atoms, often with low product selectivity and yield; more efforts need to be devoted to developing electrocatalysts that are selective, stable, and scalable to produce both C1 and multicarbon products, including some of the target molecules shown in Figure 7-2.
- Photochemical conversion of CO_2 requires further fundamental understanding in developing materials and devices that can improve light absorption, generation and separation of electron-hole pairs, and subsequent reduction of CO_2. Additional research into reactor design is also needed to optimize performance metrics and help inform scale up.
- Plasmachemical CO_2 conversion leads to unselective production of multiple products. Although the introduction of catalysts will likely improve selectivity, more in-depth understanding of plasma-catalyst interactions is needed to enable scale up for practical applications.
- Integrated capture and conversion of CO_2 offers advantages in improving overall energy efficiency and lowering capital requirements for separate steps of CO_2 capture and subsequent catalytic conversion.
- The direct utilization of CO_2 in polymerization reactions is currently limited to a narrow range of monomers. More research into catalyst design and development to enable rapid, stereoselective polymerization of a broader class of monomers with CO_2, to access polymers with properties more like those of conventional thermoplastics, could markedly expand opportunities for polymers made directly from CO_2.
- Capital costs for thermochemical reactions conducted in bulk equipment typically scale at 0.6 power with throughput, giving improved economics at larger scale. Electro-, photo(electro)-, and plasma-chemical reactors, as well as endothermic thermochemical reactors using electrified modes of heating (e.g., electromagnetic radiation, some induction heating designs) often depend on surface area rather than volume, and capital costs increase to the 1.0 power of throughput or scale. For these cases, require of smaller units and numbering up can be required to achieve large-scale production. These processes are more amenable to distributed modular production but may be challenged to deliver low costs owing to the inability to scale individual units to obtain economies of scale.

As described above, the same product(s) in some cases can be produced by multiple conversion routes, each of which come with their own advantages and disadvantages. For example, for CO_2 conversion to CO, the thermochemical conversion can be performed using existing technology, but it requires molecular H_2; the electrochemical conversion avoids molecular H_2 by using protons and electrons, but at present, the reaction rate and long-term stability cannot yet compete with the more mature thermochemical processes. When moving beyond R&D to demonstration and deployment, factors such as energy and infrastructure requirements, life cycle emissions, policy support, safety, and cost will need to be compared among different conversion routes to determine the best option for a given product or application. These considerations are discussed in more detail in Chapters 2, 3, 4, and 10.

7.3.1 Findings and Recommendations

The preceding overarching R&D needs for chemical CO_2 conversion led the committee to the following findings and recommendations:

Finding 7-1: Challenges and opportunities for thermochemical CO_2 conversion—Thermochemical CO_2 conversion typically requires high temperatures, leading to challenges in controlling catalytic selectivity and catalyst stability (including tolerance to impurities). For thermochemical CO_2 conversions to have net-zero or net-negative emissions, carbon-neutral energy, hydrogen, or other reductants are required. Alternative heating methods, such as (pulsed) electrical heating, could have potential advantages for better catalytic performance, energy savings from not needing to continually heat the entire reactor, and reduced greenhouse gas emissions by using clean electricity.

Integration of solar thermochemical hydrogen production and CO_2 conversion with thermal energy storage could improve the economic viability of high capital intensity processes (e.g., reverse water gas shift reaction, methanol synthesis, and Fischer-Tropsch synthesis).

Recommendation 7-1: Support research on catalyst development, electrical heating, and carbon-neutral reductants for thermochemical CO_2 conversion—Basic Energy Sciences within the Department of Energy's Office of Science (DOE-BES), the National Science Foundation, and the Department of Defense (DoD) should increase support for experimental and theoretical discovery research into catalysts and processes that utilize carbon-neutral and efficient methods of electrical heating to convert CO_2 to useful chemicals and chemical intermediates (e.g., targeted heating, microwave heating). DOE-BES, DoD, and DOE's Office of Energy Efficiency and Renewable Energy, Office of Fossil Energy and Carbon Management, Office of Clean Energy Demonstrations, and Advanced Research Projects Agency–Energy should continue to support research and development (R&D) that facilitates scale up of thermochemical CO_2 conversion to achieve net-zero CO_2 utilization. This includes R&D on the production of low-carbon hydrogen and other carbon-neutral reductants and the integration of solar thermochemical hydrogen production and CO_2 conversion with thermal energy storage.

Finding 7-2: Engineering and systems optimization needs for thermochemical CO_2 conversion—Thermochemical and thermocatalytic conversion of CO_2 to hydrocarbon products typically requires multiple reaction steps and is energy- and capital-intensive relative to current routes from fossil-based feedstocks. Incorporation and integration of low-carbon energy sources, such as variable renewable energy with low-cost storage, and the ability to deliver this low-carbon energy to high-temperature reaction systems, including options for dynamic operation, will require new engineering and systems optimization to provide plausible pathways for net-zero emissions chemical production.

Recommendation 7-2: Support research on integrated systems for thermochemical CO_2 conversion—The Department of Energy's Office of Fossil Energy and Carbon Management should fund applied research on integration of variable renewable energy and energy storage into efficient, heat-integrated process systems for CO_2 conversion to hydrocarbon products.

Finding 7-3: Challenges for electrochemical CO_2 conversion—Low-temperature electrochemical CO_2 conversion faces challenges in long-term catalyst stability and robustness to impurities in the CO_2 source. Copper remains the primary element that catalyzes the production of hydrocarbons and oxygenates containing two or more carbon atoms, often with low product selectivity. The cost and efficiency of electrochemical CO_2 conversion is also impacted by the materials and performance metrics of the anodic reaction and the membrane.

Recommendation 7-3: Support research on developing electrocatalysts from abundant elements and membrane materials for electrochemical CO_2 conversion technologies—Basic Energy Sciences within the Department of Energy's Office of Science (DOE-BES) and DOE's Office of Fossil Energy and Carbon Management (DOE-FECM) should devote more effort to experimental and theoretical research for discovering and developing electrocatalysts from abundant elements that are selective, stable, and scalable to produce both single- and multicarbon products for both low- and high-temperature electrochemical processes. DOE-BES and DOE-FECM should also invest in developing abundant-element electrocatalysts for water oxidation or alternative anodic reactions as well as cost-effective, scalable membrane materials that function over a wide pH range to lower the overall cost of electrochemical CO_2 conversion. Long-term stability testing should be encouraged with new electrocatalyst development, along with testing for product selectivity and current density.

Finding 7-4: Fundamental research needs for photo(electro)chemical and plasmachemical CO_2 conversion—Fundamental understanding of the sequence of processes involved in photochemical and photoelectrochemical conversion of CO_2 is incomplete. Such understanding is required to improve light absorption, generation and separation of electron-hole pairs, and subsequent reduction of CO_2. Plasmachemical CO_2 conversion also lacks in-depth understanding of plasma-catalyst interactions to improve product selectivity for practical applications. More research is needed to improve reactor design and reaction engineering for photochemical, photoelectrochemical, and plasmachemical CO_2 conversions.

Recommendation 7-4: Support research on mechanisms, materials, and reactor design for photo(electro) chemical and plasmachemical CO_2 conversion—Basic Energy Sciences within the Department of Energy's (DOE's) Office of Science should support more experimental and theoretical research into understanding fundamental mechanisms and materials discovery for photochemical, photoelectrochemical, and plasmachemical catalytic conversion of CO_2. DOE's Office of Fossil Energy and Carbon Management and Advanced Research Projects Agency–Energy should support research to enable development of improved materials, devices, and reactor design for such conversions.

Finding 7-5: Potential advantages of tandem catalysis for CO_2 conversion—There are potential advantages in using tandem catalysis that combines two or more of the chemical conversion routes: thermochemical, electrochemical, photochemical, and plasmachemical processes. In some cases, the tandem strategy can lead to products that a single process cannot achieve.

Recommendation 7-5: Increase support for research on tandem catalysis for CO_2 conversion—The Department of Energy's Office of Fossil Energy and Carbon Management and Advanced Research Projects Agency–Energy should increase support for basic and applied research into tandem catalysis, including catalyst and membrane development, tandem reactor design, and process optimization.

Finding 7-6: Potential advantages of integrated capture and conversion of CO_2—If the energy requirements and operational scales of the capture and conversion steps are matched, integrated capture and conversion of CO_2 can offer advantages in improving overall energy efficiency and lowering capital requirements compared to separate steps of CO_2 capture and subsequent catalytic conversion.

Recommendation 7-6: Increase support for research on integrated capture and conversion of CO_2—Basic Energy Sciences within the Department of Energy's (DOE's) Office of Science, the National Science Foundation, and DOE's Office of Fossil Energy and Carbon Management, Office of Energy Efficiency and Renewable Energy, and Advanced Research Projects Agency–Energy should increase support for basic and applied research into integrated CO_2 capture and conversion, including discovery of molecules and materials, catalytic mechanisms, process optimization, CO_2 stream purification, and reactor design.

Finding 7-7: Direct utilization of CO_2 in polymerization is limited—The direct utilization of CO_2 in polymerization reactions is currently limited to a narrow range of monomers.

Recommendation 7-7: Support research on catalyst development for CO_2 polymerization with a broader class of monomers—Basic Energy Sciences within the Department of Energy's (DOE's) Office of Science and DOE's Office of Fossil Energy and Carbon Management should support more experimental and theoretical research into catalyst design and development to enable rapid, stereoselective polymerization of a broader class of monomers with CO_2. Such research could lead to polymers with properties more like those of conventional thermoplastics and/or thermosets, which could markedly expand opportunities for polymers made directly from CO_2.

7.3.2 Research Agenda for Chemical CO_2 Conversion to Organic Products

Table 7-3 presents the committee's research agenda for chemical CO_2 conversion to organic products, including research needs (numbered by chapter), and related research agenda recommendations (a subset of research-related

TABLE 7-3 Research Agenda for Chemical CO_2 Conversion to Organic Products

Research, Development, and Demonstration Need	Funding Agencies or Other Actors	Basic, Applied, Demonstration, Enabling	Research Area	Product Class	Long- or Short-Lived	Research Theme	Source
7-A. Improvements in catalytic activity, selectivity, and stability (including tolerance to impurities) for thermochemical CO_2 conversion.	DOE-BES NSF DoD	Basic	Chemical—Thermochemical	Chemicals	Short-lived	Catalyst innovation and optimization Computational modeling and machine learning	Fin. 7-1 Sec. 7.2.1.3
7-B. Discovery research into catalysts and processes that use alternative heating methods, such as (pulsed) electrical heating, with goals of improving catalyst performance, yielding energy savings, and reducing GHG emissions by using clean electricity.	DOE-BES NSF DoD	Basic	Chemical—Thermochemical	Chemicals	Short-lived	Catalyst innovation and optimization Energy efficiency, electrification, and alternative heating Computational modeling and machine learning	Fin. 7-1 Rec. 7-1
7-C. Continued research and development into low-carbon hydrogen and other carbon-neutral reductants to facilitate scale up of thermochemical CO_2 conversion that can achieve net-zero CO_2 utilization.	DOE-BES DOE-EERE DOE-FECM DOE-ARPA-E DoD	Enabling	Chemical—Thermochemical	Chemicals	Short-lived	Enabling technology and infrastructure needs	Rec. 7-1
Recommendation 7-1: Support research on catalyst development, electrical heating, and carbon-neutral reductants for thermochemical CO_2 conversion—Basic Energy Sciences within the Department of Energy's Office of Science (DOE-BES), the National Science Foundation, and the Department of Defense (DoD) should increase support for experimental and theoretical discovery research into catalysts and processes that utilize carbon-neutral and efficient methods of electrical heating to convert CO_2 to useful chemicals and chemical intermediates (e.g., targeted heating, microwave heating). DOE-BES, DoD, and DOE's Office of Energy Efficiency and Renewable Energy, Office of Fossil Energy and Carbon Management, Office of Clean Energy Demonstrations, and Advanced Research Projects Agency–Energy should continue to support research and development (R&D) that facilitates scale up of thermochemical CO_2 conversion to achieve net-zero CO_2 utilization. This includes R&D on the production of low-carbon hydrogen and other carbon-neutral reductants and the integration of solar thermochemical hydrogen production and CO_2 conversion with thermal energy storage.							
7-D. Engineering and systems optimization to integrate low-carbon energy sources with high-temperature reaction systems for CO_2 conversion to hydrocarbon products.	DOE-FECM	Applied	Chemical—Thermochemical	Chemicals	Short-lived	Reactor design and reaction engineering Energy efficiency, electrification, and alternative heating	Fin. 7-2 Rec. 7-2
Recommendation 7-2: Support research on integrated systems for thermochemical CO_2 conversion—The Department of Energy's Office of Fossil Energy and Carbon Management should fund applied research on integration of variable renewable energy and energy storage into efficient, heat-integrated process systems for CO_2 conversion to hydrocarbon products.							

continued

TABLE 7-3 Continued

Research, Development, and Demonstration Need	Funding Agencies or Other Actors	Basic, Applied, Demonstration, Enabling	Research Area	Product Class	Long- or Short-Lived	Research Theme	Source
7-E. Discovery and development of electrocatalysts from abundant elements that are selective, stable, robust to impurities in CO_2 sources, and scalable, and that can produce single- and multicarbon products for both low- and high-temperature electrochemical processes.	DOE-BES DOE-FECM	Basic Applied	Chemical—Electrochemical	Chemicals	Short-lived	Catalyst innovation and optimization Computational modeling and machine learning	Fin. 7-3 Rec. 7-3
7-F. Discovery and development of abundant-element electrocatalysts for water oxidation or alternative anodic reactions to improve the cost and efficiency of electrochemical CO_2 conversion.	DOE-BES DOE-FECM	Basic Applied	Chemical—Electrochemical	Chemicals	Short-lived	Catalyst innovation and optimization Computational modeling and machine learning	Fin. 7-3 Rec. 7-3
7-G. Development of economical membrane materials that function over a wide pH range to improve the cost, efficiency, and scalability of electrochemical CO_2 conversion.	DOE-BES DOE-FECM	Basic Applied	Chemical—Electrochemical	Chemicals	Short-lived	Reactor design and reaction engineering Separations	Fin. 7-3 Rec. 7-3
Recommendation 7-3: Support research on developing electrocatalysts from abundant elements and membrane materials for electrochemical CO_2 conversion technologies—Basic Energy Sciences within the Department of Energy's Office of Science (DOE-BES) and DOE's Office of Fossil Energy and Carbon Management (DOE-FECM) should devote more effort to experimental and theoretical research for discovering and developing electrocatalysts from abundant elements that are selective, stable, and scalable to produce both single- and multicarbon products for both low- and high-temperature electrochemical processes. DOE-BES and DOE-FECM should also invest in developing abundant-element electrocatalysts for water oxidation or alternative anodic reactions as well as cost-effective, scalable membrane materials that function over a wide pH range to lower the overall cost of electrochemical CO_2 conversion. Long-term stability testing should be encouraged with new electrocatalyst development, along with testing for product selectivity and current density.							
7-H. Fundamental understanding of the sequence of processes involved in photochemical and photoelectrochemical conversion of CO_2 for light absorption, generation and separation of electron-hole pairs, and subsequent reduction of CO_2, across a variety of material types.	DOE-BES	Basic	Chemical—Photochemical Chemical—Photoelectrochemical	Chemicals	Short-lived	Fundamental knowledge Computational modeling and machine learning	Fin. 7-4 Rec. 7-4
7-I. Discovery research into materials for photochemical, photoelectrochemical, and plasmachemical catalytic conversion of CO_2.	DOE-BES	Basic	Chemical—Photochemical Chemical—Photoelectrochemical Chemical—Plasmachemical	Chemicals	Short-lived	Catalyst innovation and optimization Computational modeling and machine learning	Rec. 7-4

7-J. In-depth understanding of plasma-catalyst interactions for product selectivity.	DOE-BES	Basic	Chemical—Plasmachemical	Chemicals	Short-lived	Fundamental knowledge Computational modeling and machine learning	Fin. 7-4 Rec. 7-4
7-K. Improved devices, reactor design, and reaction engineering for photochemical, photoelectrochemical, and plasmachemical CO_2 conversions to optimize performance metrics and inform scale up.	DOE-FECM DOE-ARPA-E	Applied	Chemical—Photochemical Chemical—Photoelectrochemical Chemical—Plasmachemical	Chemicals	Short-lived	Reactor design and reaction engineering	Fin. 7-4 Rec. 7-4

Recommendation 7-4: Support research on mechanisms, materials, and reactor design for photo(electro)chemical and plasmachemical CO_2 conversion—Basic Energy Sciences within the Department of Energy's (DOE's) Office of Science should support more experimental and theoretical research into understanding fundamental mechanisms and materials discovery for photochemical, photoelectrochemical, and plasmachemical catalytic conversion of CO_2. DOE's Office of Fossil Energy and Carbon Management and Advanced Research Projects Agency–Energy should support research to enable development of improved materials, devices, and reactor design for such conversions.

7-L. Development of tandem catalysis processes that couple two or more thermochemical, electrochemical, photochemical, and plasmachemical processes, with a goal of accessing products that a single process alone cannot achieve.	DOE-FECM DOE-ARPA-E	Basic Applied	Chemical	Chemicals	Short-lived	Integrated systems Computational modeling and machine learning	Fin. 7-5 Rec. 7-5

Recommendation 7-5: Increase support for research on tandem catalysis for CO_2 conversion—The Department of Energy's Office of Fossil Energy and Carbon Management and Advanced Research Projects Agency–Energy should increase support for basic and applied research into tandem catalysis, including catalyst and membrane development, tandem reactor design, and process optimization.

7-M. Development of integrated CO_2 capture and conversion, including discovery of molecules and materials, catalytic mechanisms, process optimization, CO_2 stream purification, and reactor design.	DOE-BES NSF DOE-FECM DOE-ARPA-E DOE-EERE	Basic Applied	Chemical	Chemicals	Short-lived	Integrated systems Reactor design and reaction engineering	Fin. 7-6 Rec. 7-6

Recommendation 7-6: Increase support for research on integrated capture and conversion of CO_2—Basic Energy Sciences within the Department of Energy's (DOE's) Office of Science, the National Science Foundation, and DOE's Office of Fossil Energy and Carbon Management, Office of Energy Efficiency and Renewable Energy, and Advanced Research Projects Agency–Energy should increase support for basic and applied research into integrated CO_2 capture and conversion, including discovery of molecules and materials, catalytic mechanisms, process optimization, CO_2 stream purification, and reactor design.

7-N. Design and development of catalysts for rapid, stereoselective polymerization of a broader class of monomers with CO_2, especially those that can lead to polymers with properties more like thermoplastics and/or thermosets.	DOE-BES DOE-FECM	Basic Applied	Chemical—Thermochemical	Polymers	Long-lived Short-lived	Fundamental knowledge Catalyst innovation and optimization Computational modeling and machine learning	Fin. 7-7 Rec. 7-7

Recommendation 7-7: Support research on catalyst development for CO_2 polymerization with a broader class of monomers—Basic Energy Sciences within the Department of Energy's (DOE's) Office of Science and DOE's Office of Fossil Energy and Carbon Management should support more experimental and theoretical research into catalyst design and development to enable rapid, stereoselective polymerization of a broader class of monomers with CO_2. Such research could lead to polymers with properties more like those of conventional thermoplastics and/or thermosets, which could markedly expand opportunities for polymers made directly from CO_2.

NOTE: ARPA-E = Advanced Research Projects Agency–Energy; BES = Basic Energy Sciences; DoD = Department of Defense; DOE = Department of Energy; EERE = Office of Energy Efficiency and Renewable Energy; FECM = Office of Fossil Energy and Carbon Management; NSF = National Science Foundation.

recommendations from the chapter). The table includes the relevant funding agencies or other actors; whether the need is for basic research, applied research, technology demonstration, or enabling technologies and processes for CO_2 utilization; the research theme(s) that the research need falls into; the relevant research area and product class covered by the research need; whether the relevant product(s) are long- or short-lived; and the source of the research need (chapter section, finding, or recommendation). The committee's full research agenda can be found in Chapter 11.

7.4 REFERENCES

Al-Shankiti, I., B.D. Ehrhart, and A.W. Weimer. 2017. "Isothermal Redox for H_2O and CO_2 Splitting—A Review and Perspective." *Solar Energy* 156(November):21–29. https://doi.org/10.1016/j.solener.2017.05.028.

Alvarez-Guerra, M., J. Albo, E. Alvarez-Guerra, and A. Irabien. 2015. "Ionic Liquids in the Electrochemical Valorisation of CO_2." *Energy and Environmental Science* 8(9):2574–2599. https://doi.org/10.1039/C5EE01486G.

Ampelli, C., S. Perathoner, and G. Centi. 2015. "CO_2 Utilization: An Enabling Element to Move to a Resource- and Energy-Efficient Chemical and Fuel Production." *Philosophical Transactions of the Royal Society A: Mathematical, Physical and Engineering Sciences* 373(2037):20140177. https://doi.org/10.1098/rsta.2014.0177.

Appaturi, J.N., R.J. Ramalingam, M.K. Gnanamani, G. Periyasami, P. Arunachalam, R. Adnan, F. Adam, M.D. Wasmiah, and H.A. Al-Lohedan. 2021. "Review on Carbon Dioxide Utilization for Cycloaddition of Epoxides by Ionic Liquid-Modified Hybrid Catalysts: Effect of Influential Parameters and Mechanisms Insight." *Catalysts* 11(1). https://doi.org/10.3390/catal11010004.

Appel, A.M., and J.Y. Yang. 2024. "Maximum and Comparative Efficiency Calculations for Integrated Capture and Electrochemical Conversion of CO_2." *ACS Energy Letters* 9(2):768–770. https://doi.org/10.1021/acsenergylett.3c02489.

Ardo, S., D. Fernandez Rivas, M.A. Modestino, V. Schulze Greiving, F.F. Abdi, E. Alarcon Llado, V. Artero, et al. 2018. "Pathways to Electrochemical Solar-Hydrogen Technologies." *Energy and Environmental Science* 11(10):2768–2783. https://doi.org/10.1039/C7EE03639F.

Aresta, M., A. Dibenedetto, and A. Angelini. 2014. "Catalysis for the Valorization of Exhaust Carbon: From CO_2 to Chemicals, Materials, and Fuels. Technological Use of CO_2." *Chemical Reviews* 114(3):1709–1742. https://doi.org/10.1021/cr4002758.

Asahi Kasei. n.d. "Non-Phosgene Process for Producing Polycarbonate." *AsahiKASEI*. https://www.asahi-kasei.com/r_and_d/innovation/#anc-10.

Ashford, B., and X. Tu. 2017. "Non-Thermal Plasma Technology for the Conversion of CO_2." *CO_2 Capture and Chemistry 2017* 3(February 1):45–49. https://doi.org/10.1016/j.cogsc.2016.12.001.

Azara, A., E.-H. Benyoussef, F. Mohellebi, M. Chamoumi, F. Gitzhofer, and N. Abatzoglou. 2019. "Catalytic Dry Reforming and Cracking of Ethylene for Carbon Nanofilaments and Hydrogen Production Using a Catalyst Derived from a Mining Residue." *Catalysts* 9(12):1069. https://doi.org/10.3390/catal9121069.

Azcarate, I., C. Costentin, M. Robert, and J.-M. Savéant. 2016. "Through-Space Charge Interaction Substituent Effects in Molecular Catalysis Leading to the Design of the Most Efficient Catalyst of CO_2-to-CO Electrochemical Conversion." *Journal of the American Chemical Society* 138(51):16639–16644. https://doi.org/10.1021/jacs.6b07014.

Banerjee, S., X. Han, and V. Sara Thoi. 2019. "Modulating the Electrode–Electrolyte Interface with Cationic Surfactants in Carbon Dioxide Reduction." *ACS Catalysis* 9(6):5631–5637. https://doi.org/10.1021/acscatal.9b00449.

Bansode, A., and A. Urakawa. 2014. "Towards Full One-Pass Conversion of Carbon Dioxide to Methanol and Methanol-Derived Products." *Journal of Catalysis* 309(January 1):66–70. https://doi.org/10.1016/j.jcat.2013.09.005.

Barlow, J.M., and J.Y. Yang. 2019. "Thermodynamic Considerations for Optimizing Selective CO_2 Reduction by Molecular Catalysts." *ACS Central Science* 5(4):580–588. https://doi.org/10.1021/acscentsci.9b00095.

Barton Cole, E., P.S. Lakkaraju, D.M. Rampulla, A.J. Morris, E. Abelev, and A.B. Bocarsly. 2010. "Using a One-Electron Shuttle for the Multielectron Reduction of CO_2 to Methanol: Kinetic, Mechanistic, and Structural Insights." *Journal of the American Chemical Society* 132(33):11539–11551. https://doi.org/10.1021/ja1023496.

Behr, A., and K. Nowakowski. 2014. "Chapter Seven—Catalytic Hydrogenation of Carbon Dioxide to Formic Acid." Pp. 223–258 in *Advances in Inorganic Chemistry* 66, M. Aresta and R. van Eldik, eds. Academic Press. https://doi.org/10.1016/B978-0-12-420221-4.00007-X.

Bernatis, P.R., A. Miedaner, R.C. Haltiwanger, and D.L. DuBois. 1994. "Exclusion of Six-Coordinate Intermediates in the Electrochemical Reduction of CO_2 Catalyzed by $[Pd(Triphosphine)(CH_3CN)](BF_4)_2$ Complexes." *Organometallics* 13(12):4835–4843. https://doi.org/10.1021/om00024a029.

Bernoud, E., A. Company, and X. Ribas. 2017. "Direct Use of CO_2 for O-Arylcarbamate Synthesis via Mild Cu(II)-Catalyzed Aerobic C-H Functionalization in Pincer-Like Macrocyclic Systems." *Organometallic Chemistry of Pincer Complexes* 845(September 15):44–48. https://doi.org/10.1016/j.jorganchem.2017.02.004.

Bernskoetter, W.H., and N. Hazari. 2017. "Reversible Hydrogenation of Carbon Dioxide to Formic Acid and Methanol: Lewis Acid Enhancement of Base Metal Catalysts." *Accounts of Chemical Research* 50(4):1049–1058. https://doi.org/10.1021/acs.accounts.7b00039.

Biswas, A.N., L.R. Winter, B. Loenders, Z. Xie, A. Bogaerts and J.G. Chen. 2022a. "Oxygenate Production from Plasma-Activated Reaction of CO_2 and Ethane." *ACS Energy Letters* 7:236–241. https://pubs.acs.org/doi/10.1021/acsenergylett.1c02355.

Biswas, A.N., Z. Xie, R. Xia, S. Overa, F. Jiao, J.G. Chen. 2022b. "Tandem Electrocatalytic–Thermocatalytic Reaction Scheme for CO_2 Conversion to C3 Oxygenates." *ACS Energy Letters* 7(9):2904–2910. https://doi.org/10.1021/acsenergylett.2c01454.

Biswas, A.N., L.R. Winter, Z. Xie, and J.G. Chen. 2023. "Utilizing CO_2 as a Reactant for C3 Oxygenate Production via Tandem Reactions." *JACS Au* 3(2):293–305. https://doi.org/10.1021/jacsau.2c00533.

Bogaerts, A., C. De Bie, R. Snoeckx, and T. Kozák. 2017. "Plasma Based CO_2 and CH_4 Conversion: A Modeling Perspective." *Plasma Processes and Polymers* 14(6):1600070. https://doi.org/10.1002/ppap.201600070.

Boutin, E., and M. Robert. 2021. "Molecular Electrochemical Reduction of CO_2 Beyond Two Electrons." *Trends in Chemistry* 3(5):359–372. https://doi.org/10.1016/j.trechm.2021.02.003.

Boutin, E., M. Wang, J.C. Lin, M. Mesnage, D. Mendoza, B. Lassalle-Kaiser, C. Hahn, T.F. Jaramillo, and M. Robert. 2019. "Aqueous Electrochemical Reduction of Carbon Dioxide and Carbon Monoxide into Methanol with Cobalt Phthalocyanine." *Angewandte Chemie International Edition* 58(45):16172–16176. https://doi.org/10.1002/anie.201909257.

Bui, J.C., E.W. Lees, D.H. Marin, T.N. Stovall, L. Chen, A. Kusoglu, A.C. Nielander, et al. 2024. "Multi-Scale Physics of Bipolar Membranes in Electrochemical Processes." *Nature Chemical Engineering* 1(1):45–60. https://doi.org/10.1038/s44286-023-00009-x.

Bukur, D.B., B. Todic, and N. Elbashir. 2016. "Role of Water-Gas-Shift Reaction in Fischer–Tropsch Synthesis on Iron Catalysts: A Review." *Catalysis Today* 275(October):66–75. https://doi.org/10.1016/j.cattod.2015.11.005.

Bustamante, F., R.M. Enick, A.V. Cugini, R.P. Killmeyer, B.H. Howard, K.S. Rothenberger, M.V. Ciocco, B.D. Morreale, S. Chattopadhyay, and S. Shi. 2004. "High-Temperature Kinetics of the Homogeneous Reverse Water-Gas Shift Reaction." *AIChE Journal* 50(5):1028–1041. https://doi.org/10.1002/aic.10099.

Buysch, H.-J. 2011. "Carbonic Esters." In *Ullmann's Encyclopedia of Industrial Chemistry*, 7th edition. https://doi.org/10.1002/14356007.a05_197.

Cai, X., and Y.H. Hu. 2019. "Advances in Catalytic Conversion of Methane and Carbon Dioxide to Highly Valuable Products." *Energy Science and Engineering* 7(1):4–29. https://doi.org/10.1002/ese3.278.

Cao, G., R.M. Handler, W.L. Luyben, Y. Xiao, C.-H. Chen, and J. Baltrusaitis. 2022. "CO_2 Conversion to Syngas via Electrification of Endothermal Reactors: Process Design and Environmental Impact Analysis." *Energy Conversion and Management* 265(August 1):115763. https://doi.org/10.1016/j.enconman.2022.115763.

Carbon. 2023. "Climate Change Solution." *Carbon*. https://carboncorp.org/climate-change-solution.html.

Carbon Recycling International. n.d. "George Olah Renewable Methanol Plant: First Production of Fuel from CO_2 at Industrial Scale." *Carbon Recycling International*. https://www.carbonrecycling.is/project-goplant.

Cauwenbergh, R., V. Goyal, R. Maiti, K. Natte, and S. Das. 2022. "Challenges and Recent Advancements in the Transformation of CO2 into Carboxylic Acids: Straightforward Assembly with Homogeneous 3d Metals." *Chemical Society Reviews* 51(22):9371–9423. https://doi.org/10.1039/D1CS00921D.

Centi, G., and S. Perathoner. 2023. "Catalysis for an Electrified Chemical Production." *Catalysis Today* 423(November 1): 113935. https://doi.org/10.1016/j.cattod.2022.10.017.

Challiwala, M.S., H.A. Choudhury, D. Wang, M.M. El-Halwagi, E. Weitz, and N.O. Elbashir. 2021. "A Novel CO_2 Utilization Technology for the Synergistic Co-Production of Multi-Walled Carbon Nanotubes and Syngas." *Scientific Reports* 11(1):1417. https://doi.org/10.1038/s41598-021-80986-2.

Chang, F., G. Zhan, Z. Wu, Y. Duan, S. Shi, S. Zeng, X. Zhang, and S. Zhang. 2021. "Technoeconomic Analysis and Process Design for CO_2 Electroreduction to CO in Ionic Liquid Electrolyte." *ACS Sustainable Chemistry and Engineering* 9(27):9045–9052. https://doi.org/10.1021/acssuschemeng.1c02065.

Chang, X., T. Wang, and J. Gong. 2016. "CO_2 Photo-Reduction: Insights into CO_2 Activation and Reaction on Surfaces of Photocatalysts." *Energy and Environmental Science* 9(7):2177–2196. https://doi.org/10.1039/C6EE00383D.

Chatterjee, D., N. Jaiswal, and P. Banerjee. 2014. "Electrochemical Conversion of Bicarbonate to Formate Mediated by the Complex Ru^{III}(edta) ($edta^{4-}$ = ethylenediaminetetraacetate)." *European Journal of Inorganic Chemistry* 2014(34):5856–5859. https://doi.org/10.1002/ejic.201402831.

Chen, G., G.I.N. Waterhouse, R. Shi, J. Zhao, Z. Li, L.-Z. Wu, C.-H. Tung, and T. Zhang. 2019. "From Solar Energy to Fuels: Recent Advances in Light-Driven C1 Chemistry." *Angewandte Chemie International Edition* 58(49):17528–17551. https://doi.org/10.1002/anie.201814313.

Chen, X., Y. Chen, C. Song, P. Ji, N. Wang, W. Wang, and L. Cui. 2020. "Recent Advances in Supported Metal Catalysts and Oxide Catalysts for the Reverse Water-Gas Shift Reaction." *Frontiers in Chemistry* 8(August):709. https://doi.org/10.3389/fchem.2020.00709.

Chen, X., X. Su, H.-Y. Su, X. Liu, S. Miao, Y. Zhao, K. Sun, Y. Huang, and T. Zhang. 2017. "Theoretical Insights and the Corresponding Construction of Supported Metal Catalysts for Highly Selective CO_2 to CO Conversion." *ACS Catalysis* 7(7):4613–4620. https://doi.org/10.1021/acscatal.7b00903.

Cho, W., H. Yu, and Y. Mo. 2017. "CO_2 Conversion to Chemicals and Fuel for Carbon Utilization." Chapter 9 in *Recent Advances in Carbon Capture and Storage*, Y. Yun, ed. Rijeka: IntechOpen. https://doi.org/10.5772/67316.

Coates, G.W., and D.R. Moore. 2004. "Discrete Metal-Based Catalysts for the Copolymerization of CO_2 and Epoxides: Discovery, Reactivity, Optimization, and Mechanism." *Angewandte Chemie International Edition* 43(48):6618–6639. https://doi.org/10.1002/anie.200460442.

Cohen, K.Y., R. Evans, S. Dulovic, and A.B. Bocarsly. 2022. "Using Light and Electrons to Bend Carbon Dioxide: Developing and Understanding Catalysts for CO_2 Conversion to Fuels and Feedstocks." *Accounts of Chemical Research* 55(7):944–954. https://doi.org/10.1021/acs.accounts.1c00643.

Costentin, C., S. Drouet, M. Robert, and J.-M. Savéant. 2012. "A Local Proton Source Enhances CO_2 Electroreduction to CO by a Molecular Fe Catalyst." *Science* 338(6103):90–94. https://doi.org/10.1126/science.1224581.

Costentin, C., J.-M. Savéant, and C. Tard. 2018. "Catalysis of CO_2 Electrochemical Reduction by Protonated Pyridine and Similar Molecules. Useful Lessons from a Methodological Misadventure." *ACS Energy Letters* 3(3):695–703. https://doi.org/10.1021/acsenergylett.8b00008.

Covert, L.W., and H. Adkins. 1932. "Nickel by the Raney Process as a Catalyst of Hydrogenation." *Journal of the American Chemical Society* 54(10):4116–4117. https://doi.org/10.1021/ja01349a510.

Dabral, S., and T. Schaub. 2019. "The Use of Carbon Dioxide (CO_2) as a Building Block in Organic Synthesis from an Industrial Perspective." *Advanced Synthesis and Catalysis* 361(2):223–246. https://doi.org/10.1002/adsc.201801215.

Dalle, K.E., J. Warnan, J.J. Leung, B. Reuillard, I.S. Karmel, and E. Reisner. 2019. "Electro- and Solar-Driven Fuel Synthesis with First Row Transition Metal Complexes." *Chemical Reviews* 119(4):2752–2875. https://doi.org/10.1021/acs.chemrev.8b00392.

Dang, S., H. Yang, P. Gao, H. Wang, X. Li, W. Wei, and Y. Sun. 2019. "A Review of Research Progress on Heterogeneous Catalysts for Methanol Synthesis from Carbon Dioxide Hydrogenation." *SI:18ncc_Energy* 330(June 15):61–75. https://doi.org/10.1016/j.cattod.2018.04.021.

Darensbourg, D.J. 2007. "Making Plastics from Carbon Dioxide: Salen Metal Complexes as Catalysts for the Production of Polycarbonates from Epoxides and CO_2." *Chemical Reviews* 107(6):2388–2410. https://doi.org/10.1021/cr068363q.

Darensbourg, D.J., and M.W. Holtcamp. 1996. "Catalysts for the Reactions of Epoxides and Carbon Dioxide." *Coordination Chemistry Reviews* 153(August 1):155–174. https://doi.org/10.1016/0010-8545(95)01232-X.

Das, S., D. Nugegoda, W. Yao, F. Qu, M.T. Figgins, R.W. Lamb, C.E. Webster, J.H. Delcamp, and E.T. Papish. 2022. "Sensitized and Self-Sensitized Photocatalytic Carbon Dioxide Reduction Under Visible Light with Ruthenium Catalysts Shows Enhancements with More Conjugated Pincer Ligands." *European Journal of Inorganic Chemistry* 2022(8):e202101016. https://doi.org/10.1002/ejic.202101016.

Davies, J., J.R. Lyonnet, D.P. Zimin, and R. Martin. 2021. "The Road to Industrialization of Fine Chemical Carboxylation Reactions." *Chem* 7(11):2927–2942. https://doi.org/10.1016/j.chempr.2021.10.016.

Daza, Y.A., and J.N. Kuhn. 2016. "CO_2 Conversion by Reverse Water Gas Shift Catalysis: Comparison of Catalysts, Mechanisms and Their Consequences for CO_2 Conversion to Liquid Fuels." *RSC Advances* 6(55):49675–49691. https://doi.org/10.1039/C6RA05414E.

De Klerk, A. 2014. "Consider Technology Implications for Small-Scale Fischer-Tropsch GTL." *Gas Processing and LNG*. http://gasprocessingnews.com/articles/2014/08/consider-technology-implications-for-small-scale-fischer-tropsch-gtl.

Dieterich, V., A. Buttler, A. Hanel, H. Spliethoff, and S. Fendt. 2020. "Power-to-Liquid via Synthesis of Methanol, DME or Fischer–Tropsch-Fuels: A Review." *Energy and Environmental Science* 13(10):3207–3252. https://doi.org/10.1039/D0EE01187H.

Dimitriou, I., P. García-Gutiérrez, R.H. Elder, R.M. Cuéllar-Franca, A. Azapagic, and R.W.K. Allen. 2015. "Carbon Dioxide Utilisation for Production of Transport Fuels: Process and Economic Analysis." *Energy and Environmental Science* 8(6):1775–1789. https://doi.org/10.1039/C4EE04117H.

Dineen, J. 2023. "Jet Fuel Made with Captured CO_2 and Clean Electricity Set for Take-Off." *NewScientist* September 28. https://www.newscientist.com/article/2394108-jet-fuel-made-with-captured-co2-and-clean-electricity-set-for-take-off.

Ding, M., and H.-L. Jiang. 2018. "Incorporation of Imidazolium-Based Poly(Ionic Liquid)s into a Metal–Organic Framework for CO_2 Capture and Conversion." *ACS Catalysis* 8(4):3194–3201. https://doi.org/10.1021/acscatal.7b03404.

Dodge, H.M., B.S. Natinsky, B.J. Jolly, H. Zhang, Y. Mu, S.M. Chapp, T.V. Tran, et al. 2023. "Polyketones from Carbon Dioxide and Ethylene by Integrating Electrochemical and Organometallic Catalysis." *ACS Catalysis* 13(7):4053–4059. https://doi.org/10.1021/acscatal.3c00769.

DOE (Department of Energy). 2022. "Industrial Decarbonization Roadmap." DOE/EE-2635. Washington, DC: Department of Energy. https://www.energy.gov/sites/default/files/2022-09/Industrial%20Decarbonization%20Roadmap.pdf.

Dong, Q., Y. Yao, S. Cheng, K. Alexopoulos, J. Gao, S. Srinivas, Y. Wang, et al. 2022. "Programmable Heating and Quenching for Efficient Thermochemical Synthesis." *Nature* 605(7910):470–476. https://doi.org/10.1038/s41586-022-04568-6.

Dorner, R.W., D.R. Hardy, F.W. Williams, and H.D. Willauer. 2010. "Heterogeneous Catalytic CO_2 Conversion to Value-Added Hydrocarbons." *Energy and Environmental Science* 3(7):884–890. https://doi.org/10.1039/C001514H.

Dzuryk, S., and E. Rezaei. 2022. "Dimensionless Analysis of Reverse Water Gas Shift Packed-Bed Membrane Reactors." *Chemical Engineering Science* 250(March 15):117377. https://doi.org/10.1016/j.ces.2021.117377.

Edwards, J.P., T. Alerte, C.P. O'Brien, C.M. Gabardo, S. Liu, J. Wicks, A. Gaona, et al. 2023. "Pilot-Scale CO_2 Electrolysis Enables a Semi-Empirical Electrolyzer Model." *ACS Energy Letters* 8(6):2576–2584. https://doi.org/10.1021/acsenergylett.3c00620.

EIA (U.S. Energy Information Administration). 2023. "Annual Energy Outlook 2023 Table 19. Energy-Related Carbon Dioxide Emissions by End Use." https://www.eia.gov/outlooks/aeo/data/browser/#/?id=22-AEO2023&cases=ref2023&sourcekey=0.

Ellis, P.R., M.J. Hayes, N. Macleod, S.J. Schuyten, C.L. Tway, and C.M. Zalitis. 2023. "Chapter 10—Carbon Conversion: Opportunities in Chemical Productions." Pp. 479–524 in *Surface Process, Transportation, and Storage*, Q. Wang, ed. Vol. 4. Gulf Professional Publishing. https://doi.org/10.1016/B978-0-12-823891-2.00006-5.

Elsernagawy, O.Y.H., A. Hoadley, J. Patel, T. Bhatelia, S. Lim, N. Haque, and C. Li. 2020. "Thermo-Economic Analysis of Reverse Water-Gas Shift Process with Different Temperatures for Green Methanol Production as a Hydrogen Carrier." *Journal of CO_2 Utilization* 41(October):101280. https://doi.org/10.1016/j.jcou.2020.101280.

EPA (U.S. Environmental Protection Agency). 2018. "Biofuels and the Environment: Second Triennial Report to Congress." EPA/600/R-18/195. Washington, DC: U.S. Environmental Protection Agency. https://cfpub.epa.gov/si/si_public_record_Report.cfm?Lab=IO&dirEntryId=341491.

Eryazici, I., N. Ramesh, and C. Villa. 2021. "Electrification of the Chemical Industry—Materials Innovations for a Lower Carbon Future." *MRS Bulletin* 46(12):1197–1204. https://doi.org/10.1557/s43577-021-00243-9.

Fang, W., W. Guo, R. Lu, Y. Yan, X. Liu, D. Wu, F.M. Li, et al. 2024. "Durable CO_2 Conversion in the Proton-Exchange Membrane System." *Nature* 626(7997):86–91. https://doi.org/10.1038/s41586-023-06917-5.

Feroci, M., M. Orsini, G. Sotgiu, L. Rossi, and A. Inesi. 2005. "Electrochemically Promoted C–N Bond Formation from Acetylenic Amines and CO_2. Synthesis of 5-Methylene-1,3-Oxazolidin-2-Ones." *The Journal of Organic Chemistry* 70(19):7795–7798. https://doi.org/10.1021/jo0511804.

Filonenko, G.A., R. van Putten, E.J.M. Hensen, E.A. Pidko. 2018. "Catalytic (de)hydrogenation Promoted by Non-precious Metals—Co, Fe and Mn: Recent Advances in an Emerging Field." *Chemical Society Reviews* 47:1459–1483. https://doi.org/10.1039/C7CS00334J.

Fors, S.A., and C.A. Malapit. 2023. "Homogeneous Catalysis for the Conversion of CO_2, CO, CH_3OH, and CH_4 to C2+ Chemicals via C–C Bond Formation." *ACS Catalysis* 13(7):4231–4249. https://doi.org/10.1021/acscatal.2c05517.

Francke, R., B. Schille, and M. Roemelt. 2018. "Homogeneously Catalyzed Electroreduction of Carbon Dioxide—Methods, Mechanisms, and Catalysts." *Chemical Reviews* 118(9):4631–4701. https://doi.org/10.1021/acs.chemrev.7b00459.

Freire Ordóñez, D., T. Halfdanarson, C. Ganzer, N. Shah, N. Mac Dowell, and G. Guillén-Gosálbez. 2022. "Evaluation of the Potential Use of E-Fuels in the European Aviation Sector: A Comprehensive Economic and Environmental Assessment Including Externalities." *Sustainable Energy and Fuels* 6(20):4749–4764. https://doi.org/10.1039/D2SE00757F.

Freyman, M.C., Z. Huang, D. Ravikumar, E.B. Duoss, Y. Li, S.E. Baker, S.H. Pang, and J.A. Schaidle. 2023. "Reactive CO_2 Capture: A Path Forward for Process Integration in Carbon Management." *Joule* 7(4):631–651. https://doi.org/10.1016/j.joule.2023.03.013.

Fukuoka, S. 2012. "Non-Phosgene Polycarbonate from CO_2: Industrialization of Green Chemical Process." New York: Nova Science Publishers.

Fukuoka, S., I. Fukawa, T. Adachi, H. Fujita, N. Sugiyama, and T. Sawa. 2019. "Industrialization and Expansion of Green Sustainable Chemical Process: A Review of Non-Phosgene Polycarbonate from CO_2." *Organic Process Research and Development* 23(2):145–169. https://doi.org/10.1021/acs.oprd.8b00391.

Gambo, Y., S. Adamu, G. Tanimu, I.M. Abdullahi, R.A. Lucky, M.S. Ba-Shammakh, and M.M. Hossain. 2021. "CO_2-Mediated Oxidative Dehydrogenation of Light Alkanes to Olefins: Advances and Perspectives in Catalyst Design and Process Improvement." *Applied Catalysis A: General* 623(August 5):118273. https://doi.org/10.1016/j.apcata.2021.118273.

GAO (U.S. Government Accountability Office). 2022. "Decarbonization: Status, Challenges, and Policy Options for Carbon Capture, Utilization, and Storage." GAO-22-105274. Technology Assessment. Washington, DC: U.S. Government Accountability Office. https://www.gao.gov/assets/730/723198.pdf.

Gao, P., S. Li, X. Bu, S. Dang, Z. Liu, H. Wang, L. Zhong, et al. 2017. "Direct Conversion of CO_2 into Liquid Fuels with High Selectivity Over a Bifunctional Catalyst." *Nature Chemistry* 9(10):1019–1024. https://doi.org/10.1038/nchem.2794.

Gao, X., T. Atchimarungsri, Q. Ma, T.-S. Zhao, and N. Tsubaki. 2020. "Realizing Efficient Carbon Dioxide Hydrogenation to Liquid Hydrocarbons by Tandem Catalysis Design." *EnergyChem* 2(4):100038. https://doi.org/10.1016/j.enchem.2020.100038.

Garg, S., Z. Xie, and J.G. Chen. 2024. "Tandem Reactors and Reactions for CO_2 Conversion." *Nature Chemical Engineering* 1(2):139–148. https://doi.org/10.1038/s44286-023-00020-2.

Ghobadi, M.H., M. Firuzi, and E. Asghari-Kaljahi. 2016. "Relationships Between Geological Formations and Groundwater Chemistry and Their Effects on the Concrete Lining of Tunnels (Case Study: Tabriz Metro Line 2)." *Environmental Earth Sciences* 75(12):987. https://doi.org/10.1007/s12665-016-5785-0.

Gogate, M.R. 2019. "Methanol-to-Olefins Process Technology: Current Status and Future Prospects." *Petroleum Science and Technology* 37(5):559–565. https://doi.org/10.1080/10916466.2018.1555589.

Goguet, A., F. Meunier, J.P. Breen, R. Burch, M.I. Petch, and A. Faur Ghenciu. 2004. "Study of the Origin of the Deactivation of a Pt/CeO_2 Catalyst During Reverse Water Gas Shift (RWGS) Reaction." *Journal of Catalysis* 226(2):382–392. https://doi.org/10.1016/j.jcat.2004.06.011.

González-Castaño, M., B. Dorneanu, and H. Arellano-García. 2021. "The Reverse Water Gas Shift Reaction: A Process Systems Engineering Perspective." *Reaction Chemistry and Engineering* 6(6):954–976. https://doi.org/10.1039/D0RE00478B.

Govind Rajan, A., J.M.P. Martirez, and E.A. Carter. 2020. "Why Do We Use the Materials and Operating Conditions We Use for Heterogeneous (Photo)Electrochemical Water Splitting?" *ACS Catalysis* 10(19):11177–11234. https://doi.org/10.1021/acscatal.0c01862.

Grignard, B., S. Gennen, C. Jérôme, A.W. Kleij, and C. Detrembleur. 2019. "Advances in the Use of CO_2 as a Renewable Feedstock for the Synthesis of Polymers." *Chemical Society Reviews* 48(16):4466–4514. https://doi.org/10.1039/C9CS00047J.

Grills, D.C., and E. Fujita. 2010. "New Directions for the Photocatalytic Reduction of CO_2: Supramolecular, scCO_2 or Biphasic Ionic Liquid–scCO_2 Systems." *The Journal of Physical Chemistry Letters* 1(18):2709–2718. https://doi.org/10.1021/jz1010237.

Grills, D.C., Y. Matsubara, Y. Kuwahara, S.R. Golisz, D.A. Kurtz, and B.A. Mello. 2014. "Electrocatalytic CO_2 Reduction with a Homogeneous Catalyst in Ionic Liquid: High Catalytic Activity at Low Overpotential." *The Journal of Physical Chemistry Letters* 5(11):2033–2038. https://doi.org/10.1021/jz500759x.

Grim, R.G., J.R. Ferrell III, Z. Huang, L. Tao, and M.G. Resch. 2023. "The Feasibility of Direct CO_2 Conversion Technologies on Impacting Mid-Century Climate Goals." *Joule* 7(8):1684–1699. https://doi.org/10.1016/j.joule.2023.07.008.

Grimm, A., W.A. de Jong, and G.J. Kramer. 2020. "Renewable Hydrogen Production: A Techno-Economic Comparison of Photoelectrochemical Cells and Photovoltaic-Electrolysis." *International Journal of Hydrogen Energy* 45(43):22545–22555. https://doi.org/10.1016/j.ijhydene.2020.06.092.

Gui, M.M., W.P. Cathie Lee, L.K. Putri, X.Y. Kong, L.L. Tan, and S.-P. Chai. 2021. "Photo-Driven Reduction of Carbon Dioxide: A Sustainable Approach Towards Achieving Carbon Neutrality Goal." *Frontiers in Chemical Engineering* 3. https://www.frontiersin.org/articles/10.3389/fceng.2021.744911.

Han, B., W. Wei, L. Chang, P. Cheng, and Y.H. Hu. 2016. "Efficient Visible Light Photocatalytic CO_2 Reforming of CH_4." *ACS Catalysis* 6(2):494–497. https://doi.org/10.1021/acscatal.5b02653.

Hauch, A., R. Küngas, P. Blennow, A.B. Hansen, J.B. Hansen, B.V. Mathiesen, and M.B. Mogensen. 2020. "Recent Advances in Solid Oxide Cell Technology for Electrolysis." *Science* 370(6513):eaba6118. https://doi.org/10.1126/science.aba6118.

Hawecker, J., J.-M. Lehn, and R. Ziessel. 1986. "Photochemical and Electrochemical Reduction of Carbon Dioxide to Carbon Monoxide Mediated by (2,2'-Bipyridine)Tricarbonylchlororhenium(I) and Related Complexes as Homogeneous Catalysts." *Helvetica Chimica Acta* 69(8):1990–2012. https://doi.org/10.1002/hlca.19860690824.

He, Z., M. Cui, Q. Qian, J. Zhang, H. Liu, and B. Han. 2019. "Synthesis of Liquid Fuel via Direct Hydrogenation of CO_2." *Proceedings of the National Academy of Sciences* 116(26):12654–12659. https://doi.org/10.1073/pnas.1821231116.

Heldebrant, D.J., J. Kothandaraman, N. Mac Dowell, and L. Brickett. 2022. "Next Steps for Solvent-Based CO_2 Capture; Integration of Capture, Conversion, and Mineralisation." *Chemical Science* 13(22):6445–6456. https://doi.org/10.1039/D2SC00220E.

Hietala, J., A. Vuori, P. Johnsson, I. Pollari, W. Reutemann, and H. Kieczka. 2016. "Formic Acid." Pp. 1–22 in *Ullmann's Encyclopedia of Industrial Chemistry*. https://doi.org/10.1002/14356007.a12_013.pub3.

Ho, P.-Y., S.-C. Cheng, F. Yu, Y.-Y. Yeung, W.-X. Ni, C.-C. Ko, C.-F. Leung, T.-C. Lau, and M. Robert. 2023. "Light-Driven Reduction of CO_2 to CO in Water with a Cobalt Molecular Catalyst and an Organic Sensitizer." *ACS Catalysis* 13(9):5979–5985. https://doi.org/10.1021/acscatal.3c00036.

Hu, C., X. Chen, J. Low, Y.-W. Yang, H. Li, D. Wu, S. Chen, et al. 2023. "Near-Infrared-Featured Broadband CO_2 Reduction with Water to Hydrocarbons by Surface Plasmon." *Nature Communications* 14(1):221. https://doi.org/10.1038/s41467-023-35860-2.

Huang, J., J.C. Worch, A.P. Dove, and O. Coulembier. 2020. "Update and Challenges in Carbon Dioxide-Based Polycarbonate Synthesis." *ChemSusChem* 13(3):469–487. https://doi.org/10.1002/cssc.201902719.

Hunt, J., A. Ferrari, A. Lita, M. Crosswhite, B. Ashley, and A.E. Stiegman. 2013. "Microwave-Specific Enhancement of the Carbon–Carbon Dioxide (Boudouard) Reaction." *The Journal of Physical Chemistry C* 117(51):26871–26880. https://doi.org/10.1021/jp4076965.

Inoue, S., H. Koinuma, and T. Tsuruta. 1969a. "Copolymerization of Carbon Dioxide and Epoxide." *Journal of Polymer Science Part B: Polymer Letters* 7(4):287–292. https://doi.org/10.1002/pol.1969.110070408.

Inoue, S., H. Koinuma, and T. Tsuruta. 1969b. "Copolymerization of Carbon Dioxide and Epoxide with Organometallic Compounds." *Die Makromolekulare Chemie* 130(1):210–220. https://doi.org/10.1002/macp.1969.021300112.

Iota, V., C.S. Yoo, and H. Cynn. 1999. "Quartzlike Carbon Dioxide: An Optically Nonlinear Extended Solid at High Pressures and Temperatures." *Science* 283(5407):1510–1513. https://doi.org/10.1126/science.283.5407.1510.

Jang, W.-J., J.-O. Shim, H.-M. Kim, S.-Y. Yoo, and H.-S. Roh. 2019. "A Review on Dry Reforming of Methane in Aspect of Catalytic Properties." *SI: Green Catalysis* 324(March 1):15–26. https://doi.org/10.1016/j.cattod.2018.07.032.

Jens, C.M., L. Müller, K. Leonhard, and A. Bardow. 2019. "To Integrate or Not to Integrate—Techno-Economic and Life Cycle Assessment of CO_2 Capture and Conversion to Methyl Formate Using Methanol." *ACS Sustainable Chemistry and Engineering* 7(14):12270–12280. https://doi.org/10.1021/acssuschemeng.9b01603.

Jiang, X., X. Nie, X. Guo, C. Song, and J.G. Chen. 2020. "Recent Advances in Carbon Dioxide Hydrogenation to Methanol via Heterogeneous Catalysis." *Chemical Reviews* 120(15):7984–8034. https://doi.org/10.1021/acs.chemrev.9b00723.

Jiao, F., J. Li, X. Pan, J. Xiao, H. Li, H. Ma, M. Wei, et al. 2016. "Selective Conversion of Syngas to Light Olefins." *Science* 351(6277):1065–1068. https://www.science.org/doi/10.1126/science.aaf1835.

Joo, O.-S., K.-D. Jung, I. Moon, A.Y. Rozovskii, G.I. Lin, S.-H. Han, and S.-J. Uhm. 1999. "Carbon Dioxide Hydrogenation to Form Methanol via a Reverse-Water-Gas-Shift Reaction (the CAMERE Process)." *Industrial and Engineering Chemistry Research* 38(5):1808–1812. https://doi.org/10.1021/ie9806848.

Jouny, M., G.S. Hutchings, and F. Jiao. 2019. "Carbon Monoxide Electroreduction as an Emerging Platform for Carbon Utilization." *Nature Catalysis* 2(12):1062–1070. https://www.nature.com/articles/s41929-019-0388-2.

Kalamaras, E., M.M. Maroto-Valer, M. Shao, J. Xuan, and H. Wang. 2018. "Solar Carbon Fuel via Photoelectrochemistry." *SI: Decarbonising Fossil Fuel* 317(November 1):56–75. https://doi.org/10.1016/j.cattod.2018.02.045.

Kalck, P., C. Le Berre, and P. Serp. 2020. "Recent Advances in the Methanol Carbonylation Reaction into Acetic Acid." *Coordination Chemistry Reviews* 402(January):213078. https://doi.org/10.1016/j.ccr.2019.213078.

Kar, S., A. Goeppert, V. Galvan, R. Chowdhury, J. Olah, and G.K. Surya Prakash. 2018a. "A Carbon-Neutral CO_2 Capture, Conversion, and Utilization Cycle with Low-Temperature Regeneration of Sodium Hydroxide." *Journal of the American Chemical Society* 140(49):16873–16876. https://doi.org/10.1021/jacs.8b09325.

Kar, S., R. Sen, A. Goeppert, and G.K. Surya Prakash. 2018b. "Integrative CO_2 Capture and Hydrogenation to Methanol with Reusable Catalyst and Amine: Toward a Carbon Neutral Methanol Economy." *Journal of the American Chemical Society* 140(5):1580–1583. https://doi.org/10.1021/jacs.7b12183.

Kar, S., A. Goeppert, and G.K. Surya Prakash. 2019a. "Combined CO_2 Capture and Hydrogenation to Methanol: Amine Immobilization Enables Easy Recycling of Active Elements." *ChemSusChem* 12(13):3172–3177. https://doi.org/10.1002/cssc.201900324.

Kar, S., A. Goeppert, and G.K. Surya Prakash. 2019b. "Integrated CO_2 Capture and Conversion to Formate and Methanol: Connecting Two Threads." *Accounts of Chemical Research* 52(10):2892–2903. https://doi.org/10.1021/acs.accounts.9b00324.

Kember, M.R., A. Buchard, and C.K. Williams. 2011. "Catalysts for CO_2/Epoxide Copolymerisation." *Chemical Communications* 47(1):141–163. https://doi.org/10.1039/C0CC02207A.

Kempler, P.A., and A.C. Nielander. 2023. "Reliable Reporting of Faradaic Efficiencies for Electrocatalysis Research." *Nature Communications* 14(1):1158. https://doi.org/10.1038/s41467-023-36880-8.

Khan, A.A., and M. Tahir. 2019. "Recent Advancements in Engineering Approach Towards Design of Photo-Reactors for Selective Photocatalytic CO_2 Reduction to Renewable Fuels." *Journal of CO_2 Utilization* 29(January 1):205–239. https://doi.org/10.1016/j.jcou.2018.12.008.

Kim, S.M., P.M. Abdala, M. Broda, D. Hosseini, C. Copéret, and C. Müller. 2018. "Integrated CO_2 Capture and Conversion as an Efficient Process for Fuels from Greenhouse Gases." *ACS Catalysis* 8(4):2815–2823. https://doi.org/10.1021/acscatal.7b03063.

Klug, C.M., A.J.P. Cardenas, R.M. Bullock, M. O'Hagan, and E.S. Wiedner. 2018. "Reversing the Tradeoff Between Rate and Overpotential in Molecular Electrocatalysts for H_2 Production." *ACS Catalysis* 8(4):3286–3296. https://doi.org/10.1021/acscatal.7b04379.

Kothandaraman, J., and D.J. Heldebrant. 2020. "Catalytic Coproduction of Methanol and Glycol in One Pot from Epoxide, CO_2, and H_2." *RSC Advances* 10(69):2557–42563. https://doi.org/10.1039/D0RA09459E.

Kothandaraman, J., A. Goeppert, M. Czaun, G.A. Olah, and G.K. Surya Prakash. 2016a. "CO_2 Capture by Amines in Aqueous Media and Its Subsequent Conversion to Formate with Reusable Ruthenium and Iron Catalysts." *Green Chemistry* 18(21):5831–5838. https://doi.org/10.1039/C6GC01165A.

Kothandaraman, J., A. Goeppert, M. Czaun, G.A. Olah, and G.K. Surya Prakash. 2016b. "Conversion of CO_2 from Air into Methanol Using a Polyamine and a Homogeneous Ruthenium Catalyst." *Journal of the American Chemical Society* 138(3):778–781. https://doi.org/10.1021/jacs.5b12354.

Krishnamurty, K.V., G. McLeod Harris, and V.S. Sastri. 1970. "Chemistry of the Metal Carbonato Complexes." *Chemical Reviews* 70(2):171–197. https://doi.org/10.1021/cr60264a001.

Kumar, B., M. Llorente, J. Froehlich, T. Dang, A. Sathrum, and C.P. Kubiak. 2012. "Photochemical and Photoelectrochemical Reduction of CO_2." *Annual Review of Physical Chemistry* 63(May):541–569. https://doi.org/10.1146/annurev-physchem-032511-143759.

Küngas, R. 2020. "Review—Electrochemical CO_2 Reduction for CO Production: Comparison of Low- and High-Temperature Electrolysis Technologies." *Journal of the Electrochemical Society* 167(4):044508. https://doi.org/10.1149/1945-7111/ab7099.

La Plante, E.C., D.A. Simonetti, J. Wang, A. Al-Turki, X. Chen, D. Jassby, and G.N. Sant. 2021. "Saline Water-Based Mineralization Pathway for Gigatonne-Scale CO_2 Management." *ACS Sustainable Chemistry and Engineering* 9(3):1073–1089. https://doi.org/10.1021/acssuschemeng.0c08561.

La Plante, E.C., X. Chen, S. Bustillos, A. Bouissonnie, T. Traynor, D. Jassby, L. Corsini, D.A. Simonetti, and G.N. Sant. 2023. "Electrolytic Seawater Mineralization and the Mass Balances That Demonstrate Carbon Dioxide Removal." *ACS ES&T Engineering* 3(7):955–968. https://doi.org/10.1021/acsestengg.3c00004.

Lam, E., K. Larmier, P. Wolf, S. Tada, O.V. Safonova, and C. Copéret. 2018. "Isolated Zr Surface Sites on Silica Promote Hydrogenation of CO_2 to CH_3OH in Supported Cu Catalysts." *Journal of the American Chemical Society* 140(33):10530–10535. https://doi.org/10.1021/jacs.8b05595.

Lamprecht, D. 2007. "Fischer–Tropsch Fuel for Use by the U.S. Military as Battlefield-Use Fuel of the Future." *Energy and Fuels* 21(3):1448–1453. https://doi.org/10.1021/ef060607m.

Langanke, J., A. Wolf, and M. Peters. 2015. "Chapter 5—Polymers from CO_2—An Industrial Perspective." Pp. 59–71 in *Carbon Dioxide Utilisation*, P. Styring, E.A. Quadrelli, and K. Armstrong, eds. Amsterdam: Elsevier. https://doi.org/10.1016/B978-0-444-62746-9.00005-0.

Lazard. 2024. "Lazard's Levelized Cost of Energy+." Lazard and Roland Berger. https://www.lazard.com/media/xemfey0k/lazards-lcoeplus-june-2024-_vf.pdf.

le Saché, E., and T.R. Reina. 2022. "Analysis of Dry Reforming as Direct Route for Gas Phase CO_2 Conversion. The Past, the Present and Future of Catalytic DRM Technologies." *Progress in Energy and Combustion Science* 89(March 1):100970. https://doi.org/10.1016/j.pecs.2021.100970.

Lee, C.W., K.D. Yang, D.-H. Nam, J.H. Jang, N.H. Cho, S.W. Im, and K.T. Nam. 2018. "Defining a Materials Database for the Design of Copper Binary Alloy Catalysts for Electrochemical CO_2 Conversion." *Advanced Materials* 30(42):1704717. https://doi.org/10.1002/adma.201704717.

Lee, M.G., X.-Y. Li, A. Ozden, J. Wicks, P. Ou, Y. Li, R. Dorakhan, et al. 2023. "Selective Synthesis of Butane from Carbon Monoxide Using Cascade Electrolysis and Thermocatalysis at Ambient Conditions." *Nature Catalysis* 6(4):310–318. https://doi.org/10.1038/s41929-023-00937-0.

Lees, E.W., A. Liu, J.C. Bui, S. Ren, A.Z. Weber, and C.P. Berlinguette. 2022. "Electrolytic Methane Production from Reactive Carbon Solutions." *ACS Energy Letters* 7(5):1712–1718. https://doi.org/10.1021/acsenergylett.2c00283.

Liao, P., and E.A. Carter. 2013. "New Concepts and Modeling Strategies to Design and Evaluate Photo-Electro-Catalysts Based on Transition Metal Oxides." *Chemical Society Reviews* 42(6):2401–2422. https://doi.org/10.1039/C2CS35267B.

Lichterman, M.F., K. Sun, S. Hu, X. Zhou, M.T. McDowell, M.R. Shaner, M.H. Richter, et al. 2016. "Protection of Inorganic Semiconductors for Sustained, Efficient Photoelectrochemical Water Oxidation." *Electrocatalysis* 262(March 15):11–23. https://doi.org/10.1016/j.cattod.2015.08.017.

Lidston, C.A.L., S.M. Severson, B.A. Abel, and G.W. Coates. 2022. "Multifunctional Catalysts for Ring-Opening Copolymerizations." *ACS Catalysis* 12(18):11037–11070. https://doi.org/10.1021/acscatal.2c02524.

Lin, T., Y. An, F. Yu, K. Gong, H. Yu, C. Wang, Y. Sun, and L. Zhong. 2022. "Advances in Selectivity Control for Fischer–Tropsch Synthesis to Fuels and Chemicals with High Carbon Efficiency." *ACS Catalysis* 12(19):12092–120112. https://doi.org/10.1021/acscatal.2c03404.

Liu, A.-H., R. Ma, C. Song, Z.-Z. Yang, A. Yu, Y. Cai, L.-N. He, Y.-N. Zhao, B. Yu, and Q.-W. Song. 2012b. "Equimolar CO_2 Capture by N-Substituted Amino Acid Salts and Subsequent Conversion." *Angewandte Chemie International Edition* 51(45):11306–11310. https://doi.org/10.1002/anie.201205362.

Liu, B., L. Ma, H. Feng, Y. Zhang, J. Duan, Y. Wang, D. Liu, and Q. Li. 2023. "Photovoltaic-Powered Electrochemical CO_2 Reduction: Benchmarking Against the Theoretical Limit." *ACS Energy Letters* 8(2):981–987. https://doi.org/10.1021/acsenergylett.2c02906.

Liu, L., S.-M. Wang, Z.-B. Han, M. Ding, D.-Q. Yuan, and H.-L. Jiang. 2016a. "Exceptionally Robust In-Based Metal–Organic Framework for Highly Efficient Carbon Dioxide Capture and Conversion." *Inorganic Chemistry* 55(7):3558–3565. https://doi.org/10.1021/acs.inorgchem.6b00050.

Liu, M., L. Liang, X. Li, X. Gao, and J. Sun. 2016b. "Novel Urea Derivative-Based Ionic Liquids with Dual-Functions: CO_2 Capture and Conversion Under Metal- and Solvent-Free Conditions." *Green Chemistry* 18(9):2851–2863. https://doi.org/10.1039/C5GC02605A.

Liu, S., L.R. Winter and J.G. Chen, 2020. "Review of Plasma-Assisted Catalysis for Selective Generation of Oxygenates from CO_2 and CH_4," *ACS Catalysis* 10:2855–2871. https://pubs.acs.org/doi/abs/10.1021/acscatal.9b04811.

Liu, X., S. Zhang, and Y. Ding. 2012a. "Synthesis, Characterization and Properties of a μ–η^2: η^2-Carbonato-Bridged Bis(Phosphinoferrocenyl) Copper(I) Complex from CO_2 Fixation." *Inorganic Chemistry Communications* 18(April 1): 83–86. https://doi.org/10.1016/j.inoche.2012.01.023.

Liu, Y., and X.-B. Lu. 2023. "Current Challenges and Perspectives in CO_2-Based Polymers." *Macromolecules* 56(5):1759–1777. https://doi.org/10.1021/acs.macromol.2c02483.

Margarit, C.G., N.G. Asimow, M.I. Gonzalez, and D.G. Nocera. 2020. "Double Hangman Iron Porphyrin and the Effect of Electrostatic Nonbonding Interactions on Carbon Dioxide Reduction." *The Journal of Physical Chemistry Letters* 11(5):1890–1895. https://doi.org/10.1021/acs.jpclett.9b03897.

Marocco Stuardi, F., F. MacPherson, and J. Leclaire. 2019. "Integrated CO_2 Capture and Utilization: A Priority Research Direction." *CO_2 Capture and Chemistry* 16(April 1):71–76. https://doi.org/10.1016/j.cogsc.2019.02.003.

Martin, D.J., C.F. Wise, M.L. Pegis, and J.M. Mayer. 2020. "Developing Scaling Relationships for Molecular Electrocatalysis Through Studies of Fe-Porphyrin-Catalyzed O_2 Reduction." *Accounts of Chemical Research* 53(5):1056–1065. https://doi.org/10.1021/acs.accounts.0c00044.

Martín, M., and I.E. Grossmann. 2011. "Process Optimization of FT-Diesel Production from Lignocellulosic Switchgrass." *Industrial and Engineering Chemistry Research* 50(23):13485–13499. https://doi.org/10.1021/ie201261t.

Martirez, J.M.P., and E.A. Carter. 2023. "Solvent Dynamics Are Critical to Understanding Carbon Dioxide Dissolution and Hydration in Water." *Journal of the American Chemical Society* 145(23):12561–12575. https://doi.org/10.1021/jacs.3c01283.

Martirez, J.M.P., J. L. Bao, and E.A. Carter. 2021. "First-Principles Insights into Plasmon-Induced Catalysis." *Annual Review of Physical Chemistry* 72(1):99–119. https://doi.org/10.1146/annurev-physchem-061020-053501.

Masel, R.I., Z. Liu, H. Yang, J.J. Kaczur, D. Carrillo, S. Ren, D. Salvatore, and C.P. Berlinguette. 2021. "An Industrial Perspective on Catalysts for Low-Temperature CO_2 Electrolysis." *Nature Nanotechnology* 16(2):118–128. https://doi.org/10.1038/s41565-020-00823-x.

Matsubara, Y., D.C. Grills, and Y. Kuwahara. 2015. "Thermodynamic Aspects of Electrocatalytic CO_2 Reduction in Acetonitrile and with an Ionic Liquid as Solvent or Electrolyte." *ACS Catalysis* 5(11):6440–6452. https://doi.org/10.1021/acscatal.5b00656.

Mayer, J.M. 2023. "Bonds Over Electrons: Proton Coupled Electron Transfer at Solid–Solution Interfaces." *Journal of the American Chemical Society* 145(13):7050–7064. https://doi.org/10.1021/jacs.2c10212.

McGhee, W., D. Riley, K. Christ, Y. Pan, and B. Parnas. 1995. "Carbon Dioxide as a Phosgene Replacement: Synthesis and Mechanistic Studies of Urethanes from Amines, CO_2, and Alkyl Chlorides." *The Journal of Organic Chemistry* 60(9):2820–2830. https://doi.org/10.1021/jo00114a035.

Mondal, T., and D. Chatterjee. 2016. "Ru^{III}-edta ($edta^{4-}$ = ethylenediaminetetraacetate) Mediated Photocatalytic Conversion of Bicarbonate to Formate Over Visible Light Irradiated Non-Metal Doped TiO_2 Semiconductor Photocatalysts." *RSC Advances* 6(68):63488–63492. https://doi.org/10.1039/C6RA11464D.

NASEM (National Academies of Sciences, Engineering, and Medicine). 2019. *Gaseous Carbon Waste Streams Utilization: Status and Research Needs*. Washington, DC: The National Academies Press. https://doi.org/10.17226/25232.

NASEM. 2023. *Carbon Dioxide Utilization Markets and Infrastructure: Status and Opportunities: A First Report*. Washington, DC: The National Academies Press. https://doi.org/10.17226/26703.

Neelis, M., M. Patel, K. Blok, W. Haije, and P. Bach. 2007. "Approximation of Theoretical Energy-Saving Potentials for the Petrochemical Industry Using Energy Balances for 68 Key Processes." *Energy* 32(7):1104–1123. https://doi.org/10.1016/j.energy.2006.08.005.

NETL (National Energy Technology Laboratory). n.d.(a). "Syngas Conversion to Methanol." National Energy Technology Laboratory. https://www.netl.doe.gov/research/carbon-management/energy-systems/gasification/gasifipedia/methanol.

NETL. n.d.(b). "Conversion of Methanol to Gasoline." National Energy Technology Laboratory. https://www.netl.doe.gov/research/carbon-management/energy-systems/gasification/gasifipedia/methanol-to-gasoline.

NETL. n.d.(c). "Fischer-Tropsch Synthesis." National Energy Technology Laboratory. https://www.netl.doe.gov/research/carbon-management/energy-systems/gasification/gasifipedia/ftsynthesis.

NETL. n.d.(d). "Point Source Carbon Capture from Power Generation Sources." National Energy Technology Laboratory. https://netl.doe.gov/carbon-capture/power-generation.

Nie, W., and C.C.L. McCrory. 2022. "Strategies for Breaking Molecular Scaling Relationships for the Electrochemical CO_2 Reduction Reaction." *Dalton Transactions* 51(18):6993–7010. https://doi.org/10.1039/D2DT00333C.

Nitopi, S., E. Bertheussen, S.B. Scott, X. Liu, A.K. Engstfeld, S. Horch, B. Seger, et al. 2019. "Progress and Perspectives of Electrochemical CO_2 Reduction on Copper in Aqueous Electrolyte." *Chemical Reviews* 119(12):7610–7672. https://doi.org/10.1021/acs.chemrev.8b00705.

Nova-Institut. 2023. "The Rise of Carbon Dioxide (CO_2) as a Renewable Carbon Feedstock—More Than 1.3 Million Tonnes Capacity for CO_2-Based Products Already Exist and Are Expected to at Least Quadruple by 2030." https://nova-institute.eu/press/?id=428.

NPC (National Petroleum Council). 2024. "Chapter 2—Production at Scale." In *Harnessing Hydrogen: A Key Element of the U.S. Energy Future*. https://harnessinghydrogen.npc.org/files/H2-CH_2-Production_at_scale-2024-04-23.pdf.

NRC (National Research Council). 2006. *Sustainability in the Chemical Industry: Grand Challenges and Research Needs*. Washington, DC: The National Academies Press. https://doi.org/10.17226/11437.

Omodolor, I.S., H.O. Otor, J.A. Andonegui, B.J. Allen, and A.C. Alba-Rubio. 2020. "Dual-Function Materials for CO_2 Capture and Conversion: A Review." *Industrial and Engineering Chemistry Research* 59(40):17612–17631. https://doi.org/10.1021/acs.iecr.0c02218.

Pescarmona, P.P. 2021. "Cyclic Carbonates Synthesised from CO_2: Applications, Challenges and Recent Research Trends." *Current Opinion in Green and Sustainable Chemistry* 29(June 1):100457. https://doi.org/10.1016/j.cogsc.2021.100457.

Poltavsky, I., and A. Tkatchenko. 2021. "Machine Learning Force Fields: Recent Advances and Remaining Challenges." *The Journal of Physical Chemistry Letters* 12(28):6551–6564. https://doi.org/10.1021/acs.jpclett.1c01204.

Prajapati, A., R. Sartape, M.T. Galante, J. Xie, S.L. Leung, I. Bessa, M.H.S. Andrade, et al. 2022. "Fully-Integrated Electrochemical System That Captures CO_2 from Flue Gas to Produce Value-Added Chemicals at Ambient Conditions." *Energy and Environmental Science* 15(12):5105–5117. https://doi.org/10.1039/D2EE03396H.

Pullar, R.C., R.M. Novais, A.P.F. Caetano, M.A. Barreiros, S. Abanades, and F.A. Costa Oliveira. 2019. "A Review of Solar Thermochemical CO_2 Splitting Using Ceria-Based Ceramics with Designed Morphologies and Microstructures." *Frontiers in Chemistry* 7. https://www.frontiersin.org/articles/10.3389/fchem.2019.00601.

Queyriaux, N. 2021. "Redox-Active Ligands in Electroassisted Catalytic H^+ and CO_2 Reductions: Benefits and Risks." *ACS Catalysis* 11(7):4024–4035. https://doi.org/10.1021/acscatal.1c00237.

Rehman, A., F. Saleem, F. Javed, A. Ikhlaq, S. Waqas Ahmad, and A. Harvey. 2021. "Recent Advances in the Synthesis of Cyclic Carbonates via CO_2 Cycloaddition to Epoxides." *Journal of Environmental Chemical Engineering* 9(2):105113. https://doi.org/10.1016/j.jece.2021.105113.

Richard, A.R., and M. Fan. 2017. "Low-Pressure Hydrogenation of CO_2 to CH_3OH Using Ni-In-Al/SiO_2 Catalyst Synthesized via a Phyllosilicate Precursor." *ACS Catalysis* 7(9):5679–5692. https://doi.org/10.1021/acscatal.7b00848.

Robatjazi, H., L. Yuan, Y. Yuan, and N.J. Halas. 2021. "Heterogeneous Plasmonic Photocatalysis: Light-Driven Chemical Reactions Introduce a New Approach to Industrially-Relevant Chemistry." Pp. 363–387 in *Emerging Trends in Chemical Applications of Lasers*. ACS Symposium Series 1398. American Chemical Society. https://doi.org/10.1021/bk-2021-1398.ch016.

Rothschild, A., and H. Dotan. 2017. "Beating the Efficiency of Photovoltaics-Powered Electrolysis with Tandem Cell Photoelectrolysis." *ACS Energy Letters* 2(1):45–51. https://doi.org/10.1021/acsenergylett.6b00610.

Rungtaweevoranit, B., J. Baek, J.R. Araujo, B.S. Archanjo, K.M. Choi, O.M. Yaghi, and G.A. Somorjai. 2016. "Copper Nanocrystals Encapsulated in Zr-Based Metal–Organic Frameworks for Highly Selective CO_2 Hydrogenation to Methanol." *Nano Letters* 16(12):7645–7649. https://doi.org/10.1021/acs.nanolett.6b03637.

Saeidi, S., S. Najari, V. Hessel, K. Wilson, F.J. Keil, P. Concepción, S.L. Suib, and A.E. Rodrigues. 2021. "Recent Advances in CO_2 Hydrogenation to Value-Added Products—Current Challenges and Future Directions." *Progress in Energy and Combustion Science* 85(July 1):100905. https://doi.org/10.1016/j.pecs.2021.100905.

Samimi, F., D. Karimipourfard, and M.R. Rahimpour. 2018. "Green Methanol Synthesis Process from Carbon Dioxide via Reverse Water Gas Shift Reaction in a Membrane Reactor." *Chemical Engineering Research and Design* 140(December 1):44–67. https://doi.org/10.1016/j.cherd.2018.10.001.

Sampaio, R.N., D.C. Grills, D.E. Polyansky, D.J. Szalda, and E. Fujita. 2020. "Unexpected Roles of Triethanolamine in the Photochemical Reduction of CO_2 to Formate by Ruthenium Complexes." *Journal of the American Chemical Society* 142(5):2413–2428. https://doi.org/10.1021/jacs.9b11897.

Sandoval-Diaz, L.E., R. Schlögl, and T. Lunkenbein. 2022. "Quo Vadis Dry Reforming of Methane?–A Review on Its Chemical, Environmental, and Industrial Prospects." *Catalysts* 12(5). https://doi.org/10.3390/catal12050465.

Sarswat, A., D.S. Sholl, and R.P. Lively. 2022. "Achieving Order of Magnitude Increases in CO_2 Reduction Reaction Efficiency by Product Separations and Recycling." *Sustainable Energy and Fuels* 6(20):4598–4604. https://doi.org/10.1039/D2SE01156E.

Sattler, W., and G. Parkin. 2014. "Reduction of Bicarbonate and Carbonate to Formate in Molecular Zinc Complexes." *Catalysis Science and Technology* 4(6):1578–1584. https://doi.org/10.1039/C3CY01065A.

Saw, S.Z., and J. Nandong. 2016. "Simulation and Control of Water-Gas Shift Packed Bed Reactor with Inter-Stage Cooling." *IOP Conference Series: Materials Science and Engineering* 121(March):012022. https://doi.org/10.1088/1757-899X/121/1/012022.

Sears, W.M., and S.R. Morrison. 1985. "Carbon Dioxide Reduction on Gallium Arsenide Electrodes." *The Journal of Physical Chemistry* 89(15):3295–3298. https://doi.org/10.1021/j100261a026.

Shafaat, H.S., and J.Y. Yang. 2021. "Uniting Biological and Chemical Strategies for Selective CO_2 Reduction." *Nature Catalysis* 4(11):928–933. https://doi.org/10.1038/s41929-021-00683-1.

Shaner, M.R., H.A. Atwater, N.S. Lewis, and E.W. McFarland. 2016. "A Comparative Technoeconomic Analysis of Renewable Hydrogen Production Using Solar Energy." *Energy and Environmental Science* 9(7):2354–2371. https://doi.org/10.1039/C5EE02573G.

Shao, B., Y. Zhang, Z. Sun, J. Li, Z. Gao, Z. Xie, J. Hu, and H. Liu. 2022. "CO_2 Capture and In-Situ Conversion: Recent Progresses and Perspectives." *Green Chemical Engineering* 3(3):189–198. https://doi.org/10.1016/j.gce.2021.11.009.

Shi, L., G. Yang, K. Tao, Y. Yoneyama, Y. Tan, and N. Tsubaki. 2013. "An Introduction of CO_2 Conversion by Dry Reforming with Methane and New Route of Low-Temperature Methanol Synthesis." *Accounts of Chemical Research* 46(8):1838–1847. https://doi.org/10.1021/ar300217j.

Sibi, M.G., D. Verma and J. Kim. 2022. "Direct Conversion of CO_2 into Aromatics Over Multifunctional Heterogeneous Catalysts." *Catalysis Reviews* 1–60. https://www.tandfonline.com/doi/abs/10.1080/01614940.2022.2099058.

Snoeckx, R., and A. Bogaerts. 2017. "Plasma Technology—A Novel Solution for CO_2 Conversion?" *Chemical Society Reviews* 46:5805–5863. https://doi.org/10.1039/C6CS00066E.

Soler, A., V. Gordillo, W. Lilley, P. Schmidt, W. Werner, T. Houghton, and S. Dell-Orco. 2022. "E-Fuels: A Technoeconomic Assessment of European Domestic Production and Imports Towards 2050." Report no. 17/22. Brussels, Belgium: Concawe and Aramco. https://www.concawe.eu/wp-content/uploads/Rpt_22-17.pdf.

Song, B., A. Qin, and B.Z. Tang. 2022. "Syntheses, Properties, and Applications of CO_2-Based Functional Polymers." *Cell Reports Physical Science* 3(2):100719. https://doi.org/10.1016/j.xcrp.2021.100719.

Stephens, I.E.L., K. Chan, A. Bagger, S.W. Boettcher, J. Bonin, E. Boutin, A.K. Buckley, R. Buonsanti, E.R. Cave, and X. Chang. 2022. "2022 Roadmap on Low Temperature Electrochemical CO_2 Reduction." *JPhys Energy* 4(4):042003. https://doi.org/10.1088/2515-7655/ac7823.

Storch, H.H., N. Golumbic, and R.B. Anderson. 1961. *The Fischer-Tropsch and Related Syntheses*. 1st edition. New York: John Wiley and Sons.

Sun, S., H. Sun, P.T. Williams, and C. Wu. 2021. "Recent Advances in Integrated CO_2 Capture and Utilization: A Review." *Sustainable Energy and Fuels* 5(18):4546–4559. https://doi.org/10.1039/D1SE00797A.

Sung, S., D. Kumar, M. Gil-Sepulcre, and M. Nippe. 2017. "Electrocatalytic CO_2 Reduction by Imidazolium-Functionalized Molecular Catalysts." *Journal of the American Chemical Society* 139(40):13993–13996. https://doi.org/10.1021/jacs.7b07709.

Tabanelli, T., D. Bonincontro, S. Albonetti, and F. Cavani. 2019. "Chapter 7—Conversion of CO_2 to Valuable Chemicals: Organic Carbonate as Green Candidates for the Replacement of Noxious Reactants." Pp. 125–144 In *Studies in Surface Science and Catalysis* 178, S. Albonetti, S. Perathoner, and E.A. Quadrelli, eds. Elsevier. https://doi.org/10.1016/B978-0-444-64127-4.00007-0.

Talapaneni, S.N., O. Buyukcakir, S.H. Je, S. Srinivasan, Y. Seo, K. Polychronopoulou, and A. Coskun. 2015. "Nanoporous Polymers Incorporating Sterically Confined N-Heterocyclic Carbenes for Simultaneous CO_2 Capture and Conversion at Ambient Pressure." *Chemistry of Materials* 27(19):6818–6826. https://doi.org/10.1021/acs.chemmater.5b03104.

Tan, E.W.P., J.L. Hedrick, P.L. Arrechea, T. Erdmann, V. Kiyek, S. Lottier, Y.Y. Yang, and N.H. Park. 2021. "Overcoming Barriers in Polycarbonate Synthesis: A Streamlined Approach for the Synthesis of Cyclic Carbonate Monomers." *Macromolecules* 54(4):1767–1774. https://doi.org/10.1021/acs.macromol.0c02880.

Tang, R., Z. Zhu, C. Li, M. Xiao, Z. Wu, D. Zhang, C. Zhang, et al. 2021. "Ru-Catalyzed Reverse Water Gas Shift Reaction with Near-Unity Selectivity and Superior Stability." *ACS Materials Letters* 3(12):1652–1659. https://doi.org/10.1021/acsmaterialslett.1c00523.

Thor Wismann, S., K.-E. Larsen, and P. Mølgaard Mortensen. 2022. "Electrical Reverse Shift: Sustainable CO_2 Valorization for Industrial Scale." *Angewandte Chemie International Edition* 61(8):e202109696. https://doi.org/10.1002/anie.202109696.

Tian, P., Y. Wei, M. Ye, and Z. Liu. 2015. "Methanol to Olefins (MTO): From Fundamentals to Commercialization." *ACS Catalysis* 5(3):1922–1938. https://doi.org/10.1021/acscatal.5b00007.

Timmerhaus, K.D., and M.S. Peters. 1991. *Plant Design and Economics for Chemical Engineers*. 4th edition. New York: McGraw-Hill.

Tlili, A., E. Blondiaux, X. Frogneux, and T. Cantat. 2015. "Reductive Functionalization of CO_2 with Amines: An Entry to Formamide, Formamidine and Methylamine Derivatives." *Green Chemistry* 17(1):157–168. https://doi.org/10.1039/C4GC01614A.

Ton, T.N., R.J. Baker, and K. Manthiram. 2024. "Recent Progress in the Development of Electrode Materials for Electrochemical Carboxylation with CO_2." *Journal of Catalysis* 432(April 1):115371. https://doi.org/10.1016/j.jcat.2024.115371.

Tortajada, A., F. Juliá-Hernández, M. Börjesson, T. Moragas, and R. Martin. 2018. "Transition-Metal-Catalyzed Carboxylation Reactions with Carbon Dioxide." *Angewandte Chemie International Edition* 57(49):15948–15982. https://doi.org/10.1002/anie.201803186.

Unke, O.T., S. Chmiela, H.E. Sauceda, M. Gastegger, I. Poltavsky, K.T. Schütt, A. Tkatchenko, and K.-R. Müller. 2021. "Machine Learning Force Fields." *Chemical Reviews* 121(16):10142–10186. https://doi.org/10.1021/acs.chemrev.0c01111.

van Den Bosch, B., J. Krasovic, B. Rawls, and A.L. Jongerius. 2022. "Research Targets for Upcycling of CO_2 to Formate and Carbon Monoxide with Paired Electrolysis." *Current Opinion in Green and Sustainable Chemistry* 34(April):100592. https://doi.org/10.1016/j.cogsc.2022.100592.

Vanhoof, J.R., S. Spittaels, and D.E. De Vos. 2024. "A Comparative Overview of the Electrochemical Valorization and Incorporation of CO2 in Industrially Relevant Compounds." *EES Catalysis* 2(3):753–779. https://doi.org/10.1039/D4EY00005F.

Verma, R., R. Belgamwar, and V. Polshettiwar. 2021. "Plasmonic Photocatalysis for CO_2 Conversion to Chemicals and Fuels." *ACS Materials Letters* 3(5):574–598. https://doi.org/10.1021/acsmaterialslett.1c00081.

von der Assen, N., and A. Bardow. 2014. "Life Cycle Assessment of Polyols for Polyurethane Production Using CO_2 as Feedstock: Insights from an Industrial Case Study." *Green Chemistry* 16(6):3272–3280. https://doi.org/10.1039/C4GC00513A.

Wakerley, D., S. Lamaison, J. Wicks, A. Clemens, J. Feaster, D. Corral, S.A. Jaffer, et al. 2022. "Gas Diffusion Electrodes, Reactor Designs and Key Metrics of Low-Temperature CO_2 Electrolysers." *Nature Energy* 7(2):130–143. https://doi.org/10.1038/s41560-021-00973-9.

Wang, F., Z. Lu, H. Guo, G. Zhang, Y. Li, Y. Hu, W. Jiang, and G. Liu. 2023a. "Plasmonic Photocatalysis for CO_2 Reduction: Advances, Understanding and Possibilities." *Chemistry—A European Journal* 29(25):e202202716. https://doi.org/10.1002/chem.202202716.

Wang, J., W. Sng, G. Yi, and Y. Zhang. 2015b. "Imidazolium Salt-Modified Porous Hypercrosslinked Polymers for Synergistic CO_2 Capture and Conversion." *Chemical Communications* 51(60):12076–12079. https://doi.org/10.1039/C5CC04702A.

Wang, J.-W., X. Zhang, L. Velasco, M. Karnahl, Z. Li, Z.-M. Luo, Y. Huang, et al. 2023b. "Precious-Metal-Free CO_2 Photoreduction Boosted by Dynamic Coordinative Interaction between Pyridine-Tethered Cu(I) Sensitizers and a Co(II) Catalyst." *JACS Au* 3(7):1984–1997. https://doi.org/10.1021/jacsau.3c00218.

Wang, W.-H., Y. Himeda, J.T. Muckerman, G.F. Manbeck, and E. Fujita. 2015a. "CO_2 Hydrogenation to Formate and Methanol as an Alternative to Photo- and Electrochemical CO_2 Reduction." *Chemical Reviews* 115(23):12936–12973. https://doi.org/10.1021/acs.chemrev.5b00197.

Wang, X., H. Wang, and Y. Sun. 2017. "Synthesis of Acrylic Acid Derivatives from CO_2 and Ethylene." *Chem* 3(2):211–228. https://doi.org/10.1016/j.chempr.2017.07.006.

Wenzel, M., L. Rihko-Struckmann, and K. Sundmacher. 2016. "Thermodynamic Analysis and Optimization of RWGS Processes for Solar Syngas Production from CO_2." *AIChE Journal* 63(1):15–22. https://doi.org/10.1002/aic.15445.

Wexler, R.B., E.B. Stechel, and E.A. Carter. 2023. "Materials Design Directions for Solar Thermochemical Water Splitting." Pp. 1–63 in *Solar Fuels*. https://doi.org/10.1002/9781119752097.ch1.

White, J.L., M.F. Baruch, J.E. Pander III, Y. Hu, I.C. Fortmeyer, J.E. Park, T. Zhang, et al. 2015. "Light-Driven Heterogeneous Reduction of Carbon Dioxide: Photocatalysts and Photoelectrodes." *Chemical Reviews* 115(23):12888–12935. https://doi.org/10.1021/acs.chemrev.5b00370.

Wołosz, D., P.G. Parzuchowski, and K. Rolińska. 2022. "Environmentally Friendly Synthesis of Urea-Free Poly(Carbonate-Urethane) Elastomers." *Macromolecules* 55(12):4995–5008. https://doi.org/10.1021/acs.macromol.2c00706.

Wu, S., X. Yang, X. Zhao, Z. Li, M. Lu, X. Xie, and J. Yan. 2023. "Applications and Advances in Machine Learning Force Fields." *Journal of Chemical Information and Modeling* 63(22):6972–6985. https://doi.org/10.1021/acs.jcim.3c00889.

Xia, R., S. Overa, and F. Jiao. 2022. "Emerging Electrochemical Processes to Decarbonize the Chemical Industry." *JACS Au* 2(5):1054–1070. https://doi.org/10.1021/jacsau.2c00138.

Xie, J., and U. Olsbye. 2023. "The Oxygenate-Mediated Conversion of CO_x to Hydrocarbons–On the Role of Zeolites in Tandem Catalysis." *Chemical Reviews* 123(20):11775–11816. https://doi.org/10.1021/acs.chemrev.3c00058.

Xie, W., J. Xu, U. Md Idros, J. Katsuhira, M. Fuki, M. Hayashi, M. Yamanaka, Y. Kobori, and R. Matsubara. 2023. "Metal-Free Reduction of CO_2 to Formate Using a Photochemical Organohydride-Catalyst Recycling Strategy." *Nature Chemistry* 15(6):794–802. https://doi.org/10.1038/s41557-023-01157-6.

Xu, S., and E.A. Carter. 2019a. "Optimal Functionalization of a Molecular Electrocatalyst for Hydride Transfer." *Proceedings of the National Academy of Sciences* 116(46):22953–22958. https://doi.org/10.1073/pnas.1911948116.

Xu, S., and E.A. Carter. 2019b. "Theoretical Insights into Heterogeneous (Photo)Electrochemical CO_2 Reduction." *Chemical Reviews* 119(11):6631–6669. https://doi.org/10.1021/acs.chemrev.8b00481.

Xu, S., L. Li, and E.A. Carter. 2018. "Why and How Carbon Dioxide Conversion to Methanol Happens on Functionalized Semiconductor Photoelectrodes." *Journal of the American Chemical Society* 140(48):16749–16757. https://doi.org/10.1021/jacs.8b09946.

Xu, S., H. Chen, C. Hardacre, and X. Fan. 2021. "Non-Thermal Plasma Catalysis for CO_2 Conversion and Catalyst Design for the Process." *Journal of Physics D: Applied Physics* 54(23):233001. https://doi.org/10.1088/1361-6463/abe9e1.

Yamazaki, Y., H. Takeda, and O. Ishitani. 2015. "Photocatalytic Reduction of CO_2 Using Metal Complexes." *Journal of Photochemistry and Photobiology C: Photochemistry Reviews* 25(December 1):106–137. https://doi.org/10.1016/j.jphotochemrev.2015.09.001.

Yan, T., X. Chen, L. Kumari, J. Lin, M. Li, Q. Fan, H. Chi, T.J. Meyer, S. Zhang, and X. Ma. 2023. "Multiscale CO_2 Electrocatalysis to C2+ Products: Reaction Mechanisms, Catalyst Design, and Device Fabrication." *Chemical Reviews* 123(17):10530–10583. https://doi.org/10.1021/acs.chemrev.2c00514.

Ye, R.-P., J. Ding, W. Gong, M.D. Argyle, Q. Zhong, Y. Wang, C.K. Russell, et al. 2019. "CO_2 Hydrogenation to High-Value Products via Heterogeneous Catalysis." *Nature Communications* 10(1):5698. https://doi.org/10.1038/s41467-019-13638-9.

Yeung, C.W.S., G.E.K.K. Seah, A.Y.X. Tan, S.Y. Tee, J.Y.C. Lim, and S.S. Goh. 2023. "Chapter 5—Functional Polymers from CO_2 as Feedstock." Pp. 129–171 in *Circularity of Plastics*, L. Zibiao, J.Y.C. Lim, and C.-G. Wang, eds. Elsevier. https://doi.org/10.1016/B978-0-323-91198-6.00005-X.

Yin, Z., J. Yu, Z. Xie, S.-W. Yu, L, Zhang, T. Akauola, J.G. Chen, W. Huang, L. Qi, and S. Zhang. 2022. "Hybrid Catalyst Coupling Single-Atom Ni and Nanoscale Cu for Efficient CO_2 Electroreduction to Ethylene." *Journal of the American Chemical Society* 144(45):20931–20938. https://doi.org/10.1021/jacs.2c09773.

Yu, S., and P.K. Jain. 2019. "Plasmonic Photosynthesis of C1–C3 Hydrocarbons from Carbon Dioxide Assisted by an Ionic Liquid." *Nature Communications* 10(1):2022. https://doi.org/10.1038/s41467-019-10084-5.

Zhang, G., G. Wei, Z. Liu, S.R.J. Oliver, and H. Fei. 2016. "A Robust Sulfonate-Based Metal–Organic Framework with Permanent Porosity for Efficient CO_2 Capture and Conversion." *Chemistry of Materials* 28(17):6276–6281. https://doi.org/10.1021/acs.chemmater.6b02511.

Zhang, J., B. Guan, X. Wu, Y. Chen, J. Guo, Z. Ma, S. Bao, et al. 2023. "Research on Photocatalytic CO_2 Conversion to Renewable Synthetic Fuels Based on Localized Surface Plasmon Resonance: Current Progress and Future Perspectives." *Catalysis Science and Technology* 13(7):1932–1975. https://doi.org/10.1039/D2CY01967A.

Zhang, L., and Z. Hou. 2013. "Chapter 9—Transition-Metal-Catalyzed C-C Bond Forming Reactions with Carbon Dioxide." Pp. 253–273 in *New and Future Developments in Catalysis: Activation of Carbon Dioxide*. Amsterdam: Elsevier Science and Technology.

Zhang, Q., M. Bown, L. Pastor-Pérez, M.S. Duyar, and T.R. Reina. 2022a. "CO_2 Conversion via Reverse Water Gas Shift Reaction Using Fully Selective Mo–P Multicomponent Catalysts." *Industrial and Engineering Chemistry Research* 61(34):12857–12865. https://doi.org/10.1021/acs.iecr.2c00305.

Zhang, T., J.C. Bui, Z. Li, A.T. Bell, A.Z. Weber, J. Wu. 2022b. "Highly Selective and Productive Reduction of Carbon Dioxide to Multicarbon Products via In Situ CO Management Using Segmented Tandem Electrodes." *Nature Catalysis* 5(3):202–211. https://doi.org/10.1038/s41929-022-00751-0.

Zhang, X., A. Zhang, X. Jiang, J. Zhu, J. Liu, J. Li, G. Zhang, C. Song, and X. Guo. 2019a. "Utilization of CO_2 for Aromatics Production Over ZnO/ZrO_2-ZSM-5 Tandem Catalyst." *Journal of CO_2 Utilization* 29(January 1):140–145. https://doi.org/10.1016/j.jcou.2018.12.002.

Zhang, X., M. Cibian, A. Call, K. Yamauchi, and K. Sakai. 2019b. "Photochemical CO_2 Reduction Driven by Water-Soluble Copper(I) Photosensitizer with the Catalysis Accelerated by Multi-Electron Chargeable Cobalt Porphyrin." *ACS Catalysis* 9(12):11263–11273. https://doi.org/10.1021/acscatal.9b04023.

Zhang, X., G. Zhang, C. Song, and X. Guo. 2021. "Catalytic Conversion of Carbon Dioxide to Methanol: Current Status and Future Perspective." *Frontiers in Energy Research* 8. https://www.frontiersin.org/articles/10.3389/fenrg.2020.621119.

Zhang, X., W. Huang, L. Yu, M. García-Melchor, D. Wang, L. Zhi, and H. Zhang. 2024. "Enabling Heterogeneous Catalysis to Achieve Carbon Neutrality: Directional Catalytic Conversion of CO_2 into Carboxylic Acids." *Carbon Energy* 6(3): e362. https://doi.org/10.1002/cey2.362.

Zhang, Z., E.W. Lees, F. Habibzadeh, D.A. Salvatore, S. Ren, G.L. Simpson, D.G. Wheeler, A. Liu, and C.P. Berlinguette. 2022c. "Porous Metal Electrodes Enable Efficient Electrolysis of Carbon Capture Solutions." *Energy and Environmental Science* 15(2):705–713. https://doi.org/10.1039/D1EE02608A.

Zhao, Q., J.M.P. Martirez, and E.A. Carter. 2021. "Revisiting Understanding of Electrochemical CO_2 Reduction on Cu(111): Competing Proton-Coupled Electron Transfer Reaction Mechanisms Revealed by Embedded Correlated Wavefunction Theory." *Journal of the American Chemical Society* 143(16):6152–6164. https://doi.org/10.1021/jacs.1c00880.

Zheng, L., M. Ambrosetti, A. Beretta, G. Groppi, and E. Tronconi. 2023. "Electrified CO_2 Valorization Driven by Direct Joule Heating of Catalytic Cellular Substrates." *Chemical Engineering Journal* 466(June 15):143154. https://doi.org/10.1016/j.cej.2023.143154.

Zhou, C., J. Zhang, Y. Fu, and H. Dai. 2023. "Recent Advances in the Reverse Water–Gas Conversion Reaction." *Molecules* 28(22). https://doi.org/10.3390/molecules28227657.

Zhou, G., B. Dai, H. Xie, G. Zhang, K. Xiong, and X. Zheng. 2017. "CeCu Composite Catalyst for CO Synthesis by Reverse Water–Gas Shift Reaction: Effect of Ce/Cu Mole Ratio." *Journal of CO2 Utilization* 21(October 1):292–301. https://doi.org/10.1016/j.jcou.2017.07.004.

Zhou, L., J.M.P. Martirez, J. Finzel, C. Zhang, D.F. Swearer, S. Tian, H. Robatjazi, et al. 2020. "Light-Driven Methane Dry Reforming with Single Atomic Site Antenna-Reactor Plasmonic Photocatalysts." *Nature Energy* 5(1):61–70. https://doi.org/10.1038/s41560-019-0517-9.

Zhou, Y., A. José Martín, F. Dattila, S. Xi, N. López, J. Pérez-Ramírez, and B.S. Yeo. 2022. "Long-Chain Hydrocarbons by CO_2 Electroreduction Using Polarized Nickel Catalysts." *Nature Catalysis* 5(6):545–554. https://doi.org/10.1038/s41929-022-00803-5.

Zhu, Q., Y. Zeng, and Y. Zheng. 2023. "Overview of CO_2 Capture and Electrolysis Technology in Molten Salts: Operational Parameters and Their Effects." *Industrial Chemistry and Materials* 1(4):595–617. https://doi.org/10.1039/D3IM00011G.

8

Biological CO_2 Conversion to Fuels, Chemicals, and Polymers

8.1 OVERVIEW OF BIOLOGICAL CONVERSION ROUTES FOR CO_2 TO ORGANIC PRODUCTS

To achieve net-zero emissions, it is critical to manage carbon flows and consider both strategies for CO_2 capture and long-term storage, as well as technologies to replace the emission-intensive processes characteristic of the current petrochemical industries. Biological conversion of carbon into products—either directly from CO_2 or indirectly from sugar or other intermediates derived from CO is one such strategy.[1] Biological processes occur under ambient conditions and thus have intrinsically lower energy intensity than conventional thermochemical processes, which require high temperatures and pressures. Biological processes are also generally more robust to contaminants and fluctuations in reactant stream quality and composition. Products commonly accessible from biological CO_2 utilization via photosynthetic, nonphotosynthetic, and hybrid processes include organic chemicals, lipids, terpenoids, polymer precursors, biopolymers, and food and animal feed. Products derived from bioprocessing often have additional environmental benefits, such as being compostable or biodegradable at end of life, preventing nondegradable plastic waste pollution (Mayfield and Burkart 2023; Nduko and Taguchi 2021; Sirohi et al. 2020).

Despite its positive attributes, native biological CO_2 fixation is slow and has low energy conversion efficiency, limiting growth and production rates (Liu et al. 2016). Terrestrial plant photosynthesis in general exhibits less than 1 percent conversion efficiency of light energy into chemical product energy. Algal and cyanobacterial conversion is often limited by light penetration. (Long et al. 2022; Zhu et al. 2010). In conventional biofuel production, photosynthesis converts CO_2 into plant-based carbohydrate substrates (e.g., starch, sucrose, cellulose), which are later converted to ethanol and other chemicals through fermentation, with some CO_2 as a by-product. Considering the low efficiency of natural photosynthesis and the carbon loss to CO_2 during product formation through fermentation, bioproduction using carbohydrates has very low energy efficiency from sunlight. Consequently, substantial amounts of land would be required for bioproduction to replace emission-intensive petrochemical production (Smith et al. 2023). The Department of Energy's 2023 Billion Ton Report outlines the potential for biomass resources in the contiguous United States to meet some of this demand, including a detailed analysis of non-CO_2 routes (DOE 2024).

Direct biological conversion of CO_2—the focus of this chapter—aims to combat some of the challenges of native biological fixation. Biological CO_2 utilization is defined here as the use of concentrated CO_2 (e.g., industrial waste gas

[1] Although out of scope for this report, it is noted that recent efforts have achieved full biomass utilization through carbohydrate and lignin conversion, which represents an indirect route for CO_2 utilization in which CO_2 captured in plant biomass can be used for fiber, fuels, materials, and chemicals (Liu et al. 2022; Yuan et al. 2022).

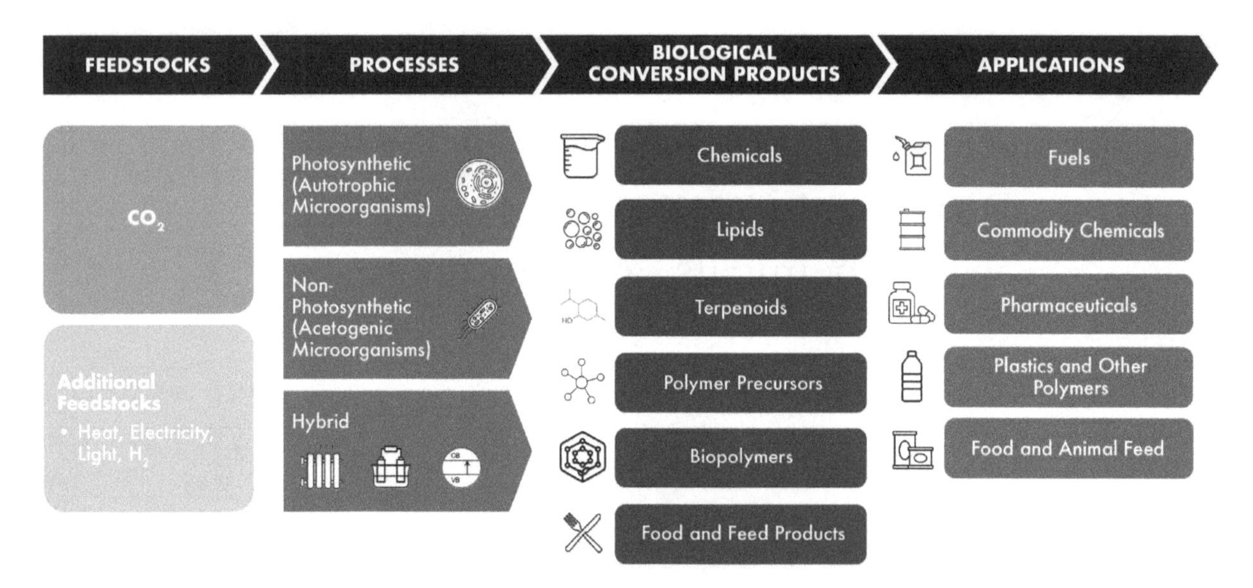

FIGURE 8-1 Summary of the inputs, processes, products, and applications for direct biological conversion of CO_2.
SOURCE: Icons from the Noun Project, https://thenounproject.com. CC BY 3.0.

streams or direct air capture CO_2) as a feedstock for biochemical production, which can be developed through autotrophic microorganisms (e.g., microalgae or cyanobacteria), acetogenic microbes, or hybrid systems (Figure 8-1). The biochemical production systems discussed in this chapter—including photosynthetic, nonphotosynthetic, and hybrid systems such as electro-bio and cell-free systems—have the potential not only to sequester CO_2 but also to possibly help replace highly polluting commodity chemical products with greener alternatives (Zhang et al. 2022), assuming many of the challenges discussed in this chapter are overcome and that adequate market incentives are met. Engineered microorganisms capable of producing commodity chemicals have gained traction as viable alternatives to traditional petrochemical approaches. Expanding feedstock pools, engineering regulatory elements of metabolism, and optimizing conditions are all methods employed to increase productivity in these microbial hosts. It is also critical to explore and advance biomanufacturing technologies that utilize CO_2 to produce a diverse range of value-added products.

This chapter addresses the strengths and challenges intrinsic to using CO_2 as a substrate for biochemical production and provides an update on biological CO_2 utilization to the 2019 National Academies' report *Gaseous Carbon Waste Streams Utilization: Status and Research Needs* (NASEM 2019). First, engineering efforts using photoautotrophs such as microalgae and cyanobacteria are discussed. Second, engineering approaches with noncanonical CO_2 fixing pathways such as within acetogenic bacteria are outlined. Third, the combination of bioconversion with electro-, thermo-, plasma-, and photo-catalysis is reviewed. These strategies demonstrate potentially more sustainable methods of making industrially relevant carbon-based products with the aid of engineered microorganisms capable of utilizing CO_2 to support a net-zero emissions future.

8.2 PHOTOSYNTHETIC PRODUCTION OF CHEMICALS FROM CO_2

8.2.1 Existing and Emerging Processes

Synthetic biology and metabolic engineering strive to establish sustainable methods for chemical production through engineered microorganisms.[2] These endeavors involve leveraging photosynthetic microorganisms capable of generating valuable chemical commodities from CO_2 and light. Microalgae and cyanobacteria—the two

[2] Genetic engineering involves the direct manipulation of an organism's genes using biochemical methods. Controversy surrounding this technology exists owing to possible ethical concerns related to unknown environmental impacts and long-term health effects; however, the applications discussed herein focus strictly on applications that enhance natural microbial processes, contained in reactors, and do not have any direct pathways to affect human, animal, or environmental health.

key categories of photosynthetic microorganisms under investigation for chemical production—exhibit potential for synthesizing a diverse range of useful compounds, including fuels, polymer precursors, and commodity chemicals.

Microalgae are photosynthetic microorganisms that can naturally fix CO_2 10 to 50 times more efficiently than other terrestrial plants (Onyeaka et al. 2021). Their carbon-fixing ability makes microalgae a promising feedstock for biofuels, bioplastics, pharmaceuticals, cosmetics, and industrial chemicals (Al-Jabri et al. 2022; Cheng et al. 2022; Sirohi et al. 2021). Successful demonstrations of microalgae-based CO_2 utilization include antioxidants, anticancer, and antimicrobial compounds, polymers, biocrude, biodiesel, biogas, and hydrogen (Cuellar-Bermudez et al. 2015; Rezvani et al. 2017; Sosa-Hernández et al. 2018). Nonetheless, industrial-scale commercialization of microalgae-based CO_2 utilization has been limited and is at low technology readiness level (TRL) (Roh et al. 2020). The predominant challenges for industrial-scale commercialization are use of microalgal biomass, evaluation of the life cycle of microalgae technologies, and development and implementation of a supportive policy and regulatory environment (Miranda et al. 2022).

Cyanobacteria are photosynthetic microorganisms found naturally in water that can fix CO_2 twice as efficiently as other plants (Hill et al. 2020). Their natural properties like high specific growth rate, abundant fatty acid and oil content, and other active metabolites make them an attractive non-food biomass source. Successful pilot demonstrations of cyanobacteria-based CO_2 utilization include the production of ethanol, butanol, biodiesel, bioplastics, and hydrogen (Agarwal et al. 2022).[3] However, differences in outdoor cultivation as compared to ideal indoor conditions, insufficient light or nutrients, contamination in open pond cultivation systems, and inefficiencies in the extraction, purification, and harvest stages have limited scaling from low TRL. Additional challenges include bioreactor design limitations, limits of CO_2 solubility in water, and land and water availability and access (Burkart et al. 2019). Photobioreactors that leverage LED lights can potentially produce high-value products and intermediates with high efficiency, but they are limited to small scales (Porto et al. 2022). The next section discusses challenges for CO_2 utilization from both types of photosynthetic microorganisms.

8.2.2 Challenges

Despite the growing interest in these photosynthetic microorganisms, challenges persist within the field. A major concern is the inefficiency of photosynthesis and CO_2 fixation. Attempts to enhance the central carbon fixation enzyme, ribulose-1,5-bisphosphate carboxylase-oxygenase (RuBisCO), have faced limited success (Cummins 2021). A related challenge is that RuBisCO cannot distinguish effectively between CO_2 and O_2, resulting in energy-intensive photorespiration caused by the oxidative reaction of RuBisCO (Hagemann and Bauwe 2016). Additionally, at high cell densities, cell shading inhibits photosynthesis, making the design of culture systems more complicated.

To address these challenges of selectivity and low activity, recent studies have explored strategies such as reviving ancestral forms of RuBisCO and drawing inspiration from natural adaptations like the carboxysome, a bacterial microcompartment concentrating RuBisCO with high CO_2 concentrations (Kerfeld and Melnicki 2016; Shih et al. 2016). Another approach to enhance chemical production capacity involves supplementing CO_2 with carbohydrates as an auxiliary carbon source, enabling photomixotrophy that utilizes carbohydrates in addition to CO_2. CO_2 fixation efficiency can be improved by redirecting carbohydrate breakdown to the RuBisCO precusor in the Calvin-Benson cycle, promoting faster growth and increased production of target compounds (Kanno et al. 2017). This strategy allows for 24-hour production periods, even in darkness, by fixing CO_2 through photomixotrophy. Although many different types of sugars are amenable to this process, xylose—a prevalent sugar in corn stover lysate—has shown promise in enhancing photomixotrophic production of various compounds (Gonzales et al. 2023; Yao et al. 2022). Additionally, cell growth is enhanced by incorporating the nonoxidative glycolysis pathway and deleting genes that elevate the intracellular concentration of acetyl-CoA (Song et al. 2021).

Large-scale culture systems for photosynthetic chemical production are still under development. Various studies are under way to establish a system that enables efficient CO_2 fixation and chemical production and is economically viable. They are discussed in detail in other reports (Sun et al. 2020).

[3] These applications are also possible using eukaryotic microalgae (Daneshvar et al. 2022).

8.2.3 Research and Development Opportunities

8.2.3.1 Exploration of Fast-Growing Cyanobacteria

The conventional focus of cyanobacterial chemical production has been on model species such as *Synechococcus elongatus* PCC 7942 and *Synechocystis* sp. PCC 6803, but efforts to discover faster-growing strains amenable to genetic manipulation are gaining traction. Cyanobacteria like *Synechococcus elongatus* UTEX 2973, *S. elongatus* PCC 11801, and *S. elongatus* PCC 11802 exhibit faster growth and efficient chemical production under specific conditions (Sengupta et al. 2020a, 2020b; Yu et al. 2015). These discoveries not only offer potential improvements for model organisms but also shed light on differences between new and traditional strains.

8.2.3.2 Advances in Genome and Metabolic Engineering Tools

Great progress has been made in photosynthetic chemical production from CO_2, but new genetic engineering strategies are needed to enhance the efficiency of CO_2 fixation and enable the establishment of economical and scalable production systems. To improve microbial CO_2 fixation, it is essential to continue to refine the tools of genetic engineering and to increase understanding of the design principles of photosynthetic metabolism.

Traditional genome modification in cyanobacteria faces limitations owing to polyploidy and antibiotic resistance markers (Griese et al. 2011). The advent of CRISPR gene editing has revolutionized cyanobacterial genome engineering, allowing markerless editing for increased efficiency (Behler et al. 2018). The CRISPR enzyme, Cas9, is toxic in some cyanobacterial species, so alternative enzymes with lower toxicity, such as Cas12a, are being investigated (Ungerer et al. 2018). CRISPR inhibition (CRISPRi) has also been established to knock down the expression of target genes in cyanobacteria (Qi et al. 2013). The Cas9 mutant without the endonuclease activity (dCas9) used in CRISPRi is functional in cyanobacteria, although Cas9 is toxic (Santos et al. 2021). These advances open new possibilities for innovative genome engineering strategies to enhance CO_2 fixation and chemical production from CO_2.

Photosynthetic carbon metabolism is an intricate process that has undergone evolutionary optimization for cell growth in natural conditions. Utilizing systems biology, including proteomics and metabolomics, is crucial for characterizing photosynthetic carbon metabolism and identifying targets to improve CO_2 fixation and product formation. Computational techniques can enable integration of complementary datasets obtained from systems biology and identify higher-level features such as regulation and network characteristics. Owing to the complex and highly interconnected nature of carbon metabolism, understanding the effects of modifications on downstream metabolism or determining necessary genetic modifications for a desired effect is often not a trivial process. Therefore, applying mathematical models becomes essential to describe, understand, and predict system behavior. Through the application of such models, one gains the ability to generate a set of testable hypotheses for system behavior. Machine learning can facilitate the training of mathematical models for better prediction.

Among many additional opportunities in photobiological research and development (R&D), some notable ideas include: light absorption for downstream metabolism (Blankenship and Chen 2013); exploiting waste dissipation processes from excess absorbed light (Niyogi and Truong 2013); and the use of smaller portions of the electromagnetic spectrum (for example, only blue or red light) to drive photosynthesis (Blankenship et al. 2011; Chen et al 2010).

8.2.3.3 Establishment of Large-Scale Cultivation

New cultivation strategies aimed at increasing productivity and CO_2 fixation efficiency need to be developed, while simultaneously minimizing the land footprint required for effective CO_2 conversion. From low-cost harvesting methods to innovative culture media formulations and robust crop protection measures, advances in cultivation technology are being made (Pittman et al. 2011). Integrating these established technologies with the latest strains that exhibit high CO_2 fixation rates and productivity is crucial to optimize overall efficiency. This integrated approach not only ensures a more sustainable and resource-efficient cultivation process, but also meets the broader goals of environmental stewardship and carbon footprint reduction. These synergistic advances in the areas of large-scale cultivation systems need to be explored and implemented, in addition to the more general effects that gas flow management and reactor pressurization may have on productivity.

8.3 NONPHOTOSYNTHETIC PRODUCTION OF CHEMICALS FROM CO_2

8.3.1 Existing and Emerging Processes

Chemolithotrophs obtain ATP by oxidizing inorganic compounds instead of relying on solar energy (Kelly 1981). These microorganisms produce NAD(P)H by reversing the electron transport chain, accepting electrons from high redox potential donors. This ability allows them to circumvent challenges such as photorespiration and cell shading faced by photosynthetic organisms.

One class of chemolithotrophs, acetogens, has attracted commercial interest owing to their unique carbon fixation strategy known as the Wood-Ljungdahl pathway (WLP) (Ragsdale 2008). Under anaerobic conditions, the WLP converts H_2, CO_2, and/or CO into acetyl-CoA, an acyl carrier, generating ATP and acetate for further carbon anabolism (Pavan et al. 2022; Ragsdale 2008; Schiel-Bengelsdorf and Dürre 2012). The WLP comprises two branches: the methyl branch, which reduces CO_2 to formate, and the carbonyl branch, driven by the carbon monoxide dehydrogenase/acetyl-CoA synthase (CODH/ACS) enzyme complex, reduces CO_2 to CO and catalyzes the condensation of CO to C2 chemicals. Although the natural WLP is not efficient, optimal enzymes for the WLP have been identified through a combination of genome mining, enzymatic characterization, omics approaches, and kinetic modeling (Liew et al. 2022). The engineered WLP can fix CO_2 and synthesize various chemicals from CO_2. The primary product of the WLP, acetate, is not a particularly desirable product; however, stoichiometric modeling suggests that the WLP is highly efficient in carbon fixation owing to its effective use of reducing power to generate ATP and the low energy cost of producing acetyl-CoA (Fast and Papoutsakis 2012). Acetogens can recapture CO_2 generated during glycolysis, establishing a closed loop.

To address the need for greater reducing power, mixotrophic fermentation that uses other substrates in addition to CO_2 emerges as a potential solution. Research efforts have targeted systems that can provide the increased reducing power and CO_2 reassimilation without carbon catabolite repression (CCR) (Fast et al. 2015; Jones et al. 2016). For example, a mixotrophic fermentation strategy combining syngas and fructose in *Clostridium ljungdahlii* that aimed to enhance acetone production (Jones et al. 2016; Otten et al. 2022) demonstrated promising results, indicating that CCR was not occurring. Another two-stage lipid biosynthesis process, involving syngas-to-acetate production in *Moorella acetecia* followed by lipid synthesis in *Yarrowia lipolytica*, showcased the potential of a gas-to-lipids production scheme (Hu et al. 2013; Ruth and Stephanopoulos 2023). Additionally, a study on co-culturing *Clostridium ljungdahlii* and *Clostridium kluyveri* demonstrated efficient synthesis of long-chain alcohols from syngas (Diender et al. 2021; Richter et al. 2016).

Despite inherent limitations and challenges in these approaches (see Section 8.3.2), acetogenic fermentation remains an attractive platform for biological CO_2 utilization owing to the efficiency of the WLP and the diverse range of products achievable through co-culturing and genetic modification of the host. The ability to operate at ambient temperatures and pressures further simplifies and makes the scale-up process more cost, environmental, and energy efficient Systems based on algae production only require sunlight for the initial conversion of CO_2, whereas hybrid systems require electricity for this same conversion step (for which the availability of renewable electrons is critical). While both types of processes require additional energy for processing and separations, the energy requirements for purely biological systems are generally considered to be lower than chemical or hybrid systems.

8.3.2 Challenges

Although they can avoid challenges faced by photosynthetic organisms, chemolithotrophs have inherent limitations, including the need for multiple substrates, complex physiochemical cellular environments, and electron donors.[4] As noted above, acetogens are commercially intriguing owing to their unique carbon fixation strategy employing the WLP. However, carbon conversion yields vary, and a strict dependence on anaerobic conditions is required, restricting the range of products that can be synthesized (Bertsch and Muller 2015; Fast et al. 2015; Kopke and Simpson 2020; Molitor et al. 2016). Photomixotrophic production would alleviate these challenges but

[4] However, the ability of chemolithotrophs to be feedstock-agnostic presents a potential opportunity in geographic flexibility.

poses other limitations like CCR and the need to optimize for multiple substrates. Co-culturing studies showcase the potential of multiorganism production strategies, but also highlight challenges with optimizing conditions for both microorganisms, maintaining co-culture health, and competition for substrates. Regardless of the cultivation strategy, the overall carbon emission impacts have to be considered. For example, acetogens use H_2 or CO as electron donors. If these electron donors are generated from natural gas, the platform will be a derivative of petrochemical platform with net-positive carbon emissions. Section 8.4 discusses the alternative option of hybrid systems, where renewable electricity can be used to generate hydrogen and CO to drive acetogen conversion.

8.3.3 R&D Opportunities

R&D opportunities in chemolithotrophic CO_2 utilization include optimizing mixotrophic fermentation to address challenges like CCR, exploring synergistic gas-to-liquids production schemes such as those involving various gas-organic substrate combinations, improvements in genetic modification tools, and further developing efficient co-culturing strategies for synthesizing diverse products. Enhancing acetogenic fermentation presents opportunities to overcome inherent limitations and challenges, improve carbon conversion yields, and broaden the range of producible compounds. Additionally, research should focus on scaling up processes and exploring new gas-to-lipids production schemes for lipid-based products. The WLP has the potential to diversify products, and research into new applications of the WLP for synthesizing a wider range of chemicals than just acetate shows further potential in this field. These initiatives collectively aim to advance the efficiency, versatility, and commercial viability of the processes under consideration. Coupling the WLP with further biological conversions that utilize acetate (or other platform chemical product) to achieve higher value-added products is a worthwhile focus for R&D. Today, the most common platform chemical for biological conversion is sugar, which can also be made via biological processes.

8.4 HYBRID BIOLOGICAL SYSTEMS

8.4.1 Existing and Emerging Processes

Recent advances have suggested multiple viable paths to convert CO_2 into valuable and emissions-neutral products by combining catalytic processes (e.g., electrocatalysis, thermocatalysis, photocatalysis, or plasmacatalysis) with bioconversion (Gassler et al. 2020; Liu et al. 2016). A variety of catalysis-bio hybrid platforms have been developed, all of which require electron donors, which can be delivered as gas (e.g., hydrogen, CO), electric current, or electron- and energy-carrying soluble molecules (e.g., formate, methanol, acetate, ethanol) (Zhang et al. 2022). Hydrogen and CO are gas intermediates often used to drive acetogen conversion in gas fermentation, as described in Section 8.3. Energy-carrying soluble molecules are further categorized as C1 intermediates (e.g., formate and methanol) and C2 intermediates (e.g., acetate and ethanol). The C2 intermediates are compatible with a broader range of microorganisms than the C1 intermediates (Zhang et al. 2022). With these intermediates, several platforms have been developed by integrating electrocatalysis or thermocatalysis with cell-based or cell-free bioconversion systems. Some of these platforms have demonstrated superior energy conversion efficiency than the natural photosynthesis (Natelson et al. 2018; Ullah et al. 2023).

8.4.1.1 Electro-Bio Hybrid Systems

The integration of electrocatalysis with bioconversion for CO_2 utilization recently has emerged as a particularly attractive pathway owing to its ambient reaction conditions, potential to achieve higher efficiency and reaction rates, and ability to manufacture diverse products that could replace emission-intensive production of fuels, chemicals, and polymers (see Figures 8-1, 8-2, and 8-3). Because both electrocatalysis and bioconversion can operate at ambient temperature, electro-bio hybrid platforms have lower energy- and carbon-intensity than some other CO_2 conversion methods. The use of CO_2 as a substrate theoretically enables a higher carbon and energy efficiency compared to sugar-based bioproduction if the system is properly designed to leverage the high efficiency of catalysis (Tan and Nielsen 2022). Additionally, the kinetics or reaction rate also can be more favorable than in natural

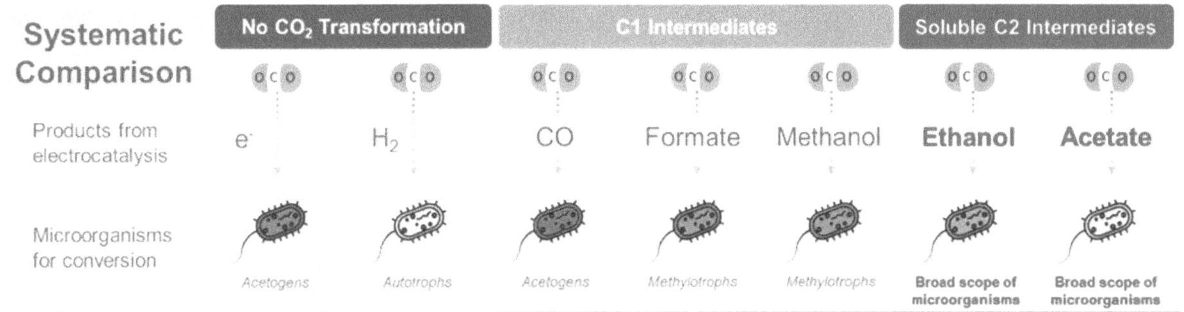

Systematic Comparison	No CO_2 Transformation		C1 Intermediates			Soluble C2 Intermediates	
Products from electrocatalysis	e^-	H_2	CO	Formate	Methanol	**Ethanol**	**Acetate**
Microorganisms for conversion	*Acetogens*	*Autotrophs*	*Acetogens*	*Methylotrophs*	*Methylotrophs*	**Broad scope of microorganisms**	**Broad scope of microorganisms**
Assimilation Pathway	W-L pathway	CBB cycle	W-L pathway	rGly pathway	RuMP cycle	Ethanol assimilation	Acetate assimilation
Net reducing equivalents	-4	-6	-2	-2	3	2	0
Net ATP per Carbon	-6.5	-9.33	-3.5	-4	2	2	-1
Steps to Acetyl-CoA	8	20	8	14	18	3	1
Electron carried	1	2	2	2	6	12	8
ΔH (kJ/mol)	/	-285.8	-283.0	-254.6	-726.5	-1367.6	-875.1
Mass transfer	★★★	★★★	★★★	★★★★★	★★★★★	★★★★★	★★★★★
Bio-compatibility	★★	★★★	★★	★★	★★★	★★★★	★★★★★

FIGURE 8-2 Comparisons of possible hybrid pathways to diverse products using integrated catalysis with bioconversion.
NOTES: The pathways in general have three classes. The first class leverages hydrogen and electrons to drive the CO_2 reduction and fixation by certain acetogens and chemolithotrophs. The second class first converts CO_2 into C1 intermediates like CO, formate, and methanol, and then converts these intermediates to other products. The second class often leverages the unique pathways of acetogens and methylotrophs. The third class utilizes the more biocompatible C2 intermediates like acetate and ethanol, which can be amenable to a much broader groups of microorganisms. Hydrogen and CO are gas intermediates. The assimilation pathways, net reducing equivalents, net ATP generation per carbon, steps to central metabolite acetyl-CoA (as molecular building block), numbers of electrons carried, reaction enthalpy, mass transfer capacity, and biocompatibility are compared.
SOURCE: Modified from Zhang et al. (2022), https://doi.org/10.1016/j.chempr.2022.09.005. CC BY 4.0.

photosynthetic processes, considering that the rapid catalytic processes bypass the slow carbon concentrating and RuBisCO carbon fixation steps (Zhang et al. 2022). Furthermore, an electro-bio platform can potentially leverage a wide array of diverse biological pathways to produce a wide range of chemicals, polymer precursors, and fuels, including longer carbon chain molecules that are more difficult or energy-intensive to access by conventional thermochemical, electrochemical, or photochemical conversion routes (Zhang et al. 2022).

Electro-bio hybrid systems have been developed using C1 intermediates, gas intermediates, and C2+ (i.e., biocompatible) intermediates (Figure 8-2). These various platforms for integrating catalysis and bioconversion have their advantages and drawbacks, as discussed in the following sections.

8.4.1.1.1 Electro-Bio Hybrid Systems with C1 Intermediates

In 2012, Li et al. first demonstrated the concept of an electro-bio conversion system, in which an electrocatalytic CO_2 reduction reaction (CO_2RR) was coupled with bioconversion using *Ralstonia eutropha* H16 to produce isobutanol (Li et al. 2012). This study showed that electrocatalytically derived formate was consumed by the microorganism and observed that hydrogen generated from electrocatalysis also might have helped to drive the CO_2 fixation and conversion in *R. eutropha*. The integration of electrocatalytic conversion of CO_2 into methanol and subsequent bioconversion also has been proposed (Guo et al. 2023). For both formate and methanol, substantial metabolic engineering has been carried out to enable the conversion of these C1 intermediates into various bioproducts (Chen et al. 2020). Despite the progress, recent work also indicates that it is very challenging to achieve high titer with the C1 soluble intermediates owing to the incompatibility with most of the industrial strains and the limited pathway kinetics for bioproducts generation (Figure 8-2). Even though gas fermentation with acetogens has been scaled up, most of the commercially relevant strains like *E. coli*, *P. putida*, and *S. cerevisiae* are not amenable to convert C1 intermediates into bioproducts at appreciable rate, efficiency, and titer. Natural methylotrophs and formatotrophs have been explored for more than half a century, yet large-scale production using these organisms remains challenging owing to a limited genetic toolbox and low fermentation titers. To overcome these challenges, recent work has focused on engineering platform industrial microorganisms into methylotrophs to achieve bioconversion (Reiter 2024). In fact, recent work has engineered *E. coli* to carry out methylotroph-type of conversion, yet the product titer remains low (Chen et al. 2020; Kim, 2020). Various pathways including a reductive glycine pathway and ribulose monophosphate (RuMP) cycle have been engineered or evolved in *E. coli*. Despite this progress, the inherent pathways for assimilation and conversion also pose challenges for C1 intermediate utilization considering the multiple steps needed to convert to central metabolic building blocks like acetyl-CoA (Figure 8-2), which limits the kinetics and rate of conversion. All these will translate into economic and scalability challenges of the platforms. Recent work has shown that it is possible to engineer and evolve *E. coli* to convert methanol into polyhydroxybutyrate (Reiter 2024), yet the titer and rate need to be further improved in future work. In order to utilize C1 intermediates more efficiently in electro-bio conversion, it is critical to study further how to engineer balanced and efficient conversion routes from these intermediates to a diverse range of products. Another option would be to leverage electrocatalysis, engineering electrocatalytic systems to produce more biocompatible and higher carbon products such as acetate, propionic acid, butyric acid, or even pyruvate, allowing the use of platform industrial microorganisms for bioconversion in proceeding steps.

8.4.1.1.2 Electro-Bio Hybrid Systems with Gas Intermediates

Hydrogen and CO have been used as the electron donors and intermediates in electro-bio hybrid systems for CO_2 conversion. In 2016, Liu et al. developed a system where electrocatalytically generated hydrogen drove the *R. eutropha* conversion of CO_2 into polyhydroxybutyrate (Liu et al. 2016). Electrocatalytic hydrogen-driven CO_2 conversion can achieve very high energy efficiency, but challenges with gas-to-liquid transfer may limit large-scale production. Other systems have been established for electro-bioconversion using CO and other intermediates (e.g., see Tan and Nielsen 2022). The CO-based platform is essentially the same as the nonphotosynthetic microorganisms discussed in Section 8.3. The CO can be derived from anaerobic digestion or CO_2 electroreduction. Overall, the proposed platforms can use CO, hydrogen, and CO_2 as substrates in various combinations with certain microorganisms, which opens various opportunities for CO_2 conversion to diverse molecules (Tan and Nielsen 2022).

8.4.1.1.3 Electro-Bio Hybrid Systems with Biocompatible Intermediates

The electro-bio platform relies heavily on the effective integration of electrocatalysis and fermentation. Recent advancements have demonstrated that two- and three-carbon (C2+) intermediates derived from a CO_2RR have much better compatibility with biological systems than gas or C1 intermediates, as more microbes, in particular, the industrial microorganisms like *Psuedomonas putida* can be used (see, e.g., Hann et al. 2022 and Figure 8-2). For example, Zhang et al. (2022) designed an integrated electro-bio conversion system, where a membrane electrode and phosphate buffer electrolyte enabled a CO_2RR to produce C2+ intermediates in a biocompatible environment. The synergistic design of catalysts, electrode, electrolyte, electro-bioconversion reactor, and microbial strains have enabled rapid microbial conversion of ethanol, acetate, and other intermediates from electrocatalytic CO_2RR into bioplastics in an integrated system (Figure 8-3).[5] This work and a recent study of a related photo-electro-bio inte-

[5] Bioplastics are a type of plastic derived from renewable biomass sources. Examples of bioplastics include polylactic acid (PLA), which is derived from corn starch or sugarcane, and polyhydroxyalkanoates (PHA), produced by bacteria through fermentation of sugar or plant oils. Another example is polyethylene derived from sugarcane ethanol, known as bio-based polyethylene. These bioplastics offer more sustainable alternatives to conventional plastics made from nonrenewable resources.

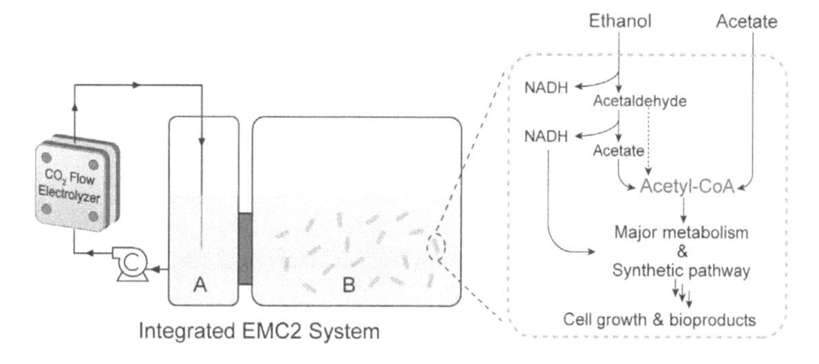

FIGURE 8-3 Example of a proof-of-concept electro-bioconversion process that utilizes ethanol and acetate.
NOTES: The figure provides a general overview of an integrated electro-bioconversion system, where the electrocatalytic CO_2RR is carried out in the CO_2 flow electrolyzer, and the produced C2 intermediates are fed into a bioreactor with two chambers (A and B). In this particular design, Chamber A contains the acetate, ethanol, and other CO_2RR products, but no bacteria. The membrane between Chambers A and B allows the CO_2RR-produced intermediates to transport freely to Chamber B, but does not allow the transport of bacteria from Chamber B to Chamber A. The microbial fermentation thus only happens in Chamber B and does not interfere with the electrocatalysis. The microbial engineering will convert ethanol and acetate to acetyl-CoA in few steps, which will then be converted to broad products. The same principle of this design can be broadly applied with different configurations, using different bacteria, intermediates, storage, and integration strategies for CO_2RR intermediates feeding into bioreactor.
SOURCE: Modified from Zhang et al. (2022), https://doi.org/10.1016/j.chempr.2022.09.005. CC BY 4.0.

gration achieved solar energy conversion efficiency to biomass of 4–4.5 percent, which is better than the terrestrial plant photosynthesis rate (Hann et al. 2022; Zhang et al. 2022). The relatively high energy conversion efficiency suggests the potential for these platforms to use low-cost renewable electrons to drive the conversion of CO_2 into various valuable products with limited land use.

8.4.1.2 Other Catalysis-Bioconversion Hybrid Systems

Beyond electro-bio conversion, thermocatalysis, plasmacatalysis, and photocatalysis all could be integrated with bioconversion. For example, Cai et al. (2021) demonstrated a route to convert CO_2 to starch by combining thermocatalytic CO_2 conversion to methanol with subsequent methanol conversion to starch through cell-free enzymatic systems. Semiconductor nanoparticles have been used to harvest sunlight to drive the bacteria *Moorella thermoacetica* to convert CO_2 to acetate (Sakimoto et al. 2016). Even though the titer and efficiency are still far from being commercially relevant, the study thus opens new avenues on how photocatalysis can be integrated with bioconversion to achieve CO_2 conversion to broad chemical products and polymer products (Hann et al. 2022). In principle, plasmacatalytic CO_2RR products can be converted into chemicals and polymers by bioconversion, too.

8.4.1.3 Cell-Free Hybrid Systems

A cell-free system is characterized by a lack of any cell walls or membranes or native DNA, potentially conferring benefits for product monitoring and purification. Cell-free systems integrate multiple enzyme steps in vitro to carry out a cascade of reactions for converting intermediates into different products. Cell-free systems also have the potential to remove competing pathways, resulting in higher efficiency (Yang et al. 2023). For the bioconversion component of a hybrid system, cell-free systems can be used in place of cell-based bioconversion, such as the abovementioned example of CO_2-to-starch conversion developed by Cai et al. (2021). Another example utilizes a cell-free system to convert electrocatalytic CO_2RR-derived ethanol to a chemical as a pharmaceutical precursor, although the yield is rather low (Jack et al. 2022). Cell-free systems can integrate with both photocatalysis and electrocatalysis, yet the photocatalytic integration mainly has yielded shorter carbon chain products. Regardless the route of integration, electron donors need to be available for CO_2RR, and these electron donors can be electrons or CO_2-derived energy dense intermediates like ethanol or CO (Jack et al. 2022).

8.4.2 Challenges

Despite the progress of hybrid CO_2 conversion systems, significant scalability, economic, and technical challenges remain in advancing them. For example, building scalable integrated reaction systems when combining bioconversion with thermo-, plasma-, and photo-catalytic processes is very challenging. In the thermocatalytic-bio hybrid system developed by Cai et al. (2021), separate reactors and steps were required for the thermocatalysis and bioconversion, and the cell-free enzyme reactions involved four different steps, without system integration. Integrated reactor and process design will be difficult for thermocatalytic and plasmachemical processes owing to the temperature constraints and incompatible reaction phases (e.g., gas/solid phase for thermochemical or plasmachemical catalysis versus liquid phase for bioconversion). To integrate photocatalysis with bioconversion, a remaining challenges is the availability of biocompatible intermediates, as photocatalytic CO_2RR generally yields C1 intermediates. Substantial R&D needs to be carried out to address these challenges in order to achieve system integration and improve the economics and scalability of hybrid conversion of CO_2.

For cell-free hybrid systems, the design of redox-balanced pathways to convert CO_2RR intermediates into targeted end products remains difficult. In particular, the reductant generated from the conversion of energy-dense intermediates like ethanol has to be sufficient to drive the downstream reactions to produce end products. Besides the reductant balance, most of the biological synthesis pathways require ATP. In biological systems, ATP is usually generated through a proton gradient across the membrane. Recent breakthroughs in membrane-free ATP generation could empower the design of a broader range of cell-free ATP generation from electricity (Luo et al. 2023). Enzyme stability and production costs also currently prevent cell-free systems from being utilized together with catalytic processes to achieve commercial relevance. Additionally, challenges related to scale up of cell-free systems include resource depletion or over-saturation, as well as system poisoning owing to accumulation of harmful by-products (Batista et al. 2021). For example, the CO_2-to-starch system developed by Cai et al. (2021) achieved bioconversion through four separate steps, each with a group of enzymes lasting only for four hours. Substantial research is needed to advance cell-free systems for the integration with catalytic processes. Computational methods and machine learning could be helpful in identifying metabolic bottlenecks (Batista et al. 2021).

8.4.3 R&D Opportunities

Hybrid systems that integrate bioconversion with various catalytic strategies have substantial potential to overcome limitations in the efficiency of natural systems. Nevertheless, substantial scientific and technology development are needed to improve efficiency, enhance scalability, decrease cost, and reduce life cycle carbon emissions. For hybrid systems using C1 intermediates, research on metabolic engineering and synthetic biology can help to improve conversion efficiency, kinetics, and titer. For gas fermentation, the systems are limited by microorganism selection—many conventional industrial strains along with their engineered functional modules cannot be used. Genetic engineering tools and pathways for gas-fermenting bacteria will need to be developed. Gas fermentation also has to overcome the fundamental limits of gas-to-liquid mass transfer, which can be accomplished by higher pressures, although this introduces new safety considerations in reactor design and higher energy costs. Pilot projects will be important to evaluate whether the current platforms can be commercially viable.[6] The technical barriers to improved economic and life cycle outcomes will have to be understood and overcome.

Among different systems, electro-bioconversion systems with biocompatible C2+ intermediates have substantial potential owing to their high efficiency, biocompatibility with the industrial strains, and improved system integration (see Figure 8-2). The C2 intermediates from electrocatalytic CO_2RR can improve electron, mass, and

[6] Lanzatech produces ethanol at commercial scale using syngas (see Kopke and Simpson 2020; Pavan et al. 2022).

energy transfer, shorten the steps to acetyl-CoA, and thus have the potential to improve the efficiency, titer, productivity, and ultimately economics and scalability of the electro-bio hybrid systems (see Figure 8-2 and Zhang et al. 2022). While recent progress has paved the way for efficient electron-to-molecule conversion, research needs to be carried out to improve performance, evaluate efficiency, and achieve commercial deployment. First, better fundamental understanding of electrocatalyst structure–function relationships is needed to improve Faradaic efficiency, product yield, and catalyst stability; reduce costs; and generate longer-carbon-chain products, particularly in biocompatible electrolytes. Current state-of-the-art systems can achieve 50–70 percent yield of acetate from CO_2, which is counted by calculating the number of carbon atoms in the acetate end products divided by the number of carbon atoms from CO_2 feeding into the system. In particularFor example, CO_2 electrolysis to CO has been demonstrated in a high-temperature solid oxide electrolysis system with nearly 100 percent yield (Hauch et al. 2020). For the CO electrolysis to acetate, multiple groups have reported a faradaic efficiency up to 80 percent. A recent report of a tandem reactor shows a CO-to-acetate faradaic efficiency of 50 percent with ethylene as the only side product of CO at about 30 percent. This selectivity can be maintained at a relatively high conversion (50–70 percent) (Overa et al. 2022). The recent one-step CO_2-to-C2 intermediate reactor can achieve over 25 percent Faradaic efficiency and high catalyst stability for CO_2 conversion to soluble C2 molecules without requiring rare earth metal catalysts in phosphate buffers (Zhang et al. 2022), yet the yield for C2+ intermediates needs to be further improved as compared to the tandem reactors.

Recent advances have shown that membrane electrodes and reactors together with bio-compatible electrolytes could achieve an integrated electro-bio system for the conversion of CO_2 to value- added chemicals, albeit at lower Faradaic efficiency than other state-of-the-art technologies in the field (Chen et al. 2020). Future catalyst design to convert CO_2 to C2 and longer carbon chain molecules instead of C1 molecules will enable better integration with bioconversion and improved efficiency, yield, stability, and scalability.

Additionally, the bioenergetic, biochemical, and metabolic limits for microbial conversion of CO_2RR products need to be better understood. For years, microbial engineering focused on carbohydrate substrates, which later expanded to industrially relevant compounds like lignin and glycerol (Lin et al. 2016). Substantial investment, particularly from the Department of Energy, recently expanded the understanding of formate and methanol metabolism, empowering the engineering of new conversion pathways and capacity in industrially relevant microbial strains like *E. coli* (Chen et al. 2020). Substantial work has been carried out to engineer methanol and formate conversion, yet the conversion efficiency and rate remain low (Chen et al. 2020). Recent research has focused on the bioconversion of electrocatalytic CO_2RR-derived C2 and C3 intermediates like acetate, ethanol, and propionic acids. These compounds need fewer steps to central metabolism and carry more energy and electrons, so the study of their use as bioconversion feedstocks for longer carbon chain products can help to advance strategies to improve conversion efficiency.

Last, bioconversion reactor design, system integration, and evaluation are needed if these systems are to reach commercial scale. It is important to build a reactor interface in which electrocatalytically derived methanol, formate, acetate, ethanol, and other products can be efficiently converted to diverse longer carbon chain products. Such integration is not trivial. For example, acetate and formate often form as salts during electrocatalysis, which could inhibit any concomitant microbial conversion. Among different possible intermediates, gas fermentation with CO has achieved commercialization. The integration of electrocatalysis-derived CO and hydrogen with anaerobic fermentation at scale also needs to be evaluated to understand the economic and life cycle impacts.

8.5 PRODUCTS ACCESSIBLE FROM BIOLOGICAL CO_2 UTILIZATION

8.5.1 Commodity Chemicals, Fuels, Food, and Pharmaceuticals

Photosynthetic and nonphotosynthetic organisms have been engineered to produce a wide range of fuels, commodity chemicals, food chemicals, and pharmaceutical chemicals. Fuels and chemicals are compelling circular economy targets, as sustainable routes to producing these valuable products will be required in a net-zero future. Most of the target products of biological systems are not durable in nature. CO_2 also can be used to produce animal

TABLE 8-1 Example Compounds Produced in Engineered Photosynthetic Organisms

Class	Compound	Reference(s)
Fuels	Ethanol	Dexter and Fu (2009)
	Butanol	Atsumi et al. (2009); Lan and Liao (2011)
	Octanol	Yunus et al. (2021)
	Isobutene	Mustila et al. (2021)
	Fatty acids	Ruffing (2014)
	Fatty acid methyl ester	Yunus et al. (2020)
	Alkanes and alkenes	Amer et al. (2020); Knoot and Pakrasi (2019)
	Limonene	Lin et al. (2021); Shinde et al. (2022); Wang et al. (2016)
	Bisabolene	Davies et al. (2014); Rodrigues and Lindberg (2021); Sebesta and Peebles (2020)
	Squalene	Choi et al. (2017)
Commodity Chemicals	2,3-butanediol	Kanno et al. (2017); Gonzales et al. (2023); Oliver et al. (2013)
	1,3-propanediol	Hirokawa et al. (2016)
	Ethylene	Mo et al. (2017); Wang et al. (2021)
	Glycogen	Aikawa et al. (2014); Ueno et al. (2017)
	Lactate	Li et al. (2015)
	3-hydroxypropanoic acid	Wang et al. (2016)
	3-hydroxybutanoic acid	Monshupanee et al. (2019); Zhang et al. (2015)
	4-hydroxybutanoic acid	Zhang et al. (2015)
	Isoprene	Gao et al. (2016); Zhou et al. (2021)
	Farnesene	Lee et al. (2017, 2021)
Food and Pharmaceutical Chemicals	Astaxanthin	Diao et al. (2020); Hasunuma et al. (2019)
	L-lysine	Dookeran and Nielsen (2021)
	L-phenylalanine	Kukil et al. (2023)
	Valencene	Dietsch et al. (2021); Sun et al. (2023)
	P-coumaric acid and Cinnamic acid	Brey et al. (2020)
	Riboflavin	Kachel and Mack (2020)
	Trehalose	Qiao et al. (2020)
	Heparosan	Sarnaik et al. (2019)
	1,8-cineole	Sakamaki et al. (2023)
	Isomaltulose	Wu et al. (2023)
Lipids and Terpenoids	Eicosapentaenoic acid and docosahexaenoic acid	Chauhan et al. (2023); Chisti (2007); Cui et al. (2021a, 2021b); Gao et al. (2022); Lu et al. (2019); Rehmanji et al. (2022)
	Ketocarotenoids and terpenoids	Amendola et al. (2023); Cazzaniga et al. (2022); Einhaus et al. (2022); Huang et al. (2021, 2023); Perozeni et al. (2020); Seger et al. (2023); Tokunaga et al. (2021); Vadrale et al. (2023); Wichmann et al. (2018); Yahya et al. (2023)

feed and food ingredients through bioconversion; however, larger scale opportunities rely mostly on microbial fermentation, algae cultivation, and nutrient recovery from waste streams (see Chapter 2).

The availability of a general platform for producing chemicals in photosynthetic organisms lags far behind that in heterotrophic model organisms such as yeast and *E. coli*. Both photosynthetic and nonphotosynthetic organisms utilize a similar core framework for essential metabolic processes, yet vast differences in substrate utilization and growth capacity exist between these microorganisms. The elucidation of factors and behaviors unique to photoautotrophic organisms would be useful in biochemical production. Table 8-1 provides a list of chemicals that have been produced in engineered photosynthetic organisms, expanded from the list in NASEM (2019). Nonphotosynthetic and hybrid systems can produce these same classes of chemicals yet can achieve a much broader product portfolio owing to extensive metabolic engineering efforts. These other products include many platform chemicals and polymer precursors, such as succinic acid and ethyl glycol (Chen et al. 2016; Gao et al. 2016).

However, scale up of many bioconversion processes remains a challenge—many of these products are produced in very small quantities at laboratory scale only. Furthermore, many of the chemicals are available at low cost from other routes, so any future processes must provide products at competitive prices.

Besides chemicals and polymers, CO_2 conversion to food and feed can be another route to reduce carbon emissions. For photosynthetic systems, some cyanobacteria and algal species have long been considered as protein sources for human consumption (e.g., Spirulina) and animal feed. Furthermore, recent advances have highlighted that hybrid systems can be used to produce food at a much higher efficiency than photosynthesis (Hann et al. 2022).

8.5.2 Polymer Precursors and Polymers

The plastics industry accounts for 4.5 percent of global carbon emissions (Cabernard 2022). Replacing emission-intensive processes and products will be critical to reduce the emissions associated with plastics manufacturing. Traditional bioplastics include products like polyhydroxyalkanoates, polyhydroxybutyrates, and polylactic acids, which are produced from biological sources and can be biodegradable or biocompostable (provided the waste infrastructure allows for the right degradation conditions, and in the case of compostables, a separate collection mechanism from recyclable plastics). These materials could replace emission-intensive plastics while also addressing daunting environmental challenges like accumulation of nondegradable waste and microplastics.

Photosynthetic, nonphotosynthetic, and hybrid microbial systems have been explored for biopolymer or polymer precursor production using CO_2 as the feedstock. Microbial conversion has been used widely for the production of polymer precursors such as 1,4-butanediol, 2,3-butanediol, succinic acid, isoprene, and others (Lee et al. 2011). Butanediol and succinic acid are precursors to polybutylene succinate. Microorganisms also have been engineered to produce polymer precursors like 2,3-butanediol (Kanno et al. 2017; Oliver et al. 2013) and succinic acid (Lan and Wei 2016; Treece et al. 2023). Considering their metabolic diversity, all three systems could be used to produce the precursors for a wide range of polymers.

For hybrid systems, an even more diverse range of microorganisms can be exploited to produce diverse polymers and polymer precursors from CO_2. R&D support could empower various hybrid platforms to be used broadly for producing polymer precursors from CO_2. Future research is needed on both fundamental and applied aspects of these systems. From a fundamental perspective, understanding the carbon flux control and bioenergetics of polymer precursor production will be helpful in identifying new pathways, improving productivity, conversion efficiency, and titer. From the applied side, advancement of new reactor designs and processes, and integration of carbon capture technologies will deliver integrated modules that directly convert CO_2 to industrially relevant precursors or polymers. The commercial deployment of these modules could have substantial impact on carbon emissions reductions, considering that they may replace current emissions intensive processes.

8.6 CONCLUSIONS

Many carbon-based products today are produced from bioconversion processes via an agriculture–fermentation route; however, terrestrial plants have limitations in CO_2 conversion owing to the low efficiency of photosynthesis and challenges with subsequent conversions to value-added chemicals. This chapter has presented processes for direct biological conversion of CO_2, whereby autotrophic microorganisms, acetogenic microbes, or hybrid systems use concentrated CO_2 sources as a feedstock for chemical production. These processes can combat some of the challenges associated with native biological fixation, but additional R&D is required to improve their performance and enable commercial-scale chemical production.

For photosynthetic systems, significant progress has been made in chemical production from CO_2. However, new engineering strategies are needed to enhance the efficiency of CO_2 fixation, enabling the more economical and scalable production systems. CO_2 fixation can be improved with continued refinement of genetic engineering tools and increased understanding of the design principles of photosynthetic metabolism. The intricate and interconnected nature of carbon metabolism makes predicting the effects of modifications challenging. Using mathematical models becomes crucial to describe and understand system behavior, aiding in the generation of testable hypotheses. Machine learning can enhance the training of these models for improved predictions. Furthermore, new cultivation strategies need to be developed to substantially improve productivity and thus reduce the land required for CO_2 conversion—for example, light-emitting diode (LED) technologies for reactor design (Porto et al. 2022). Other cultivation technologies, including low-cost harvest, media, and crop protection, have been developed but need to be integrated with the new strains with high CO_2 fixation rate. As in other systems, it is critical to carry out techno-economic assessments (TEAs) and life cycle assessments (LCAs) to evaluate the commercial potential and environmental impacts of photosynthetic CO_2 conversion platforms. This need is underscored by the many commercial challenges of systems biology-based efforts to produce commodity chemicals (Blois 2024).

Chemolithotrophs obtain ATP by oxidizing inorganic compounds and reversing the electron transport chain, offering an alternative energy source to photosynthetic organisms. The WLP in acetogens converts H_2, CO_2, and/or CO into acetyl-CoA, demonstrating efficiency in carbon fixation. Despite the limitations discussed above, acetogenic fermentation remains attractive for CO_2 utilization, with potential applications like mixotrophic fermentation and co-culturing strategies. Challenges include substrate dependency and competition. Research opportunities include optimizing mixotrophic fermentation, exploring gas-to-liquids production, and scaling up processes for broader product synthesis. Alternative options like hybrid systems using renewable electricity for electron donor generation are also possible, emphasizing the ongoing efforts to enhance efficiency, versatility, and commercial viability in chemolithotrophic CO_2 utilization.

Substantial R&D needs to be carried out to build efficient, economic, and scalable hybrid systems. For gaseous intermediates, demonstrating integration with catalytic processes at scale will be the priority. For C1 intermediates, continued microbial engineering research could lead to improvements in productivity and titer of integrated hybrid systems. For C2+ biocompatible intermediates, electro-bio hybrid systems have potential to achieve commercial and scalable production owing to their compatibility with many industrially relevant microorganisms, as well as improved electron, energy, and mass transfer, and fewer steps to acetyl-CoA. However, substantial research is still required in four primary areas: (1) development of efficient, selective, high-yield, cost-effective, and stable electrocatalysts for C2+ intermediates; (2) substantial advances in the fundamental understanding of metabolic and biochemical limits for C2+ intermediate conversion; (3) development of scalable reactor designs for system integration; and (4) development of highly efficient alternative bioconversion systems, including cell-free systems.

For all three routes (photosynthetic, nonphotosynthetic, and hybrid), TEAs and LCAs will be needed. Much work has been done on these assessments for algae biofuels, yet limited research has been carried out for hybrid and nonphotosynthetic systems (Handler et al. 2016; Liew et al. 2022). TEA and LCA carried out at the bench scale may identify the drivers and barriers for improving the system efficiency and economics. These assessments become even more important for pilot projects to determine commercial and environmental viability. For gas fermentation and photosynthetic systems, pilot and demonstration research and the relevant TEAs and LCAs are important to evaluate if these technologies could achieve commercialization and have a real impact on reducing carbon emissions.

8.6.1 Findings and Recommendations

Based on the above research needs, the committee makes the following findings and recommendations:

Finding 8-1: Opportunities to produce diverse products—Biological and hybrid systems have enormous opportunities to convert CO_2 to a variety of products for a circular carbon economy and carbon storage. However, key challenges exist, including low photosynthetic efficiency, low overall energy and carbon efficiency, and system integration and bioreactor design optimization. Both biological and hybrid systems merit further exploration because of their potential to perform selective conversions of CO_2 to a wide variety of products under mild conditions.

Recommendation 8-1: Coordination of fundamental and applied research is needed—Substantial fundamental and applied research needs to be conducted in order to understand and overcome biochemical, bioenergetic, and metabolic limits to higher reaction rates, conversion efficiency, and product titers. Various Department of Energy offices, including the Office of Science, Bioenergy Technologies Office, and Office of Fossil Energy and Carbon Management, as well as the Department of Defense, should coordinate fundamental and applied research to accelerate the advancement of efficient and implementable biochemical systems for carbon conversion.

Finding 8-2: Low productivities and titers are a barrier to commercialization—Most biochemical conversion of CO_2 work is still at an early stage, focused primarily on process optimization and not commercial products. The productivities and titers (mostly on the order of 1 mg/liter) for chemicals produced by these systems are too low to make commercialization of this technology appealing. Further improvements are needed to gain a sophisticated understanding of carbon metabolism and develop more efficient genetic manipulation tools (e.g., systems modeling with machine learning and cultivation optimization).

Recommendation 8-2: Support for advances in genetic engineering, systems modeling, and fundamental research is critical—New genetic engineering strategies must be developed to enhance the efficiency of CO_2 fixation, enabling the establishment of economical and scalable production systems. Continued refinement of genetic engineering tools and better understanding of the design principles of carbon metabolism are needed to improve CO_2 fixation. Additionally, systems modeling and machine learning can be exploited to optimize nutrient input, CO_2 delivery, light penetration, and other conditions to achieve higher productivities. Last, fundamental research to improve enzyme stability and the scalability of redox balanced systems is needed to make hybrid systems commercially viable and scalable. The Department of Energy's (DOE's) Office of Science and the National Science Foundation should continue to support fundamental research on these topics, and DOE's Bioenergy Technologies Office, Office of Fossil Energy and Carbon Management, and Advanced Research Projects Agency–Energy should support the related applied research.

Finding 8-3: Electro-bio hybrid systems require improvements in process engineering and techno-economic and life cycle assessments—Electro-bio hybrid systems for converting CO_2 to valuable chemicals have shown promise, especially to biocompatible intermediates like ethanol and acetate, but are still in their infancy. Achieving economically viable electro-bio conversion will require development of advanced technologies in catalyst design, microbial engineering, and process and reactor engineering. Besides improvements in process engineering, techno-economic and life cycle assessments are needed to evaluate the technical feasibility, commercial viability, and environmental impact of biochemical processes as compared to alternatives, such as chemical processes.

Recommendation 8-3: Explore electrocatalysts that operate under biologically amenable conditions with high activity, selectivity, and stability—The National Science Foundation and the Department of Energy's Office of Science, Bioenergy Technologies Office, Office of Fossil Energy and Carbon Management, and Advanced Research Projects Agency–Energy should support fundamental and applied research on the development of electrocatalysts that operate under biologically amenable conditions with high activity, selectivity, and stability. Systems that produce and utilize biocompatible (i.e., nontoxic, multicarbon) intermediates should be prioritized.

Recommendation 8-4: Develop microorganisms and cell-free systems that can efficiently produce target chemicals via intermediates derived from electrocatalysis—The National Science Foundation and the Department of Energy's Office of Science, Bioenergy Technologies Office, Office of Fossil Energy and Carbon Management, and Advanced Research Projects Agency–Energy should support fundamental

and applied research on the development of microorganisms and cell-free systems that can efficiently produce target chemicals from catalysis-derived intermediates under conditions amenable to electrocatalysis. These efforts should include systems biology understanding of the limitations for the conversion of various electrocatalysis-derived intermediates, in particular, biocompatible intermediates, as well as the synthetic biology engineering of microorganisms and cell-free systems for efficient conversion of these intermediates to chemicals, materials, and fuels.

Recommendation 8-5: Evaluate reactor design and system integration for hybrid systems—The Department of Energy's Bioenergy Technologies Office, Advanced Research Projects Agency–Energy, and Office of Fossil Energy and Carbon Management should support research to investigate system integration and scale up of catalysis and bioconversion. The reactor and process design need to be optimized to the specific intermediates and desired products, and techno-economic and life cycle assessments need to be carried out to evaluate the economic, environmental, and emissions impacts of hybrid systems.

Finding 8-4: Integration of thermocatalytic and photocatalytic CO_2 conversion with bioconversion—Only limited laboratory/research-scale examples have been reported of hybrid processes for CO_2 utilization that couple thermo- or photocatalytic CO_2 reduction with bioconversion. More research is needed to explore such systems—in particular, to improve and evaluate their efficiency, economics, and scalability.

Recommendation 8-6: Advance prototype hybrid systems to integrate thermocatalytic or photocatalytic CO_2 conversion with bioconversion—The Department of Energy's Bioenergy Technologies Office, Advanced Research Projects Agency–Energy, and Office of Fossil Energy and Carbon Management should support research to investigate the concept of integrating thermocatalytic or photocatalytic conversion of CO_2 into intermediates, and subsequently convert the intermediates via bioconversion to diverse chemical and polymer products. The evaluation of system efficiency and economics will help to assess whether such integrated systems are feasible.

8.6.2 Research Agenda for Biological CO_2 Conversion to Organic Products

Table 8-2 presents the committee's research agenda for biological CO_2 conversion to organic products, including research needs (numbered by chapter), and related research agenda recommendations (a subset of research-related recommendations from the chapter). The table includes the relevant funding agencies or other actors; whether the need is for basic research, applied research, technology demonstration, or enabling technologies and processes for CO_2 utilization; the research theme(s) that the research need falls into; the relevant research area and product class covered by the research need; whether the relevant product(s) are long- or short-lived; and the source of the research need (chapter section, finding, or recommendation). The committee's full research agenda can be found in Chapter 11.

TABLE 8-2 Research Agenda for Biological CO_2 Conversion to Organic Products

Research, Development, and Demonstration Need	Funding Agencies or Other Actors	Basic, Applied, Demonstration, or Enabling	Research Area	Product Class	Long- or Short-Lived	Research Themes	Source
8-A. Pathway modeling and metabolic engineering of microorganisms to overcome biochemical, bioenergetic and metabolic limits to enhance the efficiency, titer, and productivity of photosynthetic, nonphotosynthetic, and hybrid systems.	DOE-BES DOE-BER DOE-BETO DOE-FECM	Basic Applied	Biological	Chemicals Polymers	Short-lived	Metabolic understanding and engineering Reactor design and reaction engineering	Fin. 8-1 Rec. 8-1
Recommendation 8-1: Coordination of fundamental and applied research is needed—Substantial fundamental and applied research needs to be conducted in order to understand and overcome biochemical, bioenergetic, and metabolic limits to higher reaction rates, conversion efficiency, and product titers. Various Department of Energy offices, including the Office of Science, Bioenergy Technologies Office, and Office of Fossil Energy and Carbon Management, as well as the Department of Defense, should coordinate fundamental and applied research to accelerate the advancement of efficient and implementable biochemical systems for carbon conversion.							
8-B. New, more efficient genetic manipulation tools must be developed to enhance the efficiency of CO_2 fixation and improve the understanding of carbon metabolism. Computational modeling and machine learning can also be exploited to this end.	DOE-BES DOE-BER NSF	Basic	Biological	Chemicals	Short-lived	Metabolic understanding and engineering Genetic manipulation Computational modeling and machine learning	Fin. 8-2 Rec. 8-2
8-C. Improved enzyme efficiency, selectivity, and stability, along with multienzyme metabolon design to overcome biochemical limits for photosynthetic, nonphotosynthetic, and hybrid systems.	DOE-BES DOE-BER NSF DOE-BETO DOE-FECM DOE-ARPA-E	Basic Applied	Biological	Chemicals Polymers	Short-lived	Fundamental knowledge Computational modeling and machine learning Metabolic understanding and engineering	Fin. 8-1 Fin. 8-2 Rec. 8-1 Rec. 8-2
8-D. Improved enzyme stability and scalability of redox-balanced systems to facilitate demonstration and scale up of cell-free and hybrid systems.	DOE-BES DOE-BER NSF DOE-BETO DOE-FECM DOE-ARPA-E	Basic Applied	Biological	Chemicals Polymers	Short-lived	Fundamental knowledge Reactor design and reaction engineering Integrated systems	Rec. 8-2
Recommendation 8-2: Support for advances in genetic engineering, systems modeling, and fundamental research is critical—New genetic engineering strategies must be developed to enhance the efficiency of CO_2 fixation, enabling the establishment of economical and scalable production systems. Continued refinement of genetic engineering tools and better understanding of the design principles of carbon metabolism are needed to improve CO_2 fixation. Additionally, systems modeling and machine learning can be exploited to optimize nutrient input, CO_2 delivery, light penetration, and other conditions to achieve higher productivities. Last, fundamental research to improve enzyme stability and the scalability of redox balanced systems is needed to make hybrid systems commercially viable and scalable. The Department of Energy's (DOE's) Office of Science and the National Science Foundation should continue to support fundamental research on these topics, and DOE's Bioenergy Technologies Office, Office of Fossil Energy and Carbon Management, and Advanced Research Projects Agency–Energy should support the related applied research.							

continued

TABLE 8-2 Continued

Research, Development, and Demonstration Need	Funding Agencies or Other Actors	Basic, Applied, Demonstration, or Enabling	Research Area	Product Class	Long- or Short-Lived	Research Themes	Source
8-E. Improve fundamental understanding of electrocatalyst design to increase efficiency, selectivity, and product profile control under biocompatible conditions.	DOE-BES DOE-BER NSF DOE-BETO DOE-FECM DOE-ARPA-E	Basic Applied	Biological—Hybrid Electro-bio	Chemicals Polymers	Short-lived	Fundamental Knowledge Catalyst innovation and optimization	Fin. 8-3 Rec. 8-3
Recommendation 8-3: Explore electrocatalysts that operate under biologically amenable conditions with high activity, selectivity, and stability—The National Science Foundation and the Department of Energy's Office of Science, Bioenergy Technologies Office, Office of Fossil Energy and Carbon Management, and Advanced Research Projects Agency–Energy should support fundamental and applied research on the development of electrocatalysts that operate under biologically amenable conditions with high efficiency, selectivity, and stability. Systems that produce and utilize bio-compatible (i.e., nontoxic, multicarbon) intermediates should be prioritized.							
8-F. Develop microorganisms and cell-free systems compatible with intermediates derived from electrocatalysis.	DOE-BES DOE-BER NSF DOE-BETO DOE-FECM DOE-ARPA-E	Basic Applied	Biological—Hybrid Electro-bio	Chemicals Polymers	Short-lived	Microbial engineering	Rec. 8-4
Recommendation 8-4: Develop microorganisms and cell-free systems that can efficiently produce target chemicals via intermediates derived from electrocatalysis—The National Science Foundation and the Department of Energy's Office of Science, Bioenergy Technologies Office, Office of Fossil Energy and Carbon Management, and Advanced Research Projects Agency–Energy should support fundamental and applied research on the development of microorganisms and cell-free systems that can efficiently produce target chemicals from catalysis-derived intermediates under conditions amenable to electrocatalysis. These efforts should include systems biology understanding of the limitations for the conversion of various electrocatalysis-derived intermediates, in particular, biocompatible intermediates, as well as the synthetic biology engineering of microorganisms and cell-free systems for efficient conversion of these intermediates to chemicals, materials, and fuels.							
8-G. Optimization of hybrid systems via evaluation of reactor design.	DOE-BETO DOE-ARPA-E DOE-FECM	Applied Demonstration	Biological—Hybrid	Chemicals Polymers	Short-lived	Integrated Systems Reactor design and reaction engineering	Rec. 8-5
Recommendation 8-5: Evaluate reactor design and system integration for hybrid systems—The Department of Energy's Bioenergy Technologies Office, Advanced Research Projects Agency–Energy, and Office of Fossil Energy and Carbon Management should support research to investigate system integration and scale up of catalysis and bioconversion. The reactor and process design need to be optimized to the specific intermediates and desired products, and techno-economic and life cycle assessments need to be carried out to evaluate the economic, environmental, and emissions impacts of hybrid systems.							
8-H. Feasibility study for integrating thermocatalytic or photocatalytic CO_2 conversion with bioconversion to evaluate the efficiency, economics, and scalability.	DOE-ARPA-E DOE-FECM DOE-BETO	Applied	Biological—Hybrid	Chemicals Polymers	Short-lived	Reactor design and reaction engineering Integrated Systems	Rec. 8-6
Recommendation 8-6: Advance prototype hybrid systems that integrate thermocatalytic or photocatalytic CO_2 conversion with bioconversion—The Department of Energy's Bioenergy Technologies Office, Advanced Research Projects Agency–Energy, and Office of Fossil Energy and Carbon Management should support research to investigate the concept of integrating thermocatalytic or photocatalytic conversion of CO_2 into intermediates, and subsequently convert the intermediates via bioconversion to diverse chemical and polymer products. The evaluation of system efficiency and economics will help to assess whether such integrated systems are feasible.							

NOTE: ARPA-E = Advanced Research Projects Agency–Energy; BER = Biological and Environmental Research; BES = Basic Energy Sciences; BETO = Bioenergy Technologies Office; DOE = Department of Energy; FECM = Office of Fossil Energy and Carbon Management; NSF = National Science Foundation.

8.7 REFERENCES

Agarwal, P., R. Soni, P. Kaur, A. Madan, R. Mishra, J. Pandey, S. Singh, and G. Singh. 2022. "Cyanobacteria as a Promising Alternative for Sustainable Environment: Synthesis of Biofuel and Biodegradable Plastics." *Frontiers in Microbiology* 13:939347.

Aikawa, S., A. Nishida, S.-H. Ho, J.-S. Chang, T. Hasunuma, and A. Kondo. 2014. "Glycogen Production for Biofuels by the Euryhaline Cyanobacteria *Synechococcus* sp. Strain PCC 7002 from an Oceanic Environment." *Biotechnology for Biofuels* 7(1):88. https://doi.org/10.1186/1754-6834-7-88.

Al-Jabri, H., P. Das, S. Khan, M. AbdulQuadir, M.I. Thaher, K. Hoekman, and A.H. Hawari. 2022. "A Comparison of Bio-Crude Oil Production from Five Marine Microalgae—Using Life Cycle Analysis." *Energy* 251:123954.

Amendola, S., J.S. Kneip, F. Meyer, F. Perozeni, S. Cazzaniga, K.J. Lauersen, M. Ballottari, and T. Baier. 2023. "Metabolic Engineering for Efficient Ketocarotenoid Accumulation in the Green Microalga *Chlamydomonas reinhardtii*." *ACS Synthetic Biology* 12(3):820–831. https://doi.org/10.1021/acssynbio.2c00616.

Amer, M., E.Z. Wojcik, C. Sun, R. Hoeven, J.M.X. Hughes, M. Faulkner, I.S. Yunus, et al. 2020. "Low Carbon Strategies for Sustainable Bio-Alkane Gas Production and Renewable Energy." *Energy and Environmental Science* 13(6):1818–1831. https://doi.org/10.1039/D0EE00095G.

Atsumi, S., W. Higashide, and J.C. Liao. 2009. "Direct Photosynthetic Recycling of Carbon Dioxide to Isobutyraldehyde." *Nature Biotechnology* 27(12):1177–1180. https://doi.org/10.1038/nbt.1586.

Batista, A., P. Soudier, and M. Kushwaha, J.-L. 2021. "Optimizing Protein Synthesis in Cell-Free Systems, A Review." *Engineering Biology* 5(1):10–19. https://doi.org/10.1049/enb2.12004.

Behler, J., D. Vijay, W.R. Hess, and M.K. Akhtar. 2018. "CRISPR-Based Technologies for Metabolic Engineering in Cyanobacteria." *Trends in Biotechnology* 36(10):996–1010. https://doi.org/10.1016/j.tibtech.2018.05.011.

Bertsch, J., and V. Müller. 2015. "Bioenergetic Constraints for Conversion of Syngas to Biofuels in Acetogenic Bacteria." *Biotechnology for Biofuels* 8(1):210. https://doi.org/10.1186/s13068-015-0393-x.

Blankenship R.E., and M. Chen. 2013. "Spectral Expansion and Antenna Reduction Can Enhance Photosynthesis for Energy Production." *Current Opinion in Chemical Biology* 17(3):457–461. https://doi.org/10.1016/j.cbpa.2013.03.031.

Blankenship, R.E., D.M. Tiede, J. Barber, G.W. Brudvig, G. Fleming, M. Ghrardi and W. Zinth. 2011. "Comparing Photosynthetic and Photovoltaic Efficiencies and Recognizing the Potential for Improvement." *Science* 332(6031):805–809. https://www.science.org/doi/10.1126/science.1200165.

Blois, M. 2024. "Biomanufacturing Isn't Cleaning Up Chemicals: Synthetic Biology Firms Promised a Low-Carbon Industry, But So Far They Haven't Delivered." *Chemical and Engineering News* 102(12). https://cen.acs.org/business/biobased-chemicals/Biomanufacturing-isnt-cleaning-chemicals/102/i12.

Brey, L.F., A.J. Włodarczyk, J.F. Bang Thøfner, M. Burow, C. Crocoll, I. Nielsen, A.J. Zygadlo Nielsen, and P.E. Jensen. 2020. "Metabolic Engineering of *Synechococcus* sp. PCC 6803 for the Production of Aromatic Amino Acids and Derived Phenylpropanoids." *Metabolic Engineering* 57(January):129–139. https://doi.org/10.1016/j.ymben.2019.11.002.

Burkart, M.D., N. Hazari, C.L. Tway, and E.L. Zeitler. 2019. "Opportunities and Challenges for Catalysis in Carbon Dioxide Utilization." *ACS Catalysis* 9(9):7937–7956.

Cabernard, L., S. Pfister, C. Oberschelp, and S. Hellweg. 2022. "Growing Environmental Footprint of Plastics Driven by Coal Combustion." *Nature Sustainability* 5:139–148. https://doi.org/10.1038/s41893-021-00807-2.

Cai, T., H. Sun, J. Qiao, L. Zhu, F. Zhang, J. Zhang, Z. Tang, et al. 2021. "Cell-Free Chemoenzymatic Starch Synthesis from Carbon Dioxide." *Science* 373(6562):1523–1527. https://doi.org/10.1126/science.abh4049.

Cazzaniga, S., F. Perozeni, T. Baier, and M. Ballottari. 2022. "Engineering Astaxanthin Accumulation Reduces Photoinhibition and Increases Biomass Productivity Under High Light in *Chlamydomonas reinhardtii*." *Biotechnology for Biofuels and Bioproducts* 15(1):77. https://doi.org/10.1186/s13068-022-02173-3.

Chauhan, A.S., A.K. Patel, C.-W. Chen, J.-S. Chang, P. Michaud, C.-D. Dong, and R.R. Singhania. 2023. "Enhanced Production of High-Value Polyunsaturated Fatty Acids (PUFAs) from Potential Thraustochytrid *Aurantiochytrium sp*." *Bioresource Technology* 370(February):128536. https://doi.org/10.1016/j.biortech.2022.128536.

Chen, F.Y.-H., H.-W. Jung, C.-Y. Tsuei, and J.C. Liao. 2020. "Converting Escherichia Coli to a Synthetic Methylotroph Growing Solely on Methanol." *Cell* 182(4):933–946. https://doi.org/10.1016/j.cell.2020.07.010.

Chen, M., M. Schliep, R.D. Willows, Z.-L. Cai, B.A. Neilan and H. Scheer. 2010. "A Red-Shifted Chlorophyll." *Science* 329(5997):1318–1319. https://www.science.org/doi/10.1126/science.1191127.

Chen, Z., J. Haung, Y. Wu, and D. Liu. 2016 "Metabolic Engineering of Corynebacterium Glutamicum for the de Novo Production of Ethylene Glycol from Glucose." *Metabolic Engineering* 33:12–18. https://doi.org/10.1016/j.ymben.2015.10.013.

Cheng, P., Y. Li, C. Wang, J. Guo, C. Zhou, R. Zhang, Y. Ma, et al. 2022. "Integrated Marine Microalgae Biorefineries for Improved Bioactive Compounds: A Review." *Science of the Total Environment* 817:152895.

Chisti, Y. 2007. "Biodiesel from Microalgae." *Biotechnology Advances* 25(3):294–306. https://doi.org/10.1016/j.biotechadv.2007.02.001.

Choi, S.Y., J.-Y. Wang, H.S. Kwak, S.-M. Lee, Y. Um, Y. Kim, S.J. Sim, J. Choi, and H.M. Woo. 2017. "Improvement of Squalene Production from CO_2 in *Synechococcus elongatus* PCC 7942 by Metabolic Engineering and Scalable Production in a Photobioreactor." *ACS Synthetic Biology* 6(7):1289–1295. https://doi.org/10.1021/acssynbio.7b00083.

Cuellar-Bermudez, S.P., J.S. Garcia-Perez, B.E. Rittmann, and R. Parra-Saldivar. 2015. "Photosynthetic Bioenergy Utilizing CO_2: An Approach on Flue Gases Utilization for Third Generation Biofuels." *Journal of Cleaner Production* 98:53–65.

Cui, Y., S.R. Thomas-Hall, E.T. Chua, and P.M. Schenk. 2021a. "Development of High-Level Omega-3 Eicosapentaenoic Acid (EPA) Production from *Phaeodactylum tricornutum*." *Journal of Phycology* 57(1):258–268. https://doi.org/10.1111/jpy.13082.

Cui, Y., S.R. Thomas-Hall, E.T. Chua, and P.M. Schenk. 2021b. "Development of a *Phaeodactylum tricornutum* Biorefinery to Sustainably Produce Omega-3 Fatty Acids and Protein." *Journal of Cleaner Production* 300:126839. https://doi.org/10.1016/j.jclepro.2021.126839.

Cummins, P.L. 2021. "The Coevolution of RuBisCO, Photorespiration, and Carbon Concentrating Mechanisms in Higher Plants." *Frontiers in Plant Science* 12:662425. https://doi.org/10.3389/fpls.2021.662425.

Daneshvar, E., R.J. Wicker, P.-L. Show, and A. Bhatnagar. 2022. "Biologically-Mediated Carbon Capture and Utilization by Microalgae Towards Sustainable CO2 Biofixation and Biomass Valorization—A Review." *Chemical Engineering Journal* 427:130884. https://doi.org/10.1016/j.cej.2021.130884.

Davies, F.K., V.H. Work, A.S. Beliaev, and M.C. Posewitz. 2014. "Engineering Limonene and Bisabolene Production in Wild Type and a Glycogen-Deficient Mutant of *Synechococcus* sp. PCC 7002." *Frontiers in Bioengineering and Biotechnology* 2. https://www.frontiersin.org/articles/10.3389/fbioe.2014.00021.

Dexter, J., and P. Fu. 2009. "Metabolic Engineering of Cyanobacteria for Ethanol Production." *Energy and Environmental Science* 2(8):857–864. https://doi.org/10.1039/B811937F.

Diao, J., X. Song, L. Zhang, J. Cui, L. Chen, and W. Zhang. 2020. "Tailoring Cyanobacteria as a New Platform for Highly Efficient Synthesis of Astaxanthin." *Metabolic Engineering* 61(September):275–287. https://doi.org/10.1016/j.ymben.2020.07.003.

Diender, M., I. Parera Olm, and D.Z. Sousa. 2021. "Synthetic Co-Cultures: Novel Avenues for Bio-Based Processes." *Current Opinion in Biotechnology* 67(February):72–79. https://doi.org/10.1016/j.copbio.2021.01.006.

Dietsch, M., A. Behle, P. Westhoff, and I.M. Axmann. 2021. "Metabolic Engineering of *Synechocystis* sp. PCC 6803 for the Photoproduction of the Sesquiterpene Valencene." *Metabolic Engineering Communications* 13(December):e00178. https://doi.org/10.1016/j.mec.2021.e00178.

DOE (Department of Energy). 2024. "2023 Billion-Ton Report: An Assessment of U.S. Renewable Carbon Resources." M.H. Langholz (Lead). Oak Ridge, TN: Oak Ridge National Laboratory. ORNL/SPR-2024/3103. doi: 10.23720/BT2023/2316165.

Dookeran, Z.A., and D.R. Nielsen. 2021. "Systematic Engineering of *Synechococcus elongatus* UTEX 2973 for Photosynthetic Production of L-Lysine, Cadaverine, and Glutarate." *ACS Synthetic Biology* 10(12):3561–3575. https://doi.org/10.1021/acssynbio.1c00492.

Einhaus, A., J. Steube, R.A. Freudenberg, J. Barczyk, T. Baier, and O. Kruse. 2022. "Engineering a Powerful Green Cell Factory for Robust Photoautotrophic Diterpenoid Production." *Metabolic Engineering* 73(September):82–90. https://doi.org/10.1016/j.ymben.2022.06.002.

Fast, A.G., and E.T. Papoutsakis. 2012. "Stoichiometric and Energetic Analyses of Non-Photosynthetic CO2-Fixation Pathways to Support Synthetic Biology Strategies for Production of Fuels and Chemicals." *Current Opinion in Chemical Engineering* 1(4):380–395. https://doi.org/10.1016/j.coche.2012.07.005.

Gao, C., X. Yang, H. Wang, C. Perez Rivero, C. Li, Z. Cui, Q. Qi, and C.S. Lin. 2016. "Robust Succinc Acid Production from Crude Glycerol Using Engineered *Yarrowia lipolytica*." *Biotechnology for Biofuels and Bioproducts* 179. https://doi.org/10.1186/s13068-016-0597-8.

Gao, F., I.T. Dominguez Cabanelas, R.H. Wijffels, and M.J. Barbosa. 2022. "Fucoxanthin and Docosahexaenoic Acid Production by Cold-Adapted *Tisochrysis lutea*." *New Biotechnology* 66(January):16–24. https://doi.org/10.1016/j.nbt.2021.08.005.

Gao, X., F. Gao, D. Liu, H. Zhang, X. Nie, and C. Yang. 2016. "Engineering the Methylerythritol Phosphate Pathway in Cyanobacteria for Photosynthetic Isoprene Production from CO_2." *Energy and Environmental Science* 9(4):1400–1411. https://doi.org/10.1039/C5EE03102H.

Gassler, T., M. Sauer, B. Gasser, M. Egermeier, C. Troyer, T. Causon, S. Hann, D. Mattanovich, and M.G. Steiger. 2020. "The Industrial Yeast *Pichia pastoris* Is Converted from a Heterotroph into an Autotroph Capable of Growth on CO_2." *Nature Biotechnology* 38(2):210–216. https://doi.org/10.1038/s41587-019-0363-0.

Gonzales, J.N., T.R. Treece, S.P. Mayfield, R. Simkovsky, and S. Atsumi. 2023. "Utilization of Lignocellulosic Hydrolysates for Photomixotrophic Chemical Production in Synechococcus Elongatus PCC 7942." *Communications Biology* 6(1):1022. https://doi.org/10.1038/s42003-023-05394-w.

Griese, M., C. Lange, and J. Soppa. 2011. "Ploidy in Cyanobacteria." *FEMS Microbiology Letters* 323(2):124–131. https://doi.org/10.1111/j.1574-6968.2011.02368.x.

Guo, F., Y. Qiao, F. Xin, W. Zhang, and M. Jiang. 2023. "Bioconversion of C1 Feedstocks for Chemical Production Using *Pichia pastoris*." *Trends in Biotechnology* 41(8):1066–1079. https://doi.org/10.1016/j.tibtech.2023.03.006.

Hagemann, M., and H. Bauwe. 2016. "Photorespiration and the Potential to Improve Photosynthesis." *Energy Mechanistic Biology* 35(December):109–116. https://doi.org/10.1016/j.cbpa.2016.09.014.

Handler, R.M., D.R. Shonnard, E.M. Griffing, A. Lai, and I. Palou-Rivera. 2016. "Life Cycle Assessments of Ethanol Production via Gas Fermentation: Anticipated Greenhouse Gas Emissions for Cellulosic and Waste Gas Feedstocks." *Industrial & Engineering Chemistry Research* 55(12):3253–3261. https://doi.org/10.1021/acs.iecr.5b03215.

Hann, E.C., S. Overa, M. Harland-Dunaway, A.F. Narvaez, D.N. Le, M.L. Orozco-Cárdenas, F. Jiao, and R.E. Jinkerson. 2022. "A Hybrid Inorganic–Biological Artificial Photosynthesis System for Energy-Efficient Food Production." *Nature Food* 3(6):461–471. https://doi.org/10.1038/s43016-022-00530-x.

Hasunuma, T., A. Takaki, M. Matsuda, Y. Kato, C.J. Vavricka, and A. Kondo. 2019. "Single-Stage Astaxanthin Production Enhances the Nonmevalonate Pathway and Photosynthetic Central Metabolism in *Synechococcus* sp. PCC 7002." *ACS Synthetic Biology* 8(12):2701–2709. https://doi.org/10.1021/acssynbio.9b00280.

Hauch, A., R. Kungas, P. Blennow, A.B. Hansen, J.B. Hansen, B.V. Mathiesen, and M.B. Mogensen. 2020. "Recent Advances in Solid Oxide Cell Technology for Electrolysis." *Science*. https://doi.org/10.1126/science.aba6118

Hill, N.C., J.W. Tay, S. Altus, D.M. Bortz, and J.C. Cameron. 2020. "Life Cycle of a Cyanobacterial Carboxysome." *Science Advances* 6(19):eaba1269.

Hirokawa, Y., Y. Maki, Y. Tatsuke, and T. Hanai. 2016. "Cyanobacterial Production of 1,3-Propanediol Directly from Carbon Dioxide Using a Synthetic Metabolic Pathway." *Metabolic Engineering* 34:97–103. https://doi.org/10.1016/j.ymben.2015.12.008.

Hu, P., H. Rismani-Yazdi, and G. Stephanopoulos. 2013. "Anaerobic CO2 Fixation by the Acetogenic Bacterium *Moorella Thermoacetica*." *AIChE Journal* 59(9):3176–3183. https://doi.org/10.1002/aic.14127.

Huang, D., C. Liu, M. Su, Z. Zeng, C. Wang, Z. Hu, S. Lou, and H. Li. 2023. "Enhancement of β-Carotene Content in *Chlamydomonas reinhardtii* by Expressing Bacterium-Driven Lycopene β-Cyclase." *Biotechnology for Biofuels and Bioproducts* 16(1):127. https://doi.org/10.1186/s13068-023-02377-1.

Huang, P.-W., L.-R. Wang, S.-S. Geng, C. Ye, X.-M. Sun, and H. Huang. 2021. "Strategies for Enhancing Terpenoids Accumulation in Microalgae." *Applied Microbiology and Biotechnology* 105(12):4919–4930. https://doi.org/10.1007/s00253-021-11368-x.

Jack, J., H. Fu, A. Leininger, T.K. Hyster, and Z.J. Ren. 2022. "Cell-Free CO_2 Valorization to C6 Pharmaceutical Precursors via a Novel Electro-Enzymatic Process." *ACS Sustainable Chemistry and Engineering* 10(13):4114–4121.

Jones, S.W., A.G. Fast, E.D. Carlson, C.A. Wiedel, J. Au, M.R. Antoniewicz, E.T. Papoutsakis, and B.P. Tracy. 2016. "CO_2 Fixation by Anaerobic Non-Photosynthetic Mixotrophy for Improved Carbon Conversion." *Nature Communications* 7(1):12800. https://doi.org/10.1038/ncomms12800.

Kachel, B., and M. Mack. 2020. "Engineering of *Synechococcus* sp. Strain PCC 7002 for the Photoautotrophic Production of Light-Sensitive Riboflavin (Vitamin B2)." *Metabolic Engineering* 62(November):275–286. https://doi.org/10.1016/j.ymben.2020.09.010.

Kanno, M., A.L. Carroll, and S. Atsumi. 2017. "Global Metabolic Rewiring for Improved CO_2 Fixation and Chemical Production in Cyanobacteria." *Nature Communications* 8(1):14724. https://doi.org/10.1038/ncomms14724.

Kelly, D.P. 1981. "Chemolithotrophic Carbon Dioxide Fixation in Tube Worms—Symbiotic Primary Production." *Nature* 293(5834):609–609. https://doi.org/10.1038/293609b0.

Kerfeld, C.A, and M.R. Melnicki. 2016. "Assembly, Function and Evolution of Cyanobacterial Carboxysomes." *SI: 31: Physiology and Metabolism 2016* 31(June):66–75. https://doi.org/10.1016/j.pbi.2016.03.009.

Kim, S., S.N. Linder, S. Aslan, O. Yishai, S. Went, K. Schann, and A. Bar-Even. 2020. "Growth of *E. coli* on Formate and Methanol via the Reductive Glycine Pathway." *Nature chemical biology* 16:538–545. https://www.nature.com/articles/s41589-020-0473-5.

Knoot, C.J., and H.B. Pakrasi. 2019. "Diverse Hydrocarbon Biosynthetic Enzymes Can Substitute for Olefin Synthase in the Cyanobacterium Synechococcus Sp. PCC 7002." *Scientific Reports* 9(1):1360. https://doi.org/10.1038/s41598-018-38124-y.

Kopke, M., and S.D. Simpson. 2020. "Pollution to Products: Recycling of 'Above Ground' Carbon by Gas Fermentation." *Current Opinion in Biotechnology* 65:180–189. https://doi.org/10.1016/j.copbio.2020.02.017.

Kukil, K., E. Englund, N. Crang, E.P. Hudson, and P. Lindberg. 2023. "Laboratory Evolution of *Synechocystis* sp. PCC 6803 for Phenylpropanoid Production." *Metabolic Engineering* 79(September):27–37. https://doi.org/10.1016/j.ymben.2023.06.014.

Lan, E.I., and J.C. Liao. 2011. "Metabolic Engineering of Cyanobacteria for 1-Butanol Production from Carbon Dioxide." *Metabolic Engineering* 13(4):353–363. https://doi.org/10.1016/j.ymben.2011.04.004.

Lan, E.I., and C.T. Wei. 2016. "Metabolic Engineering of Cyanobacteria for the Photosynthetic Production of Succinate." *Metabolic Engineering* 38(November):483–493. https://doi.org/10.1016/j.ymben.2016.10.014.

Lee, H.J., J. Lee, S.-M. Lee, Y. Um, Y. Kim, S.J. Sim, J. Choi, and H.M. Woo. 2017. "Direct Conversion of CO_2 to α-Farnesene Using Metabolically Engineered *Synechococcus elongatus* PCC 7942." *Journal of Agricultural and Food Chemistry* 65(48):10424–10428. https://doi.org/10.1021/acs.jafc.7b03625.

Lee, H.J., J. Choi, and H.M. Woo. 2021. "Biocontainment of Engineered *Synechococcus elongatus* PCC 7942 for Photosynthetic Production of α-Farnesene from CO_2." *Journal of Agricultural and Food Chemistry* 69(2):698–703. https://doi.org/10.1021/acs.jafc.0c07020.

Lee, J.W., H.U. Kim, S. Choi, J. Yi, and S.Y. Lee. 2011. "Microbial Production of Building Block Chemicals and Polymers." *Current Opinion in Biotechnology* 22(6):758–767. https://doi.org/10.1016/j.copbio.2011.02.011.

Li, C., F. Tao, J. Ni, Y. Wang, F. Yao, and P. Xu. 2015. "Enhancing the Light-Driven Production of d-Lactate by Engineering Cyanobacterium Using a Combinational Strategy." *Scientific Reports* 5(1):9777. https://doi.org/10.1038/srep09777.

Li, H., P.H. Opgenorth, D.G. Wernick, S. Rogers, T.-Y. Wu, W. Higashide, P. Malati, Y.-X. Huo, K.M. Cho, and J.C. Liao. 2012. "Integrated Electromicrobial Conversion of CO2 to Higher Alcohols." *Science* 335(6076):1596. https://doi.org/10.1126/science.1217643.

Liew, F.E., R. Nogle, T. Abdalla, B.J. Rasor, C. Canter, R.O. Jensen, L. Wang, et al. 2022. "Carbon-Negative Production of Acetone and Isopropanol by Gas Fermentation at Industrial Pilot Scale." *Nature Biotechnology* 40:335–344. https://www.nature.com/articles/s41587-021-01195.

Lin, L., Y. Cheng, Y. Pu, S. Sun, X. Li, M. Jin, E.A. Pierson, et al. 2016. "Systems Biology-Guided Biodesign of Consolidated Lignin Conversion." *Green Chemistry* 18(20):5536–5547. https://doi.org/10.1039/C6GC01131D.

Lin, P.-C., F. Zhang, and H.B. Pakrasi. 2021. "Enhanced Limonene Production in a Fast-Growing Cyanobacterium Through Combinatorial Metabolic Engineering." *Metabolic Engineering Communications* 12(June):e00164. https://doi.org/10.1016/j.mec.2021.e00164.

Liu, C., B.C. Colón, M. Ziesack, P.A. Silver, and D.G. Nocera. 2016. "Water Splitting–Biosynthetic System with CO2 Reduction Efficiencies Exceeding Photosynthesis." *Science* 352(6290):1210–1213. https://doi.org/10.1126/science.aaf5039.

Liu, Z.-H., B.-Z. Li, J.S. Yuan, and Y.-J. Yuan. 2022. "Creative Biological Lignin Conversion Routes Toward Lignin Valorization." *Trends in Biotechnology* 40(12):1550–1566. https://doi.org/10.1016/j.tibtech.2022.09.014.

Long, B., B. Fischer, Y. Zeng, Z. Amerigian, Q. Li, H. Bryant, M. Li, S.Y. Dai, and J.S. Yuan. 2022. "Machine Learning-Informed and Synthetic Biology-Enabled Semi-Continuous Algal Cultivation to Unleash Renewable Fuel Productivity." *Nature Communications* 13(1):541. https://doi.org/10.1038/s41467-021-27665-y.

Lu, Z., B. Liu, Y. He, B. Guo, H. Sun, and F. Chen. 2019. "Novel Insights into Mixotrophic Cultivation of *Nitzschia laevis* for Co-Production of Fucoxanthin and Eicosapentaenoic Acid." *Bioresource Technology* 294(December):122145. https://doi.org/10.1016/j.biortech.2019.122145.

Luo, S., D. Adam, S. Giaveri, S. Barthel, S. Cestellos-Blanco, D. Hege, N. Paczia, et al. 2023. "ATP Production from Electricity with a New-to-Nature Electrobiological Module." *Joule* 7(8):1745–1758. https://doi.org/10.1016/j.joule.2023.07.012.

Mayfield, S.P., and M.D. Burkart. 2023. "Chapter 12—The Future of Biobased Polymers from Algae." Pp. 281–291 in *Rethinking Polyester Polyurethanes*, R.S. Pomeroy, ed. Elsevier. https://doi.org/10.1016/B978-0-323-99982-3.00012-2.

Miranda, A.M., F. Hernandez-Tenorio, D. Ocampo, G.J. Vargas, and A.A. Sáez. 2022. "Trends on CO_2 Capture with Microalgae: A Bibliometric Analysis." Molecules 27(15):4669.

Mo, H., X. Xie, T. Zhu, and X. Lu. 2017. "Effects of Global Transcription Factor NtcA on Photosynthetic Production of Ethylene in Recombinant *Synechocystis* sp. PCC 6803." *Biotechnology for Biofuels* 10(1):145. https://doi.org/10.1186/s13068-017-0832-y.

Molitor, B., H. Richter, M.E. Martin, R.O. Jensen, A. Juminaga, C. Mihalcea, and L.T. Angenent. 2016. "Carbon Recovery by Fermentation of CO-Rich off Gases—Turning Steel Mills into Biorefineries." *Bioresource Technology* 215(September):386–396. https://doi.org/10.1016/j.biortech.2016.03.094.

Monshupanee, T., C. Chairattanawat, and A. Incharoensakdi. 2019. "Disruption of Cyanobacterial γ-Aminobutyric Acid Shunt Pathway Reduces Metabolites Levels in Tricarboxylic Acid Cycle, But Enhances Pyruvate and Poly(3-Hydroxybutyrate) Accumulation." *Scientific Reports* 9(1):8184. https://doi.org/10.1038/s41598-019-44729-8.

Mustila, H., A. Kugler, and K. Stensjö. 2021. "Isobutene Production in Synechocystis Sp. PCC 6803 by Introducing α-Ketoisocaproate Dioxygenase from Rattus Norvegicus." *Metabolic Engineering Communications* 12(June):e00163. https://doi.org/10.1016/j.mec.2021.e00163.

NASEM (National Academies of Sciences, Engineering, and Medicine). 2019. *Gaseous Carbon Waste Streams Utilization: Status and Research Needs*. Washington, DC: The National Academies Press. https://doi.org/10.17226/25232.

Natelson, R., M. Resch, J. Fitzgerald, B. Hoffman, I. Rowe, and D. Babson. 2018. "Cell-Free Synthetic Biology and Biocatalysis (Listening Day Summary Report)." DOE/EE-1860. Washington, DC: Department of Energy Office of Energy Efficiency and Renewable Energy.

Nduko, J.M., and S. Taguchi. 2021. "Microbial Production of Biodegradable Lactate-Based Polymers and Oligomeric Building Blocks from Renewable and Waste Resources." *Frontiers in Bioengineering and Biotechnology* 8. https://www.frontiersin.org/articles/10.3389/fbioe.2020.618077.

Niyogi, K.K., and T.B. Truong. 2013. "Evolution of Flexible Non-Photochemical Quenching Mechanisms That Regulate Light Harvesting in Oxygenic Photosynthesis." *Current Opinion in Plant Biology* 16(3):307–314. https://doi.org/10.1016/j.pbi.2013.03.011.

Oliver, J.W.K., I.M.P. Machado, H. Yoneda, and S. Atsumi. 2013. "Cyanobacterial Conversion of Carbon Dioxide to 2,3-Butanediol." *Proceedings of the National Academy of Sciences* 110(4):1249–1254. https://doi.org/10.1073/pnas.1213024110.

Onyeaka, H., T. Miri, K. Obileke, A. Hart, C. Anumudu, and Z.T. Al-Sharify. 2021. "Minimizing Carbon Footprint via Microalgae as a Biological Capture." *Carbon Capture Science and Technology* 1:100007.

Otten, J.K., Y. Zou, and E.T. Papoutsakis. 2022. "The Potential of Caproate (Hexanoate) Production Using Clostridium Kluyveri Syntrophic Cocultures with Clostridium Acetobutylicum or Clostridium Saccharolyticum." *Frontiers in Bioengineering and Biotechnology* 10(August):965614. https://doi.org/10.3389/fbioe.2022.965614.

Overa, S., B.S. Crandall, B. Shrimant, D. Tian, B.H. Ko, H. Shin, C. Bae and F. Jiao. 2022. "Enhancing Acetate Selectivity by Coupling Anodic Oxidation to Carbon Monoxide Electroreduction." *Nature Catalysis* 5:738–745. https://www.nature.com/articles/s41929-022-00828-w.

Pavan, M., K. Reinmets, S. Garg, A.P. Mueller, E. Marcellin, M. Kopke, K. Valgepea. 2022. "Advances in Systems Metabolic Engineering of Autotrophic Carbon Oxide-Fixing Biocatalysts Towards a Circular Economy." *Metabolic Engineering* 71:117–141. https://doi.org/10.1016/j.ymben.2022.01.015

Pittman, J.K., A.P. Dean, and O. Osundeko. 2011. "The Potential of Sustainable Algal Biofuel Production Using Wastewater Resources." *Bioresource Technology* 102(1):17–25.

Perozeni, F., S. Cazzaniga, T. Baier, F. Zanoni, G. Zoccatelli, K.J. Lauersen, L. Wobbe, and M. Ballottari. 2020. "Turning a Green Alga Red: Engineering Astaxanthin Biosynthesis by Intragenic Pseudogene Revival in *Chlmydomonas reinhardtii*." *Plant Biotechnology Journal* 18(10):2053–2067. https://doi.org/10.1111/pbi.13364.

Porto, B., T.F.C.V. Silva, A.L. Goncalves, A.F. Esteves, S.M.A. Guelli U. de Souza, A.A.U. de Souza, J.C.M. Pires, and V.J. Vilar. 2022. "Tubular Photobioreactors Illuminated with LEDs to Boost Microalgal Biomass Production." *Chemical Engineering Journal* 435:134747. https://doi.org/10.1016/j.cej.2022.134747.

Qi, L.S., M.H. Larson, L.A. Gilbert, J.A. Doudna, J.S. Weissman, A.P. Arkin, and W.A. Lim. 2013. "Repurposing CRISPR as an RNA-Guided Platform for Sequence-Specific Control of Gene Expression." *Cell* 152(5):1173–1183. https://doi.org/10.1016/j.cell.2013.02.022.

Qiao, Y., W. Wang, and X. Lu. 2020. "Engineering Cyanobacteria as Cell Factories for Direct Trehalose Production from CO2." *Metabolic Engineering* 62(November):161–171. https://doi.org/10.1016/j.ymben.2020.08.014.

Ragsdale, S.W. 2008. *Enzymology of the Wood–Ljungdahl Pathway of Acetogenesis. Annals of the New York Academy of Sciences* 1125(1):129–136. https://doi.org/10.1196/annals.1419.015.

Rehmanji, M., A.A. Nesamma, N.J. Khan, T. Fatma, and P.P. Jutur. 2022. "Media Engineering in Marine Diatom *Phaeodactylum tricornutum* Employing Cost-Effective Substrates for Sustainable Production of High-Value Renewables." *Biotechnology Journal* 17(10):2100684. https://doi.org/10.1002/biot.202100684.

Reiter, M.A., T. Bradley, L.A. Buchel, P. Keller, E. Hegedis, T. Gassler and J.A. Vorholt. 2024. "A Synthetic Methylotrophic *Escherichia coli* as a Chassis for Bioproduction from Methanol." *Nature Catalysis* 7:560–573. https://doi.org/10.1038/s41929-024-01137-0.

Rezvani, F., M.-H. Sarrafzadeh, and H.-M. Oh. 2020. "Hydrogen Producer Microalgae in Interaction with Hydrogen Consumer Denitrifiers as a Novel Strategy for Nitrate Removal from Groundwater and Biomass Production." *Algal Research* 45:101747.

Richter, H., B. Molitor, M. Diender, D.Z. Sousa, and L.T. Angenent. 2016. "A Narrow pH Range Supports Butanol, Hexanol, and Octanol Production from Syngas in a Continuous Co-Culture of Clostridium Ljungdahlii and Clostridium Kluyveri with In-Line Product Extraction." *Frontiers in Microbiology* 7(November). https://doi.org/10.3389/fmicb.2016.01773.

Rodrigues, J.S., and P. Lindberg. 2021. "Metabolic Engineering of *Synechocystis* sp. PCC 6803 for Improved Bisabolene Production." *Metabolic Engineering Communications* 12(June):e00159. https://doi.org/10.1016/j.mec.2020.e00159.

Roh, K., A. Bardow, D. Bongartz, J. Burre, W. Chung, S. Deutz, D. Han et al. 2020. "Early-Stage Evaluation of Emerging CO_2 Utilization Technologies at Low Technology Readiness Levels." *Green Chemistry* 22(12):3842–3859.

Ruffing, A.M. 2014. "Improved Free Fatty Acid Production in Cyanobacteria with *Synechocystis* sp. PCC 7002 as Host." *Frontiers in Bioengineering and Biotechnology* 2. https://www.frontiersin.org/articles/10.3389/fbioe.2014.00017.

Ruth, J.C., and G. Stephanopoulos. 2023. "Synthetic Fuels: What Are They and Where Do They Come From?" *Current Opinion in Biotechnology* 81(June):102919. https://doi.org/10.1016/j.copbio.2023.102919.

Sakamaki, Y., M. Ono, N. Shigenari, T. Chibazakura, K. Shimomura, and S. Watanabe. 2023. "Photosynthetic 1,8-Cineole Production Using Cyanobacteria." *Bioscience, Biotechnology, and Biochemistry* 87(5):563–568. https://doi.org/10.1093/bbb/zbad012.

Sakimoto, K.K., A.B. Wong, and P. Yang. 2016. "Self-Photosensitization of Nonphotosynthetic Bacteria for Solar-to-Chemical Production." *Science* 351(6268):74–77.

Santos, L., K. Mention, K. Cavusoglu-Doran, D.J. Sanz, M. Bacalhau, M. Lopes-Pacheco, P.T. Harrison, and C.M. Farinha. 2021. "Comparison of Cas9 and Cas12a CRISPR Editing Methods to Correct the W1282X-CFTR Mutation." *Journal of Cystic Fibrosis* 21(1):181–187. https://doi.org/10.1016/j.jcf.2021.05.014.

Sarnaik, A., M.H. Abernathy, X. Han, Y. Ouyang, K. Xia, Y. Chen, B. Cress, et al. 2019. "Metabolic Engineering of Cyanobacteria for Photoautotrophic Production of Heparosan, a Pharmaceutical Precursor of Heparin." *Algal Research* 37(January):57–63. https://doi.org/10.1016/j.algal.2018.11.010.

Sebesta, J., and C.A.M. Peebles. 2020. "Improving Heterologous Protein Expression in *Synechocystis* sp. PCC 6803 for Alpha-Bisabolene Production." *Metabolic Engineering Communications* 10(June):e00117. https://doi.org/10.1016/j.mec.2019.e00117.

Seger, M., F. Mammadova, M. Villegas-Valencia, B. Bastos de Freitas, C. Chang, I. Isachsen, H. Hemstreet, et al. 2023. "Engineered Ketocarotenoid Biosynthesis in the Polyextremophilic Red Microalga *Cyanidioschyzon merolae* 10D." *Metabolic Engineering Communications* 17(December):e00226. https://doi.org/10.1016/j.mec.2023.e00226.

Sengupta, S., D. Jaiswal, A. Sengupta, S. Shah, S. Gadagkar, and P.P. Wangikar. 2020a. "Metabolic Engineering of a Fast-Growing Cyanobacterium Synechococcus Elongatus PCC 11801 for Photoautotrophic Production of Succinic Acid." *Biotechnology for Biofuels* 13(1):89. https://doi.org/10.1186/s13068-020-01727-7.

Sengupta, A., P. Pritam, D. Jaiswal, A. Bandyopadhyay, H.B. Pakrasi, and P.P. Wangikar. 2020b. "Photosynthetic Co-Production of Succinate and Ethylene in a Fast-Growing Cyanobacterium, Synechococcus Elongatus PCC 11801." *Metabolites* 10(6). https://doi.org/10.3390/metabo10060250.

Schiel-Bengelsdorf, B., and P. Dürre. 2012. "Pathway Engineering and Synthetic Biology Using Acetogens." *FEBS Letters* 586(15):2191–2198. https://doi.org/10.1016/j.febslet.2012.04.043.

Shih, P.M., A. Occhialini, J.C. Cameron, P.J. Andralojc, M.A.J. Parry, and C.A. Kerfeld. 2016. "Biochemical Characterization of Predicted Precambrian RuBisCO." *Nature Communications* 7(1):10382. https://doi.org/10.1038/ncomms10382.

Shinde, S., S. Singapuri, Z. Jiang, B. Long, D. Wilcox, C. Klatt, J.A. Jones, J.S. Yuan, and X. Wang. 2022. "Thermodynamics Contributes to High Limonene Productivity in Cyanobacteria." *Metabolic Engineering Communications* 14(June):e00193. https://doi.org/10.1016/j.mec.2022.e00193.

Sirohi, R., J.P. Pandey, V.K. Gaur, E. Gnansounou, and R. Sindhu. 2020. "Critical Overview of Biomass Feedstocks as Sustainable Substrates for the Production of Polyhydroxybutyrate (PHB)." *Bioresource Technology* 311(September):123536. https://doi.org/10.1016/j.biortech.2020.123536.

Sirohi, R., J.S. Lee, B.S. Yu, H. Roh, and S.J. Sim. 2021. "Sustainable Production of Polyhydroxybutyrate from Autotrophs Using CO_2 as Feedstock: Challenges and Opportunities." *Bioresource Technology* 341:125751.

Smith, J.P., B.J. Limb, C.M. Beal, K.R. Banta, J.L. Field, S.J. Simske, and J.C. Quinn. 2023. "Evaluating the Sustainability of the 2017 US Biofuel Industry with an Integrated Techno-Economic Analysis and Life Cycle Assessment." *Journal of Cleaner Production* 413(August):137364. https://doi.org/10.1016/j.jclepro.2023.137364.

Song, X., J. Diao, J. Yao, J. Cui, T. Sun, L. Chen, and W. Zhang. 2021. "Engineering a Central Carbon Metabolism Pathway to Increase the Intracellular Acetyl-CoA Pool in Synechocystis Sp. PCC 6803 Grown Under Photomixotrophic Conditions." *ACS Synthetic Biology* 10(4):836–846. https://doi.org/10.1021/acssynbio.0c00629.

Sosa-Hernández, J.E., Z. Escobedo-Avellaneda, H.M.N. Iqbal, and J. Welti-Chanes. 2018. "State-of-the-Art Extraction Methodologies for Bioactive Compounds from Algal Biome to Meet Bio-Economy Challenges and Opportunities." Molecules 23(11):2953.

Sun, J., X. Xu, Y. Wu, H. Sun, G. Luan, and X. Lu. 2023. "Conversion of Carbon Dioxide into Valencene and Other Sesquiterpenes with Metabolic Engineered *Synechocystis* sp. PCC 6803 Cell Factories." *GCB Bioenergy* 15(9):1154–1165. https://doi.org/10.1111/gcbb.13086.

Sun, N., R. Skaggs, M. Wignosta, A. Coleman, M. Huesemann, and S. Edmundson. 2020. "Growth Modeling to Evaluate Alternative Cultivation Strategies to Enhance National Microalgal Biomass Production." Pacific Northwest National Laboratory. https://www.osti.gov/biblio/1638493.

Tan, X., and J. Nielsen. 2022. "The Integration of Bio-Catalysis and Electrocatalysis to Produce Fuels and Chemicals from Carbon Dioxide." *Chemical Society Reviews* 51(11):4763–4785. https://doi.org/10.1039/D2CS00309K.

Tokunaga, S., D. Morimoto, T. Koyama, Y. Kubo, M. Shiroi, K. Ohara, T. Higashine, Y. Mori, S. Nakagawa, and S. Sawayama. 2021. "Enhanced Lutein Production in *Chlamydomonas reinhardtii* by Overexpression of the Lycopene Epsilon Cyclase Gene." *Applied Biochemistry and Biotechnology* 193(6):1967–1978. https://doi.org/10.1007/s12010-021-03524-w.

Treece, T.R., M. Tessman, R.S. Pomeroy, S.P. Mayfield, R. Simkovsky, and S. Atsumi. 2023. "Fluctuating pH for Efficient Photomixotrophic Succinate Production." *Metabolic Engineering* 79(September):118–129. https://doi.org/10.1016/j.ymben.2023.07.008.

Ueno, K., Y. Sakai, C. Shono, I. Sakamoto, K. Tsukakoshi, Y. Hihara, K. Sode, and K. Ikebukuro. 2017. "Applying a Riboregulator as a New Chromosomal Gene Regulation Tool for Higher Glycogen Production in *Synechocystis* sp. PCC 6803." *Applied Microbiology and Biotechnology* 101(23):8465–8474. https://doi.org/10.1007/s00253-017-8570-4.

Ullah, M.W., S. Manan, M. Ul Islam, W.A. Khattak, K.A. Khan, J. Liu, G. Yang, and J Sun. 2023. "Cell-free Systems for Biosynthesis: Towards a Sustainable and Economical Approach." *Green Chemistry*.

Ungerer, J., H.-Y. Chen, and H.B. Pakrasi. 2018. "Adjustments to Photosystem Stoichiometry and Electron Transfer Proteins Are Key to the Remarkably Fast Growth of the Cyanobacterium Synechococcus Elongatus UTEX 2973." *mBio* 9(1):10.1128/mbio.02327-17. https://doi.org/10.1128/mbio.02327-17.

Vadrale, A.P., C.-D. Dong, D. Haldar, C.-H. Wu, C.-W. Chen, R.R. Singhania, and A.K. Patel. 2023. "Bioprocess Development to Enhance Biomass and Lutein Production from *Chlorella sorokiniana* Kh12." *Bioresource Technology* 370(February):128583. https://doi.org/10.1016/j.biortech.2023.128583.

Wang, B., Y. Xu, X. Wang, J.S. Yuan, C.H. Johnson, J.D. Young, and J. Yu. 2021. "A Guanidine-Degrading Enzyme Controls Genomic Stability of Ethylene-Producing Cyanobacteria." *Nature Communications* 12(1):5150. https://doi.org/10.1038/s41467-021-25369-x.

Wang, X., W. Liu, C. Xin, Y. Zheng, Y. Cheng, S. Sun, R. Li, et al. 2016. "Enhanced Limonene Production in Cyanobacteria Reveals Photosynthesis Limitations." *Proceedings of the National Academy of Sciences* 113(50):14225–14230. https://doi.org/10.1073/pnas.1613340113.

Wang, Y., T. Sun, X. Gao, M. Shi, L. Wu, L. Chen, and W. Zhang. 2016. "Biosynthesis of Platform Chemical 3-Hydroxypropionic Acid (3-HP) Directly from CO2 in Cyanobacterium *Synechocystis* sp. PCC 6803." *Metabolic Engineering* 34(March):60–70. https://doi.org/10.1016/j.ymben.2015.10.008.

Wichmann, J., T. Baier, E. Wentnagel, K.J. Lauersen, and O. Kruse. 2018. "Tailored Carbon Partitioning for Phototrophic Production of (E)-α-Bisabolene from the Green Microalga *Chlamydomonas reinhardtii*." *Metabolic Engineering* 45(January):211–222. https://doi.org/10.1016/j.ymben.2017.12.010.

Wu, Y., J. Sun, X. Xu, S. Mao, G. Luan, and X. Lu. 2023. "Engineering Cyanobacteria for Converting Carbon Dioxide into Isomaltulose." *Journal of Biotechnology* 364(February):1–4. https://doi.org/10.1016/j.jbiotec.2023.01.007.

Yahya, R.Z., G.B. Wellman, S. Overmans, and K.J. Lauersen. 2023. "Engineered Production of Isoprene from the Model Green Microalga *Chlamydomonas reinhardtii*." *Metabolic Engineering Communications* 16(June):e00221. https://doi.org/10.1016/j.mec.2023.e00221.

Yang, J., C.-K. Sou, and Y. Lu. 2023. "Cell-Free Biocatalysis Coupled with Photo-Catalysis and Electro-Catalysis: Efficient CO_2-to-Chemical Conversion." *Green Energy and Environment*.

Yao, J., J. Wang, Y. Ju, Z. Dong, X. Song, L. Chen, and W. Zhang. 2022. "Engineering a Xylose-Utilizing Synechococcus Elongatus UTEX 2973 Chassis for 3-Hydroxypropionic Acid Biosynthesis Under Photomixotrophic Conditions." *ACS Synthetic Biology* 11(2):678–688. https://doi.org/10.1021/acssynbio.1c00364.

Yu, J., M. Liberton, P.F. Cliften, R.D. Head, J.M. Jacobs, R.D. Smith, D.W. Koppenaal, J.J. Brand, and H.B. Pakrasi. 2015. "Synechococcus Elongatus UTEX 2973, a Fast Growing Cyanobacterial Chassis for Biosynthesis Using Light and CO_2." *Scientific Reports* 5(1):8132. https://doi.org/10.1038/srep08132.

Yuan, J.S., M.J. Pavlovich, A.J. Ragauskas, and B. Han. 2022. "Biotechnology for a Sustainable Future: Biomass and Beyond." *Trends in Biotechnology* 40(12):1395–1398. https://doi.org/10.1016/j.tibtech.2022.09.020.

Yunus, I.S., A. Palma, D.L. Trudeau, D.S. Tawfik, and P.R. Jones. 2020. "Methanol-Free Biosynthesis of Fatty Acid Methyl Ester (FAME) in Synechocystis Sp. PCC 6803." *Metabolic Engineering* 57(January):217–227. https://doi.org/10.1016/j.ymben.2019.12.001.

Yunus, I.S., Z. Wang, P. Sattayawat, J. Muller, F.W. Zemichael, K. Hellgardt, and P.R. Jones. 2021. "Improved Bioproduction of 1-Octanol Using Engineered Synechocystis Sp. PCC 6803." *ACS Synthetic Biology* 10(6):1417–1428. https://doi.org/10.1021/acssynbio.1c00029.

Zhang, P., K. Chen, B. Xu, J. Li, C. Hu, J.S. Yuan, and S.Y. Dai. 2022. "Chem-Bio Interface Design for Rapid Conversion of CO_2 to Bioplastics in an Integrated System." *Chem* 8(12):3363–3381. https://doi.org/10.1016/j.chempr.2022.09.005.

Zhang, S., Y. Liu, and D.A. Bryant. 2015. "Metabolic Engineering of *Synechococcus* sp. PCC 7002 to Produce Poly-3-Hydroxybutyrate and Poly-3-Hydroxybutyrate-Co-4-Hydroxybutyrate." *Metabolic Engineering* 32(November):174–183. https://doi.org/10.1016/j.ymben.2015.10.001.

Zhou, J., F. Yang, F. Zhang, H. Meng, Y. Zhang, and Y. Li. 2021. "Impairing Photorespiration Increases Photosynthetic Conversion of CO_2 to Isoprene in Engineered Cyanobacteria." *Bioresources and Bioprocessing* 8(1):42. https://doi.org/10.1186/s40643-021-00398-y.

Zhu, X.-G., S.P. Long, and D.R. Ort. 2010. "Improving Photosynthetic Efficiency for Greater Yield." *Annual Review of Plant Biology* 61(1):235–261. https://doi.org/10.1146/annurev-arplant-042809-112206.

9

Products from Coal Waste

This chapter discusses opportunities for the beneficial repurposing of wastes generated from the coal supply chain (i.e., coal waste). It first describes the sources, compositions, and locations of coal wastes in the United States and then outlines the potential market opportunities for carbon-based products and critical minerals and materials that could be derived from coal waste. Considerations for repurposing coal waste are examined, including separations, existing and emerging applications, and product safety.

9.1 COAL WASTE COMPOSITION

In 2022, the United States consumed 513 million short tons of coal, with 92 percent dedicated to generating electricity (EIA 2023b). Of the remainder, industrial uses predominated, notably coke production, which accounted for 3.1 percent. This contrasts with the peak coal usage in 2007, when nearly 1130 million short tons of coal were used (EIA 2023b). While coal production and use are expected to continue declining in the United States over the coming years (NASEM 2024), the significant quantities of legacy waste streams contain valuable components that could be repurposed for societal benefit.

Coal waste streams considered in this chapter as potential feedstocks for carbon-based products and/or sources of critical minerals and materials include acid mine drainage (AMD), coal impoundment wastes, and coal combustion residuals (CCRs).[1] As illustrated in Figure 9-1, these wastes are generated throughout the coal supply chain, from mining, to processing and preparation, to combustion at a power plant or another industrial facility. AMD is characterized by the release of acidic and metal-laden water from abandoned coal mines and is a significant environmental problem primarily associated with historical coal mining, particularly in the Appalachian region. This acidic runoff can harm aquatic ecosystems, corrode infrastructure, and contaminate drinking water sources, posing serious environmental and public health challenges. Coal impoundment waste, often referred to as coal slurry, coal refuse, or coal sludge, is a by-product generated during the processing and cleaning of coal. Impoundment waste is a mixture of water, coal fines (small particles of coal), and other substances generated during coal mining and processing activities. This waste material is typically stored in large containment structures called impoundments,

[1] Mineral-dominated portions of coal beds (e.g., underclays, partings) also have been identified as potential sources of rare earth elements (Kolker et al. 2024) but are out of scope for this report.

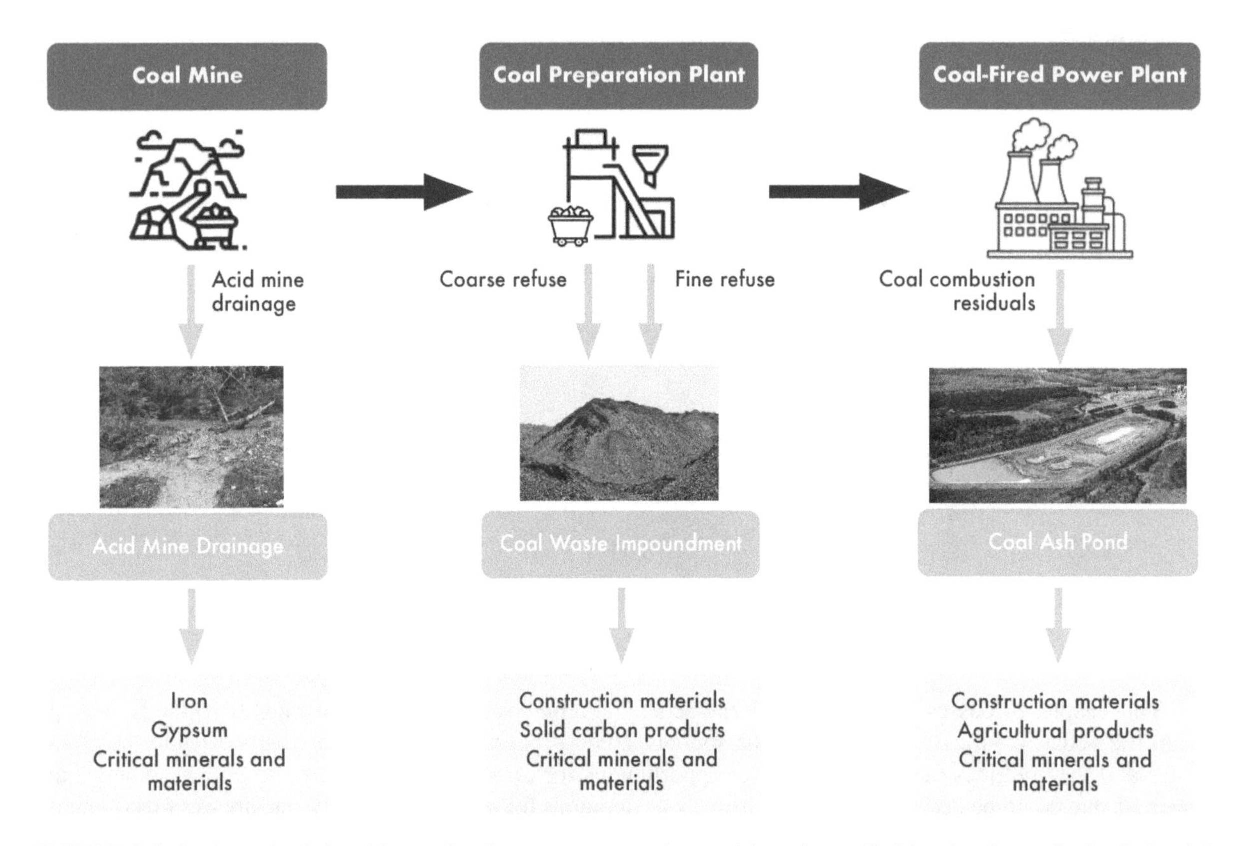

FIGURE 9-1 Coal supply chain with associated waste streams and potential products. Coal combustion and other industrial applications, excluding electricity generation, are not depicted.
SOURCES: Icons from the Noun Project, https://thenounproject.com. CC BY 3.0. Photos: (*acid mine*) Dr. Matthew Kirk, https://upload.wikimedia.org/wikipedia/commons/8/8a/AMD_at_the_Davis_Mine.jpg, CC BY-SA 4.0; (*coal waste*) Jakec, https://commons.wikimedia.org/wiki/File:Coal_waste_pile_west_of_Trevorton,_Pennsylvania_detail_5.JPG. CC BY-SA 4.0; (*coal ash pond*) Waterkeeper Alliance Inc., https://www.flickr.com/photos/waterkeeperalliance/13183371303/in/photostream. CC BY-NC-ND 2.0.

which are typically located near coal mines or coal processing facilities. Approximately 30 percent of the mined product is rejected as waste at the preparation plant (Karfakis et al. 1996), and the United States generates 70–90 million tons of impoundment waste annually (Gassenheimer and Shaynak 2023), with several billion tons stored in nearly 600 slurry impoundments across the country (Environmental Integrity Project 2019). CCRs are the by-products generated from burning coal and its associated environmental controls primarily to produce electricity in power plants. CCRs consist of fly ash, boiler slag, and flue gas desulfurization (FGD) products and are contained in nearly 750 coal ash impoundments located across the United States (Earthjustice 2022).

9.1.1 Classification, Definitions, and Characteristics of Coal Wastes

Specific definitions and characteristics of AMD,[2] impoundment waste (both coarse and fine refuse), and CCRs, including fly ash, bottom ash, boiler slag, and FGD products, are provided in Table 9-1. Each waste stream has its own unique characteristics that influence potential beneficial reuse applications.

[2] Note that for this report, acid mine drainage is considered only as a potential source of critical minerals and materials.

TABLE 9-1 Coal Waste Types, Definitions, and Characteristics

Waste Type	Definition	Characteristics
Acid Mine Drainage		
Acid Mine Drainage (AMD)	Acidic water with pH <6.0, discharged from mining operations and formed from the chemical reaction of surface water and subsurface water with rocks containing sulfur-bearing minerals. The sulfuric acid generated in this reaction can leach heavy metals from other rocks, yielding highly toxic wastewater.[a]	1. The pyrite quantities in rocks and the oxidation degree of the pyrite, as well as the $Fe^{2+} \leftrightarrow Fe^{3+}$ conversion chemistry determine the pH value of AMD. 2. The geochemical properties of the remnant mine strata influence AMD chemistry (e.g., cation and anion concentrations). 3. The concentrations of total rare earth elements (REEs, including lanthanide series, and scandium and yttrium that are not always accounted as REEs by some researchers) in AMD vary significantly. Most acid mine drainage contains <1 ppm total REEs and very few streams carry >1 ppm total REEs.[b] 4. Heavy rare earth elements (HREEs) account for less than 50 wt% of total REE content.[b]
Impoundment Waste		
Coarse Refuse	Large-particle waste product from the coal cleaning process. Typically used to construct impoundment-retaining embankments.[c]	1. Size: >2 mm. 2. High inorganic content (rock, shale, slate, clay). 3. Low carbon content.
Fine Refuse	Fine-grained particle waste from the coal cleaning process. Pumped via slurry and stored in impoundments.	1. Size: <2 mm. 2. Can contain 30–80 wt% coal, with clay, shale, and other mineral matter.[d]
Coal Combustion Residuals		
Fly Ash	A finely ground, powdery substance primarily consisting of silica, produced through the combustion of finely pulverized coal in a boiler. Its composition may include small carbon particles, varying based on the conditions of combustion.[e]	1. Fly ash generally accounts for ~80% of the total ash generated from coal combustion, except for combustion in slag-tap or cyclone boilers (see bottom ash and boiler slag below).[f] 2. Inorganic matter accounts for the majority of fly ash by mass. 3. Inorganic matter consists of amorphous and crystalline phases. 4. Inorganic matter consists of Si, O, Ca, Al, Fe, Mg, K, Ti, Na, P, N, Ba, rare earth, and other trace elements. 5. REEs in fly ash are found primarily in aluminosilicate glasses.[g] 6. To a lesser extent, REEs also are found in carbon grains surrounding aluminosilicate phases.[h] 7. The concentrations of REEs can be >1500 ppm.[i] 8. Light rare earth elements account for less than 50 wt% of total REE content.
Bottom Ash	A rough, sharply angled ash particle, which is too sizable to be carried into the smokestacks and thus accumulates at the base of the coal furnace.[e]	1. Bottom ash generally accounts for about 20 wt% of the total ash generated from coal combustion in a dry-bottom boiler.[f] 2. Specific gravity: 2.1–2.7, with respect to H_2O.[f] 3. Si and Al, as SiO_2 and Al_2O_3, respectively, are the major elements in bottom ash, although the concentrations of these two elements and other elements in bottom ash vary with the coal types used for combustion. 4. REEs are typically more enriched in bottom ash compared to flue gas desulfurization sludge.[j]

continued

TABLE 9-1 Continued

Waste Type	Definition	Characteristics
Boiler Slag	Molten ash formed in cyclone furnaces. Forms smooth pellets flowing through slag tap at furnace bottom, glass-like appearance once cooled in water.[e]	1. Boiler slag is generated when coal is burned in a slag-tap or cyclone boiler. 2. Boiler slag accounts for ~50 percent of the total ash produced by a slag-tap furnace.[f] 3. Boiler slag accounts for 70–80 percent of the total ash produced by a cyclone furnace.[e,f] 4. Specific gravity of boiler slag from both slag-tap and cyclone boilers: 2.3–2.9.[f] 5. Si and Al, as SiO_2 and Al_2O_3, respectively, are the major elements in slag, although the concentrations of these two elements and other elements in slag vary with the coal types used for combustion.
Flue Gas Desulfurization (FGD) Products	A product from the process used to decrease sulfur dioxide emissions from coal-fired boilers. It can be a moist sludge made up of calcium sulfite or calcium sulfate, or a dry, powdery mix of sulfites and sulfates.[e]	1. The quantity and quality of FGD products mainly depend on the quality of the coal used in a coal-fired power plant, primarily depending on the chemical composition of the coal. 2. The major compound in dry FGD products is $CaSO_4$. 3. The major heavy metals in FGD products are As, Cu, Cr, Cd, Hg, Pb, and Zn.

NOTES: REEs: Scandium (Sc), yttrium (Y), lanthanum (La), cerium (Ce), praseodynium (Pr), neodynium (Nd), promethium (Pm), samarium (Sm), europium (Eu), gadolinium (Gd), terbium (Tb), dysproium (Dy), holium (Ho), erbium (Er), thulium (Tm), ytterbium (Yb), lutetium (Lu). Heavy REEs: Dysprosium (Dy), terbium (Tb), erbium (Er), thulium (Tm), ytterbium (Yb), lutetium (Lu), yttrium (Y), holmium (Ho).

[a] EPA (2023a) and 30 CFR § 710.5.
[b] Vass et al. (2019).
[c] Luttrell and Honaker (2012).
[d] MSHA (2009); Rezaee and Honaker (2020).
[e] EPA (2023b); Luttrell and Honaker (2012).
[f] FHWA (2016).
[g] Kolker et al. (2017).
[h] Hower and Groppo (2021).
[i] Scott and Kolker (2019).
[j] Ekmann (2012); Wewerka and Williams (1978).

9.1.2 Locations of Coal Wastes

As illustrated in Figures 9-2, 9-3, and 9-4, coal wastes are located across the United States, although largely concentrated in Appalachia and the Intermountain West. Figure 9-2 shows the locations of surface and underground coal mines, thus indicating the approximate locations of impoundment wastes. Mining impoundments holding coal wastes can be categorized into three distinct states: (1) active, where they are currently in use; (2) in the process of reclamation, where restoration or rehabilitation efforts are under way; and (3) released, where erosion control, earth stabilization, topsoil replacement, and revegetation measures are complete, and the lands have been cleared from active or reclamation status. Coal wastes may also be located on abandoned mine lands. Figure 9-3 shows the locations of coal ash impoundments with color coding based on whether the ponds are regulated and/or are legacy units.[3] Figure 9-4 shows locations of potential unconventional and secondary sources of critical minerals, which include coal ash ponds, coal-fired power plants, and abandoned coal mines.

[3] Regulated ponds are those required by the U.S. Environmental Protection Agency (EPA) to safely dispose of coal combustion residuals by following a set of technical and reporting criteria outlined in EPA's 2015 rule "Disposal of Coal Combustion Residuals from Electric Utilities." Legacy units are inactive coal ash dump sites at inactive electric utilities; they are currently exempt from EPA regulation, although a proposed rule for their regulation was issued in May 2023 (Earthjustice 2023; EPA 2024b).

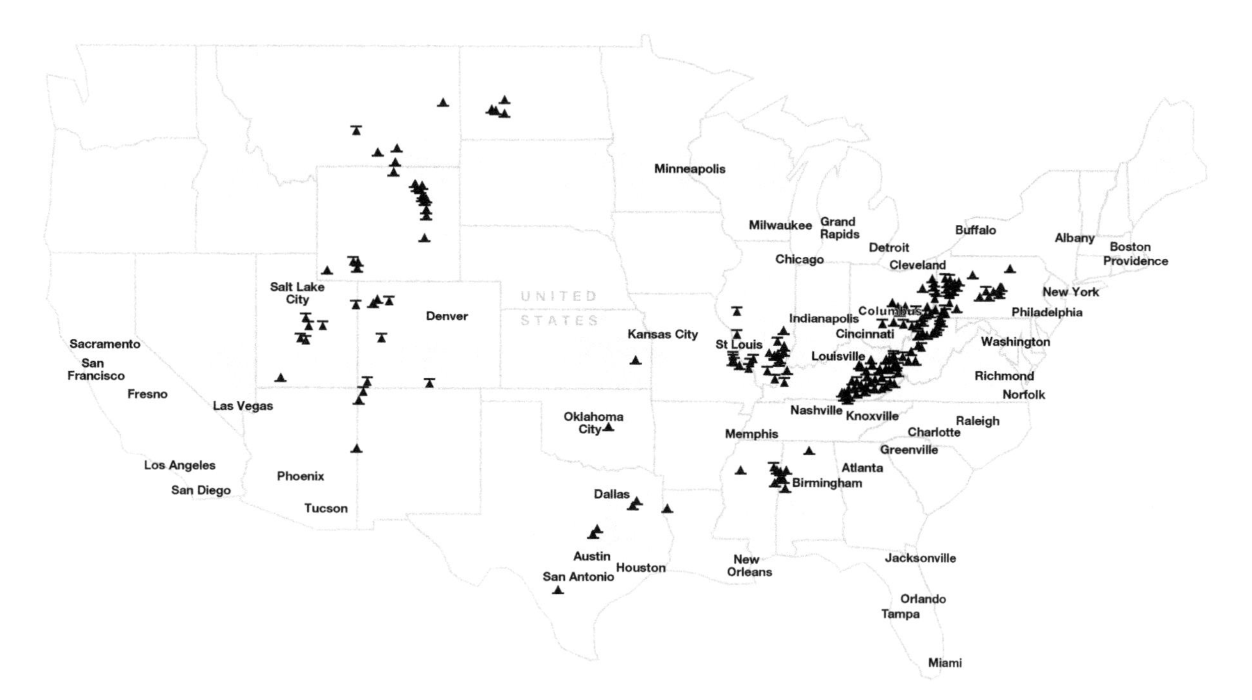

FIGURE 9-2 Locations of operating surface and underground coal mines in the United States, indicating approximate locations of impoundment wastes.
SOURCE: U.S. Energy Information Administration (2023b).

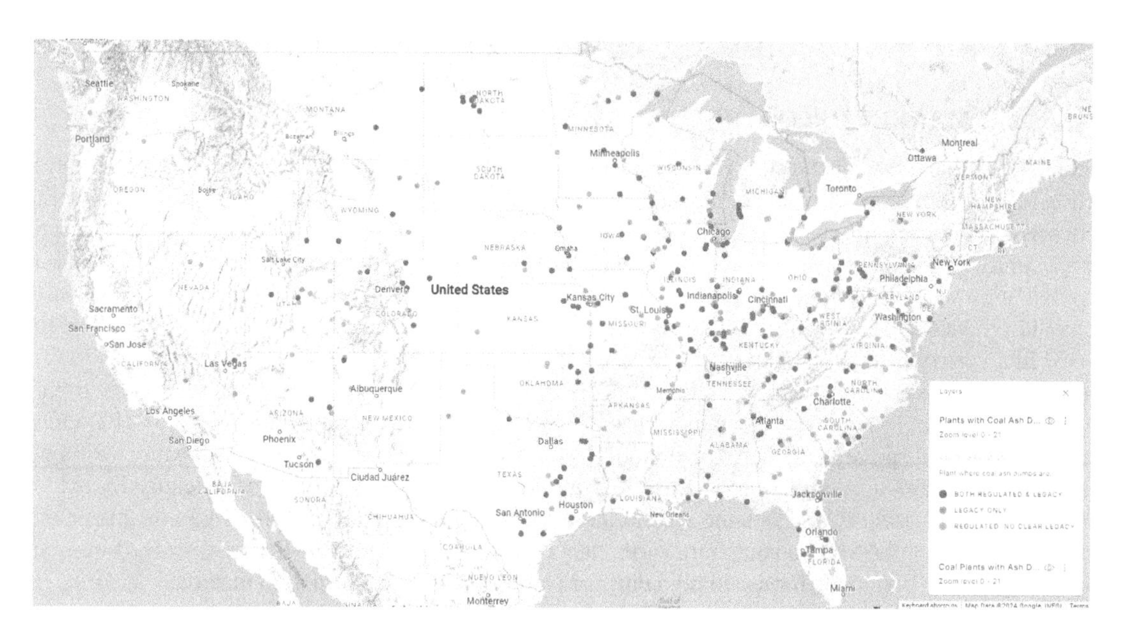

FIGURE 9-3 Locations of coal ash in the contiguous United States.
SOURCES: Earthjustice (2023); Basemap: (c) 2024 Google, Instituto Nacional de Estadística y Geografia.

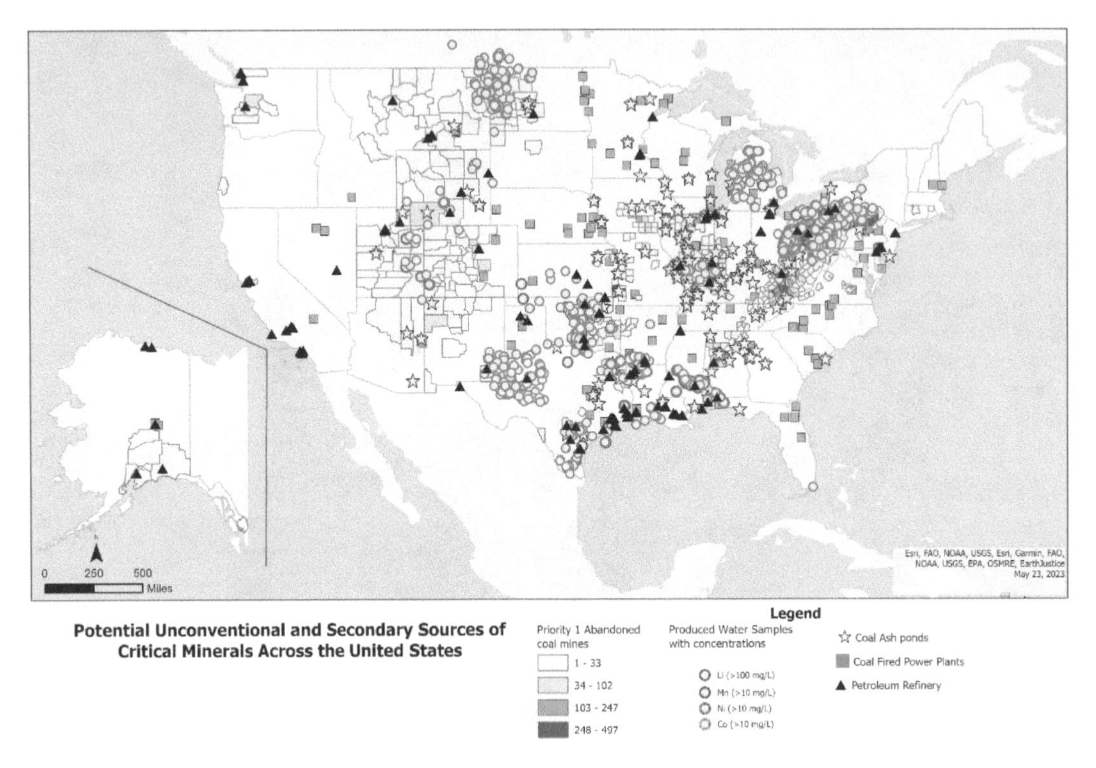

FIGURE 9-4 Locations of potential unconventional and secondary sources of critical minerals in the United States, including several sources relevant for this report: coal ash ponds (yellow stars), coal-fired power plants (gray squares), and abandoned coal mines (green shading).
SOURCE: Granite et al. (2023).

9.2 MARKET OPPORTUNITIES FOR COAL WASTE–DERIVED PRODUCTS

Table 9-2 outlines existing and emerging market opportunities for all three categories of coal wastes introduced in Section 9.1. By replacing virgin materials that are extracted or harvested from the Earth, use of coal wastes can help conserve natural resources. This chapter prioritizes coal waste applications that maximize product yield and also offer improvements in properties, reductions in manufacturing costs, improvements in environmental impact, or a combination of these benefits. Additional applications such as the gasification or pyrolysis of coal waste, which could produce syngas for the manufacture of liquid fuels or chemicals, were not considered. CCRs have been used in a variety of existing market applications (ACAA n.d.). The U.S. Environmental Protection Agency (EPA) encourages the beneficial use of CCRs in a safe and responsible manner, as it can lead to environmental, economic, and product benefits such as reduced use of virgin resources, lower greenhouse gas (GHG) emissions, reduced cost of CCR disposal, and improved strength and durability of materials. According to a survey conducted by the American Coal Ash Association, at least 35.2 million tons of coal ash were beneficially reused in 2021 (EPA 2024a). FGD gypsum ($CaSO_4$) is the second most widely used coal waste, primarily in the manufacture of wall board (EPA 2014a). AMD contains rare earth elements (REEs), is a rich source of dissolved iron, and has been used to generate iron oxide for use in pigments for paints, coatings, construction materials, and inks (Riefler 2021; Riefler et al. 2023). AMD is also a potentially attractive source of critical minerals and materials (CMMs), which are crucial to U.S. energy security and reduction of GHG emissions through their use in sustainable energy technologies.

TABLE 9-2 Market Opportunities for Beneficial Coal Waste Reutilization

Market Segment	Market Value (billion $)[a]	Compound Annual Growth Rate (%)	Applications
Acid Mine Drainage			
Pigments (iron oxide)	2.2[b]	4.6	Paints, cement, polymers, inks, ceramics
Critical minerals and materials	325[c]	8–30[d]	Catalysts, clean electricity, magnets, batteries, metallurgy
Impoundment Waste			
Construction materials	49.9[e]	6.7	Engineered composites, roofing tiles, building materials
Energy storage materials (graphite, graphene)	37.9[e]	14.4	Lithium-ion battery anodes, supercapacitors
3D-printing materials	4.6[e]	4.5	Electronics, touch screens
Carbon fiber	4.3[e]	11.2	Aerospace, composites, vehicles, reinforced concrete
Carbon foam	0.11[e]	14	Aerospace tooling, engineered components for military applications
Coal Combustion Residuals			
Cement	405[f]	4.3	Buildings, roads, infrastructure
Concrete bricks and blocks	370[g]	6.3	Buildings, construction, walkways
Asphalt	3.8[h]	5.1	Roads, roofs
Drywall	55.9[i]	12.7	Buildings
Critical minerals and materials	325[c]	8-30[d]	Catalysts, clean electricity, magnets, batteries, metallurgy

[a] Total market value, not just potential market value for coal waste derived–product.
[b] Straits Research (2022).
[c] IEA (2024).
[d] Represents demand for lithium, nickel, cobalt, and graphite.
[e] Stoffa (2023).
[f] *Fortune Business Insights* (2023a).
[g] *Fortune Business Insights* (2023b).
[h] SkyQuest (2023).
[i] Grand View Research (2024).

9.3 EXISTING MARKET APPLICATIONS AND RESEARCH, DEVELOPMENT, AND DEMONSTRATION NEEDS FOR COMMERCIAL USES OF COAL WASTES AND COAL WASTE–DERIVED MATERIALS

Existing market applications and research, development, and demonstration (RD&D) needs for commercializing coal waste utilization are presented below, including those for (1) direct solid waste utilization; (2) coal waste separations; (3) coal waste conversions to solid carbon products; and (4) CMM recovery. Figure 9-5 shows the major features of coal waste utilization to produce long-lived, solid carbon products and extract CMMs, including feedstock inputs, processes, products, and applications. Sections 9.3.3 and 9.3.4 describe these methods of coal waste utilization and the resulting products in more detail, noting relevant applications where appropriate. Safety considerations for using coal waste in commercial products also are discussed, emphasizing the necessary environmental testing, human health concerns, and product performance requirements. The Department of Energy's (DOE's) National Energy Technology Laboratory requires federally supported research and development (R&D) projects aiming to develop materials from coal waste to evaluate the safety and performance of these materials in accordance with their intended application (Stoffa 2023). The performance and safety of coal waste–derived products, when available, are reported in the following sections.

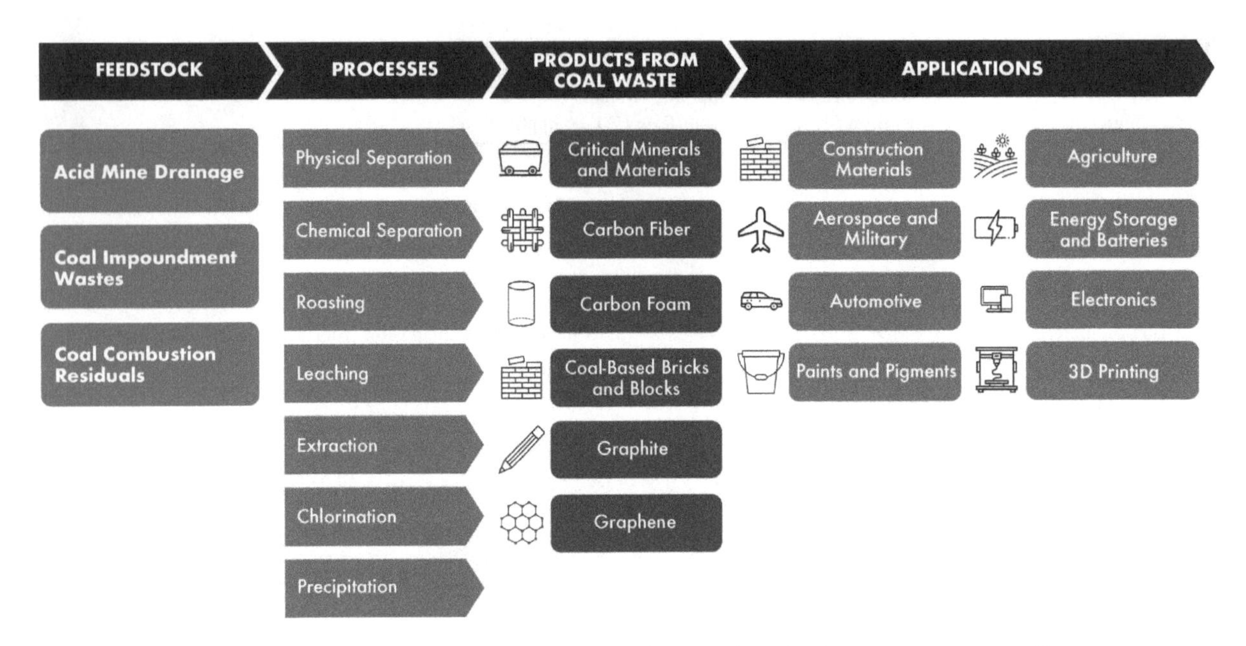

FIGURE 9-5 Summary of the feedstock inputs, processes, products, and applications for coal waste utilization to produce long-lived, solid carbon products and extract critical minerals and materials.
SOURCE: Icons from the Noun Project, https://thenounproject.com. CC BY 3.0.

9.3.1 Direct Solid Waste Utilization

This subsection provides an overview of direct reuse applications for solid coal wastes. The primary existing solid coal utilization markets include agriculture (fly ash and FGD gypsum), pavement and concrete applications, and other building product applications.

9.3.1.1 Agriculture Applications

Coal waste, in particular fly ash, has been used widely to improve soil textures in developing countries. Previous studies have shown that coal waste soil amendment could even double the crop yield in certain cases (Elseewi et al. 1980; Yunusa et al. 2012). Traditionally, it is believed that soil improvement mainly comes from the physical improvement of soil texture (Tejasvi and Kumar 2012); however, some studies have highlighted that the choices of coal waste for a particular type of soil improvement could maximize the benefits (Yunusa et al. 2012). In particular, high pH fly ash like Class C and Class can be used to modify soil acidity,[4] with Class C apparently more effective owing to its high calcium oxide content (Phung et al. 1978; Yunusa et al. 2012). Besides adjusting acidity, fly ash can be used to mitigate soil salinity, and the minerals and trace elements contained in fly ash are believed to be able to enrich fertile soils (Yunusa et al. 2012). The incorporation of fly ash modified by low-temperature roasting and hydro-thermal synthesis into contaminated soils has been shown to stabilize the migration of lead and cadmium (Xu et al. 2021).

[4] Fly ash classifications are based on chemical composition. Fly ash that meets the requirements of ASTM C618, which is necessary for its use in Portland cement concrete, is classified as either Class C or Class F (FHWA 2016). Class C fly ashes "are generally derived from sub-bituminous coals and consist primarily of calcium alumino-sulfate glass, as well as quartz, tricalcium aluminate, and free lime (CaO)" (FHWA 2017). Class F fly ashes "are typically derived from bituminous and anthracite coals and consist primarily of an alumino-silicate glass, with quartz, mullite, and magnetite also present" (FHWA 2017). A key difference between the two classes is the amount of calcium, with Class C ashes containing >20 percent CaO and Class F ashes containing <10 percent CaO.

As mentioned above, a broader agricultural application for fly ash is to ameliorate physical constraints in soils. Overall, the application of fly ash to soil improvement and agriculture depends on the type of soil and the type of fly ash. Fly ash can be used to mitigate particular types of soil property deficiencies. Fly ash does contain heavy metals (arsenic, chromium, lead, mercury, and others), and routine use in agriculture could contaminate surface and ground water and allow uptake into plants and animals (Carlson and Adriano 1991; Ishak et al. 2002; Izquierdo and Querol 2012; Kukier et al. 2003; Taylor and Schuman 1988). Previous research has shown that using FGD gypsum in agriculture can provide crops with essential nutrients and reduce the amount of phosphorus runoff into nearby water bodies. An analysis performed by EPA, the Department of Agriculture, and RTI International found that, in all scenarios evaluated, agricultural use of FGD gypsum did not lead to accumulation of inorganic constituents (e.g., arsenic, cadmium, mercury, thallium) in soil, crops, livestock, air, or groundwater at levels harmful for human or environmental health (EPA, USDA, and RTI International 2023).

9.3.1.2 Pavement and Concrete Applications

While both fly ash and bottom ash can be used for construction and pavement materials, other processing waste, like coal tailings, also can be mixed with rejuvenated asphalt for pavement (Mohanty et al. 2023). The American Coal Ash Association reports that more than half of the concrete produced in the United States today uses fly ash in some quantity as a substitute for traditional cement (ACAA Educational Foundation n.d.). The use of up to 40 percent fly ash in concrete can improve the durability, workability, and strength of concrete, while reducing the amount of required cement (Bentz et al. 2013; Bouaissi et al. 2020). Both fly ash and bottom ash can be mixed with limestone to partially replace cement in pavement materials while fulfilling the mechanical strength needs (Indian Roads Congress 2010). In certain cases, the coal waste mixtures can also improve the pavement temperature resistance (Cao et al. 2011). Utilizing fly ash for pavement projects could lead to the leaching of heavy metals into surface and groundwater, a factor that must be taken into account in engineering risk assessments (Kang et al. 2011; McCallister et al. 2002).

Sand is a primary component of concrete building materials, with an annual global consumption of nearly 50 billion tons, and current extraction rates exceed natural replenishment rates, creating an impending supply shortage (Advincula et al. 2023; UNEP 2022). Fine refuse could be used directly in a host of low-cost construction applications, such as a sand substitute in concrete applications or road base (Jahandari et al. 2023; Leininger et al. 1987). Studies of different types of coal wastes are needed to determine various optimal strategies to apply for pavement improvements (Mohanty et al. 2023) or other construction applications. Comprehensive analyses of both environmental impacts and the mechanical performance of coal waste enhanced pavement materials need to be carried out.

9.3.1.3 Other Building Product Applications

FGD gypsum is commonly used in building product applications, primarily wallboard and similar products. Almost half of all U.S. wall board is manufactured using FGD gypsum generated at coal-fired power plants (Gypsum Association 2024). According to an EPA study, the release of constituents of potential concern from FGD gypsum wallboard during use by the consumer is comparable to or lower than that from analogous non-CCR products (EPA 2014a).

9.3.2 Coal Waste Separations

Coal waste streams may require processing to separate and refine useful organic and inorganic materials, using either physical or chemical methods. This subsection reviews existing and emerging coal waste separation technologies.

9.3.2.1 Physical Methods

Fine refuse generated from coal preparation plants represents a primary feedstock opportunity for repurposing into carbon products or extracting CMMs. Fine refuse streams can contain upward of 60 wt% of coal in the form of ultrafine particles (<50 μm). Advances in ultrafine particle recovery from fine refuse are being developed, and key technologies being considered include selective flocculation-flotation (Liang et al. 2019), hydrophobic flocculation flotation (Song and Trass 1997), carrier flotation (Ateşok et al. 2001), micro/nano-bubble flotation (Sobhy and Tao 2013), nanoparticle flotation (Li et al. 2019), oil agglomeration (Özer et al. 2017), and two-liquid flotation (Pires and Solari 1988).

Reducing air bubble size in the flotation process is effective in improving the recovery of fine coal particles. Nanobubbles (<1 μm) increase recovery of fine coal particles by preferable adherence to coal's surface over inorganic slurry components (Fan et al. 2010). Nucleation of nanobubbles on the coal's surface enhances its hydrophobicity, stability of bubble-particle aggregates, and increased aggregation of fine particles. Disadvantages of utilizing smaller bubbles for recovery include increased residence time for separation, increased energy consumption to generate micro/nanobubbles, and the need to use heat to dry the recovered product (Li 2019). The micro/nanobubble technology is being used to process coal refuse at 14 tons per hour for 20 hours per day at a plant in Pennsylvania (Gassenheimer and Shaynak 2023).

Recent development of a two-liquid flotation technology termed hydrophobic-hydrophilic separation (HHS) has shown promise. In the HHS process, a hydrocarbon liquid (e.g., pentane, heptane) rather than an air bubble is used to collect hydrophobic particles from aqueous slurry waste, as shown in Figure 9-6. Pilot-scale testing of the HHS process has demonstrated effective recovery of fine coal from a range of Appalachian fine refuse streams consisting of upward of 67.5 wt% inorganic content, resulting in ultrafine coal product streams with <1.0 wt% inorganic content and >99 percent liquid hydrocarbon recovery for reuse (Yoon et al. 2022). The first commercial-scale installation using HHS technology is in the commissioning phase at a plant in Tuscaloosa, Alabama, which can process up to 20 tons of coal per hour (Troutman 2023).

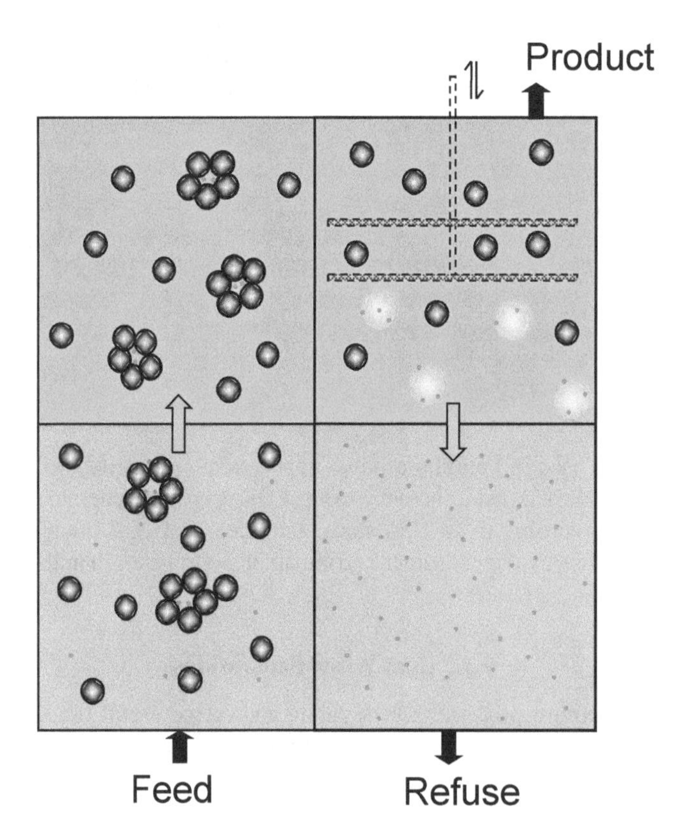

FIGURE 9-6 Conceptual representation of the HHS process, where a hydrophobic particle is transferred from water (blue) to a hydrocarbon/water interface and subsequently to a hydrocarbon phase (gold). Water-in-oil emulsions are broken mechanically, liberating coal particles (white) and entrapped water droplets (blue) that contain hydrophilic particles (gray).
SOURCE: Yoon et al. (2022).

9.3.2.2 Chemical Methods

The majority of coal waste separation is performed using physical methods, and only a few methods for chemical separations have been described. Tailings often require dewatering to separate the solid and liquid phases, but the presence of fine clays makes traditional filtration challenging. One approach uses chemical additives to facilitate this dewatering step. Coal tailings often are composed primarily of clay minerals such as kaolinite and montmorillonite, which have negative surface charges that inhibit their settlement during physical separation methods. Various inorganic salts with cations such as Na^+, Ca^{2+}, Mg^{2+}, and Al^{3+} neutralize these surface charges and facilitate agglomeration through centrifugation, with Al^{3+} having the greatest efficacy (Nguyen et al. 2021). Another approach uses oil and water phases to induce HHS in coal gasification fine slag. This method provides enrichment of carbon products compared to inorganic elements, which contain silicates and metals. (Xue et al. 2022). Nanofiltration and reverse osmosis membranes have been proposed to purify acid mine wastewater; they could potentially generate concentrated brine streams for further processing and purified water, thereby dewatering such streams (Ighalo et al. 2022; Xia et al. 2023).

9.3.3 Waste Coal Utilization

The separations described in Section 9.3.2 produce a waste coal product with low ash content, which can be converted into a variety of carbon-containing products using methods that directly or indirectly utilize waste coal. This subsection provides an overview of ongoing R&D focused on repurposing the carbon fraction (i.e., waste coal) of coal waste into solid carbon products. Because information on the use of waste coal is limited to date, several products that have been obtained from coal are discussed for illustrative purposes, as similar methods could be developed for using waste coal as the starting feedstock. Product categories include construction materials, energy storage materials, carbon fiber, carbon foam, and three-dimensional (3D) printing applications. With DOE support, technologies spanning several areas are being developed that convert coal or waste coal into value-added solid products (Stoffa 2022).

9.3.3.1 Construction Materials

The mass of construction materials used in the global built environment to date, estimated at approximately 1.1 teratonnes, is equal to living biomass on Earth (Elhacham et al. 2020). Construction material demand has been roughly doubling every 20 years over the past century and is expected to continue increasing (Elhacham et al. 2020), providing market growth opportunities for inclusion of new construction materials and feedstocks. Long-lived construction materials offer a significant beneficial utilization opportunity, as they sequester the carbon content of waste coal. Emerging research, detailed below, has been analyzing the use of coal or coal waste in construction materials, primarily as filler in engineered composites for construction applications (e.g., blocks, decking, piping). Coal has been introduced to a wide range of composite materials consisting of thermoplastic, thermoset, ceramic, and fiber-reinforced cement composites.

Coal-based bricks and blocks (CBBs), made from composites composed of coal mixed with thermoplastic resin, offer an alternative to traditional clay materials. CBB materials containing up to 70 wt% coal have been investigated, with cross-linked high-density polyethylene and polyvinyl chloride materials exceeding the 5000-psi compressive strength code requirement (Vander Wal and Heim 2023). CBBs made with thermosetting epoxies have been shown to possess compressive strengths exceeding 5,000 psi (Vander Wal and Heim 2023). The manufacture of thermoplastic and thermoset based CBBs will have very limited to no CO_2 from waste coal content. Thermoplastic- and thermoset-based CBBs that utilize mined coal are currently at a bench-scale level of development. Further R&D is needed to incorporate waste coal, scale up manufacturing processes, and perform necessary building code testing. Coal-derived chars, generated via pyrolysis at 850°C, also have been studied for use in CBB applications, with materials containing 40 wt% char content possessing higher compressive strength (49.5–52.5 MPa) than conventional clay bricks (10–20 MPa). Char-based CBBs also possess lower density and water absorption (Yu et al. 2023). In addition, CBBs have been made from pyrolyzing pressed mixtures of coal and preceramic polymer resin (PCR) at temperatures of 180°C (plastic/ceramic bricks) or 1000°C (ceramic bricks) (Sherwood 2022b). The coal-PCR bricks and blocks exhibit compressive strength (4465 and 4863 psi, respectively) and density (1.57 and 1.55 g/cm^3, respectively) advantages over conventional brick and block materials (2845 and 4400 psi, respectively, and 2.3 g/cm^3 [block]) (Sherwood 2022b).

Pyrolysis processing will generally retain approximately 60 wt% of coal's carbon content when generating char (Chen et al. 2006; Seo et al. 2011), with the remaining carbon forming carbon oxides and condensable and noncondensable hydrocarbons; similar emissions are anticipated from pyrolysis processing of waste coals. Pyrolyzed coal-PCR composites also have been evaluated for façade and roofing tile applications. PC-PCR roofing tiles possess flexural strength of 3317 psi with 35 percent lower weight than clay roofing tiles and passed ASTM specifications for hail impact, water absorption, and flexural strength (Sherwood 2022a). In addition, testing of a composite material made from phenolic resin and coal char derived from pyrolyzed subbituminous coal indicates the material has significant potential for load-bearing building applications (Wang et al. 2023). Additional R&D is required to determine whether coal-char can be produced from waste coal and its performance in such applications. Carbon aggregate made via flash Joule heating (FJH) of coal-derived metallurgical coke has been tested as a replacement for sand in concrete applications. The replacement of sand with FJH-derived carbon aggregate reduced concrete density by 25 percent while increasing toughness, peak strain, and specific compressive strength by 32 percent, 33 percent, and 21 percent, respectively (Advincula et al. 2023).

Composite decking materials developed from high-density polyethylene with coal and waste coal have reached an advanced stage (technology readiness level [TRL] 8). These coal-based decking boards meet ASTM specifications and with projected pricing equivalent or lower than typical wood-plastic composites when manufactured at scale (Al-Majali et al. 2023b). Coal plastic composites (CPCs) also potentially offer greater resistance to oxidation (typically associated with product service life) than commercial wood-plastic composites without antioxidant additives owing to the primary and secondary antioxidant components of coal (Al-Majali et al. 2022). A recent analysis, using finite element analysis modeling with material properties for CPCs made with bituminous waste coal recovered from an active impoundment, suggested that decking boards made with CPC should meet building code requirements (Al-Majali et al. 2023a). Commercially manufactured CPC product has been demonstrated to meet ASTM D7032 specifications for composite decking, including passing respirable dust (NIOSH 600), leaching (EPA 1311), and fire rating (ASTM E84, Class B) tests (Al-Majali et al. 2023b). Preliminary life cycle assessments (LCAs) indicate that coal-based composite decking has less embodied energy and emissions than its wood-plastic counterparts (Al-Majali et al. 2019). CPC materials also offer opportunities to incorporate significant recycled plastic content, similar to currently offered commercial composite building products (AZEK 2024; MoistureShield, Inc. 2022; Trex 2024). Furthermore, thermoplastic composites made from polyvinyl chloride and waste coal have been evaluated for drainage and vent pipe applications. These composites were shown to meet ASTM D1784 specifications for piping compounds, and 2-inch schedule 40 pipe has been successfully manufactured (Trembly et al. 2023a).

9.3.3.2 Energy Storage Materials

Carbon materials for energy storage applications are categorized into two main categories, graphitizable soft carbons and nongraphitizable hard carbons. Two primary soft carbon materials being targeted from waste coal are graphite and graphene, with graphite further categorized as flake or amorphous. Flake graphite is primarily used in batteries, foundries, refractory materials, and lubricants. Amorphous graphite is used in foundries, refractories, recarburization processes, and lubricants. The United States does not produce natural graphite, relying on imports from China (33 percent), Mexico (18 percent), Canada (17 percent), Madagascar (10 percent), and others (22 percent) (percentages for 2018–2021; USGS 2023), which creates an energy security risk for the U.S. electric vehicle market—the primary factor increasing graphite demand. Production of lithium-ion battery (LIB)-grade graphite is an attractive option for waste coal. Spherical graphite is the preferred shape for many commercial battery applications because spherical particles pack well, leading to high tap density, which can increase the overall energy density of the battery, and because the spherical shape has a shorter diffusion path for lithium ions, which can enhance charging/discharging rates.

Traditionally, synthetic graphite is manufactured by treating a carbon precursor at high temperature (up to 3000°C) over a long time. A portion of the synthetic graphite supply chain is manufactured from mesophase pitch, a collection of aromatic hydrocarbons that exhibit optical anisotropy. While the majority of mesophase pitch is derived from petroleum processing, the carbonaceous precursor also can be synthesized from coal tar pitch, a by-product of coke production. Mesophase pitch with a significant amount of polycyclic aromatic hydrocarbons and polynuclear aromatic structural units has the potential to be used as a precursor for producing soft carbon material with a high degree of graphitization (Zhang et al. 2022a). Recent research has demonstrated the synthesis of mesocarbon microbeads at 2350°C from coal-derived mesophase pitch, with good performance in LIB coin cells (Prakash et al. 2022). Another study showed that a low-temperature solvothermal preparation method can produce carbonaceous mesophase materials at 230°C with promising electrochemical properties, offering a less energy-intensive preparation method (Wu et al. 2023).

Direct preparation of graphite via thermal treatment of highly volatile bituminous and anthracite coals from 2000–2800°C has been reported, but LIB performance data to assess commercial viability is limited (Han et al. 2021; Shi et al. 2021; Xing et al. 2018). Similar preparation methods could be applied to waste coal. To assess maximum performance potential, waste coal-derived graphite intended for LIB applications would have to undergo spheroidization and amorphous carbon coating before testing. Processes that use high temperature to synthesize graphite or related carbon materials, in particular, will require LCA to analyze process GHG emissions. (See Chapter 3 for more detail on LCA.) Another method involving the thermal treatment of lignite up to 3000°C after mineral acid treatment has been explored; subsequent LIB tests suggest the resulting soft graphitizable carbon is not suitable for electric vehicle applications (Azenkeng 2022). A bituminous coal graphitized at 2850°C was compared to synthetic commercial graphite possessing a higher d-spacing[5] (3.396±0.001 Å versus 3.389±0.006 Å), lower I_D:I_G6[6] ratio (0.78 versus 0.97), lower degree of graphitization (51±1 percent versus 60±7 percent), and higher surface area (7.34±0.29 m^2/g versus 1.35±0.06 m^2/g) (Paul et al. 2023, 2024). The bituminous coal graphite and commercial graphite were evaluated in LIB half-cells demonstrating capacities of 270±4 $mAh·g^{-1}$ for coal graphite (theoretical capacity is 372 $mAh·g^{-1}$) and 329±7 $mAh·g^{-1}$ for commercial material and coulombic efficiencies of 99.19±0.24 percent for coal graphite and 99.51±1.49 percent for commercial material (Paul et al. 2024). The bituminous coal graphite was not spheroidized or coated with amorphous carbon, typical battery-grade graphite processing techniques. Novel synthetic graphite preparation methods from coal and waste coal include laser irradiation and molten salt synthesis. Graphite derived from lignite through laser irradiation has been studied (Banek et al. 2018), but the material did not attain the performance metrics of commercial graphite (Wagner 2022). Highly crystalline nano-graphite has been synthesized via molten calcium/magnesium chloride-assisted electrocatalytic graphitization of coal chars (Thapaliya et al. 2021). Related work on beneficial reuse of coal for LIBs evaluates the use of silicon oxycarbide (SiOC) polymer-derived ceramic with 25 wt% bituminous coal as an anode material and reports a specific capacity of about 700 mAh/g in coin half cells (Marcus 2022).

In addition to energy storage, graphite has many additional industrial applications that are possible opportunities for waste coal reutilization. Graphite electrodes are used in the production of steel, aluminum, and silicon (Jäger and Frohs 2021). Other applications include high-temperature refractories, lubricants, conductivity additives, gaskets, fire extinguishing agents, and lubrication of industrial manufacturing and machining processes (Al-Samarai et al. 2020; Chung 1987; Jäger and Frohs 2021).

Recent studies have reported the use of ab initio molecular dynamics simulations to evaluate the formation of amorphous graphite and carbon nanotubes from carbonaceous material such as waste coal (Thapa et al. 2022; Ugwumadu et al. 2023). Atomic-scale modeling via—for example, molecular dynamics or kinetic Monte Carlo simulations are useful tools for fundamental research, offering the potential to enhance understanding of complex carbon chemistry involved in transforming waste coal structures into useful carbon products.

Graphene is a soft carbon nanomaterial that can be manufactured from waste coal. Applications of graphene are extensive and include energy storage, ultraconductors, composites, separations, biomedical, and electronics. Various direct and indirect methods for producing graphene from coal have been investigated, such as exfoliation (Leandro et al. 2021; Yan et al. 2020), chemical vapor deposition (Vijapur et al. 2017), electric arc (plasma)

[5] D-spacing is a measure of the distance between parallel planes of atoms in a crystal structure.

[6] I_D:I_G is the ratio of the intensity of D to G peaks in Raman spectroscopy.

(Awasthi et al. 2015), and flash Joule heating (Du et al. 2023; Tour 2022). Carbon quantum dots have also been prepared from coal and coal-derived materials using electrochemical exfoliation (He et al. 2018), chemical oxidation (Ye et al. 2015), plasma arc discharge (Xu et al. 2004), hydrothermal (Fei et al. 2014), and laser ablation (Kumar Thiyagarajan et al. 2016) synthesis methods. Carbon nanotubes also can be synthesized from coal or waste coal utilizing arc plasma jet (Tian et al. 2004), arc discharge (Awasthi et al. 2015), laser ablation (Kumar Thiyagarajan et al. 2016), chemical vapor deposition (Tian et al. 2004), or chemical pyrolysis (Moothi et al. 2015) methods. Most methods for producing graphene from waste coal are energy-intensive (thermal, plasma, arc discharge, and laser ablation) or require energy processing agents with high embodied energy (exfoliation). Electrochemical methods operating at near ambient temperature are attractive if supplied by renewable energy. Research that focuses on reducing the energy intensity of graphene synthesis from waste coal is needed and will require LCA to analyze process GHG emissions.

Hard carbon, which can be produced from both mined and waste coal, serves various applications such as sodium-ion batteries, potassium-ion batteries, catalyst supports (Lu et al. 2019; Wang et al. 2020a; Xiao et al. 2019; Yang et al. 2011), and adsorption media. Manufacture of soft carbons requires a precursor with a high aromatic content (i.e., coal tar pitch; Alvira et al. 2022) and processing temperature (2200°C–3000°C; Marsh and Rodríguez-Reinoso 2006), making nonfusing coal or waste coals better suited precursors for hard carbon manufacturing at lower temperatures (1000°C–1800°C; Marsh 1989). Sodium-ion batteries are an attractive alternative to LIB technology, owing to sodium's greater abundance and lower cost (Abraham 2020). Studies have shown that hard carbon derived from pyrolyzed samples of anthracite, subbituminous, and bituminous coal possess energy storage capacities of 252 (Wang et al. 2020b), 291 (Lu et al. 2019), and 270 (Kong et al. 2022) $mAh{\cdot}g^{-1}$, respectively. These values are within the range of 250 to 350 $mAh{\cdot}g^{-1}$ reported in literature for hard carbons from various sources (Kong et al. 2022; Liu et al. 2022; Lu et al. 2019; Wang et al. 2020b). Additionally, activated carbons made from all types of coal are commercially produced and used in the purification of water, air, chemicals, and food, among other applications (Carbon Activated Corporation 2024). Hard carbon applications are an attractive option for waste coal utilization owing to less stringent aromatic content requirements.

9.3.3.3 Carbon Fiber

Carbon fiber is a high-strength, lightweight material composed of polycrystalline carbon atoms largely aligned parallel to the long axis of the fiber, which results in a corrosion-resistant high tensile strength material. Carbon fiber is used in advanced materials as a reinforcing material in composite structures, including carbon fiber reinforced polymers and carbon-carbon composites. Current applications of carbon fiber reinforced polymers span many sectors, including aviation (they make up 80 percent by volume of the Boeing 787; Giurgiutiu 2022), automotive, energy, infrastructure, marine, and electronics (Belarbi et al. 2016).

More than 95 percent of carbon fiber is prepared from polyacrylonitrile (Grand View Research 2022), but carbon fiber also can be manufactured using mesophase pitch derived from petroleum or coal resources. Automotive industry recommendations for carbon fiber for car frames are that the tensile strength, elongation ratio, and Young's modulus be at least 1.7 GPa, 1.5 percent, and 170 GPa, respectively, at a cost of less than $11 per kg (Yang et al. 2014). Although polyacrylonitrile-based carbon fibers exceed the mechanical performance requirements, their cost (>$20/kg) inhibits their adoption, so alternative carbon fiber preparation methods are needed. Das and Nagapurkar (2021) conducted LCA and technoeconomic assessment (TEA) to compare carbon fiber production from polyacrylonitrile with that from coal pitch derived from mined coal. The embodied energy for coal pitch carbon fiber manufacturing was estimated to be 2.4–2.5 times lower than that for polyacrylonitrile carbon fiber manufacturing, estimated at 1188 MJ/kg. The LCA considered nine environmental impact categories, indicating that coal pitch fiber manufacturing would result in lower emissions across all categories owing to higher manufacturing yield, generally producing less than 50 percent of the emissions of the conventional polyacrylonitrile carbon fiber process. TEA estimates for a 3750 tonne/year carbon fiber manufacturing facility indicated costs of $10.29/kg for coal pitch sourcing, compared to $18/kg to $22/kg for polyacrylonitrile sourcing.

Owing to the low costs of coal and waste coal, and the higher conversion yield associated with coal pitch carbon fiber manufacturing, their use as feedstocks for carbon fiber production could provide value to this market if mechanical performance requirements can be achieved. Carbon fiber manufacturing and performance are highly susceptible to impurities in coal pitch, such as quinoline insolubles and ash, which create stress points and lead to breakage (Banerjee et al. 2021a; Cao et al. 2012). It is essential to understand the impurities introduced during coal pitch manufacturing when using reclaimed coal as feedstock. Recent research has shown that carbon fiber made from mesophase pitch derived from coal fractionation product possesses tensile strength of 1.8 and 3.0 GPa, elongation of 1.4 percent and 0.7 percent, and Young's moduli of 140 and 450 GPa, after carbonization at 1000°C for 30 minutes and graphitization at 2800°C for 10 minutes, respectively (Shimanoe et al. 2020). Another study reported a tensile strength of 3.86 GPa, elongation of 0.62 percent, and Young's modulus of 620 GPa (Guo et al. 2020). These results indicate that the elongation ratio of coal mesophase pitch carbon fiber needs to be improved to meet industry standards, but the standards for tensile strength and Young's moduli have been achieved. Encouragingly, isotropic pitch-based carbon fibers derived from waste coal have been shown recently to possess mechanical properties similar to those of general-purpose carbon fibers (Craddock et al. 2024).

9.3.3.4 Carbon Foam

Carbon foam is a low-density, porous material made predominantly of carbon. It has unique tunable properties including strength and conductivity (both electrical and thermal), with high temperature resistance and chemical inertness. Although commercially manufactured, carbon foam is still an emerging material owing to its cost. Applications for carbon foam include tooling to produce carbon fiber composites, thermal insulation, fireproofing, aerospace, heat sinks and exchangers, electrodes, electromagnetic interference shielding, acoustic insulation, energy absorption, and contaminant adsorption.

Carbon foam currently is made from three primary sources: phenolic resin, petroleum-derived pitch or coal-derived pitch, and caking coals. The foaming process involves controlled heating of the precursors (pitch or coal) (up to 500°C) under pressure (up to 3.5 MPa) in an inert atmosphere to form an amorphous carbon (Chen et al. 2006). Further thermal treatment of the amorphous product is completed to control product properties, up to graphitization. During heating, the evolving volatiles from the decomposing light fractions serve as bubble agents to create a foam cell in the highly viscous precursor material. Further heating results in solidification of the precursor, which fixes the foam matrix (Calvo et al. 2008; Chen et al. 2006). Waste coal can serve as a direct or indirect precursor for carbon foam synthesis. Carbon foam products directly synthesized from coal with varying density (20–30 lb/ft^3) are commercially available (CFOAM LLC n.d.(a)). Cost is the primary hindrance to adoption of carbon foam for use in commodity applications such as construction and building products, which is associated with the high-pressure batch nature of current commercial manufacturing methods.[7] Unique market opportunities that warrant carbon foam's price premium include tooling and defense applications owing to its unique thermal and physical properties, such as high thermal stability, resistance to melting, and low coefficient of thermal expansion (CFOAM LLC n.d.(b)). As volumes increase in more successful applications, opportunities will arise for cost reductions through automation and process improvements. Recent research has focused on developing methods to synthesize carbon foam continuously, to perform the foaming step at atmospheric pressure, or both. Carbon foams made from strongly caking coals in a batch-wise process at atmospheric pressure possess compressive strengths of 2.7–18.1 MPa (Yang et al. 2022), compared to 6.0–16.0 MPa for commercially available carbon foams (CFOAM LLC n.d.(c)). Continuous production of carbon foam panels using a continuous atmospheric belt kiln has recently been demonstrated, with products possessing densities of 25–32 lb/ft^3 and compressive strength of 9.6 to 16.5 MPa (Olson 2022).

Façades made from waste-coal-derived carbon foam and carbon-foam-enhanced fiber-reinforced cement composites also have been demonstrated. Carbon foam materials with a density of 35 lb/ft^3 were shown to meet ASTM C1186 Grade I specifications, while carbon foam (27 lb/ft^3) with backing material exceeded ASTM C1186 Grade II specifications (Trembly et al. 2023b). Fiber-reinforced cement composites made by replacing sand filler with carbon foam particulate demonstrated equivalent strength as a conventionally prepared fiber-reinforced cement composite control, with up to 30 percent lower density (Trembly et al. 2023b).

[7] Dr. Rudolph Olson, personal communication with the committee.

9.3.3.5 3D Printing

Additive manufacturing, or 3D printing, is an emerging technique with industrial applications already developed. Fused deposition modeling (FDM) and fused granulate fabrication are the most widely used 3D printing techniques, involving the extrusion of a thermoplastic filament or pellet to deposit material layer by layer. Commercial FDM applications include prototyping, tooling and jigs, furniture, automotive components, parts (e.g., gears, bumpers, valves, covers), and architecture.

Anthracite and lignite have been incorporated in polyamide-12 (PA 12) resin to form a composite and printed using the FDM procedure. The addition of lignite improved Young's modulus and thermal conductivity compared to unmodified PA 12 (Veley et al. 2023a, 2023b). The incorporation of bituminous coal into polylactic acid, polyethylene terephthalate glycol, high-density polyethylene, and PA 12 resulted in FDM filaments with similar glass and melt transition temperatures, lower heat capacity and thermal conductivity, and, notably, a reduced thermal expansion coefficient for high-density polyethylene (Veley et al. 2023a). PA 12 filaments made with waste coal demonstrated greater maximum tensile and flexural strengths than unfilled plastic, likely owing to beneficial hydrogen bonding between the waste coal filler and the matrix (Veley et al. 2023b). DOE has active projects investigating the use of waste coal in 3D printing applications (DOE-FECM 2021). The addition of fly ash to cement-based 3D printing formulations improved flowability (Yu et al. 2021). Coal waste, produced from both mining and combustion processes, holds potential for use in 3D printing applications. Further research is essential to assess the performance and safety of additive manufacturing materials developed using coal waste.

9.3.4 Critical Minerals and Materials Recovery

Currently, the United States imports most minerals deemed "critical" by the U.S. Geological Survey (USGS; see Figure 9-7), which can lead to supply chain vulnerabilities. In 2022, imports comprised over 50 percent of the demand for 43 critical minerals, with 12 of those being 100 percent imported (USGS 2023). Coal has diverse and

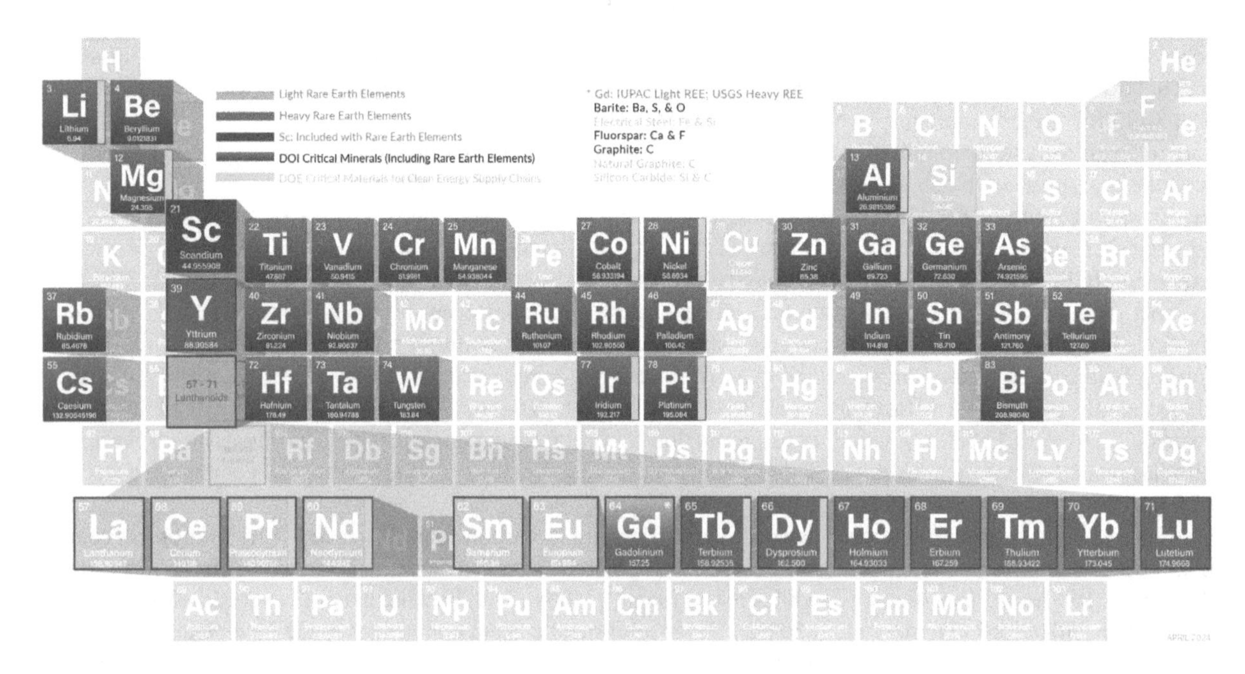

FIGURE 9-7 Critical minerals, including rare earth elements, that are found in coal wastes. Critical minerals as defined by the Department of the Interior (DOI) are indicated in dark gray, light rare earth elements in light blue, heavy rare earth elements in dark blue, and those elements that DOE deems critical for clean energy supply chains are in green or, if they are also a DOI critical mineral, designated with a green bar. Scandium is typically considered as a light rare earth element.
SOURCE: NETL (n.d.).

TABLE 9-3 Advantages and Disadvantages of REE Extraction from Different Coal Wastes

	REE Recovery	
Coal Waste Type	Advantages	Disadvantages
Acid mine drainage	REEs that exist as precipitates are near 300 ppm concentration (DOE interest level).	REEs that exist in solution are at much lower concentration than those in coal-related solids; geographically limited to areas with past mining.
Impoundment waste	Relative enrichment in REEs and lithophile elements (e.g., Li, Al, Ti, Sc, Rb, Y, Zr, Cs, Ba) compared to raw coal.	Enriched in harmful chalcophile elements (e.g., Hg, As, Sb, Pb).
Fly ash	Highest REE enrichment of coal sources because REEs strongly retained in smaller mass.	Difficult to extract the significant fraction of REEs contained in aluminosilicate glasses.

SOURCE: Based on data from Kolker (2023).

complex chemistry, containing all the elements existing in nature, and coal wastes contain vast amounts of critical minerals (CMs), including REEs (Kolker 2023; Kolker et al. 2024; McNulty et al. 2022). Thus, increasing attention is being paid to the development of potential methods for recovering CMs, including REEs from coal wastes. Table 9-2 above provides market information on critical materials from coal wastes. This section reviews R&D on extraction of REEs, typically in the form of rare earth oxides (REOs), from coal wastes and discusses research on the extraction of two other CMs, lithium and nickel. The chapter does not discuss the subsequent separation of individual REEs from REOs, which is a distinct separations issue not affected considerably by the characteristics of coal wastes.

REEs can be found in solid and liquid coal wastes. Solid coal wastes containing REEs include impoundment refuse and CCRs. Recovery of REEs from solid coal wastes requires dissolving REOs via a leaching process into an extractable aqueous phase. AMD, a liquid coal waste stream, offers potentially direct extraction opportunities when containing a sufficient REE concentration. REEs in coal wastes can be recovered by combining physical beneficiation and subsequent hydrometallurgical processes. Table 9-3 presents advantages and disadvantages of different coal waste types with respect to REE extraction. Physical beneficiation methods, including gravity, magnetic, and flotation separations, are used to enrich the concentrations of REEs in solid coal wastes to obtain REE-preconcentrated solid coal wastes as a feedstock for the next process—hydrometallurgical treatment (see, e.g., Eterigho-Ikelegbe et al. 2021; Fu et al. 2022; Mwewa et al. 2022; Zhang et al. 2020a). Physical beneficiation reduces the amount of unburned carbon (that would consume chemicals used during hydrometallurgical treatment), magnetic materials that typically contain fewer REEs than nonmagnetic fractions, and large-size particles that lower the leaching efficiency in hydrometallurgical processing.

Hydrometallurgical processes are used to obtain high-purity CM mixtures from REE-enriched solid coal wastes or preconcentrates (see, e.g., Dodbiba and Fujita 2023; Eterigho-Ikelegbe et al. 2021; Fu et al. 2022; Mwewa et al. 2022; Zhang et al. 2020a). Typically, leaching is the first step in hydrometallurgical processes, followed by various extraction and precipitation steps. To increase leaching efficiency, a pretreatment of roasting preconcentrates obtained with chemicals, such as Na_2CO_3, is first completed. Alternatively, mixing with sodium hydroxide (NaOH) or other strong alkalis decomposes the glassy aluminum-silicate matrix to form soluble species—for example, $H_2Si_2O_6^{2-}$ and $Al(OH)_4^-$, thereby liberating the REEs captured in the glassy matrix. REOs or rare earth hydroxides ($RE(OH)_3$), are subsequently generated via precipitation.

Membrane-based separation processes are also proposed for recovery of critical metals, including nanofiltration, reverse osmosis, and electrically driven separation processes such as electrodialysis, bipolar electrodialysis, and selective bipolar electrodialysis (Chen et al. 2022; Elbashier et al. 2021; Huang and Xu 2006; Sarker et al. 2022; Zhou et al. 2018). Such membranes were originally designed to remove relatively low concentrations of ions from aqueous solutions, and much of the current research is aimed at understanding required modifications to membrane design to improve their utility in these new applications.

With an eye toward securing U.S. supplies of REEs, DOE's Office of Fossil Energy and Carbon Management and National Energy Technology Laboratory assessed the feasibilities of extracting REEs from coal ore and coal combustion by-products at a request from Congress in 2014 and submitted the findings of the assessments to Congress in January 2017 (DOE 2017b). Subsequently, DOE began funding many RD&D projects to demonstrate

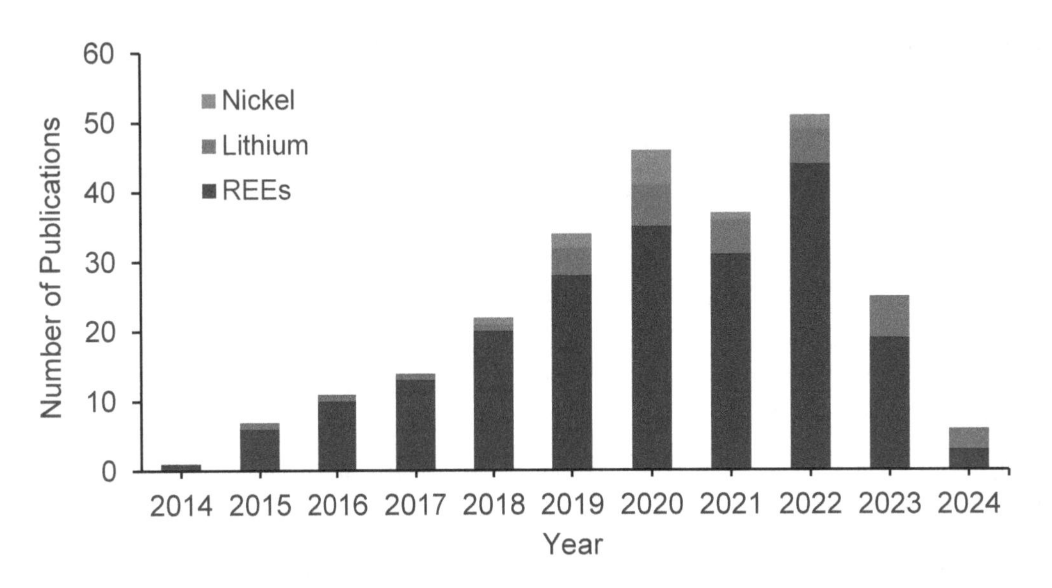

FIGURE 9-8 The quantities of annually published work since 2014 involving REE, lithium (Li), and nickel (Ni) extractions. NOTE: The 2024 data include publications through March.

the feasibility of producing REEs from coals and coal wastes. The committee performed a literature review of publications on this topic since 2015, following the initial request from Congress for a feasibility assessment. The literature review, which can be found in Appendix L, analyzed the extraction method and leaching agent employed in each publication, as well as the resultant leaching efficiencies of REEs, lithium, and nickel. Since 2015, 211 journal articles, book chapters, conference materials, theses/dissertations, and DOE project reports have been published globally in the areas of REE, and/or lithium, and/or nickel extractions from coal wastes. Figure 9-8 shows the number of publications covering REE, lithium, and nickel by year, illustrating that most of the work has been performed on REE extractions. Table 9-4 shows the total number of publications by element/element group, by type of publications (for REE extraction), and by feedstock source (for research articles about REE extraction).

9.3.4.1 Current Technologies for Recovering Rare Earth Elements from Coal Wastes

Light rare earth elements (LREEs) are found in bastnaesite [(La,Ce,Y)CO_3F], monazite [(Ce,La,Th)PO_4], allanite [(Ce,Ca,Y,La)$_2$(Al,Fe^{+3})$_3$(SiO_4)$_3$(OH)], ancylite [Sr(Ce,La)(CO_3)$_2$(OH)·H_2O], cerite [(Ce,La,Ca)$_9$(Mg,Fe^{3+})(SiO_4)$_6$(SiO_3OH)(OH)$_3$], cerianite [(Ce,Th)O_2], fluocerite [(Ce,La)F_3], lanthanite [(REY)$_2$(CO_3)$_3$·8(H_2O)], loparite [(Ce,Na,Ca)(Ti,Nb)O_3], parisite [Ca(Ce,La)$_2$(CO_3)$_3F_2$], and stillwellite [(Ce,La,Ca)BSiO_5]. Bastnaesite is the primary source for praseodymium and neodymium (Omodara et al. 2019). The majority of heavy rare earth elements (HREEs) are found in xenotime (YPO_4), yttrotungstite [$YW_2O_6(OH)_3$], samarskite [(Y$Fe^{3+}Fe^{2+}$U,Th,Ca)$_2$(Nb,Ta)$_2O_8$], euxenite [(Y,Ca,Ce,U,Th)(Nb,Ta,Ti)$_2O_6$], gadolinite [(Ce,La,Nd,Y)$_2$Fe$Be_2Si_2O_{10}$], yttrotantalite [(Y,U,Fe^{2+})(Ta,Nb)O_4], yttrialite [(Y,Th)$_2Si_2O_7$], and fergusonite (REY,NbO_4). Ion adsorption clays are the dominant sources for dysprosium (Zapp et al. 2018) and holmium (Sanz et al. 2022) Monazite and bastnaesite are the major sources for erbium (RSC 2024). Terbium can be found in monazite, bastnaesite, and xenotime, as well as ion adsorption clays, which are its richest marketable sources (Sinha et al. 2023).

In solid coal wastes, REEs primarily exist in monazite (containing light rare earth elements) and xenotime (containing an HREE) in addition to organic matter/clays. The REE-containing minerals are exceedingly fine-grained with particle sizes ranging from less than 1 μm to 5 μm (Hedin et al. 2020; Li and Zhang 2022). In addition, some common REE-bearing minerals in coals do not contain REE structural constituents. These include zircon ($ZrSiO_4$), REE-bearing phosphates such as apatite ($Ca_5(PO_4)_3$(OH,F,Cl)) and crandallite ($CaAl_3(PO_{3.5}(OH)_{0.5})_2(OH)_6$), and rhabdophane (REE$PO_4$·$H_2O$) (Dai et al. 2016; Finkelman et al. 2019; Kolker et al. 2024; Ward 2016).

TABLE 9-4 Number of Publications on REE, Li, and Ni Extraction from Coal Waste Since 2015 by Target Element, Publication Type, and Feedstock Source

Category	Publications, 2015–2024[a]
Target Element(s)	
REEs	210
Lithium	33
Nickel	11
REE Extraction by Publication Type	
Research Article	128
Review Article	21
Book	12
Conference Paper	12
Project Report	18
Thesis/Dissertation	9
Patent	8
Source of REE in Research Articles	
Coal	18
Impoundment Waste	20
Coal Fly Ash	66
Coal Bottom Ash	2
Mixture of Coal Ash	12
Acid Mine Drainage	10

[a] 2024 data include publications through March.

Although solid coal wastes from mining and combustion contain much lower concentrations of REEs than REE ores, they represent a significant potential resource for REE production, given the large volumes of these wastes (Das et al. 2018; DOE 2022; Huang et al. 2020; Jha et al. 2016; Opare et al. 2021; Peiravi et al. 2021; Wu et al. 2018; Zhang et al. 2020a). DOE's "interest level" for extracting REEs from coal waste is 300 ppm, based on demonstration of technical feasibility for obtaining high-purity REE from coal wastes at these concentrations (DOE 2017a, 2022). REE-bearing coal refuse contains REE concentrations as high as 300 ppm (Zhang and Honaker 2020a), and coal slags and ashes typically have higher concentrations of REEs than coal refuses (in some cases more than 1500 ppm; Scott and Kolker 2019). Total REE concentrations in AMD solids, on the other hand, can be as high as 2000 ppm (Hedin et al. 2024).

The total concentration of critical REEs (neodymium, europium, terbium, dysprosium, yttrium, and erbium)[8] in coal ashes is a key factor that determines their potential economic values. Globally, the average fraction of critical REEs in coal ashes is 36 percent (Hower et al. In press), which is higher than some conventional REE ores (Fu et al. 2022). Thus, coal ashes could be important resources for providing critical REEs.

REEs can be extracted directly from sedimentary rocks including coal refuse or overburden (a carbonaceous rock with less organic matter than coal). Figure 9-9 shows a conventional process for extracting REEs from solid coal wastes, which includes both physical and chemical treatments. The first few operation steps are physical processes, and the remaining steps are chemical processes. Physical separation methods include air classification (Shapiro and Galperin 2005), gravity separation, magnetic separation, and flotation. Chemical methods include

[6] Critical REEs are those for which supply (production) is projected to be less than demand (industrial consumption) (Seredin 2010; Seredin and Dai 2012). They are a subset of the critical minerals defined by the USGS.

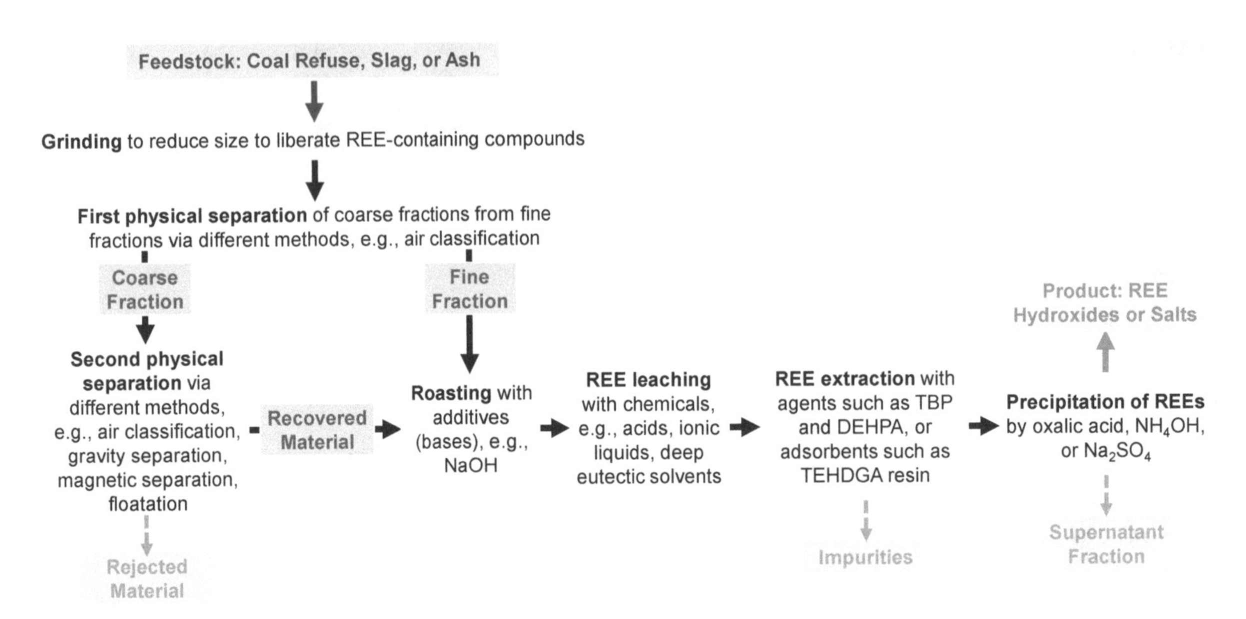

FIGURE 9-9 Schematic of a sample processes of REE separation from solid coal wastes.
NOTES: DEHPA = di-(2-ethyl hexyl) phosphoric acid; Na_2SO_4 = sodium sulfate; NaOH = sodium hydroxide; NH_4OH = ammonium hydroxide; TBP = tributyl phosphate; TEHDGA = N,N,N',N' tetra (2-ethylhexyl) diglycolamide. Processes are discussed in the text. Feedstock indicated in purple text, intermediates in pink text, and product in blue text.

roasting with additives, leaching with chemicals (e.g., acids, ionic liquids, deep eutectic solvents), extraction, and precipitation. The physical treatment steps are similar for all types of solid coal wastes, although the order and operation conditions may differ slightly depending on the physical properties of the solid wastes. In contrast, the materials used in each chemical treatment step can differ significantly depending on the chemical structure and elemental compositions of the treated solid coal wastes.

Understanding the original structure and composition of REE-containing compounds in coal and their changes during coal combustion is important to the choices of physical and chemical treatment methods in Figure 9-9. Good progress has been made in this area. For example, a collaborative survey by the USGS, University of Kentucky Center for Applied Energy Research, and a utility company in Indiana on the ashes of Wyodak-Anderson (WY) coal indicated that the REEs mainly exist as inorganic phosphates (monazite) with trace silicates and carbonates (Brownfield et al. 2005). More recently, Fu et al. reviewed the content and occurrence mode of REEs in coal ashes from a variety of different countries, including the United States, finding that the average total REE concentration in the 257 tested U.S. samples from coal-fired power plants is 459.6 $\mu g \cdot g^{-1}$ or ppm, which is higher than the average value of samples from six European countries (Spain, England, Poland, Bulgaria, Romania, and Finland; 278.7 $\mu g \cdot g^{-1}$) and four other countries (South Africa, India, South Korea, Indonesia; 298 $\mu g \cdot g^{-1}$) (Fu et al. 2022). The average concentrations of light (lanthanum to samarium, without the presences of scandium and promethium in this study); middle (europium to dysprosium, and yttrium); and heavy (holmium to lutetium) REEs in the U.S. samples are 340.8 $\mu g \cdot g^{-1}$, 100.5 $\mu g \cdot g^{-1}$, 18.3 $\mu g \cdot g^{-1}$, respectively (Fu et al. 2022). The average fraction of critical REEs in U.S. coal ashes was found to be 37 percent, slightly higher than the average fraction of critical REEs in the 581 tested coal ash samples from around the world (36 percent) (Fu et al. 2022). Fu et al. (2022) also found that REE-bearing phases in coal can undergo complex chemical transformation during coal combustion, as shown in Table 9-5. However, more studies are needed to characterize coal wastes to facilitate selection of appropriate chemicals for leaching REEs—for example, the choice of inorganic or organic acids.

An important chemical treatment step in the overall procedure of REE extraction from solid coal wastes is to leach REEs from the solid materials that contain many compounds, including a number of non-REE cations. Among the cations in the solid materials are Na^+, Mg^{2+}, Ca^{2+}, Fe^{3+}, Al^{3+}, and Si^{4+}, of which all except Si^{4+} can be

TABLE 9-5 Overview of Reported Thermal Decomposition and Transformation of Common REE Phases During Coal Combustion

REE Speciation in Coal	Reference Compound	Phase Transformation During Thermal Conversion	Change in Oxidation State
Organic associations	REE-lignin	REE-lignin → REE-oxides	Ce(III) → Ce(IV)
Carbonates	$Y_2(CO_3)_3$	$Y_2(CO_3)_3 \rightarrow Y_2O_3$	No change
	$Ce_2(CO_3)_3$	$Ce_2(CO_3)_3 \rightarrow CeO_2$	Ce(III) → Ce(IV)
	$(Ce,La)CO_3(F,OH)$	$(Ce,La)CO_3(F,OH) \rightarrow (Ce,La)O_2$	Ce(III) → Ce(IV)
	REE-doped calcite	$CaCO_3 \rightarrow CaO$	No change
Phosphates	Hydrated YPO_4	$YPO_4{\cdot}2H_2O \rightarrow YPO_4$	No change
	Hydrated $CePO_4$	$CePO_4{\cdot}2H_2O \rightarrow CePO_4$	No change
	Monazite	Size reduction via fragmentation (>1400°C)	
	Xenotime	No Change	
	Calcium Apatite	Fluorapatite melts at 1644°C; Chlorapatite structure change begins 200°C, with melting at 1530°C; Hydroxy apatite dehydroxylates at 900°C and decomposes above 1200°C	No change Partial oxidation
Silicates	Zircon	No change below 1000°C Melting above 1000°C $ZrSiO_4 \rightarrow ZrO_2\text{-t} + SiO_2$ (>1285°C)	No change

SOURCES: Based on data from Fu et al. (2022); Hood et al. (2017); Liu et al. (2020); and Tõnsuaadu et al. (2012).

easily leached from the source material. The concentrations of each of these cations can be more than 1000 times higher than the total concentration of REEs in solid coal wastes. Therefore, the choice of solvent to maximally leach REE cations and minimally leach non-REE cations is important for cost and efficiency. In general, organic acids are better than mineral acids for selectively leaching REEs from solid coal wastes (Banerjee et al. 2021b). The use of organic acids to leach REEs also maximally maintains the chemical structure of coal fly ash, which is important to its major applications in cement or brick industries.

In recent years, new leaching approaches have been developed to overcome safety and environmental issues resulting from strong chemicals—for example, corrosion, toxicity, and explosion, by using ionic liquids and deep eutectic solvents for solvent extraction (Alguacil and Robla 2023; Danso et al. 2021; Karan et al. 2022). Owing to their very low vapor pressure and intrinsic electric conductivity, ionic liquids and deep eutectic solvents could replace the organic phase in liquid-liquid extraction processes, resulting in safer operation systems. Stoy et al. (2021) achieved 83 percent and >90 percent leaching efficiencies with a fly ash without and with pretreatments, respectively, when using an ionic liquid [Hbet][Tf_2N] (Nockemann et al. 2006, 2008), to separate REEs from a coal fly ash sample. Karan et al. (2022) used a choline-chloride-based deep eutectic solvent system for REE extraction from a coal fly ash sample and achieved 85–95 percent leachability. Despite their potential benefits, ionic liquids tend to be significantly more costly than conventional solvents. Most ongoing research on ionic liquids and deep eutectic solvents is still at bench scale, and their feasibility at larger scales continues to be evaluated.

Supercritical fluids, including supercritical CO_2, have been explored for leaching REEs from coal by-products. Supercritical-fluid-based REE extraction technology is promising owing to the high diffusivities and low viscosities of supercritical fluids and ease of scale up. High extraction efficiencies, especially for HREEs, can be achieved with supercritical fluids, particularly supercritical CO_2. REE extraction with supercritical CO_2 is environmentally friendly and has been demonstrated to concentrate REEs at 312 ppm in fly ash to 99.4 percent in the final product in form of REOs, with the five critically important HREEs (dysprosium, europium, neodymium, terbium, and yttrium) accounting for up to ~63 percent of the total weight of the final REO product (Fan and Huang 2023; Huang et al. 2018).

Another approach to REE separation and recovery from coal by-products uses molecular recognition technology, which has the capability to perform selective separations at various stages in metal life cycles (Bentzen et al. 2013; Gielen and Lyons 2022; Oberhaus 2023). High REE selectivity, both as a group or as individual elements, can be obtained using a predesigned ligand bonded chemically by a tether to a silica gel solid support. Separations are performed in column mode using feed solutions containing the target REE in a matrix of acid and/or other metals. The target REE is selectively separated by the silica gel–bound ligand, leaving other solution components to go to the raffinate, where individual components can be further recovered. The high selectivity means metal impurities do not have to be removed downstream, which simplifies the process. It has been claimed that a molecular recognition technology plant will offer lower capital and operating costs, as well as a lower physical and environmental footprint, than an equivalent conventional solvent extraction plant (Ucore 2022). The process could be superior to alternatives in capital and operating costs and with potential environmental benefits.

Different REE extraction methods have their advantages and disadvantages from different perspectives. Many factors, including the type of coal waste, REE extraction method, and conditions (e.g., type of leaching agent, leaching temperature, and leaching time) affect the overall REE recovery efficiency. Furthermore, the form and content of REEs differs between coal ashes from the western and eastern United States. The majority of REEs in low-rank western coals are complexed with organic compounds rather than inorganic materials, and thus are more readily extractable than higher-rank eastern coals, despite having a lower overall REE content.

As mentioned earlier, the committee performed a literature review of REE extractions from coal waste streams to examine the state of the field and understand some of these factors impacting recovery efficiency. The full literature review is presented in Appendix L, while results of the committee's subsequent analysis are depicted in Figures 9-10, 9-11, and 9-12. Based on data reported from 2019 to 2023, the average REE recovery efficiency obtained with AMD is higher than those achieved with other coal waste types (Figure 9-10). Figure 9-11 indicates that chemical leaching is, on average, more efficient than bioleaching when coal waste pretreatment is used to enhance REE leaching with chemicals. Much longer leaching times are needed for bioleaching processes.

Challenges for extraction of REEs from coal wastes are significant, which could be a major reason why no considerable increase in average leaching efficiency has been observed in recent years, as indicated in Figure 9-12. One challenge is the use of strong mineral acids in leaching processes, which can lead to serious environmental pollution. Thus, R&D opportunities for extraction of REEs from coal waste center around developing improved physical and chemical separations processes. As noted above, a better understanding of the structure or morphology of REEs in coal by-products would facilitate the choice of physical beneficiation and operation steps. Integrating different physical separation methods could improve the efficiency of physical separation processes, with a goal of rejecting the CM-containing fractions and preconcentrating the CMs. The low concentration of REEs in coal wastes and coal combustion by-products (typically <100 μg g^{-1} for an individual element) makes leaching processes expensive. Development of more selective solvents could yield cost reductions.

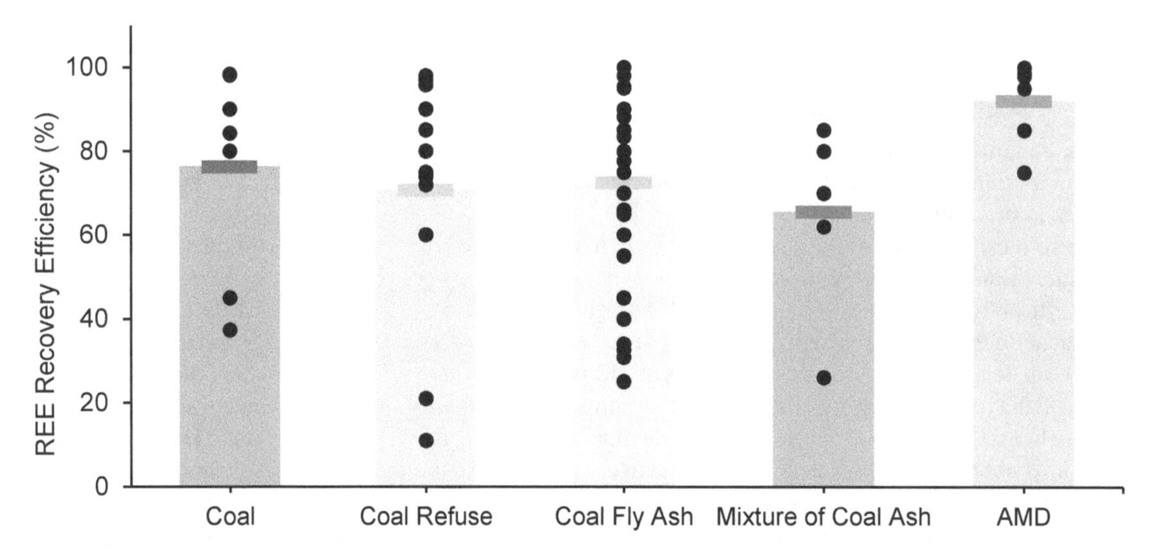

FIGURE 9-10 The effect of coal waste source type on recovery efficiencies.
NOTES: See Table L-1 in Appendix L for underlying data. Bar height indicates average recovery efficiency value.

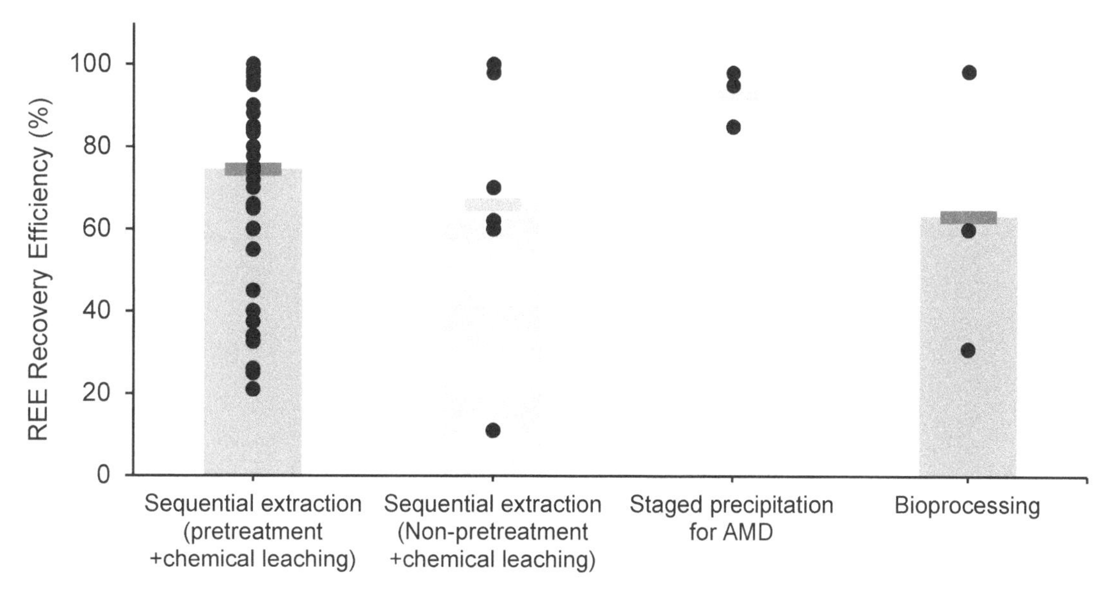

FIGURE 9-11 The effect of extraction method on efficiency of REE recovery from coal wastes.
NOTES: See Table L-1 in Appendix L for underlying data. Bar height indicates average recovery efficiency value.

9.3.4.2 Current Technologies for Recovering Energy-Relevant Critical Minerals from Coal Wastes

Deploying clean energy technologies at scale will require substantial amounts of CMMs, and supply chain risks or bottlenecks in obtaining these minerals and materials could inhibit the net-zero transition (DOE 2023). Table 9-6 lists the clean energy technology applications of 12 elements and materials of particular interest: neodymium, praseodymium, dysprosium, lithium, cobalt, nickel, manganese, graphite, iridium, platinum, gallium, and germanium. Coal waste streams represent one potential domestic source of these energy-relevant minerals and materials, with estimates of tens to thousands of years of supply for certain elements at current rates of consumption (see Table 9-6). Methods to obtain graphite from coal wastes were discussed in Section 9.3.3.2 above. In the following sections, extraction of lithium and nickel are described as case examples because most literature on CM recovery from coal wastes (mainly) focuses on these elements, and the estimated masses of lithium and nickel

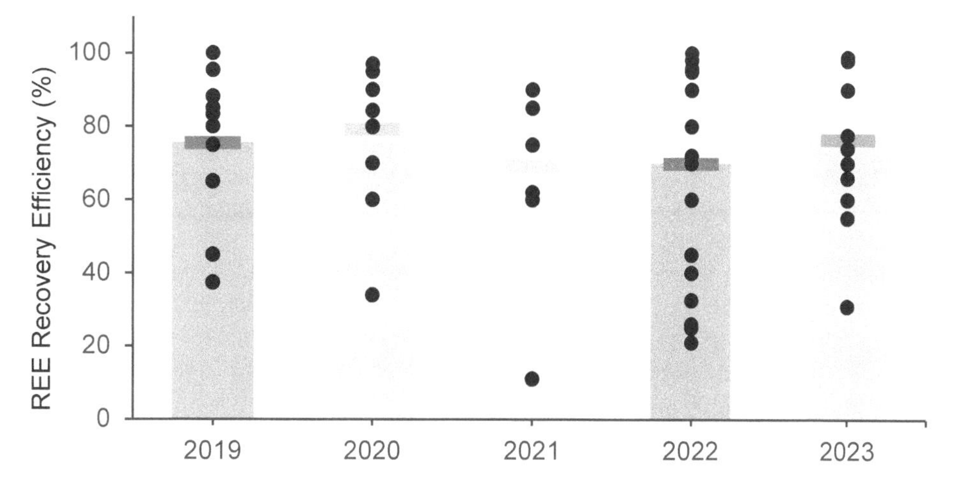

FIGURE 9-12 Average REE recovery efficiencies achieved during 2019–2023 across all coal waste types and extraction methods.
NOTES: See Table L-1 in Appendix L for underlying data. Bar height indicates average recovery efficiency value.

TABLE 9-6 Key Elements and Materials for Clean Energy Technologies and Estimated Domestic Supply from Coal Waste Streams

Metal/Material	Technology Application	Estimated Mass (tons)	Estimated Supply (years)
Neodymium	Magnets for wind energy generators, electric and fuel cell vehicle motors, and industrial motors	172,000	40
Praseodymium		Not reported	Not reported
Dysprosium		62,000	14
Lithium	Batteries for electric vehicles and electricity storage	288,000	130
Cobalt		110,000	15
Nickel		252,000	1.1
Manganese		Not reported	Not reported
Graphite		Not reported	Not reported
Iridium	Electrolyzers for hydrogen production; fuel cells for transportation, stationary energy storage	40	15
Platinum		600	15
Gallium	Wide bandgap power electronics for connecting high voltage power generation to the grid	20,000	1100
Germanium	Semiconductors; fiber and infrared optics for sensors, data, and control	30,000	3900

SOURCE: Adapted from Wilcox (2023).

in coal wastes are higher than those of other elements (Table 9-6). Both lithium and nickel exist in coal (Zhang et al. 2020b), AMD (DOE 2022; Fritz et al. 2021; Hedin et al. 2020; Li and Zhang 2022; Stuckman 2022), and coal ash (Gupta et al. 2023; Hamidi et al. 2023) at varying concentrations, depending on the structure and chemical composition of the coal, including sulfur content and metal concentration. A final section briefly describes processes to extract other CMMs relevant to energy supply chains from coal wastes.

9.3.4.2.1 Characteristics of Lithium in Coal Wastes and Separation of Lithium from Coal Wastes

Lithium is a widely used critical mineral that has been playing an increasingly important role in rechargeable batteries for electric vehicles and various electronic devices, in addition to its use in the production of pharmaceuticals, ceramics, glass, metallurgy, and polymers. Between 2010 and 2022, the world's production of lithium, in the form of lithium carbonate or lithium hydroxide, nearly quadrupled (Energy Institute 2023). Lithium markets are anticipated to increase by 13 times by 2040 in a base case scenario and could grow by 40 times by 2040 in a sustainable development scenario (IEA 2022). The well-known lithium deposits are peralkaline and peraluminous pegmatite deposits and their associated metasomatic rocks, lithium-rich hectorite clays derived from volcanic deposits, and Salar evaporites and geothermal deposits (Bowell et al. 2020). Lithium-rich rocks include spodumene (6–9 wt% Li_2O), petalite (3.0–4.73 wt% Li_2O), lepidolite (3.0–4.19 wt% Li_2O), zinnwaldite (2–5 wt% Li_2O), amblygonite (7.4–9.5 wt% Li_2O), montebrasite (7.4 wt% Li_2O), eucryptite (4.5–9.7 wt% Li_2O), triphylite (9.47 wt% Li_2O), jadarite (7.3 wt% Li_2O), and hectorite (<1–3 wt% Li_2O) (de los Hoyos 2022; Evans 2014; Garrett 2004; London 2008). However, these mining resources of lithium are being exploited steadily and quickly depleted. Thus, other lithium-containing resources, such as brines, seawater, and waters from oilfields, geothermal fields, and mining of coal and other ores, need to be developed.

Coal contains lithium at global average concentrations of about 12 ppm (Ketris and Yudovich 2009), while the average concentration of lithium in U.S. coals is 16 ppm (Orem and Finkelman 2003). Coal fly ash contains a global average concentration of 66 ppm lithium (Ketris and Yudovich 2009). More than 90 percent of the lithium in U.S. coals is in clays and micas, and the remaining amount is associated with either organics or tosudite [$Na_{0.5}(Li,Al,Mg)_6((Si,Al)_8O_{18})(OH)_{12}\cdot 5H_2O$] and cookeite [$LiAl_4(AlSi_3O_{10})(OH)_8$] (Finkelman et al. 2018; Seredin et al. 2013; Zhang et al. 2020b). The concentrations of lithium in high-rank coals are significantly higher than those in low-rank coals owing to the association of lithium with detrital silicates in the former (Dai et al. 2021). Accordingly, the concentrations of lithium in AMD, solid coal wastes, and coal combustion wastes vary significantly with the types and locations of coals.

AMD contains lithium that can dissolve in water at low pH (Griswold 2022; Kolker et al. 2024). Thus, lithium can be separated from AMD using reported methods for recovering lithium from water (Baudino et al. 2022), including evaporation, precipitation, lithium-ion sieves (e.g., aluminum hydroxide ion sieves, lithium manganese oxide ion sieves, and lithium titanium oxide ion sieves), membranes, supramolecular chemistry, ionic liquids, and electrochemistry. The principles, advantages, and disadvantages of different methods are shown in Table 9-7. Different factors affect the technological performance, economic feasibility, and carbon and overall environmental footprint of various lithium extraction technologies. Some approaches contain multiple operation steps, with different factors affecting these same metrics for each step. For example, the conventional precipitation approach, shown in Figure 9-13, involves six steps. The operational performance of Steps 1, 3, and 6 is controlled by the quantities of the added chemicals (i.e., $CaO/Ca(OH)_2$, $Na_2C_2O_4$, Na_2CO_3) and the pH values of the liquids, while the operational performance of Steps 2 and 4 is determined by temperature and characteristics of the filters, as well as filtration operation conditions. Multiple technologies can be integrated to achieve high lithium recovery efficiency and product purity.

TABLE 9-7 Technologies for Separating Lithium from Acid Mine Drainage and Other Lithium-Containing Aqueous Solutions

	Technology	Advantages	Disadvantages
Conventional	**Vaporization** Based on concentrating Li^+ in solution via the loss of H_2O, realized by using solar energy (i.e., solar-heated evaporation)	1. Environmentally friendly 2. Cost-effective	1. Need favorable weather conditions 2. Slow
Conventional	**Precipitation** Based on the formation of insoluble or low solubility lithium salts • **Method 1:** Carbonate formation, e.g., $2Li^+ + Na_2CO_3 \rightarrow 2Na^+ + Li_2CO_3\downarrow$ • **Method 2:** Redox and coagulation, e.g., $Li^+ + 2Al + Cl^- + (x + 6)\ H_2O \rightarrow LiCl{\cdot}2[Al(OH)_3]\ xH_2O\downarrow + 3H_2\uparrow$ • **Method 3:** Coagulation, e.g., $Li^+ + 2AlCl_3 + (x + 6)\ H_2O \rightarrow LiCl{\cdot}2[Al(OH)_3]{\cdot}xH_2O\downarrow + 6H^+ + 5Cl^-$	1. Efficient for low Mg^{2+}/Li^+ mass ratio solutions 2. Simple 3. Easy scale up and commercialization 4. Low cost owing to use of inexpensive precipitation agents and processes	1. Necessary to preremove major non-Li^+ ions (e.g., Mg^{2+}, Ca^{2+}, SO_4^{2-}) 2. Difficult to manage co-precipitation of other cations, especially Mg^{2+}
Conventional	**Ion exchange** Based on sorption and desorption • **Method 1:** Manganese oxide sorbents (Spinel form or H-form) • **Method 2:** Titanium dioxide sorbents • **Method 3:** Other sorbents (e.g., aluminum hydroxide)	1. Fast Li^+ adsorption and desorption kinetics 2. High Li^+ adsorption selectivity 3. Generates high-quality lithium products 4. Environmentally friendly	1. Sorbent fouling owing to the presence of silica 2. High cost owing to the need for expensive raw materials and small sorbent porosity, sorbent dissolution, and acids for Li^+ desorption
Conventional	**Solvent extraction** Based on (1) the difference of Li^+ solubilities in aqueous solution and organic solvents (e.g., tributyl phosphate [TBP]) and (2) the formation of Li^+-containing complex structures dissolvable in organic materials, and stripping Li^+ out of the organic phase with acids (e.g., HCl)	1. High Li^+ separation efficiency 2. High selectivity 3. Fast Li^+ stripping from the organic phase owing to use of strong acids (e.g., HCl)	1. Need for expensive organic extraction agents 2. Stabilities of extraction agents 3. Pollution from organic extractant
Nonconventional	**Membrane-based electrodialysis** Based on the selective diffusion of lithium via a specific Li^+-ion-selective membrane set between anode and cathode under the stimulus of electricity, which is determined by membrane's size and surface charge	1. Potentially applicable to high Mg^{2+}/Li^+ mass ratio solutions 2. Versatile 3. Relatively easy to industrialize	1. Sensitive to silica that damages ionic membranes and electrodialysis 2. Durability challenges 3. Short lifetime
Nonconventional	**Electrochemical process** Based on the combination of a working electrode in which Li^+ is captured and consequently released into a recovery solution and an opposing electrode onto which anions are deposited	1. High Li^+ selectivity 2. Good reversibility 3. Low energy usage (and thus cost)	1. Mainly applicable for solutions with high Li^+ concentration 2. Difficult to adjust pH

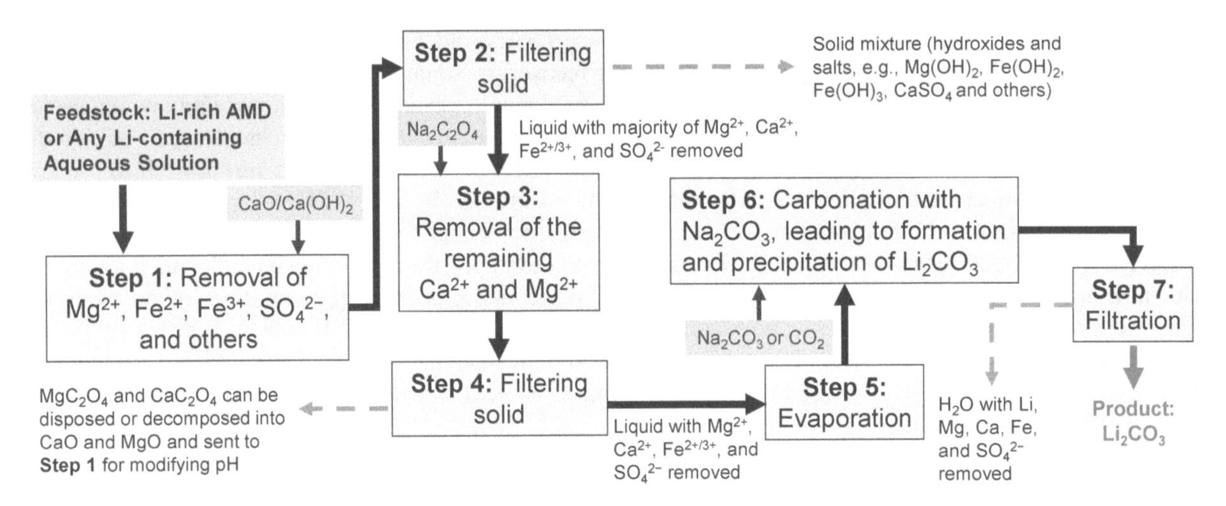

FIGURE 9-13 Schematic drawing of conventional precipitation-based lithium extraction from AMD.
NOTES: Feedstock indicated in purple text, additives in pink text, and product in blue text. Dashed arrows indicated rejected fractions.

Separating lithium from solid waste CCRs differs from separating lithium from AMD because pretreatment steps and leaching processes are required, as shown in Figure 9-14. Pretreatments are used to select the fractions with higher lithium concentrations, which reduces the overall cost of lithium extraction from coal wastes and coal combustion products. Calcination and roasting, frequently performed with alkaline materials and salts, increase the leaching or chlorination reaction kinetics. Temperatures for both calcination and roasting typically exceed 700°C. Higher temperatures and longer operation times can lead to more decomposition reactions and stronger phase transformation for calcination and roasting processes. Leaching conditions, including the concentration of sulfuric acid/hydrochloric acid (H_2SO_4/HCl) and the times of H_2SO_4 leaching and chlorination, need to be optimized to maximize lithium dissolution and minimize dissolution of undesired elements. Then different technologies, including precipitation, can be used to separate lithium ions (Li^+) from the aqueous solution obtained via filtration, as discussed above and shown in Figure 9-13. Given the requirement for additional front-end processing, lithium extraction and purification from solid coal wastes and coal combustion products is likely more expensive and less environmentally friendly than lithium extraction and purification from AMD.

9.3.4.2.2 Characteristics of Nickel in Coal Wastes and Separation of Nickel from Coal Wastes

The world's nickel resources are estimated to be ~350 million tons (Nickel Institute 2024). New nickel reserves and resources continue to rise steadily owing to increased knowledge of nickel-containing mineral deposits and innovative mineral exploration technologies. However, the distribution of nickel resources in the world is highly uneven; Indonesia alone accounted for about half of global mine production of nickel in 2023 (USGS 2024). Moreover, nickel will continue to play an essential role in many industries, including stainless steel, high-temperature alloys, nonferrous alloys, superalloys, and batteries (Li et al. 2023; Mohammadi and Saif 2023; Pandey et al. 2023; Staszuk et al. 2023; Torres et al. 2023).

Nickel mainly exists in sulfide and oxide ores in nature. It is typically found in the +2 oxidation state in water, although it also can be in the +3 and +4 oxidation states as Ni_2O_3 and NiO_2, respectively. Pentlandite [Ni_9S_8 or $(Fe,Ni)_9S_8$] and laterites (nickeliferous limonite [$(Fe,Ni)O(OH)$], and garnierite [$(NiMg)_6Si_4O_{10}(OH)_8$]) have been the major nickel-containing ores for nickel production (USGS n.d.). Unlike natural nickel ore, the existing form of nickel in coal is not fully clear yet. Nickel might be chemically bonded with organic or inorganic compounds, such as sulfide minerals (O'Keefe 1996). The concentrations of nickel in coal wastes vary from undetectable levels (0 ppm) to >10,000 ppm, determined by the coal resource used for coal combustion (Hower et al. 2021; Ilander and Väisänen 2009; Kalderis et al. 2008; Salgansky et al. 2022; Sun et al. 2021; Wierońska et al. 2019).

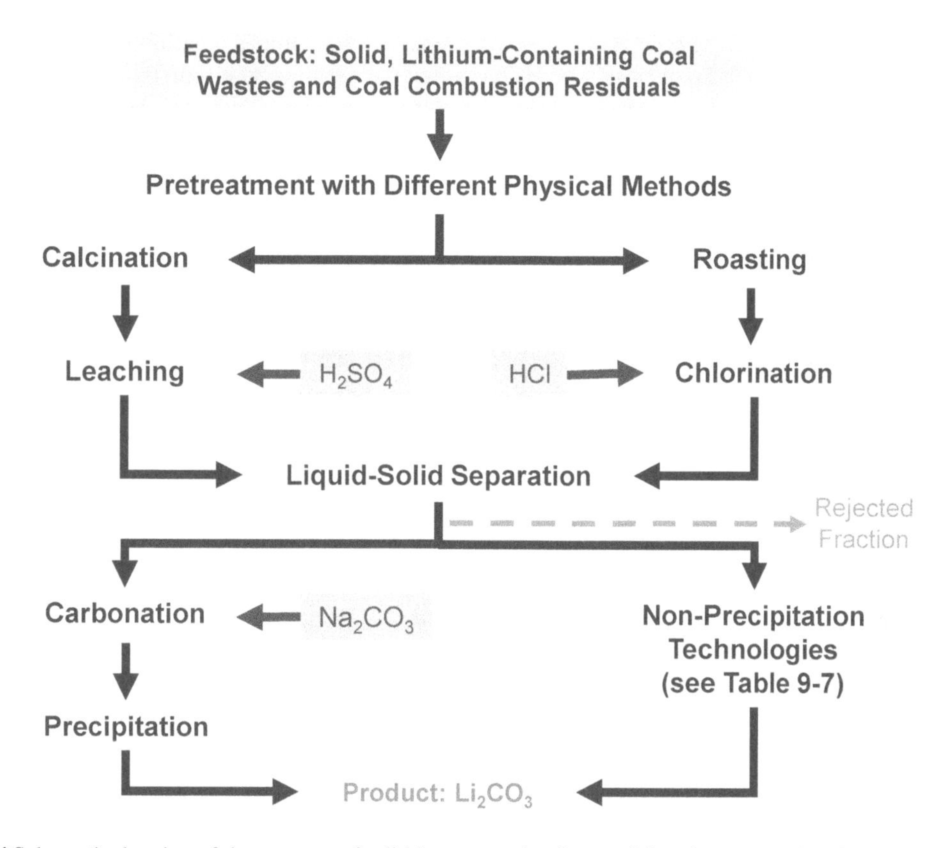

FIGURE 9-14 Schematic drawing of the processes for lithium extraction from solid coal wastes and coal combustion residues. NOTE: Feedstock indicated in purple text, additives in pink text, and product in blue text.

The concentrations of nickel in AMD vary significantly with coal source and can be as high as 6720 ppm (Berger et al. 2011; Hedin et al. 2020; Park et al. 2013; Sahoo et al. 2012; Toropitsyna et al. 2023).

The United States is not rich in nickel-containing ore deposits, but it is rich in coals, which contain substantial amounts of potentially economically recoverable nickel resources. Therefore, developing processes for recovering nickel from coal wastes (e.g., coal overburdens) could benefit the U.S. economy. Moreover, like other heavy metals, nickel is harmful to the environment and ecosystem. Thus, extracting nickel from coal, coal wastes, and coal combustion by-products could provide both economic development and environmental protection.

Several studies on the recovery of nickel from AMD and fly ashes have been performed (Park et al. 2013; Toropitsyna et al. 2023). The methods for separating nickel from AMD, other coal wastes (e.g., overburdens), and combustion wastes (e.g., fly ashes) are essentially the same. The only difference is that the nickel in solid coal wastes and coal combustion wastes needs to be leached out with acids, such as hydrochloric acid (HCl) and sulfuric acid (H_2SO_4). Prior to separating nickel from aqueous solutions, including AMD and the solutions that result from leaching of nickel-containing solid coal wastes and combustion products, other critical materials (e.g., REEs, lithium, manganese) and undesired elements (e.g., iron, magnesium, calcium) typically need to be removed with various other methods. For example, staged precipitation can be used to separate iron, aluminum, and others from solution by tuning the pH. Sulfides can be used to separate copper, zinc, and others from solution via fractional precipitation, but they cannot be used to separate nickel from cobalt owing to their very similar precipitation properties. Nickel/cobalt separation can be achieved using solvent extraction with bis-2-ethylhexyl phosphoric acid (DEHPA, D2EHPA, HDEHP) (Nadimi et al. 2017), but this is challenging owing to the high similarity of the extraction pH isotherms of nickel and cobalt, and thus the small difference between pH values corresponding to 50 percent extraction of Ni and Co, expressed as $\Delta pH_{0.5}^{Ni-Co}$. Agents for shifting the extraction pH isotherms of nickel and cobalt, such as citrate, can be used to help overcome this challenge. The extraction isotherms of cobalt and nickel with DEHPA in the presence of citrate ion as a carboxylate ligand are different from the isotherms of cobalt and nickel with DEHPA without the presence of citrate (Nadimi et al. 2017). Given the importance of both nickel and cobalt to clean energy technology supply chains, more efficient methods for their separation would be valuable.

9.3.4.2.3 Other Materials Critical to Energy Supply Chains

In addition to the four REEs (dysprosium, neodymium, praseodymium, and terbium), lithium, and nickel previously discussed in this chapter, other minerals and materials deemed by DOE as critical for clean energy supply chains include natural graphite, electrical steel, and silicon carbide (SiC), aluminum, cobalt, copper, fluorine, gallium, iridium, magnesium, platinum, and silicon (DOE 2023). All the materials either exist—that is, natural graphite, or can be manufactured by extraction, subsequent separation, and reactions with other materials. Their concentrations in coal wastes vary from one element to another, and the differences can be significant. For example, the concentrations of cobalt in coal ashes are typically lower than 50 ppm (Talan and Huang 2022), while those of gallium in coal ashes can be as high as ~300 ppm (Arroyo et al. 2014).

Natural graphite is a product of regional metamorphism or the interaction of larger metamorphosis agents with a seam of anthracite coal.[9] It can be retained in the resulting coal wastes (Quan et al. 2022), especially in coal waste impoundments, and even in fly ashes. The natural coaly graphite separated from coal wastes before REE extraction can be purified using different physical and chemical methods, including flotation, high-temperature treatment (Ling et al. 2020), pressure microwave-assisted treatment, chlorination roasting (Quan et al. 2022), and electrochemical processing (Quan et al. 2022).

Electrical steel mainly contains iron and low percentages of silicon. Coal wastes are rich in both iron and silicon. Some coal wastes contain high percentages of iron oxides, which can be treated as iron ores, and technologies to convert iron ores to iron are mature. Also, silicon dioxide (SiO_2) can be produced along with REE extraction from coal wastes via alkali-acid and water-leaching processes (Pan et al. 2021). Then, the iron and stoichiometrically small amount of SiO_2 produced from coal wastes can react to form electrical steel (Vaish 1994).

SiC can be prepared using SiO_2 generated from coal wastes during REE extraction processes, as just mentioned above (Pan et al. 2021), and the carbon separated from coal wastes (e.g., unburned carbon from fly ashes; Li et al. 2018; Maroto-Valer et al. 2001; Xiao et al. 2018). Technologies to produce SiC from SiO_2 and carbon have been developed, and the associated reaction mechanisms are well understood (Sun et al. 2019).

Methods for extracting aluminum (Cui et al. 2024; Li et al. 2020; Pan et al. 2021), cobalt (Zhang and Honaker 2020a), copper (Zhang and Honaker 2020b), fluorine (Geng et al. 2007; Luo et al. 2004), gallium (Lisowyj et al. 1987a, 1987b; Xu et al. 2022), iridium (Minter 2004, 2005), magnesium (Pan et al. 2021; Zhang et al. 2022b), platinum (Minter 2004, 2005), and silicon (Pan et al. 2021; Zhang et al. 2022b) from coal wastes have been explored. Some of them are based on integrating their separations with REE extractions, which simplifies the overall critical minerals separation process and increases the recovery of REEs, thus reducing coal waste valorization costs.[10] For example, increases in the separations of aluminum and silicon (two materials critical to energy supply chains) from fly ashes can lead to >300 percent improvement of the recovery efficiencies of REEs owing to the breaking of aluminosilicate glassy phase in fly ashes with the addition of the strong alkali NaOH to roasting processes (Pan et al. 2021). Most of the non-REE CMs (e.g., lithium, copper, cobalt, and nickel cations) stay in solution after the REEs are dissolved with mineral acids (e.g., nitric acid or hydrochloric acid) or oxalic acid (Pan et al. 2021) and can be separated with different methods. For example, lithium (as Li^+) can be precipitated by adding CO_2 or sodium carbonate (Na_2CO_3). Copper (as Cu^{2+}) can be precipitated by adding sodium sulfide (Na_2S). Then, cobalt (as Co^{2+}) and nickel (as Ni^{2+}) separation can be realized by using their difference in solvation capabilities in the same solution adjusted with lactic acid (Yuan et al. 2022). Electrodeposition or electrowinning also can be used to separate CMs from solution or from other non-CMs.

---

[7] Metamorphosis agents include temperature, pressure, and chemical fluids in different forms (e.g., tectonic stress, magma). Larger metamorphosis agents, such as larger hot fluids or semifluid materials, can come into contact with anthracite coal, resulting in metamorphism and structural changes of carbon materials via chemical reactions (GeologyHub 2023; Superior Graphite n.d.).

[8] Similar opportunities exist with recovery of scandium, which, although not directly involved in energy supply chains, is a high-price product. Its co-recovery with REEs could significantly improve the economic feasibility of REE extraction from coal or fly ashes.

9.3.5 Safety of Coal Waste Applications

Coal wastes can contain hazardous components such as heavy metals and volatile organic compounds. Thus, it is critical to evaluate the safety of coal waste–derived products and processes with regard to public health, the environment, and field performance. Appropriate analyses of products derived from coal wastes are necessary to ensure the safety and confidence of consumers. While most of the minerals in coal waste have been oxidized, the complex outdoor environment and potential exposure to acid rain could lead to the release of these heavy metals in various forms. Studies have evaluated the environmental safety of coal waste applications in soil and pavement improvement (Carlson and Adriano 1991; Ishak et al. 2002; Izquierdo and Querol 2012; Kang et al. 2011; Kukier et al. 2003; McCallister et al. 2002; Taylor and Schuman 1988), but more research is needed on different types of coal waste, and standards need to be established for using coal waste in applications with exposure.

9.3.5.1 Environmental Testing

A range of analyses have to be performed to assess the environmental safety of processes and products involving coal waste (EPA 2014b). These include identifying the fate of heavy metals and other hazardous components during coal waste processing and assessing the leaching characteristics and vapor emissions that may emerge from potential products. Such evaluations help to determine the suitability of coal waste–derived products for specific applications, as well as requirements for their final disposal. In cases where leaching potential is indicated, ecological risk assessments are needed to gauge the potential impact of these products on local ecosystems (Chen et al. 2024).

9.3.5.2 Human Health

When evaluating coal waste–derived products, exposure studies are required to safeguard the well-being of workers and communities (EPA 2014b). These studies serve as a vital component in ensuring the safety of individuals who may come into contact with coal waste or the products derived from it. Where relevant, assessments need to be carried out to evaluate the probability of inhalation of airborne coal waste or particles originating from coal waste-derived products or the processes that produced them, as well as potential exposure through dermatological contact and oral ingestion, by workers and consumers.

9.3.5.3 Product Performance

Products derived from coal waste often find application in construction, manufacturing, and industrial settings. The functionality and performance of these products has to be evaluated thoroughly for their intended applications. Such an evaluation may involve examination of mechanical, thermal, electrical, or chemical properties. It is important to align product development efforts with performance codes specific to their intended applications and to implement associated testing requirements accordingly.

9.4 CONCLUSIONS

This section summarizes the potential and practicality of repurposing coal waste for beneficial uses detailed in this chapter. Three main types of coal waste are generated along the coal supply chain: acid mine drainage, coal refuse or impoundment waste, and coal combustion residuals. This supply chain produces a substantial volume of waste, encompassing active mining operations, waste in impoundments either under reclamation or already rehabilitated, and waste from abandoned mine sites (e.g., AMD). While rigorous volume estimates are challenging to ascertain, significant quantities of coal waste exist in major coal-producing areas in the United States, notably in Appalachia and the Intermountain West.

Opportunities for the beneficial reuse of coal waste, both existing and emerging, have been identified across all three categories of waste. Figure 9-15 offers a summary of these reuse options, along with the maturity of the associated technologies. Key existing opportunities for beneficial reuse include the use of fly ash or bottom

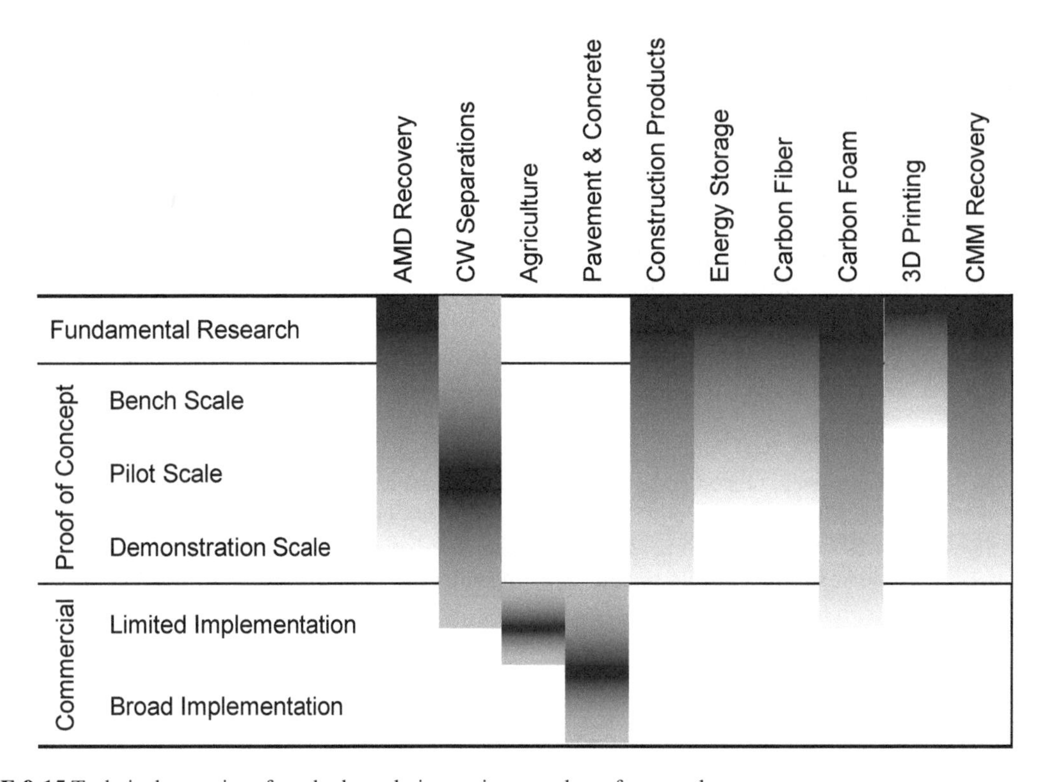

FIGURE 9-15 Technical maturity of methods to derive various products from coal waste.
NOTES: Shading gradient indicates level of R&D activity in each category, blank areas indicate little or no activity in that stage. CW = coal waste.

ash for agricultural soil amendment, the incorporation of fly ash in pavement and concrete, and the use of FGD gypsum in construction products.

AMD offers promising near-term opportunities for extracting economically valuable minerals, including both critical and noncritical elements like iron. Emerging technologies based on physical methods show potential in reclaiming waste coal and minerals, offering substantial prospects for repurposing these immense volumes of waste. Carbon-based materials such as engineered composites, graphite, graphene, fiber, and foam show promise for production from reclaimed waste coal. Current research indicates that building products, fibers, and foams made from waste coal can match or surpass the performance of existing market materials. Producing graphite or graphene from waste coal is an especially appealing prospect for supporting the U.S. electric vehicle industry, grid-scale electricity storage, and other key industrial sectors. The development of 3D printing media from waste coal or coal-derived materials is an emerging field with significant potential in tooling and building product applications. Furthermore, repurposing coal waste, including both coal and mineral fractions, has the potential to bolster a domestic critical mineral supply chain. Technologies in this area span from fundamental research to demonstration-scale projects.

9.4.1 Findings and Recommendations

Finding 9-1: Impoundment waste—Significant amounts of coal impoundment waste are found in three regions: Appalachia, the Illinois Basin, and the Intermountain West. Quantifying impoundment waste location, volume, and composition will be important for assessing potential beneficial waste utilization opportunities and applications.

Recommendation 9-1: Fund mapping of coal waste resources and infrastructure development to link coal waste sites with product markets—The Department of Energy's Office of Fossil Energy and Carbon Management (DOE-FECM), in collaboration with the Department of the Interior's Office of Surface Mining Reclamation and Enforcement, the U.S. Geological Survey, and the U.S. Environmental Protection Agency's Office of Land and Emergency Management, should fund an effort to evaluate and map coal waste resources, including acid mine drainage, impoundment wastes, and coal combustion residuals, to facilitate their potential use in producing solid carbon products and/or critical minerals and materials. Additionally, DOE-FECM and Basic Energy Sciences within DOE's Office of Science should fund translational and basic research of coal waste separations. DOE-FECM, jointly with the Department of Transportation, should support research and development focused on linking coal waste sites to solid carbon and critical minerals and materials markets and creating infrastructure to process and efficiently transport large amounts of coal wastes (both liquid and solid).

Finding 9-2: Beneficial utilization of impoundment wastes—Research conducted to date has shown that waste coal can be used to produce a host of materials for construction, electricity storage, and transportation applications, although these processes are at varying technology readiness levels. Additionally, the information necessary to conduct robust life cycle and techno-economic assessments of these processes is largely unavailable.

Finding 9-3: Long-lived carbon product market potential—The global utilization potential for long-lived carbon products is significant. The volume of waste coal utilized in a net-zero economy will be driven by the market value of long-lived carbon products and competitiveness of waste coal as a feedstock, both of which will be determined by a variety of factors, including demand for long-lived carbon-based products; relative cost compared to traditional fossil, renewable, or recycled carbon-based alternatives; life cycle implications of waste coal–derived, long-lived carbon products versus alternatives; and policy incentives and regulatory frameworks, including reclamation incentives.

Recommendation 9-2: Increase research on conversion of coal waste to long-lived carbon products, including techno-economic, environmental, and safety assessments—The Department of Energy's Office of Fossil Energy and Carbon Management (DOE-FECM) and the Department of Defense should fund research that focuses on the efficient transformation of waste coal through thermochemical, electrochemical, and plasmachemical processes into long-lived solid carbon products, for example, carbon fibers, graphite, graphene, carbon foam, 3D-printing materials, and engineered composites. Specifically, low-cost and environmentally friendly conversions of coal waste to long-lived carbon products should be developed, and approaches to valorize the by-products of conversions of coal waste to long-lived carbon products should be explored. As part of these efforts, DOE should increase support for their national laboratories to develop databases that can assist researchers in completing robust life cycle and techno-economic assessments of coal waste utilization processes. DOE-FECM should evaluate the functionality and performance of coal waste–derived products to ensure that they conform with codes specific to their intended application and, in collaboration with ASTM International, establish standards for using coal waste in applications with environmental exposure to ensure product safety.

Recommendation 9-3: Appropriate funds for the Carbon Materials Science Initiative—The U.S. Congress should appropriate funds for the "Carbon Materials Science Initiative" as authorized in the CHIPS and Science Act of 2022 "to expand knowledge of coal, coal wastes, and carbon ore chemistry useful for understanding the conversion of carbon to material products." Funding for basic research into low-carbon-emission pathways (e.g., clean electricity-driven heating, electrolysis, and plasma) for coal waste utilization should be prioritized by the Department of Energy's (DOE's) Office of Basic Energy Sciences, along with associated applied research and demonstration supported by DOE's Office of Fossil Energy and Carbon Management.

Finding 9-4: Characterization of coal waste streams for rare earth element extraction—Information about the structure and morphology of solid coal wastes and coal combustion by-products is incomplete, which is an obstacle for choosing effective physical beneficiation methods that can enrich the concentration of rare earth elements (REEs) in these waste streams. Roasting solid coal wastes and coal combustion by-products with chemicals or additives is important for destroying the glass structure in coal combustion by-products and thus realizing high leaching efficiency for REEs. Transformative, sustainable extraction agents, systems, and processes for REE recovery are still lacking.

Recommendation 9-4: Support characterization of coal wastes to facilitate development of new physical beneficiation methods—Basic Energy Sciences within the Department of Energy's (DOE's) Office of Science and the U.S. Geological Survey should support the characterization of solid coal wastes and coal combustion by-products. DOE's Office of Fossil Energy and Carbon Management and Advanced Research Projects Agency–Energy should fund the development of innovative physical beneficiation methods, including separation of unburned carbon and magnetic fractions in fly ashes and removal of oversized particles, novel roasting methods, and highly efficient, sustainable leaching agents and extractants for rare earth elements.

Finding 9-5: Improvements are needed in critical material extraction from solid wastes and recovery from liquid wastes to support a domestic supply chain—Extraction of critical materials (CMs) from solid coal wastes, such as impoundment fines and fly ash, has been demonstrated using acids, ionic liquids, and deep eutectic solvents, but requires significant improvement in overall recovery, sustainability, and selectivity over non-CM components. Recovery of CMs from acid mine drainage and extraction of CMs from solid coal wastes is complicated and costly owing to complex multivalent cation solution chemistry, and these processes are currently too expensive compared to production of CMs from their mineral ores to be commercially viable.

Recommendation 9-5: Develop novel technologies to extract critical minerals from liquid and solid coal waste streams—The Department of Defense and the Department of Energy's (DOE's) Office of Fossil Energy and Carbon Management, Office of Energy Efficiency and Renewable Energy, and Advanced Research Projects Agency–Energy should fund the development of novel technologies for extracting critical materials, including rare earth and energy-relevant elements, from solid (e.g., coal combustion residuals, waste coal) and liquid (e.g., acid mine drainage) coal waste streams, as well as the separation of individual elements once extracted. In parallel, DOE's Office of Basic Energy Sciences and the National Science Foundation should support research on separations from complex mixtures, including to understand species partitioning, ion pairing, and ion dissociation. Technologies and processes to be explored include development of novel acids and electrified methods (e.g., microwave, inductive Joule heating) for extractions; development of novel, sustainable sorbents for separations; and multiphysics simulations and artificial intelligence to analyze and understand the composition of complex waste streams.

9.4.2 Research Agenda for Coal Waste Utilization

Table 9-8 presents the committee's research agenda for coal waste utilization, including research needs (numbered by chapter), and related research agenda recommendations (a subset of research-related recommendations from the chapter). The table includes the relevant funding agencies or other actors; whether the need is for basic research, applied research, technology demonstration, or enabling technologies and processes for CO_2 utilization; the research theme(s) that the research need falls into; the relevant research area and product class covered by the research need; whether the relevant product(s) are long- or short-lived; and the source of the research need (chapter section, finding, or recommendation). The committee's full research agenda can be found in Chapter 11.

TABLE 9-8 Research Agenda for Coal Waste Utilization

Research, Development, and Demonstration Need	Funding Agencies or Other Actors	Basic, Applied, Demonstration, or Enabling	Research Area	Product Class	Long- or Short-Lived	Research Theme	Source
9-A. Evaluation and mapping of coal waste resources, including composition, volume, and locations.	DOE-FECM OSMRE EPA USGS	Enabling	Coal waste utilization	Coal waste–derived carbon products Metal coal waste by-products	Long-lived	Resource mapping	Fin. 9-1 Rec. 9-1
9-B. Strategies for linking coal waste sites to markets for solid carbon and critical minerals and materials and creating infrastructure to process and efficiently transport large amounts of coal wastes (both liquid and solid).	DOE-FECM DOT	Applied Enabling	Coal waste utilization	Coal waste–derived carbon products Metal coal waste by-products	Long-lived	Enabling technology and infrastructure needs Market opportunities	Rec. 9-1
9-C. Physical and chemical methods for separating mineral matter from carbon in coal wastes.	DOE-BES DOE-FECM	Basic Applied	Coal waste utilization	Coal waste–derived carbon products Metal coal waste by-products	Long-lived	Separations	Rec. 9-1
Recommendation 9-1: Fund mapping of coal waste resources and infrastructure development to link coal waste sites with product markets—The Department of Energy's Office of Fossil Energy and Carbon Management (DOE-FECM), in collaboration with the Department of the Interior's Office of Surface Mining Reclamation and Enforcement, the U.S. Geological Survey, and the U.S. Environmental Protection Agency's Office of Land and Emergency Management, should fund an effort to evaluate and map coal waste resources, including acid mine drainage, impoundment wastes, and coal combustion residuals, to facilitate their potential use in producing solid carbon products and/or critical minerals and materials. Additionally, DOE-FECM and Basic Energy Sciences within DOE's Office of Science should fund translational and basic research of coal waste separations. DOE-FECM, jointly with the Department of Transportation, should support research and development focused on linking coal waste sites to solid carbon and critical minerals and materials markets and creating infrastructure to process and efficiently transport large amounts of coal wastes (both liquid and solid).							
9-D. Efficient transformation of waste coal into long-lived carbon products with lower embodied carbon than existing products—including engineered composites, graphite, graphene, fiber, and foam—for construction, energy storage technologies, transportation, and defense applications.	DOE-FECM DoD	Applied Demonstration	Coal waste utilization—long-lived carbon products	Coal waste–derived carbon products	Long-lived	Reactor design and reaction engineering	Fin. 9-2 Rec. 9-2 Rec. 9-3
9-E. Evaluation of different types of coal waste to determine their ability to enhance pavement performance and to understand the fate and transfer of heavy metals over long time periods.	DOE-FECM DOT-FHWA DOT-OST-R State departments of transportation	Applied	Coal waste utilization	Coal waste–derived carbon products	Long-lived	Certification and standards	Sec. 9.3.1.2
9-F. Atomic- and multi-scale computer simulations to better understand the conversion of coal waste carbon into various solid-carbon products and solid-carbon product precursors.	NSF DOE-BES	Basic	Coal waste utilization—long-lived carbon products	Coal waste–derived carbon products	Long-lived	Fundamental knowledge Computational modeling and machine learning	Sec. 9.3.3.2 Rec. 9-3

continued

TABLE 9-8 Continued

Research, Development, and Demonstration Need	Funding Agencies or Other Actors	Basic, Applied, Demonstration, or Enabling	Research Area	Product Class	Long- or Short-Lived	Research Theme	Source
9-G. Develop 3D-printing media from waste coal or coal-derived materials for tooling and building product applications.	DOE-FECM	Applied	Coal waste utilization—long-lived carbon products	Coal waste–derived carbon products	Long-lived	Certification and standards	Rec. 9-2
9-H. Evaluations of functionality and performance of coal waste–derived products to ensure that they conform with codes specific to their intended application. These evaluations could include examination of mechanical, thermal, electrical, and/or chemical properties.	DOE-FECM	Applied	Coal waste utilization—long-lived carbon products	Coal waste–derived carbon products	Long-lived	Certification and standards	Rec. 9-2
9-I. Establish standards for using coal waste in applications with environmental exposure to ensure product safety.	ASTM International	Enabling	Coal waste utilization—long-lived carbon products	Coal waste–derived carbon products	Long-lived	Certification and standards	Rec. 9-2
9-J. Data and tools to conduct life cycle and techno-economic assessments of coal waste utilization processes.	DOE-FECM	Applied	Coal waste utilization	Coal waste–derived carbon products Metal coal waste by-products	Long-lived	Certification and standards Understanding environmental and societal impacts of CO_2 and coal waste utilization technologies	Fin. 9-2 Rec. 9-2

Recommendation 9-2: Increase research on conversion of coal waste to long-lived carbon products, including techno-economic, environmental, and safety assessments—The Department of Energy's Office of Fossil Energy and Carbon Management (DOE-FECM) and the Department of Defense should fund research that focuses on the efficient transformation of waste coal through thermochemical, electrochemical, and plasma-chemical processes into long-lived solid carbon products, for example, carbon fibers, graphite, graphene, carbon foam, 3D-printing materials, and engineered composites. Specifically, low-cost and environmentally friendly conversions of coal waste to long-lived carbon products should be developed, and approaches to valorize the by-products of conversions of coal waste to long-lived carbon products should be explored. As part of these efforts, DOE should increase support for their national laboratories to develop databases that can assist researchers in completing robust life cycle and techno-economic assessments of coal waste utilization processes. DOE-FECM should evaluate the functionality and performance of coal waste–derived products to ensure that they conform with codes specific to their intended application and, in collaboration with ASTM International, establish standards for using coal waste in applications with environmental exposure to ensure product safety.

Recommendation 9-3: Appropriate funds for the Carbon Materials Science Initiative—The U.S. Congress should appropriate funds for the "Carbon Materials Science Initiative" as authorized in the CHIPS and Science Act of 2022 "to expand knowledge of coal, coal wastes, and carbon ore chemistry useful for understanding the conversion of carbon to material products." Funding for basic research into low-carbon-emission pathways (e.g., clean electricity-driven heating, electrolysis, and plasma) for coal waste utilization should be prioritized by the Department of Energy's (DOE's) Office of Basic Energy Sciences, along with associated applied research and demonstration supported by DOE's Office of Fossil Energy and Carbon Management.

9-K. Characterization of the structure and morphology of solid coal wastes and coal combustion by-products.	DOE-BES USGS	Basic	Coal waste utilization—metal extraction	Metal coal waste by-products	Long-lived	Fundamental knowledge	Fin. 9-4 Rec. 9-4

Recommendation 9-4: Support characterization of coal wastes to facilitate development of new physical beneficiation methods—Basic Energy Sciences within the Department of Energy's (DOE's) Office of Science and the U.S. Geological Survey should support the characterization of solid coal wastes and coal combustion by-products. DOE's Office of Fossil Energy and Carbon Management and Advanced Research Projects Agency–Energy should fund the development of innovative physical beneficiation methods, including separation of unburned carbon and magnetic fractions in fly ashes and removal of oversized particles, novel roasting methods, and highly efficient, sustainable leaching agents and extractants for rare earth elements.

9-L. Discovery and development of more selective, sustainable leaching agents, membranes, and processes for extracting rare earth elements from coal waste.	DOE-BES NSF	Basic	Coal waste utilization—metal extraction	Metal coal waste by-products	Long-lived	Separations	Fin. 9-5 Rec. 9-5
9-M. Technologies for extracting rare earth elements, lithium, and other energy-relevant critical materials from solid wastes (e.g., waste coal and coal combustion residuals).	DOE-FECM DOE-EERE DOE-ARPA-E DoD	Applied	Coal waste utilization—metal extraction	Metal coal waste by-products	Long-lived	Separations	Fin. 9-5 Rec. 9-5
9-N. Technologies for extracting rare earth elements, lithium, and other energy-relevant critical minerals from liquid wastes (e.g., acid mine drainage).	DOE-FECM DOE-EERE DOE-ARPA-E DoD	Applied	Coal waste utilization—metal extraction	Metal coal waste by-products	Long-lived	Separations	Fin. 9-5 Rec. 9-5
9-O. Separation of individual elements once extracted from coal wastes, especially separation of nickel from cobalt.	DOE-BES NSF DOE-FECM DOE-EERE DOE-ARPA-E DoD	Basic Applied	Coal waste utilization—metal extraction	Metal coal waste by-products	Long-lived	Separations	Fin. 9-5 Rec. 9-5

Recommendation 9-5: Develop novel technologies to extract critical minerals from liquid and solid coal waste streams—The Department of Defense and the Department of Energy's (DOE's) Office of Fossil Energy and Carbon Management, Office of Energy Efficiency and Renewable Energy, and Advanced Research Projects Agency–Energy should fund the development of novel technologies for extracting critical materials, including rare earth and energy-relevant elements, from solid (e.g., coal combustion residuals, waste coal) and liquid (e.g., acid mine drainage) coal waste streams, as well as the separation of individual elements once extracted. In parallel, DOE's Office of Basic Energy Sciences and the National Science Foundation should support research on separations from complex mixtures, including to understand species partitioning, ion pairing, and ion dissociation. Technologies and processes to be explored include development of novel acids and electrified methods (e.g., microwave, inductive Joule heating) for extractions; development of novel, sustainable sorbents for separations; and multiphysics simulations and artificial intelligence to analyze and understand the composition of complex waste streams.

NOTE: ARPA-E = Advanced Research Projects Agency–Energy; BES = Basic Energy Sciences; DoD = Department of Defense; DOE = Department of Energy; DOT = Department of Transportation; EERE = Office of Energy Efficiency and Renewable Energy; EPA = U.S. Environmental Protection Agency; FECM = Office of Fossil Energy and Carbon Management; FHWA = Federal Highway Administration; NSF = National Science Foundation; OSMRE = Office of Surface Mining Reclamation and Enforcement; OST-R = Office of the Assistant Secretary for Research and Technology; USGS = U.S. Geological Survey.

9.5 REFERENCES

30 CFR § 710.5. "Initial Regulatory Program." Office of Surface Mining Reclamation and Enforcement, Department of the Interior. https://www.ecfr.gov/current/title-30/chapter-VII/subchapter-B/part-710/section-710.5.

Abraham, K.M. 2020. "How Comparable Are Sodium-Ion Batteries to Lithium-Ion Counterparts?" *ACS Energy Letters* 5(11):3544–3547. https://doi.org/10.1021/acsenergylett.0c02181.

ACAA (American Coal Ash Association). n.d. "An American Recycling Success Story: Beneficial Use of Coal Combustion Products." Sandy, Utah: American Coal Ash Association. https://acaa-usa.org/wp-content/uploads/2022/12/22-ACAA-Brochure_FINAL.pdf.

ACAA Educational Foundation (American Coal Ash Association Educational Foundation). n.d. "Sustainable Construction with Coal Combustion Products." Farmington Hills, MI: American Coal Ash Association Educational Foundation. https://acaa-usa.org/wp-content/uploads/free-publications/Sustainability_Construction_w_CCPs(Consolidated).pdf.

Advincula, P.A., W. Meng, L. J. Eddy, P.Z. Scotland, J.L. Beckham, S. Nagarajaiah, and J.M. Tour. 2023. "Replacement of Concrete Aggregates with Coal-Derived Flash Graphene." *ACS Applied Materials and Interfaces* 16(1):1474–1481. https://doi.org/10.1021/acsami.3c15156.

Alguacil, F.J., and J.I. Robla. 2023. "Recent Work on the Recovery of Rare Earths Using Ionic Liquids and Deep Eutectic Solvents." *Minerals* 13(10). https://doi.org/10.3390/min13101288.

Al-Majali, Y.A., C.T. Chirume, E.P. Marcum, D.A. Daramola, K.S. Kappagantula, and J.P. Trembly. 2019. "Coal-Filler-Based Thermoplastic Composites as Construction Materials: A New Sustainable End-Use Application." *ACS Sustainable Chemistry and Engineering* 7(19):16870–16878. https://doi.org/10.1021/acssuschemeng.9b04453.

Al-Majali, Y.T., S. Forshey, and J.P. Trembly. 2022. "Effect of Natural Carbon Filler on Thermo-Oxidative Degradation of Thermoplastic-Based Composites." *Thermochimica Acta* 713(July 1):179226. https://doi.org/10.1016/j.tca.2022.179226.

Al-Majali, Y.T., E.S. Alamiri, B. Wisner, and J.P. Trembly. 2023a. "Mechanical Performance Assessment of Sustainable Coal Plastic Composite Building Materials." *Journal of Building Engineering* 80:108089. https://doi.org/10.1016/j.jobe.2023.108089.

Al-Majali, Y., D. Daramola, L. Veley, R. Statnick, V. Dhanapal, D. Connell, E. Shereda, and J. Trembly. 2023b. "Direct Utilization of U.S. Coal as Feedstock for the Manufacture of High-Value Coal Plastic Composites." Final Technical Report. Athens, OH: Ohio University. https://doi.org/10.2172/1907170.

Al-Samarai, R.A., A.S. Mahmood, and Y. Al-Douri. 2020. "Chapter 5—Surface Modification, Including Polymerization, Nanocoating, and Microencapsulation." Pp. 83–99 in *Metal Oxide Power Technologies*, Y. Al-Douri, ed. Elsevier. https://doi.org/10.1016/B978-0-12-817505-7.00005-1.

Alvira, D., D. Antorán, and J.J. Manyà. 2022. "Plant-Derived Hard Carbon as Anode for Sodium-Ion Batteries: A Comprehensive Review to Guide Interdisciplinary Research." *Chemical Engineering Journal* 447(November 1):137468. https://doi.org/10.1016/j.cej.2022.137468.

Arroyo, F., O. Font, J.M. Chimenos, C. Fernández-Pereira, X. Querol, and P. Coca. 2014. "IGCC Fly Ash Valorisation. Optimisation of Ge and Ga Recovery for an Industrial Application." *Fuel Processing Technology* 124(August 1):222–227. https://doi.org/10.1016/j.fuproc.2014.03.004.

Ateşok, G., F. Boylu, and M.S. Çelik. 2001. "Carrier Flotation for Desulfurization and Deashing of Difficult-to-Float Coals." *Minerals Engineering* 14(6):661–670. https://doi.org/10.1016/S0892-6875(01)00058-9.

Awasthi, S., K. Awasthi, A.K. Ghosh, S.K. Srivastava, and O.N. Srivastava. 2015. "Formation of Single and Multi-Walled Carbon Nanotubes and Graphene from Indian Bituminous Coal." *Fuel* 147(May 1):35–42. https://doi.org/10.1016/j.fuel.2015.01.043.

AZEK. 2024. "Environmental, Social and Governance." *The AZEK Company.* https://investors.azekco.com/environmental-social-and-governance/environmental-social-and-governance/default.aspx.

Azenkeng, A. 2022. "Advanced Processing of Coal and Coal Waste to Produce Graphite for Fast-Charging Lithium-Ion Battery Anode." Presented at the Resource Sustainability Project Review Meeting. October 25–27. Pittsburgh, PA. https://netl.doe.gov/sites/default/files/netl-file/22RS-25_Azenkeng_A.pdf.

Banek, N.A., D.T. Abele, K.R. Jr. McKenzie, and M.J. Wagner. 2018. "Sustainable Conversion of Lignocellulose to High-Purity, Highly Crystalline Flake Potato Graphite." *ACS Sustainable Chemistry and Engineering* 6(10):13199–13207. https://doi.org/10.1021/acssuschemeng.8b02799.

Banerjee, C., V.K. Chandaliya, and P.S. Dash. 2021a. "Recent Advancement in Coal Tar Pitch-Based Carbon Fiber Precursor Development and Fiber Manufacturing Process." *Journal of Analytical and Applied Pyrolysis* 158(September 1):105272. https://doi.org/10.1016/j.jaap.2021.105272.

Banerjee, R., A. Mohanty, S. Chakravarty, S. Chakladar, and P. Biswas. 2021b. "A Single-Step Process to Leach Out Rare Earth Elements from Coal Ash Using Organic Carboxylic Acids." *Hydrometallurgy* 201(May 1):105575.

Baudino, L., C. Santos, C.F. Pirri, F. La Mantia, and A. Lamberti. 2022. "Recent Advances in the Lithium Recovery from Water Resources: From Passive to Electrochemical Methods." *Advanced Science* 9(27):2201380. https://doi.org/10.1002/advs.202201380.

Belarbi, A., M. Dawood, and B. Acun. 2016. "Sustainability of Fiber-Reinforced Polymers (FRPs) as a Construction Material." Pp. 521–538 in *Sustainability of Construction Materials*. Woodhead Publishing. https://doi.org/10.1016/B978-0-08-100370-1.00020-2.

Bentz, D.P., C.F. Ferraris, and K.A. Snyder. 2013. "Best Practices Guide for High-Volume Fly Ash Concretes: Assuring Properties and Performance." NIST Technical Note 1812. Gaithersburg, MD: National Institute of Standards and Technology, September. https://tsapps.nist.gov/publication/get_pdf.cfm?pub_id=914225.

Bentzen III, E.H., H. Gaffari, L. Galbraith, R.F. Hammen, R.J. Robinson, S.A. Hafez, and S. Annavarapu. 2013. "Preliminary Economic Assessment on the Bokan Mountain Rare Earth Element Project, Near Ketchikan, Alaska: Prepared for UCORE Rare Metals." Document No. 1196000100-REP-R0001-02. Vancouver, BC: Tetra Tech. https://ucore.com/PEA.pdf.

Berger, V.I., D.A. Singer, J.D. Bliss, and B.C. Moring. 2011. "Ni-Co Laterite Deposites of the World—Database and Grade and Tonnage Models." Open-File Report 2011-1058. Reston, VA: U.S. Geological Survey. https://pubs.usgs.gov/of/2011/1058.

Bouaissi, A., L.Y. Li, M.M. Al Bakri Abdullah, R. Ahmad, R.A. Razak, and Z. Yahya. 2020. "Fly Ash as a Cementitious Material for Concrete." Chapter 7 in *Zero-Energy Buildings*, J.A.P. Arcas, C. Rubio-Bellido, A. Pérez-Fargallo, and I. Oropeza-Perez, eds. Rijeka: IntechOpen. https://doi.org/10.5772/intechopen.90466.

Bowell, R.J., L. Lagos, C.R. De Low Hoyos, and J. Declercq. 2020. "Classification and Characteristics of Natural Lithium Resources." *Lithium Less Is More* (August). https://www.elementsmagazine.org/classification-and-characteristics-of-natural-lithium-resources.

Brownfield, M.E., J.D. Cathcart, R.H. Affolter, I.K. Brownfield, C.A. Rice, J.T. O'Connor, R.A. Zielinski, J.H. Bullock, Jr., J.C. Hower, and G.P. Meeker. 2005. "Characterization and Modes of Occurrence of Elements in Feed Coal and Coal Combustion Products from a Power Plant Utilizing Low-Sulfur Coal from the Powder River Basin, Wyoming." 2004–5271. Scientific Investigations Report. Reston, VA: U.S. Geological Survey. https://doi.org/10.3133/sir20045271.

Calvo, M., R. García, and S.R. Moinelo. 2008. "Carbon Foams from Different Coals." *Energy and Fuels* 22(5):3376–3383. https://doi.org/10.1021/ef8000778.

Cao, D., J. Ji, Q. Liu, Z. He, H. Wang, and Z. You. 2011. "Coal Gangue Applied to Low-Volume Roads in China." *Transportation Research Record* 2204(1):258–266. https://doi.org/10.3141/2204-32.

Cao, Q., X. Xie, J. Li, J. Dong, and L. Jin. 2012. "A Novel Method for Removing Quinoline Insolubles and Ash in Coal Tar Pitch Using Electrostatic Fields." *Fuel* 96(June 1):314–318. https://doi.org/10.1016/j.fuel.2011.12.061.

Carbon Activated Corporation. 2024. "Coal Based Activated Carbon." Carbon Activated Corp. https://activatedcarbon.com/products/coal-based-activated-carbon.

Carlson, C.L., and D.C. Adriano. 1991. "Growth and Elemental Content of Two Tree Species Growing on Abandoned Coal Fly Ash Basins." *Journal of Environmental Quality* 20(3):581–587. https://doi.org/10.2134/jeq1991.00472425002000030013x.

CFOAM LLC. n.d.(a). "CFOAM® Carbon Foam." CFOAM. https://www.cfoam.com/whatis.

CFOAM LLC. n.d.(b). "CFOAM® 35 HTC." CFOAM. https://www.cfoam.com/cfoam-35-htc.

CFOAM LLC. n.d.(c). "Spec Sheets." CFOAM. https://www.cfoam.com/spec-sheets.

Chen, C., E.B. Kennel, A.H. Stiller, P.G. Stansberry, and J.W. Zondlo. 2006. "Carbon Foam Derived from Various Precursors." *Carbon* 44(8):1535–1543. https://doi.org/10.1016/j.carbon.2005.12.021.

Chen, Y., Y. Fan, Y. Huang, X. Liao, W. Xu, and T. Zhang. 2024. "A Comprehensive Review of Toxicity of Coal Fly Ash and Its Leachate in the Ecosystem." *Ecotoxicology and Environmental Safety* 269(January 1):115905. https://doi.org/10.1016/j.ecoenv.2023.115905.

Chen, Z., Z. Li, J. Chen, P. Kallem, F. Banat, and H. Qiu. 2022. "Recent Advances in Selective Separation Technologies of Rare Earth Elements: A Review." *Journal of Environmental Chemical Engineering* 10(1):107104. https://doi.org/10.1016/j.jece.2021.107104.

Chung, D.D.L. 1987. "Exfoliation of Graphite." *Journal of Materials Science* 22(12):4190–4198. https://doi.org/10.1007/BF01132008.

Craddock, J.D., G. Frank, M. Martinelli, J. Lacy, V. Edwards, A. Vego, C. Thompson, R. Andrews, and M.C. Weisenberger. 2024. "Isotropic Pitch-Derived Carbon Fiber from Waste Coal." *Carbon* 216(January 5):118590. https://doi.org/10.1016/j.carbon.2023.118590.

Cui, L., L. Feng, H. Yuan, H. Cheng, and F. Cheng. 2024. "Efficient Recovery of Aluminum, Lithium, Iron and Gallium from Coal Fly Ash Leachate via Coextraction and Stepwise Stripping." *Resources, Conservation and Recycling* 202(March 1):107380. https://doi.org/10.1016/j.resconrec.2023.107380.

Dai, S., R.B. Finkelman, D. French, J.C. Hower, I.T. Graham, and F. Zhao. 2021. "Modes of Occurrence of Elements in Coal: A Critical Evaluation." *Earth-Science Reviews* 222(November 1):103815. https://doi.org/10.1016/j.earscirev.2021.103815.

Dai, S., I.T. Graham, and C.R. Ward. 2016. "A Review of Anomalous Rare Earth Elements and Yttrium in Coal." *International Journal of Coal Geology* 159(April):82–95. https://doi.org/10.1016/j.coal.2016.04.005.

Danso, I.K., A.B. Cueva-Sola, Z. Masaud, J.-Y. Lee, and R.K. Jyothi. 2021. "Ionic Liquids for the Recovery of Rare Earth Elements from Coal Combustion Products." Pp. 617–638 in *Clean Coal Technologies: Beneficiation, Utilization, Transport Phenomena and Prospective*, R.K. Jyothi and P.K. Parhi, eds. Cham: Springer International Publishing. https://doi.org/10.1007/978-3-030-68502-7_25.

Das, S., and P. Nagapurkar. 2021. "Sustainable Coal Tar Pitch Carbon Fiber Manufacturing." ORNL/TM-2021/1889. Oak Ridge, TN: Oak Ridge National Laboratory, May 1. https://doi.org/10.2172/1784125.

Das, S., G. Gaustad, A. Sekar, and E. Williams. 2018. "Techno-Economic Analysis of Supercritical Extraction of Rare Earth Elements from Coal Ash." *Journal of Cleaner Production* 189(July 10):539–551. https://doi.org/10.1016/j.jclepro.2018.03.252.

de los Hoyos, C. 2022. "Geology of Economic Natural Lithium Deposits." Presented at the AAPG International Conference and Exhibition. April 19. Cartagena, Colombia. https://cdn-web-content.srk.com/upload/user/image/ICE_Colombia-2022-CamilodelosHoyos_SRK20220426133846909.pdf.

Dodbiba, G., and T. Fujita. 2023. "Trends in Extraction of Rare Earth Elements from Coal Ashes: A Review." *Recycling* 8(1):17. https://doi.org/10.3390/recycling8010017.

DOE (Department of Energy). 2017a. "High Concentrations of Rare Earth Elements Found in American Coal Basins." https://www.energy.gov/articles/high-concentrations-rare-earth-elements-found-american-coal-basins.

DOE. 2017b. "Report on Rare Earth Elements from Coal and Coal Byproducts." Report to Congress. Washington, DC: Department of Energy. https://www.energy.gov/sites/prod/files/2018/01/f47/EXEC-2014-000442%20-%20for%20Conrad%20Regis%202.2.17.pdf.

DOE. 2022. "Recovery of Rare Earth Elements and Critical Materials from Coal and Coal Byproducts." Report to Congress. Washington, DC: Department of Energy. https://www.energy.gov/sites/default/files/2022-05/Report%20to%20Congress%20on%20Recovery%20of%20Rare%20Earth%20Elements%20and%20Critical%20Minerals%20from%20Coal%20and%20Coal%20By-Products.pdf.

DOE. 2023. "Critical Materials Assessment." Washington, DC: Department of Energy. https://www.energy.gov/sites/default/files/2023-07/doe-critical-material-assessment_07312023.pdf.

DOE-FECM (Office of Fossil Energy and Carbon Management). 2021. "DOE Invests Nearly $7 Million to Put Coal Wastes to Work, Creating Products for a Clean Energy Economy." *Energy.gov*. https://www.energy.gov/fecm/articles/doe-invests-nearly-7-million-put-coal-wastes-work-creating-products-clean-energy.

Du, M., P.A. Advincula, X. Ding, J.M. Tour, and C. Xiang. 2023. "Coal-Based Carbon Nanomaterials: En Route to Clean Coal Conversion toward Net Zero CO_2." *Advanced Materials* 35(25):2300129. https://doi.org/10.1002/adma.202300129.

Earthjustice. 2022. "Mapping the Coal Ash Contamination." *Earthjustice*. https://earthjustice.org/feature/coal-ash-contaminated-sites-map.

Earthjustice. 2023. "Where Are Coal Ash Dump Sites?" *Earthjustice*. https://earthjustice.org/feature/coal-ash-map-sites-legacy-inactive-regulated.

EIA (U.S. Energy Information Administration). 2023a. "All Energy Infrastructure and Resources." *U.S. Energy Atlas*. https://atlas.eia.gov/apps/all-energy-infrastructure-and-resources/explore.

EIA. 2023b. "Coal Explained: Use of Coal." *EIA*. https://www.eia.gov/energyexplained/coal/use-of-coal.php.

Ekmann, J.M. 2012. "Rare Earth Elements in Coal Deposits—A Prospectivity Analysis." Cleveland, OH. https://www.searchanddiscovery.com/documents/2012/80270ekmann/ndx_ekmann.pdf.

Elbashier, E., A. Mussa, M. Hafiz, and A.H. Hawari. 2021. "Recovery of Rare Earth Elements from Waste Streams Using Membrane Processes: An Overview." *Hydrometallurgy* 204(September 1):105706. https://doi.org/10.1016/j.hydromet.2021.105706.

Elhacham, E., L. Ben-Uri, J. Grozovski, Y.M. Bar-On, and R. Milo. 2020. "Global Human-Made Mass Exceeds All Living Biomass." *Nature* 588(7838):442–444. https://doi.org/10.1038/s41586-020-3010-5.

Elseewi, A.A., I.R. Straughan, and A.L. Page. 1980. "Sequential Cropping of Fly Ash-Amended Soils: Effects on Soil Chemical Properties and Yield and Elemental Composition of Plants." *Science of the Total Environment* 15(3):247–259. https://doi.org/10.1016/0048-9697(80)90053-4.

Energy Institute (Energy Institute—Statistical Review of World Energy—with major processing by Our World in Data). 2023. "Lithium production." [2010 to 2022].

Environmental Integrity Project. 2019. "Coal's Poisonous Legacy-Groundwater Contaminated by Coal Ash Across the U.S." https://environmentalintegrity.org/wp-content/uploads/2019/03/National-Coal-Ash-Report-Revised-7.11.19.pdf.

EPA (U.S. Environmental Protection Agency). 2014a. "Coal Combustion Residual Beneficial Use Evaluation: Fly Ash Concrete and FGD Gypsum Wallboard." Washington, DC: U.S. Environmental Protection Agency. https://www.epa.gov/sites/default/files/2014-12/documents/ccr_bu_eval.pdf.

EPA. 2014b. "Human and Ecological Risk Assessment of Coal Combustion Residuals." Washington, DC: U.S. Environmental Protection Agency. Regulation Identifier Number: 2050-AE81. https://www.regulations.gov/document/EPA-HQ-OLEM-2019-0173-0008.

EPA. 2015. "Hazardous and Solid Waste Management System; Disposal of Coal Combustion Residuals from Electric Utilities." *Federal Register* 80(74):21302–213501. https://www.govinfo.gov/content/pkg/FR-2015-04-17/pdf/2015-00257.pdf.

EPA. 2023a. "Abandoned Mine Drainage." EPA. December 4. https://www.epa.gov/nps/abandoned-mine-drainage.

EPA. 2023b. "Coal Ash Basics." EPA. https://www.epa.gov/coalash/coal-ash-basics.

EPA. 2024a. "Coal Ash Reuse." EPA. https://www.epa.gov/coalash/coal-ash-reuse.

EPA. 2024b. "Disposal of Coal Combustion Residuals from Electric Utilities Rulemakings." EPA. https://www.epa.gov/coalash/coal-ash-rule.

EPA, USDA (Department of Agriculture), and RTI International. 2023. "Beneficial Use Evaluation: Flue Gas Desulfurization Gypsum as an Agricultural Amendment." EPA 530-R-23-004. Washington, DC: U.S. Environmental Protection Agency, Department of Agriculture, and RTI International. https://www.epa.gov/system/files/documents/2023-03/FGD_Ben_Use_Eval_with_Appendices_March_2023_508.pdf.

Eterigho-Ikelegbe, O., H. Harrar, and S. Bada. 2021. "Rare Earth Elements from Coal and Coal Discard—A Review." *Minerals Engineering* 173(November):107187. https://doi.org/10.1016/j.mineng.2021.107187.

Evans, K. 2014. "Lithium." Pp. 230–260 in *Critical Metals Handbook*. https://doi.org/10.1002/9781118755341.ch10.

Fan, M., and Z. Huang. 2023. "Method for Separation of Rare Earth Elements from Coal Ash Using Supercritical Carbon Dioxide." U.S. Patent No. 11814299, filed May 10, 2021, and issued November 14, 2023.

Fan, M., D. Tao, R. Honaker, and Z. Luo. 2010. "Nanobubble Generation and Its Applications in Froth Flotation (Part II): Fundamental Study and Theoretical Analysis." *Mining Science and Technology (China)* 20(2):159–177. https://doi.org/10.1016/S1674-5264(09)60179-4.

Fei, H., R. Ye, G. Ye, Y. Gong, Z. Peng, X. Fan, E.L.G. Samuel, P.M. Ajayan, and J.M. Tour. 2014. "Boron- and Nitrogen-Doped Graphene Quantum Dots/Graphene Hybrid Nanoplatelets as Efficient Electrocatalysts for Oxygen Reduction." *ACS Nano* 8(10):10837–10843. https://doi.org/10.1021/nn504637y.

FHWA (Federal Highway Administration). 2016. "User Guidelines for Waste and Byproduct Materials in Pavement Construction." FHWA-RD-97-148. Washington, DC: Department of Transportation. https://www.fhwa.dot.gov/publications/research/infrastructure/structures/97148/cfa51.cfm.

FHWA. 2017. "Fly Ash Facts for Highway Engineers." Department of Transportation. https://www.fhwa.dot.gov/pavement/recycling/fach01.cfm.

Finkelman, R.B., C.A. Palmer, and P. Wang. 2018. "Quantification of the Modes of Occurrence of 42 Elements in Coal." *International Journal of Coal Geology* 185(January 2):138–160. https://doi.org/10.1016/j.coal.2017.09.005.

Finkelman, R.B., S. Dai, and D. French. 2019. "The Importance of Minerals in Coal as the Hosts of Chemical Elements: A Review." *International Journal of Coal Geology* 212:103251. https://doi.org/10.1016/j.coal.2019.103251.

Fortune Business Insights. 2023a. "Cement Market Size, Share and COVID-19 Impact Analysis, by Type (Portland, Blended, and Others), by Application (Residential and Non-Residential), and Regional Forecast, 2023–2030." *Fortune Business Insights*. https://www.fortunebusinessinsights.com/industry-reports/cement-market-101825.

Fortune Business Insights. 2023b. "Concrete Blocks and Bricks Market Size, Share and COVID-19 Impact Analysis, by Type (Cement Block (Hollow, Fully Solid, and Cellular) and Brick (Clay, Fly Ash Clay, Sand Lime, and Others)), by Application (Residential, Commercial, and Industrial), and Regional Forecast, 2022–2029." *Fortune Business Insights*. https://www.fortunebusinessinsights.com/concrete-blocks-and-bricks-market-103784.

Fritz, A.G., T.J. Tarka, and M.S. Mauter. 2021. "Technoeconomic Assessment of a Sequential Step-leaching Process for Rare Earth Element Extraction from Acid Mine Drainage Precipitates." *ACS Sustainable Chemistry and Engineering* 9(28):9308–9316. https://doi.org/10.1021/acssuschemeng.1c02069.

Fu, B., J.C. Hower, W. Zhang, G. Luo, H. Hu, and H. Yao. 2022. "A Review of Rare Earth Elements and Yttrium in Coal Ash: Content, Modes of Occurrences, Combustion Behavior, and Extraction Methods." *Progress in Energy and Combustion Science* 88(January 1):100954. https://doi.org/10.1016/j.pecs.2021.100954.

Garrett, D.E. 2004. Handbook of Lithium and Natural Calcium Chloride: Their Deposits, Processing, Uses and Properties. Elsevier. https://doi.org/10.1016/B978-0-12-276152-2.X5035-X.

Gassenheimer, C., and C. Shaynak. 2023. "Coal Waste Recovery Presentation." Presentation to the committee. November 3. Washington, DC: National Academies of Sciences, Engineering, and Medicine.

Geng, W., T. Nakajima, H. Takanashi, and A. Ohki. 2007. "Determination of Total Fluorine in Coal by Use of Oxygen Flask Combustion Method with Catalyst." *Fuel* 86(5):715–721. https://doi.org/10.1016/j.fuel.2006.08.025.

GeologyHub. 2023. "What Are Agents of Metamorphism." *GeologyHub: Digital Education*. https://geologyhub.com/agents-of-metamorphism.

Gielen, D., and M. Lyons. 2022. "Critical Materials for the Energy Transition: Rare Earth Elements." Technical Paper 2/2022. Abu Dhabi: International Renewable Energy Agency. https://atf.asso.fr/media/technews/39/tnf39-prof3-irena-rare-earth-elements-2022.pdf.

Giurgiutiu, V. 2022. "Chapter 1—Introduction." Pp. 1–27 in *Stress, Vibration, and Wave Analysis in Aerospace Composites*, V. Giurgiutiu, ed. Academic Press. https://doi.org/10.1016/B978-0-12-813308-8.00006-5.

Grand View Research. 2022. "Carbon Fiber Market Size, Share and Trends Analysis Report by Raw Material (PAN-Based, Pitch-Based), by Tow Size, by Application (Automotive, Aerospace and Defense), by Region, and Segment Forecasts, 2023–2030." GVR-1-68038-523-6. San Francisco, CA: Grand View Research. https://www.grandviewresearch.com/industry-analysis/carbon-fiber-market-analysis.

Grand View Research. 2024. "Gypsum Board Market Size, Share and Trends Analysis Report by Product (Wallboard, Ceiling Board, Pre-Decorated), by Application (Pre-Engineered Metal Building, Residential, Isndustrial), by Region, and Segment Forecasts, 2024–2030." San Francisco, CA: Grand View Research. https://www.grandviewresearch.com/industry-analysis/gypsum-board-market.

Granite, E.J., G. Bromhal, J. Wilcox, and M.A. Alvin. 2023. "Domestic Wastes and Byproducts: A Resource for Critical Material Supply Chains." *The Bridge* 53(3):59–66. https://www.nae.edu/File.aspx?id=300392.

Griswold, E. 2022. "Could Coal Waste Be Used to Make Sustainable Batteries?" *The New Yorker*, August 26. https://www.newyorker.com/news/us-journal/could-coal-waste-be-used-to-make-sustainable-batteries.

Guo, J., X. Li, H. Xu, H. Zhu, B. Li, and A. Westwood. 2020. "Molecular Structure Control in Mesophase Pitch via Co-Carbonization of Coal Tar Pitch and Petroleum Pitch for Production of Carbon Fibers with Both High Mechanical Properties and Thermal Conductivity." *Energy and Fuels* 34(5):6474–6482. https://doi.org/10.1021/acs.energyfuels.0c00196.

Gupta, S., K. Chapman, S. Praneeth, P.M. Stemmer, M.J. Allen, T.M. Dittrich, and J.J. Kodanko. 2023. "Semi-Synthetic Proteins as Metal Ion Capture Agents: Catch and Release of Ni(II) and Cu(II) with Myoglobin Bioconjugates." *ACS Sustainable Chemistry and Engineering* 11(30):11305–11312. https://doi.org/10.1021/acssuschemeng.3c03148.

Gypsum Association. 2024. "FGD Gypsum and Sustainability." *Gypsum Association*. https://gypsum.org/press-roomfgd-gypsum-board.

Hamidi, A., P. Nazari, S. Shakibania, and F. Rashchi. 2023. "Microwave Irradiation for the Recovery Enhancement of Fly Ash Components: Thermodynamic and Kinetic Aspects." *Chemical Engineering and Processing—Process Intensification* 191(September 1):109472. https://doi.org/10.1016/j.cep.2023.109472.

Han, L., X. Zhu, F. Yang, Q. Liu, and X. Jia. 2021. "Eco-Conversion of Coal into a Nonporous Graphite for High-Performance Anodes of Lithium-Ion Batteries." *Powder Technology* 382(April 1):40–47. https://doi.org/10.1016/j.powtec.2020.12.052.

He, M., X. Guo, J. Huang, H. Shen, Q. Zeng, and L. Wang. 2018. "Mass Production of Tunable Multicolor Graphene Quantum Dots from an Energy Resource of Coke by a One-Step Electrochemical Exfoliation." *Carbon* 140(December 1):508–520. https://doi.org/10.1016/j.carbon.2018.08.067.

Hedin, B.C., R.S. Hedin, R.C. Capo, and B.W. Stewart. 2020. "Critical Metal Recovery Potential of Appalachian Acid Mine Drainage Treatment Solids." *International Journal of Coal Geology* 231(November 1):103610. https://doi.org/10.1016/j.coal.2020.103610.

Hedin, B.C., M.Y. Stuckman, C.A. Cravotta, C.L. Lopano, and R.C. Capo. 2024. "Determination and Prediction of Micro Scale Rare Earth Element Geochemical Associations in Mine Drainage Treatment Wastes." *Chemosphere* 346(January 1):140475. https://doi.org/10.1016/j.chemosphere.2023.140475.

Hood, M.M., R.K. Taggart, R.C. Smith, H. Hsu-Kim, K.R. Henke, U. Graham, J.G. Groppo, J.M. Unrine, and J.C. Hower. 2017. "Rare Earth Element Distribution in Fly Ash Derived from the Fire Clay Coal, Kentucky." *Coal Combustion and Gasification Products* 9(1):22–33. https://doi.org/10.4177/CCGP-D-17-00002.1.

Hower, J.C., and J.G. Groppo. 2021. "Rare Earth-Bearing Particles in Fly Ash Carbons: Examples from the Combustion of Eastern Kentucky Coals." *Coal Energy and Environmental Impacts* 2(2):90–98. https://doi.org/10.1016/j.engeos.2020.09.003.

Hower, J.C., J.G. Groppo, H. Hsu-Kim, and R.K. Taggart. 2021. "Distribution of Rare Earth Elements in Fly Ash Derived from the Combustion of Illinois Basin Coals." *Fuel* 289(April 1):119990. https://doi.org/10.1016/j.fuel.2020.119990.

Hower, J.C., A. Kolker, H. Hsu-Kim, and D.L. Plata. In press. "Rare Earth Elements in Coal Fly Ash and Their Potential Recovery," Chapter 2 in *Rare Earth Elements: Sustainable Processing, Purification, and Recovery*, A.K. Karamalidis and R. Eggert, eds. American Geophysical Union Special Publication 79. https://doi.org/10.1002/9781119515005.ch2.

Huang, C., and T. Xu. 2006. "Electrodialysis with Bipolar Membranes for Sustainable Development." *Environmental Science and Technology* 40(17):5233–5243. https://doi.org/10.1021/es060039p.

Huang, Z., M. Fan, and H. Tian. 2018. "Coal and Coal Byproducts: A Large and Developable Unconventional Resource for Critical Materials—Rare Earth Elements." *Journal of Rare Earths* 36(4):337–338. https://doi.org/10.1016/j.jre.2018.01.002.

Huang, Z., M. Fan, and H. Tian. 2020. "Rare Earth Elements of Fly Ash from Wyoming's Powder River Basin Coal." *Journal of Rare Earths* 38(2):219–226. https://doi.org/10.1016/j.jre.2019.05.004.

IEA (International Energy Agency). 2022. "The Role of Critical Minerals in Clean Energy Transitions." World Energy Outlook Special Report. Paris, France: International Energy Agency. https://iea.blob.core.windows.net/assets/ffd2a83b-8c30-4e9d-980a-52b6d9a86fdc/TheRoleofCriticalMineralsinCleanEnergyTransitions.pdf.

IEA. 2024. "Global Critical Minerals Outlook 2024." Paris, France: International Energy Agency. https://www.iea.org/reports/global-critical-minerals-outlook-2024.

Ighalo, J.O., S.B. Kurniawan, K.O. Iwuozor, C.O. Aniagor, O.J. Ajala, S.N. Oba, F.U. Iwuchukwu, S. Ahmadi, and C.A. Igwegbe. 2022. "A Review of Treatment Technologies for the Mitigation of the Toxic Environmental Effects of Acid Mine Drainage (AMD)." *Process Safety and Environmental Protection* 157(January 1):37–58. https://doi.org/10.1016/j.psep.2021.11.008.

Ilander, A., and A. Väisänen. 2009. "The Determination of Trace Element Concentrations in Fly Ash Samples Using Ultrasound-Assisted Digestion Followed with Inductively Coupled Plasma Optical Emission Spectrometry." *Ultrasonics Sonochemistry* 16(6):763–768. https://doi.org/10.1016/j.ultsonch.2009.03.001.

Indian Roads Congress. 2010. "Guidelines for Soil and Granular Material Stabilization Using Cement, Lime, and Fly Ash." IRC:SP:89-2010. New Delhi: Indian Roads Congress. https://law.resource.org/pub/in/bis/irc/irc.gov.in.sp.089-1.2010.pdf.

Ishak, C.F., J.C. Seaman, W.P. Miller, and M. Sumner. 2002. "Contaminant Mobility in Soils Amended with Fly Ash and Flue-Gas Gypsum: Intact Soil Cores and Repacked Columns." *Water, Air, and Soil Pollution* 134(1):285–303. https://doi.org/10.1023/A:1014101217340.

Izquierdo, M., and X. Querol. 2012. "Leaching Behaviour of Elements from Coal Combustion Fly Ash: An Overview." *Minerals and Trace Elements in Coal* 94(May 1):54–66. https://doi.org/10.1016/j.coal.2011.10.006.

Jäger, H., and W. Frohs, eds. 2021. *Industrial Carbon and Graphite Materials, Volume I: Raw Materials, Production and Applications*. Weinheim, Germany: Wiley-VCH. https://onlinelibrary.wiley.com/doi/book/10.1002/9783527674046.

Jahandari, S., Z. Tao, Z. Chen, D. Osborne, and M. Rahme. 2023. "4—Coal Wastes: Handling, Pollution, Impacts, and Utilization." Pp. 97–163 in *The Coal Handbook*, 2nd edition, D. Osborne, ed. Woodhead Publishing Series in Energy. Woodhead Publishing. https://doi.org/10.1016/B978-0-12-824327-5.00001-6.

Jha, M.K., A. Kumari, R. Panda, J.R. Kumar, K. Yoo, and J.Y. Lee. 2016. "Review on Hydrometallurgical Recovery of Rare Earth Metals." *SI: IC-LGO 2015* 165(October 1):2–26. https://doi.org/10.1016/j.hydromet.2016.01.035.

Kalderis, D., E. Tsolaki, C. Antoniou, and E. Diamadopoulos. 2008. "Characterization and Treatment of Wastewater Produced During the Hydro-Metallurgical Extraction of Germanium from Fly Ash." *Desalination* 230(1):162–174. https://doi.org/10.1016/j.desal.2007.11.023.

Kang, D.-H., S.C. Gupta, P.R. Bloom, A.Z. Ranaivoson, R. Roberson, and J. Siekmeier. 2011. "Recycled Materials as Substitutes for Virgin Aggregates in Road Construction: II. Inorganic Contaminant Leaching." *Soil Science Society of America Journal* 75(4):1276–1284. https://doi.org/10.2136/sssaj2010.0296.

Karan, R., T. Sreenivas, M. Ajay Kumar, and D.K. Singh. 2022. "Recovery of Rare Earth Elements from Coal Flyash Using Deep Eutectic Solvents as Leachants and Precipitating as Oxalate or Fluoride." *Hydrometallurgy* 214(October 1):105952. https://doi.org/10.1016/j.hydromet.2022.105952.

Karfakis, M.G., C.H. Bowman, and E. Topuz. 1996. "Characterization of Coal-Mine Refuse as Backfilling Material." *Geotechnical and Geological Engineering* 14(2):129–150. https://doi.org/10.1007/BF00430273.

Ketris, M.P., and Y.E. Yudovich. 2009. "Estimations of Clarkes for Carbonaceous Biolithes: World Averages for Trace Element Contents in Black Shales and Coals." *International Journal of Coal Geology* 78(2):135–148. https://doi.org/10.1016/j.coal.2009.01.002.

Kolker, A. 2023. "Rare Earth Elements and Critical Minerals in Coal and Coal Byproducts." 2023. Presentation to the committee. June 28. Washington DC: National Academies of Sciences, Engineering, and Medicine.

Kolker, A., L. Lefticariu, and S.T. Anderson. 2024. "Energy-Related Rare Earth Element Sources." Pp. 57–102 in *Rare Earth Metals and Minerals Industries: Status and Prospects*, Y.V. Murty, M.A. Alvin, and J.P. Lifton, eds. Cham: Springer International Publishing. https://doi.org/10.1007/978-3-031-31867-2_3.

Kolker, A., C. Scott, J.C. Hower, J.A. Vazquez, C.L. Lopano, and S. Dai. 2017. "Distribution of Rare Earth Elements in Coal Combustion Fly Ash, Determined by SHRIMP-RG Ion Microprobe." *International Journal of Coal Geology* 184(November 1):1–10. https://doi.org/10.1016/j.coal.2017.10.002.

Kong, J., G. Pan, and Z. Su. 2022. "The Coal as High Performance and Low Cost Anodes for Sodium-Ion Batteries." *Materials Letters: X* 13(March 1):100123. https://doi.org/10.1016/j.mlblux.2022.100123.

Kukier, U., C.F. Ishak, M.E. Sumner, and W.P. Miller. 2003. "Composition and Element Solubility of Magnetic and Non-Magnetic Fly Ash Fractions." *Environmental Pollution* 123(2):255–266. https://doi.org/10.1016/S0269-7491(02)00376-7.

Kumar Thiyagarajan, S., S. Raghupathy, D. Palanivel, K. Raji, and P. Ramamurthy. 2016. "Fluorescent Carbon Nano Dots from Lignite: Unveiling the Impeccable Evidence for Quantum Confinement." *Physical Chemistry Chemical Physics* 18(17):12065–12073. https://doi.org/10.1039/C6CP00867D.

Leandro, A.P.M., M.A. Seas, K. Vap, A.S. Tyrrell, V. Jain, H. Wahab, and P.A. Johnson. 2021. "Evolution of Structural and Electrical Properties in Coal-Derived Graphene Oxide Nanomaterials During High-Temperature Annealing." *Diamond and Related Materials* 112(February 1):108244. https://doi.org/10.1016/j.diamond.2021.108244.

Leininger, D., J. Leonhard, W. Erdmann, and T. Schieder. 1987. "Research on Suitability of Coal Preparation Refuse in Civil Engineering in the Federal Republic of Germany." Pp. 55–67 in *Advances in Mining Science and Technology*, Vol. 2, D. Leininger, J. Leonhard, W. Erdmann, and T. Schieder, eds. Reclamation, Treatment and Utilization of Coal Mining Wastes. Elsevier. https://doi.org/10.1016/B978-0-444-42876-9.50010-5.

Li, B. 2019. "Hydrophobic-Hydrophilic Separation Process for the Recovery of Ultrafine Particles." Doctor of Philosophy in Mineral Engineering, Virginia Polytechnic Institute and State University. https://vtechworks.lib.vt.edu/handle/10919/103278.

Li, F., Z. Guo, G. Su, C. Guo, G. Sun, and X. Zhu. 2018. "Preparation of SiC from Acid-Leached Coal Gangue by Carbothermal Reduction." *International Journal of Applied Ceramic Technology* 15(3):625–632. https://doi.org/10.1111/ijac.12856.

Li, J., D. Zhan, Z. Jiang, H. Zhang, Y. Yang, and Y. Zhang. 2023. "Progress on Improving Strength-Toughness of Ultra-High Strength Martensitic Steels for Aerospace Applications: A Review." *Journal of Materials Research and Technology* 23(March 1):172–190. https://doi.org/10.1016/j.jmrt.2022.12.177.

Li, Q., and W. Zhang. 2022. "Process Development for Recovering Critical Elements from Acid Mine Drainage." *Resources, Conservation and Recycling* 180(May 1):106214. https://doi.org/10.1016/j.resconrec.2022.106214.

Li, S., P. Bo, L. Kang, H. Guo, W. Gao, and S. Qin. 2020. "Activation Pretreatment and Leaching Process of High-Alumina Coal Fly Ash to Extract Lithium and Aluminum." *Metals* 10(7). https://doi.org/10.3390/met10070893.

Li, Z., F. Rao, M.A. Corona-Arroyo, A. Bedolla-Jacuinde, and S. Song. 2019. "Comminution Effect on Surface Roughness and Flotation Behavior of Malachite Particles." *Minerals Engineering* 132(March 1):1–7. https://doi.org/10.1016/j.mineng.2018.11.056.

Liang, L., J. Tan, B. Li, and G. Xie. 2019. "Reducing Quartz Entrainment in Fine Coal Flotation by Polyaluminum Chloride." *Fuel* 235(January 1):150–157. https://doi.org/10.1016/j.fuel.2018.07.106.

Ling, F., Y.E., K.J. Garriga Francis, B. Zhang, and X.-C. Zhang. 2020. "Investigation on THz Generation from Influences of Gold Nanoparticles in Water Solution," Proc. SPIE 11559, Infrared, Millimeter-Wave, and Terahertz Technologies VII, 1155911. https://doi.org/10.1117/12.2575244.

Lisowyj, B., D. Hitchcock, and H. Epstein. 1987a. "Direct Fuel-Fired Furnace Arrangement for the Recovery of Gallium and Germanium from Coal Fly Ash." US4643110A, filed July 7, 1986, and issued February 17, 1987. https://patents.google.com/patent/US4643110A/en.

Lisowyj, B., D. Hitchcock, and H. Epstein. 1987b. "Process for the Recovery of Gallium and Germanium from Coal Fly Ash." US4678647A, filed May 12, 1986, and issued July 7, 1987. https://patents.google.com/patent/US4678647A/en.

Liu, J.L., T.-H. Yan, and S. Bashir. 2022. *Advanced Nanomaterials and Their Applications in Renewable Energy*, 2nd edition. Elsevier. https://doi.org/10.1016/C2021-0-00377-1.

Liu, P., L. Yang, Q. Wang, B. Wan, Q. Ma, H. Chen, and Y. Tang. 2020. "Speciation Transformation of Rare Earth Elements (REEs) during Heating and Implications for REE Behaviors during Coal Combustion." *International Journal of Coal Geology* 219(February 15):103371. https://doi.org/10.1016/j.coal.2019.103371.

London, D. 2008. *Pegmatites*. The Canadian Mineralogist Special Publication. The Mineralogical Association.

Lu, H., S. Sun, L. Xiao, J. Qian, X. Ai, H. Yang, A.-H. Lu, and Y. Cao. 2019. "High-Capacity Hard Carbon Pyrolyzed from Subbituminous Coal as Anode for Sodium-Ion Batteries." *ACS Applied Energy Materials* 2(1):729–735. https://doi.org/10.1021/acsaem.8b01784.

Luo, K., D. Ren, L. Xu, S. Dai, D. Cao, F. Feng, and J. Tan. 2004. "Fluorine Content and Distribution Pattern in Chinese Coals." *International Journal of Coal Geology* 57(2):143–149. https://doi.org/10.1016/j.coal.2003.10.003.

Luttrell, G.H., and R.Q. Honaker. 2012. "Coal Preparation." Pp. 2194–2222 in *Encyclopedia of Sustainability Science and Technology*, R.A. Meyers, ed. New York: Springer New York. https://doi.org/10.1007/978-1-4419-0851-3_431.

Marcus, K. 2022. "Coal as Value-Added for Lithium Battery Anodes." Presented at the Department of Energy National Energy Technology Laboratory Resource Sustainability Project Review Meeting. October 25. Pittsburgh, PA. https://netl.doe.gov/sites/default/files/netl-file/22RS-25_Marcus.pdf.

Maroto-Valer, M.M., D.N. Taulbee, and J.C. Hower. 2001. "Characterization of Differing Forms of Unburned Carbon Present in Fly Ash Separated by Density Gradient Centrifugation." *Fuel* 80(6):795–800. https://doi.org/10.1016/S0016-2361(00)00154-X.

Marsh, H., ed. 1989. *Introduction to Carbon Science*. London: Butterworth & Co (Publishers) Ltd. https://doi.org/10.1016/C2013-0-04111-4.

Marsh, H., and F. Rodríguez-Reinoso. 2006. "Chapter 9—Production and Reference Material." Pp. 454–508 in *Activated Carbon*, H. Marsh and F. Rodríguez-Reinoso, eds. Oxford: Elsevier Science Ltd. https://doi.org/10.1016/B978-008044463-5/50023-6.

McCallister, D.L., K.D. Frank, W.B. Stevens, G.W. Hergert, R.R. Renken, and D.B. Marx. 2002. "Coal Fly Ash as an Acid-Reducing Soil Amendment and Its Side Effects 1." *Soil Science* 167(12). https://journals.lww.com/soilsci/fulltext/2002/12000/coal_fly_ash_as_an_acid_reducing_soil_amendment.5.aspx.

McNulty, T., N. Hazen, and S. Park. 2022. "Processing the Ores of Rare-Earth Elements." *MRS Bulletin* 47(3):258–266. https://doi.org/10.1557/s43577-022-00288-4.

Minter, B. 2004. "Method for Recovering Trace Elements from Coal." US6827837B2, filed November 22, 2002, and issued December 7, 2004. https://patents.google.com/patent/US6827837B2/en?oq=US6827837B2.

Minter, B. 2005. "Method for Recovering Trace Elements from Coal." US20050056548A1, filed October 27, 2004, and issued March 17, 2005. https://patents.google.com/patent/US20050056548A1/en.

Mohammadi, F., and M. Saif. 2023. "A Comprehensive Overview of Electric Vehicle Batteries Market." *E-prime—Advances in Electrical Engineering, Electronics and Energy* 3(March 1):100127. https://doi.org/10.1016/j.prime.2023.100127.

Mohanty, M., D.R. Biswal, and S.S. Mohapatra. 2023. "A Systematic Review Exploring the Utilization of Coal Mining and Processing Wastes as Secondary Aggregate in Sub-Base and Base Layers of Pavement." *Construction and Building Materials* 368(March 3):130408. https://doi.org/10.1016/j.conbuildmat.2023.130408.

MoistureShield, Inc. 2022. "ICC-ES VAR Environmental Report." VAR-1015. ICC Evaluation Service, LLC. https://www.moistureshield.com/wp-content/uploads/2021/03/VAR-1015-Recycled-Content.pdf.

Moothi, K., G.S. Simate, R. Falcon, S.E. Iyuke, and M. Meyyappan. 2015. "Carbon Nanotube Synthesis Using Coal Pyrolysis." *Langmuir* 31(34):9464–9472. https://doi.org/10.1021/acs.langmuir.5b01894.

MSHA (Mine Safety and Health Administration). 2009. "Engineering and Design Manual: Coal Refuse Disposal Facilities." 2nd edition. Arlington, VA: Department of Labor, May (Rev. Aug. 2010). https://arlweb.msha.gov/Impoundments/DesignManual/2009ImpoundmentDesignManual.pdf.

Mwewa, B., M. Tadie, S. Ndlovu, G.S. Simate, and E. Matinde. 2022. "Recovery of Rare Earth Elements from Acid Mine Drainage: A Review of the Extraction Methods." *Journal of Environmental Chemical Engineering* 10(3):107704. https://doi.org/10.1016/j.jece.2022.107704.

Nadimi, H., D. Haghshenas Fatmehsari, and S. Firoozi. 2017. "Separation of Ni and Co by D2EHPA in the Presence of Citrate Ion." *Metallurgical and Materials Transactions B* 48(5):2751–2758. https://doi.org/10.1007/s11663-017-1008-7.

NASEM (National Academies of Sciences, Engineering, and Medicine). 2024. *Accelerating Decarbonization in the United States: Technology, Policy, and Societal Dimensions*. Washington, DC: National Academies Press. https://doi.org/10.17226/25931.

NETL (National Energy Technology Laboratory). n.d. "Rare Earth Elements—A Subset of Critical Minerals." https://www.netl.doe.gov/resource-sustainability/critical-minerals-and-materials/rare-earth-elements.

Nguyen, C.V., A.V. Nguyen, A. Doi, E. Dinh, Thuong V. Nguyen, M. Ejtemaei, and D. Osborne. 2021. "Advanced Solid-Liquid Separation for Dewatering Fine Coal Tailings by Combining Chemical Reagents and Solid Bowl Centrifugation." *Separation and Purification Technology* 259(March 15):118172. https://doi.org/10.1016/j.seppur.2020.118172.

Nickel Institute. 2024. "About Nickel." https://nickelinstitute.org/en/about-nickel-and-its-applications.

Nockemann, P., B. Thijs, S. Pittois, J. Thoen, C. Glorieux, K. Van Hecke, L. Van Meervelt, B. Kirchner, and K. Binnemans. 2006. "Task-Specific Ionic Liquid for Solubilizing Metal Oxides." *The Journal of Physical Chemistry B* 110(42):20978–20992. https://doi.org/10.1021/jp0642995.

Nockemann, P., B. Thijs, T.N. Parac-Vogt, K. Van Hecke, L. Van Meervelt, B. Tinant, I. Hartenbach, et al. 2008. "Carboxyl-Functionalized Task-Specific Ionic Liquids for Solubilizing Metal Oxides." *Inorganic Chemistry* 47(21):9987–9999. https://doi.org/10.1021/ic801213z.

Oberhaus, D. 2023. "Rare Earths for America's Future." Washington, DC: Innovation Frontier Project. https://www.progressivepolicy.org/wp-content/uploads/2023/01/IFP_Rare-Earths_.pdf.

O'Keefe, C.A. 1996. "Selective Leaching of Coal and Coal Combustion Solid Residues." In *Conversion and Utilization of Waste Materials*, 1st edition. Routledge. https://doi.org/10.1201/9781315140360.

Olson III, R. 2022. "Manufacture of Carbon Foam in Continuous Process at Atmospheric Pressure." Presented at the National Energy Technology Laboratory Resource Sustainability Project Review Meeting. October 25–27. Pittsburgh, PA. https://netl.doe.gov/sites/default/files/netl-file/22RS-26_Olson.pdf.

Omodara, L., S. Pitkäaho, E.-M. Turpeinen, P. Saavalainen, K. Oravisjärvi, and R.L. Keiski. 2019. "Recycling and Substitution of Light Rare Earth Elements, Cerium, Lanthanum, Neodymium, and Praseodymium from End-of-Life Applications—A Review." *Journal of Cleaner Production* 236(November 1):117573. https://doi.org/10.1016/j.jclepro.2019.07.048.

Opare, E.O., E. Struhs, and A. Mirkouei. 2021. "A Comparative State-of-Technology Review and Future Directions for Rare Earth Element Separation." *Renewable and Sustainable Energy Reviews* 143(June 1):110917. https://doi.org/10.1016/j.rser.2021.110917.

Orem, W.H., and R.B. Finkelman. 2003. "7.08—Coal Formation and Geochemistry." Pp. 191–222 in *Treatise on Geochemistry*, H.D. Holland and K.K. Turekian, eds. Oxford: Pergamon. https://doi.org/10.1016/B0-08-043751-6/07097-3.

Özer, M., O.M. Basha, and B. Morsi. 2017. "Coal-Agglomeration Processes: A Review." *International Journal of Coal Preparation and Utilization* 37(3):131–167. https://doi.org/10.1080/19392699.2016.1142443.

Pan, J., B.V. Hassas, M. Rezaee, C. Zhou, and S.V. Pisupati. 2021. "Recovery of Rare Earth Elements from Coal Fly Ash Through Sequential Chemical Roasting, Water Leaching, and Acid Leaching Processes." *Journal of Cleaner Production* 284(February 15):124725. https://doi.org/10.1016/j.jclepro.2020.124725.

Pandey, N., S.K. Tripathy, S.K. Patra, and G. Jha. 2023. "Recent Progress in Hydrometallurgical Processing of Nickel Lateritic Ore." *Transactions of the Indian Institute of Metals* 76(1):11–30. https://doi.org/10.1007/s12666-022-02706-2.

Park, S.-M., J.-C. Yoo, S.-W. Ji, J.-S. Yang, and K. Baek. 2013. "Selective Recovery of Cu, Zn, and Ni from Acid Mine Drainage." *Environmental Geochemistry and Health* 35(6):735–743. https://doi.org/10.1007/s10653-013-9531-1.

Paul, A., R. Magee, N. Wichert, W. Wilczewski, J. Trembly, T.R. Garrick, and J.A. Staser. 2023. "Carbon-Based Porous Materials for Energy Storage Applications." *ECS Meeting Abstracts* MA2023-01, no. 24: 1621. https://doi.org/10.1149/MA2023-01241621mtgabs.

Paul, A., R. Magee, W. Wilczewski, N. Wichert, C. Gula, R. Olson, E. Shereda, et al. 2024. "Application of the Multi-Species, Multi-Reaction Model to Coal-Derived Graphite for Lithium-Ion Batteries." *Journal of the Electrochemical Society* 171(2):023501. https://doi.org/10.1149/1945-7111/ad2061.

Peiravi, M., F. Dehghani, L. Ackah, A. Baharlouei, J. Godbold, J. Liu, M. Mohanty, and T. Ghosh. 2021. "A Review of Rare-Earth Elements Extraction with Emphasis on Non-Conventional Sources: Coal and Coal Byproducts, Iron Ore Tailings, Apatite, and Phosphate Byproducts." *Mining, Metallurgy and Exploration* 38(1):1–26. https://doi.org/10.1007/s42461-020-00307-5.

Phung, H.T., L.J. Lund, and A.L. Page. 1978. "Potential Use of Fly Ash as a Liming Material." Pp. 504–515 in *Environmental Chemistry and Cycling Processes*, D.C. Adrinao and L. Brisbin, eds. Tech. Inform. Cent. Publ. CONF-760429. Department of Commerce, Springfield, VA.

Pires, M., and J.A. Solari. 1988. "Ultrafine Coal Beneficiation by Liquid—Liquid Extraction." Pp. 363–372 in *Production and Processing of Fine Particles*, A.J. Plumpton, ed. Amsterdam: Pergamon. https://doi.org/10.1016/B978-0-08-036448-3.50042-6.

Prakash, S., R. Kumar, A. Gupta, A. Chaudhary, V.K. Chandaliya, P.S. Dash, P. Gurunathan, K. Ramesha, S. Kumari, and S.R. Dhakate. 2022. "A Process for Developing Spherical Graphite from Coal Tar as High Performing Carbon Anode for Li-Ion Batteries." *Materials Chemistry and Physics* 281(April 1):125836. https://doi.org/10.1016/j.matchemphys.2022.125836.

Quan, Y., Q. Liu, K. Li, H. Zhang, and L. Yuan. 2022. "Highly Efficient Purification of Natural Coaly Graphite via an Electrochemical Method." *Separation and Purification Technology* 281(January 15):119931. https://doi.org/10.1016/j.seppur.2021.119931.

Rezaee, M., and R.Q. Honaker. 2020. "Long-Term Leaching Characteristic Study of Coal Processing Waste Streams." *Chemosphere* 249(June 1):126081. https://doi.org/10.1016/j.chemosphere.2020.126081.

Riefler, R.G. 2021. "Method for Actively Treating Mining Wastewater for Pigment Production." US11198629B2. Ohio University filed November 17, 2017, and issued December 14, 2021. https://patents.google.com/patent/US11198629B2/en.

Riefler, G., F. Reshma, L. Myers, J. Stanley, M. Shively MacIver, K. Shaw, and J. Sabraw. 2023. "Dewatering of Iron Sludge for Pigment Production." Presented at the National Association of Abandoned Mine Land Programs Conference. September 25–26. Chicago, IL.

RSC (Royal Society of Chemistry). 2024. "Erbium." Periodic Table. https://www.rsc.org/periodic-table/element/68/erbium.

Sahoo, P.K., S. Tripathy, Sk. Md. Equeenuddin, and M.K. Panigrahi. 2012. "Geochemical Characteristics of Coal Mine Discharge Vis-à-Vis Behavior of Rare Earth Elements at Jaintia Hills Coalfield, Northeastern India." *Journal of Geochemical Exploration* 112(January 1):235–243. https://doi.org/10.1016/j.gexplo.2011.09.001.

Salgansky, E.A., V.M. Kislov, M.V. Tsvetkov, A.Yu. Zaichenko, D.N. Podlesniy, M.V. Salganskaya, K.M. Kadiev, M.Ya. Visaliev, and L.A. Zekel. 2022. "Energy Production and Recovery of Rare Metals from Ash Residue During Coal Filtration Combustion." *Russian Journal of Physical Chemistry B* 16(2):268–277. https://doi.org/10.1134/S1990793122020105.

Sanz, J., O. Tomasa, A. Jimenez-Franco, and N. Sidki-Rius. 2022. "Holmium (Ho) [Z = 67]." Pp. 285–287 in *Elements and Mineral Resources*. Springer Textbooks in Earth Sciences, Geography and Environment. Cham: Springer. https://link.springer.com/chapter/10.1007/978-3-030-85889-6_71.

Sarker, S.K., N. Haque, M. Bhuiyan, W. Bruckard, and B.K. Pramanik. 2022. "Recovery of Strategically Important Critical Minerals from Mine Tailings." *Journal of Environmental Chemical Engineering* 10(3):107622. https://doi.org/10.1016/j.jece.2022.107622.

Scott, C., and A. Kolker. 2019. "Rare Earth Elements in Coal and Coal Fly Ash." Reston, VA: U.S. Geological Survey. https://pubs.usgs.gov/fs/2019/3048/fs20193048.pdf.

Seo, D.K., S.S. Park, Y.T. Kim, J. Hwang, and T.-U, Yu. 2011. "Study of Coal Pyrolysis by Thermo-Gravimetric Analysis (TGA) and Concentration Measurements of the Evolved Species." *Journal of Analytical and Applied Pyrolysis* 92(1):209–216. https://doi.org/10.1016/j.jaap.2011.05.012.

Seredin, V.V. 2010. "A New Method for Primary Evaluation of the Outlook for Rare Earth Element Ores." *Geology of Ore Deposits* 52(5):428–433. https://doi.org/10.1134/S1075701510050077.

Seredin, V.V., and S. Dai. 2012. "Coal Deposits as Potential Alternative Sources for Lanthanides and Yttrium." *International Journal of Coal Geology, Minerals and Trace Elements in Coal* 94(May 1):67–93. https://doi.org/10.1016/j.coal.2011.11.001.

Seredin, V.V., S. Dai, Y. Sun, and I.Y. Chekryzhov. 2013. "Coal Deposits as Promising Sources of Rare Metals for Alternative Power and Energy-Efficient Technologies." *Applied Geochemistry* 31(April 1):1–11. https://doi.org/10.1016/j.apgeochem.2013.01.009.

Shapiro, M., and V. Galperin. 2005. "Air Classification of Solid Particles: A Review." *Pneumatic Conveying and Handling of Particulate Solids* 44(2):279–285. https://doi.org/10.1016/j.cep.2004.02.022.

Sherwood, W. 2022a. "Coal Core Composites for Low Cost, Light Weight, Fire Resistant Panels and Roofing Materials." Presented at the Department of Energy National Energy Technology Laboratory Resource Sustainability Project Review Meeting. October 26. Pittsburgh, PA. https://netl.doe.gov/sites/default/files/netl-file/22RS-26_Sherwood_C.pdf.

Sherwood, W. 2022b. "Low Weight, High Strength Coal-Based Building Materials for Infrastructure Products." Presented at the Department of Energy National Energy Technology Laboratory Resource Sustainability Project Review Meeting. October 26. Pittsburgh, PA. https://netl.doe.gov/sites/default/files/netl-file/22RS-26_Sherwood_A.pdf.

Shi, M., C. Song, Z. Tai, K. Zou, Y. Duan, X. Dai, J. Sun, Y. Chen, and Y. Liu. 2021. "Coal-Derived Synthetic Graphite with High Specific Capacity and Excellent Cyclic Stability as Anode Material for Lithium-Ion Batteries." *Fuel* 292(May 15):120250. https://doi.org/10.1016/j.fuel.2021.120250.

Shimanoe, H., T. Mashio, K. Nakabayashi, T. Inoue, M. Hamaguchi, J. Miyawaki, I. Mochida, and S.-H. Yoon. 2020. "Manufacturing Spinnable Mesophase Pitch Using Direct Coal Extracted Fraction and Its Derived Mesophase Pitch Based Carbon Fiber." *Carbon* 158(March 1):922–929. https://doi.org/10.1016/j.carbon.2019.11.082.

Sinha, M.K., H. Tanvar, S.K. Sahu, and B. Mishra. 2023. "A Review on Recovery of Terbium from Primary and Secondary Resources: Current State and Future Perspective." *Mineral Processing and Extractive Metallurgy Review* 1–24. https://doi.org/10.1080/08827508.2023.2253490.

SkyQuest. 2023. "Global Asphalt Market." SkyQuest Technology Consulting Pvt. Ltd. https://www.skyquestt.com/report/asphalt-market.

Sobhy, A., and D. Tao. 2013. "Nanobubble Column Flotation of Fine Coal Particles and Associated Fundamentals." *International Journal of Mineral Processing* 124(November 14):109–116. https://doi.org/10.1016/j.minpro.2013.04.016.

Song, S., and O. Trass. 1997. "Floc Flotation of Prince Coal with Simultaneous Grinding and Hydrophobic Flocculation in a Szego Mill." *Fuel* 76(9):839–844. https://doi.org/10.1016/S0016-2361(97)00068-9.

Staszuk, M., D. Pakuła, Ł. Reimann, A. Kloc-Ptaszna, and K. Lukaszkowicz. 2023. "Structure and Properties of the TiN/ZnO Coating Obtained by the Hybrid Method Combining PVD and ALD Technologies on Austenitic Cr-Ni-Mo Steel Substrate." *Surfaces and Interfaces* 37(April 1):102693. https://doi.org/10.1016/j.surfin.2023.102693.

Stoffa, J. 2022. "Carbon Ore Processing Program." Presented at the Resource Sustainability Project Review Meeting. October 25. Pittsburgh, PA. https://netl.doe.gov/sites/default/files/netl-file/22RS-25_Stoffa.pdf.

Stoffa, J. 2023. "Carbon Conversion Program Overview." Presentation to the committee. June 28. Washington DC: National Academies of Sciences, Engineering, and Medicine.

Stoy, L., V. Diaz, and C.-H. Huang. 2021. "Preferential Recovery of Rare-Earth Elements from Coal Fly Ash Using a Recyclable Ionic Liquid." *Environmental Science and Technology* 55(13):9209–9220. https://doi.org/10.1021/acs.est.1c00630.

Straits Research. 2022. "Iron Oxide Pigments Market." Straits Research. https://straitsresearch.com/report/iron-oxide-pigments-market.

Stuckman, M. 2022. "Rare Earth Element/Critical Mineral (REE/CM) Recovery from Coal Byproducts and Acid Mine Drainage." June 17. Pittsburgh, PA. https://www.osti.gov/servlets/purl/1893527.

Sun, K., T. Wang, Z. Chen, W. Lu, X. He, W. Gong, M. Tang, et al. 2019. "Clean and Low-Cost Synthesis of High Purity Beta-Silicon Carbide with Carbon Fiber Production Residual and a Sandstone." *Journal of Cleaner Production* 238:117875. https://doi.org/10.1016/j.jclepro.2019.117875.

Sun, S., K. Yang, C. Liu, G. Tu, and F. Xiao. 2021. "Recovery of Nickel and Preparation of Ferronickel Alloy from Spent Petroleum Catalyst via Cooperative Smelting–Vitrification Process with Coal Fly Ash." *Environmental Technology* 1–11. https://doi.org/10.1080/09593330.2021.2002421.

Superior Graphite. n.d. "Natural Graphite." https://superiorgraphite.com/about-us/natural-graphite.

Talan, D., and Q. Huang. 2022. "A Review Study of Rare Earth, Cobalt, Lithium, and Manganese in Coal-Based Sources and Process Development for Their Recovery." *Minerals Engineering* 189(November 1):107897. https://doi.org/10.1016/j.mineng.2022.107897.

Taylor, Jr., E.M., and G.E. Schuman. 1988. "Fly Ash and Lime Amendment of Acidic Coal Spoil to Aid Revegetation." *Journal of Environmental Quality* 17(1):120–124. https://doi.org/10.2134/jeq1988.00472425001700010018x.

Tejasvi, A., and S. Kumar. 2012. "Impact of Fly Ash on Soil Properties." *National Academy Science Letters* 35(1):13–16. https://doi.org/10.1007/s40009-011-0002-x.

Thapa, R., C. Ugwumadu, K. Nepal, J. Trembly, and D. A. Drabold. 2022. "*Ab Initio* Simulation of Amorphous Graphite." *Physical Review Letters* 128(23):236402. https://doi.org/10.1103/PhysRevLett.128.236402.

Thapaliya, B.P., H. Luo, M. Li, W.-Y. Tsai, H.M. Meyer, J.R. Dunlap, J. Nanda, I. Belharouak, and S. Dai. 2021. "Molten Salt Assisted Low-Temperature Electro-Catalytic Graphitization of Coal Chars." *Journal of The Electrochemical Society* 168(4):046504. https://doi.org/10.1149/1945-7111/abf219.

Tian, Y., Y. Zhang, B. Wang, W. Ji, Y. Zhang, and K. Xie. 2004. "Coal-Derived Carbon Nanotubes by Thermal Plasma Jet." *Carbon* 42(12):2597–2601. https://doi.org/10.1016/j.carbon.2004.05.042.

Tõnsuaadu, K., K.A. Gross, L. Plūduma, and M. Veiderma. 2012. "A Review on the Thermal Stability of Calcium Apatites." *Journal of Thermal Analysis and Calorimetry* 110(2):647–659. https://doi.org/10.1007/s10973-011-1877-y.

Toropitsyna, J., L. Jelinek, R. Wilson, and M. Paidar. 2023. "Selective Removal of Transient Metal Ions from Acid Mine Drainage and the Possibility of Metallic Copper Recovery with Electrolysis." *Solvent Extraction and Ion Exchange* 41(2):176–204. https://doi.org/10.1080/07366299.2023.2181090.

Torres, H., K. Pichelbauer, S. Budnyk, T. Schachinger, C. Gachot, and M. Rodríguez Ripoll. 2023. "A Ni-Bi Self-Lubricating Ti6Al4V Alloy for High Temperature Sliding Contacts." *Journal of Alloys and Compounds* 944(May 25):169216. https://doi.org/10.1016/j.jallcom.2023.169216.

Tour, J.M. 2022. "Conversion of Domestic US Coal into Exceedingly High-Quality Graphene." Houston, TX: Rice University. https://www.osti.gov/servlets/purl/1842469.

Trembly, J., Y. Al-Majali, L. Veley, D. Daramola, V. Dhanapal, R. Statnick, and E. Shereda. 2023a. "Coal Plastic Composite Piping Infrastructure Components." DOE Cooperative Agreement No. DE-FE0031982. Athens, OH: Ohio University. https://www.osti.gov/servlets/purl/1970232.

Trembly, J., Y. Al-Majali, L. Veley, M. Zhang, R. Olson III, and E. Shereda. 2023b. "Coal-Derived Alternatives to Fiber-Cementitious Building Materials Final Technical Report." DOE Cooperative Agreement No. DE-FE0031981. Athens, OH: Ohio University. https://www.osti.gov/servlets/purl/1970234.

Trex. 2024. "Trex® Environmental, Social and Governance Report." https://www.trex.com/why-trex/esg.

Troutman, T. 2023. "HHS Technology Presentation." Presented at the Carbon Utilization Infrastructure, Markets, Research and Development Information Gathering: Coal Waste Separations. Virtually. October 23, 2023. https://vimeo.com/event/3765010.

Ucore. 2022. "RapidSX™." Ucore. https://ucore.com/rapidsx.

Ugwumadu, C., R. Thapa, Y. Al-Majali, J. Trembly, and D.A. Drabold. 2023. "Formation of Amorphous Carbon Multi-Walled Nanotubes from Random Initial Configurations." *Physica Status Solidi (b)* 260(3):2200527. https://doi.org/10.1002/pssb.202200527.

UNEP (United Nations Environment Programme). 2022. "Sand and Sustainability: 10 Strategic Recommendations to Avert a Crisis." GRID-Geneva. Geneva, Switzerland: United Nations Environment Programme. https://www.unep.org/resources/report/sand-and-sustainability-10-strategic-recommendations-avert-crisis.

USGS (U.S. Geological Survey). 2023. "Mineral Commodity Summaries 2023." Reston, VA: U.S. Geological Survey. https://pubs.usgs.gov/periodicals/mcs2023/mcs2023.pdf.

USGS. 2024. "Mineral Commodity Summaries 2024." Mineral Commodity Summaries. Reston, VA: U.S. Geological Survey. https://doi.org/10.3133/mcs2024.

USGS. n.d. "Nickel Statistics and Information." USGS. https://www.usgs.gov/centers/national-minerals-information-center/nickel-statistics-and-information.

Vaish, A.K. 1994. "Production of Ferro-Silicon and Calcium Silicon Alloys." Jamshedpur: National Metallurgical Laboratory. https://eprints.nmlindia.org/5785/1/6.01-6.17.PDF.

Vander Wal, R. 2023. "Coal-Based Bricks and Blocks (CBBs): Process Development to Prototype Fabrication Coupled with Techno-Economic Analysis and Market Survey." Presented at the TechConnect WORLD INNOVATION Conference and Expo. June 19. Washington, DC. https://www.osti.gov/servlets/purl/1993726.

Vass, C.R., A. Noble, and P.F. Ziemkiewicz. 2019. "The Occurrence and Concentration of Rare Earth Elements in Acid Mine Drainage and Treatment Byproducts. Part 2: Regional Survey of Northern and Central Appalachian Coal Basins." *Mining, Metallurgy and Exploration* 36(5):917–929. https://doi.org/10.1007/s42461-019-00112-9.

Veley, L., J. Trembly, and Y. Al-Majali. 2023a. "3D Printing of Sustainable Coal Polymer Composites: Thermo-physical Characteristics." *Materials Today Communications* 37(December 1):106989. https://doi.org/10.1016/j.mtcomm.2023.106989.

Veley, L.E., C. Ugwumadu, J.P. Trembly, D.A. Drabold, and Y. Al-Majali. 2023b. "3D Printing of Sustainable Coal Polymer Composites: Study of Processing, Mechanical Performance, and Atomistic Matrix–Filler Interaction." *ACS Applied Polymer Materials* 5(11):9286–9296. https://doi.org/10.1021/acsapm.3c01784.

Vijapur, S.H., D. Wang, D.C. Ingram, and G.G. Botte. 2017. "An Investigation of Growth Mechanism of Coal Derived Graphene Films." *Materials Today Communications* 11(June 1):147–155. https://doi.org/10.1016/j.mtcomm.2017.04.003.

Wagner, M.J. 2022. "Conversion of Coal to Li-Ion Battery Grade 'Potato' Graphite." Presented at the Resource Sustainability Project Review Meeting. October 25. Pittsburgh, PA. https://netl.doe.gov/sites/default/files/netl-file/22RS-25_Wagner.pdf.

Wang, B.-Y., J.-L. Xia, X.-L. Dong, X.-S. Wu, L.-J. Jin, and W.-C. Li. 2020b. "Highly Purified Carbon Derived from Deashed Anthracite for Sodium-Ion Storage with Enhanced Capacity and Rate Performance." *Energy and Fuels* 34(12):16831–16837. https://doi.org/10.1021/acs.energyfuels.0c03138.

Wang, X., A.R. Zanjanijam, S. Holberg, H.C. Thomas, and P.A. Johnson. 2023. "Coal Char as an Economical Filler for Phenolic Composites." *Composites Part B: Engineering* 264(September 1):110923. https://doi.org/10.1016/j.compositesb.2023.110923.

Wang, Y., Y.-J. Hu, X. Hao, P. Peng, J.-Y. Shi, F. Peng, and R.-C. Sun. 2020a. "Hydrothermal Synthesis and Applications of Advanced Carbonaceous Materials from Biomass: A Review." *Advanced Composites and Hybrid Materials* 3(3):267–284. https://doi.org/10.1007/s42114-020-00158-0.

Ward, C.R. 2016. "Analysis, Origin and Significance of Mineral Matter in Coal: An Updated Review." *International Journal of Coal Geology* 165(August 1):1–27. https://doi.org/10.1016/j.coal.2016.07.014.

Wewerka, E.M., and J.M. Williams. 1978. "Trace Element Characterization of Coal Wastes—First Annual Report." LA-6835-PR. EPA-600/7-79-028. Los Alamos, New Mexico: Los Alamos Scientific Laboratory. https://nepis.epa.gov/Exe/ZyPDF.cgi?Dockey=20012W2I.PDF.

Wierońska, F., D. Makowska, A. Strugała, and K. Bytnar. 2019. "Analysis of the Content of Nickel, Chromium, Lead and Zinc in Solid Products of Coal Combustion (CCPs) Coming from Polish Power Plants." *IOP Conference Series: Earth and Environmental Science* 214(1):012029. https://doi.org/10.1088/1755-1315/214/1/012029.

Wilcox, J. 2023. "Coal Waste Separations—Critical Mineral Recovery." Presentation to the committee. October 23. Washington, DC: National Academies of Sciences, Engineering, and Medicine.

Wu, P., C.-R. Zhou, X.-Y. Xu, Z.-B. Zhang, T.-Y. Han, and C.-A. Xiong. 2023. "Low-Temperature Solvothermal Method for Coal Tar-Based Carbonaceous Mesophase Preparation and Its Excellent Performance." *Energy and Fuels* 37(16):11683–11693. https://doi.org/10.1021/acs.energyfuels.3c01225.

Wu, S., L. Wang, L. Zhao, P. Zhang, H. El-Shall, B. Moudgil, X. Huang, and L. Zhang. 2018. "Recovery of Rare Earth Elements from Phosphate Rock by Hydrometallurgical Processes—A Critical Review." *Chemical Engineering Journal* 335(March 1):774–800. https://doi.org/10.1016/j.cej.2017.10.143.

Xia, S., Z. Song, X. Zhao, and J. Li. 2023. "Review of the Recent Advances in the Prevention, Treatment, and Resource Recovery of Acid Mine Wastewater Discharged in Coal Mines." *Journal of Water Process Engineering* 52(April 1):103555. https://doi.org/10.1016/j.jwpe.2023.103555.

Xiao, J., L. Zhang, J. Yuan, Z. Yao, L. Tang, Z. Wang, and Z. Zhang. 2018. "Co-Utilization of Spent Pot-Lining and Coal Gangue by Hydrothermal Acid-Leaching Method to Prepare Silicon Carbide Powder." *Journal of Cleaner Production* 204(December 10):848–860. https://doi.org/10.1016/j.jclepro.2018.08.331.

Xiao, N., X. Zhang, C. Liu, Y. Wang, H. Li, and J. Qiu. 2019. "Coal-Based Carbon Anodes for High-Performance Potassium-Ion Batteries." *Carbon* 147(June 1):574–581. https://doi.org/10.1016/j.carbon.2019.03.020.

Xing, B., C. Zhang, Y. Cao, G. Huang, Q. Liu, C. Zhang, Z. Chen, G. Yi, L. Chen, and J. Yu. 2018. "Preparation of Synthetic Graphite from Bituminous Coal as Anode Materials for High Performance Lithium-Ion Batteries." *Fuel Processing Technology* 172(April 1):162–171. https://doi.org/10.1016/j.fuproc.2017.12.018.

Xu, D., P. Ji, L. Wang, X. Zhao, X. Hu, X. Huang, H. Zhao, and F. Liu. 2021. "Effect of Modified Fly Ash on Environmental Safety of Two Soils Contaminated with Cadmium and Lead." *Ecotoxicology and Environmental Safety* 215(June 1):112175. https://doi.org/10.1016/j.ecoenv.2021.112175.

Xu, F., S. Qin, S. Li, J. Wang, D. Qi, Q. Lu, and J. Xing. 2022. "Distribution, Occurrence Mode, and Extraction Potential of Critical Elements in Coal Ashes of the Chongqing Power Plant." *Journal of Cleaner Production* 342(March 15):130910. https://doi.org/10.1016/j.jclepro.2022.130910.

Xu, X., R. Ray, Y. Gu, H.J. Ploehn, L. Gearheart, K. Raker, and W.A. Scrivens. 2004. "Electrophoretic Analysis and Purification of Fluorescent Single-Walled Carbon Nanotube Fragments." *Journal of the American Chemical Society* 126(40): 12736–12737. https://doi.org/10.1021/ja040082h.

Xue, Z., L. Dong, X. Fan, Z. Ren, X. Liu, P. Fan, M. Fan, W. Bao, and J. Wang. 2022. "Physical and Chemical Properties of Coal Gasification Fine Slag and Its Carbon Products by Hydrophobic–Hydrophilic Separation." *ACS Omega* 7(19): 16484–16493. https://doi.org/10.1021/acsomega.2c00484.

Yan, Y., F.Z. Nashath, S. Chen, S. Manickam, S.S. Lim, H. Zhao, E. Lester, T. Wu, and C.H. Pang. 2020. "Synthesis of Graphene: Potential Carbon Precursors and Approaches," *Nanotechnology Reviews*, 9(1):1284–1314. https://doi.org/10.1515/ntrev-2020-0100.

Yang, K.S., B.H. Kim, and S.H. Yoon. 2014. "Pitch Based Carbon Fibers for Automotive Body and Electrodes." *Carbon Letters* 15(3):162–170. https://doi.org/10.5714/CL.2014.15.3.162.

Yang, N., X. Gao, Y. Shen, M. Wang, L. Chang, and Y. Lv. 2022. "Effects of Coal Characteristics on the Structure and Performance of Coal-Based Carbon Foam Prepared by Self-Foaming Technique Under Atmospheric Pressure." *Journal of Analytical and Applied Pyrolysis* 164(June 1):105516. https://doi.org/10.1016/j.jaap.2022.105516.

Yang, Y., K. Chiang, and N. Burke. 2011. "Porous Carbon-Supported Catalysts for Energy and Environmental Applications: A Short Review." *Catalysis Today, Catalysis for Energy and a Clean Environment* 178(1):197–205. https://doi.org/10.1016/j.cattod.2011.08.028.

Ye, R., Z. Peng, A. Metzger, J. Lin, J.A. Mann, K. Huang, C. Xiang, et al. 2015. "Bandgap Engineering of Coal-Derived Graphene Quantum Dots." *ACS Applied Materials and Interfaces* 7(12):7041–7048. https://doi.org/10.1021/acsami.5b01419.

Yoon, R.-H., A. Noble, and S. Suboleski. 2022. "Pilot-Scale Testing of the Hydrophobic-Hydrophilic Separation (HHS) Process to Produce Value-Added Products from Waste Coals." Presented at the Department of Energy National Energy Technology Laboratory Resource Sustainability Project Review Meeting. October 27. Pittsburgh, PA. https://netl.doe.gov/sites/default/files/netl-file/22RS-27_Yoon.pdf.

Yu, H., S. Kharel, C. Lau, and K. Ng. 2023. "Development of High-Strength and Durable Coal Char-Based Building Bricks." *Journal of Building Engineering* 74:106908. https://doi.org/10.1016/j.jobe.2023.106908.

Yu, K., W. McGee, T.Y. Ng, H. Zhu, and V.C. Li. 2021. "3D-Printable Engineered Cementitious Composites (3DP-ECC): Fresh and Hardened Properties." *Cement and Concrete Research* 143:106388. https://doi.org/10.1016/j.cemconres.2021.106388.

Yuan, L., J. Wen, P. Ning, H. Yang, Z. Sun, and H. Cao. 2022. "Inhibition Role of Solvation on the Selective Extraction of Co(II): Toward Eco-Friendly Separation of Ni and Co." *ACS Sustainable Chemistry and Engineering* 10(3):1160–1171. https://doi.org/10.1021/acssuschemeng.1c06307.

Yunusa, I.A.M., P. Loganathan, S.P. Nissanka, V. Manoharan, M.D. Burchett, C.G. Skilbeck, and D. Eamus. 2012. "Application of Coal Fly Ash in Agriculture: A Strategic Perspective." *Critical Reviews in Environmental Science and Technology* 42(6):559–600. https://doi.org/10.1080/10643389.2010.520236.

Zapp, P., J. Marx, A. Schreiber, B. Friedrich, and D. Voβenkaul. 2018. "Comparison of Dysprosium Production from Different Resources by Life Cycle Assessment." *Resources, Conservation and Recycling* 130:248–259. https://doi.org/10.1016/j.resconrec.2017.12.006.

Zhang, J., Y. Qi, J. Yang, K. Shi, J. Li, and X. Zhang. 2022a. "Molecular Structure Effects of Mesophase Pitch and Isotropic Pitch on Morphology and Properties of Carbon Nanofibers by Electrospinning." *Diamond and Related Materials* 126:109079. https://doi.org/10.1016/j.diamond.2022.109079.

Zhang, R., C. Zhang, Y. Cao, et al. 2022b. "The Enhanced Extraction of Rare Earth Elements from Coal Gangue and Coal Fly Ash." PREPRINT (Version 1) available at Research Square. https://doi.org/10.21203/rs.3.rs-1438617/v1.

Zhang, W., and R. Honaker. 2020a. "Characterization and Recovery of Rare Earth Elements and Other Critical Metals (Co, Cr, Li, Mn, Sr, and V) from the Calcination Products of a Coal Refuse Sample." *Fuel* 267:117236. https://doi.org/10.1016/j.fuel.2020.117236.

Zhang, W., and R. Honaker. 2020b. "Process Development for the Recovery of Rare Earth Elements and Critical Metals from an Acid Mine Leachate." *Minerals Engineering* 153:106382. https://doi.org/10.1016/j.mineng.2020.106382.

Zhang, W., A. Noble, X. Yang, and R. Honaker. 2020a. "A Comprehensive Review of Rare Earth Elements Recovery from Coal-Related Materials." *Minerals* 10(5). https://doi.org/10.3390/min10050451.

Zhang, W., A. Noble, X. Yang, and R. Honaker. 2020b. "Lithium Leaching Recovery and Mechanisms from Density Fractions of an Illinois Basin Bituminous Coal." *Fuel* 268:117319. https://doi.org/10.1016/j.fuel.2020.117319.

Zhou, Y., H. Yan, X. Wang, L. Wu, Y. Wang, and T. Xu. 2018. "Electrodialytic Concentrating Lithium Salt from Primary Resource." *Desalination* 425:30–36. https://doi.org/10.1016/j.desal.2017.10.013.

10

CO_2 Utilization Infrastructure

10.1 STATUS AND GOALS OF CO_2 UTILIZATION INFRASTRUCTURE

As noted in the preceding chapters, CO_2 utilization (i.e., conversion) can play a role in developing a circular carbon economy, in storing carbon dioxide, and in enabling a net-zero-emissions future. Expanding CO_2 utilization requires developing or repurposing infrastructure for CO_2 capture, transportation of CO_2 and other feedstocks and inputs, CO_2 conversion, and transportation of products and wastes. Such infrastructure for the full life cycle of CO_2 utilization is of limited extent today. Congress and the Department of Energy (DOE) requested this study to assess infrastructure needs to enable CO_2 utilization, focused on a future where carbon wastes are fundamental participants in a circular carbon economy with net-zero carbon emissions to the atmosphere (U.S. Congress 2020). Over two reports, the committee was instructed to analyze challenges in expanding carbon utilization infrastructure, mitigating environmental impacts, accessing capital, overcoming technical hurdles, and addressing geographic, community, and equity issues. The first report's analysis, summarized below, focused on the current state of CO_2 transportation, use, and storage infrastructure, and identified priority opportunities to develop, improve, and expand that infrastructure to enable utilization (NASEM 2023a). This final report builds off the first report and assesses infrastructure updates needed to ensure safe and reliable CO_2 transportation, use, and storage for carbon utilization purposes. The committee considers how carbon utilization fits into larger needs and opportunities for carbon capture and storage (CCS) infrastructure and describes the economic, climate, and environmental impacts of a well-integrated CO_2 pipeline system as applied to carbon utilization. The committee's analysis includes suggestions for policies that could improve the economic impact of the system and mitigate its climate and environmental impacts.

10.1.1 Summary of First Report's Infrastructure Analysis

The committee's first report (NASEM 2023a) assessed the state of existing infrastructure for CO_2 transportation, use, and storage; outlined considerations for developing new CO_2 utilization infrastructure; and identified priority opportunities for the development of such infrastructure. Currently, CO_2 utilization[1] occurs on a commercial scale for the synthesis of urea, and to a lesser extent salicylic acid, methanol, and organic and inorganic carbon-

[1] CO_2 utilization, for the purposes of this report, includes chemical conversion of CO_2 into products, and excludes uses of CO_2 that do not result in a chemical transformation, such as use of CO_2 as a working fluid for enhanced oil recovery, or in beverage carbonation or fire suppression.

ates. Other CO_2-derived products, such as hydrocarbon fuels, are the target of research or pilot-scale activities, but expanded market opportunities for CO_2 conversion are limited by the high energy requirements and lack of financial incentives and policy mechanisms to use CO_2 as a feedstock in place of fossil carbon. Most existing CO_2 capture and transportation infrastructure has been developed for enhanced oil recovery (EOR), connecting geologic or fossil CO_2 sources with depleted oil reservoirs. Limited opportunities exist to use this infrastructure for sustainable (i.e., net-zero or net-negative emissions) CO_2 utilization processes. However, the anticipated build-out of additional infrastructure for CCS over the coming decades potentially could enable sustainable CO_2 utilization, depending on the CO_2 source, utilization product, and other energy and feedstock requirements (NASEM 2023a). Expanding CO_2 utilization to produce net-zero- or net-negative-emissions materials, chemicals, and fuels would necessitate significant expansion of infrastructure for clean electricity and clean hydrogen in a safe, environmentally benign, and sustainable manner.

In its first report, the committee laid out considerations for developing CO_2 utilization infrastructure to serve a net-zero future. It noted that "the economics of infrastructure placement will be dictated in part by the ease of transporting CO_2, hydrogen, electricity, and other inputs, versus the ease of transporting the carbon-based products" (Finding 4.9, NASEM 2023a, p. 100) and "the optimal CO_2 transport and delivery infrastructure to enable utilization depends on the product type" (Finding 4.10, NASEM 2023a, p. 100). For example, chemicals and fuels production might benefit from centralized CO_2 capture, transportation, and conversion infrastructure, potentially taking advantage of existing chemical production facilities, while concrete and aggregate production preferably may occur in a distributed manner, with smaller-scale CO_2 capture and distribution networks, to serve localized needs for these products. Decisions about CO_2 transportation method(s) for a given project should consider the location and type of CO_2 source as well as the site of utilization and estimated product volumes, with a goal of minimizing cost, environmental, and justice impacts, and addressing safety concerns.

Overarching recommendations from the committee's first report provided opportunities to integrate CO_2 utilization infrastructure with infrastructure for CCS, clean electricity, clean hydrogen, and other enabling inputs. For example, the committee recommended that CCS infrastructure be designed with the flexibility to connect to CO_2 utilization processes and technologies in the future (Recommendation 6.2, NASEM 2023a, p. 134) and that studies to identify the most promising CO_2 utilization opportunities "determine the value of co-locating specific CO_2 utilization activities with specific source types of CO_2, as well as the value of minimizing transport, identifying those that maximize climate benefits" (Recommendation 6.5, NASEM 2023a, p. 134). Additionally, DOE, as part of its industrial decarbonization efforts, should provide technical and financial support for development of carbon capture, utilization, and storage (CCUS) clusters, which "should involve best practices for community engagement and allow for flexibility in utilization scenarios over the long term, for example, by incorporating hydrogen production, chemical and fuel manufacturing, and low-carbon electricity generation" (Recommendation 6.3, NASEM 2023a).

This chapter expands on the committee's prior findings and recommendations by providing an update on CCUS infrastructure under development in the United States, discussing opportunities and challenges for CO_2 utilization infrastructure planning at the regional or national scale, including multimodal transport of the captured CO_2, and evaluating potential economic, climate, environmental, health, safety, justice, and societal impacts of CO_2 utilization infrastructure.

10.1.2 What Are the Infrastructure Needs for CO_2 Utilization to Contribute to a Net-Zero Energy System?

As CO_2 utilization is developed to enable 2050 net-zero goals, it is imperative to chart a course toward the scale of infrastructure required to meet these targets. Several analyses have examined the infrastructure capacity required, including for capture and removal of CO_2, expansion of transportation networks, establishment of secure storage sites, and development of CO_2 conversion facilities. Although momentum has been growing, the disparity between present capacity and the required infrastructure is evident from a comparison of different scenarios (see Table 10-1).

The largest disparity remains in CO_2 capture. The current and announced capacity of point source CO_2 capture facilities is 161 million tonnes per annum (MTPA), significantly lower than—and likely not on track to meet—the amount estimated to be required by 2050, which ranges from a few hundred to a few thousand MTPA in the United States (Table 10-1). Moreover, current and announced capacities for direct air capture (DAC) are less than 8 MTPA, while the

TABLE 10-1 Summary of Published Modeling and Planning for National- or Regional-Scale CCUS Infrastructure

Infrastructure Category		Current U.S. Capacity	Announced U.S. Capacity	Required by 2050		
				Scale	Reference	Scope of Analysis
CO_2 Capture	Point source CO_2 capture	20 MTPA[a]	141 MTPA[a]	230 (by 2030) MTPA	McKinsey Sustainability[b]	United States
				380–610 MTPA	National Resources Defense Council (NRDC)[c]	United States
				700–1800 MTPA	Princeton Net-Zero America[d]	United States
				669 MTPA	Great Plains Institute[e]	Midwest, Gulf Coast, and Rockies
				640–1063 MTPA	Williams et al. (2021)	United States
				300–2400 MTPA	Decarb America[f]	United States
				14 MTPA	Net-Zero Northwest[g,h]	Montana, Idaho, Washington, and Oregon
				304.7 MTPA	Great Plains Institute[i]	Mid-Atlantic
	Direct Air Capture	2000 TPA[j]	7.5 MTPA[k]	40–320 MTPA	NRDC[c]	United States
				90–600 MTPA	Decarb America[f]	United States
				690–2260 MTPA	Rhodium Group[l]	United States
				24.5 MTPA	Net-Zero Northwest[g]	Montana, Idaho, Washington, and Oregon
CO_2 Transport	CO_2 Pipelines	5354 miles[m]	2280 miles[n]	65,000–70,000 miles	Princeton Net-Zero America[d]	United States
				29,0000 miles	Great Plains Institute[e]	Midwest, Gulf Coast and Rockies
				96,000 miles	Pathways to Commercial Liftoff: Carbon Management[o]	United States (stress case)
				6719 miles	Great Plains Institute[i]	Mid-Atlantic
Enabling Inputs for CO_2 Utilization	Carbon-free electricity	434.3 GW[p]	1250 GW[q]	1600–6300 GW	Princeton Net-Zero America[d]	Overall clean electricity generating capacity, not specific to CO_2 utilization
				1160–5000 GW	Decarb America[f,r]	
				9264–15,190 TWh	Williams et al. (2021)	
				564 TWh	Net-Zero Northwest	Montana, Idaho, Washington, and Oregon
	Low-carbon hydrogen	~0.5 MTPA[s]	12 MTPA[t,u]	50 MTPA	U.S. National Clean Hydrogen Strategy and Roadmap[u]	Overall clean hydrogen production, not specific to CO_2 utilization
				58–136 MTPA	Princeton Net-Zero America[d]	
				68–190 MTPA	Decarb America[f]	
				3.8 MTPA	Net-Zero Northwest[g]	Montana, Idaho, Washington, and Oregon

continued

TABLE 10-1 Continued

Infrastructure Category		Current U.S. Capacity	Announced U.S. Capacity	Required by 2050 Scale	Reference	Scope of Analysis
CO_2 Storage	Class VI wells	1.68 MTPA[v]	23.75MTPA[w]	1860 MTPA	Princeton Net-Zero America[d,x]	United States

[a] From Figures 5 and 16 of DOE (2023a).
[b] Clune et al. (2022).
[c] Ennis and Levin (2023).
[d] Larson et al. (2021).
[e] Abramson et al. (2020).
[f] Decarb America Research Initiative (2021); range reflects all modeled net-zero scenarios in 2050.
[g] CETI (2023).
[h] Includes carbon capture from biogasification, cement and lime, and power generation.
[i] Kammer et al. (2023); Mid-Atlantic region includes Delaware, Kentucky, Maryland, New Jersey, Ohio, Pennsylvania, Virginia, West Virginia, and the District of Columbia.
[j] As of November 2023; includes Heirloom (Heirloom 2023) and Global Thermostat facilities (Global Thermostat 2023).
[k] As of November 2023; includes Project Bison, Stratos, Project Cypress, and the South Texas DAC Hub.
[l] Larsen et al. (2019).
[m] PHMSA (2023a).
[n] As of November 2023; includes Summit Carbon Solutions and Wolf Carbon Solutions projects (Summit Carbon Solutions n.d.(b); Wolf Carbon Solutions n.d.).
[o] DOE (2023a).
[p] From Table 4.3 of EIA (2023a).
[q] Total capacity from nuclear, hydro, wind (onshore and offshore, and paired with storage), and solar (including solar paired with storage) in the interconnection queue as of the end of 2022 from LBNL (n.d.).
[r] Includes onshore wind, offshore wind, solar, and nuclear.
[s] Estimated as <5 percent of 2022 hydrogen production, per Figure 2 of DOE (2023b).
[t] As of end of year 2022.
[u] DOE (2023c).
[v] Includes Archer Daniels Midland (EPA 2023a), Blue Flint (Harvestone Group 2023), and Red Trail (Red Trail Energy LLC 2022) sequestration sites.
[w] Includes Wabash Carbon Services (EPA 2023a), Carbon TerraVault JV Storage Company Sub 1, LLC (EPA 2023a), Eastern Wyoming Sequestration Hub (Tallgrass 2023), Sweetwater Carbon Storage Hub (Frontier Carbon Solutions LLC 2022), DCC West Center Broom Creek (DCC West Project LLC 2023), Project Tundra (Minnkota Power Cooperative 2022a), Minnkota Center MRYS Deadwood (Minnkota Power Cooperative 2022b), and Great Plains CO_2 Sequestration Project (Dakota Gasification Company 2023).
[x] Base case storage capacity, per Annex I of Larson et al. (2021).

required capacity by 2050 is estimated to be as much as 300 times these announcements, per Larsen et al. (2019). Carbon capture continues to represent a significant cost component of CO_2 utilization, as well as a large cost reduction potential. Various challenges limit widespread deployment of CO_2 capture infrastructure, as described in NASEM (2023a).

Another significant gap in the CO_2 utilization supply chain is CO_2 transportation infrastructure, which is dominated by pipeline transport. The current capacity is just over 5000 miles, mostly supporting the transport of CO_2 from natural reservoirs, power plants, and industrial sources to nearby oil fields for EOR applications in sparsely populated areas (NASEM 2023a; NPC 2019). The scale and geographic distribution of present transport capacity does not support the creation of a nationwide CO_2 utilization network. Adopting mixed modes of CO_2 transport, such as pipelines, ships, trains, and trucks, can be an efficient solution to address the transportation gap, and this approach is explored in Section 10.3.1.

Enabling infrastructure in the form of clean electricity and clean hydrogen production capacity must scale up significantly to achieve net-zero goals. Although most models do not establish specific electricity and hydrogen targets for CO_2 utilization, these inputs are vital for converting CO_2 into valuable products such as synthetic fuels and chemicals. Furthermore, the United States has significant capacity to store CO_2 underground that is currently underutilized, and the infrastructure built for CO_2 storage can be leveraged to support CO_2 utilization projects.

Substantial amounts of capital investment are necessary to close the gap in CO_2 infrastructure development. DOE has estimated that between $300 billion and $600 billion of total investment along the carbon management value chain is required until 2050 to meet net-zero decarbonization goals (DOE 2023a). The private sector's long-term commercial investments require supportive policy measures and regulatory frameworks that can provide market certainty and encourage investments in CCUS infrastructure. Additionally, public–private partnerships offer a vital approach to leveraging the expertise, resources, and funding of both sectors to scale up CO_2 utilization infrastructure to meet the net-zero targets indicated by different modeling scenarios. Section 10.2 describes numerous CCUS infrastructure announcements since the committee's first report, indicating that plans are gaining momentum. Nonetheless, the current pace of development remains insufficient to achieve most of the referenced analysis models' targets.

10.2 CCUS INFRASTRUCTURE UNDER DEVELOPMENT

As noted above, the committee's first report described the status of existing infrastructure for CCUS in the United States (Chapter 2 of NASEM 2023a). This section briefly discusses new developments in infrastructure for CO_2 capture, transport, utilization, and storage, as well as hydrogen production, since the first report was published.

10.2.1 Direct Air Capture Hubs

In August 2023, DOE selected two (of four total to be chosen) DAC projects for award negotiations as part of the Regional Direct Air Capture Hubs program authorized and appropriated in the Infrastructure Investment and Jobs Act (IIJA) (DOE-OCED 2023a). Both projects—Project Cypress, which will be located in Calcasieu Parish, Louisiana, and the South Texas DAC Hub, which will be located in Kleberg County, Texas—plan to capture 1 million tonnes of CO_2 per year for sequestration in a saline aquifer (DOE-OCED n.d.(a)). DOE is also supporting feasibility and design studies for DAC projects located throughout the United States (DOE-FECM 2023a). Figure 10-1 shows the locations and relative funding amounts of DOE-funded DAC projects.

In addition to the DOE selections, private companies are developing commercial-scale DAC facilities in the United States. Examples include Occidental and 1PointFive's Stratos plant under construction in the Texas Permian Basin (1PointFive 2023b; Oxy 2022), Heirloom's operational DAC-to-concrete facility in California (Heirloom 2023), CarbonCapture, Inc.'s announced Project Bison project in Wyoming (CarbonCapture Inc. 2023), and Global Thermostat's demonstration plant in Colorado (Global Thermostat 2023).

10.2.2 Hydrogen Hubs

In October 2023, DOE announced the selection of seven Regional Clean Hydrogen Hubs for award negotiations, as authorized in the IIJA. Figure 10-2 shows the locations of these projects, and Table 10-2 summarizes their plans for hydrogen generation and use, expected job creation, and estimated emissions reductions. The seven hubs aim collectively to produce more than 3 million metric tons (megatonnes, Mt) of hydrogen annually and reduce CO_2 emissions from end uses by 25 Mt per year (DOE-OCED 2023b; White House 2023). Given the importance of hydrogen for many CO_2 utilization applications, project developers may consider proximity to hydrogen producers in their siting decisions. The committee's first report recommended that project planners consider co-locating hydrogen generation with facilities that capture and use CO_2 to reduce the need for additional costly and complex infrastructure for hydrogen storage and transport (Recommendation 4.5, NASEM 2023a). The Gulf Coast Hydrogen Hub, which is targeting the use of hydrogen as a feedstock in refineries and petrochemicals, and the South Texas DAC Hub could be considered co-located, in the sense that only short trunklines of H_2 or CO_2 would be required to connect facilities for H_2 generation, CO_2 capture, and CO_2 utilization. The Midwest Hydrogen Hub lists sustainable aviation fuels as one of its use cases and is located near proposed sites for CO_2 pipelines and geologic storage, so that location may be a prime opportunity for coordination of H_2 and CO_2 utilization infrastructure.

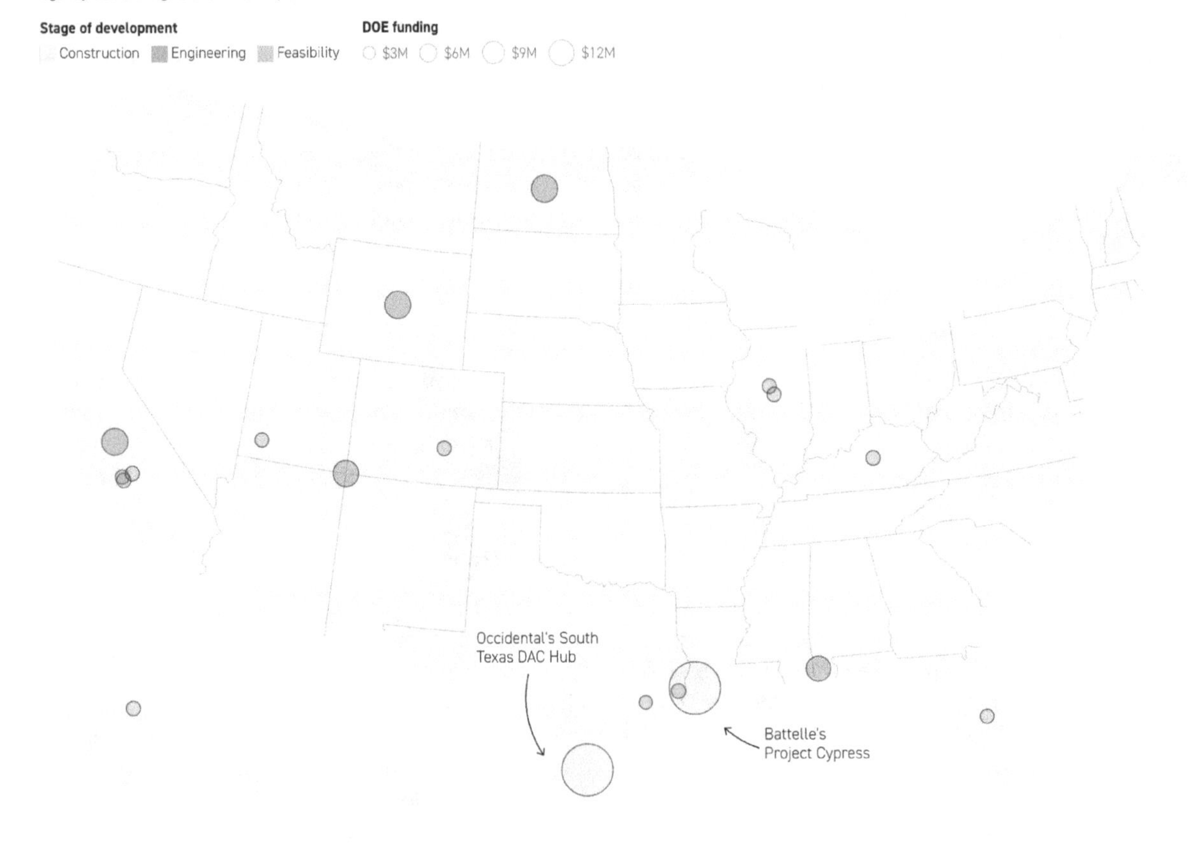

FIGURE 10-1 Locations of DOE-funded direct air capture projects, where yellow circles indicate DAC hubs, orange circles indicate engineering design studies, and blue circles indicate feasibility studies. The size of the circle denotes the relative amount of DOE funding that each project will receive.
SOURCE: Hiar (2023).

FIGURE 10-2 Location of DOE-funded regional clean hydrogen hubs.
SOURCE: DOE-OCED (n.d.(b)).

10.2.3 CO_2 Pipelines

The United States currently has 5354 miles of CO_2 pipeline infrastructure[2] concentrated in the Gulf Coast, Permian Basin, and Wyoming, which primarily transports supercritical CO_2 from geological sources to depleted oil reservoirs for EOR (PHMSA 2023a). At the start of the committee's writing, three major new CO_2 pipeline projects were under way in the Midwest, being developed by Summit Carbon Solutions, Navigator CO_2 Ventures, and Wolf Carbon Solutions. Summit Carbon Solutions plans to build about 2000 miles of pipeline to transport CO_2 captured from 57 ethanol plants to permanent geological storage in North Dakota (Summit Carbon Solutions n.d.(a)). The pipeline network would traverse five states—Iowa, Minnesota, Nebraska, North Dakota, and South Dakota—and transport about 18 million tons of CO_2 annually for storage (Summit Carbon Solutions n.d.(b)). Navigator CO_2 Ventures' Heartland Greenway project, now canceled, would have been a 1300-mile pipeline network across Illinois, Iowa, Minnesota, Nebraska, and South Dakota that would have captured 15 Mt of CO_2 per year from ethanol and fertilizer facilities for permanent geologic storage in Illinois, with the option of off-takes for CO_2 use (Navigator Heartland Greenway LLC 2021; Voegele 2023). The Mt. Simon Hub project being developed by Wolf Carbon Solutions would transport 12 million tons of CO_2 per year from two ethanol plants in Iowa to permanent geological storage in Illinois via 280 miles of pipeline (Wolf Carbon Solutions n.d.).

All three projects have faced legal and regulatory challenges and have seen pushback from local communities concerned about safety and use of eminent domain for pipeline siting (see, e.g., Ahmed 2023; Peterson 2023; Ramos

[2] For scale and context, the United States has about 190,000 miles of petroleum pipelines (API 2021) and 3 million miles of natural gas pipelines (EIA 2024).

TABLE 10-2 Summary of Hydrogen Hub Selections

Hub Name	State(s) Involved	Source of Hydrogen	Use of Hydrogen	Estimated Job Creation	Estimated Emissions Reductions
Appalachian Hydrogen Hub	Ohio, Pennsylvania, West Virginia	Natural gas with carbon capture	Not reported	21,000 (18,000 construction; 3000 permanent)	9 Mt/yr
California Hydrogen Hub	California	Renewable electricity and biomass	Public transportation, heavy-duty trucking, port operations	220,000 (130,000 construction; 90,000 permanent)	2 Mt/yr
Gulf Coast Hydrogen Hub	Texas	Natural gas with carbon capture and renewable electricity	Fuel cell electric trucks, industrial processes, ammonia production, marine fuel, refineries and petrochemicals	45,000 (35,000 construction; 10,000 permanent)	7 Mt/yr
Heartland Hydrogen Hub	Minnesota, North Dakota, South Dakota	Not reported	Fertilizer production, power generation, cold climate space heating	3880 (3067 construction; 703 permanent)	1 Mt/yr
Mid-Atlantic Hydrogen Hub	Delaware, New Jersey, Pennsylvania	Renewable and nuclear electricity	Heavy-duty transportation, manufacturing and industrial processes, combined heat and power	20,800 (14,400 construction; 6400 permanent)	1 Mt/yr
Midwest Hydrogen Hub	Illinois, Indiana, Michigan	Renewable and nuclear electricity, natural gas with carbon capture	Steel and glass production, power generation, heavy-duty transportation, refining, sustainable aviation fuel	13,600 (12,100 construction; 1500 permanent)	3.9 Mt/yr
Pacific Northwest Hydrogen Hub	Montana, Oregon, Washington	Renewable electricity	Heavy-duty transportation, industry, ports, aviation, fertilizer production	10,000 (8050 construction; 350 permanent)	1 Mt/yr

SOURCES: Based on data from DOE-OCED (n.d.(b)) and White House (2023).

2023; Soraghan 2023). For example, regulators in North and South Dakota denied permits to Summit Carbon Solutions and Navigator CO_2 Ventures, respectively, although North Dakota plans to reconsider the permit request (Dura 2023a, 2023b; Dura and Karnowski 2023). In October 2023, Navigator announced its cancellation of the Heartland Greenway project (Douglas 2023), Summit announced a delay in its pipeline start-up date from 2024 to early 2026 (Anchondo 2023), and the Illinois Commerce Commission staff recommended that state regulators reject Wolf Carbon Solutions' pipeline application (Tomich 2023).

At a public meeting on CO_2 pipeline safety hosted by the Pipeline and Hazardous Materials Safety Administration (PHMSA) in Des Moines, Iowa, from May 31 to June 1, 2023, public comments centered on a lack of trust in pipeline companies, the limited capacity of volunteer emergency response teams in rural areas, appropriate setback distances, the potential for induced seismicity from CO_2 injection for sequestration, and the durability of pipeline materials in Iowa's geographic and environmental conditions (e.g., freeze/thaw cycles, soil composition, ground vibrations from farm equipment) (PHMSA 2023b). During the same meeting, members of the public made several requests of PHMSA as it develops updated safety regulations for CO_2 pipelines, including (1) place a national moratorium on CO_2 pipelines until the updated safety regulations are in place, (2) provide guidance on regulatory jurisdictions and clear definitions of CO_2 (supercritical, gaseous, and liquid phases) in regulations, (3) establish clear guidelines on disclosure of emergency response plans and mandate that pipeline companies pay for the necessary emergency response equipment, and (4) in the case of a pipeline emergency, require that companies notify all customers and emergency planning departments within a reasonable distance of the route (PHMSA 2023b).

In addition to opposition by individuals and communities directly affected by pipeline projects, some groups also oppose CO_2 pipelines as a means to block the development of CCUS projects, which they consider a moral hazard owing to their potential to perpetuate fossil fuel development and use. Opposition is not unique to CO_2 pipelines. Other net-zero energy system infrastructure projects have been opposed owing to the concerns of host or neighboring communities (e.g., solar and wind developments and transmission lines), or owing to the moral hazard of fossil fuel development, or both (e.g., H_2 hubs) (Christol et al. 2021; Gordon et al. 2023; Romero-Lankao 2023). See Chapter 4 for more detail on the policy, regulatory, and societal aspects of CO_2 utilization infrastructure development.

10.2.4 Update on Commercial CO_2 Capture, Utilization, and Storage Projects

Since the committee's first report was released in December 2022, additional CCUS projects have been and continue to be announced. As this is a rapidly evolving space, producing a comprehensive list of projects and facilities is impractical. Rather, the committee points the reader to resources that continually update information on carbon management projects (see Table 10-3).

In addition to individual projects, several CCS hub developments have been announced. Examples include the Midland Basin hub (Milestone Carbon 2023), Bayou Bend CCS hub (OGCI n.d.), and Bluebonnet Hub (1PointFive 2023a) in Texas; the Cameron Parish CO_2 Hub offshore of Louisiana (Carbonvert and Castex 2023);

TABLE 10-3 Resources with Information on Commercial Carbon Capture, Utilization, and Storage Projects

Resource Name	Coordinating Organization	Project Type(s) Included	Description	Link
Carbon Matchmaker	U.S. DOE Office of Fossil Energy and Carbon Management	Source Capture Utilization Storage Removal Transport	Map of self-reported and DOE-supported carbon capture, utilization, and storage activities. Includes brief description and status of each project.	https://www.energy.gov/fecm/carbon-matchmaker
CCU Activity Hub	Global CO_2 Initiative, University of Michigan	Capture and Utilization	Map of carbon capture and utilization start-up companies worldwide. Ability to layer with locations of publications and research centers focused on CCU, as well as states, provinces, or countries with carbon tax or emissions trading system.	https://www.globalco2initiative.org/evaluation/carbon-capture-activity-hub
U.S. Carbon Capture Activity and Project Map	Clean Air Task Force	Capture	Map of carbon capture projects that are operational or in development, differentiated by capture capacity and subsector. Also shows locations of CO_2 storage potential.	https://www.catf.us/ccsmapus
Innovator Index	Circular Carbon Network	Capture Utilization Removal	List and brief description of companies involved in CO_2 capture, utilization, and removal. Also includes companies that provide circular carbon market infrastructure.	https://circularcarbon.org/innovator-index
CCUS Companies	Carbon Utilization Alliance	Utilization	List and brief description of carbon utilization companies.	https://www.cua.earth/ccus-companies
CO_2RE Facilities Database	Global CCS Institute	Capture Storage	Map and brief description of CCS facilities worldwide. Ability to filter by region, country, category (commercial, pilot/demonstration), and status (e.g., early development, advanced development, operational).	https://co2re.co/FacilityData
CCUS Projects Explorer	International Energy Agency	Capture Utilization Transport Storage	Database of CO_2 capture, utilization, transport, and storage projects worldwide. Includes both planned and operational projects, with information about location, project partners, anticipated capacity, sector, and fate of CO_2.	https://www.iea.org/data-and-statistics/data-tools/ccus-projects-explorer

the Central Louisiana Regional Carbon Storage Hub ("CENLA Hub," CapturePoint 2024), and the Eastern Wyoming Sequestration Hub (Tallgrass 2022). These could be prime locations for future CO_2 utilization infrastructure development, in line with the committee's recommendations in its first report to support development of industrial clusters for CCUS and identify opportunities to co-locate utilization with existing CO_2 transport infrastructure (Recommendations 6.3 and 6.5, NASEM 2023a).

10.3 INTEGRATED CO_2 UTILIZATION INFRASTRUCTURE PLANNING AND DEVELOPMENT AT THE REGIONAL OR NATIONAL SCALE

10.3.1 Optimal Multimodal, Regional CO_2 Transportation Infrastructure

As part of the CCUS chain, pressurized pipelines are generally considered to be the most economical and safest method for large-scale CO_2 transport. However, as indicated in Section 10.2.3, CO_2 pipeline development faces regulatory challenges and public opposition. The recent cancellation of a 1300-mile CO_2 pipeline project aiming to transport 15 Mt of CO_2 annually from Midwest ethanol plants for geological storage (Tomich et al. 2023) owing to opposition of residents along its route highlights such barriers. As a result, ships, river barges, trucks, and trains are becoming increasingly attractive, as they have shorter timelines for implementation and likely face fewer regulatory and public acceptance barriers, especially when passing through or near more densely populated areas. Such forms of transportation also will likely be needed to collect and distribute CO_2 at the origin and destination of pipelines, or for smaller, distributed sources or conversion facilities, even if a larger system of CO_2 pipelines is built (Pett-Ridge et al. 2023).

The combination of different CO_2 transport modes can, in many cases, represent a better alternative to a single transport mode, especially for small, dispersed emitters that are not within easy reach of industrial CCUS clusters enjoying shared transport infrastructure. From a societal perspective, the optimal design of multimodal CO_2 transport infrastructure for utilization and storage involves a cost-benefit analysis of the transport network with the aim of minimizing costs and environmental impact and reducing the risk of failure, while maximizing the CO_2 utilization potential in the region.

Meeting these complex challenges requires the development of dedicated mathematical optimization models that can determine key impact indices for each of the above factors, which in turn can be employed as a valuable decision-making tool for the design of the optimal CO_2 transport network. Parolin et al. (2022) propose a similar analytical tool for hydrogen delivery infrastructure, but the proposed methodology does not include factors related to safety and financial risks, nor environmental impacts during transportation, and only considers land-based transport of hydrogen. Lawrence Livermore National Laboratory's "Roads to Removal" report presents a model aimed at identifying the most economical route of multimodal CO_2 transportation from the CO_2 source to a storage location using established cost models and literature data (Pett-Ridge et al. 2023). However, it assumes that fluid transportation conditions (e.g., temperature, pressure) are the same across the modular transport chain and thus adopts only one levelized cost associated with liquefaction. In practice, CO_2 may be transported in liquid form at different temperatures and pressures depending on transport mode, resulting in different conditioning costs and further conditioning stations when transferring across different transport modes (e.g., from trucks to barges or trains). Such requirements may significantly impact the optimal multimodal transport solution. Furthermore, although Pett-Ridge et al. (2023) discuss the need to evaluate the selected routes under different criteria—societal impact, in particular—they do not include calculations of risks and emissions associated with the different options. A more rigorous, all-encompassing, multiobjective approach is required to determine the optimal multimodal transport solution for transporting captured CO_2 gathered from small, dispersed emitters for utilization. The following text describes the main steps in developing such a tool.

Step 1. Mixed-Integer Linear Programming (MILP): Minimizing the Cost of Transport

First, a set of transport configurations for determining the minimum costs is selected via the construction of a MILP problem incorporating several considerations and constraints (Lee et al. 2017). These may include

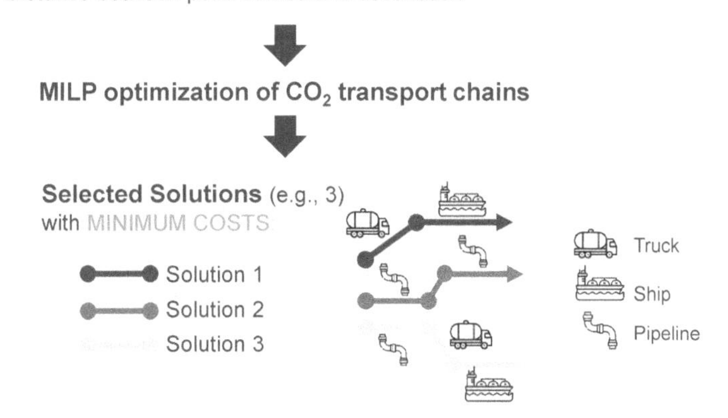

FIGURE 10-3 Diagram showing an example of expected results of the MILP for the multimodal transport of CO_2.
SOURCE: Icons from the Noun Project, https://thenounproject.com. CC BY 3.0.

the geographical locations of the emitters and utilization sites, the availability or accessibility of infrastructure for different modes of transport, along with the respective amounts of CO_2 and corresponding temperatures and pressures. Established techno-economic models are employed to determine the CO_2 transport and conditioning unit costs for the various transport modes, such as pipelines (Knoope et al. 2014), ships (Element Energy 2018; Roussanaly et al. 2021), and trucks (Stolaroff et al. 2021). The solutions proposed by MILP are expected to fall in a range of costs that are considered reasonable and might highlight different routes and a combination of different modes. A graphical representation is presented in Figure 10-3.

Step 2. Multimodal Transport of CO_2: Incorporating Safety and Environmental Impact Costs

Step 2 involves developing tools to quantify the safety and environmental impact costs associated with different modes of transportation. These tools can be in the form of a Safety and Reliability Index and an Environmental Index, as previously developed in several studies for the transportation of different goods (e.g., H_2) (Bevrani et al. 2020; d'Amore et al. 2018; Kim et al. 2011; Lee et al. 2017). The safety index considers factors such as failure rates for the various transportation modes, and failure consequence analysis takes account of the population density along the selected route. The environmental index considers emissions generated from the construction and implementation of the different transport modes, which also depends on distance covered and selected routes.

A multicriteria optimization model then has to be developed, which simultaneously analyzes the risk and environmental impact for each of the transport solutions identified by the MILP, as illustrated in Figure 10-4, and returns a set of possible solutions, indicating the necessary trade-off between the different selected criteria, covering cost, safety, and environmental impacts.

The following subsections describe a few case studies of CO_2 transportation infrastructure development at different locations and CO_2 emission scales, taking into account the optimization methodologies for multimodal transport described above. For each case study, implications and opportunities for CO_2 utilization are discussed.

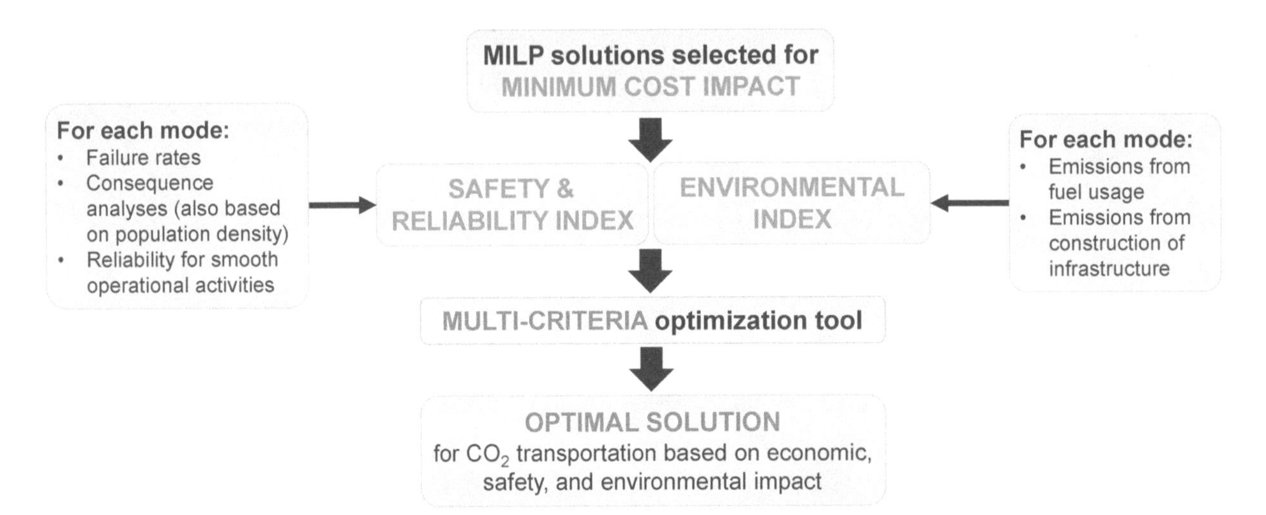

FIGURE 10-4 Illustration of the process and parameters involved in applying the multicriteria optimization tool.

10.3.1.1 Distributed Small- to Medium-Scale CO_2 Emitters

Small- to medium-scale CO_2 emitters, or those that emit less than 1 Mt CO_2 equivalents per year, span power plants, chemicals, minerals, breweries, paper and pulp, food, commercial, and public administration sectors. Such emitters are widely distributed across the United States, many in remote locations, and collectively contribute a significant proportion of overall U.S. CO_2 emissions. As shown in Figure 10-5, 92 percent of U.S. industrial and power plant facilities reporting emissions to the U.S. Environmental Protection Agency (EPA) in 2022[3] emitted less than 1 Mt CO_2e, and these 6955 facilities accounted for 33 percent of reported emissions (EPA 2023b). The decarbonization of small- to medium-scale emitters is imperative in successfully meeting the 2050 net-zero-emission target and could involve CO_2 capture and utilization in addition to other decarbonization strategies like electrification and improvements in energy efficiency. Considering only the industrial sector, small- to medium-scale industrial emitters, defined as emitting between 12,500 and 60,000 tCO_2 per year, comprise 25 percent of all U.S. industrial point-source emissions (Moniz et al. 2023). Moniz et al. (2023) identified ten regional targets for clusters of small-to-midsize emitters that could share resources and risks, and develop economies of scale and effort. Three of that report's identified clusters are in the same regions (Midwest and Gulf Coast) discussed as possible industrial clusters in Section 10.3.1.2.

For some small- to medium-scale emitters, deployment of renewable electricity and onsite utilization of captured CO_2 using modular technologies may be more cost-effective than process modification. New capture technologies based on membranes (Etxeberria-Benavides et al. 2018), enhanced adsorption processes (Crake et al. 2017; Wang et al. 2018), and molten carbonate fuel cells (FuelCell Energy 2023) are promising at small to medium scale because they are modular, able to be retrofitted on existing infrastructure, easy to scale up, and cost-competitive. They are also relatively simpler than conventional capture technologies, such as solvent-based post-combustion capture, which is not expected to be feasible at medium scale (50 $ktCO_2$ per year) owing to its large physical footprint (Sharma et al. 2019). The Advanced Research Projects Agency–Energy's (ARPA-E's) GREENWELLS program is exploring the feasibility of producing chemicals and fuels from CO_2 using intermittent renewable electricity and hydrogen (ARPA-E 2023), which, if successful, could be a valuable opportunity for onsite CO_2 utilization at small- to medium-scale emitters.

In cases where onsite utilization of captured CO_2 may not be a feasible option, deployment of optimal multimodal CO_2 transport solutions and "right-size" infrastructure is an attractive addition to the mitigation portfolio

[3] EPA's Greenhouse Gas Reporting Program requires facilities that emit greater than 25,000 metric tons CO_2e per year to report their emissions annually.

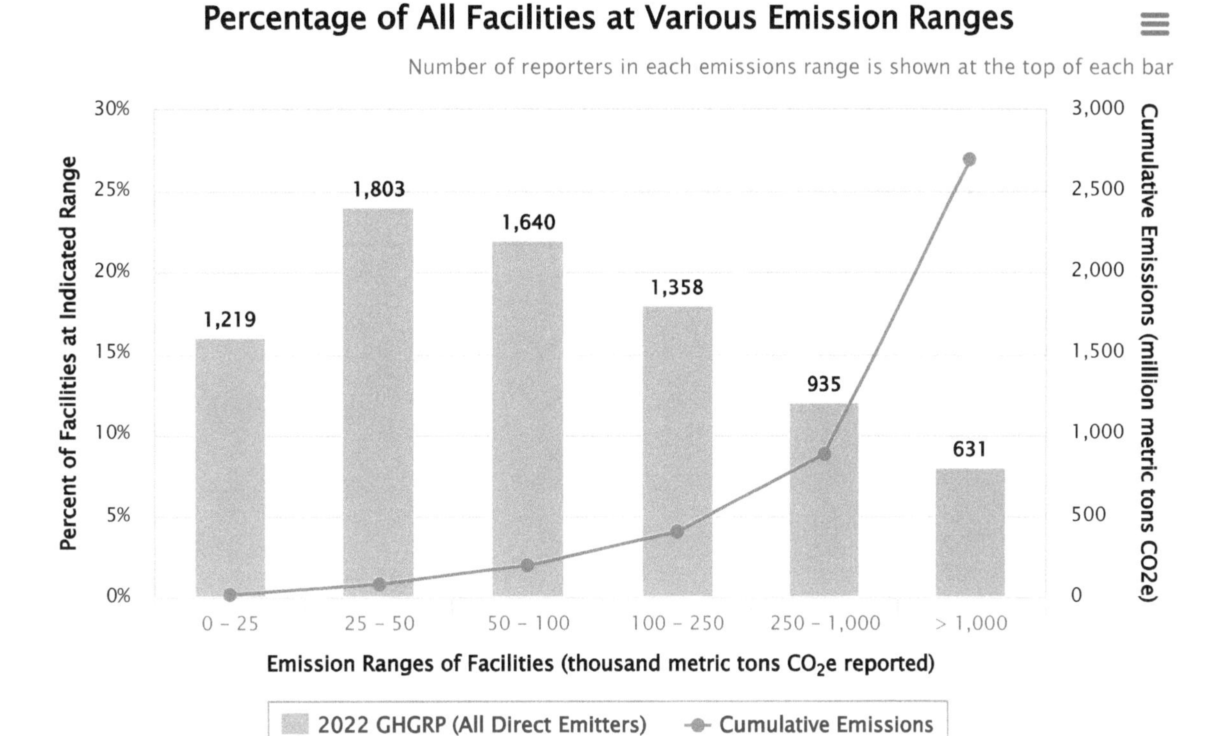

FIGURE 10-5 Emissions from U.S. industrial and power plant facilities in 2022, reported via EPA's Greenhouse Gas Reporting Program (GHGRP). Bars show percent of facilities in different ranges of emissions, indicating that the majority of facilities emit less than 1,000,000 metric tons of CO_2e per year. Blue line shows cumulative emissions, demonstrating that about two-thirds of emissions come from the largest facilities (>1 million metric tons CO_2e per year) and about one-third comes from small- to medium-scale emitters.
SOURCE: EPA (2023b).

for small-to-medium emitters, allowing CO_2 emissions to be substantially reduced and holistically integrated with electrification, hydrogen, and biomass technologies. This is particularly so as planned CCUS industrial clusters comprising large CO_2 emitters (e.g., cement and steel production) often cannot embrace distant small-to-medium-scale emitters, given the additional CO_2 transport infrastructure costs which may become unrealistic (Moniz et al. 2023). Moreover, when building pipelines for small sources is too costly, smaller companies are unable to take advantage of the 45Q tax credit, placing them at a financial disadvantage relative to larger companies. In these instances, transport of CO_2 by truck, rail, barge, or ship may offer a solution and be particularly important for early CCUS adopters.

10.3.1.2 Industrial Clusters for Large Volumes of CO_2

Given the multiple utility and feedstock needs for CO_2 capture and utilization, and the expense, challenge, and public concern over new pipelines, sites where all needed components are co-located are distinctly advantageous for CO_2 utilization deployment. Two promising opportunities for co-locating CO_2 capture and utilization are around bioethanol facilities in the Midwest and heavy industry and refining in the Gulf Coast. Midwest bioethanol facilities are small- to medium-scale emitters located in close enough proximity that shared infrastructure can aggregate CO_2 to obtain volumes suitable for conversion to products at scale. On the other hand, many industrial facilities in the Gulf Coast region are large emitters, each generating substantial volumes of CO_2 for utilization, but likewise

FIGURE 10-6 Bioethanol CO_2 capture projects in development and operational (blue circles, with bold outlines indicating those that are operational). Gray shaded areas indicate potential CO_2 storage capacity.
SOURCE: CATF (2023).

could benefit from shared infrastructure to reduce costs. Such infrastructure could also serve the numerous small- and medium-size emitters also located in the Gulf region, functioning as an "anchor tenant" around which larger CCUS networks could develop in the longer term.

Biogenic CO_2 from ethanol plants is advantaged for use in synthetic fuels and chemicals, as it is a sustainable CO_2 source with relatively low capture cost, around \$0–\$55 per tonne of CO_2 (Bennett et al. 2023; GAO 2022; Hughes et al. 2022; Moniz et al. 2023; NPC 2019). Most of the planned and operational CO_2 capture projects at bioethanol facilities are located in the Midwest, primarily in Iowa, Minnesota, Nebraska, and South Dakota (Figure 10-6). As discussed in Section 10.2.3, CO_2 pipeline projects are under development to transport this captured CO_2 to geologic sequestration sites, although they have experienced setbacks and delays. Some of this captured CO_2 instead could be diverted for utilization; however, given the small scale of individual bioethanol plants, captured CO_2 will likely need to be collected in a single location to enable conversion to chemicals and fuels at economies of scale. This could be done using local pipeline networks, along with other modes of transport—especially when passing through or close to populated areas—per the methodologies described above. CO_2 conversion to fuels will require hydrogen, which could be produced at the CO_2 collection site using electrolysis powered by clean electricity.

The U.S. Gulf Coast region is home to nearly 50 percent of U.S. refining and petrochemicals manufacturing (EIA 2023b), making it a prime opportunity for deployment of point-source carbon capture. These refining and petrochemicals facilities are co-located with an existing array of CO_2 and hydrogen pipelines; storage sites in well-characterized depleted hydrocarbon reservoirs, including offshore storage in federal waters where the U.S. government can have long-term ownership; and salt domes for low-cost hydrogen storage for use with renewable power, and are in close proximity to ports to allow export to markets (e.g., European Union, Japan) where low-carbon products are given market incentives (Bayer and Aklin 2020; Datta et al. 2020; LSU Center for Energy Studies 2023). Figure 10-7 shows planned and operational CO_2 capture projects in the Gulf Coast region from a variety of point sources, which could be aggregated and utilized or transported for geological storage.

FIGURE 10-7 CO_2 capture projects in development and operational (colored circles, with bold outlines indicating those that are operational) in the U.S. Gulf Coast Region. Gray shaded areas indicate potential CO_2 storage capacity.
SOURCE: CATF (2023).

Because most of the captured CO_2 from this region is of fossil origin, utilization products would have to be long-lived (e.g., mineral carbonates, solid carbon products) to ensure durable carbon sequestration. This could present a challenge, as mineral products are typically low value, and end-use of hydrocarbon products such as plastics typically cannot guarantee sequestration. On the other hand, carbon fibers, graphite, and other elemental carbon forms offer higher value (see Chapters 2 and 6). Use of fossil CO_2 to make short-lived products from large point sources could be considered in the near term, wherein fossil CO_2-based production replaces similar production from oil or natural gas in demand-limited market scenarios (see Section 10.4.5). Full market life cycle assessment (LCA) would be needed to ensure net fossil CO_2 mitigation. In any case, given that there will be a mix of CO_2 sources aggregated for utilization and storage, a rigorous accounting method will be required to determine the carbon intensity of utilization products. Such accounting methods for common carrier CO_2 infrastructure are discussed in Chapter 4.

10.3.1.3 Shared CO_2 Transport Pipeline Networks for CO_2 Storage and Utilization in CCUS Industrial Clusters

As the committee recommended in its first report, DOE should consider favorably the ability of CO_2 capture, transport, and storage demonstration projects to connect to future CO_2 utilization opportunities because allowing for shared use of CO_2 pipelines for both utilization and storage could take advantage of economies of scale (Recommendation 6.2, NASEM 2023a). In general, wider availability of CO_2 through improved transportation and storage infrastructure could open the CO_2 marketplace to traditional market demand dynamics and enable CO_2 pull from the market where its conversion is most affordable (e.g., near low-cost clean electricity and/or hydrogen) and where the resulting product can be used. For CO_2 transport pipeline networks in CCUS industrial clusters, depending on the CO_2 purity and market demand, some of the CO_2 stream destined for geological storage could be diverted for utilization. In addition, as the CO_2 emission rates in the cluster substantially decrease because of a transition to clean energy or electrification, it may be more economical to divert the entire CO_2 stream for utilization, taking account of the costs associated with any upstream CO_2 purification that may be required. Such plans already exist in Europe (see, e.g., C^4U Project 2020) that could set the scene for the United States.

For example, as part of a CCUS industrial cluster commencing operation in 2025, the planned Fluxys CO_2 pipeline network in Belgium is expected to handle eventually more than 50 percent of the 40 MTPA total CO_2 emissions captured from several major industries along its route (Fluxys Belgium 2022). Several of these emitters, such as the steel and cement industries, plan to divert some of the CO_2 for utilization to produce fuels such as e-methanol (see, e.g., the North CCU Hub; CO_2 Value Europe n.d.), with future potential opportunities for producing methane, e-kerosene, and polymers. The Belgian CO_2 CCUS value chain encompasses CO_2 capture and purification; multimodal transport involving pipeline, ship, train, and trucks; CO_2 liquefaction; CO_2 utilization; and CO_2 storage in disused gas fields under the North Sea seabed (Fluxys Belgium 2022).

As another example of shared CO_2 transport and utilization pipelines, OCAP (Organic CO_2 for Assimilation in Plants) currently supplies about 500,000 tons of CO_2 per year to enhance crop growth[4] for approximately 600 greenhouse companies in the western part of the Netherlands via a 97-kilometer transport pipeline and distribution network of 250 kilometers (OCAP n.d.). This CO_2 is produced during the production of hydrogen at Shell in the Botlek area and during the production of bioethanol at Alco in Europoort Rotterdam. In situations where CO_2 supply exceeds demand, the surplus CO_2 may in future be diverted for nearby geological storage sites by joining existing CO_2 pipeline infrastructure.

To this end, the Porthos (Port of Rotterdam CO_2 Transport Hub and Offshore Storage) project intends to provide transport and storage infrastructure to energy-intensive industries in the Port of Rotterdam and, possibly, to industries in the Antwerp and North Rhine Westphalia areas at a later stage (Porthos 2023). The project will link CO_2 capture facilities and the existing OCAP pipeline with a new onshore pipeline, which will transport the aggregated CO_2 in a CO_2 hub in the Port of Rotterdam and subsequently via an offshore pipeline to a depleted gas field 20 kilometers off the coast for permanent storage. The final investment decision for Porthos was made in October 2023, construction of the Porthos infrastructure will start in 2024, and the system is expected to be operational starting in 2026.

10.3.2 Retrofitting Existing Infrastructure for CO_2 Capture, Transport, and Utilization

10.3.2.1 Addition of Carbon Capture to Existing Industrial Facilities to Enable CO_2 Utilization

DOE's Industrial Decarbonization Roadmap (DOE 2022b) details research needs and challenges for decarbonizing the industrial sector, addressing CO_2 footprints for scope-1 and -2 emissions, which includes the process energy, heat, and utilities required for manufacturing products, but not emissions associated with use of products (e.g., as fuel) or production of feedstock. As shown in Figure 10-8, the U.S. industries with the largest energy-related CO_2 emissions are chemical manufacturing and petroleum refining. One opportunity for mitigating emissions from these industries is incorporating CO_2 capture paired with sequestration or utilization to form a long-lived product. Existing facilities in principle can be retrofitted for CO_2 capture, but achieving significant decarbonization may require capture from several places within the process, which is one reason the industrial sector can be considered "difficult to decarbonize."

The cost of retrofitting an industrial facility with CO_2 capture is an important consideration for the feasibility of subsequent CO_2 utilization or storage. The National Energy Technology Laboratory and National Petroleum Council have performed rigorous cost estimates for CO_2 capture from industrial processes (Hughes et al. 2022; NPC 2019). A few processes, such as ethanol fermentation or petrochemical production of ethylene oxide, produce a relatively pure CO_2 stream, and capturing that CO_2 to produce a purified product has fairly low cost. For example, as mentioned earlier, costs to capture CO_2 offgas from ethanol fermentation range from \$0–\$55 per tonne[5] (Bennett et al. 2023; GAO 2022; Hughes et al. 2022; Moniz et al. 2023; NPC 2019). The CO_2 stream from hydrogen production via steam reforming of natural gas, which accounts for approximately half of the CO_2 footprint of the process, has a capture cost on the order of \$60–\$115 per tonne CO_2 (DOE 2023a; NPC 2019). Capture of CO_2 from process furnaces used to provide heat and power comes at an even higher cost. Commercial CO_2 capture projects often only pursue capture from the most economical, high-concentration, high-pressure streams. For example, many current CO_2 capture demonstration projects for hydrogen production by steam methane

---

[4] CO_2 utilization for enhanced crop growth is out of scope for this report but is included here as a case example of shared utilization and storage infrastructure.

[5] Range includes first-of-a-kind and *n*th-of-a-kind facilities.

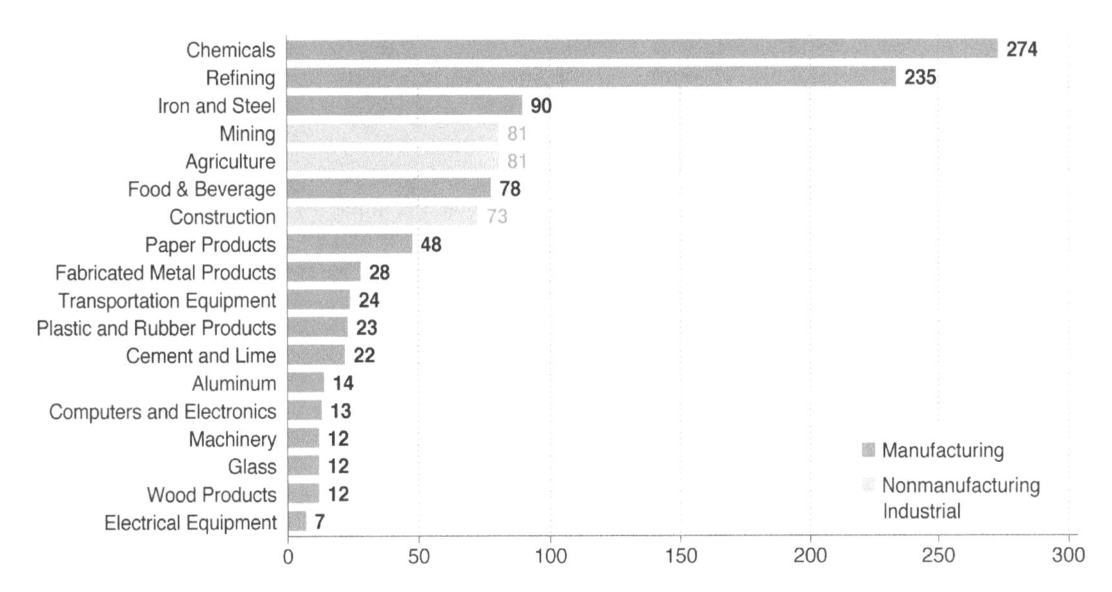

FIGURE 10-8 Energy-related CO_2 emissions from industrial subsectors in 2020 (Mt CO_2).
SOURCE: DOE (2022a).

reforming install a single capture unit at a point in the process that provides the lowest unit costs, and therefore only capture 40–60 percent of the overall CO_2 emissions. Many industrial processes entail CO_2 emissions from both process and utility streams, thus requiring capture from multiple point sources, which drives capture costs to as high as $200 per tonne CO_2 (NPC 2019).

In some cases, industry has constructed facilities that are "CO_2 capture ready" to facilitate tie-ins and space for CO_2 capture systems. Where CO_2 capture and mitigation is targeted, new technologies can be more efficient; for example, production of hydrogen from natural gas by autothermal reforming or partial oxidation can result in virtually complete decarbonization via capture from a single stream in the process unit. These technologies often require a complete rebuild of the production unit (Liu 2021; NPC 2019) or a substantial upgrade including addition of a partial oxidation reactor (Mahabir et al. 2022).

Until passage of the Inflation Reduction Act (IRA), the 45Q tax credits—at $50/tonne for geologic storage and $35/tonne for use in EOR by 2026 (Beck 2020)—were generally too low for industry to invest in CO_2 capture, except where the CO_2 had a coproduct value for utilization in EOR. (EOR can be economic at higher oil prices found prior to the discovery of unconventional shale oil production methods.) Recent increases in 45Q tax credits to $85 per tonne for storage and $60 per ton for utilization allow technologies with mid-range CO_2 capture costs, such as those involved in hydrogen production, to be considered, but they are still too low for the $100–$200 per tonne capture costs of many of the processes in the "difficult to decarbonize" industrial sector (NPC 2019). For capturing these industrial CO_2 streams for utilization or storage to be economical, there would need to be a further increase in the tax credits and/or additional research and development (R&D) on CO_2 capture technologies, such solid sorbents, to reduce costs. See Chapter 4 for more detail on policy options to support CO_2 utilization and Chapter 11 for more information about R&D needs for CO_2 capture.

10.3.2.2 Conversion of Existing Natural Gas and Oil Pipelines for Transporting CO_2, Hydrogen, or CO_2-Derived Products

The committee's first report examined the feasibility of retrofitting natural gas pipelines for transporting CO_2, finding that this would have to be determined using rigorous systems analysis on a case-by-case basis given the large number of parameters involved (Finding 4.8, NASEM 2023a). Retrofit of natural gas pipelines for CO_2

service can be considered over shorter distances (trunklines, see, e.g., Tallgrass 2024), but pressure capabilities of existing pipelines will not be sufficient for large-scale, long-distance transport (Kenton and Silton n.d.; NPC 2019). CO_2 is normally transported as a supercritical fluid exhibiting the high density of liquid CO_2 but the low viscosity of a gas. Natural gas is generally transported in pipelines in gaseous form at pressures between 800 and 1160 psi. The critical point of CO_2 is at 30.9°C and 1070 psi, such that the pressure for CO_2 transportation must be at least 1200 psi to avoid phase changes from temperature fluctuation, which is much higher than the standard operating parameters for existing natural gas pipelines (Kenton and Silton n.d.). All major CO_2 pipelines today transport at pressures above 1900 psi. For retrofit, a dehydration system would be required to minimize water content, because wet CO_2 forms carbonic acid, which offers a high risk of corrosion. High-pressure CO_2 pipelines require crack arrestors to prevent catastrophic failure in the event of corrosion or external forces such as subsidence or collision damage. Modifications to the gaskets and nonferrous materials of the original pipeline may be required to prevent deterioration in the presence of concentrated CO_2 (Kenton and Silton n.d.).

Pipe-in-pipe technologies may be considered for laying new CO_2-compatible pipe within existing pipelines (Enbridge 2022). In principle, land used for pipeline rights of way can be used to lay new CO_2-compatible pipe; however, a formal right of way for transport of a given gas (e.g., natural gas) does not translate into a right of way for transport of a new gas (e.g., CO_2). CO_2 pipelines and rights of way have to be approved for the new CO_2 service on a case-by-case basis, including scenario modeling for release and risk of asphyxiation from release of a vapor that is heavier than air. Compatibility of pipeline metals and wetted components also has to be approved on a case-by-case basis, especially for retrofitted systems. CO_2 transportation challenges, including repurposing of existing pipelines to service CO_2, are described in more detail in a workshop report from DOE (DOE-FECM 2023b).

Retrofitting existing natural gas networks for hydrogen transport is also of interest, as hydrogen is an enabling input for many CO_2 utilization processes. However, there are challenges associated with doing so. For example, the existing natural gas network may not be able to handle the high pressures required for hydrogen transport. Additionally, hydrogen tends to embrittle metals, which may require upgrades to existing pipelines, such as adding a copper or polymer coating by retrofit pigging operations or installing pipe-in-pipe technologies. Coating technologies present concerns about long-term robustness and safety, while pipe-in-pipe technologies allow the preferred metallurgy to be installed but reduce capacity. For both options, the cost may be greater than new pipe installation. However, studies have shown that, in some cases, converting existing natural gas pipelines into dedicated hydrogen pipelines could reduce hydrogen transmission costs by 20–60 percent compared to constructing new hydrogen pipelines because of savings across the entire value chain of materials, the permitting and time expense, land use acquisition costs, construction costs, and costs of additional infrastructure (e.g., compression, power) (Cerniauskus et al. 2020). Thus, decisions will have to be made on a site-specific basis. Despite these challenges, there has been some progress in using existing gas networks for hydrogen transport. In the United Kingdom, for example, the H21 project is exploring the feasibility of converting the gas network in northern England to run entirely on hydrogen (Northern Gas Networks 2016). In the United States, unlike in Europe, there often is enough space to lay down more pipe in existing rights of way, perhaps reducing the need to retrofit existing infrastructure.

As infrastructure is built out to transport CO_2 for storage and utilization and to use hydrogen for decarbonization (including via its reaction with CO_2 to make hydrocarbon products), there may be competition between CO_2 and hydrogen for natural gas pipeline repurposing. Optimization of the existing gas pipeline network for future use by CO_2 and/or H_2 will require integrated coordination among usage options. Owing to this competition and other factors, there will likely be a need for new pipelines to transport CO_2 or H_2 beyond what can be accommodated by existing natural gas pipelines.

In addition to natural gas pipelines, over 190,000 miles of liquid petroleum pipelines traverse the United States (API 2021), and with the transition to renewable energy and electrification, many of these pipelines may become obsolete as demand for fossil fuels decreases. Given their extensive geographical spread, it is very likely that some will pass near CO_2 utilization facilities. Depending on their locations and taking account of any additional safety concerns that might arise in the event of an accidental release, there may be opportunities to use some of these pipelines to transport CO_2-derived fuels or chemicals. As with any pipeline retrofit, questions of safety, environmental impacts, technical feasibility, and economics would have to be addressed adequately before moving forward with the project.

10.3.2.3 Converting Fossil Facilities and Chemical Plants for CO_2 Utilization

The conversion of existing fossil facilities and chemical plants to accommodate CO_2 utilization depends on the specific project economics and thus has to be considered on a case-by-case basis. For example, converting a conventional methanol synthesis plant using syngas to one using CO_2 and H_2 as feedstocks would require a CO_2 purification unit and redesign of the reactor and methanol distillation column to separate excess water. Project developers would have to determine if such a conversion is cost-effective compared to construction of a new facility. Converting or rebuilding on existing facility sites could allow for reuse of the connected power, feedstock, and product offtake infrastructure, as well as retain the existing workforce associated with the facility.

10.3.3 Enabling Infrastructure Needs for Water, Hydrogen, and Electricity

CO_2 is a fully oxidized form of carbon, thermodynamically degraded and devoid of energy except when used for some mineralization reactions to form carbonates. Therefore, activating CO_2 for conversion requires inputs of energy in the form of electricity, hydrogen, and/or heat. As described in previous chapters, there are multiple pathways for converting CO_2 into useful products and chemical intermediates. For example, CO_2 can be electrochemically or thermochemically reduced to CO, which can be further reacted with H_2 to form hydrocarbon products, effectively reproducing the current hydrocarbon economy. Given finite conversion efficiency, using renewable wind or solar energy to power hydrogen production requires at least twice the amount of energy as current commercial processes for making hydrogen, while further conversion into hydrocarbon products requires two-fold more energy (Adolf et al. 2018, 2020). Land use can also be a significant issue for CO_2 utilization in systems where renewable electricity is used to provide both the energy and hydrogen required to upgrade CO_2 to valuable products (Gabrielli et al. 2023; Merrill 2021). At a national scale, water requirements for CO_2 utilization do not represent a significant increase over current usage, but local impacts need to be evaluated. The committee's first report detailed these enabling infrastructure requirements to supply clean electricity, clean hydrogen, water, land, and energy storage for CO_2 utilization projects (Chapter 4 of NASEM 2023a), and the committee refers readers to that discussion for more information. This section covers additional aspects of CO_2 utilization enabling infrastructure not discussed in depth in the first report (e.g., transportation of hydrogen) and highlights regional considerations for electricity, hydrogen, and water infrastructure when developing CO_2 utilization facilities.

10.3.3.1 Hydrogen Pipelines

Clean hydrogen[6] is a required feedstock for many approaches to convert captured CO_2 into synthetic fuels and chemicals. For dispersed CO_2 emitters, onsite production of clean hydrogen for CO_2 utilization may not always be a viable option. In such circumstances, to take advantage of economies of scale, hydrogen may be produced at a central facility for distribution to the various emitters. Given the large volumes involved, transportation of gas-phase hydrogen using high-pressure pipelines in combination with other modes of transport may be the most viable option. Developing infrastructure to supply clean hydrogen at large scale (e.g., for use in vehicles and power generation) would require an expansive hydrogen pipeline network (Parfomak 2021). Clean hydrogen for fuel cells or combustion will compete with battery electrification of vehicles and renewable power generation technologies, so it is not yet clear how much hydrogen fuel will be needed, and hence if any extensive hydrogen fuel infrastructure will be built. If hydrogen pipelines are built, some of them may also supply hydrogen to dispersed emitters for CO_2 utilization depending on their proximity.

To facilitate hydrogen infrastructure development, the Consolidated Appropriations Act, 2021 provided funding to DOE's Hydrogen and Fuel Cell Technologies Office to support R&D for topics including hydrogen pipeline research. The IIJA appropriated $9.5 billion for clean hydrogen (DOE 2022a), including to develop the Regional Clean Hydrogen Hubs discussed in Section 10.2.2. The IRA provided additional beneficial policies and incentives for the U.S. hydrogen industry to take center stage in the clean energy transition (Webster 2022).

[6] DOE's Clean Hydrogen Production Standard considers low-carbon (i.e., clean) hydrogen that which has "well-to-gate lifecycle greenhouse gas emissions of ≤ 4.0 $kgCO_2e/kgH_2$" (DOE 2023c, p. 2).

Experience with high-pressure transportation of hydrogen is relatively limited. It is currently done on a much smaller scale than other methods of transport, with only 1600 miles of pipelines in operation in the United States, mainly located in the Gulf Coast region (DOE-HFTO n.d.). More than 80 percent of these pipelines are in areas of low population density, defined as a class location unit 1 under current federal pipeline safety regulations (Kuprewicz 2022) (see Figure 10-9).

The anticipated increase in demand for hydrogen—for CO_2 utilization and other fuel and feedstock applications—could require the development of a national high-pressure hydrogen transport pipeline network. NPC (2024) provides an in-depth analysis of the technology, policy, and partnerships needed to build out hydrogen infrastructure that is safe, integrated, flexible, scalable, and resilient. In a large-scale system, some hydrogen pipelines would need to pass through or nearby populated areas, so their safe operation is of paramount importance. Hydrogen has a unique hazard profile, substantially different than those for CO_2 or hydrocarbons, which requires important and stringent modifications to minimum federal and state pipeline safety regulations (DOE 2023d; Kuprewicz 2022). Relevant risk factors to consider when drafting hydrogen pipeline regulations include the following (Kuprewicz 2022):

- Hydrogen has a much greater flammability range than natural gas and hence is more likely to combust.
- Because hydrogen is the smallest chemical element, it readily diffuses through most materials, and thus hydrogen pipelines are more susceptible to leaks than CO_2 or natural gas pipelines.
- On a weight-for-weight basis, hydrogen has more than double the energy intensity of natural gas.
- Hydrogen has a much lower autoignition temperature and faster burn velocity than natural gas, meaning that its accidental release is much more likely to lead to detonation and explosion as compared to natural gas.
- Over time, hydrogen can cause metal embrittlement, increasing the probability of pipeline failure.
- Hydrogen is an indirect greenhouse gas, with potentially 33 times the warming power of CO_2 in the first 20 years.

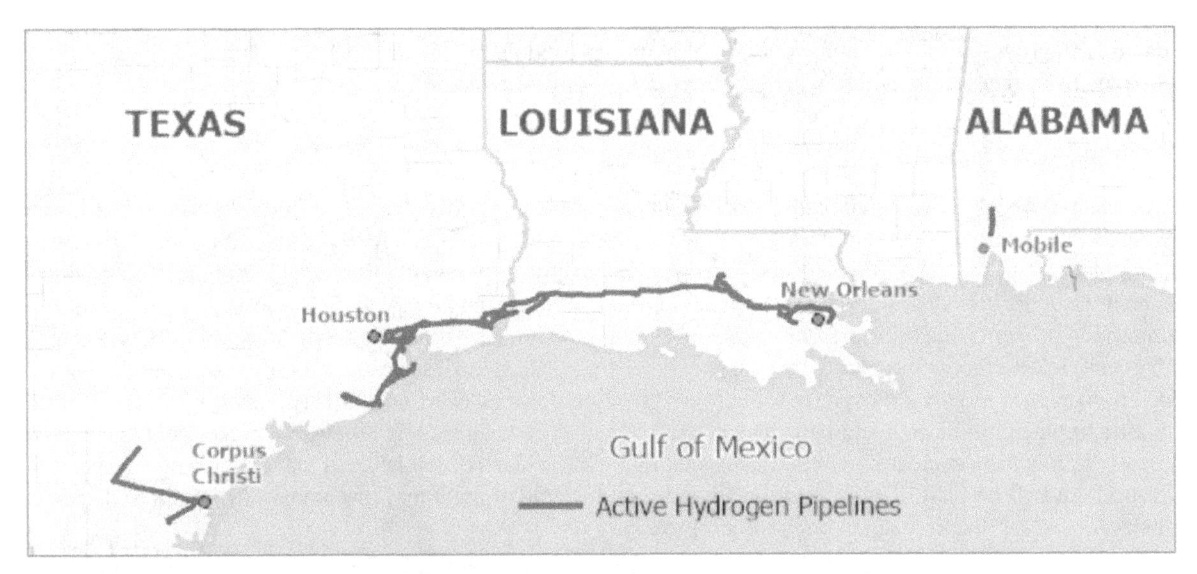

Source: CRS using data from Texas Railroad Commission, Public GIS Viewer, 2020, https://www.rrc.state.tx.us/about-us/resource-center/research/gis-viewers/; PHMSA, National Pipeline Mapping System, Public Map Viewer, 2020, https://pvnpms.phmsa.dot.gov/PublicViewer/; PHMSA, Gas Transmission and Gathering Annual, online database, 2020, https://www.phmsa.dot.gov/data-and-statistics/pipeline/gas-distribution-gas-gathering-gas-transmission-hazardous-liquids; and Esri Data and Map, 2019.

FIGURE 10-9 Hydrogen pipeline infrastructure in the United States as of 2020.
SOURCE: Parfomak (2021).

- Odorants are routinely added to natural gas to detect accidental leaks. Special odorants will need to be developed for hydrogen that do not lead to mixtures that adversely interact with pipeline materials, have minimal health and environmental impacts, and do not require costly separation depending on the end use (Murugan et al. 2019).

10.3.3.2 Regional Considerations for Clean Electricity, Hydrogen, and Water Infrastructure for CO_2 Utilization

Enabling infrastructure requirements for CO_2 utilization—in particular the needs for clean electricity, clean hydrogen, and water—are likely to impact siting decisions for CO_2 utilization facilities. For example, clean electricity is needed to power CO_2 capture and conversion, electrolytic hydrogen generation, and other processes to ensure that CO_2-derived products have lower emissions than incumbent products on a lifecycle basis. Grid emissions intensity varies regionally, and such variations are projected to continue through 2030 and 2050, as illustrated in Figure 10-10. Thus, in the absence of dedicated, onsite clean energy generation for a CO_2 utilization project, developers may preferentially site facilities in regions with lower average grid emissions. Alternatively, developers could contract for emissions-free electricity through a power purchase agreement or work with a utility to set up a tariff structure to obtain clean electricity for their project. In either case, robust LCA would be required to determine eligibility for renewable energy tax credits.

If a CO_2 utilization project developer decides to deploy dedicated renewable resources (e.g., onsite wind or solar) to obtain clean electricity, the varying resource potential across the country (see NREL n.d.) would need to be considered in site selection. Furthermore, as discussed in the committee's first report, many regions with abundant renewable resources are water stressed (Finding 4.14, NASEM 2023a), which could limit deployment of CO_2 utilization projects that require water (e.g., some carbon capture technologies, algae cultivation, and hydrogen production).

The locations of planned hydrogen hubs (see Figure 10-2 and Section 10.2.2) could also impact CO_2 utilization infrastructure siting, given that many CO_2 utilization processes require clean hydrogen. To that end, the committee's first report recommended that DOE consider co-locating hydrogen and DAC hubs (Recommendation 6.4, NASEM 2023a), which it could still consider for the two DAC hubs yet to be selected. As shown in Figure 10-11, a 2023 Great Plains Institute analysis identified promising locations for DAC technology deployment based on proximity to low-carbon electricity and heat, geological carbon storage, and existing CO_2 transport, as well as appropriate climate and atmospheric conditions for DAC operation (Abramson et al. 2023). An analysis by Cai et al. (2024) examined the effects of meteorological conditions (e.g., temperature, humidity, atmospheric pressure, and local

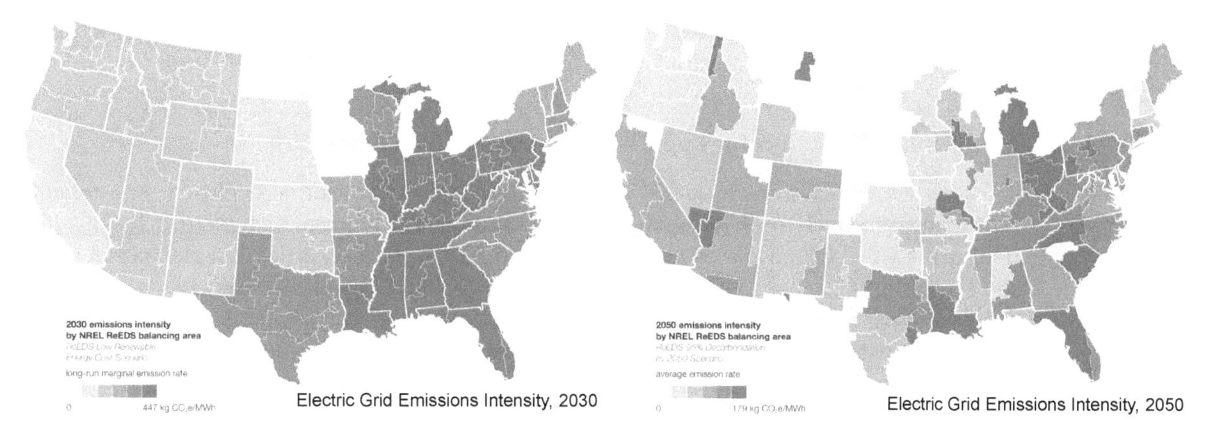

FIGURE 10-10 Projected emissions intensity of the U.S. electricity grid in 2030 (left) and 2050 (right), where darker blue indicates a higher average emissions rate. In 2030, the darkest blue shading represents an emissions intensity of 447 kg CO_2e/MWh, while in 2050, the darkest blue shading represents an emissions intensity of 179 kg CO_2e/MWh.
SOURCE: Adapted from Abramson et al. (2023).

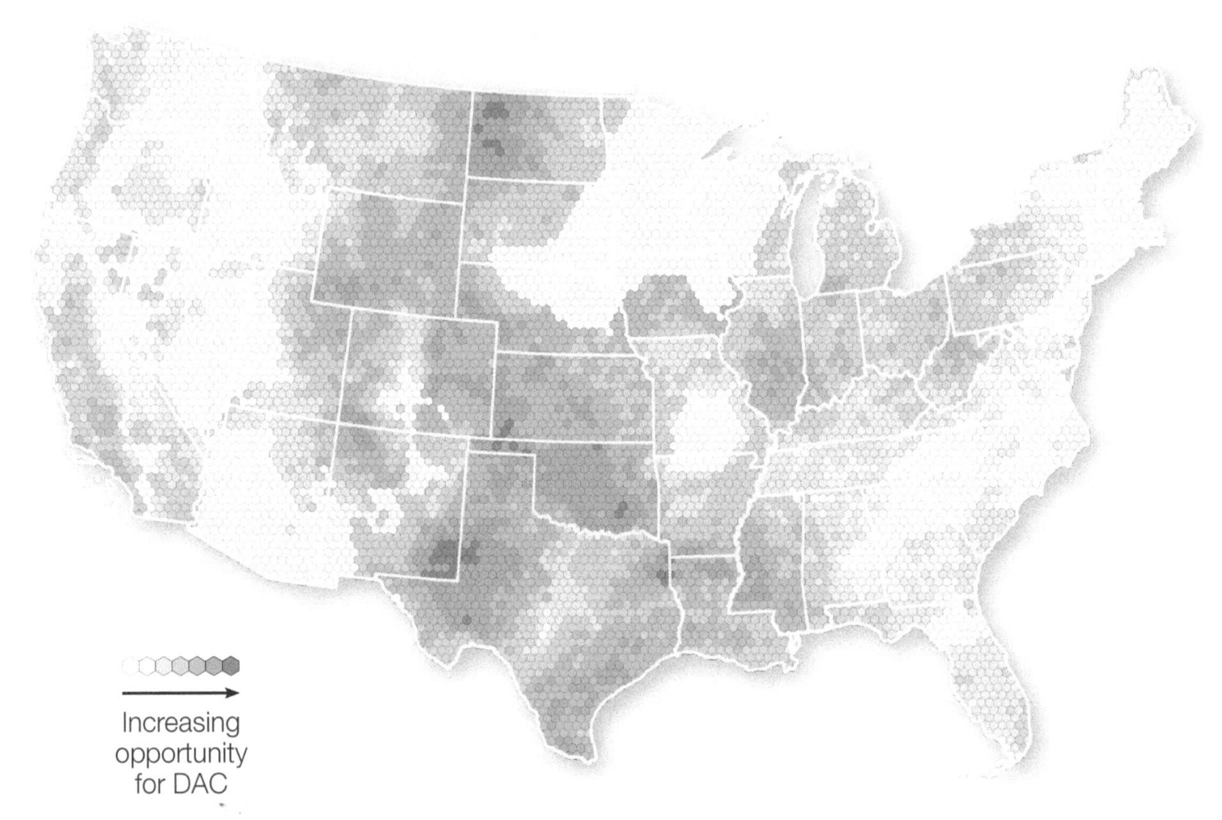

FIGURE 10-11 Promising locations for DAC technology deployment based on six categories: proximity to geological carbon storage and transport infrastructure, regional resources for low-carbon electricity, regional resources for low-carbon heat, electric power and industrial facilities with potential for waste heat supply, natural gas availability for heat with carbon capture, and optimal climate and atmospheric conditions for DAC. Darker coloring indicates greater opportunity for DAC deployment. SOURCE: Abramson et al. (2023).

CO_2 concentration) on the performance of amine-based DAC systems, concluding that process optimization at a specific location can significantly improve system performance and that consideration of local atmospheric conditions may impact siting decisions for DAC facilities. Overlaying these results with the hydrogen hub locations could identify promising sites to develop infrastructure for production of chemicals and fuels from CO_2.

10.3.4 Timescale for Implementation and Potential Barriers

A number of factors will influence whether CO_2 utilization infrastructure can be deployed on a timeline that allows it to contribute effectively to midcentury decarbonization goals. Primary barriers to CO_2 utilization development are the cost of CO_2 capture and, in cases where onsite CO_2 utilization is not feasible, transportation infrastructure costs. The cost of CO_2 capture is inversely proportional to the CO_2 concentration in the gas stream, ranging from less than \$20 per tonne CO_2 for some high-purity streams (e.g., ethanol fermentation off-gas and natural gas processing) to upward of \$1000 per tonne CO_2 for low-concentration streams (e.g., DAC) (Budinis and Lo Re 2023; DOE 2023a). Transportation of captured CO_2 for utilization and/or storage also adds cost, estimated at \$5–\$25 per tonne CO_2 for pipeline transport, \$14–\$25 per tonne CO_2 for ship transport, and \$35–\$60 per tonne CO_2 for rail and truck transport (DOE 2023a). Purification requirements for different transport modes (see Table H-3 in Appendix H) could further increase costs.

The long lead times for developing and deploying CO_2 capture, transport, and storage infrastructure, as depicted in Figure 10-12, also could slow large-scale CO_2 utilization rollout. For example, a lack of operational

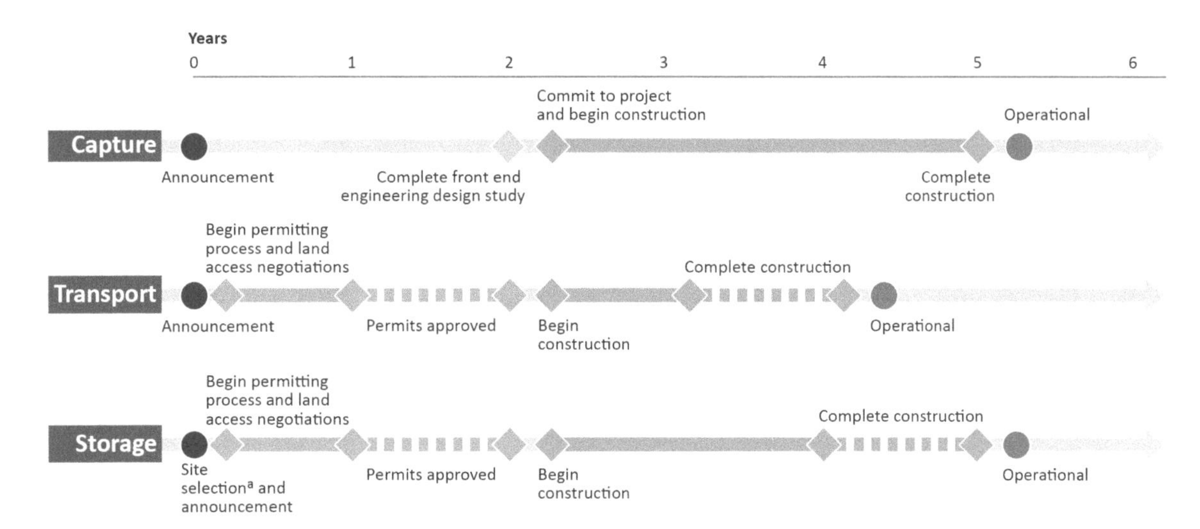

FIGURE 10-12 Approximate timelines for development and deployment of CO_2 capture, transport, and storage infrastructure.
SOURCE: GAO (2022).

storage capacity may deter investments in pipeline infrastructure, which in turn could delay CO_2 utilization projects that plan to obtain CO_2 from a larger pipeline network. As another example, CO_2 utilization projects aiming to produce hydrocarbon fuels from low-cost, high-purity, biogenic CO_2 from ethanol fermentation facilities may not be feasible without a pipeline network in place that can aggregate CO_2 sources to achieve economies of scale for fuel production. In addition to having long lead times, pipeline development also faces challenges with public acceptance, as discussed in Section 10.2.3. Regional or national CO_2 infrastructure planning is also challenged by uncertain policy and business environments that will impact future availability and cost of CO_2 and other inputs to CO_2 utilization like clean electricity and hydrogen. For instance, legal challenges to EPA's power plant emissions rule raise questions about the future availability, quantity, and location of CO_2 sourced from power plants, which affects business cases and decisionmaking on how, when, and where to build CCUS infrastructure. Uncertainty in the durability of state and national policies, regulations, and incentives for CO_2 pipelines, CO_2 use and storage, clean electricity, and clean hydrogen also impedes long-term investment decisions for infrastructure.

10.4 INFRASTRUCTURE IMPACTS

The committee's statement of task asks it to "[d]escribe the economic, climate, and environmental impacts of any well-integrated national carbon dioxide pipeline system as applied for carbon utilization purposes." Recognizing that pipelines are only one component of the infrastructure needed to support CO_2 utilization—and, in fact, may not be required for all CO_2 utilization opportunities—this section also considers impacts of carbon capture, enabling inputs like hydrogen and water, and industrial facility siting and development.

10.4.1 Economic and Cost Impacts of Infrastructure Development for CO_2 Utilization

Potential infrastructure deployment scenarios for utilization, such as those described in Section 10.3.1 above, are speculative. However, existing analyses of U.S. sources of CO_2 and geological storage locations (e.g., Abramson et al. 2022; Larson et al. 2021) can inform assumptions about where these projects and their needed infrastructure might be built first. As described in previous chapters, CO_2 utilization technologies have an inherent challenge of scale, making co-location with sites that are amenable to geological storage (see USGS 2013) an attractive option. Additionally, CO_2 pipeline infrastructure is expensive (capital costs of recently announced projects in Table 10-4 range from $1.8 million to $2.7 million per mile), which adds incentive to develop utilization projects in close proximity to existing projects that will capture, transport, and store large quantities of CO_2. As noted above, locating

CO_2 utilization projects near enabling infrastructure for clean electricity and hydrogen production also could be beneficial. Multiobjective optimization models can assist project developers in determining the best options for infrastructure co-location.

In the coming months and years, continued implementation of the IIJA and IRA will provide a better sense of where and how this infrastructure will be deployed. Federal incentives will likely shape future infrastructure development, as many early movers may use federal funding programs, loans, or tax incentives to begin their projects. By one estimate, incentives in the IRA will spur $90 billion–$126 billion of investments in CO_2 transport and storage infrastructure between 2023 and 2035 (Jenkins et al. 2023). Because IRA incentives improve the economic viability of carbon capture for steel, cement, refineries, and natural gas- and coal-fired power generation, around 200 million tons of CO_2 could be captured from industry and power generation per year by the mid-2030s (Jenkins et al. 2023). In a net-zero future, the CO_2 captured from these facilities will have to either be stored or used in long-lived products, as its use in short-lived products would result in net-positive CO_2 emissions to the atmosphere. In the near term, before net-zero is reached, the lower cost of CO_2 capture from fossil point sources versus DAC may enable more fossil CO_2 removal per unit of investment, given that substitution of synthetic fuels or chemicals derived from CO_2 reduce demand for petroleum or natural gas production. This possibility can be evaluated with a levelized cost of CO_2 abatement combined with a full LCA that accounts for the initial energy services that result in fossil CO_2 emissions, followed by capture and reuse.

When making net-zero infrastructure investment decisions, the levelized cost of CO_2 abatement is a valuable metric. It measures the cost of a specific policy, technology, or investment per amount of CO_2 reduced or removed, where the cost (including both capital and operating expenses) is annualized over the lifetime of the project. This accounting for cost of capital, amortization, and net present value across the project lifetime distinguishes the levelized cost of CO_2 abatement from marginal abatement calculations and enables more relevant comparisons across options (Friedmann et al. 2020). Additionally, unlike many marginal abatement cost calculations, levelized abatement costs take into account the emissions source(s) being displaced upon implementation of the new policy, technology, or investment, which adds more local or regional specificity to the calculation (Friedmann et al. 2020). Any comparisons between levelized costs of CO_2 abatement for different policy, technology, or investment options must use the same financing metrics and assumptions. Emissions reductions are estimated as the difference between the baseline condition and the "new" condition (i.e., the result of the policy, technology, or investment). As described in Friedmann et al. (2020), a simplified formula for calculating levelized cost (L) of CO_2 abatement is as follows:

$$L = \frac{C}{(E_0 - E_1)}$$

where C is the levelized cost of the investment or change in policy or technology, E_0 is the emissions associated with the baseline condition, and E_1 is the emissions upon making the investment or change in policy or technology. A smaller positive value for levelized cost of CO_2 abatement indicates a more cost-effective action, although other factors beyond lowest cost (e.g., jobs, national security, equity implications) may also be important in making policy and technology decisions (Friedmann et al. 2020).

Public–private partnerships can also play a role in the development of large infrastructure such as a CCUS hub, as they facilitate sharing of resources. These partnerships can take different forms based on which entity owns and operates the various project components (e.g., capture infrastructure, utilization infrastructure, transport and storage infrastructure, possession of stored carbon, and title of stored carbon) (EFI Foundation and Horizon Climate Group 2023). As noted by the EFI Foundation and Horizon Climate Group, public–private partnerships for CCUS hubs could include "binding community benefits agreements for financial compensation, public authority or utility models for CO_2 transport and storage management, or government entities assuming long-term liability and postinjection site care responsibilities" (EFI Foundation and Horizon Climate Group 2023, p. 68).

The potential economic contributions from a regional CO_2 network for both storage and utilization can be substantial. CO_2 infrastructure development could contribute to the U.S. economy via an expansion of the local workforce (jobs), purchase of goods and services from local businesses (direct spend), and payment of federal, state, and local taxes (taxes). These benefits can be compounded through indirect and induced impacts. Indirect impacts refer to the secondary effects of infrastructure deployment, such as the changes in output, employment, or labor earnings of industries that support CO_2 utilization. Induced impacts emerge from the interaction between the CO_2 utilization sector and other sectors. While development of a regional CO_2 pipeline network in the United States has been challenging (see Section 10.2.3), the experience of those project developers provides insight into the estimated economic benefits. Table 10-4 summarizes publicly available information from specific projects.

Despite these potential benefits, development of CO_2 utilization infrastructure also could have some negative impacts for local communities, project developers, and the existing CO_2 market. For instance, CO_2 infrastructure development could result in land use changes that affect local economies owing to changes in property values and/or tax base. Project developers wanting to obtain CO_2 from pipelines destined for storage could see high initial costs as pipeline capacity ramps up, which could deter investment or result in higher costs being passed along to the first customers. A nascent but growing CO_2 utilization industry could also raise the cost of CO_2 for direct use markets, such as food and beverage, where potential negative social impacts include an increase in cost of those products.

10.4.2 Climate and Environmental Impacts of Infrastructure for CO_2 Utilization

Regional- and national-scale infrastructure for CO_2 utilization will have benefits and impacts concurrent to those outlined in the committee's first report (NASEM 2023a), which focused on enabling infrastructure. This section emphasizes the combined impacts of both CO_2 utilization and enabling infrastructure, and outlines gaps.

Decisions about the design and siting of CO_2 utilization facilities impact air emissions. LCA results for CO_2 utilization demonstrate a large range of climate and air pollution impacts, depending on the product and the way in which the LCA was conducted (see Chapter 3 for more detail). For CO_2 capture infrastructure, air pollution impacts may not be displaced unless operations are powered by renewable energy and the systems are designed

TABLE 10-4 Estimated Economic Benefits from Proposed CO_2 Pipeline Projects

				Costs			
Project	Location	Pipeline Distance	Estimated CO_2 Captured	Total Capital	Annual Operating	Estimated Job Creation	Estimated Tax Impact
Summit Carbon Solutions[a]	Iowa, Minnesota, Nebraska, North Dakota, South Dakota	1991 miles	16 MTPA	\$4.8 billion	\$213 million	Construction phase (2021–2027): 6921/yr Operations phase: 989	Construction phase: \$493.6 million/phase Operations phase: \$94.7 million/yr
Economic Impacts of CO_2 Pipelines in South Dakota[b]	South Dakota	361 miles (Navigator) 474 miles (Summit)	Not reported	\$1.5 billion	\$792 million	Construction phase (2024–2025): 5353/yr Operations phase: 436	Not reported
Wolf Carbon Solutions[c]	Illinois, Iowa	280 miles	12 MTPA	Not reported	Not reported	Construction phase (first year): 2780 Operations phase: 342	Not reported
Navigator CO_2 Solutions[d,e]	Illinois, Iowa, Minnesota, Nebraska, South Dakota	1300 miles	15 MTPA	\$3.5 billion	Not reported	Construction phase: 8000 Operations phase: 80	\$43 million/yr

[a] Summit Carbon Solutions (2023, 2024).
[b] Dakota Institute (2023).
[c] Wolf Carbon Solutions U.S. LLC (2023).
[d] Eller (2023); Navigator Heartland Greenway LLC (2021).
[e] Despite project cancellation, included for reference of potential economic impacts of CO_2 pipeline projects.

to reduce or eliminate co-pollutants (Jacobson 2019). Systems designed to reduce both CO_2 and co-pollutant emissions could remove at least 75 percent of NO_x emissions, 98 percent of SO_2 emissions, and all condensable particulate matter (PM), alongside 90 percent capture of CO_2 emissions (Bennett et al. 2023). Figure 10-13 shows potential regional emissions reductions for CO_2, NO_x, SO_2, and $PM_{2.5}$ that could be achieved by deploying carbon capture on cement plants, providing a representative example of carbon capture co-benefits.

The transport of CO_2 through pipelines often requires compression for gas phase and pumping for dense phase, often involving intermediate boosting in the case of long distances (Doctor et al. 2005), which in turn requires energy. For natural gas pipeline transport, such demand is typically fueled using combined cycle gas turbines, which result in CO_2 emissions and other pollutants (e.g., volatile organic carbons [VOCs]); however, in the case of CO_2 pipelines, electricity may serve as a less carbon-intensive source of power, depending on the localized generation mix (Doctor et al. 2005). While metering has been recommended to account for any lost and unaccounted for gas, leak detection systems for CO_2 pipelines can reduce potential risks (Han et al. 2019; Santos 2012).

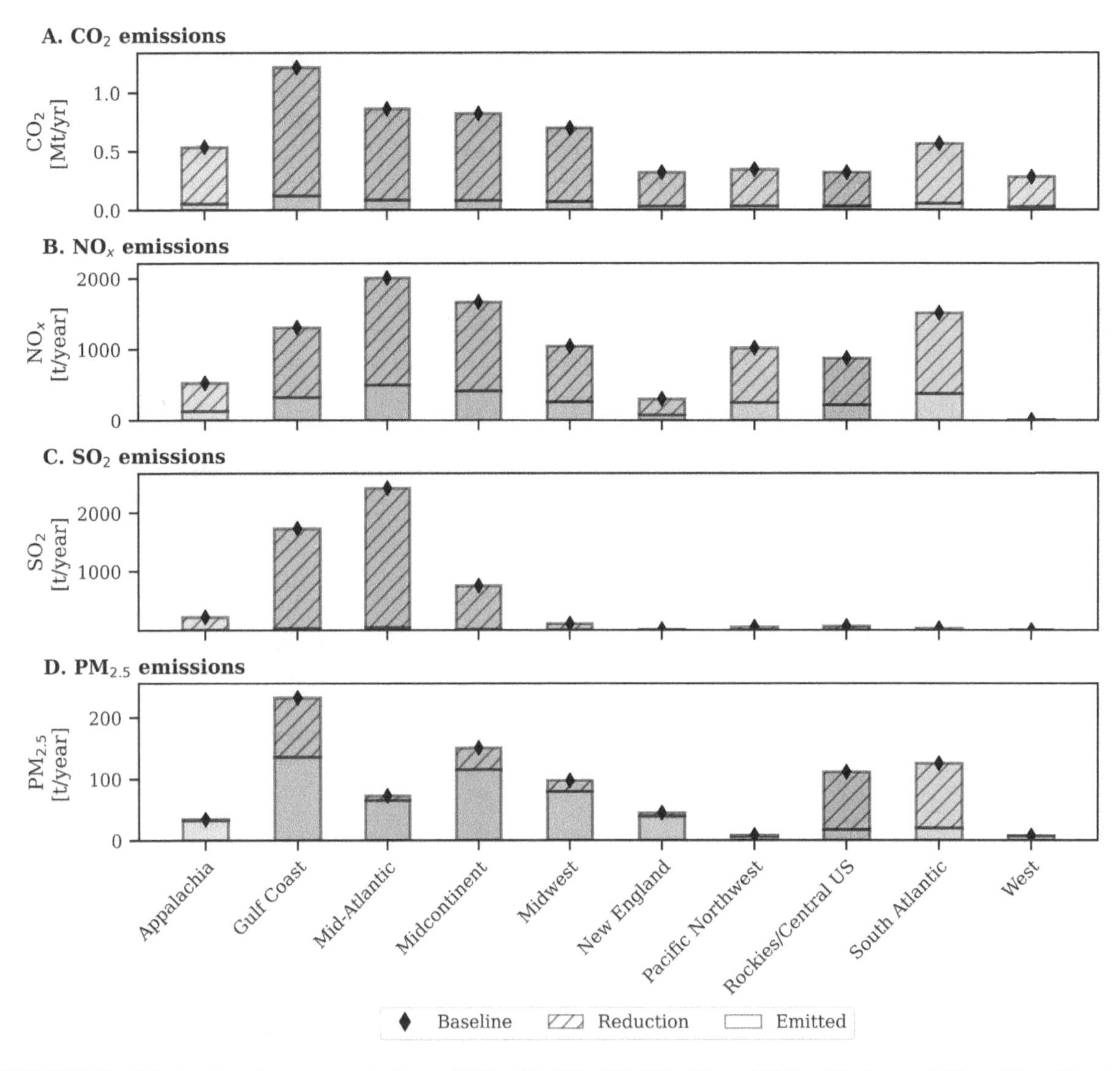

FIGURE 10-13 Reductions in annual emissions of CO_2 (A), NO_x (B), SO_2 (C), and $PM_{2.5}$ (D) that could be achieved by deploying carbon capture on cement plants in different regions of the United States. Hatched bars represent mitigated emissions, and solid bars indicate remaining emissions.
SOURCE: Bennett et al. (2023).

Hydrogen, comprising part of the enabling infrastructure for CO_2 utilization, has indirect global warming impacts, with an estimated global warming potential over a 100-year time horizon (GWP100) of 11.6 ± 2.8 (one standard deviation) (Sand et al. 2023). A recent review found that documented leaks across the present and future hydrogen value chain range widely; for example, from 0.5–1.0 percent for hydrogen production by steam methane reforming, 0.0–1.5 percent for hydrogen production by steam methane reforming with carbon capture, and 0.03–9.2 percent for hydrogen production by water electrolysis (Esquivel-Elizondo et al. 2023). If hydrogen production relies on natural gas, there may be additional upstream leaks that could further confound climate benefits, but these could be avoided if hydrogen is produced using renewable energy through electrolysis or other nonfossil methods (Bertagni et al. 2022). These findings present particular significance for developing measurement programs for enabling infrastructure involving hydrogen value chains.

Different types of CO_2 utilization operations can increase specific impacts other than climate change as well. For example, for CO_2 capture from thermal power plants, water requirements may increase owing to additional cooling needs and process water make-up (Magneschi et al. 2017; Ou et al. 2018). Net-zero scenarios that rely on CO_2 utilization may face land constraints without adequate planning, incurring ecosystem costs (Williams et al. 2021). Enabling infrastructure such as renewable power plants and transmission capacity will incur new land requirements and possibly damage to ecosystems without diligent siting analysis for new facilities (Hernandez et al. 2020). Furthermore, if the CO_2 supply is sourced from biomass, development may face additional land and water constraints.

10.4.3 Health and Safety Impacts of CO_2 Utilization Infrastructure

As more CO_2 utilization infrastructure is deployed, care will have to be taken to ensure safe operation of all technologies throughout the value chain, from CO_2 capture, to transport, to use in industry. The potential health hazards need to be understood and measures put in place to mitigate them. The sections below briefly describe health and safety considerations for CO_2 capture technologies, CO_2 transport in pipelines, and CO_2 use in industry.

10.4.3.1 CO_2 Capture Technologies

Solvent-based technologies remain one of the leading methods for large-scale post-combustion CO_2 capture. Research conducted at the Technology Centre Mongstad (TCM) has shed light on the potential harmful effects of amines used in carbon capture processes (Flø et al. 2017; Morken et al. 2014, 2017). During test campaigns with monoethanolamine between 2015 and 2018, solvent degradation and associated air emissions were observed. These degradation products include nitrosamine and nitramines, which are carcinogenic, toxic, and mutagenic compounds. The same studies at TCM also investigated strategies to reduce air emissions by addressing the main mechanisms for solvent degradation, namely exposure to heat (thermal degradation), oxygen (oxidative degradation), and reactions of the amine with flue gas contaminants such as SO_x, NO_x, halogenated compounds, and other impurities.

Advanced approaches to solvent management, flue gas pretreatment, and optimized plant process configuration have been shown to reduce solvent degradation. The initial concentration of solvent also has been shown to have a significant impact on degradation (Azarpour and Zendehboudi 2023). It is important to keep the solvent fresh by setting strict threshold limits for key indicators, such as ammonia concentration, and performing regular solvent exchange. Advanced novel process concepts have been developed to reduce and control amine emissions. A water wash section in the absorber and Brownian diffusion filter upstream from the absorber have been shown to remove greater than 95 percent aerosol contaminants that contribute to solvent degradation (Shah et al. 2018).

Other complications associated with amine-based CO_2 capture technology include the occupational hygiene risks associated with exposure of plant operators to amines, waste handling, and disposal of spent solvent and the risks associated with materials management, such as potential contamination of groundwater or surface water from unintended spills or leaks in the plant or in the chemical inventory. The absence of large-scale CCS facilities or reported data in the literature from pilot facilities make it difficult to estimate potential amine exposures and predict formation of and exposure to degradation products. Moreover, technology providers make use of proprietary amine blends that could have different impacts. However, with proper worker training and curbed containment areas surrounding the facility, these risks can be properly mitigated. Although amine solvents are the most commercially prevalent technology today, other technologies are in various stages of development (NASEM 2023a; Pett-Ridge et al. 2023). Next-generation innovations in sorbents, membranes, looping cycles, and other novel concepts can mitigate health and safety risks associated with amine-based carbon capture systems.

A report from the Great Plains Institute highlighted health benefits of carbon capture facilities from the removal of co-pollutants and subsequent improvement of air quality (Bennett et al. 2023). As previously mentioned, amine-based capture systems require flue gas pretreatment to remove NO_x, SO_2, and $PM_{2.5}$ to reduce solvent degradation and improve capture performance. Bennett et al. (2023) developed a methodology to evaluate the co-benefits of applying pretreatment for these co-pollutants to amine-based carbon capture for seven industries. The analysis used the EPA CO-Benefits Risk Assessment (COBRA) Health Impacts Screening and Mapping tool to quantify changes in air quality from removal of these co-pollutants and calculate the changes in 12 different health outcomes, including adult and infant mortality, asthma exacerbations, and respiratory and cardiovascular-related hospital admissions. It then calculated the total monetary value of all impacts in millions of dollars per year. The analysis showed that co-pollutant removal resulted in health benefits across all industries and regions, with an economic value ranging from $6.8 million to $481.2 million per year, with the highest potential impact in the Mid-Atlantic and Appalachia regions (Bennett et al. 2023).

10.4.3.2 CO_2 Pipelines

Safety considerations for CO_2 transport were discussed in detail in the committee's first report (Section 4.3.3, NASEM 2023a). The risks and hazards differ somewhat by mode of transport owing to pressure and temperature differences but are the same whether the CO_2 is being transported for utilization or storage. As noted above, for large quantities of CO_2, transport by pipelines is the safest and most cost-effective approach. The United States has been operating CO_2 pipelines safely since the 1970s, primarily to transport CO_2 for EOR (Wallace et al. 2015). CO_2 pipeline safety is overseen by PHMSA, which requires operators to follow specific guidelines for operations and maintenance, control room management, public awareness, damage prevention, operator qualification, and drug and alcohol testing (Thomley and Kammer 2023). For example, to ensure safe operation, pipeline operators have to pay attention to pipeline design, monitor for leaks, protect against corrosion, and safeguard against overpressure (Carbon Capture Coalition 2023).

While CO_2 pipelines have a strong safety record overall, incidents can be and have been significant. A primary example is the rupture of a CO_2 pipeline near Satartia, Mississippi, in February 2020, which required 45 individuals to seek medical attention and 200 to evacuate (Thomley and Kammer 2023). This rupture resulted in the formation of a crater owing to a high momentum CO_2 jet, as illustrated in Figure 10-14. The failure investigation of this incident, led by PHMSA, found that the plume dispersion model used as part of the risk assessment underestimated the safety distances (PHMSA 2022). PHMSA's investigation indicated needs for improvements in (1) mitigation efforts to address "integrity threats owing to changing climate, geohazards, and soil stability issues" and (2) public and emergency responder awareness of CO_2 pipeline safety (PHMSA 2022, p. 2). As a result of this incident, PHMSA initiated a rulemaking process to update its safety standards for CO_2 pipelines, with plans to release a Notice of Proposed Rulemaking in 2024 (Parfomak 2023). In April 2024, pipeline developer Tallgrass established a community benefits agreement with Bold Alliance that includes $600,000 in funding for emergency response equipment and training in communities along the pipeline route, among other provisions (Hammel 2024). (See Chapter 4 for more on community benefits agreements and community engagement around CO_2 pipeline siting.) The Carbon Capture Coalition, a nonpartisan group working to facilitate commercial deployment of carbon management, proposed the following measures to increase CO_2 pipeline safety while supporting build-out of this infrastructure to meet net-zero emissions goals: "expand first responder training for CO_2 pipeline safety incidents; request that PHMSA conduct additional reporting on the public safety record of CO_2 pipelines; require that project proponents more rigorously consider potential geohazard impacts on CO_2 pipelines during design, siting, construction, and maintenance; carry out a national assessment of the CO_2 network necessary to meet net-zero emissions" (Carbon Capture Coalition 2023, p. 2).

Additional R&D on dispersion modeling for CO_2 and its typical stream impurities and on propagating brittle and ductile fractures in CO_2 pipelines also could help improve CO_2 pipeline safety. For example, while there are relatively accurate computer simulations that model CO_2 dispersion behavior following accidental release from a pipeline, more research is needed to model the dispersion of CO_2 and its impurities from a crater formed owing to the rupture of a

FIGURE 10-14 Photograph of the crater that formed following the rupture of a CO_2 pipeline in Satartia, Mississippi, in 2020.
SOURCE: Adapted from PHMSA (2022).

buried pipeline. Such research could inform the minimum safety distance required. Additionally, as highlighted in the committee's first report (NASEM 2023a), the risk of propagating ductile and brittle fractures in CO_2 pipelines warrants significant research attention. Rigorous mathematical models need to be validated using realistic-scale test facilities to determine a pipeline's susceptibility to these types of fractures and to identify the appropriate pipeline materials to mitigate these risks. Given the multidisciplinary nature of the expertise needed in understanding and modeling the behavior of CO_2 pipelines, coupled with the significant capital investment needed to conduct realistic field tests, international collaboration of the type funded by the European Commission's Horizon Europe program (European Commission n.d.) can be a key enabler in accelerating the safe, large-scale implementation of CO_2 pipelines.

10.4.3.3 Industrial Use of CO_2

Carbon dioxide is a colorless, odorless, and tasteless gas at normal temperatures and atmospheric pressures, making it undetectable without monitoring devices (Airgas 2018; FSIS Environmental Health and Safety Group 2020; Spitzenberger and Flechas 2023). With a density of 1.98 kg/m^3 at standard temperature and pressure, CO_2 is approximately 1.5 times heavier than air and can cause low-lying vapor clouds along the ground, or in depressed areas such as pits and cellars. CO_2 is typically transported, stored, and handled in its liquid form, either in cylinders or noninsulated storage tanks at ambient temperature, or under pressure in insulated tankers and storage tanks at temperatures between −35°C and −15°C (CO_2 Gas Company 2017). Many industries also utilize CO_2 in its solid form (dry ice) for chilling and packing product.

Workers that handle CO_2 storage or portable containers, dry ice, or operate near areas where CO_2 is produced or stored, risk CO_2 exposure. A CO_2 cylinder may rupture (or burst) if heated or if the container is overfilled.

Compressed CO_2 gas that is released from a cylinder can injure skin or eyes. Contact with solid CO_2 (dry ice) and refrigerated liquid CO_2 will cause frostbite upon contact owing to the extremely cold temperatures. Dry ice rapidly sublimes and off-gases CO_2, and levels as high as 30,000 ppm have been measured in rooms where dry ice is handled without adequate ventilation (FSIS Environmental Health and Safety Group 2020). A potential asphyxiation hazard exists when CO_2 is dispensed from transportation vessels to stationary, low-pressure storage tanks if there are inadequate transfer seals (OSHA 1996). Exposure to CO_2 gas can lead to negative health effects, which range from headaches and dizziness with minimal CO_2 exposure to difficulty breathing, malaise, increased heart rate, elevated blood pressure, asphyxia, convulsions, unconsciousness, and even death.

CO_2 generally is not found at hazardous levels, unless there is limited ventilation where gas is being stored or produced, or in confined or enclosed spaces. The industry has developed standards and procedures to minimize the development of hazardous conditions (OSHA 1996). These include adequate employee training; CO_2 safety warning signs around stored CO_2; gas detection and alarm systems that can alert building occupants of a CO_2 gas release; proper personal protective equipment such as cryogenic gloves and eye protection; as well as regular maintenance and inspection of CO_2 equipment by qualified personnel.

Owing to its potential to create a hazardous atmosphere, the transport, storage, and handling of CO_2 are regulated by several agencies, including the National Fire Protection Association, the National Board Inspection Code, the International Fire Code, the Occupational Safety and Health Administration (OSHA), and the National Institute for Occupational Safety and Health (Department of Labor 1974; EPA 2000). These organizations have agreed to workplace exposure limits for CO_2 set by OSHA that would likely be applicable to future CO_2 utilization markets (FSIS Environmental Health and Safety Group 2020).

10.4.4 Environmental Justice Considerations and Societal Support for CO_2 Utilization Infrastructure Development and Use

As discussed in more detail in Chapter 4, CO_2 utilization infrastructure development, siting, and selection will play impactful roles in determining the overall benefits of utilization projects, as well as broader societal support for them. Incorporating environmental and social justice tenets in the development of this infrastructure is not a guarantee for support but remains a foundational premise if the sector is to play an integral role in a more equitable circular economy and to avoid past and ongoing harms of linear economies that rely on extractive and disenfranchising principles.

Until the CO_2 utilization sector establishes itself in a consumer market in a tangible way, it will continue to be viewed primarily in the context of other carbon management infrastructure, especially CCS. Many individuals of historically marginalized backgrounds and disenfranchised communities consider carbon management to be a part of the extractive, carbon-based industries that concentrate economic and political power and harm their communities and the global environment (Chemnick 2023; Climate Justice Alliance 2019; Earthjustice and Clean Energy Program 2023; Just Transition Alliance 2020; New Energy Economy n.d.). The opposition to emerging carbon management technologies like carbon capture and carbon removal mirrors the themes seen in opposition to other emerging industries such as advanced nuclear and hydrogen (NASEM 2023b, 2024). However, most societal acceptance research has determined that the general public lacks information about CO_2 utilization and the carbon management sphere as a whole (Offermann-van Heek et al. 2018), which provides an opportunity both to embed environmental justice and community-centered principles into the build-out of the utilization sector and shape further discourse on the subject.

For infrastructure that has yet to be built out, there are opportunities for developers to consider incorporating and prioritizing specific environmental justice principles that may impact how the work gets done, including considerations around shared resources like energy, land, and water use. Best practices for community engagement around project development will differ across communities, but core tenets include avoiding past mistakes, building trust and maintaining relationships, centering enfranchising frameworks, providing resources for engagement, employing alternative ownership models, and sharing information transparently, all of which are elaborated upon further in Chapter 4.

Project siting has to take into account community history and perceptions around new infrastructure versus existing infrastructure, as well as aspects like pipelines that could substantially increase in number as the sector builds out. The centering of strategies developed by minority and overburdened communities—especially those that have had significant intersections with resource management and pipeline development like U.S. Indigenous communities and Tribal nations—and lessons learned from past community experiences can play significant roles in ensuring that community concerns are adequately addressed. Similarly, opportunities to incorporate justice principles into permitting and workforce development in the utilization space must be considered. Lastly, the siting of utilization infrastructure needs to consider the role the end products will play in creating a more just circular economy; understand impacts from infrastructure and use frameworks, such as Community Benefit Agreements and Free, Prior, and Informed Consent, to encourage their just distribution; and incorporate these frameworks into decision-making processes.

10.4.5 CO_2 Utilization Infrastructure Considerations and Trade-Offs

Infrastructure development is a critical bottleneck for CO_2 utilization. Infrastructure decisions require considerations across the value chain to ensure safety and economic viability, minimize environmental impacts, engage with communities, and enable sustainable production of the chemicals and materials needed in a net-zero future. These decisions necessarily involve trade-offs—for example, where one solution might have lower cost, another might yield lower emissions—and decision makers will need to determine which factor(s) are more critical for a particular project. In this section, the committee highlights some considerations and trade-offs for CO_2 utilization infrastructure deployment.

The source of CO_2 utilized influences the life cycle carbon intensity and public acceptance of, and corporate investment in, different conversion processes, as well as the infrastructure required. For example, the largest point sources of CO_2 today are of fossil origin. They can enable economies of scale for CO_2-derived products and can avoid extensive use of CO_2 pipelines via co-locating CO_2 capture and production of CO_2-derived products. However, using such sources for production of short-lived products like hydrocarbon fuels would not be sustainable in a net-zero future, as the fossil-derived carbon would be reemitted quickly to the atmosphere, adding net-positive emissions. As discussed in the committee's first report, atmospheric or biogenic CO_2 is a more sustainable feedstock for hydrocarbon products (NASEM 2023a). However, costs are high for CO_2 produced from DAC, and supply is limited from biogenic sources, which are orders of magnitude smaller than from petroleum refineries and require extensive agricultural land (and water) for corn-derived bioethanol fuel production, which simultaneously forms the CO_2 off-gas. CO_2 from fossil point sources at 4–25 percent by volume will always be less expensive to collect and purify than CO_2 from DAC at ~450 ppm CO_2. The cost of widely available fossil CO_2 from power plants or industrial heating and power applications ranges from $40–$290 per tonne (Bennett et al. 2023; GAO 2022; Hughes et al. 2022; Moniz et al. 2023; NPC 2019), compared to costs of $90–$1000 per ton CO_2 for DAC (Budinis and Lo Re 2023; DOE 2023a; NASEM 2019).

Additional questions surround the long-term viability of investments in CO_2 capture from fossil point sources and the related CO_2 transport, use, and storage infrastructure. For example, what, if any, fossil-powered electricity generation or industrial heat and power will remain in the net-zero future, requiring CO_2 capture and utilization or sequestration infrastructure? Which fossil-powered facilities will be last to be retired, thus making them better candidates for CO_2 capture retrofits? Will small emitters continue to emit and capture their CO_2, or change their processes to no longer emit? The answers to these questions might differ for emitters that produce pure streams of CO_2 versus those with CO_2 streams requiring purification prior to use or storage, or for utilization processes that can be done economically on a small scale (e.g., concrete and fertilizer production) versus those requiring larger scales for viability.

One scenario to consider is the development of a CO_2-to-fuels commercial plant, where the developer has a choice of using fossil point source CO_2 or DAC CO_2 (assuming that low-cost biogenic CO_2 is unavailable owing to volume or location constraints). The CO_2-derived liquid fuel product would reduce the use of petroleum to supply final global demand for liquid fuels.[7] For a given investment, more CO_2-derived fuel can be made from fossil

[7] In the current policy environment, independent oil producers would respond to an increase in CO_2-derived fuels by reducing drilling operations, which would cause oil prices to decrease owing to oversupply until existing reservoirs decline to match the lower demand. Such actions would make it even harder for additional biofuels or CO_2-to-fuels projects to compete, may reduce the impetus for electrification of transportation, and could lead to enhanced liquid fuel use (assuming consumers limit gasoline or diesel use when prices are high). While relevant to note these nuances, this scenario presents investment considerations only between the two CO_2 sources.

point source CO_2 owing to the lower capital costs, so the lower-cost point source CO_2 reduces more petroleum than DAC CO_2 would. There are several risks to choosing fossil point source CO_2, however:

- Climate, energy, or tax policy may lead to early retirement of assets producing fossil CO_2.
- Public opinion and global policy[8] favor use of DAC and biogenic CO_2; thus, these sources may be incentivized over fossil point source CO_2.

The net emissions reductions resulting from using either fossil or DAC CO_2 also need to be considered. Use of fossil point source CO_2 to produce fuel results in net positive fossil carbon emissions to the atmosphere, while producing at most twice the energy output.[9] On the other hand, use of DAC CO_2 for hydrocarbon fuel production can result in a net-zero emissions fuel if all other inputs are also net-zero carbon emissions. Another complication with DAC CO_2-to-fuels plants is that, given the current high costs of DAC, investments in DAC CO_2-to-fuels facilities may only support a small plant capacity, which may also be less efficient than a larger fuels production facility. The levelized cost of CO_2 abatement (see Section 10.4.1) from fossil point sources may remain less than from DAC sources during the time when most fuel is still being produced from petroleum. Levelized cost of CO_2 abatement calculations can facilitate infrastructure investment decisions by indicating which facilities are likely to abate the most CO_2 per investment cost over their lifetime. Government policy and corporate climate goals may not be designed to use this relationship, and hence risk in the fossil point-source investment is higher than its levelized cost of abatement.

In general, CCUS infrastructure is best positioned in proximity to both CO_2 point sources and low-cost geologic storage sites, to enable sharing of infrastructure for CO_2 supply and transport (De Luna et al. 2023). Nonetheless, CO_2 utilization facilities have to be sited on a case-by-case basis, considering infrastructure needs, product type, market acceptance for the carbon-mitigation potential, and community acceptance of new facilities and infrastructure, all of which will vary across the nation and globally. Thus, infrastructure siting is best decided on a project basis, after development of a fully integrated technology and market plan, and engagement with local stakeholders. Public–private partnership funding can be provided to facilitate infrastructure costs for project development, as has been seen with some CCUS hubs under development in Europe (EFI Foundation and Horizon Climate Group 2023). Infrastructure investment in CCUS can contribute to local employment and support global CO_2 emissions reduction goals, although local communities may face burdens of construction, land use changes, and potential pipeline development that need to be addressed. As discussed in this committee's first report, regulators need to assess distributional impacts, particularly for historically disadvantaged groups, in collaboration with affected communities and CO_2 utilization project developers and can consider compensation for those communities negatively impacted by the project or choose not to invest in the project if there are unavoidable and unacceptable equity implications (NASEM 2023a). Section 4.4 in Chapter 4 provides more information about considerations for selecting and siting projects with community input.

10.5 CONCLUSIONS

Building on the committee's first report, this chapter provides an update on existing CO_2 utilization infrastructure and considers economic, climate, environmental, health, safety, and environmental justice impacts of further CO_2 utilization infrastructure development. Government and private sector investments in carbon management infrastructure continue to increase, notably through DOE's Regional Direct Air Capture Hubs and Regional Clean Hydrogen Hubs and private development of CO_2 pipelines and CCS hubs. Nonetheless, for CO_2 capture, transport, storage, and utilization infrastructure, as well as for the enabling infrastructure of clean electricity and clean

[8] For example, the European Union requires that greenhouse gas (GHG) emissions savings for recycled carbon fuels (including CO_2-derived fuels) be at least 70 percent compared to a fossil-carbon-based fuel and notes that "capturing of emissions from nonsustainable fuels should not be considered as avoiding emissions indefinitely when determining the greenhouse gas emissions savings from the use of renewable liquid and gaseous transport fuels of nonbiological origin and recycled carbon fuels" (European Commission 2023).

[9] As described in the committee's first report, "The combustion of a hydrocarbon fuel yields the lower heating value of energy release and the associated CO_2 emissions. If renewable energy is used to capture the emitted CO_2 and convert it back into a hydrocarbon fuel, and then the fuel is combusted again, the same net emissions result but with twice the energy output. However, that 50 percent reduction in CO_2 footprint assumes that capture and conversion are 100 percent efficient, so in reality, the CO_2 footprint will be reduced by less than 50 percent compared to the 'no recycling' case" (NASEM 2023a, p. 132).

hydrogen generation, there are significant gaps between the current and announced capacity and the projected needs to meet midcentury net-zero targets.

Development of infrastructure to transport CO_2, H_2, and CO_2-derived products could be the limiting factor in scaling up CO_2 utilization given the significant costs, long lead times, and public concern. There may be opportunities to retrofit existing pipeline infrastructure to transport CO_2, H_2, or CO_2-derived products, but these determinations will have to be made on a case-by-case basis, as such retrofitting can be technically challenging and costly. Furthermore, additional research is needed to understand and mitigate risks with H_2 and CO_2 transport. This includes, for CO_2 pipelines, development of mathematical models and realistic-scale experiments to analyze issues with propagating brittle and ductile fractures and to examine CO_2 dispersion following accidental rupture of buried CO_2 pipelines that results in formation of a crater owing to the high-momentum CO_2 jet. For H_2 pipelines, safety concerns requiring attention include H_2 embrittlement and its propensity to leak.

Deployment of CO_2 utilization technologies can be facilitated by integrated infrastructure planning and development that balances cost, safety, and environmental impacts and prioritizes public engagement. Use of mathematical models to optimize multimodal CO_2 transportation networks, development of shared CO_2 pipelines for utilization and storage, establishment of public–private partnerships to facilitate resource sharing, and close coordination between developers of CO_2 capture, transport, utilization, and storage projects may accelerate the build-out of CO_2 utilization infrastructure. At the same time, the impacts of CO_2 utilization infrastructure development on pollutant emissions, human health and safety, and local communities must be taken into account. Clear, durable government policies are needed to facilitate long-term investment decisions by industry that meet these societal needs, align with net-zero objectives, and avoid stranded assets.

10.5.1 Findings and Recommendations

Finding 10-1: CO_2 utilization infrastructure development for midcentury net-zero targets—Although infrastructure to support CO_2 utilization (e.g., for CO_2 capture, transport, and storage, and generation of clean electricity and clean hydrogen) is expanding, the existing and announced capacities are not on track to meet the projected needs for achieving midcentury net-zero emissions targets. In cases where CO_2 utilization provides a competitive option to achieve net zero, CO_2 infrastructure development would require substantial amounts of capital investment—for example, across the carbon management value chain encompassing CO_2 capture, transport, utilization, and storage, the Department of Energy estimates upward of \$300 billion–\$600 billion through 2050.

Finding 10-2: Expansion of CO_2 pipeline infrastructure—The successful large-scale build-out of CO_2 capture and storage would require a significant increase in the current stock of CO_2 pipelines. Such pipelines could also enable CO_2 utilization projects. A main challenge to expanding this infrastructure is recent public concerns regarding CO_2 pipeline safety and siting for host communities. Opposition to pipelines is also seen by some groups as a means to block the development of CO_2 capture, utilization, and storage projects, which they consider a moral hazard that could perpetuate use of fossil fuels.

Recommendation 10-1: Enhance public engagement on CO_2 pipeline safety—In updating its safety standards for CO_2 pipelines, the Pipeline and Hazardous Materials Safety Administration should highlight the importance of proactively addressing public concerns regarding pipeline safety at the earliest opportunity through open forum consultations as part of the process for obtaining planning consent.

Finding 10-3: CO_2 utilization as a decarbonization option for small- to medium-scale emitters—Small- to medium-scale CO_2 emitters, estimated to number around 7000, are distributed throughout the United States and collectively contribute a significant amount (approximately 33 percent) of overall U.S. emissions. Given the unique challenges of small- to medium-scale emitters compared to large emitters, dedicated strategies for their decarbonization will need to be developed. In addition to other decarbonization approaches, like electrification and improvements in energy efficiency, such strategies could involve CO_2 capture and utilization using modular technologies that can accommodate smaller amounts of CO_2 to match the emissions of the facility.

Recommendation 10-2: Evaluate CO_2 capture and utilization for small- to medium-scale emitters—Small- to medium-scale emitters that cannot eliminate emissions through energy efficiency, electrification, and

other decarbonization strategies should evaluate the economic feasibility of performing CO_2 capture and utilization on site. This may involve deploying renewable electricity generation, clean hydrogen production, modular carbon capture technologies, and utilization processes suited to small-scale conversion. In cases where this is not possible, taking account of their emissions rates and geographical location, such emitters should seek multimodal transport solutions for the CO_2 and, if relevant, CO_2-derived product, striking a balance between cost, safety, and environmental impacts using established methodologies that also have been proposed for hydrogen delivery infrastructure development. Additionally, as recommended in this committee's first report (Recommendation 4.2), the Department of Energy should develop dedicated methodologies for optimizing multimodal CO_2 transport to assist in these infrastructure planning efforts.

Finding 10-4: Shared CO_2 pipelines for utilization and storage—The development of shared CO_2 pipeline transportation infrastructure for utilization and storage with the flexibility of seamless switching between the two modes of application can significantly reduce infrastructure costs and environmental impacts. Such configurations are already being implemented in several carbon capture, utilization, and storage projects in Europe.

Recommendation 10-3: Support dual use of CO_2 pipelines for utilization and storage—Congress should direct the Department of Transportation to develop policies and regulations that support the dual use of CO_2 pipelines for both CO_2 sequestration and utilization.

Finding 10-5: Deployment of carbon capture on industrial facilities—Current industrial processes were designed without consideration of carbon capture, often leading to the need to capture CO_2 from multiple points in the process, or at low concentrations or pressures, which drives up the cost of carbon capture. Future commercial deployments that redesign and reconfigure processes to improve efficiency of heat integration and CO_2 capture (e.g., oxy-fuel combustion) could reduce CO_2 capture costs.

Finding 10-6: Hydrogen transport to support CO_2 utilization—Given the unique hazard profile of hydrogen, further research and development is needed before pressurized pipelines could play a major role as part of the hydrogen transport infrastructure for making chemicals, direct use as a fuel, or use as a CO_2 utilization feedstock. If such pipelines are developed, they could supply hydrogen to CO_2 utilization facilities. Examples of such planned dual purpose hydrogen pipeline transport infrastructure have already been approved in Europe.

Recommendation 10-4: Support flexible offtake agreements for hydrogen to facilitate CO_2 utilization—Congress should direct the Department of Transportation and Department of Energy to develop policies and regulations that would support flexible hydrogen pipeline offtake agreements to facilitate CO_2 utilization where co-location with hydrogen generation is impractical, thus avoiding the need to build dedicated hydrogen transport infrastructure for distributed CO_2 utilization.

Finding 10-7: Development of transportation infrastructure may limit the scale up of CO_2 utilization—Transportation infrastructure—for CO_2, hydrogen, CO_2-derived products, and other enabling inputs—may be the limiting factor in the scale up of the CO_2 utilization industry and consequently impact siting opportunities for CO_2 capture and utilization facilities. In particular, in the event of limited build-out of CO_2 pipelines, given cost, environmental, and safety considerations, opportunities for CO_2 utilization that do not require significant CO_2 transport infrastructure, such as distributed, onsite conversion of captured CO_2, would be favored.

Finding 10-8: Partnerships and coordination to accelerate build-out of CO_2 infrastructure—Establishing public–private partnerships could accelerate the build-out of a shared regional or national CO_2 transportation infrastructure network as a public good to achieve economies of scale for CO_2 capture, utilization, and storage. Close communication and coordination between developers of CO_2 capture, transport, utilization, and storage projects is necessary to develop such a regional- or national-scale network because of different costs, risks, and technical requirements across the value chain.

Finding 10-9: Modeling and experimentation needs to improve CO_2 pipeline safety—Relatively accurate computer simulations are already available for modeling the dispersion behavior of CO_2 following accidental releases. However, in the case of failure of buried CO_2 pipelines resulting in the formation of a crater owing to the high-momentum jet, more research is needed to model the subsequent dispersion behavior of the CO_2 plume exiting the crater. Preliminary evidence suggests that in such cases, the minimum safety distance is increased. A further issue related to CO_2 pipeline safety is the risk of propagating ductile and brittle fractures. Here rigorous mathematical models validated using realistic-scale test facilities are needed to determine pipeline susceptibility to such types of failure and hence enable the selection of pipeline materials with sufficient mechanical strength to avoid them.

Recommendation 10-5: Fund research on modeling and testing aimed at improving CO_2 pipeline safety—The Pipeline and Hazardous Materials Safety Administration should collaborate with national laboratories, university researchers, industry, and international partners to co-fund and implement research to develop rigorous mathematical models, which in turn should be extensively validated using realistic-scale experiments, to

a. **Simulate the fluid/structure interaction and subsequent atmospheric dispersion of the escaping overground CO_2 plume following accidental rupture of buried CO_2 pipelines that results in the formation of a crater owing to the high-momentum jet impingement in order to determine minimum safe distances to populated areas and emergency response planning.**
b. **Understand and mitigate issues with propagating brittle and ductile fractures in CO_2 pipelines, as described in detail in Recommendation 4.3 of the committee's first report, in order to select pipeline materials capable of resisting such types of failures.**
c. **Implement the validated mathematical models developed in (a) and (b) into robust and easy-to-use computer programs to be routinely used as a design and decision-making tool for pipeline developers in order to reduce and mitigate the risks associated with pipeline failures to as low as reasonably practicable.**

Finding 10-10: Use of fossil CO_2 for short-lived products in the near, mid-, and long term—Production of short-lived products from fossil CO_2 rather than direct air capture (DAC) CO_2 could provide greater emissions reductions per dollar investment during the near- to mid-term decarbonization transition (i.e., a lower levelized cost of CO_2 abatement). However, in a net-zero future, short-lived products like fuels will need to be produced from biogenic, DAC, or direct ocean capture CO_2 sources. Current science, policy, and public opinion favor use of nonfossil CO_2 for products in which CO_2 is not permanently sequestered, placing investment risk on the development of a CO_2-to-fuels plant using fossil point source CO_2.

Recommendation 10-6: Establish durable policies to facilitate long-term decisions by industry—To facilitate long-term investment decisions by industry that are aligned with net-zero goals and that avoid stranded assets, Congress should establish clear, durable policies and direct federal agencies to develop metrics for the use of different CO_2 sources in the production of short- and long-lived products, the incentives available for enabling technologies like clean electricity and clean hydrogen, and the methodologies used to determine compliance with regulations and incentives.

10.5.2 Research Agenda for CO_2 Utilization Infrastructure

Table 10-5 presents the committee's research agenda for CO_2 utilization infrastructure, including research needs (numbered by chapter), and related research agenda recommendations (a subset of research-related recommendations from the chapter). The table includes the relevant funding agencies or other actors; whether the need is for basic research, applied research, technology demonstration, or enabling technologies and processes for CO_2 utilization; the research theme(s) that the research need falls into; the relevant research area and product class covered by the research need; whether the relevant product(s) are long- or short-lived; and the source of the research need (chapter section, finding, or recommendation). The committee's full research agenda can be found in Chapter 11.

TABLE 10-5 Research Agenda for CO_2 Utilization Infrastructure

Research, Development, and Demonstration Need	Funding Agencies or Other Actors	Basic, Applied, Demonstration, or Enabling	Research Area	Product Class	Long- or Short-Lived	Research Theme	Source
10-A. Development of robust computational tools for optimal multimodal transportation of CO_2 captured from stranded emitters for centralized utilization.	DOE-FECM	Enabling	Infrastructure	All	Long-lived Short-lived	Enabling technology and infrastructure needs Computational modeling and machine learning	Rec. 10-2
10-B. Techno-economic assessment of centralized versus distributed/onsite utilization of CO_2 for small- to medium-scale emitters.	Small- to medium-scale emitters	Enabling	Infrastructure	All	Long-lived Short-lived	Enabling technology and infrastructure needs	Rec. 10-2
Recommendation 10-2: Evaluate CO_2 capture and utilization for small- to medium-scale emitters—Small- to medium-scale emitters that cannot eliminate emissions through energy efficiency, electrification, and other decarbonization strategies should evaluate the economic feasibility of performing CO_2 capture and utilization on site. This may involve deploying renewable electricity generation, clean hydrogen production, modular carbon capture technologies, and utilization processes suited to small-scale conversion. In cases where this is not possible, taking account of their emissions rates and geographical location, such emitters should seek multimodal transport solutions for the CO_2 and, if relevant, CO_2-derived product, striking a balance between cost, safety, and environmental impacts using established methodologies that also have been proposed for hydrogen delivery infrastructure development. Additionally, as recommended in this committee's first report (Recommendation 4.2), the Department of Energy should develop dedicated methodologies for optimizing multimodal CO_2 transport to assist in these infrastructure planning efforts.							
10-C. Better understanding and development of approaches to mitigate issues with propagating brittle and ductile fractures in CO_2 pipelines.	PHMSA National laboratories University researchers Industry	Enabling	Infrastructure	All	Long-lived Short-lived	Enabling technology and infrastructure needs	Rec 10-5
10-D. CO_2 dispersion modeling calculations for the case of accidental rupture of buried CO_2 pipelines that results in formation of a crater owing to the high-momentum CO_2 jet.	PHMSA National laboratories University researchers Industry	Enabling	Infrastructure	All	Long-lived Short-lived	Enabling technology and infrastructure needs Computational modeling and machine learning	Rec 10-5
10-E. Computer programs that implement validated mathematical models on dispersion modeling and propagating fractures to serve as decision-making tools for pipeline developers.	PHMSA National laboratories University researchers Industry	Enabling	Infrastructure	All	Long-lived Short-lived	Enabling technology and infrastructure needs	Rec 10-5
Recommendation 10-5: Fund research on modeling and testing aimed at improving CO_2 pipeline safety—The Pipeline and Hazardous Materials Safety Administration should collaborate with national laboratories, university researchers, industry, and international partners to co-fund and implement research to develop rigorous mathematical models, which in turn should be extensively validated using realistic-scale experiments, to **a. Simulate the fluid/structure interaction and subsequent atmospheric dispersion of the escaping overground CO_2 plume following accidental rupture of buried CO_2 pipelines that results in the formation of a crater owing to the high-momentum jet impingement in order to determine minimum safe distances to populated areas and emergency response planning.** **b. Understand and mitigate issues with propagating brittle and ductile fractures in CO_2 pipelines, as described in detail in Recommendation 4.3 of the committee's first report, in order to select pipeline materials capable of resisting such types of failures.** **c. Implement the validated mathematical models developed in (a) and (b) into robust and easy-to-use computer programs to be routinely used as a design and decision-making tool for pipeline developers in order to reduce and mitigate the risks associated with pipeline failures to as low as reasonably practicable.**							

NOTE: FECM = Office of Fossil Energy and Carbon Management; PHMSA = Pipeline and Hazardous Materials Safety Administration.

10.6 REFERENCES

1PointFive. 2023a. "1PointFive Announces Plan to Develop a Carbon Capture and Sequestration Hub in Southeast Texas." *Oxy*. https://www.oxy.com/news/news-releases/1pointfive-announces-plan-to-develop-a-carbon-capture-and-sequestration-hub-in-southeast-texas.

1PointFive. 2023b. "Stratos." 1PointFive. https://www.1pointfive.com/ector-county-tx.

Abramson, E., D. McFarlane, and J. Brown. 2020. "Transport Infrastructure for Carbon Capture and Storage." Minneapolis, MN: Great Plains Institute. https://betterenergy.org/wp-content/uploads/2020/06/GPI_RegionalCO2Whitepaper.pdf.

Abramson, E., E. Thomley, and D. McFarlane. 2022. "An Atlas of Carbon and Hydrogen Hubs for United States Decarbonization." Minneapolis, MN: Great Plains Institute. https://scripts.betterenergy.org/CarbonCaptureReady/GPI_Carbon_and_Hydrogen_Hubs_Atlas.pdf.

Abramson, E., D. McFarlane, A. Jordan, D. Rodriguez, J. Ogland-Hand, N. Holwerda, M. Fry, R. Kammer, and E. Thomley. 2023. "An Atlas of Direct Air Capture Opportunities for Negative Emissions in the United States." *Carbon Solutions LLC and Great Plains Institute*. https://carboncaptureready.betterenergy.org/wp-content/uploads/2023/03/DAC-Hubs-Atlas-2023.pdf.

Adolf, J., W. Warnecke, C. Balzer, X. Fu, G. Hagenow, Y. Hemberger, A. Janssen, M. Klokkenburg, A. Made, and T. McKnight. 2018. "The Road to Sustainable Fuels for Zero-Emissions Mobility. Status of and Perspectives for Power-to-Liquids (PTL) Fuels." Presented at the 39th International Vienna Motor Symposium.

Adolf, J., W. Warnecke, P. Karzel, A. Kolbeck, A. Made, J. Müller-Belau, J. Powell, et al. 2020. "On Route to CO_2-Free Fuels. Hydrogen-Latest Developments in Its Supply Chain and Applications in Transport." Presented at 41st Vienna Motor Symposium. https://doi.org/10.13140/RG.2.2.21744.58889.

Ahmed, T. 2023. "North Dakota Landowners at Odds in Carbon Pipeline Plans." *AP News*, January 27. https://apnews.com/article/business-north-dakota-climate-and-environment-5db9f9e49f3a68687b5ecb78c1900c0f.

Airgas. 2018. "Safety Data Sheet: Carbon Dioxide." https://www.airgas.com/msds/001013.pdf.

Anchondo, C. 2023. "CO_2 Pipeline Company Delays Major Midwest Project." *Energywire*, October 19. https://subscriber.politicopro.com/article/eenews/2023/10/19/co2-pipeline-company-delays-major-midwest-project-00122318.

API (American Petroleum Institute). 2021. "Where Are the Pipelines?" *American Petroleum Institute*. https://www.api.org/oil-and-natural-gas/wells-to-consumer/transporting-oil-natural-gas/pipeline/where-are-the-pipelines.

ARPA-E (Advanced Research Projects Agency–Energy). 2023. "Grid-Free Renewable Energy Enabling New Ways to Economical Liquids and Long-Term Storage." *ARPA-E*, December 12. https://arpa-e.energy.gov/technologies/programs/greenwells.

Azarpour, A., and S. Zendehboudi. 2023. "Hybrid Smart Strategies to Predict Amine Thermal Degradation in Industrial CO_2 Capture Processes." *ACS Omega* 8(30):26850–26870. https://doi.org/10.1021/acsomega.3c01475.

Bayer, P., and M. Aklin. 2020. "The European Union Emissions Trading System Reduced CO_2 Emissions Despite Low Prices." *Proceedings of the National Academy of Sciences* 117(16):8804–8812.

Beck, L. 2020. "The US Section of 45Q Tax Credit for Carbon Oxide Sequestration: An Update." Brief. Global CCS Institute. https://www.globalccsinstitute.com/wp-content/uploads/2020/04/45Q_Brief_in_template_LLB.pdf.

Bennett, J., R. Kammer, J. Eidbo, M. Ford, S. Henao, N. Holwerda, E. Middleton, et al. 2023. "Carbon Capture Co-Benefits." Great Plains Institute. https://carboncaptureready.betterenergy.org/wp-content/uploads/2023/08/Carbon-Capture-Co-Benefits.pdf.

Bertagni, M.B., S.W. Pacala, F. Paulot, and A. Porporato. 2022. "Risk of the Hydrogen Economy for Atmospheric Methane." *Nature Communications* 13(1):7706. https://doi.org/10.1038/s41467-022-35419-7.

Bevrani, B., R. Burdett, A. Bhaskar, and P.K.D.V. Yarlagadda. 2020. "A Multi-Criteria Multi-Commodity Flow Model for Analysing Transportation Networks." *Operations Research Perspectives* 7(January 1):100159. https://doi.org/10.1016/j.orp.2020.100159.

Budinis, S., and L.Lo Re. 2023. "Unlocking the Potential of Direct Air Capture: Is Scaling Up Through Carbon Markets Possible?" *IEA*, May 11. https://www.iea.org/commentaries/unlocking-the-potential-of-direct-air-capture-is-scaling-up-through-carbon-markets-possible.

C^4U Project. 2020. "The C^4U Project: Advanced Carbon Capture for Steel Industries Integrated in CCUS Clusters." C^4U. https://c4u-project.eu.

Cai, X., M.A. Coletti, D.S. Sholl, and M.R. Allen-Dumas. 2024. "Assessing Impacts of Atmospheric Conditions on Efficiency and Siting of Large-Scale Direct Air Capture Facilities." *JACS Au* 4(5):1883–1891. https://doi.org/10.1021/jacsau.4c00082.

CapturePoint. 2024. "Our Projects." *CapturePoint*. https://www.capturepointllc.com/projects.

Carbon Capture Coalition. 2023. "PHMSA/Pipeline Safety Fact Sheet." https://carboncapturecoalition.org/wp-content/uploads/2023/11/Pipeline-Safety-Fact-Sheet.pdf.

CarbonCapture Inc. 2023. "Project Bison." https://www.carboncapture.com/project-bison.

Carbonvert and Castex. 2023. "Carbonvert, Castex Joint Venture Executes Operating Agreement for Offshore Carbon Storage Hub in Louisiana." *Businesswire*, September 19. https://www.businesswire.com/news/home/20230919740569/en/Carbonvert-Castex-Joint-Venture-Executes-Operating-Agreement-for-Offshore-Carbon-Storage-Hub-in-Louisiana.

CATF (Clean Air Task Force). 2023. "US Carbon Capture Activity and Project Map." https://www.catf.us/ccsmapus.

Cerniauskas, S., A.J. Chavez Junco, T. Grube, M. Robinius, and D. Stolten. 2020. "Options of Natural Gas Pipeline Reassignment for Hydrogen: Cost Assessment for a Germany Case Study." *International Journal of Hydrogen Energy* 45(21):12095–12107. https://doi.org/10.1016/j.ijhydene.2020.02.121.

CETI (Clean Energy Transition Institute). 2023. "Net-Zero Northwest: Technical and Economic Pathways to 2050." Net-Zero Northwest. https://www.nznw.org.

Chemnick, J. 2023. "The Carbon Removal Project That Puts Communities in the Driver's Seat." *E&E News: Climatewire*. October 26. https://www.eenews.net/articles/the-carbon-removal-project-that-puts-communities-in-the-drivers-seat.

Christol, C., F. Oteri, and M. Laurienti. 2021. "Land-Based Wind Energy Siting: A Foundational and Technical Resource." NREL/TP-5000-78591; DOE/GO-102021-5608. National Renewable Energy Laboratory (NREL), Golden, CO.

Climate Justice Alliance. 2019. "Just Transition Principles." Berkeley, CA: Climate Justice Alliance. https://climatejusticealliance.org/wp-content/uploads/2019/11/CJA_JustTransition_highres.pdf.

Clune, R., L. Corb, W. Glazener, K. Henderson, D. Pinner, and D. Walter. 2022. "Navigating America's Net-Zero Frontier: A Guide for Business Leaders." *McKinsey Sustainability*, May 5. https://www.mckinsey.com/capabilities/sustainability/our-insights/navigating-americas-net-zero-frontier-a-guide-for-business-leaders.

CO_2 Gas Company. 2017. "What Is Liquid CO_2 and What Can It Be Useful for?" https://co2gas.co.uk/what-is-liquid-co2.

CO_2 Value Europe. n.d. "North CCU Hub." CO_2 Value Europe. https://database.co2value.eu/projects/218.

Crake, A., K.C. Christoforidis, A. Kafizas, S. Zafeiratos, and C. Petit. 2017. "CO_2 Capture and Photocatalytic Reduction Using Bifunctional TiO_2/MOF Nanocomposites under UV–Vis Irradiation." *Applied Catalysis B: Environmental* 210:131–140. https://doi.org/10.1016/j.apcatb.2017.03.039.

Dakota Gasification Company. 2023. "In the Matter of a Hearing Called on a Motion of the Commission to Consider the Application of Dakota Gasification Company Requesting Consideration for the Geologic Storage of Carbon Dioxide from the Great Plains Synfuels Plant Located in Sections 5, 6, 7, 8, 17, 18, 19, Township 145 North, Range 87 West, Sections 1, 2, 3, 4, 9, 10, 11, 12, 13, 14, 15, 16, 22, 23, 24, Township 145 North, Range 88 West, Sections 30, 31, 32, Township 146 North, Range 87 West, Sections 25, 26, 27, 33, 34, 35, 36, Township 146 North, Range 88 West, Mercer County, North Dakota Pursuant to North Dakota Administrative Code Section 43-05-01." Industrial Commission of the State of North Dakota, Case No. 29450, Order No. 32250. https://www.dmr.nd.gov/dmr/sites/www/files/documents/Oil%20and%20Gas/Class%20VI/DGC/C29450.pdf.

Dakota Institute. 2023. "Economic Impacts of CO_2 Pipelines in South Dakota." Sioux Falls, SD: Dakota Institute. https://smartcarbonnetwork.com/wp-content/uploads/2023/05/Economic-Impacts-of-CO2-Pipelines-Final-Report1.pdf.

d'Amore, F., P. Mocellin, C. Vianello, G. Maschio, and F. Bezzo. 2018. "Economic Optimisation of European Supply Chains for CO_2 Capture, Transport and Sequestration, Including Societal Risk Analysis and Risk Mitigation Measures." *Applied Energy* 223:401–415. https://doi.org/10.1016/j.apenergy.2018.04.043.

Datta, A., R. De Leon, and R. Krishnamoorti. 2020. "Advancing Carbon Management Through the Global Commoditization of CO_2: The Case for Dual-Use LNG-CO_2 Shipping." *Carbon Management* 11(6):611–630. https://doi.org/10.1080/17583004.2020.1840871.

DCC West Project LLC. 2023. "In the Matter of a Hearing Called on a Motion of the Commission to Consider the Application of DCC West Project LLC Requesting Consideration for the Geologic Storage of Carbon Dioxide in the Broom Creek Formation from the Milton R. Young Station and Other Sources Location in Sections 2, 3, 4, 5, 6, 7, 8, 9, 10, 11, 14, 15, 16, 17, 18, 19, 20, 21, 29, 30, 31, and 32, Township 141 North, Range 84 West, Section 1, 2, 3, 4, 9, 10, 11, 12, 13, 14, 15, 16, 22, 23, 24, 25, 26, 27, and 36, Township 141 North, Range 85 West, Sections 19, 20, 21, 28, 29, 30, 31, 32, 33, and 34, Township 142 North, Range 84 West, and Sections 24, 25, 33, 34, 35, and 36, Township 142 North, Range 85 West, Oliver County, North Dakota, Pursuant to North Dakota Administrative Code Chapter 43-05-01." Industrial Commission of the State of North Dakota, Case No. 30122. Order No. 32806. https://www.dmr.nd.gov/dmr/sites/www/files/documents/Oil%20and%20Gas/Class%20VI/DCC%20West/C30122.pdf.

De Luna, P., L. Di Fiori, Y. Li, A. Nojek, and B. Stackhouse. 2023. "The World Needs to Capture, Use, and Store Gigatons of CO_2: Where and How?" *McKinsey & Company*, April 5. https://www.mckinsey.com/industries/oil-and-gas/our-insights/the-world-needs-to-capture-use-and-store-gigatons-of-co2-where-and-how.

Decarb America Research Initiative. 2021. "Interactive Maps: Energy Infrastructure Needs for a Net-Zero Economy." Decarb America. https://decarbamerica.org/interactive-maps.

Department of Labor. 1974. Occupational Safety and Health Standards, 1910 Subpart H Hazardous Materials. https://www.osha.gov/lawsregs/regulations/standardnumber/1910/1910.101.

Doctor, R., A. Palmer, D. Coleman, J. Davison, C. Hendriks, O. Kaarstad, M. Ozaki, and M. Austell. 2005. "4—Transport of CO_2." In *IPCC Special Report on Carbon Dioxide Capture and Storage*. Cambridge, UK: Cambridge University Press. https://www.ipcc.ch/site/assets/uploads/2018/03/srccs_chapter4-1.pdf.

DOE (Department of Energy). 2022a. "DOE Establishes Bipartisan Infrastructure Law's $9.5 Billion Clean Hydrogen Initiatives." *Energy.gov*, February 15. https://www.energy.gov/articles/doe-establishes-bipartisan-infrastructure-laws-95-billion-clean-hydrogen-initiatives.

DOE. 2022b. "Industrial Decarbonization Roadmap." DOE/EE-2635. Washington, DC: Department of Energy. https://www.energy.gov/sites/default/files/2022-09/Industrial%20Decarbonization%20Roadmap.pdf.

DOE. 2023a. "Pathways to Commercial Liftoff: Carbon Management." Washington, DC: Department of Energy. https://liftoff.energy.gov/wp-content/uploads/2023/04/20230424-Liftoff-Carbon-Management-vPUB_update.pdf.

DOE. 2023b. "Pathways to Commercial Liftoff: Clean Hydrogen." Washington, DC: Department of Energy. https://liftoff.energy.gov/wp-content/uploads/2023/05/20230523-Pathways-to-Commercial-Liftoff-Clean-Hydrogen.pdf.

DOE. 2023c. "U.S. Department of Energy Clean Hydrogen Production Standard (CHPS) Guidance." Washington, DC: Department of Energy. https://www.hydrogen.energy.gov/docs/hydrogenprogramlibraries/pdfs/clean-hydrogen-production-standard-guidance.pdf.

DOE. 2023d. "U.S. National Clean Hydrogen Strategy and Roadmap." Washington, DC: Department of Energy. https://www.hydrogen.energy.gov/docs/hydrogenprogramlibraries/pdfs/us-national-clean-hydrogen-strategy-roadmap.pdf.

DOE-FECM (Office of Fossil Energy and Carbon Management). 2023a. "Project Selections for FOA 2735: Regional Direct Air Capture Hubs—Topic Area 1 (Feasibility) and Topic Area 2 (Design)." https://www.energy.gov/fecm/project-selections-foa-2735-regional-direct-air-capture-hubs-topic-area-1-feasibility-and.

DOE-FECM. 2023b. "Workshop on Applied Research for CO_2 Transport." Washington, DC: Department of Energy. https://www.energy.gov/sites/default/files/2023-08/Workshop-on-Applied-Research-for-CO2-Transport-Summary-Report-2023_0.pdf.

DOE-HFTO (Hydrogen and Fuel Cell Technologies Office). n.d. "Hydrogen Pipelines." https://www.energy.gov/eere/fuelcells/hydrogen-pipelines.

DOE-OCED (Office of Clean Energy Demonstrations). 2023a. "Biden-Harris Administration Announces $7 Billion for America's First Clean Hydrogen Hubs, Driving Clean Manufacturing and Delivering New Economic Opportunities Nationwide." *Energy.gov*, October 13. https://www.energy.gov/articles/biden-harris-administration-announces-7-billion-americas-first-clean-hydrogen-hubs-driving?utm_medium=email&utm_source=govdelivery.

DOE-OCED. 2023b. "Biden-Harris Administration Announces Up to $1.2 Billion for Nation's First Direct Air Capture Demonstrations in Texas and Louisiana." *Energy.gov*, August 11. https://www.energy.gov/articles/biden-harris-administration-announces-12-billion-nations-first-direct-air-capture.

DOE-OCED. n.d.(a). "Regional Clean Hydrogen Hubs Selections for Award Negotiations." https://www.energy.gov/oced/regional-clean-hydrogen-hubs-selections-award-negotiations.

DOE-OCED. n.d.(b). "Regional Direct Air Capture Hubs Selections for Award Negotiations." https://www.energy.gov/oced/regional-direct-air-capture-hubs-selections-award-negotiations.

Douglas, L. 2023. "Navigator CO_2 Ventures Cancels Carbon-Capture Pipeline Project in U.S. Midwest." *Reuters*, October 20. https://www.reuters.com/sustainability/climate-energy/navigator-co2-ventures-cancels-carbon-capture-pipeline-project-us-midwest-2023-10-20.

Dura, J. 2023a. "North Dakota Panel Will Reconsider Denying Permit for Summit CO_2 Pipeline." *AP News*, September 15. https://apnews.com/article/north-dakota-midwest-co2-pipeline-summit-carbon-dioxide-13c4c1524f96e8fbf52d0dcd6ab4abd0.

Dura, J. 2023b. "North Dakota Regulators Deny Siting Permit for Summit Carbon Dioxide Pipeline; Company Will Reapply." *AP News*, August 4. https://apnews.com/article/north-dakota-carbon-dioxide-pipeline-29d15d0d29782f9f28b7907b6bb1896e.

Dura, J., and S. Karnowski. 2023. "CO_2 Pipeline Project Denied Key Permit in South Dakota; Another Seeks Second Chance in North Dakota." *AP News*, September 6. https://apnews.com/article/south-dakota-carbon-dioxide-pipeline-665a20269d43d4aff5944969f869bb03.

Earthjustice and Clean Energy Program. 2023. "Carbon Capture: The Fossil Fuel Industry's False Climate Solution." Earthjustice. https://earthjustice.org/article/carbon-capture-the-fossil-fuel-industrys-false-climate-solution.

EFI Foundation and Horizon Climate Group. 2023. "A New U.S. Industrial Backbone: Exploring Regional CCUS Hubs for Small-to-Midsize Industrial Emitters." Washington, DC: EFI Foundation and Horizon Climate Group. https://efifoundation.org/wp-content/uploads/sites/3/2023/12/CCUS-Hubs-FINAL-12142023-WITH-COVER-1.pdf.

EIA (U.S. Energy Information Administration). 2023a. "Electric Power Annual 2022." Washington, DC: U.S. Energy Information Administration. https://www.eia.gov/electricity/annual/pdf/epa.pdf.

EIA. 2023b. "Gulf of Mexico Fact Sheet." U.S. Energy Information Administration. https://www.eia.gov/special/gulf_of_mexico/#gulf_coast_refinery.

EIA. 2024. "Natural Gas Explained: Natural Gas Pipelines." U.S. Energy Information Administration. https://www.eia.gov/energyexplained/natural-gas/natural-gas-pipelines.php.

Element Energy. 2018. "Shipping Carbon Dioxide (CO_2): UK Cost Estimation Study." https://www.gov.uk/government/publications/shipping-carbon-dioxide-co2-uk-cost-estimation-study.

Eller, D. 2023. "Navigator Kills Its $3.5B Carbon Capture Pipeline Across Iowa, South Dakota, Other States." *Des Moines Register*, October 20, sec. Business. https://www.desmoinesregister.com/story/money/business/2023/10/20/navigator-kills-its-carbon-capture-pipeline-in-iowa-other-states-ethanol-poet/71253882007.

Enbridge. 2022. "A Novel Retrofit Solution for Existing Pipelines." *Enbridge Blog* (blog), May 3. https://www.enbridge.com/stories/2022/may/enbridge-invests-in-smartpipe-pipeline-retrofit-enabling-co2-hydrogen-transportation.

Ennis, J., and A. Levin. 2023. "Clean Energy Now for a Safer Climate Future." New York: Natural Resources Defense Council. https://www.nrdc.org/sites/default/files/2023-04/clean-energy-pathways-net-zero-2050-report.pdf.

EPA (U.S. Environmental Protection Agency). 2000. "Office of Air and Radiation. Carbon Dioxide as a Fire Suppressant: Examining The Risks." Air and Radiation, U.S. Environmental Protection Agency.

EPA. 2023a. "GHGRP Emissions Ranges." https://www.epa.gov/ghgreporting/ghgrp-emissions-ranges.

EPA. 2023b. "Table of EPA's Draft and Final Class VI Well Permits." https://www.epa.gov/uic/table-epas-draft-and-final-class-vi-well-permits.

Esquivel-Elizondo, S., A. Hormaza Mejia, T. Sun, E. Shrestha, S.P. Hamburg, and I.B. Ocko. 2023. "Wide Range in Estimates of Hydrogen Emissions from Infrastructure." *Frontiers in Energy Research* 11. https://www.frontiersin.org/articles/10.3389/fenrg.2023.1207208.

Etxeberria-Benavides, M., O. David, T. Johnson, M.M. Łozińska, A. Orsi, P.A. Wright, S. Mastel, R. Hillenbrand, F. Kapteijn, and J. Gascon. 2018. "High Performance Mixed Matrix Membranes (MMMs) Composed of ZIF-94 Filler and 6FDA-DAM Polymer." *Journal of Membrane Science* 550(March 15):198–207. https://doi.org/10.1016/j.memsci.2017.12.033.

European Commission. 2023. "COMMISSION DELEGATED REGULATION (EU) 2023/1185 of 10 February 2023 Supplementing Directive (EU) 2018/2001 of the European Parliament and of the Council by Establishing a Minimum Threshold for Greenhouse Gas Emissions Savings of Recycled Carbon Fuels and by Specifying a Methodology for Assessing Greenhouse Gas Emissions Savings from Renewable Liquid and Gaseous Transport Fuels of Non-Biological Origin and from Recycled Carbon Fuels." Commission Delegated Regulation (EU) 2023/1185. *Brussels, Belgium: European Commission*, February 10. https://eur-lex.europa.eu/eli/reg_del/2023/1185/oj.

European Commission. n.d. "Horizon Europe." European Commission: Research and Innovation. https://research-and-innovation.ec.europa.eu/funding/funding-opportunities/funding-programmes-and-open-calls/horizon-europe_en.

Flø, N.E., L. Faramarzi, T. de Cazenove, O.A. Hvidsten, A. Kolstad Morken, E. Steinseth Hamborg, K. Vernstad, et al. 2017. "Results from MEA Degradation and Reclaiming Processes at the CO_2 Technology Centre Mongstad." 13th International Conference on Greenhouse Gas Control Technologies, GHGT-13, 14–18 November 2016, Lausanne, Switzerland 114(July 1):1307–1324. https://doi.org/10.1016/j.egypro.2017.03.1899.

Fluxys Belgium. 2022. "Information Memorandum for CO_2 Infrastructure." Belgium: Fluxys. https://www.fluxys.com/en/about-us/energy-transition/hydrogen-carbon-infrastructure/carbon_preparing-to-build-the-network.

Friedmann, S.J., Z. Fan, Z. Byrum, E. Ochu, A. Bhardwaj, and H. Sheerazi. 2020. "Levelized Cost of Carbon Abatement: An Improved Cost-Assessment Methodology for a Net-Zero Emissions World." New York: Columbia University Center on Global Energy Policy. https://www.eesi.org/files/Levelized_Cost_of_Carbon_Abatement.pdf.

Frontier Carbon Solutions LLC. 2022. "Underground Injection Control Carbon Sequestration Class VI Permit Application Sweetwater Carbon Storage Hub." https://drive.google.com/file/d/10f7cqI85wYpOg6Q6HmSw72qYiEObtDjI/view.

FSIS Environmental Safety and Health Group. 2020. "Carbon Dioxide Health Hazard Information Sheet." https://fsis.usda.gov/sites/default/files/media_file/2020-08/Carbon-Dioxide.pdf.

FuelCell Energy. 2023. "Carbonate Fuel Cell Power Plants." https://www.fuelcellenergy.com/platform/fuel-cell-power-plants.

Gabrielli, P., L. Rosa, M. Gazzani, R. Meys, A. Bardow, M. Mazzotti, and G. Sansavini. 2023. "Net-Zero Emissions Chemical Industry in a World of Limited Resources." *One Earth* 6(6):682–704. https://doi.org/10.1016/j.oneear.2023.05.006.

GAO (U.S. Government Accountability Office). 2022. "Status, Challenges, and Policy Options for Carbon Capture, Utilization, and Storage." Technology Assessment. Washington, DC: U.S. Government Accountability Office. https://www.gao.gov/assets/730/723198.pdf.

Global Thermostat. 2023. "This Is What Confronting Climate Change Looks Like!" *Global Thermostat*, April 14. https://www.globalthermostat.com/news-and-updates/this-is-what-confronting-climate-change-looks-like.

Gordon, J.A., N. Balta-Ozkan, and S.A. Nabavi. 2023. "Price Promises, Trust Deficits and Energy Justice: Public Perceptions of Hydrogen Homes." *Renewable and Sustainable Energy Reviews* 188:113810.

Hammel, P. 2024. "Pipeline Company, Environmental Group Strike Unique 'Community Benefits' Agreement." *Nebraska Examiner*, April 9. https://nebraskaexaminer.com/2024/04/09/pipeline-company-environmental-group-strike-unique-community-benefits-agreement.

Han, X., S. Zhao, X. Cui, and Y. Yan. 2019. "Localization of CO_2 Gas Leakages Through Acoustic Emission Multi-Sensor Fusion Based on Wavelet-RBFN Modeling." *Measurement Science and Technology* 30(8):085007. https://doi.org/10.1088/1361-6501/ab1025.

Harvestone Group. 2023. "Harvestone: CO_2 Injection Commences at Blue Flint Ethanol CCS Project." *Ethanol Producer Magazine*. November 6. https://ethanolproducer.com/articles/harvestone-co2-injection-commences-at-blue-flint-ethanol-ccs-project.

Heirloom. 2023. "Heirloom Unveils America's First Commercial Direct Air Capture Facility." *Heirloom*, November 9. https://www.heirloomcarbon.com/news/heirloom-unveils-americas-first-commercial-direct-air-capture-facility.

Hernandez, R.R., S.M. Jordaan, B. Kaldunski, and N. Kumar. 2020. "Aligning Climate Change and Sustainable Development Goals with an Innovation Systems Roadmap for Renewable Power." *Frontiers in Sustainability* 1. https://www.frontiersin.org/articles/10.3389/frsus.2020.583090.

Hiar, C. 2023. "What's Next for Direct Air Capture?" *E&E News: Climatewire*, October 11. https://www.eenews.net/articles/whats-next-for-direct-air-capture.

Hughes, S., A. Zoelle, M. Woods, S. Henry, S. Homsy, S. Pidaparti, N. Kuehn, et al. 2022. "Cost of Capturing CO_2 from Industrial Sources." DOE/NETL-2022/3319. Pittsburgh, PA: National Energy Technology Laboratory. https://www.netl.doe.gov/projects/files/CostofCapturingCO2fromIndustrialSources_071522.pdf.

Jacobson, M.Z. 2019. "The Health and Climate Impacts of Carbon Capture and Direct Air Capture." *Energy and Environmental Science* 12(12):3567–3574. https://doi.org/10.1039/C9EE02709B.

Jenkins, J.D., E.N. Mayfield, J. Farbes, G. Schivley, N. Patankar, and R. Jones. 2023. "Climate Progress and the 117th Congress: The Impacts of the Inflation Reduction Act and Infrastructure Investment and Jobs Act." Princeton, NJ: Princeton University. https://repeatproject.org/docs/REPEAT_Climate_Progress_and_the_117th_Congress.pdf.

Just Transition Alliance. 2020. "False Solutions to Address Climate Change." https://jtalliance.org/wp-content/uploads/2020/02/False-Solutions.pdf.

Kammer, R., D. Rodriguez, and D. McFarlane. 2023. "Carbon Capture and Storage Opportunities in the Mid-Atlantic." Minneapolis, MN: Great Plains Institute. https://carboncaptureready.betterenergy.org/wp-content/uploads/2023/11/Mid-Atlantic-Report.pdf.

Kenton, C., and B. Silton. n.d. "Repurposing Natural Gas Lines: The CO_2 Opportunity." *ADL Ventures* (blog). https://www.adlventures.com/blogs/repurposing-natural-gas-lines-the-co2-opportunity.

Kim, J., Y. Lee, and I. Moon. 2011. "An Index-Based Risk Assessment Model for Hydrogen Infrastructure." *International Journal of Hydrogen Energy* 36(11):6387–6398. https://doi.org/10.1016/j.ijhydene.2011.02.127.

Knoope, M.M.J., W. Guijt, A. Ramírez, and A.P.C. Faaij. 2014. "Improved Cost Models for Optimizing CO_2 Pipeline Configuration for Point-to-Point Pipelines and Simple Networks." *International Journal of Greenhouse Gas Control* 22(March 1):25–46. https://doi.org/10.1016/j.ijggc.2013.12.016.

Kuprewicz, R.B. 2022. "Safety of Hydrogen Transportation by Gas Pipelines." Redmond, WA: Accufacts Inc. https://pstrust.org/wp-content/uploads/2022/11/11-28-22-Final-Accufacts-Hydrogen-Pipeline-Report.pdf.

Larsen, J., W. Herndon, M. Grant, and P. Marsters. 2019. "Capturing Leadership: Policies for the US to Advance Direct Air Capture Technology." New York: Rhodium Group, LLC. https://rhg.com/wp-content/uploads/2019/05/Rhodium_CapturingLeadership_May2019-1.pdf.

Larson, E., C. Greig, J. Jenkins, E. Mayfield, A. Pascale, C. Zhang, J. Drossman, et al. 2021. "Net-Zero America: Potential Pathways, Infrastructure, and Impacts." Princeton, NJ: Princeton University. https://www.dropbox.com/s/ptp92f65lgds5n2/Princeton%20NZA%20FINAL%20REPORT%20%2829Oct2021%29.pdf?dl=0.

LBNL (Lawrence Berkeley National Laboratory). n.d. "Generation, Storage, and Hybrid Capacity in Interconnection Queues." https://emp.lbl.gov/generation-storage-and-hybrid-capacity.

Lee, S.-Y., J.-U. Lee, I.-B. Lee, and J. Han. 2017. "Design Under Uncertainty of Carbon Capture and Storage Infrastructure Considering Cost, Environmental Impact, and Preference on Risk." *Applied Energy* 189(March 1):725–738. https://doi.org/10.1016/j.apenergy.2016.12.066.

Liu, N. 2021. "Increasing Blue Hydrogen Production Affordability." *Gulf Energy*, June 1. https://gulfenergyinfo.com/h2tech/articles/2021/q2-2021/increasing-blue-hydrogen-production-affordability.

LSU (Louisiana State University) Center for Energy Studies. 2023. "The Economic Implications of Carbon Capture and Sequestration for the Gulf Coast Economy." https://www.lsu.edu/ces/publications/2023/economic-implications-carbon-capture-sequestration-2022-online-updated.pdf.

Magneschi, G., T. Zhang, and R. Munson. 2017. "The Impact of CO_2 Capture on Water Requirements of Power Plants." 13th International Conference on Greenhouse Gas Control Technologies, GHGT-13, 14–18 November 2016, Lausanne, Switzerland 114(July 1):6337–6347. https://doi.org/10.1016/j.egypro.2017.03.1770.

Mahabir, J., N. Samaroo, M. Janardhanan, and K. Ward. 2022. "Pathways to Sustainable Methanol Operations Using Gas-Heated Reforming (GHR) Technologies." *Journal of CO_2 Utilization* 66(December 1):102302. https://doi.org/10.1016/j.jcou.2022.102302.

Merrill, D. 2021. "The U.S. Will Need a Lot of Land for a Zero-Carbon Economy." *Bloomberg*, June 3. https://www.bloomberg.com/graphics/2021-energy-land-use-economy/#xj4y7vzkg.

Milestone Carbon. 2023. "Milestone Carbon Announces Development of Carbon Sequestration Hub in Midland Basin." *Businesswire*, September 26. https://www.businesswire.com/news/home/20230926373201/en.

Minnkota Power Cooperative. 2022a. "In the Matter of a Hearing Called on a Motion of the Commission to Consider the Application of Minnkota Power Cooperative, Inc. Requesting Consideration for the Geologic Storage of Carbon Dioxide in the Broom Creek Formation from the Milton R. Young Station Location in Sections 35 and 36, T.142N, R.84W, Sections 19, 20, 21, 22, 26, 27, 28, 29, 30, 31, 32, 33, 34, and 35, T.142N, R.83W, Sections 1, 2, 12, and 13, T.141N, R.84W, Sections 1, 2, 3, 4, 5, 6, 7, 8, 9, 10, 11, 12, 14, 15, 16, 17, 18, 19, 20, and 21, T.141N, R.83W, Oliver County, ND Pursuant to NDAC Section 43-05-01." Industrial Commission of the State of North Dakota, Case No. 29029, Order No. 31583. October 14. https://www.dmr.nd.gov/dmr/sites/www/files/documents/Oil%20and%20Gas/Class%20VI/Minnkota/BC/C29029.pdf.

Minnkota Power Cooperative. 2022b. "In the Matter of a Hearing Called on a Motion of the Commission to Consider the Application of Minnkota Power Cooperative, Inc. Requesting Consideration for the Geologic Storage of Carbon Dioxide in the Deadwood Formation from the Milton R. Young Station Located in Sections 35 and 36, T.142N, R.84W, Sections 19, 20, 21, 22, 26, 27, 28, 29, 30, 31, 32, 33, 34, and 35, T.142N, R.83W, Sections 1, 2, 12, and 13, T.141N, R.84W, Sections 1, 2, 3, 4, 5, 6, 7, 8, 9, 10, 11, 12, 14, 15, 16, 17, 18, 19, 20 and 21, T.141N, R.83W, Oliver County, ND Pursuant to NDAC Section 43-05-01." Industrial Commission of the State of North Dakota, Case No. 29032, Order No. 31586. October 14. https://www.dmr.nd.gov/dmr/sites/www/files/documents/Oil%20and%20Gas/Class%20VI/Minnkota/DW/C29032.pdf.

Moniz, E.J., J.D. Brown, S.D. Comello, M. Jeong, M. Downey, and M.I. Cohen. 2023. "Turning CCS Projects in Heavy Industry and Power into Blue Chip Financial Investments." Energy Future Initiative. https://efifoundation.org/wp-content/uploads/sites/3/2023/02/20230212-CCS-Final_Full-copy.pdf.

Morken, A.K., B. Nenseter, S. Pedersen, M. Chhaganlal, J.K. Feste, R. Bøe Tyborgnes, Ø. Ullestad, et al. 2014. "Emission Results of Amine Plant Operations from MEA Testing at the CO_2 Technology Centre Mongstad." 12th International Conference on Greenhouse Gas Control Technologies, GHGT-12 63(January 1):6023–6038. https://doi.org/10.1016/j.egypro.2014.11.636.

Morken, A.K., S. Pedersen, E.R. Kleppe, A. Wisthaler, K. Vernstad, Ø. Ullestad, N. Enaasen Flø, L. Faramarzi, and E. Steinseth Hamborg. 2017. "Degradation and Emission Results of Amine Plant Operations from MEA Testing at the CO_2 Technology Centre Mongstad." *13th International Conference on Greenhouse Gas Control Technologies, GHGT-13, 14–18 November 2016, Lausanne, Switzerland* 114(July 1):1245–1262. https://doi.org/10.1016/j.egypro.2017.03.1379.

Murugan, A., S. Bartlett, J. Hesketh, H. Becker, and G. Hinds. 2019. "Project Closure Report: Hydrogen Odorant and Leak Detection Part 1, Hydrogen Odorant." United Kingdom: Scotland Gas Networks. https://sgn.co.uk/sites/default/files/media-entities/documents/2020-09/Hydrogen_Odorant_and_Leak_Detection_Project_Closure_Report_SGN.pdf.

NASEM (National Academies of Sciences, Engineering, and Medicine). 2019. *Negative Emissions Technologies and Reliable Sequestration: A Research Agenda*. Washington, DC: The National Academies Press. https://doi.org/10.17226/25259.

NASEM. 2023a. *Carbon Dioxide Utilization Markets and Infrastructure: Status and Opportunities: A First Report*. Washington, DC: The National Academies Press. https://doi.org/10.17226/26703.

NASEM. 2023b. *Laying the Foundation for New and Advanced Nuclear Reactors in the United States*. Washington, DC: The National Academies Press. https://doi.org/10.17226/26630.

NASEM. 2024. *Accelerating Decarbonization in the United States: Technology, Policy, and Societal Dimensions*. Washington, DC: The National Academies Press. https://doi.org/10.17226/25931.

Navigator Heartland Greenway LLC. 2021. "Heartland Greenway: A Navigator CO_2 Ventures LLC Project." Iowa Utilities Board, HLP-2021-0003. https://wcc.efs.iowa.gov/cs/idcplg?IdcService=GET_FILE&allowInterrupt=1&RevisionSelectionMethod=latest&dDocName=2076639&noSaveAs=1.

New Energy Economy. n.d. "Opposing False Solutions." New Energy Economy. https://www.newenergyeconomy.org/opposing-false-solutions.

Northern Gas Networks. 2016. "H21 Leeds City Gate." United Kingdom: Northern Gas Networks. https://www.northerngasnetworks.co.uk/wp-content/uploads/2017/04/H21-Executive-Summary-Interactive-PDF-July-2016-V2.pdf.

NPC (National Petroleum Council). 2019. "Meeting the Dual Challenge: A Roadmap to At-Scale Deployment of Carbon Capture, Use, and Storage." National Petroleum Council. https://dualchallenge.npc.org/downloads.php.

NPC. 2024. "Chapter 3—LCI Hydrogen—Connecting Infrastructure." In *Harnessing Hydrogen: A Key Element of the U.S. Energy Future*. https://harnessinghydrogen.npc.org/files/H2-CH_3-Connecting_Infra-2024-04-23.pdf.

NREL (National Renewable Energy Laboratory). n.d. "RE Data Explorer." RE Data Explorer. https://data.re-explorer.org.

OCAP (Organic CO_2 for Assimilation by Plants). n.d. "Pure CO_2 for Greenhouses." https://www.ocap.nl/nl/images/OCAP_Factsheet_English_tcm978-561158.pdf;%20https://ec.europa.eu/competition/state_aid/cases/270589/270589_1965702_106_2.pdf;%20https://www.youtube.com/watch?v=0rC37N4Bbo4.

Offermann-van Heek, J., K. Arning, A. Linzenich, and M. Ziefle. 2018. "Trust and Distrust in Carbon Capture and Utilization Industry as Relevant Factors for the Acceptance of Carbon-Based Products." *Frontiers in Energy Research* 6. https://www.frontiersin.org/articles/10.3389/fenrg.2018.00073.

OGCI (Oil and Gas Climate Initiative). n.d. "Bayou Bend CCS." The CCUS Hub. https://ccushub.ogci.com/focus_hubs/bayou-bend-ccs.

OSHA (Occupational Safety and Health Administration). 1996. "OSHA Hazard Information Bulletins Potential Carbon Dioxide (CO_2) Asphyxiation Hazard When Filling Stationary Low Pressure CO_2 Supply Systems." Department of Labor. https://www.osha.gov/publications/hib19960605.

Ou, Y., W. Shi, S.J. Smith, C.M. Ledna, J.J. West, C.G. Nolte, and D.H. Loughlin. 2018. "Estimating Environmental Co-Benefits of U.S. Low-Carbon Pathways Using an Integrated Assessment Model with State-Level Resolution." *Applied Energy* 216(April 15):482–493. https://doi.org/10.1016/j.apenergy.2018.02.122.

Oxy. 2022. "Occidental, 1PointFive to Begin Construction of World's Largest Direct Air Capture Plant in Texas Permian Basin." *Oxy*, August 25. https://www.oxy.com/news/news-releases/occidental-1pointfive-to-begin-construction-of-worlds-largest-direct-air-capture-plant-in-the-texas-permian-basin.

Parfomak, P.W. 2021. "Pipeline Transportation of Hydrogen: Regulation, Research, and Policy." R46700. Washington, DC: Congressional Research Service. https://www.everycrsreport.com/files/2021-03-02_R46700_294547743ff4516b1d562f7c4dae166186f1833e.pdf.

Parfomak, P.W. 2023. "Carbon Dioxide (CO_2) Pipeline Development: Federal Initiatives." Washington, DC: Congressional Research Service. https://crsreports.congress.gov/product/pdf/IN/IN12169.

Parolin, F., P. Colbertaldo, and S. Campanari. 2022. "Development of a Multi-Modality Hydrogen Delivery Infrastructure: An Optimization Model for Design and Operation." *Energy Conversion and Management* 266(August 15):115650. https://doi.org/10.1016/j.enconman.2022.115650.

Peterson, T. 2023. "New Midwest Battles Brew over CO_2 Pipelines." *Stateline*, June 13. https://stateline.org/2023/06/13/new-midwest-battles-brew-over-co2-pipelines.

Pett-Ridge, J., H.Z. Ammad, A. Aui, M. Ashton, S.E. Baker, B. Basso, M. Bradford, et al. 2023. "Roads to Removal: Options for Carbon Dioxide Removal in the United States." LLNL-TR-852901. Lawrence Livermore National Laboratory. https://roads2removal.org.

PHMSA (Pipeline and Hazardous Materials Safety Administration). 2022. "Failure Investigation Report—Denbury Gulf Coast Pipelines, LLC—Pipeline Rupture/Natural Force Damage." Washington, DC: Department of Transportation. https://www.phmsa.dot.gov/sites/phmsa.dot.gov/files/2022-05/Failure%20Investigation%20Report%20-%20Denbury%20Gulf%20Coast%20Pipeline.pdf.

PHMSA. 2023a. "Annual Report Mileage for Hazardous Liquid or Carbon Dioxide Systems." https://www.phmsa.dot.gov/data-and-statistics/pipeline/annual-report-mileage-hazardous-liquid-or-carbon-dioxide-systems.

PHMSA. 2023b. "Meeting Archive: CO_2 Safety Public Meeting 2023." https://www.onlinevideoservice.com/clients/PHMSA/053123.

Porthos. 2023. "CO_2 Reduction Through Storage Under the North Sea." Porthos: CO_2 Transport and Storage. https://www.porthosco2.nl/en.

Ramos, M. 2023. "'Carbon-Capture' Pipeline Plans Across Central Illinois Worry Land Owners." *Chicago Sun Times*, May 6. https://chicago.suntimes.com/2023/5/6/23709936/carbon-capture-pipeline-central-illinois-navigator-co2-heartland-greenway.

Red Trail Energy LLC. 2022. "Red Trail Energy Begins Carbon Capture and Storage." *Ethanol Producer Magazine*, July 17. https://ethanolproducer.com/articles/red-trail-energy-begins-carbon-capture-and-storage-19447.

Romero-Lankao, P., N. Rosner, R.A. Efroymson, E.S. Parisch, L. Blanco, S. Smolinski, and K. Kline. 2023. "Community Engagement and Equity in Renewable Energy Projects: A Literature Review."

Roussanaly, S., N. Berghout, T. Fout, M. Garcia, S. Gardarsdottir, S.M. Nazir, A. Ramirez, and E.S. Rubin. 2021. "Towards Improved Cost Evaluation of Carbon Capture and Storage from Industry." *International Journal of Greenhouse Gas Control* 106(March 1):103263. https://doi.org/10.1016/j.ijggc.2021.103263.

Sand, M., R.B. Skeie, M. Sandstad, S. Krishnan, G. Myhre, H. Bryant, R. Derwent, et al. 2023. "A Multi-Model Assessment of the Global Warming Potential of Hydrogen." *Communications Earth and Environment* 4(1):203. https://doi.org/10.1038/s43247-023-00857-8.

Santos, S. 2012. "CO_2 Transport via Pipeline and Ship." Presented at the CCS Opportunities in CCOP Region. September. Indonesia. https://www.ccop.or.th/eppm/projects/42/docs/(4.)%20CO2%20Transport%20-%20Overview%20(IEAGHG).pdf.

Shah, M.I., G. Lombardo, B. Fostås, C. Benquet, A. Kolstad Morken, and T. de Cazenove. 2018. "CO_2 Capture from RFCC Flue Gas with 30w% MEA at Technology Centre Mongstad, Process Optimization and Performance Comparison." Melbourne, Australia. https://dx.doi.org/10.2139/ssrn.3366149.

Sharma, I., D. Friedrich, T. Golden, and S. Brandani. 2019. "Exploring the Opportunities for Carbon Capture in Modular, Small-Scale Steam Methane Reforming: An Energetic Perspective." *International Journal of Hydrogen Energy* 44(29):14732–14743. https://doi.org/10.1016/j.ijhydene.2019.04.080.

Soraghan, M. 2023. "Midwest CO_2 Pipeline Rush Creates Regulatory Chaos." *E&E News*, March 3. https://www.eenews.net/articles/midwest-co2-pipeline-rush-creates-regulatory-chaos.

Spitzenberger, C., and T. Flechas. 2023. "Carbon Dioxide Major Accident Hazards Awareness." AIChE. https://www.aiche.org/resources/publications/cep/2023/june/carbon-dioxide-major-accident-hazards-awareness.

Stolaroff, J.K., S.H. Pang, W. Li, W.G. Kirkendall, H.M. Goldstein, R.D. Aines, and S.E. Baker. 2021. "Transport Cost for Carbon Removal Projects with Biomass and CO_2 Storage." *Frontiers in Energy Research* 9. https://www.frontiersin.org/articles/10.3389/fenrg.2021.639943.

Summit Carbon Solutions. 2023. "New Study Shows Summit Carbon Solutions Will Drive Economic Growth Across the Midwest." https://summitcarbonsolutions.com/driving-economic-growth.

Summit Carbon Solutions. 2024. "Summit Carbon Solutions Announces New Shipper for its Carbon Capture Project." https://summitcarbonsolutions.com/summit-carbon-solutions-announces-new-shipper-for-its-carbon-capture-project.

Summit Carbon Solutions. n.d.(a). "Project Benefits." https://summitcarbonsolutions.com/project-benefits.

Summit Carbon Solutions. n.d.(b). "Project Footprint." https://summitcarbonsolutions.com/project-footprint.

Tallgrass. 2022. "Tallgrass to Develop a Commercial-Scale CO_2 Sequestration Hub in Wyoming." *Tallgrass*, January 20. https://www.tallgrass.com/newsroom/press-releases/tallgrass-to-develop-a-commercial-scale-co2-sequestration-hub-in-wyoming.

Tallgrass. 2023. "Underground Injection Control Carbon Sequestration Class VI Permit Application." https://drive.google.com/file/d/1DX6U1qKS809G5FnQoYFybYh8rnZc8V26/view.

Tallgrass. 2024. "CO2." Tallgrass. https://tallgrass.com/energy-solutions/co2.

Thomley, E., and R. Kammer. 2023. "Carbon Dioxide (CO_2) Pipeline Safety: Safely Building a CO_2 Transportation and Storage Network." Great Plains Institute. https://carbonactionalliance.org/wp-content/uploads/CO2-Pipeline-Safey-Factsheet_1_30_2023.pdf.

Tomich, J. 2023. "Another CO_2 Pipeline Hurdle Surfaces in Illinois." *Energywire*, October 27. https://subscriber.politicopro.com/article/eenews/2023/10/27/latest-co2-pipeline-hurdle-surfaces-in-illinois-00123747.

Tomich, J., J. Plautz, and N.H. Farah. 2023. "Scuttled CO_2 Pipeline Renews Debate About State Hurdles." *Energywire*, October 23. https://www.eenews.net/articles/scuttled-co2-pipeline-renews-debate-about-state-hurdles.

U.S. Congress. 2020. "Consolidated Appropriations Act, 2021." Public Law 116-260. 116th Congress (2019–2020). https://www.congress.gov/116/plaws/publ260/PLAW-116publ260.pdf.

USGS (U.S. Geological Survey). 2013. "National Assessment of Geologic Carbon Dioxide Storage Resources—Results." U.S. Geological Survey Circular 1386. Reston, VA: U.S. Geological Survey. https://pubs.usgs.gov/circ/1386.

Voegele, E. 2023. "Navigator Cancels Heartland Greenway CO_2 Pipeline Project." *Ethanol Producer Magazine*, October 20. https://ethanolproducer.com/articles/navigator-cancels-heartland-greenway-co2-pipeline-project.

Wallace, M., L. Goudarzi, K. Callahan, and R. Wallace. 2015. "A Review of the CO_2 Pipeline Infrastructure in the U.S." DOE/NETL-2014/1681. Department of Energy Office of Energy Policy and Systems Analysis and National Energy Technology Laboratory. https://www.energy.gov/policy/articles/review-co2-pipeline-infrastructure-us.

Wang, S., R.J. Farrauto, S. Karp, J.H. Jeon, and E.T. Schrunk. 2018. "Parametric, Cyclic Aging and Characterization Studies for CO_2 Capture from Flue Gas and Catalytic Conversion to Synthetic Natural Gas Using a Dual Functional Material (DFM)." *Journal of CO_2 Utilization* 27(October 1):390–397. https://doi.org/10.1016/j.jcou.2018.08.012.

Webster, J. 2022. "The Inflation Reduction Act Will Accelerate Clean Hydrogen Adoption." *Atlantic Council* (blog), September 20. https://www.atlanticcouncil.org/blogs/energysource/the-inflation-reduction-act-will-accelerate-clean-hydrogen-adoption.

White House. 2023. "Biden-Harris Administration Announces Regional Clean Hydrogen Hubs to Drive Clean Manufacturing and Jobs." https://www.whitehouse.gov/briefing-room/statements-releases/2023/10/13/biden-harris-administration-announces-regional-clean-hydrogen-hubs-to-drive-clean-manufacturing-and-jobs.

Williams, J.H., R.A. Jones, B. Haley, G. Kwok, J. Hargreaves, J. Farbes, and M.S. Torn. 2021. "Carbon-Neutral Pathways for the United States." *AGU Advances* 2(1):e2020AV000284. https://doi.org/10.1029/2020AV000284.

Wolf Carbon Solutions. n.d. "Mt. Simon Hub." https://wolfcarbonsolutions.com/mt-simon-hub.

Wolf Carbon Solutions US LLC. 2023. "Wolf Carbon Solutions US LLC Mt. Simon Hub Pipeline System Application for Certificate of Authority." https://www.icc.illinois.gov/docket/P2023-0475/documents/339033.

11

A Comprehensive Research Agenda for CO_2 and Coal Waste Utilization

In its analysis of the status and challenges for CO_2 and coal waste utilization, the committee identified key research, development, and demonstration (RD&D) needs to enable future utilization opportunities. For CO_2 utilization, research needs exist for all conversion processes (mineralization, chemical, biological) and all product classes that can be derived from CO_2 (construction materials, elemental carbon materials, fuels and chemicals, polymers). For coal waste utilization, there are research needs across conversion and separation processes to generate carbon-based products (e.g., construction materials, elemental carbon materials) and extraction of critical minerals and rare earth elements. Beyond science, engineering, and technology development, research is needed on CO_2 utilization infrastructure, tools to assess economic and environmental impacts of CO_2 utilization processes, public perception of CO_2 utilization, and market opportunities for CO_2- and coal waste–derived products. These research needs are summarized in Table 11-1, along with the committee's corresponding recommendation on how to address the need.[1] This research agenda updates and builds on the one in the 2019 National Academies' report *Gaseous Carbon Waste Streams Utilization: Status and Research Needs*.

For each research need, Table 11-1 identifies the relevant funding agencies or other actors; specifies whether the need is for basic research, applied research, technology demonstration, or enabling technologies and processes; denotes the research theme(s) that the research need falls into; and, where applicable, indicates the relevant research area and product class covered by the research need, as well as whether the product is long- or short-lived. While Table 11-1 does not explicitly include recommendations to industry, the committee recognizes that industrial companies perform RD&D in these areas, and any recommendation of federal funding should be available to all eligible entities, including industrial researchers. See Appendix E for a more detailed description of the research agenda and tables of the research needs grouped by the Department of Energy (DOE) office to which they are directed.

[1] The findings corresponding to each recommendation and more detail about the research needs can be found in the individual R&D chapters: Chapter 5: Mineralization of CO_2 to Inorganic Carbonates; Chapter 6: Chemical CO_2 Conversion to Elemental Carbon Materials; Chapter 7: Chemical CO_2 Conversion to Fuels, Chemicals, and Polymers; Chapter 8: Biological CO_2 Conversion to Fuels, Chemicals, and Polymers; and Chapter 9: Products from Coal Waste.

TABLE 11-1 A Comprehensive Research Agenda for CO_2 and Coal Waste Utilization, Covering CO_2 Mineralization, Chemical CO_2 Conversion, Biological CO_2 Conversion, Coal Waste Utilization, CO_2 Utilization Markets, LCA, TEA, and Societal/Equity Assessments, Policy and Regulatory Needs for CO_2 Utilization, and CO_2 Utilization Infrastructure

Research, Development, and Demonstration Need	Funding Agencies or Other Actors	Basic, Applied, Demonstration, or Enabling	Research Area	Product Class	Long- or Short-Lived	Research Themes	Source
CO_2 Mineralization to Inorganic Carbonates							
5-A. Evaluation and expansion of mapping of alkaline resources, including minerals and industrial wastes, as well as their chemical and physical properties.	DOE-FECM USGS EPA	Enabling	Mineralization	Construction materials	Long-lived	Resource mapping	Fin. 5-1 Rec. 5-1
5-B. Multimodal optimization of existing and new infrastructure to link feedstocks, including CO_2 and reactant minerals, to sites of carbon mineralization and product markets.	DOE-FECM DOT	Enabling	Mineralization	Construction materials	Long-lived	Enabling technology and infrastructure needs Market opportunities	Fin. 5-1 Rec. 5-1
5-C. Fundamental and translational research on emerging approaches to carbon mineralization to improve energy efficiency, process efficiency, product selectivity, and ability to scale to the gigatonne level.	NSF DOE-BES DOE-ARPA-E DoD USGS	Basic Applied Demonstration	Mineralization	Construction materials	Long-lived	Reactor design and reaction engineering Energy efficiency, electrification, and alternative heating	Fin. 5-1 Rec. 5-2

Recommendation 5-1: Support research and development to link alkaline resources to carbon mineralization sites—The availability of alkaline resources, including minerals and industrial wastes, and their chemical and physical properties should be carefully evaluated and mapped for carbon mineralization integrated with different CO_2 sources to create a carbon mineralization atlas. This effort should be funded by the Department of Energy (DOE) and the U.S. Geological Survey. DOE and the Department of Transportation should support research and development efforts focused on how to link CO_2 and mineral sources to carbon mineralization sites and product markets by designing multimodal solutions based on new and existing infrastructure to process and transport efficiently large amounts of solids (both feedstock and carbonate products).

Recommendation 5-2: Support research and development to scale carbon mineralization technologies—The National Science Foundation (NSF), Department of Energy's (DOE's) Office of Basic Energy Sciences and Advanced Research Projects Agency–Energy, and U.S. Geological Survey, should support fundamental and translational research into emerging carbon mineralization approaches (e.g., electrochemically driven carbon mineralization). Funding should be available for university–industry–national laboratory collaborations to rapidly scale up and deploy carbon mineralization technologies and address challenges associated with energy requirements for large-scale mining and mineral processing (e.g., grinding), process integration, chemical recycling, environmental challenges (e.g., handling asbestos), and water requirements, among others. Additionally, DOE and NSF should provide more fundamental and applied research funding for new materials discovery and characterization that would enable new processing such as 3D-printed concrete.

5-D. Understanding of local environmental and ecological impacts of ocean-based CO_2 utilization and development of an environmental protocol to assess and mitigate unexpected environmental and ecological impacts from pH changes.	DOE-ARPA-E DOE-EERE DOE-FECM DoD-ONR NOAA EPA	Applied	Mineralization—ocean-based CO_2 utilization	Construction materials	Long-lived	Environmental and societal considerations for CO_2 and coal waste utilization technologies	Fin. 5-2 Rec. 5-3
5-E. Development of a testing facility platform, similar to the National Carbon Capture Center, where various ocean-based carbon mineralization concepts and technologies can be evaluated in real ocean conditions with minimal environmental impacts.	NOAA DOE-FECM	Demonstration	Mineralization—ocean-based CO_2 utilization	Construction materials	Long-lived	Research centers and facilities	Fin. 5-2 Rec. 5-3
Recommendation 5-3: Support research and development for ocean-based carbon utilization technologies—The Department of Energy (DOE), along with other relevant agencies (Department of Defense [DoD], U.S. Environmental Protection Agency [EPA], and National Oceanic and Atmospheric Administration [NOAA]) should support research and development to understand the local environmental and ecological impacts of ocean-based CO_2 utilization solutions, which are still largely at early stages of development. A testing facility should be developed and installed (similar to the National Carbon Capture Center) to provide a platform where various ocean-based carbon mineralization concepts and technologies can be evaluated in actual ocean conditions but with minimal environmental impacts. DOE, DoD, EPA, and NOAA should also develop an environmental protocol to assess and mitigate unexpected environmental and ecological impacts because a local pH spike or alkalinity change could significantly impact the ocean environment.							
5-F. Full spectrum of research, development, and demonstration (RD&D) activities for electrochemically driven CO_2 mineralization under a wide range of electrolyte conditions, including seawater and brine. RD&D needs to include catalyst development, electrochemical cell design, membrane materials, and overall systems engineering and integration with carbon mineralization.	DOE-BES DOE-EERE DOE-FECM	Basic Applied Demonstration	Mineralization—Electrochemical	Construction materials	Long-lived	Catalyst innovation and optimization Reactor design and reaction engineering Integrated systems Computational modeling and machine learning	Fin. 5-3 Rec. 5-4

continued

TABLE 11-1 Continued

Research, Development, and Demonstration Need	Funding Agencies or Other Actors	Basic, Applied, Demonstration, or Enabling	Research Area	Product Class	Long- or Short-Lived	Research Themes	Source
5-G. Monitoring of side reactions (e.g., seawater oxidation at the anode producing chlorine gas) and membrane fouling issues (e.g., precipitation of undesired solid phases). Development of tools to monitor local reaction environments at electrode/electrolyte interfaces and evaluation of recyclability of process water with spent acids and bases as well as dissolved ions.	DOE-EERE DOE-FECM DOE-BES	Basic Applied	Mineralization—Electrochemical	Construction materials	Long-lived	Fundamental knowledge Reactor design and reaction engineering Environmental and societal considerations for CO_2 and coal waste utilization technologies	Fin. 5-3 Rec. 5-4
Recommendation 5-4: Support research, development, and demonstration for electrochemically driven CO_2 mineralization under a wide range of electrolyte conditions—The Department of Energy's (DOE's) Office of Basic Energy Sciences, Office of Energy Efficiency and Renewable Energy, and Office of Fossil Energy and Carbon Management should support a full range of research, development, and deployment (RD&D) activities for electrochemically driven CO_2 mineralization under a wide range of electrolyte conditions, including seawater and brine. Specifically, DOE should increase support for fundamental experimental and theoretical research into catalyst development, electrochemical cell design, membrane materials, and overall systems engineering and integration with carbon mineralization. As part of this effort, DOE should also fund RD&D on monitoring side reactions (e.g., seawater oxidation at the anode producing chlorine gas) and membrane fouling issues (e.g., precipitation of undesired solid phases) that could impact the robustness of electrochemically driven carbon mineralization; developing tools to monitor local reaction environments at electrode/electrolyte interfaces; and evaluating the possibility of recycling process water with spent acids, bases, and dissolved ions.							
5-H. Fundamental to translational research of carbon mineralization integrated with metal recovery, focused on energy-efficient grinding/comminution, selective separation, improved recycling, reduced emissions, and systems integration and optimization.	DOE-BES DOE-EERE DOE-FECM DOE-ARPA-E	Basic Applied	Mineralization	Construction materials	Long-lived	Reactor design and reaction engineering Integrated systems Energy efficiency, electrification, and alternative heating Separations	Rec. 5-5
5-I. University–industry–national laboratory collaborations to rapidly scale up and deploy carbon-negative mining technologies with large CO_2 utilization potential.	DOE-EERE DOE-FECM DOE-ARPA-E	Applied Demonstration	Mineralization	Construction materials	Long-lived	Research centers and facilities	Rec. 5-5

Recommendation 5-5: Increase support for research into CO_2 mineralization integrated with metal recovery and separation—The Department of Energy's Offices of Basic Energy Sciences, Energy Efficiency and Renewable Energy, Fossil Energy and Carbon Management, and Advanced Research Projects Agency–Energy (ARPA-E) should increase support for basic and applied research into carbon mineralization integrated with metal recovery (e.g., the MINER program at ARPA-E). The research should focus on, but not be limited to, the development of energy-efficient grinding/comminution of minerals and rocks, selective separation of metals in the presence of large amounts of competing ions (e.g., Ca and Mg ions), improved recycling of chemicals (e.g., ligands), reduced emissions (e.g., mine tailings) and systems integration and optimization. Funding should also be available for university–industry–national laboratory collaborations to rapidly scale up and deploy carbon-negative mining technologies with large CO_2 utilization potential, producing solid carbonates.

5-J. Testing, standardization, and certification systems for replacement construction materials produced from CO_2.	NIST Industrial associations	Enabling	Mineralization	Construction materials	Long-lived	Certification and standards	Rec. 5-6
5-K. Materials discovery and characterization of new forms of mineral carbonates to enable new processing like 3D-printed concrete.	DOE-BES DOE-EERE-AMMTO NSF	Basic Applied	Mineralization	Construction materials	Long-lived	Fundamental knowledge	Rec. 5-6

Recommendation 5-6: Develop performance-based standards for construction materials to enable and encourage use of innovative materials—The National Institute of Standards and Technology should collaborate with industrial associations (e.g., Portland Cement Association, National Concrete Masonry Association) to develop testing, standardization, and certification systems for replacement building materials such as low-carbon or carbon-negative cement and aggregates from CO_2 mineralization in terms of the product performance (e.g., compressive strength) and carbon content. This work should result in an industry standard and certification process to provide to carbon mineralization companies.

Chemical CO_2 Conversion to Elemental Carbon Materials

6-A. Foundational knowledge of thermochemical, electrochemical, photochemical, and plasma processes to make elemental carbon products from CO_2.	DOE-BES NSF	Basic	Chemical	Elemental Carbon Materials	Long-lived Short-lived	Fundamental knowledge	Rec. 6-1

Recommendation 6-1: Support basic research to advance CO_2 to elemental carbon technologies—Basic Energy Sciences within the Department of Energy's Office of Science and the National Science Foundation should invest in building the knowledge foundation and accelerating the maturities of the four CO_2 to elemental carbon technology areas: thermochemical, electrochemical, photochemical, and plasmachemical.

6-B. Novel and improved catalysts and low-energy reaction processes to produce elemental carbon products from CO_2.	DOE-BES DOE-EERE DOE-FECM	Basic Applied	Chemical	Elemental Carbon Materials	Long-lived Short-lived	Catalyst innovation and optimization Reactor design and reaction engineering Energy efficiency, electrification, and alternative heating	Fin. 6-2 Rec. 6-2

continued

TABLE 11-1 Continued

Research, Development, and Demonstration Need	Funding Agencies or Other Actors	Basic, Applied, Demonstration, or Enabling	Research Area	Product Class	Long- or Short-Lived	Research Themes	Source
6-C. Catalysts and processes that are selective for particular material morphologies.	DOE-BES DOE-EERE DOE-FECM	Basic Applied	Chemical	Elemental Carbon Materials	Long-lived Short-lived	Catalyst innovation and optimization	Fin. 6-2 Rec. 6-2
6-D. Enhanced activity, selectivity, and stability of catalysts to achieve high performance of reactions transforming CO_2 to elemental carbon products.	DOE-BES NSF	Basic	Chemical	Elemental Carbon Materials	Long-lived Short-lived	Catalyst innovation and optimization	Fin. 6-3 Rec. 6-2
6-E. Understanding and control of processes that produce CO_2-derived elemental carbon products.	DOE-BES NSF	Basic	Chemical	Elemental Carbon Materials	Long-lived Short-lived	Fundamental knowledge	Fin. 6-3 Rec. 6-2
6-F. Reaction electrification and heat integration including plasma processes (thermochemical, plasmachemical, etc.).	DOE-BES DOE-EERE DOE-FECM	Basic Applied	Chemical	Elemental Carbon Materials	Long-lived Short-lived	Reactor design and reaction engineering Energy efficiency, electrification, and alternative heating	Fin. 6-3 Rec. 6-2
6-G. Separation of catalyst from solid carbon products, and different elemental carbon materials from each other.	DOE-BES DOE-EERE DOE-FECM	Basic Applied	Chemical	Elemental Carbon Materials	Long-lived Short-lived	Separations	Fin. 6-3 Rec. 6-2
Recommendation 6-2: Fund research into catalysts and materials for CO_2 to elemental carbon conversion—Basic Energy Sciences within the Department of Energy's Office of Science (DOE-BES) and the National Science Foundation should fund basic research into the discovery of high-performance catalysts that are active, morphologically selective, and robust for low-cost CO_2 to elemental carbon (CTEC) conversion. DOE-BES, DOE's Office of Energy Efficiency and Renewable Energy, and DOE's Office of Fossil Energy and Carbon Management should, jointly or independently, fund research on materials (e.g., catalysts, reducing agents) used in CO_2 to elemental carbon processes. These investigations should aim to discover new, optimal materials for catalysis and separation; understand how to control CTEC reactions to increase the diversity of products and selectively generate desired morphologies; and increase energy efficiency of CTEC reaction processes that can be powered by clean energy.							
6-H. Development of tandem processes to produce elemental carbon products from CO_2.	DOE-BES DOE-EERE DOE-ARPA-E	Basic Applied	Chemical	Elemental Carbon Materials	Long-lived Short-lived	Integrated systems Reactor design and reaction engineering	Fin. 6-4 Rec. 6-3
Recommendation 6-3: Fund the development of tandem CO_2 to elemental carbon technologies to maximize economic and environmental benefits—Basic Energy Sciences within the Department of Energy's (DOE's) Office of Science, DOE's Office of Energy Efficiency and Renewable Energy, and the Advanced Research Projects Agency–Energy should fund independently and/or collectively the development of tandem CO_2 to elemental carbon (CTEC) technologies that can combine the advantages of different types of CTEC processes to maximize the economic and environmental benefits of the converted carbon materials.							

6-I. Integrated CO_2 capture and conversion to elemental carbon materials including improved technology integration and enhanced economic and/or environmental benefits.	DOE-FECM DOE-EERE DOE-ARPA-E	Applied	Chemical	Elemental Carbon Materials	Long-lived Short-lived	Integrated systems	Fin. 6-5 Rec. 6-4

Recommendation 6-4: Fund research on integrated CO_2 capture and conversion to elemental carbon materials to maximize economic and environmental benefits—The Department of Energy's Office of Fossil Energy and Carbon Management, Office of Energy Efficiency and Renewable Energy, and the Advanced Research Projects Agency–Energy should fund research on integrated CO_2 capture and conversion to elemental carbon materials, with particular consideration of technology integration and economic and environmental benefit enhancement.

Chemical CO_2 Conversion to Organic Products

7-A. Improvements in catalytic activity, selectivity, and stability (including tolerance to impurities) for thermochemical CO_2 conversion.	DOE-BES NSF DoD	Basic	Chemical—Thermochemical	Chemicals	Short-lived	Catalyst innovation and optimization Computational modeling and machine learning	Fin. 7-1 Sec. 7.2.1.3
7-B. Discovery research into catalysts and processes that use alternative heating methods, such as (pulsed) electrical heating, with goals of improving catalyst performance, yielding energy savings, and reducing GHG emissions by using clean electricity.	DOE-BES NSF DoD	Basic	Chemical—Thermochemical	Chemicals	Short-lived	Catalyst innovation and optimization Energy efficiency, electrification, and alternative heating Computational modeling and machine learning	Fin. 7-1 Rec. 7-1
7-C. Continued research and development into low-carbon hydrogen and other carbon-neutral reductants to facilitate scale up of thermochemical CO_2 conversion that can achieve net-zero CO_2 utilization.	DOE-BES DOE-EERE DOE-FECM DOE-ARPA-E DoD	Enabling	Chemical—Thermochemical	Chemicals	Short-lived	Enabling technology and infrastructure needs	Rec. 7-1

Recommendation 7-1: Support research on catalyst development, electrical heating, and carbon-neutral reductants for thermochemical CO_2 conversion—Basic Energy Sciences within the Department of Energy's Office of Science (DOE-BES), the National Science Foundation, and the Department of Defense (DoD) should increase support for experimental and theoretical discovery research into catalysts and processes that utilize carbon-neutral and efficient methods of electrical heating to convert CO_2 to useful chemicals and chemical intermediates (e.g., targeted heating, microwave heating). DOE-BES, DoD, and DOE's Office of Energy Efficiency and Renewable Energy, Office of Fossil Energy and Carbon Management, Office of Clean Energy Demonstrations, and Advanced Research Projects Agency–Energy should continue to support research and development (R&D) that facilitates scale up of thermochemical CO_2 conversion to achieve net-zero CO_2 utilization. This includes R&D on the production of low-carbon hydrogen and other carbon-neutral reductants and the integration of solar thermochemical hydrogen production and CO_2 conversion with thermal energy storage.

continued

TABLE 11-1 Continued

Research, Development, and Demonstration Need	Funding Agencies or Other Actors	Basic, Applied, Demonstration, or Enabling	Research Area	Product Class	Long- or Short-Lived	Research Themes	Source
7-D. Engineering and systems optimization to integrate low-carbon energy sources with high-temperature reaction systems for CO_2 conversion to hydrocarbon products.	DOE-FECM	Applied	Chemical—Thermochemical	Chemicals	Short-lived	Reactor design and reaction engineering Energy efficiency, electrification, and alternative heating	Fin. 7-2 Rec. 7-2
Recommendation 7-2: Support research on integrated systems for thermochemical CO_2 conversion—The Department of Energy's Office of Fossil Energy and Carbon Management should fund applied research on integration of variable renewable energy and energy storage into efficient, heat-integrated process systems for CO_2 conversion to hydrocarbon products.							
7-E. Discovery and development of electrocatalysts from abundant elements that are selective, stable, robust to impurities in CO_2 sources, and scalable, and that can produce single- and multicarbon products for both low- and high-temperature electrochemical processes.	DOE-BES DOE-FECM	Basic Applied	Chemical—Electrochemical	Chemicals	Short-lived	Catalyst innovation and optimization Computational modeling and machine learning	Fin. 7-3 Rec. 7-3
7-F. Discovery and development of abundant-element electrocatalysts for water oxidation or alternative anodic reactions to improve the cost and efficiency of electrochemical CO_2 conversion.	DOE-BES DOE-FECM	Basic Applied	Chemical—Electrochemical	Chemicals	Short-lived	Catalyst innovation and optimization Computational modeling and machine learning	Fin. 7-3 Rec. 7-3
7-G. Development of economical membrane materials that function over a wide pH range to improve the cost, efficiency, and scalability of electrochemical CO_2 conversion.	DOE-BES DOE-FECM	Basic Applied	Chemical—Electrochemical	Chemicals	Short-lived	Reactor design and reaction engineering Separations	Fin. 7-3 Rec. 7-3
Recommendation 7-3: Support research on developing electrocatalysts from abundant elements and membrane materials for electrochemical CO_2 conversion technologies—Basic Energy Sciences within the Department of Energy's Office of Science (DOE-BES) and DOE's Office of Fossil Energy and Carbon Management (DOE-FECM) should devote more effort to experimental and theoretical research for discovering and developing electrocatalysts from abundant elements that are selective, stable, and scalable to produce both single- and multicarbon products for both low- and high-temperature electrochemical processes. DOE-BES and DOE-FECM should also invest in developing abundant-element electrocatalysts for water oxidation or alternative anodic reactions as well as cost-effective, scalable membrane materials that function over a wide pH range to lower the overall cost of electrochemical CO_2 conversion. Long-term stability testing should be encouraged with new electrocatalyst development, along with testing for product selectivity and current density.							

7-H. Fundamental understanding of the sequence of processes involved in photochemical and photoelectrochemical conversion of CO_2 for light absorption, generation and separation of electron-hole pairs, and subsequent reduction of CO_2, across a variety of material types.	DOE-BES	Basic	Chemical—Photochemical Chemical—Photo-electrochemical	Chemicals	Short-lived	Fundamental knowledge Computational modeling and machine learning	Fin. 7-4 Rec. 7-4
7-I. Discovery research into materials for photochemical, photoelectrochemical, and plasmachemical catalytic conversion of CO_2.	DOE-BES	Basic	Chemical—Photochemical Chemical—Photo-electrochemical Chemical—Plasmachemical	Chemicals	Short-lived	Catalyst innovation and optimization Computational modeling and machine learning	Rec. 7-4
7-J. In-depth understanding of plasma-catalyst interactions for product selectivity.	DOE-BES	Basic	Chemical—Plasmachemical	Chemicals	Short-lived	Fundamental knowledge Computational modeling and machine learning	Fin. 7-4 Rec. 7-4
7-K. Improved devices, reactor design, and reaction engineering for photochemical, photoelectrochemical, and plasmachemical CO_2 conversions to optimize performance metrics and inform scale up.	DOE-FECM DOE-ARPA-E	Applied	Chemical—Photochemical Chemical—Photo-electrochemical Chemical—Plasmachemical	Chemicals	Short-lived	Reactor design and reaction engineering	Fin. 7-4 Rec. 7-4
Recommendation 7-4: Support research on mechanisms, materials, and reactor design for photo(electro)chemical and plasmachemical CO_2 conversion—Basic Energy Sciences within the Department of Energy's (DOE's) Office of Science should support more experimental and theoretical research into understanding fundamental mechanisms and materials discovery for photochemical, photoelectrochemical, and plasmachemical catalytic conversion of CO_2. DOE's Office of Fossil Energy and Carbon Management and Advanced Research Projects Agency–Energy should support research to enable development of improved materials, devices, and reactor design for such conversions.							
7-L. Development of tandem catalysis processes that couple two or more thermochemical, electrochemical, photochemical, and plasmachemical processes, with a goal of accessing products that a single process alone cannot achieve.	DOE-FECM DOE-ARPA-E	Basic Applied	Chemical	Chemicals	Short-lived	Integrated systems Computational modeling and machine learning	Fin. 7-5 Rec. 7-5

continued

TABLE 11-1 Continued

Research, Development, and Demonstration Need	Funding Agencies or Other Actors	Basic, Applied, Demonstration, or Enabling	Research Area	Product Class	Long- or Short-Lived	Research Themes	Source
Recommendation 7-5: Increase support for research on tandem catalysis for CO_2 conversion—The Department of Energy's Office of Fossil Energy and Carbon Management and Advanced Research Projects Agency–Energy should increase support for basic and applied research into tandem catalysis, including catalyst and membrane development, tandem reactor design, and process optimization.							
7-M. Development of integrated CO_2 capture and conversion, including discovery of molecules and materials, catalytic mechanisms, process optimization, CO_2 stream purification, and reactor design.	DOE-BES NSF DOE-FECM DOE-ARPA-E DOE-EERE	Basic Applied	Chemical	Chemicals	Short-lived	Integrated systems Reactor design and reaction engineering	Fin. 7-6 Rec. 7-6
Recommendation 7-6: Increase support for research on integrated capture and conversion of CO_2—Basic Energy Sciences within the Department of Energy's (DOE's) Office of Science, the National Science Foundation, and DOE's Office of Fossil Energy and Carbon Management, Office of Energy Efficiency and Renewable Energy, and Advanced Research Projects Agency–Energy should increase support for basic and applied research into integrated CO_2 capture and conversion, including discovery of molecules and materials, catalytic mechanisms, process optimization, CO_2 stream purification, and reactor design.							
7-N. Design and development of catalysts for rapid, stereoselective polymerization of a broader class of monomers with CO_2, especially those that can lead to polymers with properties more like thermoplastics and/or thermosets.	DOE-BES DOE-FECM	Basic Applied	Chemical—Thermochemical	Polymers	Long-lived Short-lived	Fundamental knowledge Catalyst innovation and optimization Computational modeling and machine learning	Fin. 7-7 Rec. 7-7
Recommendation 7-7: Support research on catalyst development for CO_2 polymerization with a broader class of monomers—Basic Energy Sciences within the Department of Energy's (DOE's) Office of Science and DOE's Office of Fossil Energy and Carbon Management should support more experimental and theoretical research into catalyst design and development to enable rapid, stereoselective polymerization of a broader class of monomers with CO_2. Such research could lead to polymers with properties more like those of conventional thermoplastics and/or thermosets, which could markedly expand opportunities for polymers made directly from CO_2.							
Biological CO_2 Conversion to Organic Products							
8-A. Pathway modeling and metabolic engineering of microorganisms to overcome biochemical, bioenergetic and metabolic limits to enhance the efficiency, titer, and productivity of photosynthetic, nonphotosynthetic, and hybrid systems.	DOE-BES DOE-BER DOE-BETO DOE-FECM	Basic Applied	Biological	Chemicals Polymers	Short-lived	Metabolic understanding and engineering Reactor design and reaction engineering	Fin. 8-1 Rec. 8-1

Recommendation 8-1: Coordination of fundamental and applied research is needed—Substantial fundamental and applied research needs to be conducted in order to understand and overcome biochemical, bioenergetic, and metabolic limits to higher reaction rates, conversion efficiency, and product titers. Various Department of Energy offices, including the Office of Science, Bioenergy Technologies Office, and Office of Fossil Energy and Carbon Management, as well as the Department of Defense, should coordinate fundamental and applied research to accelerate the advancement of efficient and implementable biochemical systems for carbon conversion.							
8-B. New, more efficient genetic manipulation tools must be developed to enhance the efficiency of CO_2 fixation and improve the understanding of carbon metabolism. Computational modeling and machine learning can also be exploited to this end.	DOE-BES DOE-BER NSF	Basic	Biological	Chemicals	Short-lived	Metabolic understanding and engineering Genetic manipulation Computational modeling and machine learning	Fin. 8-2 Rec. 8-2
8-C. Improved enzyme efficiency, selectivity, and stability, along with multienzyme metabolon design to overcome biochemical limits for photosynthetic, nonphotosynthetic, and hybrid systems.	DOE-BES DOE-BER NSF DOE- BETO DOE-FECM DOE-ARPA-E	Basic Applied	Biological	Chemicals Polymers	Short-lived	Fundamental knowledge Computational modeling and machine learning Metabolic understanding and engineering	Fin. 8-1 Fin. 8-2 Rec. 8-1 Rec. 8-2
8-D. Improved enzyme stability and scalability of redox-balanced systems to facilitate demonstration and scale up of cell-free and hybrid systems.	DOE-BES DOE-BER NSF DOE-BETO DOE-FECM DOE-ARPA-E	Basic Applied	Biological	Chemicals Polymers	Short-lived	Fundamental knowledge Reactor design and reaction engineering Integrated systems	Rec. 8-2
Recommendation 8-2: Support for advances in genetic engineering, systems modeling, and fundamental research is critical—New genetic engineering strategies must be developed to enhance the efficiency of CO_2 fixation, enabling the establishment of economical and scalable production systems. Continued refinement of genetic engineering tools and better understanding of the design principles of carbon metabolism are needed to improve CO_2 fixation. Additionally, systems modeling and machine learning can be exploited to optimize nutrient input, CO_2 delivery, light penetration, and other conditions to achieve higher productivities. Last, fundamental research to improve enzyme stability and the scalability of redox balanced systems is needed to make hybrid systems commercially viable and scalable. The Department of Energy's (DOE's) Office of Science and the National Science Foundation should continue to support fundamental research on these topics, and DOE's Bioenergy Technologies Office, Office of Fossil Energy and Carbon Management, and Advanced Research Projects Agency–Energy should support the related applied research.							

continued

TABLE 11-1 Continued

Research, Development, and Demonstration Need	Funding Agencies or Other Actors	Basic, Applied, Demonstration, or Enabling	Research Area	Product Class	Long- or Short-Lived	Research Themes	Source
8-E. Improve fundamental understanding of electrocatalyst design to increase efficiency, selectivity, and product profile control under biocompatible conditions.	DOE-BES DOE-BER NSF DOE-BETO DOE-FECM DOE-ARPA-E	Basic Applied	Biological—Hybrid Electro-bio	Chemicals Polymers	Short-lived	Fundamental Knowledge Catalyst innovation and optimization	Fin. 8-3 Rec. 8-3
Recommendation 8-3: Explore electrocatalysts that operate under biologically amenable conditions with high activity, selectivity, and stability—The National Science Foundation and the Department of Energy's Office of Science, Bioenergy Technologies Office, Office of Fossil Energy and Carbon Management, and Advanced Research Projects Agency–Energy should support fundamental and applied research on the development of electrocatalysts that operate under biologically amenable conditions with high efficiency, selectivity, and stability. Systems that produce and utilize bio-compatible (i.e., nontoxic, multicarbon) intermediates should be prioritized.							
8-F. Develop microorganisms and cell-free systems compatible with intermediates derived from electrocatalysis.	DOE-BES DOE-BER NSF DOE-BETO DOE-FECM DOE-ARPA-E	Basic Applied	Biological—Hybrid Electro-bio	Chemicals Polymers	Short-lived	Microbial engineering	Rec. 8-4
Recommendation 8-4: Develop microorganisms and cell-free systems that can efficiently produce target chemicals via intermediates derived from electrocatalysis—The National Science Foundation and the Department of Energy's Office of Science, Bioenergy Technologies Office, Office of Fossil Energy and Carbon Management, and Advanced Research Projects Agency–Energy should support fundamental and applied research on the development of microorganisms and cell-free systems that can efficiently produce target chemicals from catalysis-derived intermediates under conditions amenable to electrocatalysis. These efforts should include systems biology understanding of the limitations for the conversion of various electrocatalysis-derived intermediates, in particular, biocompatible intermediates, as well as the synthetic biology engineering of microorganisms and cell-free systems for efficient conversion of these intermediates to chemicals, materials, and fuels.							
8-G. Optimization of hybrid systems via evaluation of reactor design.	DOE-BETO DOE-ARPA-E DOE-FECM	Applied Demonstration	Biological—Hybrid	Chemicals Polymers	Short-lived	Integrated systems Reactor design and reaction engineering	Rec. 8-5
Recommendation 8-5: Evaluate reactor design and system integration for hybrid systems—The Department of Energy's Bioenergy Technologies Office, Advanced Research Projects Agency–Energy, and Office of Fossil Energy and Carbon Management should support research to investigate system integration and scale up of catalysis and bioconversion. The reactor and process design need to be optimized to the specific intermediates and desired products, and techno-economic and life cycle assessments need to be carried out to evaluate the economic, environmental, and emissions impacts of hybrid systems.							
8-H. Feasibility study for integrating thermocatalytic or photocatalytic CO_2 conversion with bioconversion to evaluate the efficiency, economics, and scalability.	DOE-ARPA-E DOE-FECM DOE-BETO	Applied	Biological-Hybrid	Chemicals Polymers	Short-lived	Reactor design and reaction engineering Integrated systems	Rec. 8-6

Recommendation 8-6: Advance prototype hybrid systems to integrate thermocatalytic or photocatalytic CO_2 conversion with bioconversion—The Department of Energy's Bioenergy Technologies Office, Advanced Research Projects Agency–Energy, and Office of Fossil Energy and Carbon Management should support research to investigate the concept of integrating thermocatalytic or photocatalytic conversion of CO_2 into intermediates, and subsequently convert the intermediates via bioconversion to diverse chemical and polymer products. The evaluation of system efficiency and economics will help to assess whether such integrated systems are feasible.

Coal Waste Utilization							
9-A. Evaluation and mapping of coal waste resources, including composition, volume, and locations.	DOE-FECM OSMRE EPA USGS	Enabling	Coal waste utilization	Coal waste-derived carbon products Metal coal waste by-products	Long-lived	Resource mapping	Fin. 9-1 Rec. 9-1
9-B. Strategies for linking coal waste sites to markets for solid carbon and critical minerals and materials and creating infrastructure to process and efficiently transport large amounts of coal wastes (both liquid and solid).	DOE-FECM DOT	Applied Enabling	Coal waste utilization	Coal waste-derived carbon products Metal coal waste by-products	Long-lived	Enabling technology and infrastructure needs Market opportunities	Rec. 9-1
9-C. Physical and chemical methods for separating mineral matter from carbon in coal wastes.	DOE-BES DOE-FECM	Basic Applied	Coal waste utilization	Coal waste-derived carbon products Metal coal waste by-products	Long-lived	Separations	Rec. 9-1

Recommendation 9-1: Fund mapping of coal waste resources and infrastructure development to link coal waste sites with product markets—The Department of Energy's Office of Fossil Energy and Carbon Management (DOE-FECM), in collaboration with the Department of the Interior's Office of Surface Mining Reclamation and Enforcement, the U.S. Geological Survey, and the U.S. Environmental Protection Agency's Office of Land and Emergency Management, should fund an effort to evaluate and map coal waste resources, including acid mine drainage, impoundment wastes, and coal combustion residuals, to facilitate their potential use in producing solid carbon products and/or critical minerals and materials. Additionally, DOE-FECM and Basic Energy Sciences within DOE's Office of Science should fund translational and basic research of coal waste separations. DOE-FECM, jointly with the Department of Transportation, should support research and development focused on linking coal waste sites to solid carbon and critical minerals and materials markets and creating infrastructure to process and efficiently transport large amounts of coal wastes (both liquid and solid).

9-D. Efficient transformation of waste coal into long-lived carbon products with lower embodied carbon than existing products—including engineered composites, graphite, graphene, fiber, and foam—for construction, energy storage technologies, transportation, and defense applications.	DOE-FECM DoD	Applied Demonstration	Coal waste utilization—long-lived carbon products	Coal waste-derived carbon products	Long-lived	Reactor design and reaction engineering	Fin. 9-2 Rec. 9-2 Rec. 9-3

continued

TABLE 11-1 Continued

Research, Development, and Demonstration Need	Funding Agencies or Other Actors	Basic, Applied, Demonstration, or Enabling	Research Area	Product Class	Long- or Short-Lived	Research Themes	Source
9-E. Evaluation of different types of coal waste to determine their ability to enhance pavement performance and to understand the fate and transfer of heavy metals over long time periods.	DOE-FECM DOT-FHWA DOT-OST-R State departments of transportation	Applied	Coal waste utilization	Coal waste-derived carbon products	Long-lived	Certification and standards	Sec. 9.3.1.2
9-F. Atomic- and multi-scale computer simulations to better understand the conversion of coal waste carbon into various solid-carbon products and solid-carbon product precursors.	NSF DOE-BES	Basic	Coal waste utilization—long-lived carbon products	Coal waste-derived carbon products	Long-lived	Fundamental knowledge Computational modeling and machine learning	Sec. 9.3.3.2 Rec. 9-3
9-G. Develop 3D printing media from waste coal or coal-derived materials for tooling and building product applications.	DOE-FECM	Applied	Coal waste utilization—long-lived carbon products	Coal waste-derived carbon products	Long-lived	Certification and standards	Rec. 9-2
9-H. Evaluations of functionality and performance of coal waste–derived products to ensure they conform with codes specific to their intended application. These evaluations could include examination of mechanical, thermal, electrical, and/or chemical properties.	DOE-FECM	Applied	Coal waste utilization—long-lived carbon products	Coal waste-derived carbon products	Long-lived	Certification and standards	Rec. 9-2
9-I. Establish standards for using coal waste in applications with environmental exposure to ensure product safety.	ASTM International	Enabling	Coal waste utilization—long-lived carbon products	Coal waste-derived carbon products	Long-lived	Certification and standards	Rec. 9-2
9-J. Data and tools to conduct life cycle and techno-economic assessments of coal waste utilization processes.	DOE-FECM	Applied	Coal waste utilization	Coal waste-derived carbon products Metal coal waste by-products	Long-lived	Certification and standards Environmental and societal considerations for CO_2 and coal waste utilization technologies	Fin. 9-2 Rec. 9-2

continued

Recommendation 9-2: Increase research on conversion of coal waste to long-lived carbon products, including techno-economic, environmental, and safety assessments—The Department of Energy's Office of Fossil Energy and Carbon Management (DOE-FECM) and the Department of Defense should fund research that focuses on the efficient transformation of waste coal through thermochemical, electrochemical, and plasma-chemical processes into long-lived solid carbon products, for example, carbon fibers, graphite, graphene, carbon foam, 3D-printing materials, and engineered composites. Specifically, low-cost and environmentally friendly conversions of coal waste to long-lived carbon products should be developed, and approaches to valorize the by-products of conversions of coal waste to long-lived carbon products should be explored. As part of these efforts, DOE should increase support for their national laboratories to develop databases that can assist researchers in completing robust life cycle and techno-economic assessments of coal waste utilization processes. DOE-FECM should evaluate the functionality and performance of coal waste–derived products to ensure that they conform with codes specific to their intended application and, in collaboration with ASTM International, establish standards for using coal waste in applications with environmental exposure to ensure product safety.

Recommendation 9-3: Appropriate funds for the Carbon Materials Science Initiative—The U.S. Congress should appropriate funds for the "Carbon Materials Science Initiative" as authorized in the CHIPS and Science Act of 2022 "to expand knowledge of coal, coal wastes, and carbon ore chemistry useful for understanding the conversion of carbon to material products." Funding for basic research into low-carbon-emission pathways (e.g., clean electricity-driven heating, electrolysis, and plasma) for coal waste utilization should be prioritized by the Department of Energy's (DOE's) Office of Basic Energy Sciences, along with associated applied research and demonstration supported by DOE's Office of Fossil Energy and Carbon Management.

9-K. Characterization of the structure and morphology of solid coal wastes and coal combustion by-products.	DOE-BES USGS	Basic	Coal waste utilization—metal extraction	Metal coal waste by-products	Long-lived	Fundamental knowledge	Fin. 9-4 Rec. 9-4

Recommendation 9-4: Support characterization of coal wastes to facilitate development of new physical beneficiation methods—Basic Energy Sciences within the Department of Energy's (DOE's) Office of Science and the U.S. Geological Survey should support the characterization of solid coal wastes and coal combustion by-products. DOE's Office of Fossil Energy and Carbon Management and Advanced Research Projects Agency–Energy should fund the development of innovative physical beneficiation methods, including separation of unburned carbon and magnetic fractions in fly ashes and removal of oversized particles, novel roasting methods, and highly efficient, sustainable leaching agents and extractants for rare earth elements.

9-L. Discovery and development of more selective, sustainable leaching agents, membranes, and processes for extracting rare earth elements from coal waste.	DOE-BES NSF	Basic	Coal waste utilization—metal extraction	Metal coal waste by-products	Long-lived	Separations	Fin. 9-5 Rec. 9-5
9-M. Technologies for extracting rare earth elements, lithium, and other energy-relevant critical materials from solid wastes (e.g., waste coal and coal combustion residuals).	DOE-FECM DOE-EERE DOE-ARPA-E DoD	Applied	Coal waste utilization—metal extraction	Metal coal waste by-products.	Long-lived	Separations	Fin. 9-5 Rec. 9-5
9-N. Technologies for extracting rare earth elements, lithium, and other energy-relevant critical minerals from liquid wastes (e.g., acid mine drainage).	DOE-FECM DOE-EERE DOE-ARPA-E DoD	Applied	Coal waste utilization—metal extraction	Metal coal waste by-products.	Long-lived	Separations	Fin. 9-5 Rec. 9-5

continued

TABLE 11-1 Continued

Research, Development, and Demonstration Need	Funding Agencies or Other Actors	Basic, Applied, Demonstration, or Enabling	Research Area	Product Class	Long- or Short-Lived	Research Themes	Source
9-O. Separation of individual elements once extracted from coal wastes, especially separation of nickel from cobalt.	DOE-BES NSF DOE-FECM DOE-EERE DOE-ARPA-E DoD	Basic Applied	Coal waste utilization—metal extraction	Metal coal waste by-products.	Long-lived	Separations	Fin. 9-5 Rec. 9-5
Recommendation 9-5: Develop novel technologies to extract critical minerals from liquid and solid coal waste streams—The Department of Defense and the Department of Energy's (DOE's) Office of Fossil Energy and Carbon Management, Office of Energy Efficiency and Renewable Energy, and Advanced Research Projects Agency–Energy should fund the development of novel technologies for extracting critical materials, including rare earth and energy-relevant elements, from solid (e.g., coal combustion residuals, waste coal) and liquid (e.g., acid mine drainage) coal waste streams, as well as the separation of individual elements once extracted. In parallel, DOE's Office of Basic Energy Sciences and the National Science Foundation should support research on separations from complex mixtures, including to understand species partitioning, ion pairing, and ion dissociation. Technologies and processes to be explored include development of novel acids and electrified methods (e.g., microwave, inductive Joule heating) for extractions; development of novel, sustainable sorbents for separations; and multiphysics simulations and artificial intelligence to analyze and understand the composition of complex waste streams.							
CO_2 Utilization Markets							
2-A. Understand broader impacts of CO_2 conversion on the environment, resource (re-)allocation, and jobs gains and/or losses.	DOE	Enabling	Societal Impacts	All	Long-lived Short-lived	Environmental and societal considerations for CO_2 and coal waste utilization technologies	Rec. 2-2
2-B. Understand broader impact of CO_2 conversion to (a) meet national needs for carbon products, (b) meet national targets for the transition to carbon neutrality, and (c) evaluate effectiveness of incentives and other policies.	DOE GSA State-level actors	Enabling	Markets Societal Impacts	All	Long-lived Short-lived	Market opportunities	Rec. 2-2
Recommendation 2-2: Close information gaps—The Department of Energy should support system-level studies to understand the broader impact of CO_2 conversion on the environment, markets, resource (re-)allocation, and jobs gains and/or losses. Related studies should be conducted to close information gaps to realize market opportunities for CO_2 conversion to (a) meet national needs for carbon products, (b) meet national targets for the transition to carbon neutrality, and (c) evaluate incentives and other policies for effectiveness.							

Research, Development, and Demonstration Need	Funding Agencies or Other Actors	Basic, Applied, Demonstration, or Enabling	Research Area	Product Class	Long- or Short-Lived	Research Themes	Source
LCA, TEA, and Societal/Equity Assessments for CO_2 Utilization							
3-A. Understanding the impact of fluctuations in CO_2 purity in the life cycle and techno-economic assessment of CO_2 utilization technologies.	DOE-EERE DOE-FECM	Enabling	LCA/TEA	All	Long-lived Short-lived	Environmental and societal considerations for CO_2 and coal waste utilization technologies	Fin. 3-4 Rec. 3-2
3-B. Development of improved CO_2 purification technologies that are more flexible, modular, and less energy-intensive.	DOE-EERE DOE-FECM DOE-BES	Basic Applied	Chemical	All	Long-lived Short-lived	Separations	Fin. 3-4 Rec. 3-2 Sec. 11.1.2
Recommendation 3-2: Research needs for CO_2 purity in techno-economic and life cycle assessments—The Department of Energy (DOE) and other relevant funding agencies should fund projects that examine the robustness of CO_2 utilization technologies to different CO_2 purities as well as fund further research and development of CO_2 purification technologies. Insights from these projects should be disseminated to the larger community by DOE. DOE should require awardees of applied research and development funding for CO_2 utilization technologies to perform techno-economic and life cycle assessments that explicitly address the purity requirements of the CO_2 streams.							
3-C. Understanding of non-CO_2-emissions impacts of CO_2 utilization technologies within life cycle assessments (e.g., impacts on chemical toxicity, water requirements, and air quality of carbon mineralization at the gigatonne scale).	DOE-EERE DOE-FECM EPA USGS	Enabling	LCA/TEA	All	Long-lived Short-lived	Environmental and societal considerations for CO_2 and coal waste utilization technologies	Fin. 3-6 Rec. 3-4
Recommendation 3-4: Non-CO_2-emission impacts within life cycle assessments—The Department of Energy and other relevant funding agencies such as the U.S. Environmental Protection Agency and the National Institute of Standards and Technology should support research into improving evaluation of non-CO_2-emissions impacts within life cycle assessments (LCAs) of CO_2 utilization technologies, including: **a. Evaluating the appropriate but differentiated applications for global and local impact categories, as the latter generally involves data and information with high spatial and temporal granularity (e.g., processes versus facilities, technology readiness level of various components of the technology).** **b. Evaluating appropriate applications of social LCA (s-LCA) and further developing s-LCA tools and their potential integration with environmental LCA and techno-economic assessments.**							
3-D. Development of life cycle assessment approaches that can address circularity of CO_2-derived products over time.	DOE-FECM National Laboratories NIST	Enabling	LCA/TEA	Chemicals Polymers	Short-lived	Environmental and societal considerations for CO_2 and coal waste utilization technologies	Rec. 3-6

continued

TABLE 11-1 Continued

Research, Development, and Demonstration Need	Funding Agencies or Other Actors	Basic, Applied, Demonstration, or Enabling	Research Area	Product Class	Long- or Short-Lived	Research Themes	Source
3-E. Understanding the flows of carbon through product life cycles to enable a circular carbon system, including identifying leakage potential from circular systems, the fate of products under different end of life conditions, and how processes and demand may evolve through multiple cycles of use and reuse.	DOE-FECM National Laboratories NIST	Enabling	LCA/TEA	All	Long-lived Short-lived	Environmental and societal considerations for CO_2 and coal waste utilization technologies	Fin. 3-9 Rec. 3-6
3-F. Development of approaches and tools to trace carbon across value chains over time, including mapping of value chains, identification of tracking methods, and data collection and validation protocols.	DOE-FECM NIST	Enabling	LCA/TEA	All	Long-lived Short-lived	Environmental and societal considerations for CO_2 and coal waste utilization technologies	Rec. 3-6

Recommendation 3-6: Implications of circularity on carbon storage—The Department of Energy and the National Institute of Standards and Technology should support research that examines the feasibility and impacts of extending the duration of carbon storage through circularity strategies of short-lived products. This includes

a. **Building on state-of-the-art life cycle assessment approaches that are able to address circularity of CO_2 based products over time.**
b. **Development of approaches and tools that allow the traceability and custody of carbon across value chains over time, including mapping of value chains, identification of tracking methods, and data collection and validation protocols.**

Policy and Regulatory Needs for CO_2 Utilization

Research, Development, and Demonstration Need	Funding Agencies or Other Actors	Basic, Applied, Demonstration, or Enabling	Research Area	Product Class	Long- or Short-Lived	Research Themes	Source
4-A. Knowledge gaps in public perception of carbon utilization technologies, and factors that influence community acceptance.	NGOs Universities National laboratories Other research-conducting entities DOE	Enabling	Societal Impacts	All	Long-lived Short-lived	Environmental and societal considerations for CO_2 and coal waste utilization technologies	Fin. 4-7 Rec. 4-6

Recommendation 4-6: Educational material development—Nongovernmental organizations and research-conducting entities—such as national laboratories, think tanks, and universities—should identify gaps in knowledge, sharing their data and findings about societal acceptance of or opposition to the CO_2 utilization sector through the following actions:

a. **Nongovernmental organizations (including universities) and national laboratories should conduct targeted research to develop transparent resources to communicate findings and support improved education related to the direct impact of the CO_2 utilization sector on climate change mitigation.**
b. **Research-conducting entities should continue to conduct social analyses to determine what consumers and communities think about the CO_2 utilization sector, both in relation to and separate from the carbon management value chain, filling the gaps in knowledge about societal acceptance and potential opposition, to better understand and address the concerns of the public. These data and conclusions of this research should be shared through a centralized and broadly accessible framework.**
c. **Research-conducting entities should use analyses that combine techno-economic and life cycle assessment objectives to determine a levelized cost of CO_2 abatement to be used to assess the desirability of projects as a function of CO_2 source (fossil or biogenic point source, direct air capture, or direct ocean capture) and product durability. The results of these analyses and assessments should be communicated to the Department of Energy and other entities funding carbon utilization projects to inform them of factors that can influence community acceptance of projects and expected outcomes.**

CO_2 Utilization Infrastructure

10-A. Development of robust computational tools for optimal multimodal transportation of CO_2 captured from stranded emitters for centralized utilization.	DOE-FECM	Enabling	Infrastructure	All	Long-lived Short-lived	Enabling technology and infrastructure needs Computational modeling and machine learning	Rec. 10-2
10-B. Techno-economic assessment of centralized versus distributed/ onsite utilization of CO_2 for small- to medium-scale emitters.	Small- to medium-scale emitters	Enabling	Infrastructure	All	Long-lived Short-lived	Enabling technology and infrastructure needs	Rec. 10-2

Recommendation 10-2: Evaluate CO_2 capture and utilization for small- to medium-scale emitters—Small- to medium-scale emitters that cannot eliminate emissions through energy efficiency, electrification, and other decarbonization strategies should evaluate the economic feasibility of performing CO_2 capture and utilization onsite. This may involve deploying renewable electricity generation, clean hydrogen production, modular carbon capture technologies, and utilization processes suited to small-scale conversion. In cases where this is not possible, taking account of their emissions rates and geographical location, such emitters should seek multimodal transport solutions for the CO_2 and, if relevant, CO_2-derived product, striking a balance between cost, safety, and environmental impacts using established methodologies that also have been proposed for hydrogen delivery infrastructure development. Additionally, as recommended in this committee's first report (Recommendation 4.2), the Department of Energy should develop dedicated methodologies for optimizing multimodal CO_2 transport to assist in these infrastructure planning efforts.

10-C. Better understanding and development of approaches to mitigate issues with propagating brittle and ductile fractures in CO_2 pipelines.	PHMSA National laboratories University researchers Industry	Enabling	Infrastructure	All	Long-lived Short-lived	Enabling technology and infrastructure needs	Rec. 10-5

continued

TABLE 11-1 Continued

Research, Development, and Demonstration Need	Funding Agencies or Other Actors	Basic, Applied, Demonstration, or Enabling	Research Area	Product Class	Long- or Short-Lived	Research Themes	Source
10-D. CO_2 dispersion modeling calculations for the case of accidental rupture of buried CO_2 pipelines that results in formation of a crater owing to the high momentum CO_2 jet.	PHMSA National laboratories University researchers Industry	Enabling	Infrastructure	All	Long-lived Short-lived	Enabling technology and infrastructure needs Computational modeling and machine learning	Rec. 10-5
10-E. Computer programs that implement validated mathematical models on dispersion modeling and propagating fractures to serve as decision-making tools for pipeline developers.	PHMSA National laboratories University researchers Industry	Enabling	Infrastructure	All	Long-lived Short-lived	Enabling technology and infrastructure needs	Rec. 10-5

Recommendation 10-5: Fund research on modeling and testing aimed at improving CO_2 pipeline safety—The Pipeline and Hazardous Materials Safety Administration should collaborate with national laboratories, university researchers, industry, and international partners to co-fund and implement research to develop rigorous mathematical models, which in turn should be extensively validated using realistic scale experiments, to

a. Simulate the fluid/structure interaction and subsequent atmospheric dispersion of the escaping overground CO_2 plume following accidental rupture of buried CO_2 pipelines that results in the formation of a crater owing to the high-momentum jet impingement in order to determine minimum safe distances to populated areas and emergency response planning.

b. Understand and mitigate issues with propagating brittle and ductile fractures in CO_2 pipelines, as described in detail in Recommendation 4.3 of the committee's first report, in order to select pipeline materials capable of resisting such types of failures.

c. Implement the validated mathematical models developed in (a) and (b) into robust and easy to use computer programs to be routinely used as a design and decision-making tool for pipeline developers in order to reduce and mitigate the risks associated with pipeline failures to as low as reasonably practicable.

NOTE: 3D = three-dimensional; AMMTO = Advanced Materials and Manufacturing Technologies Office; ARPA-E = Advanced Research Projects Agency–Energy; BER = Biological and Environmental Research; BES = Basic Energy Sciences; BETO = Bioenergy Technologies Office; DoD = Department of Defense; DOE = Department of Energy; DOT = Department of Transportation; EERE = Office of Energy Efficiency and Renewable Energy; EPA = U.S. Environmental Protection Agency; FECM = Office of Fossil Energy and Carbon Management; FHWA = Federal Highway Administration; GHG = Greenhouse Gas; GSA = General Services Administration; NSF = National Science Foundation; LCA = Life Cycle Assessment; ONR = Office of Naval Research; OSMRE = Office of Surface Mining Reclamation and Enforcement; OST-R = Office of the Assistant Secretary for Research and Technology; PHMSA = Pipeline and Hazardous Materials Safety Administration; TEA = Techno-Economic Assessment; USGS = U.S. Geological Survey.

Appendixes

A

Committee Member Biographical Information

EMILY A. CARTER (*Chair*) is the Gerhard R. Andlinger Professor in Energy and the Environment and a professor of mechanical and aerospace engineering and applied and computational mathematics at Princeton University. She is also a senior strategic advisor and the associate laboratory director for applied materials and sustainability sciences at the Department of Energy's (DOE's) Princeton Plasma Physics Laboratory. Until the end of 2021, Dr. Carter served as the executive vice chancellor and provost (EVCP) and the Distinguished Professor of Chemical and Biomolecular Engineering at the University of California, Los Angeles (UCLA). Dr. Carter earned a BS in chemistry from the University of California, Berkeley (graduating Phi Beta Kappa) and a PhD in chemistry from the California Institute of Technology (Caltech), followed by a brief postdoctorate at the University of Colorado Boulder, before spending 16 years on UCLA's chemistry and biochemistry faculty. She moved to Princeton University in 2004, where she spent 15 years as a jointly appointed faculty in mechanical and aerospace engineering and applied and computational mathematics. From 2010 to 2016, she was Princeton's founding director of the Andlinger Center for Energy and the Environment and from 2016–2019 she was Princeton's dean of engineering and applied science before returning to UCLA as its EVCP. Dr. Carter has pioneered the development and application of quantum mechanics–based simulation techniques to enable the discovery and design of molecules, materials, and processes for sustainable energy, fuels, and chemicals, supported by grants from the Department of Defense, the National Science Foundation (NSF), and DOE. She has received numerous honors, including election to the National Academy of Sciences, the American Academy of Arts and Sciences, the National Academy of Inventors, the National Academy of Engineering, the European Academy of Sciences, and as a Foreign Member of the Royal Society of London.

SHOTA ATSUMI is a professor in the Department of Chemistry at the University of California, Davis. He was a postdoctoral researcher with Dr. John W. Little at the University of Arizona and with Dr. James C. Liao at UCLA. Dr. Atsumi's current research focuses on the use of synthetic biology and metabolic engineering approaches to engineer microorganisms to convert CO_2 to valuable chemicals. The primary research goals of his group are to develop a platform for valuable chemical production from carbon dioxide using photosynthetic microorganisms and to develop novel biosynthetic pathways to produce chemical compounds that microbes naturally produce in trace amounts or not at all. Dr. Atsumi received the Hellman Fellowship in 2021, the NSF CAREER award in 2014, and the Chancellor's Fellowship in 2018. He received his PhD in biological chemistry in 2002, his MS in biological chemistry in 1998, and his BS in 1996, all from Kyoto University.

MAKINI BYRON is a director of Clean Energy at Linde, a leading industrial gas and engineering company. Ms. Byron has a diverse background in research and development (R&D) and business development within clean energy and innovation. In her current role, Ms. Byron focuses on supporting the company's growth ambitions in hydrogen and carbon solutions through new clean energy business opportunities and strategic partnerships. Ms. Byron has managed or participated in several DOE-funded projects for the commercial engineering design and demonstration of post-combustion and oxy-combustion carbon capture technologies as well as several CO_2 utilization technologies. She has an MS in chemical engineering and a certificate in science, technology, and energy policy from Princeton University. Ms. Byron is a registered project management professional and a member of the Project Management Institute and the American Institute of Chemical Engineers (AIChE).

JINGGUANG CHEN is the Thayer Lindsley Professor of Chemical Engineering at Columbia University, with a joint appointment at Brookhaven National Laboratory. Dr. Chen's research interests include the thermocatalytic and electrocatalytic conversion of carbon dioxide, utilizing both experimental and theoretical approaches. He is currently the president of the North American Catalysis Society and the director of the Synchrotron Catalysis Consortium. Dr. Chen received the George Olah Award from the American Chemical Society (ACS) and the R.H. Wilhelm Award from AIChE, and he has been elected to the National Academy of Engineering. He received his PhD from the University of Pittsburgh and carried out his Alexander von Humboldt postdoctoral research at KFA-Julich in Germany.

STEPHEN COMELLO is the senior vice president of strategic initiatives at the Energy Futures Initiative (EFI) Foundation and the co-managing director of its Energy Futures Finance Forum. Previously, he served as a faculty member at the Stanford Graduate School of Business for more than a decade, co-founding and co-leading the Rapid Decarbonization Initiative. With a 23-year career dedicated to scaling emerging energy and environmental technologies, Dr. Comello specializes in policy and business model innovations. His expertise spans technoeconomic analysis, policy and project finance, corporate strategy in energy transition, and open innovation. At Stanford, he held leadership roles in various research initiatives and industrial affiliate programs. He has authored numerous publications in energy policy, industrial organization, development economics, innovation management, and carbon accounting. Dr. Comello holds bachelor's and master's degrees in mechanical and industrial engineering from the University of Toronto and a PhD in civil and environmental engineering from Stanford University. Originally from Canada, he now resides in Washington, DC.

MAOHONG FAN is a School of Energy Resources professor in chemical and petroleum engineering at the University of Wyoming and an adjunct professor in environmental engineering at the Georgia Institute of Technology (Georgia Tech). Dr. Fan has led and worked on many projects in chemical production, clean energy generation, and environmental protection. The projects have been supported by various domestic and international funding agencies such as NSF, DOE, the U.S. Environmental Protection Agency, the U.S. Geological Survey, and the Department of Agriculture in the United States; the New Energy and Industrial Technology Development Organization in Japan; the United Nations Development Programme; and industrial companies such as Siemens and Caterpillar. Dr. Fan has helped various chemical, environmental, and energy companies overcome their technical challenges. He has published many refereed papers in chemical and environmental engineering, energy, and chemistry journals. He is one of the most highly cited researchers according to Web of Science. Dr. Fan's recent NSF and DOE projects cover the areas of carbon capture, utilization, and storage (CCUS); catalyzed solar energy–driven biomass conversion; rare earth oxide extraction and reduction of rare earth oxides to rare earth metals; carbon fuel cells; and production of chemicals, materials, and fuels from fossil resources.

BENNY FREEMAN is the William J. (Bill) Murray Jr. Endowed Chair of Engineering in the McKetta Department of Chemical Engineering at The University of Texas at Austin. Dr. Freeman's research is in polymer science and engineering and, more specifically, in mass transport of small molecules in solid polymers. His research group focuses on structure and property correlation development for desalination and gas separation membrane materials, new materials for hydrogen separation, natural gas purification, carbon capture, and new materials for improving fouling resistance in liquid separation membranes. Dr. Freeman leads the Center for Materials for Water and Energy Systems, a DOE Energy Frontier Research Center, and he serves as challenge area leader for membranes in the National Alliance for Water Innovation, a 5-year, DOE-sponsored Energy–Water Desalination Hub to address critical technical barriers needed to radically reduce the cost and energy of water purification. He is a fellow of the American Association for the Advancement of Science (AAAS), AIChE, ACS, and the Polymeric Materials: Science and Engineering (PMSE) and Industrial and Engineering Chemistry Research (IECR) Divisions of ACS. He has served as the chair of the PMSE Division of ACS; chair of the Gordon Research Conference on Membranes: Materials and Processes; president of the North American Membrane Society; chair of the Membranes Area of the Separations Division of AIChE; and chair of the Separations Division of AIChE. His research has served as the basis for several start-up companies, including Energy-X and NALA Systems. Dr. Freeman is a member of the National Academy of Engineering. He completed graduate training in chemical engineering at the University of California, Berkeley, earning a PhD in 1988. In 1988 and 1989, he was a postdoctoral fellow at the Ecole Supérieure de Physique et de Chimie Industrielles de la Ville de Paris, Laboratoire Physico-Chimie Structurale et Macromoléculaire, in Paris. Dr. Freeman was a member of the chemical engineering faculty at North Carolina State University from 1989–2002, and he has been a professor of chemical engineering at The University of Texas at Austin since 2002.

MATTHEW FRY joined the Great Plains Institute in August 2021 as the state and regional policy manager, supporting the Carbon Management program. Mr. Fry has more than 20 years of experience in natural resource management, regulation, and policy in both the public and private sectors. Mr. Fry served as a senior policy advisor to Wyoming Governor Matt Mead, where he focused on natural resource, energy, and CCUS policy. Additionally, he developed and managed the Wyoming Pipeline Corridor Initiative, a project that authorized a statewide network of pipeline corridors in Wyoming that aimed to establish corridors on public lands dedicated for future use of pipelines associated with CCUS, enhanced oil recovery, and delivery of associated petroleum products. Mr. Fry earned a BS in biology and chemistry from Davis & Elkins College and a master's in natural resource law from the University of Denver Sturm College of Law.

SARAH M. JORDAAN is an associate professor of industrial ecology and life cycle assessment (LCA) at the Department of Civil Engineering and the Trottier Institute for Sustainability in Engineering and Design at McGill University. Prior to joining McGill, Dr. Jordaan held positions at Johns Hopkins University, Harvard University, the Electric Power Research Institute, Shell, the University of Calgary, and the Laboratory on International Law and Regulation at the University of California, San Diego. Her research focuses on improving LCA, techno-economic analysis, and technology innovation in support of a sustainable, low-carbon energy future. Her articles examine CCUS technologies in early R&D stages and also large-scale deployment in Paris-compliant scenarios and have been published in *Nature Climate Change*, *Nature Catalysis*, and *Energy & Environmental Science*. Dr. Jordaan won the 2022 Educational Leadership Award from the American Center for Life Cycle Assessment, where she has been a member since 2008. She is co-chairing a subgroup on a task commissioned by the Secretary of Energy for the National Petroleum Council to reduce emissions across natural gas supply chains in line with the Global Methane Pledge. Her postdoctoral fellowship at Harvard University was focused on energy technology innovation with the Belfer Center for Science and International Affairs at the Kennedy School of Government and on climate impacts of energy with the Department of Earth and Planetary Sciences. Dr. Jordaan earned her doctorate in environmental design from the University of Calgary in 2010 and her BS in physics with a computer science minor from Memorial University in 2003.

HAROUN MAHGEREFTEH is a professor of chemical engineering at University College London and a fellow of the Institution of Chemical Engineers. Dr. Mahgerefteh's research spans all aspects of CCUS, particularly CO_2 pipeline safety and operational issues. He is the coordinator of several national and multinational collaborative projects, including the European Commission FP7 and H2020 projects, CO2PipeHaz, CO2QUEST, and C4U. Project highlights include the development of best-practice guidelines for injection of CO_2 into highly depleted gas fields and the construction of the world's longest fully instrumented CO_2 pipeline rupture test facility, producing first-of-its-kind fundamentally important field data for the development and validation of source term and dispersion models for the accurate determination of minimum safety distances. He is also a key partner in the three Horizon Europe projects, CaLBy2030, ENCASE, and EMPHATICAL, working on technology readiness level 7 development of calcium looping capture technologies for the iron, steel, and cement industries; e-methanol production using CO_2; and renewable hydrogen and multi-modal transport of CO_2. Dr. Mahgerefteh is one of the two lead authors of the Zero Emission Platform report titled *A Trans-European CO_2 Transportation Infrastructure for CCUS: Opportunities & Challenges*. The report is aimed at facilitating the development of a pipeline and shipping infrastructure for transporting several Mt/yr of CO_2 captured from major regional industrial emitters for permanent offshore geological storage—considered a key enabler for meeting the net-zero emission target by 2030. The author of more than 190 publications, Dr. Mahgerefteh's professional engagements include membership on the UK Carbon Capture and Storage Research Council. He is the recipient of several prizes, including the Institution of Chemical Engineers' Global Process Safety award for his CO2QUEST project.

AH-HYUNG (ALISSA) PARK is the Ronald and Valerie Sugar Dean of the UCLA Samueli School of Engineering. Previously, Dr. Park was the Lenfest Earth Institute Professor of Climate Change in the Departments of Earth and Environmental Engineering and Chemical Engineering at Columbia University. She was also the director of the Lenfest Center for Sustainable Energy. Her research focuses on sustainable energy and materials conversion pathways with an emphasis on integrated CCUS technologies addressing climate change. Dr. Park's group is also working on direct air capture of CO_2 and negative emission technologies, including bioenergy with carbon capture and storage and sustainable construction materials with low carbon intensity. Dr. Park has received a number of professional awards and honors including the Shell Thomas Baron Award in Fluid-Particle Systems and PSRI Lectureship Award from AIChE PTF, the U.S. C3E Research Award, PSRI Lectureship Award at AIChE, ACS Energy and Fuels Division—Mid-Career and Emerging Researcher Award, ACS WCC Rising Star Award, and NSF CAREER Award. Dr. Park has also led global and national discussions on CCUS, including the Mission Innovation Workshop on CCUS in 2017 and the National Petroleum Council CCUS Report in 2019. Dr. Park received her PhD in chemical engineering from The Ohio State University and her BS from The University of British Columbia. She is an elected fellow of AIChE, ACS, the Royal Society of Chemistry, and AAAS.

JOSEPH B. POWELL is a fellow and the former director of AIChE and served as Shell's first chief scientist in chemical engineering from 2006 until retiring at the end of 2020, culminating a 36-year industry career in which he led R&D programs in new chemical processes, biofuels, and enhanced oil recovery and advised on R&D for the energy transition to a net-zero carbon economy. Dr. Powell is the co-inventor on more than 125 patent applications (60 granted); has received AIChE/ACS/R&D Magazine awards for Innovation, Service, and Practice; and is the co-author of *Sustainable Development in the Process Industries: Cases and Impact* (2010). He chaired the DOE Hydrogen and Fuel Cell Technical Advisory Committee and was elected to the National Academy of Engineering (2021) after serving two terms on the National Academies' Board on Chemical Sciences and Technology. Other roles include guest editor for *Catalysis Today Natural Gas Utilization*, editorial board for *Annual Review of Chemical and Biological Engineering*, and crosscutting technologies area lead and author for *Mission Innovation Carbon Capture Utilization and Storage* (2017). Dr. Powell currently advises on energy and chemicals and process development (ChemePD, LLC). He received a PhD from the University of Wisconsin–Madison (1984) and a BS from the University of Virginia (1978), both in chemical engineering.

CLAUDIA A. RAMÍREZ RAMÍREZ is a professor of low-carbon systems and technologies at Delft University of Technology, the Netherlands. Her research focuses on the evaluation of novel low-carbon technologies and the design of methodologies and tools to assess their potential contribution to sustainable industrial systems. Dr. Ramírez currently coordinates the research line on system integration and fair governance of the Dutch project RELEASE, aiming to develop reversible large-scale energy storage based on electrochemical conversion of CO_2 to molecules. In 2018, she was awarded one of the largest scientific grants for individuals in the Netherlands to investigate the system impacts of using alternative raw materials such as CO_2, biomass, and waste in petrochemical industrial clusters. In the past 10 years, Dr. Ramírez co-coordinated the European project

"Environmental Due Diligence of Novel CO_2 Capture and Utilization Technologies (EDDICCUT)"; led the research line Technoeconomic and Environmental Analysis of the Dutch R&D program "Catalysis for Sustainable Chemicals from Biomass (CATCHBIO)"; and coordinated the program line Transport and Chain Integration of the Dutch R&D program for CO_2 capture, "Transport and Storage (CATO)." Dr. Ramírez holds a bachelor's degree in chemical engineering, a master's in human ecology, and a PhD in industrial energy efficiency. She has authored or co-authored more than 115 publications and is the editor-in-chief of the *International Journal of Greenhouse Gas Control*.

VOLKER SICK is the DTE Energy Professor of Advanced Energy Research, the faculty director at the Center for Entrepreneurship, and an Arthur F. Thurnau Professor at the University of Michigan, Ann Arbor. Dr. Sick leads the Global CO_2 Initiative at the University of Michigan, which aims to reduce atmospheric CO_2 levels by transforming CO_2 into commercially successful products using technology assessment, technology development, and commercialization. His research focuses on accelerating deployments of CO_2-utilization technologies that will innovate existing infrastructure and manufacturing processes, thereby finding sustainable solutions and continued access to required carbon-based products to help address the climate crisis. Dr. Sick is the author of numerous publications in both peer-reviewed and popular periodicals. His most recent awards and honors include the Royal Society of Chemistry Spiers Memorial Lecture Award (2021), DTE Energy Professor of Advanced Energy Research (2019), and the President's Award for Distinguished Service in International Education (2018). He is a fellow of SAE International (2007) and the Combustion Institute (2018). Dr. Sick received his doctorate in chemistry and habilitation in physical chemistry from the University of Heidelberg, Germany. He joined the University of Michigan as a professor of mechanical engineering in 1997.

SIMONE H. STEWART currently works as the senior industrial policy specialist on the Climate and Energy Policy team at the National Wildlife Federation, with a portfolio focused on CCUS; carbon dioxide removal technologies; as well as other strategies to aid in a just green transition for difficult to decarbonize sectors such as energy and industry. Dr. Stewart's work covers the intersections of emerging technologies and environmental justice across areas such as policy, industry, nongovernmental organizations, and public education, and collaboration with government agencies. Dr. Stewart joined the National Wildlife Federation in 2021 after receiving her PhD in mechanical engineering from the University of California, Santa Barbara (UCSB), where she was an NSF Graduate Research Fellow. At UCSB, her research focused on investigating the fluid mechanics of fluxes over rough surfaces, with applications in large-scale direct air carbon capture and clean energy architecture. During that time, Dr. Stewart also worked as the graduate assistant for the UCSB Blum Center on Poverty, Inequality, and Democracy, where she led a variety of programming, created detailed information campaigns centered around justice and community enfranchisement, and helped develop a comprehensive *People's Guide to the Green New Deal* rooted in the tenets of environmental and economic justice. Prior to graduate school, she received dual bachelor's degrees in physics and Spanish language, literature, and history from William Jewell College.

JASON PATRICK TREMBLY is the Russ Professor of Mechanical Engineering and a graduate faculty member in the Department of Chemical and Biomolecular Engineering at Ohio University. He is also the director of Ohio University's Institute for Sustainable Energy and the Environment. Prior to joining Ohio University as an assistant professor, Dr. Trembly was a leading young researcher at internationally recognized energy research centers. From 2007 to 2011, he was a research chemical engineer and team leader for syngas and CO_2 conversion at RTI International's Energy Technology Division. There, he was responsible for ideation and development of processes and catalysts for conversion of syngas and CO_2 to chemicals and fuels. He is also a former ORISE Fellow at DOE's National Energy Technology Laboratory, where he completed his graduate research focused on solid oxide fuel cell development. Dr. Trembly's research group utilizes process simulation with materials R&D to develop intensified process designs to address energy and environmental issues. His main research interests include solid oxide fuel cells and electrolyzers, electrochemical capture of nutrients from waste streams, produced water remediation, and sustainable building materials. He received his PhD in chemical engineering from Ohio University in 2007.

JENNY Y. YANG is a professor of chemistry at the University of California, Irvine (UCI) and a joint appointee at the Pacific Northwest National Laboratory (PNNL). Dr. Yang worked as a senior scientist at PNNL and at the Joint Center for Artificial Photosynthesis at Caltech prior to starting her faculty position. She has been working in the area of CO_2 capture and/or utilization for more than 10 years, with a focus on electrochemical methods. Dr. Yang is on the Executive Advisory Board of the Solutions That Scale Institute at UCI and is the director of the DOE-funded Energy Frontier Research Center Closing the Carbon Cycle. She has received the Chancellor's Fellowship at UCI, the Presidential Early Career Award for Scientists and Engineers, a Sloan Foundation Fellowship, the Camille Dreyfus Teacher-Scholar Award, and an Inorganic Chemistry Lectureship. Dr. Yang received her PhD in chemistry from the Massachusetts Institute of Technology and her BS from the University of California, Berkeley. She is an elected fellow of AAAS.

JOSHUA S. YUAN joined the faculty at Washington University in St. Louis in 2022 and serves as the chair of DOE, Environmental, and Chemical Engineering. Previously, Dr. Yuan was a faculty member at Texas A&M University since 2008 and was appointed the chair for Synthetic Biology and Renewable Products in 2018. He was a Sungrant Fellow at the University of Tennessee, Knoxville, and the National Renewable Energy Laboratory before joining Texas A&M. He has served in various leadership and management positions at Texas A&M University, the University of Tennessee, and the University of California, San Francisco, from 2001. Dr. Yuan also worked at BASF from 2000 to 2001. He has been awarded three U.S. patents and has two pending. He has written more than 100 peer-reviewed journal articles, published in *Nature Communications*, *Green Chemistry*, *Chem*, *Advanced Sciences*, *ChemSusChem*, and the *Proceedings of the National Academy of Sciences*, among others. He has won numerous awards and honors, including the Regional Solid Waste Planning Award and the Environmental Educator Award in 2018; the Excellence in Innovation Award from the Texas A&M University System in 2017; and the Gamma Sigma Delta Outstanding Graduate Student Award in 2007. Dr. Yuan earned his PhD from the University of Tennessee in 2007.

B

Disclosure of Unavoidable Conflicts of Interest

The conflict of interest policy of the National Academies of Sciences, Engineering, and Medicine (http://www.nationalacademies.org/coi) prohibits the appointment of an individual to a committee authoring a Consensus Study Report if the individual has a conflict of interest that is relevant to the task to be performed. An exception to this prohibition is permitted if the National Academies determine that the conflict is unavoidable and the conflict is publicly disclosed. A determination of a conflict of interest for an individual is not an assessment of that individual's actual behavior or character or ability to act objectively despite the conflicting interest.

Ms. Makini Byron has a conflict of interest in relation to her service on the Committee on Carbon Utilization Infrastructure, Markets, Research and Development because of her employment at Linde, an industrial gas company that separates and purifies CO_2 and sells it to other companies for conversion to valuable products. The National Academies have concluded that the committee must include a member with current industry experience in managing the link between the commercial sources of the carbon dioxide—including the processes for capturing the carbon dioxide and the costs and quality of the carbon dioxide obtained—and the current and emerging markets for this carbon dioxide, including the quality requirements, the cost requirements, and the potential quantities that might be utilized. As described in her biographical summary, Ms. Byron has extensive industry experience in understanding innovation and costs of carbon dioxide capture and the use of this carbon dioxide in products. Ms. Byron has managed or participated in several Department of Energy (DOE)-funded projects for both the commercial engineering design and the scale-up demonstration of Linde's carbon capture technology developed with BASF. Her project-based knowledge also extends to biological conversion of CO_2 to valuable products, mineralization of CO_2 to cementitious material, as well as the application of supercritical CO_2 for lubrication and cooling. The National Academies have determined that the experience and expertise of Ms. Byron is needed for the committee to accomplish the task for which it has been established. The National Academies could not find another available individual with the equivalent expertise and breadth of experience who does not have a conflict of interest. Therefore, the National Academies have concluded that the conflict is unavoidable. The National Academies believe that Ms. Byron can serve effectively as a member of the committee, and the committee can produce an objective report, taking into account the composition of the committee, the work to be performed, and the procedures to be followed in completing the study.

Dr. Stephen Comello has a conflict of interest in relation to his service on the Committee on Carbon Utilization Infrastructure, Markets, Research and Development because of his technical consulting with Carbon Direct, a company that invests in carbon removal and utilization technologies, and his role as an external advisor to energy practice at the consulting firm Bain & Company. The National Academies have concluded that, given the study's focus on market opportunities for carbon dioxide–derived products and carbon utilization technologies, it is essential to have a committee member with current experience in financing methods, business models, and decision-making strategies that enable the development and deployment of clean energy technologies. As described in his biographical summary, Dr. Comello possesses a unique combination of technology and economic expertise. Dr. Comello integrates tools and approaches from engineering, finance, and systems analysis to develop methodologies for analyzing investments and innovations in low-carbon energy solutions. His expertise spans an array of business analytical skills, including environmental economics, decision analysis, life cycle analysis, and techno-economic evaluation for advanced clean energy technologies. As a technical adviser to Carbon Direct, Dr. Comello brings an understanding of the technological and organizational capabilities of start-up companies in carbon utilization, which is critical for addressing the committee's task of determining how federal agencies can support small business to further the development and deployment of carbon dioxide–based products. The National Academies have determined that the experience and expertise of Dr. Comello is needed for the committee to accomplish the task for which it has been established. The National Academies could not find another available individual with the equivalent expertise and breadth of experience who does not have a conflict of interest. Therefore, the National Academies have concluded that the conflict is unavoidable. The National Academies believe that Dr. Comello can serve effectively as a member of the committee, and the committee can produce an objective report, taking into account the composition of the committee, the work to be performed, and the procedures to be followed in completing the study.

Dr. Ah-Hyung (Alissa) Park has a conflict of interest in relation to her service on the Committee on Carbon Utilization Infrastructure, Markets, Research and Development because of her equity in the start-up company GreenOre CleanTech, LLC. Dr. Park is a co-founder of GreenOre, which focuses on carbon capture and process design, using carbon dioxide and other waste streams to generate valuable products. The National Academies have concluded that, given the rapidly accelerating developments in the science, engineering, and commercialization of carbon utilization technologies, it is essential to have a committee member with current experience in basic research activities and knowledge of the opportunities and processes for technology scale up in this field. As described in her biographical summary, Dr. Park has an active research program spanning many topics relevant to the study, including CO_2 mineralization, materials for CO_2 capture and gas separations, chemical CO_2 conversion, and clean hydrogen production. In addition to her experience as an expert and leader in carbon capture and utilization research, her experience with GreenOre translating academic research into a start-up company makes her expertise a critical addition to this committee. The National Academies have determined that the experience and expertise of Dr. Park is needed for the committee to accomplish the task for which it has been established. The National Academies could not find another available individual with the equivalent expertise and breadth of experience who does not have a conflict of interest. Therefore, the National Academies have concluded that the conflict is unavoidable. The National Academies believe that Dr. Park can serve effectively as a member of the committee, and the committee can produce an objective report, taking into account the composition of the committee, the work to be performed, and the procedures to be followed in completing the study.

Dr. Joseph B. Powell has a conflict of interest in relation to his service on the Committee on Carbon Utilization Infrastructure, Markets, Research and Development because of his stock in Shell plc. The National Academies have concluded that the committee must include a member with recent experience and expertise in the chemical fuels industry with an understanding of the industrial and process engineering involved in producing such fuels and potentially adapting existing infrastructure for utilizing captured carbon dioxide in products. As described in his biographical summary, Dr. Powell has had extensive experience in development, scale up, and commercialization of existing and new technologies. He also has industrial systems expertise that is vital to address the committee's task of assessing infrastructure and research and development needs to support a future circular carbon economy. The National Academies have determined that the experience and expertise of Dr. Powell is needed for the committee to accomplish the task for which it has been established. The National Academies could not find another available individual with the equivalent expertise and breadth of experience who does not have a conflict of interest. Therefore, the National Academies have concluded that the conflict is unavoidable. The National Academies believe that Dr. Powell can serve effectively as a member of the committee, and the committee can produce an objective report, taking into account the composition of the committee, the work to be performed, and the procedures to be followed in completing the study.

Dr. Jenny Y. Yang has a conflict of interest in relation to her service on the Committee on Carbon Utilization Infrastructure, Markets, Research and Development because of patent applications for carbon dioxide capture and utilization technologies. The National Academies have concluded that it is essential to have a committee member with current experience in basic research activities for electrochemical carbon dioxide conversion, especially those relevant for commercialization. As described in her biographical summary, Dr. Yang has extensive research experience and publications in electrochemical and thermochemical carbon dioxide capture and utilization in academic and national laboratory settings and currently leads a DOE-funded center on recycling carbon dioxide into fuels, chemicals, and materials. Her experience patenting technologies derived from academic electrochemical research will bring valuable insights to this committee as it evaluates needs and challenges to commercialize electrochemical carbon dioxide utilization processes. The National Academies have determined that the experience and expertise of Dr. Yang is needed for the committee to accomplish the task for which it has been established. The National Academies could not find another available individual with the equivalent expertise and breadth of experience who does not have a conflict of interest. Therefore, the National Academies have concluded that the conflict is unavoidable. The National Academies believe that Dr. Yang can serve effectively as a member of the committee, and the committee can produce an objective report, taking into account the composition of the committee, the work to be performed, and the procedures to be followed in completing the study.

Dr. Joshua S. Yuan has a conflict of interest in relation to his service on the Committee on Carbon Utilization Infrastructure, Markets, Research and Development because of his equity in the start-up company Renewuel, LLC. Dr. Yuan is a co-founder and director of Renewuel, which focuses on algae production using CO_2. The National Academies have concluded that, to conduct an evaluation of barriers to commercializing biological carbon dioxide utilization processes, it is essential to have a committee member with current experience and knowledge translating research into commercially viable technologies for the unique infrastructure and scale-up requirements of biological-based systems. As his biographical summary makes clear, Dr. Yuan has extensive experience developing high yielding, commercially relevant, biological-based technologies, including pursuing commercialization of those technologies. His experience spans many stages of biological carbon capture and utilization, including basic science and engineering through applied technologies. The National Academies have determined that the experience and expertise of Dr. Yuan is needed for the committee to accomplish the task for which it has been established. The National Academies could not find another available individual with the equivalent expertise and breadth of experience who does not have a conflict of interest. Therefore, the National Academies have concluded that the conflict is unavoidable. The National Academies believe that Dr. Yuan can serve effectively as a member of the committee, and the committee can produce an objective report, taking into account the composition of the committee, the work to be performed, and the procedures to be followed in completing the study.

C

Committee Information-Gathering Sessions

FEBRUARY 27–28, 2023: OPEN SESSION WITH THE DEPARTMENT OF ENERGY (DOE) AND CONGRESSIONAL SPONSORS

DOE Office of Fossil Energy and Carbon Management
- Amishi Claros, Acting Division Director for Carbon Dioxide Removal and Conversion

DOE Bioenergy Technologies Office
- Christy Sterner, Technology Manager for Advanced Algal Systems Program

DOE Biological and Environmental Research Program
- Todd Anderson, Acting Associate Director

DOE Basic Energy Sciences Program
- Gail McLean, Acting Director of the Chemical Sciences, Geosciences, and Biosciences
- Raul Miranda, Team Lead for Chemical Transformations

House Committee on Science, Space, and Technology
- Hillary O'Brien, Staff Director
- Adam Rosenberg, Subcommittee Staff Director

Senate Committee on Energy and Natural Resources
- C.J. Osman, Counsel
- Tripp Parks, Deputy Chief Counsel
- Valerie Manak, Professional Staff Member

DOE Advanced Research Projects Agency–Energy
- Marina Sofos, Program Director
- Jack Lewnard, Program Director
- James Seaba, Program Director

DOE Office of Clean Energy Demonstrations

- Todd Shrader, Director for Project Management

MAY 3, 2023: OPEN SESSION WITH THE DEPARTMENT OF ENERGY

DOE Loan Programs Office

- Harry Warren, Senior Consultant

DOE Office of Clean Energy Demonstrations

- Catherine Clark, Energy Justice Liaison

JUNE 27–29, 2023: OPEN SESSIONS ON CO_2 UTILIZATION MARKETS, INVESTMENTS, NATIONAL ENERGY TECHNOLOGY LABORATORY (NETL) RESEARCH AND DEVELOPMENT, AND ESTABLISHING CARBON CAPTURE, UTILIZATION, AND STORAGE (CCUS) INDUSTRIAL CLUSTERS

Day 1, June 27—CO_2 Potential

- Josh Schaidle, National Renewable Energy Laboratory
- Frederic Clerc, New York University CO_2toValue
- Jonathan Goldberg, Carbon Direct Capital
- Nancy Gillis, World Economic Forum

Day 2, June 28—NETL Research and Development, Coal Waste Opportunities, and Carbon Utilization Policy

- Joseph Stoffa, NETL
- Michelle Krynock, NETL
- John Thompson, Clean Air Task Force
- Allan Kolker, U.S. Geological Survey

Day 3, June 29—Establishing CCUS Clusters and Small Business Challenges and Opportunities

- Martijn Verwoerd, Carbon Connect Delta/North Sea Port Cluster
- Cathy Tway, Johnson Matthey
- Erik Mayer, CF Industries
- Matthew Rhodes, Camirus/Black Country Cluster
- William Swetra, Oxy Low Carbon Ventures
- Michele Schimpp, Small Business Administration
- Duncan Turner, HAX

SEPTEMBER 13, 2023: OPEN SESSION ON SUSTAINABLE AVIATION FUELS

- Michael Köpke, LanzaTech

OCTOBER 23, 2023: OPEN SESSION ON COAL WASTE SEPARATIONS

- Jennifer Wilcox, DOE
- Tony Troutman, Minerals Refining Company

OCTOBER 25, 2023: OPEN SESSION ON CO_2 UTILIZATION RESEARCH AND DEVELOPMENT

- Eric De Coninck, ArcelorMittal
- Patricia Ansems Bancroft, Dow Chemical Company
- Sukaran Arora, Dow Chemical Company
- Marcius Extavour, TimeCO_2
- Ian Robinson, Khosla Ventures

NOVEMBER 3, 2023: OPEN SESSION ON COAL WASTE SEPARATIONS

Omnis Energy

- Charles Gassenheimer
- Chuck Shaynak

D

Acronyms and Abbreviations

AMD	acid mine drainage
AMMTO	Advanced Materials and Manufacturing Technologies Office
ARPA-E	Advanced Research Projects Agency–Energy
BER	Biological and Environmental Research
BES	Basic Energy Sciences
BETO	Bioenergy Technologies Office
BIPOC	Black, Indigenous, and other people of color
BisA-PC	bisphenol-A polycarbonate
BOF slag	carbonated iron-making (basic oxygen furnace) slag
CBA	community benefits agreement
CBB	coal-based bricks and block
CBP	community benefits plan
CCC	carbon-carbon composite
CCR	coal combustion residual
CCS	carbon capture and storage
CCUS	carbon capture, utilization, and storage
CDR	carbon dioxide removal
CF	carbon fiber
CI	carbon intensity
CIFIA	Carbon Dioxide Transportation Infrastructure Finance and Innovation program
CM	critical mineral
CMM	critical minerals and materials
CMU	concrete masonry unit
CNF	carbon nanofiber
CNFM	carbon nanofilm
CNL	carbon nanolayer
CNMO	carbon nanomodifier

CNR	carbon nanorod
CNT	carbon nanotube
CNW	carbon nanowire
CO_2RR	CO_2 reduction reaction
COBRA	Co-Benefits Risk Assessment Health Impacts Screening and Mapping Tool (EPA)
CPC	coal plastic composite
CQD	carbon quantum dot
CTC	carbon tubular cluster
CTEC	carbon dioxide to elemental carbon
CVD	chemical vapor deposition
DAC	direct air capture
DEHPA	bis-2-ethylhexyl phosphoric acid
DMC	dimethyl carbonate
DME	dimethyl ether
DMF	dimethylformamide
DOC	direct ocean capture
DoD	Department of Defense
DOE	Department of Energy
DOE-LPO	DOE Loan Programs Office
DOT	Department of Transportation
EERE	Office of Energy Efficiency and Renewable Energy
EIA	environmental impact assessment
EJ	environmental justice
e-LCA	environmental life cycle assessment
EOR	enhanced oil recovery
EPA	U.S. Environmental Protection Agency
EPD	environmental product declaration
ERA	environmental risk assessment
EV	electric vehicle
FDM	fused deposition modeling
FECM	Office of Fossil Energy and Carbon Management
FGD	flue gas desulfurization
FOA	funding opportunity announcement
FOAK	first-of-a-kind
FPIC	free, prior, and informed consent
GAIN	Gateway for Accelerated Innovation in Nuclear program
GHG	greenhouse gas
GPI	Great Plains Institute
GQD	graphene quantum dot
HCS	hollow carbon sphere
HHS	hydrophobic-hydrophilic separation
IEA	International Energy Agency
IEDO	Industrial Efficiency and Decarbonization Office
IFC	International Fire Code

IIJA	Infrastructure Investment and Jobs Act
IL	ionic liquid
IRA	Inflation Reduction Act
LCA	life cycle assessment
LCC	life cycle costing
LIB	lithium-ion battery
MEA	monoethanolamide
MILP	mixed integer linear programming
MMT	million metric tons
MRV	monitoring, reporting, and verification
MTPA	million tonnes per annum
MWNT	multi-walled nanotube
NBIC	National Board Inspection Code
NETL	National Energy Technology Laboratory
NFPA	National Fire Protection Association
NGP	nanographite platelet
NIOSH	National Institute for Occupational Safety and Health
NMR	nuclear magnetic resonance
NOAA	National Oceanic and Atmospheric Administration
NOAK	*n*th-of-a-kind
NPC	National Petroleum Council
NSF	National Science Foundation
OPC	ordinary Portland cement
OSDBU	Office of Small and Disadvantaged Business Utilization
OSHA	Occupational Safety and Health Administration
PC	photocatalytic
PCR	preceramic polymer resin
PEC	photoelectrochemical
PHMSA	Pipeline and Hazardous Materials Safety Administration
PM	particulate matter
PV	photovoltaic
PVC	polyvinyl chloride
PV-EC	photovoltaic and electrolyzer configuration
R&D	research and development
RD&D	research, development, and demonstration
REE	rare earth element
REO	rare earth oxide
RFI	request for information
RWGS	reverse water gas shift
SA	sensitivity analysis
SAF	sustainable aviation fuel
SBIR	Small Business Innovation Research program
SC	Office of Science

SCM	supplementary cementitious material
s-LCA	social life cycle assessment
SOEC	solid oxide electrolyzer cell
STTR	Small Business Technology Transfer program
SWNT	single-walled nanotube
TEA	techno-economic assessment
TRL	technology readiness level
UPGrants	Utilization Procurement Grants program
USGS	U.S. Geological Survey
UV	ultraviolet
WGS	water gas shift
WLP	Wood-Ljungdahl pathway
WRI	World Resources Institute
XPS	x-ray photoelectron spectroscopy

E

Supplemental Material to the Comprehensive Research Agenda for CO_2 and Coal Waste Utilization

As presented in Chapter 11, the committee developed a comprehensive research agenda for CO_2 and coal waste utilization that identifies key research, development, and demonstration (RD&D) needs to enable future utilization opportunities. This report's research agenda updates and expands on the one from the 2019 National Academies' report *Gaseous Carbon Waste Streams Utilization: Status and Research Needs*, differing in three key ways: a change in scope of carbon feedstocks, the focus on products needed for a net-zero future, and advances in the field over the last 5 years. As noted above, the committee was charged with identifying research needs for CO_2 and coal waste utilization, specifically for making products that could contribute to a net-zero, circular carbon economy. The 2019 report committee examined research needs for CO_2, methane, and biogas utilization, and it did not explicitly consider product needs in a net-zero future. This difference in framing, a desire to be additive rather than duplicative, and technological advances since 2019 resulted in the 2024 research agenda that places more emphasis on applied and enabling research needs for CO_2 utilization, building on the more basic science research needs covered in 2019, along with highlighting crosscutting aspects like process integration. This committee also examined in more depth research needs for producing long-lived, elemental carbon products from CO_2, which were largely not covered in the 2019 report but could play a role in durably storing carbon in a net-zero future.

This appendix provides additional context about the development of the research agenda, including details about the research agenda descriptors and information about federal funders of CO_2 and coal utilization and enabling technologies. It also elaborates on the overarching research themes identified across CO_2 and coal waste utilization RD&D, illustrating connections among research needs for various technologies and processes. Finally, this appendix presents subsets of the full research agenda table with the research needs directed to each of the Department of Energy (DOE) offices sponsoring this study: Fossil Energy and Carbon Management, Basic Energy Sciences, Biological and Environmental Research, and Bioenergy Technologies.

RESEARCH AGENDA DESCRIPTORS

The research agenda items are classified as basic research, applied research, demonstration, or enabling, or some combination of the four. The committee uses the Office of Management and Budget's definitions of basic and applied research (EOP 2023, p. 280):

- **Basic research**—"Experimental or theoretical work undertaken primarily to acquire new knowledge of the underlying foundations of phenomena and observable facts. Basic research may include activities with broad or general applications in mind, such as the study of how plant genomes change, but should exclude research directed towards a specific application or requirement, such as the optimization of the genome of a specific crop species."

- **Applied research**—"Original investigation undertaken in order to acquire new knowledge. Applied research is, however, directed primarily towards a specific practical aim or objective."

Research agenda items classified as demonstration are those that call for projects to test CO_2 utilization technologies in real-world conditions to facilitate scale up and indicate market viability. Research agenda items classified as enabling are those that are necessary for CO_2 utilization to contribute to a net-zero future but are not research on CO_2 utilization technologies or processes themselves. Examples include basic and applied research on non-CO_2 feedstocks (e.g., clean hydrogen), life cycle assessment (LCA) and techno-economic assessment (TEA) tools, infrastructure development (e.g., CO_2 pipelines, clean hydrogen generation), resource mapping, and understanding of public perception of CO_2 utilization technologies.

Carbon dioxide and coal waste conversion processes span across technology readiness levels (TRLs), with differing needs for basic research, applied research, technology demonstrations, and research on enabling technologies and processes. Sorting the research needs for CO_2 and coal waste utilization processes by research type, as shown in Figure E-1, can provide insight into the general state of the field and where future research and development (R&D) investment might be best placed. For example, the majority of the research needs identified for chemical CO_2 conversion to elemental carbon and organic products are classified as basic research. The biological CO_2 utilization research needs are more evenly split across basic and applied research. For coal waste utilization, applied research is the primary need whereas for mineralization, there is a more even distribution across basic, applied, demonstration, and enabling research.

The research agenda (Table 11-1) also indicates the relevant research area, product class, and product lifetime for each research agenda item. Research area categories are mineralization, chemical, biological, coal waste utilization, LCA/TEA, markets, infrastructure, and societal impacts. For conversions of CO_2 and coal waste,

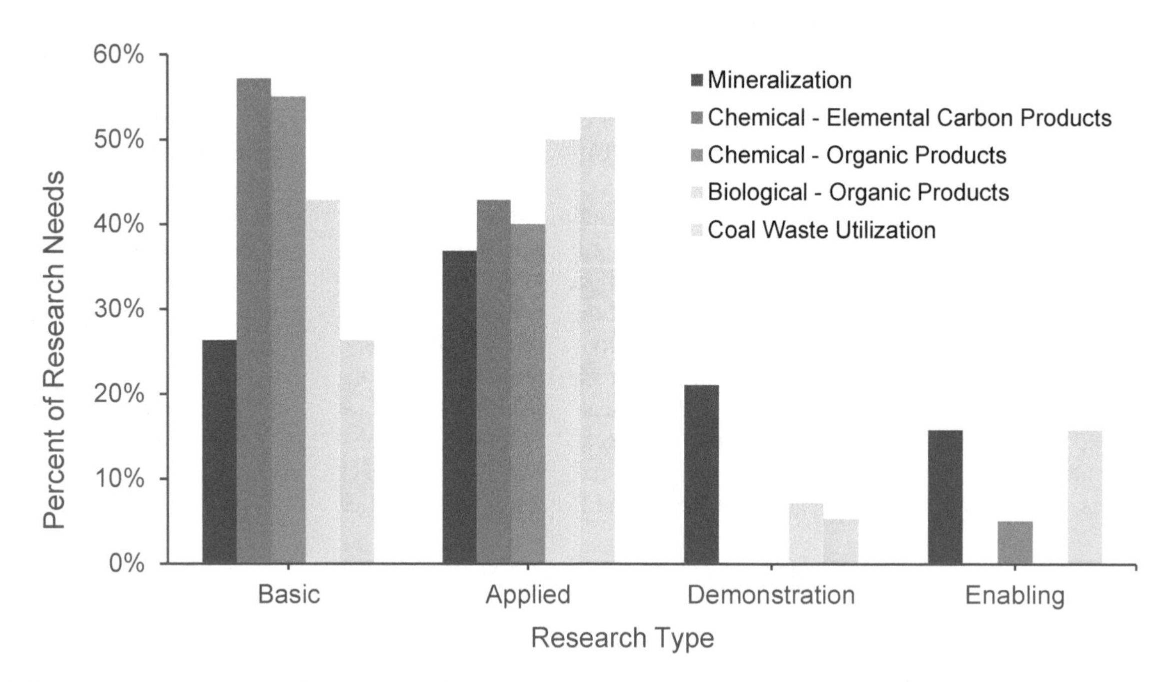

FIGURE E-1 Percent of research agenda items from Table 11-3 by research type (basic, applied, demonstration, enabling) for CO_2 mineralization, chemical CO_2 conversion to elemental carbon products, chemical CO_2 conversion to organic products, biological CO_2 conversion to organic products, and coal waste utilization, including conversion to long-lived carbon products and extraction of rare earth elements and critical minerals.

a subprocess is sometimes included—for example, chemical–electrochemical, biological–photosynthetic. Based on the research need and research area, product classes are identified; these include construction materials, elemental carbon materials, chemicals, polymers, coal waste–derived carbon products, and metal coal waste by-products. The average product lifetime is also noted, where "short-lived" indicates a lifetime of less than 100 years and "long-lived" indicates a lifetime of greater than 100 years. For research needs classified as LCA/TEA, markets, infrastructure, and societal impacts, the product class is listed as "All," with both "long-lived" and "short-lived" products being possible, as these research items would support development of any CO_2 utilization technology.

FEDERAL FUNDERS OF CO_2 UTILIZATION AND ENABLING TECHNOLOGIES

To determine the relevant funding agencies or other actors to include in the research agenda (Table 11-1), the committee reviewed current research portfolios of federal agencies that work on topics related to this study's scope, including CO_2 capture, conversion, and transport; coal waste utilization; critical minerals recovery; LCA and TEA; materials discovery and development; separations; reactor design and engineering; resource mapping; product testing and certification; and environmental and health impacts of technologies. This assessment is summarized in Table E-1.

DOE is the primary funder of CO_2 and coal waste utilization research, via the Office of Fossil Energy and Carbon Management (FECM), the Bioenergy Technologies Office (BETO), the Office of Science (SC), the Advanced Research Projects Agency–Energy (ARPA-E), the Industrial Efficiency and Decarbonization Office (IEDO), and the Advanced Materials and Manufacturing Technologies Office (AMMTO). Related to this study's scope, DOE-FECM supports applied research on point source carbon capture, carbon dioxide removal, CO_2 transport, CO_2 and coal waste conversion, critical mineral and materials extraction from coal wastes, and development of LCA and TEA tools for carbon conversion (Claros 2023; Krynock 2023; Stoffa 2023).[1] DOE-BETO supports applied research on converting CO_2 to fuels and chemicals via photosynthetic and non-photosynthetic biological routes (Sterner 2023). Basic Energy Sciences (BES) within DOE-SC supports basic research on thermo-, electro-, plasma-, and photo-catalytic CO_2 conversion; biomimetic systems for CO_2 conversion; CO_2 mineralization; catalyst and materials discovery; separations; and direct air capture (McLean and Miranda 2023). Biological and Environmental Research (BER) within DOE-SC supports research on the discovery and design of novel metabolic processes and development of cell/cell-free systems for CO_2 utilization (Anderson 2023). ARPA-E funds high-risk, high-reward energy technologies, with current and past programs related to carbon capture and storage, CO_2 utilization in building materials, CO_2 conversion to fuels and chemicals, CO_2 mineralization for metal extraction, and "exploratory topics" on direct air and ocean capture of CO_2 and electricity system models for carbon capture and storage (Sofos et al. 2023). DOE-IEDO supports applied research on carbon capture, use of low-carbon fuels and feedstocks in industry (including those produced from CO_2), and CO_2 mineralization. DOE-AMMTO supports applied research on materials and manufacturing processes that can support clean energy technologies and a circular carbon economy, as well as research on establishing domestic critical minerals supply chains.

Additional work on topics relevant for this report is supported by the National Science Foundation (NSF), the U.S. Environmental Protection Agency (EPA), the Department of Transportation (DOT), the Department of Defense (DoD), the National Oceanic and Atmospheric Administration (NOAA), and the U.S. Geological Survey (USGS). NSF funds basic research across a wide variety of topics, including chemical catalysis and mechanisms, chemical and biological separations, materials discovery and development, polymers, electrochemical systems, reaction engineering, and design and operation of civil infrastructure (NSF n.d.). EPA supports research on human health risks from environmental stressors, environmental impacts of clean energy technologies, embodied carbon in products, LCA tools, and chemical product safety (EPA 2024). DOT supports research on pipeline safety (PHMSA 2017), sustainable fuels (FAA 2024), and pavement materials like concrete and aggregates (FHWA 2021). DoD funds CO_2 conversion to aviation fuels from air or water through the Air Force Office of Scientific Research (AFOSR n.d.); basic and applied research on (electro)chemical interactions in marine environments through the Office of Naval Research (ONR n.d.); basic research on electrochemistry, polymer chemistry, microbiology, materials design and behavior, and complex systems modeling through the Army Research Laboratory (ARL 2024);

[1] This includes research performed and supported by the National Energy Technology Laboratory.

TABLE E-1 Carbon Capture and Utilization Research Across Federal Agencies

		DOE												
				EERE			SC							
Research Category	Research Topic	FECM*	ARPA-E	AMMTO	IEDO	BETO	BES	BER	NSF	EPA	DOT	DoD	USGS	NOAA
Carbon Capture	Point source capture	•	•		•				•					
	Direct air capture	•	•				•		•					
	Direct ocean capture		•				•		•			•		•
CO_2 Utilization	CO_2 conversion	•	•		•	•	•	•	•			•		
	Mineralization	•	•		•		•		•			•		•
	Integrated capture and conversion	•	•		•		•		•					
	Algae capture and conversion	•				•			•					
Coal Waste Utilization	Coal waste conversion	•							•					
	Critical materials recovery	•	•	•			•		•			•	•	
Crosscutting Basic and Applied Research	Materials discovery and design		•	•			•		•			•		
	Materials development at scale	•		•							•			
	Separations						•		•			•		
	Metabolic understanding							•						
	Reactor design	•	•						•					
Feedstock, Technology, and Product Assessments	Resource mapping	•								•			•	
	LCA data and tools	•			•	•			•	•			•	
	TEA data and tools	•			•	•			•	•			•	
	Environmental impacts at scale									•				•
	Human health risks								•	•				
	Product testing and certification	•								•	•			
Infrastructure	CO_2 transport	•							•		•			

* Including research performed through the National Energy Technology Laboratory.

NOTE: IEDO = Industrial Efficiency and Decarbonization Office; LCA = life cycle assessment; NOAA = National Oceanic and Atmospheric Administration; TEA = techno-economic assessment.

and CO_2 conversion, critical minerals recovery, and extraction and separation of rare earth elements through the Defense Advanced Research Projects Agency (DARPA n.d.). To meet its 2030 emissions reduction target, DoD funds research for advanced technological solutions to support the commercialization of early-stage carbon capture, utilization, and storage (CCUS) projects, first-of-a-kind demonstration projects, and early markets for low-carbon concrete and other construction materials. NOAA funds research on ocean-based carbon capture and mineralization (NOAA 2024). USGS conducts research on energy and mineral resources, including resource mapping (USGS n.d.).

As discussed in Chapter 4, DOE's Small Business Innovation Research (SBIR) and Small Business Technology Transfer (STTR) programs provide funding for small businesses to conduct R&D with an ultimate goal of commercialization (DOE n.d.). The SBIR and STTR programs coordinate with offices across DOE, including those that support research on CO_2 and coal waste utilization: FECM, BES, BER, and Energy Efficiency and Renewable Energy (EERE). Thus, grants through the SBIR/STTR programs could be used to address some of the research needs described in Table 11-1.

RESEARCH THEMES

Looking across the specific research needs for CO_2 and coal waste utilization, and the technologies and processes that will facilitate their deployment at scale, 16 overarching research themes emerged. Some research needs fit under multiple themes. The 16 themes can be classified into three broad categories of (1) reaction-level understanding, (2) systems-level understanding, and (3) demonstration and deployment needs (see Table E-2). Reaction-level understanding encompasses research at the atomic or molecular level that focuses on a specific reaction or process. Systems-level understanding includes research on process design and integration, modeling of complex systems, and environmental and societal impacts of technologies. Demonstration and deployment needs are items that will facilitate commercialization and scale up of CO_2 and coal waste utilization technologies, such as certification of products, deployment of required infrastructure, and development of test facilities. The research themes within each of these categories are discussed in the following sections, highlighting research that could benefit multiple approaches to CO_2 and coal waste utilization. Two research themes are classified under multiple categories. "Computational modeling and machine learning" is categorized as both reaction-level understanding and systems-level understanding because such research is needed at both scales. Similarly, "reactor design and reaction engineering" includes research needs related to systems-level understanding as well as demonstration and deployment. Figure E-2 illustrates the overlap of research themes across various CO_2 and coal waste utilization processes.

Reaction-Level Understanding

Fundamental Knowledge

A better fundamental understanding of materials and chemical processes is needed for CO_2 utilization approaches at low TRL, such as CO_2 conversion to elemental carbon products, photo(electro)chemical and

TABLE E-2 Classification of Research Themes for CO_2 and Coal Waste Utilization

Reaction-Level Understanding	Systems-Level Understanding	Demonstration and Deployment Needs
• Fundamental knowledge • Catalyst innovation and optimization • Genetic manipulation • Metabolic understanding and engineering • Microbial engineering • Computational modeling and machine learning • Separations	• Reactor design and reaction engineering • Integrated systems • Energy efficiency, electrification, and alternative heating • Environmental and societal considerations for CO_2 and coal waste utilization technologies • Computational modeling and machine learning	• Reactor design and reaction engineering • Certification and standards • Enabling technology and infrastructure needs • Resource mapping • Research centers and facilities • Market opportunities

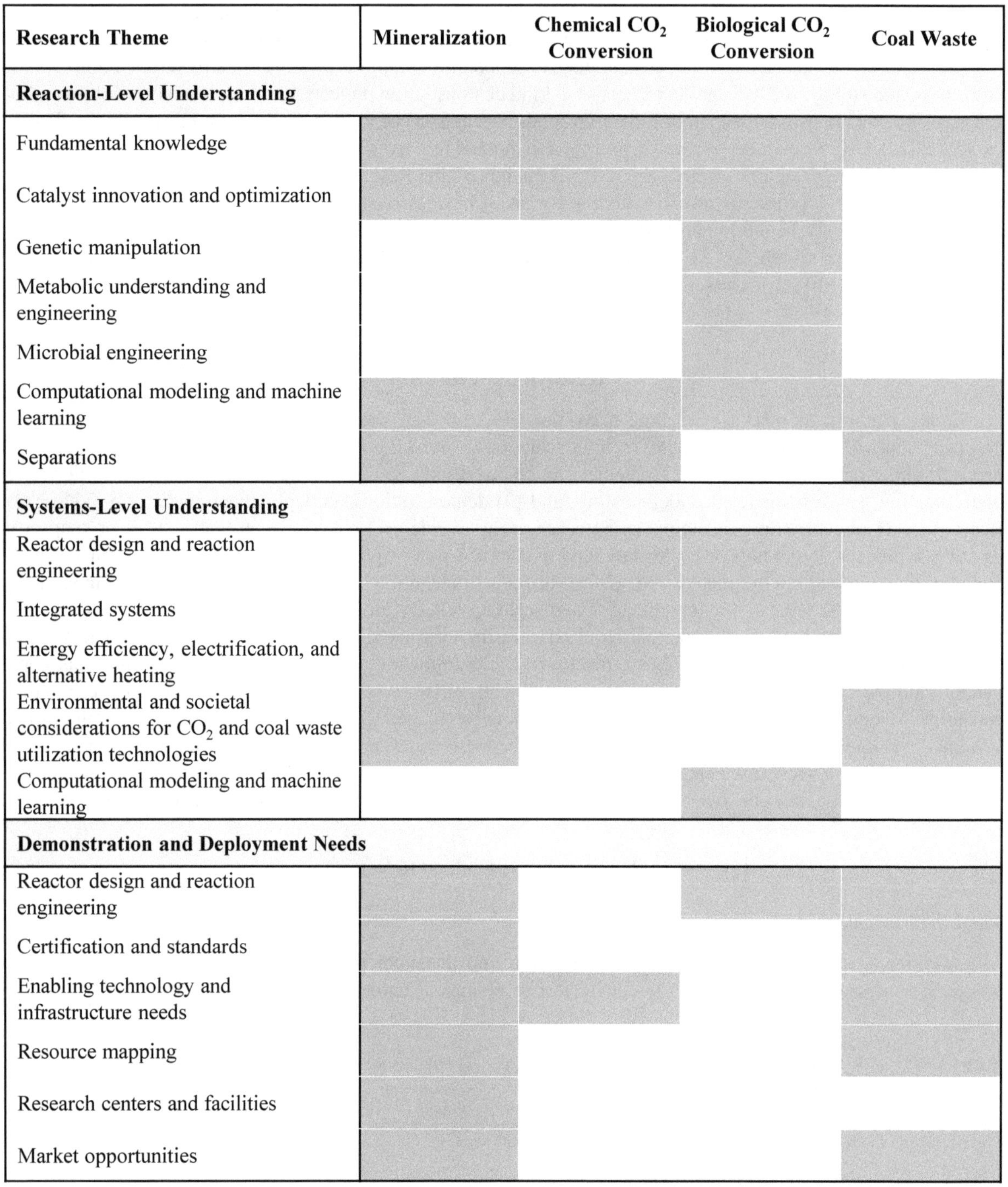

Research Theme	**Mineralization**	**Chemical CO_2 Conversion**	**Biological CO_2 Conversion**	**Coal Waste**
Reaction-Level Understanding				
Fundamental knowledge	■	■	■	■
Catalyst innovation and optimization	■	■	■	
Genetic manipulation			■	
Metabolic understanding and engineering			■	
Microbial engineering			■	
Computational modeling and machine learning	■	■	■	■
Separations	■	■		■
Systems-Level Understanding				
Reactor design and reaction engineering	■	■	■	■
Integrated systems	■	■	■	
Energy efficiency, electrification, and alternative heating	■	■		
Environmental and societal considerations for CO_2 and coal waste utilization technologies	■			■
Computational modeling and machine learning			■	
Demonstration and Deployment Needs				
Reactor design and reaction engineering	■		■	■
Certification and standards	■			■
Enabling technology and infrastructure needs	■	■		■
Resource mapping	■			■
Research centers and facilities	■			
Market opportunities	■			■

FIGURE E-2 Areas of research needed for CO_2 and coal waste utilization, indicated by gray boxes where a pathway has one or more research needs falling into research themes in categories of reaction-level understanding, systems-level understanding, and demonstration and deployment needs. The patterns show the distribution of research themes across pathways.

plasmachemical CO_2 conversion to organic products, and for some biological CO_2 and coal waste conversions. This challenge could be addressed through increased support for materials discovery and characterization, studies of mechanism and selectivity for enzymes and chemical reactions, modeling and simulations, and development of tools to monitor local reaction environments. Research agenda items 5-G, 5-K, 6-A, 6-E, 7-H, 7-J, 7-N, 8-C, 8-D, 8-E, 9-F, and 9-K describe needs related to improving fundamental knowledge.

Catalyst Innovation and Optimization

Discovery, development, and improvement of catalysts is a key research need for many CO_2 utilization approaches, including conversion of CO_2 to elemental carbon materials, fuels, chemicals, and polymers via thermochemical, electrochemical, photo(electro)chemical, and plasmachemical catalytic routes as well as electrochemically driven CO_2 mineralization. Specifically, R&D is needed to identify catalysts that are more active, selective, stable, and robust to impurities, and ideally derived from abundant elements for improved scalability. For thermochemical CO_2 conversion, discovery research into catalysts that can accommodate alternative heating methods like (pulsed) electrical heating could yield performance, energy, and efficiency improvements. Improved electrocatalysts are needed for low- and high-temperature electrochemical CO_2 conversions, for operation in biologically amenable conditions and production of bio-compatible intermediates to incorporate into electro-bio hybrid systems, and for water oxidation or alternative anodic reactions to improve the cost and efficiency of electrochemical CO_2 conversions. Catalysts for rapid, stereoselective co-polymerization of a broader class of monomers with CO_2, especially those that can lead to polymers with properties more like thermoplastics and/or thermosets, would help to further the development of CO_2-derived polymers. Research agenda items 5-F, 6-B, 6-C, 6-D, 7-A, 7-B, 7-E, 7-F, 7-I, 7-N, and 8-E describe needs related to catalyst innovation and optimization.

Genetic Manipulation

The development of new, more efficient genetic manipulation tools could enhance the efficiency of biological CO_2 fixation and improve understanding of carbon metabolism. Research agenda item 8-B describes research needs related to genetic manipulation.

Metabolic Understanding and Engineering

Biological CO_2 conversion would benefit from improved understanding of carbon metabolism, which, as noted above, could be achieved with better genetic manipulation tools. Using this knowledge, metabolic engineering of microorganisms can help to overcome biochemical, bioenergetic, and metabolic limits to improve efficiency, titer, and productivity of biological CO_2 utilization systems. Research agenda items 8-A, 8-B, and 8-C describe research needs related to metabolic understanding and engineering.

Microbial Engineering

Advances in microbial engineering could improve the productivities and titer of electro-bio hybrid systems. Research agenda item 8-F describes research needs related to microbial engineering.

Separations

Separations are a key enabling technology for CO_2 and coal waste utilization. Efficient separations are required for both the feedstock streams (e.g., CO_2 purification, separation of mineral matter from carbon in coal wastes) and the product streams (e.g., separation of catalyst from solid carbon product, separation of multiple products from each other). For electrochemical systems, development of cost-effective, scalable membrane materials that can function over a wide pH range could decrease overall costs of CO_2 conversion. Improving separations of critical minerals and rare earth elements from coal waste will require more selective, sustainable solvents and transformative systems for extractions both from solid (waste coal and coal combustion residuals) and liquid (e.g., acid mine drainage) waste streams. Research into more selective, sustainable separations is similarly needed for carbon mineralization integrated with metal recovery. Research agenda items 3-B, 5-H, 6-G, 7-G, 9-C, 9-L, 9-M, 9-N, and 9-O describe research needs related to separations.

Computational Modeling and Machine Learning

Computational modeling and machine learning at the atomic or molecular scale can increase understanding of CO_2 or coal waste conversions and direct the discovery and development of improved catalysts and materials. For example, for biological CO_2 conversion, machine learning can be used to improve CO_2 fixation efficiency and optimize nutrient input, CO_2 delivery, and light penetration to achieve higher productivities in photosynthetic organisms. For electro- and photo(electro)-chemical CO_2 conversion, computational modeling tools—including quantum methods, ab initio molecular dynamics, and machine-learned force field molecular dynamics—can provide insights into charge transfer processes, solvent configurations, and structural characteristics of electrode/electrolyte interfaces. Atomic- and multi-scale computer simulations can also improve understanding of the complex carbon chemistry involved in transforming waste coal into useful solid-carbon products. Research agenda items 5-F, 7-A, 7-B, 7-E, 7-F, 7-H, 7-I, 7-J, 7-L, 7-N, 8-B, 8-C, and 9-F describe research needs related to computational modeling and machine learning at the reaction level.

Systems-Level Understanding

Reactor Design and Reaction Engineering

Research on reactor design and reaction engineering will provide critical systems-level understanding to help advance a technology along the technology readiness scale. CO_2 utilization reactors and reactions can require incorporation of different forms of energy, facilitation of multiple reaction steps, and maintenance of reactions, including in biological systems. For example, RD&D is needed to improve electrochemical cell design, develop cost-effective, scalable membrane materials, and monitor side reactions and membrane and electrode fouling issues for electrochemically driven CO_2 mineralization and electrochemical CO_2 conversion to chemicals and fuels. RD&D also is needed on devices, reactor design, and reaction engineering for photochemical, photoelectrochemical, plasmachemical, and biological CO_2 conversions to optimize performance metrics and help inform scale up. For thermochemical processes to convert CO_2 to chemicals and elemental carbon materials, improvements in reactor design and reaction engineering can facilitate the integration of low-carbon electricity and/or heat. Tandem processes and hybrid systems (see definitions in the "Integrated Systems" section below) that combine multiple CO_2 utilization approaches could be improved with increased research into process efficiency and systems optimization. Such research could also yield more efficient transformations of waste coal into long-lived solid carbon products with lower embodied carbon than existing products. Research agenda items 5-C, 5-F, 5-G, 5-H, 6-B, 6-F, 6-H, 7-D, 7-G, 7-K, 7-M, 8-A, 8-D, 8-G, 8-H, and 9-D describe needs related to reactor design and reaction engineering at the systems level.

Integrated Systems

Integrated systems for CO_2 utilization refer to the combination of two or more conversion approaches (mineralization, thermochemical, electrochemical, photo(electro)chemical, plasmachemical, biological), or the integration of capture and conversion, to produce any of the product classes within the scope of this report (inorganic carbonates, elemental carbon materials, chemicals, fuels, polymers). This term encompasses both tandem processes (a subclass of integrated systems involving the combination of two or more conversion routes in sequence) and hybrid systems (a subclass of integrated systems involving a combination of biological and nonbiological components). A significant opportunity is integrated CO_2 capture and conversion, which removes the need for separation and purification of captured CO_2 before its conversion to product. For these systems, more research is required into molecules and materials discovery, catalytic mechanisms, process optimization, CO_2 stream purification, and reactor design. Tandem processes can yield products that a single process alone cannot access and may have economic, energy savings, and/or environmental benefits, thus warranting increased research attention. Another promising opportunity is the integration of carbon mineralization with metal recovery, for which more research is needed in energy-efficient grinding/comminution, selective separation, improved recycling, reduced emissions, and systems integration and optimization. Hybrid systems need improvements in enzyme efficiency, stability, and selectivity; scalability of redox-balanced systems; and reactor design to optimize for specific intermediates and desired products. Research agenda items 5-F, 5-H, 6-H, 6-I, 7-L, 7-M, 8-D, 8-G, and 8-H describe needs related to integrated systems.

Energy Efficiency, Electrification, and Alternative Heating

Efficiency improvements and incorporation of clean electricity or other alternative heating methods will facilitate CO_2 utilization deployment under the energy and emissions constraints of a net-zero future. Improvements in energy efficiency, particularly in grinding/comminution, would be beneficial for carbon mineralization processes. Applied research is needed on engineering and systems optimization to integrate variable renewable energy and energy storage with reaction systems for thermochemical CO_2 conversion to chemicals. R&D on reaction electrification and heat integration, including electrolytic and plasma processes, and (pulsed) electrical heating, could yield energy savings and reduced greenhouse gas (GHG) emissions (if clean electricity is used) for CO_2 conversion to chemicals, fuels, and elemental carbon products. Discovery research into catalysts that can best take advantage of these alternative electrically driven methods is also needed. Research agenda items 5-C, 5-H, 6-B, 6-F, 7-B, and 7-D describe research needs related to energy efficiency, electrification, and alternative heating.

Environmental and Societal Considerations for CO_2 and Coal Waste Utilization Technologies

To evaluate the potential for CO_2 and coal waste utilization to contribute to a net-zero future, their environmental impacts must be better understood. A greater understanding of emissions impacts of CO_2 utilization technologies within LCAs is needed, especially for non-CO_2 emissions and circular carbon processes. For circular processes, this includes understanding the leakage potential, the fate of products under different end of life conditions, and evolution of processes and demand through multiple cycles of use and reuse. More data and tools need to be developed to conduct LCAs and TEAs of coal waste utilization processes and to trace carbon across value chains over time. A better understanding the effect of CO_2 purity on the results of LCAs and TEAs could guide future R&D on CO_2 utilization technologies. Development of a protocol to assess the net environmental impacts of CO_2 mineralization at the gigatonne scale, including chemical toxicity, water requirements, and air quality, would help inform scale-up efforts. Ocean-based CO_2 mineralization requires better understanding of local environmental and ecological impacts, development of an environmental protocol to assess and mitigate unexpected impacts from pH changes, and evaluation of the recyclability of process water with spent acids, bases, and dissolved ions. In addition to environmental impacts, the societal impacts of carbon utilization need to be assessed across temporal and spatial scales, including the broader impacts of CO_2 conversion on the environment, resource (re-)allocation, distributional effects among regions, demographic groups, and communities, job gains and/or losses, and safeguards for disadvantaged communities. More information is needed about public perception of carbon utilization technologies and factors that influence community acceptance. Research agenda items 2-A, 3-A, 3-C, 3-D, 3-E, 3-F, 4-A, 5-D, 5-G, and 9-J describe research needs related to environmental and societal considerations for CO_2 and coal waste utilization technologies.

Computational Modeling and Machine Learning

Computational modeling and machine learning can help to understand and predict behavior of complex systems, providing knowledge that can be exploited to improve system integration, efficiency, safety, environmental impacts, cost-effectiveness, and other factors. For example, in hybrid biological CO_2 conversion systems, computational modeling and machine learning can be used to guide more efficient conversion of various intermediates into bioproducts. Related to infrastructure development, computational tools are needed to design optimal multimodal transportation networks to collect CO_2 captured from stranded emitters for centralized utilization. Additionally, CO_2 dispersion modeling is used to determine minimum safe distances to populated areas and for emergency response planning. More work is needed on simulating the fluid/structure interaction and subsequent atmospheric dispersion for the case of accidental rupture of buried CO_2 pipelines that results in formation of a crater owing to the high momentum CO_2 jet. Research agenda items 8-B, 8-C, 10-A, and 10-D describe research needs related to computational modeling and machine learning at the systems level.

Demonstration and Deployment Needs

Reactor Design and Reaction Engineering

Improvements to reactor design and reaction engineering will be critical in moving from basic and applied research to technology demonstrations and commercially viable systems. Building on needs for systems-level understanding of reactor/reaction systems and processes (see section 11.2.3.2), the committee identified several research needs for demonstration-scale projects. For emerging carbon mineralization systems, demonstration projects can help to identify and address challenges that may arise when moving toward gigatonne-scale production, including energy requirements for large-scale mining and mineral processing, process integration, chemical recycling, and water requirements, including for electrochemically driven systems. In biological hybrid systems, demonstration projects will inform integration and scale up of catalysis and bioconversion and facilitate evaluation of economic and environmental impacts. Demonstrations of coal waste conversions into long-lived carbon products, such as engineered composites, graphite, graphene, carbon fiber, and carbon foam, can similarly indicate the feasibility to generate products for the construction, energy storage, transportation, and defense industries. Research agenda items 5-C, 5-F, 8-G, and 9-D describe research needs–related demonstrations of reactor design and reaction engineering.

Certification and Standards

In some cases, CO_2 and coal waste utilization generate products that are not chemically identical to current products for the same application, so standards and certification methods will need to be developed to ensure that these products meet technical and safety requirements. This is particularly relevant for CO_2- and coal waste–derived construction materials, where the mechanical, thermal, electrical, and/or chemical properties need to be evaluated to ensure that the materials conform with codes specific to their intended application. Additionally, standards will need to be established for using coal waste in applications with environmental exposure to ensure product safety, given the potential presence of toxic heavy metals in coal waste streams. Research agenda items 5-J, 9-E, 9-G, 9-H, 9-I, and 9-J describe research needs related to certification and standards.

Enabling Technology and Infrastructure Needs

A common research need across CO_2 and coal waste utilization approaches is optimizing multimodal transportation networks to move feedstocks (e.g., CO_2, coal wastes, reactant minerals, hydrogen) to sites of production and products (e.g., inorganic carbonates, chemicals, solid carbon materials, critical minerals) to markets. This will require developing robust computational tools that analyze cost, safety, and environmental impact to determine possible transportation infrastructure solutions. In a similar vein, TEAs are needed to determine whether, for small- to medium-scale emitters, it is preferable to perform CO_2 utilization on site and transport the products or transport captured CO_2 from the facility for utilization or storage elsewhere. To improve the safety of CO_2 transportation by pipeline, more research is needed on (1) understanding and developing approaches to mitigate issues with propagating brittle and ductile fractures and (2) dispersion modeling calculations for the accidental rupture of buried

CO_2 pipelines that results in formation of a crater owing to the high-momentum CO_2 jet. With this knowledge, software could be developed for use as a design and decision-making tool for pipeline developers to mitigate risks associated with pipeline failures. For thermochemical CO_2 conversion to achieve net-zero emissions at commercial scale, continued research and development into low-carbon hydrogen and other carbon-neutral reductants will be required. Research agenda items 5-B, 7-C, 9-B, 10-A, 10-B, 10-C, 10-D, and 10-E describe research needs related to enabling technologies and infrastructure.

Resource Mapping

Understanding the full potential for CO_2 mineralization and coal waste utilization will require evaluation and mapping of resources used in those processes—that is, minerals, industrial wastes, and coal waste streams. More information is needed about the composition, volume, and locations of these resources, as well as their chemical and physical properties. Research agenda items 5-A and 9-A describe research needs related to resource mapping.

Research Centers and Facilities

Research centers and facilities can play a role in technology development and scale up, as they can enable testing under real-world conditions. For example, the DOE-funded direct air capture (DAC) and hydrogen hubs could incorporate CO_2 utilization research and provide testing platforms for CO_2 utilization technologies. This opportunity would be especially beneficial for co-located hubs, which could demonstrate and scale up production of net-zero fuels and chemicals using CO_2 from DAC combined with clean hydrogen. The committee also sees a need to develop two centers for carbon mineralization research: (1) a testing facility platform, similar to the National Carbon Capture Center, where various ocean-based carbon mineralization concepts and technologies can be evaluated in real ocean conditions with minimal environmental impacts and (2) university–industry–national laboratory collaborations to rapidly scale up and deploy carbon-negative mining technologies with large CO_2 utilization potential. Research agenda items 5-E and 5-I describe these needs for research centers and facilities.

Market Opportunities

For CO_2 and coal waste utilization to meet their full potential, there needs to be a better understanding of market projections for carbon-based products and critical minerals that take into consideration national targets for the transition to net-zero emissions. The development of strategies to link feedstocks to production sites to product markets, discussed under "enabling infrastructure needs" above, will facilitate market development. Research agenda items 2-B, 5-C, and 9-B describe research needs related to market opportunities.

RESEARCH NEEDS DIRECTED TO DOE OFFICES

As discussed above, DOE is the primary funder of CO_2 and coal waste utilization research, in particular through the sponsoring offices of this study: Fossil Energy and Carbon Management, Basic Energy Sciences, Biological and Environmental Research, and Bioenergy Technologies. Tables E-3 to E-6 present the research agenda items directed to each of these DOE offices.

TABLE E-3 Research Agenda Items Directed to DOE-FECM

Research Agenda Item	Basic, Applied, Demonstration, or Enabling	Research Area	Product Class	Research Barrier Addressed
3-A. Understanding impact of CO_2 purity in life cycle and techno-economic assessments.	Enabling	LCA/TEA	All	Environmental and societal considerations for CO_2 and coal waste utilization technologies
3-B. Improved CO_2 purification technologies.	Basic Applied	LCA/TEA	All	Separations
3-C. Understanding of non-CO_2-emissions impacts of CO_2 utilization technologies.	Enabling	LCA/TEA	All	Environmental and societal considerations for CO_2 and coal waste utilization technologies
3-D. LCA approaches to address circularity of CO_2-derived products.	Enabling	LCA/TEA	Chemicals Polymers	Environmental and societal considerations for CO_2 and coal waste utilization technologies
3-E. Understanding the flows of carbon through product life cycles	Enabling	LCA/TEA	All	Environmental and societal considerations for CO_2 and coal waste utilization technologies
3-F. Tools to trace carbon across value chains over time.	Enabling	LCA/TEA	All	Environmental and societal considerations for CO_2 and coal waste utilization technologies
5-A. Mapping of alkaline resources.	Enabling	Mineralization	Construction materials	Resource mapping
5-B. Optimization of infrastructure to connect feedstocks, facilities, and product markets.	Enabling	Mineralization	Construction materials	Enabling technology and infrastructure needs Market opportunities
5-D. Environmental and ecological impacts of ocean-based CO_2 mineralization.	Applied	Mineralization – ocean-based CO_2 utilization	Construction materials	Environmental and societal considerations for CO_2 and coal waste utilization technologies
5-E. Testing facility for ocean-based CO_2 mineralization.	Demonstration	Mineralization – ocean-based CO_2 utilization	Construction materials	Research centers and facilities
5-F. Catalyst, materials, and engineering design for electrochemically driven CO_2 mineralization.	Basic Applied Demonstration	Mineralization – electrochemical	Construction materials	Catalyst innovation and optimization Reactor design and reaction engineering Integrated systems Computational modeling and machine learning
5-G. Monitoring and evaluating impacts of electrochemically driven CO_2 mineralization.	Basic Applied	Mineralization – electrochemical	Construction materials	Fundamental knowledge Reactor design and reaction engineering Environmental and societal considerations for CO_2 and coal waste utilization technologies

5-H. CO_2 mineralization integrated with metal recovery.	Basic Applied	Mineralization	Construction materials	Reactor design and reaction engineering Integrated systems Energy efficiency, electrification, and alternative heating Separations
5-I. University–industry–national laboratory collaborations.	Applied Demonstration	Mineralization	Construction materials	Research centers and facilities
6-B. New catalysts and reaction processes for CO_2 conversion to elemental carbon materials.	Basic Applied	Chemical	Elemental carbon materials	Catalyst innovation and optimization Reactor design and reaction engineering Energy efficiency, electrification, and alternative heating
6-C. Selectivity for particular material morphologies.	Basic Applied	Chemical	Elemental carbon materials	Catalyst innovation and optimization
6-F. Reaction electrification and heat integration for CO_2 conversion to elemental carbon materials.	Basic Applied	Chemical	Elemental carbon materials	Reactor design and reaction engineering Energy efficiency, electrification, and alternative heating
6-G. Separations of catalyst and solid carbon product(s).	Basic Applied	Chemical	Elemental carbon materials	Separations
6-I. Integrated CO_2 capture and conversion to elemental carbon materials.	Applied	Chemical	Elemental carbon materials	Integrated systems
7-C. Carbon-neutral reductants for thermochemical CO_2 conversion.	Enabling	Chemical – thermochemical	Chemicals	Enabling technology and infrastructure needs
7-D. Engineering and systems optimization to integrate low-carbon energy with CO_2 conversion to hydrocarbons.	Applied	Chemical – thermochemical	Chemicals	Reactor design and reaction engineering Integrated systems Energy efficiency, electrification, and alternative heating
7-E. Abundant metal electrocatalysts for CO_2 conversion that are stable, selective, robust, and scalable.	Basic Applied	Chemical – electrochemical	Chemicals	Catalyst innovation and optimization Computational modeling and machine learning
7-F. Stable, abundant metal electrocatalysts for andic reactions of electrochemical CO_2 conversion.	Basic Applied	Chemical – electrochemical	Chemicals	Catalyst innovation and optimization Computational modeling and machine learning
7-G. Membrane materials that function over wide pH range.	Basic Applied	Chemical – electrochemical	Chemicals	Reactor design and reaction engineering Separations
7-K. Reactor design and engineering for photo(electro)chemical and plasmachemical CO_2 conversion.	Applied	Chemical – photochemical Chemical – photoelectrochemical Chemical – plasma	Chemicals	Reactor design and reaction engineering

continued

TABLE E-3 Continued

Research Agenda Item	Basic, Applied, Demonstration, or Enabling	Research Area	Product Class	Research Barrier Addressed
7-L. Tandem catalysis to access new products from CO_2.	Basic Applied	Chemical	Chemicals	Integrated systems Computational modeling and machine learning
7-M. Integrated CO_2 capture and conversion.	Basic Applied	Chemical	Chemicals	Integrated systems Reactor design and reaction engineering
7-N. Catalysts for rapid, stereoselective polymerization of CO_2 with broader class of monomers.	Basic Applied	Chemical – thermochemical	Polymers	Fundamental knowledge Catalyst innovation and optimization Computational modeling and machine learning
8-A. Pathway modeling and metabolic engineering of microorganisms.	Basic Applied	Biological	Chemicals Polymers	Metabolic understanding and engineering Reactor design and reaction engineering
8-C. Improvements to enzyme efficiency, stability, and selectivity and multi-enzyme metabolon design.	Basic Applied	Biological	Chemicals Polymers	Fundamental knowledge Computational modeling and machine learning Metabolic understanding and engineering
8-D. Improvements to enzyme stability and scalability of redox-balanced systems.	Basic Applied	Biological	Chemicals Polymers	Fundamental knowledge Reactor design and reaction engineering Integrated systems
8-E. Fundamental understanding of electrocatalyst design under biocompatible conditions.	Basic Applied	Biological – hybrid electro-bio	Chemicals Polymers	Fundamental knowledge Catalyst innovation and optimization
8-F. Microorganisms and cell-free systems compatible with electrochemically derived intermediates.	Basic Applied	Biological – hybrid electro-bio	Chemicals Polymers	Microbial engineering
8-G. Optimization of reactor design for hybrid systems.	Applied Demonstration	Biological – hybrid	Chemicals Polymers	Integrated systems Reactor design and reaction engineering
8-H. Feasibility studies for thermo- and photo-catalytic CO_2 reduction integrated with bioconversion	Applied	Biological – hybrid	Chemicals Polymers	Integrated systems Reactor design and reaction engineering
9-A. Mapping of coal waste resources.	Enabling	Coal waste utilization	Coal waste–derived carbon products Metal coal waste by-products	Resource mapping

9-B. Linking coal waste sites to product markets.	Applied Enabling	Coal waste utilization	Coal waste-derived carbon products Metal coal waste by-products	Enabling technology and infrastructure needs Market opportunities
9-C. Separating mineral matter from carbon in coal wastes.	Basic Applied	Coal waste utilization	Coal waste-derived carbon products Metal coal waste by-products	Separations
9-D. Efficient transformation of waste coal into long-lived solid carbon products.	Applied Demonstration	Coal waste utilization —long-lived carbon products	Coal waste-derived carbon products	Reactor design and reaction engineering
9-E. Evaluation of coal wastes for pavement applications.	Applied	Coal waste utilization	Coal waste-derived carbon products	Certification and standards
9-F. Atomic- and multi-scale computer simulations to better understand the conversion of coal waste carbon to solid-carbon products.	Basic	Coal waste utilization	Coal waste-derived carbon products	Fundamental knowledge Computational modeling and machine learning
9-G. 3D printing media from waste coal.	Applied	Coal waste utilization —long-lived carbon products	Coal waste-derived carbon products	Certification and standards
9-H. Evaluation of functionality and performance of coal-waste-derived products.	Applied	Coal waste utilization —long-lived carbon products	Coal waste-derived carbon products	Certification and standards
9-J. Life cycle and techno-economic assessments of coal waste utilization.	Applied	Coal waste utilization	Coal waste-derived carbon products Metal coal waste by-products	Certification and standards Environmental and societal considerations for CO_2 and coal waste utilization technologies
9-M. Extraction of lithium, critical minerals, and rare earth elements from solid coal wastes.	Applied	Coal waste utilization —metal extractions	Metal coal waste by-products	Separations
9-N. Extraction of lithium, critical minerals, and rare earth elements from liquid coal wastes.	Applied	Coal waste utilization —metal extractions	Metal coal waste by-products	Separations
9-O. Separation of individual elements upon extraction from coal wastes.	Basic Applied	Coal waste utilization —metal extractions	Metal coal waste by-products	Separations
10-A. Computational tools for optimal multi-modal transport of CO_2 from stranded emitters.	Enabling	Infrastructure	All	Enabling technology and infrastructure needs Computational modeling and machine learning

TABLE E-4 Research Agenda Items Directed to DOE-BES

Research Agenda Item	Basic, Applied, Demonstration, or Enabling	Research Area	Product Class	Research Barrier Addressed
3-B. Improved CO_2 purification technologies.	Basic Applied	Chemical	All	Separations
5-C. Improvements in efficiency, selectivity, and scalability.	Basic Applied Demonstration	Mineralization	Construction materials	• Reactor design and reaction engineering • Energy efficiency, electrification, and alternative heating
5-F. Catalyst, materials, and engineering design for electrochemically driven CO_2 mineralization.	Basic Applied Demonstration	Mineralization—Electrochemical	Construction materials	• Catalyst innovation and optimization • Reactor design and reaction engineering • Integrated systems • Computational modeling and machine learning
5-G. Monitoring and evaluating impacts of electrochemically driven CO_2 mineralization.	Basic Applied	Mineralization—Electrochemical	Construction materials	• Fundamental knowledge • Reactor design and reaction engineering • Environmental and societal considerations for CO_2 and coal waste utilization technologies
5-H. CO_2 mineralization integrated with metal recovery.	Basic Applied	Mineralization	Construction materials	• Reactor design and reaction engineering • Integrated systems • Energy efficiency, electrification, and alternative heating • Separations
5-K. Mineral carbonates for 3D-printed concrete.	Basic Applied	Mineralization	Construction materials	• Fundamental knowledge
6-A. Foundational knowledge of CO_2 conversion to elemental carbon materials.	Basic	Chemical	Elemental carbon materials	• Fundamental knowledge
6-B. New catalysts and reaction processes for CO_2 conversion to elemental carbon materials.	Basic Applied	Chemical	Elemental carbon materials	• Catalyst innovation and optimization • Reactor design and reaction engineering • Energy efficiency, electrification, and alternative heating
6-C. Selectivity for particular material morphologies.	Basic Applied	Chemical	Elemental carbon materials	• Catalyst innovation and optimization
6-D. Improved stability, activity, and selectivity of catalysts to convert CO_2 to elemental carbon materials.	Basic	Chemical	Elemental carbon materials	• Catalyst innovation and optimization
6-E. Understanding of reaction processes for CO_2 conversion to elemental carbon materials.	Basic	Chemical	Elemental carbon materials	• Fundamental knowledge

TABLE E-4 Continued

Research Agenda Item	Basic, Applied, Demonstration, or Enabling	Research Area	Product Class	Research Barrier Addressed
6-F. Reaction electrification and heat integration for CO_2 conversion to elemental carbon materials.	Basic Applied	Chemical	Elemental carbon materials	• Reactor design and reaction engineering • Energy efficiency, electrification, and alternative heating
6-G. Separations of catalyst and solid carbon product(s).	Basic Applied	Chemical	Elemental carbon materials	• Separations
6-H. Development of tandem processes for CO_2 conversion to elemental carbon materials.	Basic Applied	Chemical	Elemental carbon materials	• Integrated systems • Reactor design and reaction engineering
7-A. Improved catalyst selectivity and stability at high temperature.	Basic	Chemical—thermochemical	Chemicals	• Catalyst innovation and optimization • Computational modeling and machine learning
7-B. Catalysts and processes using alternative heating methods.	Basic	Chemical—thermochemical	Chemicals	• Catalyst innovation and optimization • Energy efficiency, electrification, and alternative heating • Computational modeling and machine learning
7-C. Carbon-neutral reductants for thermochemical CO_2 conversion.	Enabling	Chemical—thermochemical	Chemicals	• Enabling technology and infrastructure needs
7-E. Abundant-element electrocatalysts for CO_2 conversion that are stable, selective, robust, and scalable.	Basic Applied	Chemical—electrochemical	Chemicals	• Catalyst innovation and optimization • Computational modeling and machine learning
7-F. Stable abundant-element electrocatalysts for anodic reactions of electrochemical CO_2 conversion.	Basic Applied	Chemical—electrochemical	Chemicals	• Catalyst innovation and optimization • Computational modeling and machine learning
7-G. Cost-effective, scalable membrane materials that function over wide pH range.	Basic Applied	Chemical—electrochemical	Chemicals	• Reactor design and reaction engineering • Separations
7-H. Understanding of processes involved in photo(electro) chemical CO_2 conversion.	Basic	Chemical—photochemical Chemical—photoelectrochemical	Chemicals	• Fundamental knowledge • Computational modeling and machine learning
7-I. Materials discovery for photo(electro)chemical and plasmachemical CO_2 conversion.	Basic	Chemical—photochemical Chemical—photoelectrochemical Chemical—plasmachemical	Chemicals	• Catalyst innovation and optimization • Computational modeling and machine learning
7-J. Understanding of plasma-catalyst interactions.	Basic	Chemical—plasmachemical	Chemicals	• Fundamental knowledge • Computational modeling and machine learning

continued

TABLE E-4 Continued

Research Agenda Item	Basic, Applied, Demonstration, or Enabling	Research Area	Product Class	Research Barrier Addressed
7-M. Integrated CO_2 capture and conversion.	Basic Applied	Chemical	Chemicals	• Integrated systems • Reactor design and reaction engineering
7-N. Catalysts for rapid, stereoselective polymerization of CO_2 with broader class of monomers.	Basic Applied	Chemical—thermochemical	Polymers	• Fundamental knowledge • Catalyst innovation and optimization • Computational modeling and machine learning
8-A. Pathway modeling and metabolic engineering of microorganisms.	Basic Applied	Biological	Chemicals Polymers	• Metabolic understanding and engineering • Reactor design and reaction engineering
8-B. Tools to improve CO_2 fixation efficiency and understanding of carbon metabolism.	Basic	Biological	Chemicals	• Metabolic understanding and engineering • Genetic manipulation • Computational modeling and machine learning
8-C. Improvements to enzyme efficiency, stability, and selectivity and multi-enzyme metabolon design.	Basic Applied	Biological	Chemicals Polymers	• Fundamental knowledge • Computational modeling and machine learning • Metabolic understanding and engineering
8-D. Improvements to enzyme stability and scalability of redox-balanced systems.	Basic Applied	Biological	Chemicals Polymers	• Fundamental knowledge • Reactor design and reaction engineering • Integrated systems
8-E. Fundamental understanding of electrocatalyst design under biocompatible conditions.	Basic Applied	Biological—hybrid electro-bio	Chemicals Polymers	• Fundamental knowledge • Catalyst innovation and optimization
8-F. Microorganisms and cell-free systems compatible with electrochemically derived intermediates.	Basic Applied	Biological—hybrid electro-bio	Chemicals Polymers	• Microbial engineering
9-C. Separating mineral matter from carbon in coal wastes.	Basic Applied	Coal waste utilization	Coal waste–derived carbon products Metal coal waste by-products	• Separations
9-F. Atomic- and multi-scale computer simulations of coal waste conversions.	Basic	Coal waste utilization—long-lived carbon products	Coal waste–derived carbon products	• Fundamental knowledge • Computational modeling and machine learning
9-K. Characterization of coal waste structure and morphology.	Basic	Coal waste utilization—metal extraction	Metal coal waste by-products	• Fundamental knowledge
9-L. Sustainable leaching agents, membranes, and processes for rare earth element extraction.	Basic	Coal waste utilization—metal extraction	Metal coal waste by-products	• Separations
9-O. Separation of individual elements upon extraction from coal wastes.	Basic Applied	Coal waste utilization—metal extraction	Metal coal waste by-products	• Separations

TABLE E-5 Research Agenda Items Directed to DOE-BER

Research Agenda Item	Basic, Applied, Demonstration, or Enabling	Research Area	Product Class	Research Barrier Addressed
8-A. Pathway modeling and metabolic engineering of microorganisms.	Basic Applied	Biological	Chemicals Polymers	• Metabolic understanding and engineering • Reactor design and reaction engineering
8-B. Tools to improve CO_2 fixation efficiency and understanding of carbon metabolism.	Basic	Biological	Chemicals	• Metabolic understanding and engineering • Genetic manipulation • Computational modeling and machine learning
8-C. Improvements to enzyme efficiency, stability, and selectivity and multi-enzyme metabolon design.	Basic Applied	Biological	Chemicals Polymers	• Fundamental knowledge • Computational modeling and machine learning • Metabolic understanding and engineering
8-D. Improvements to enzyme stability and scalability of redox-balanced systems.	Basic Applied	Biological	Chemicals Polymers	• Fundamental knowledge • Reactor design and reaction engineering • Integrated systems
8-E. Fundamental understanding of electrocatalyst design under biocompatible conditions.	Basic Applied	Biological – hybrid electro-bio	Chemicals Polymers	• Fundamental knowledge • Catalyst innovation and optimization
8-F. Microorganisms and cell-free systems compatible with electrochemically derived intermediates.	Basic Applied	Biological – hybrid electro-bio	Chemicals Polymers	• Microbial engineering

TABLE E-6 Research Agenda Items Directed to DOE-BETO

Research Agenda Item	Basic, Applied, Demonstration, or Enabling	Research Area	Product Class	Research Barrier Addressed
8-A. Pathway modeling and metabolic engineering of microorganisms.	Basic Applied	Biological	Chemicals Polymers	• Metabolic understanding and engineering • Reactor design and reaction engineering
8-C. Improvements to enzyme efficiency, stability, and selectivity and multi-enzyme metabolon design.	Basic Applied	Biological	Chemicals Polymers	• Fundamental knowledge • Computational modeling and machine learning • Metabolic understanding and engineering
8-D. Improvements to enzyme stability and scalability of redox-balanced systems.	Basic Applied	Biological	Chemicals Polymers	• Fundamental knowledge • Reactor design and reaction engineering • Integrated systems
8-E. Fundamental understanding of electrocatalyst design under biocompatible conditions.	Basic Applied	Biological – hybrid electro-bio	Chemicals Polymers	• Fundamental knowledge • Catalyst innovation and optimization
8-F. Microorganisms and cell-free systems compatible with electrochemically derived intermediates.	Basic Applied	Biological – hybrid electro-bio	Chemicals Polymers	• Microbial engineering
8-G. Optimization of reactor design for hybrid systems.	Applied Demonstration	Biological – hybrid	Chemicals Polymers	• Integrated systems • Reactor design and reaction engineering
8-H. Feasibility studies for thermo- and photo-catalytic CO_2 reduction integrated with bioconversion.	Applied	Biological – hybrid	Chemicals Polymers	• Integrated systems • Reactor design and reaction engineering

REFERENCES

AFOSR (Air Force Office of Scientific Research). n.d. "AFOSR—Research Areas." https://www.afrl.af.mil/About-Us/Fact-Sheets/Fact-Sheet-Display/Article/2282138/afosr-research-areas.

Anderson, T. 2023. "Biological Systems Science." Presented at Meeting #2 of the Committee on Carbon Utilization Infrastructure, Markets, Research and Development, Washington, DC, February 27. https://www.nationalacademies.org/event/02-27-2023/carbon-utilization-infrastructure-markets-research-and-development-report-2-meeting-2.

ARL (Army Research Laboratory). 2024. "Collaborate with Us." DEVCOM. https://arl.devcom.army.mil/collaborate-with-us.

Claros, A.K. 2023. "Office of Carbon Management Technologies Overview." Presentation to the committee. February 27. Washington, DC: National Academies of Sciences, Engineering, and Medicine.

DARPA (Defense Advanced Research Projects Agency). n.d. "Our Research." https://www.darpa.mil/our-research.

DOE (Department of Energy). n.d. "DOE SBIR/STTR Programs Office." https://www.energy.gov/science/sbir/small-business-innovation-research-and-small-business-technology-transfer.

EOP (Executive Office of the President). 2023. "Circular No. A-11 Preparation, Submission, and Execution of the Budget." Washington, DC: Office of Management and Budget. https://www.whitehouse.gov/wp-content/uploads/2018/06/a11.pdf.

EPA (U.S. Environmental Protection Agency). 2024. "Research Areas." https://www.epa.gov/research/research-areas.

FAA (Federal Aviation Administration). 2024. "Sustainable Aviation Fuels (SAF)." Department of Transportation. https://www.faa.gov/about/officeorg/headquartersoffices/apl/sustainable-aviation-fuels-saf.

FHWA (Federal Highway Administration). 2021. "Pavement and Materials." Department of Transportation. https://highways.dot.gov/research/infrastructure/pavements-materials/pavement-materials.

Krynock, M. 2023. "Life Cycle Analysis Resources for Carbon Conversion and Carbon Production at NETL." Presentation to the committee. June 28. Washington, DC: National Academies of Sciences, Engineering, and Medicine.

McLean, G., and R. Miranda. "Basic Energy Sciences: Briefing for NASEM Carbon Utilization Committee." Presentation to the committee. February 27. Washington, DC: National Academies of Sciences, Engineering, and Medicine.

NOAA (National Oceanic and Atmospheric Administration). 2024. "Funding Announcements." OAP: NOAA Ocean Acidification Program. https://oceanacidification.noaa.gov/funding-opportunities.

NSF (National Science Foundation). n.d. "Our Focus Areas." https://new.nsf.gov/focus-areas.

ONR (Office of Naval Research). n.d. "Chemical Physics." https://www.nre.navy.mil/organization/departments/code-33/naval-engineering-focus-area/chemical-physics.

PHMSA (Pipeline and Hazardous Materials Safety Administration). 2017. "PHMSA Research and Development." Department of Transportation. https://www.phmsa.dot.gov/research-and-development/phmsa-research-and-development.

Sofos, M., K. Liu, and J. Seaba. 2023. "ARPA-E Efforts in Carbon Dioxide Utilization: Briefing to National Academies Committee on Carbon Utilization Infrastructure, Markets, Research and Development." Presentation to the committee. February 27. Washington, DC: National Academies of Sciences, Engineering, and Medicine.

Sterner, C. 2023. "Carbon Utilization Infrastructure, Markets, Research & Development Bioenergy Technologies Office Perspective on the NASEM Study." Presentation to the committee. February 27. Washington, DC: National Academies of Sciences, Engineering, and Medicine.

Stoffa, J. 2023. "Carbon Conversion Program Overview." Presentation to the committee. June 28. Washington, DC: National Academies of Sciences, Engineering, and Medicine.

USGS (U.S. Geological Survey). n.d. "Earth Mapping Resources Initiative (Earth MRI)." https://www.usgs.gov/earth-mapping-resources-initiative-earth-mri.

F

CO_2 Capture and Purification Technology Research, Development, and Demonstration Needs

This appendix describes research, development, and demonstration (RD&D) needs for the crosscutting technologies of CO_2 capture and CO_2 purification. Additional crosscutting research needs for markets, life cycle assessment and techno-economic assessment, policy, and infrastructure are covered in Chapters 2, 3, 4, and 10, respectively.

CO_2 CAPTURE

Common to all CO_2 utilization processes is the need first to capture CO_2 from a point source, the air, or the ocean.[1] As discussed in the committee's first report, there are a variety of carbon capture technologies at different technology readiness levels (TRLs) (see NASEM 2023, Table 4.1), and the choice of capture technology will depend on "the initial and final desired CO_2 concentration (i.e., the percentage of CO_2 to be removed), scale of CO_2 capture, operating pressure and temperature, composition and flow rate of the gas stream, integration with the original facility, and cost considerations" (NASEM 2023, p. 75). The cost and energy requirements for CO_2 pressurization are important considerations for the viability of a CO_2 utilization project, and capture systems that release CO_2 at the pressures required for transportation or downstream processing (i.e., conversion) are preferred to reduce or eliminate compression costs. Table F-1 shows the approximate concentration of CO_2 from different sources, the total annual U.S. emissions from each source type (where applicable), and estimated capture costs.

Government-supported demonstration of CO_2 capture projects, integrated with transport and storage technologies, can further the deployment and replication of CO_2 capture technologies. These demonstrations enable modularization of equipment at scale that may reduce construction costs, increase capture efficiency as improved technologies become available to substitute for current ones, and provide operational data that can build confidence with project investors and reduce financing risks. Beyond cost reductions achieved through learnings from technology scale up, additional RD&D on capture technologies will also be important for decreasing CO_2 capture costs. Table 4.2 from the committee's first report (NASEM 2023), which is reproduced and expanded upon in Table F-2 below, outlines RD&D targets to reduce costs from different classes of capture technologies. RD&D needs for enabling technologies, which could benefit multiple capture systems, include mitigating aerosol emis-

[1] In some systems, capture and conversion of CO_2 are integrated, as described in Chapter 7, Section 7.2.5.

TABLE F-1 CO_2 Concentration, Annual U.S. Emissions, and Estimated Capture Costs from Different Sources

Source	CO_2 Concentration (percent by volume)	Annual U.S. Emissions	Estimated Capture Cost (\$/t CO_2)
Power Generation	Natural gas-fired: 3–6[a,b] Coal-fired: 11–15[a,b]	1500 MMT[c]	40–290[c,d] 73–167[d,e] (coal) 82–166[d,e] (natural gas) 100–123[d,f,g] (coal) 82–98[d,f,h] (coal) 104–133[d,f,g] (natural gas) 84–105[d,f,h] (natural gas) 83–268[i,j] (coal) 93–290[i,j] (natural gas) 53–86[i,k] (coal)
Cement	Process emissions: ~14–33[c]	66 MMT[c]	45–120[c,d] 87–131[d,e] 89–109[d,f,g] 76–90[d,f,h] 64–95[i,j] 61–94[i,k] 61–64[i,l]
Iron and Steel	~17–27[c,l,m]	62 MMT[c]	40–130[c,d] 54–69[d,e] 108–121[d,f,g] 90–109[d,f,h] 75–113[i,j] 75–119[i,k] 65–67[i,l]
Hydrogen Production	14–45[b]	100 MMT[n]	103–129[d,f,g] (SMR 90 percent capture) 83–102[d,f,h] (SMR 90 percent capture) 61–88[i,j] 68–114[i,k] (SMR and steam production, 90 percent capture)
Ethanol Production	≥95[b,c]	45 MMT[c]	0–35[c,d] 42–59[d,e] 36–41[d,f,g] 33–37[d,f,h] 24–34[i,j] 18–26[i,k] 32[i,l]
Natural Gas Processing	CO_2 vent: 99[b]	26.1 MMT[o]	32–35[d,f,g] 29–32[d,f,h] 23–35[i,j] 14–20[i,k] 16[i,l]
Direct Ocean Capture	~6, varies with pH and temperature	N/A	150–2500[d,p]

TABLE F-1 Continued

Source	CO_2 Concentration (percent by volume)	Annual U.S. Emissions	Estimated Capture Cost ($/t CO_2)
Direct Air Capture	0.04[b]	N/A	600–1000[i,q] 225–600[i,k] 89–877[i,r,s] (sorbent-based) 156–506[i,r,s] (solvent-based)

[a] NETL (n.d.(a)).
[b] Claros (2023).
[c] GAO (2022).
[d] Measured in $/ton of CO_2.
[e] Bennett et al. (2023).
[f] Moniz et al. (2023).
[g] Estimated for first-of-a-kind facility.
[h] Estimated for *n*th-of-a-kind facility.
[i] Measured in $/tonne of CO_2.
[j] Table 2-7 of NPC (2019).
[k] DOE (2023a).
[l] Hughes et al. (2022).
[m] Emission streams in this range are blast furnace stove, power plant stack, and coke oven gas.
[n] DOE (2023b).
[o] Table 3-73 of EPA (2023).
[p] Capture costs for electrochemical processes; NASEM (2022).
[q] Budinis and Lo Re (2023).
[r] NASEM (2019).
[s] Cost of net CO_2 removed, accounting for any CO_2 emissions from powering the direct air capture system.

TABLE F-2 Research, Development, and Demonstration Targets to Improve Carbon Capture Systems

CO_2 Capture Technology	Research Trends for Reducing Carbon Capture Costs
Solvents	• Fast sorption and desorption kinetics • High CO_2 capacity • Lower regeneration energy requirements • Lower degradation rates • Water-lean solvent • Process intensification • Mitigation of aerosol formation and corrosion • Heat integration
Sorbents	• Low-cost materials with high CO_2 adsorption rate and capacity • Fast-spent sorbent regeneration rates • Improved durability over multiple regeneration cycles with little to no attrition • Low heats of adsorption • Adequately hydrophobic • Process intensification, novel reactor designs, enhanced process configurations
Membranes	• High CO_2 permeability and selectivity • Low-cost materials • Improved durability determined by mechanical strength, chemical resistance, and thermal stability • Integration into low-pressure drop modules • Hydrophilic (for post-combustion capture) • Tolerance to gas contaminants • Ability to be processed into thin (i.e., high flux), defect-free structures at large (>10,000 m^2) scale
Novel concepts	• Electrochemical capture • Crystallization • Microwave enhancement

SOURCES: NASEM (2023, Table 4-2); NETL (2020, n.d.(b)).

sions; improving stability, compatibility, and corrosion resistance; and reducing viscosity and degradation products (Bostick et al. 2021; NETL n.d.(b)). Basic science and applied research needs to advance solvent- and sorbent-based direct air capture (DAC) technologies, including synthesizing and testing new materials; designing and testing new equipment and system concepts; performing independent materials testing, characterization, and validation; and establishing and managing a public materials database (NASEM 2019b). For electrochemical direct ocean capture (DOC) technologies, research needs include new designs for electrochemical reactors; novel electrode and membrane materials; systems integration with rock dissolution; and demonstration projects to verify carbon removal, monitor environmental impacts, and investigate scale-up strategies (NASEM 2022).

Direct conversion of impure CO_2 streams to products can reduce the net energetic and capital costs of CO_2 utilization, as an intermediary purification step is not required. This process, termed "integrated CO_2 capture and conversion" in this report, is described in detail in Chapter 7. Primary research needs for integrated capture and conversion include discovery of relevant molecules and materials, understanding of catalytic mechanisms, process optimization, and reactor design.

CO_2 PURIFICATION

Most CO_2 transport, utilization, and storage applications require the removal of at least some impurities present in the CO_2 gas streams. The committee's first report (NASEM 2023) contained a robust discussion of typical impurities present in CO_2 gas streams (Tables 4.3 and 4.4), impurity thresholds for different CO_2 transport modes (Table 4.5), and impurities of concern for CO_2 utilization routes (Table 4.6). Appendix H of this report reproduces Tables 4.3 and 4.4 (as Tables H-1 and H-2, respectively) and updates Table 4.5 (as Table H-3) based on more recent information. Table 4.6 is reproduced as Table 3-1 in Chapter 3. From this assessment, the committee concluded that the impurities present in CO_2 streams may influence the viability of utilization processes, with mineralization and biological conversion being the most resilient to impurities[2] and electro- and thermochemical conversions being the most sensitive (Finding 4.3, NASEM 2023). The lack of standard specifications for CO_2 purity across capture, transport, utilization, and storage could result in increased energy and operational costs for some stakeholders, as well as efficiency losses throughout the value chain (Neerup et al. 2022). Common separation technologies used to reach very pure streams (akin to food-grade CO_2) are capital- and energy-intensive. Examples are cryogenic distillation and pressure-swing adsorption. These types of purification technologies, therefore, can become a key bottleneck to achieve cost-effective CO_2 utilization, let alone to run such CO_2 utilization processes flexibly. Ongoing research is investigating CO_2 conversion catalysts and technologies that can tolerate impurities (see, e.g., Harmon and Wang 2022; Ho et al. 2019), which might improve the economic feasibility of some CO_2 utilization systems. As discussed in Chapter 7, more research is needed to develop CO_2 conversion catalysts that are impervious to impurities. Additionally, developing standards and methodologies for measuring ppm or ppb levels of impurities in CO_2 streams will help to inform RD&D on CO_2 capture and utilization technologies.

The specific separation technology to be used must be selected with the contaminants and desired CO_2 purity in mind. For example, membranes today would be exceptionally effective at dehydrating CO_2 streams and could remove contaminants such as hydrogen sulfide (H_2S), but they still require significant development to increase their cost-effectiveness and performance in harsh feed conditions. Furthermore, there is a need to develop performance-tuned membrane materials that can handle gas flow rates and compositions that change with time. This will especially be important for CO_2 utilization technologies that target CO_2 from (bio)waste feedstocks. Membranes tuned to separate H_2S would not remove contaminants such as N_2, O_2, or argon. Monoethanolamine-based approaches necessarily leave the purified CO_2 stream saturated with water. Sorbents could be rapidly saturated with even low levels of condensable components. Developing more efficient CO_2 separation methods and hybrid separation technologies, rather than relying on a single method or approach, will likely yield the most promising separation results.

[2] Notably, however, anaerobic biological conversion is sensitive to oxygen, and purification to remove oxygen can be energy-intensive.

REFERENCES

Bennett, J., R. Kammer, J. Eidbo, M. Ford, S. Henao, N. Holwerda, E. Middleton, et al. 2023. "Carbon Capture Co-Benefits." Great Plains Institute. https://carboncaptureready.betterenergy.org/wp-content/uploads/2023/08/Carbon-Capture-Co-Benefits.pdf.

Bostick, D., D. Dhanraj, R. Renhui, B. Kumfer, P. Biswas, W. Sherlock, C. Lehmann, R. Jain, and N. Lemcoff. 2021. "Final Report on Aerosol Pretreatment Technology Performance and Benchmarking." DE-FE0031592. Stewartsville, NJ: Linde Gas North America. https://www.osti.gov/servlets/purl/1814890.

Budinis, S., and L.L. Re. 2023. "Unlocking the Potential of Direct Air Capture: Is Scaling Up Through Carbon Markets Possible?" IEA. https://www.iea.org/commentaries/unlocking-the-potential-of-direct-air-capture-is-scaling-up-through-carbon-markets-possible.

Claros, A.K. 2023. "Office of Carbon Management Technologies Overview." Presentation to the committee. February 27. Washington, DC: National Academies of Sciences, Engineering, and Medicine.

DOE (Department of Energy). 2023a. "Pathways to Commercial Liftoff: Carbon Management." Washington, DC: Department of Energy. https://liftoff.energy.gov/wp-content/uploads/2023/04/20230424-Liftoff-Carbon-Management-vPUB_update.pdf.

DOE. 2023b. "U.S. National Clean Hydrogen Strategy and Roadmap." Washington, DC: Department of Energy. https://www.hydrogen.energy.gov/docs/hydrogenprogramlibraries/pdfs/us-national-clean-hydrogen-strategy-roadmap.pdf.

EPA (U.S. Environmental Protection Agency). 2023. "Inventory of U.S. Greenhouse Gas Emissions and Sinks, 1990–2021." EPA 430-R-23-002. Washington, DC: U.S. Environmental Protection Agency. https://www.epa.gov/system/files/documents/2023-04/US-GHG-Inventory-2023-Main-Text.pdf.

GAO (U.S. Government Accountability Office). 2022. "Decarbonization: Status, Challenges, and Policy Options for Carbon Capture, Utilization, and Storage." GAO-22-105274. Technology Assessment. Washington, DC: U.S. Government Accountability Office. https://www.gao.gov/assets/730/723198.pdf.

Harmon, N.J., and H. Wang. 2022. "Electrochemical CO2 Reduction in the Presence of Impurities: Influences and Mitigation Strategies." *Angewandte Chemie International Edition* 61(52):e202213782. https://doi.org/10.1002/anie.202213782.

Ho, H.-J., A. Iizuka, and E. Shibata. 2019. "Carbon Capture and Utilization Technology Without Carbon Dioxide Purification and Pressurization: A Review on Its Necessity and Available Technologies." *Industrial and Engineering Chemistry Research* 58(21):8941–8954. https://doi.org/10.1021/acs.iecr.9b01213.

Hughes, S., A. Zoelle, M. Woods, S. Henry, S. Homsy, S. Pidaparti, N. Kuehn, et al. 2022. "Cost of Capturing CO2 from Industrial Sources." DOE/NETL-2022/3319. Pittsburgh, PA: National Energy Technology Laboratory. https://www.netl.doe.gov/projects/files/CostofCapturingCO2fromIndustrialSources_071522.pdf.

Moniz, E.J., J.D. Brown, S.D. Comello, M. Jeong, M. Downey, and M.I. Cohen. 2023. "Turning CCS Projects in Heavy Industry & Power into Blue Chip Financial Investments." Washington, DC: Energy Futures Initiative. https://energyfuturesinitiative.org/wp-content/uploads/sites/2/2023/02/20230212-CCS-Final_Full-copy.pdf.

NASEM (National Academies of Sciences, Engineering, and Medicine). 2019. *Negative Emissions Technologies and Reliable Sequestration: A Research Agenda*. Washington, DC: The National Academies Press. https://doi.org/10.17226/25259.

NASEM. 2022. *A Research Strategy for Ocean-Based Carbon Dioxide Removal and Sequestration*. Washington, DC: The National Academies Press. https://doi.org/10.17226/26278.

NASEM. 2023. *Carbon Dioxide Utilization Markets and Infrastructure: Status and Opportunities: A First Report*. Washington, DC: The National Academies Press. https://doi.org/10.17226/26703.

Neerup, R., I.A. Løge, K. Helgason, S.Ó. Snæbjörnsdóttir, B. Sigfússon, J.B. Svendsen, N.T. Rosted, et al. 2022. "A Call for Standards in the CO2 Value Chain." *Environmental Science and Technology* 56(24):17502–17505. https://doi.org/10.1021/acs.est.2c08119.

NETL (National Energy Technology Laboratory). 2020. *Compendium of Carbon Capture Technology*. 2020 Carbon Capture Program R&D. Pittsburgh, PA: National Energy Technology Laboratory. https://www.netl.doe.gov/sites/default/files/2020-07/Carbon-Capture-Technology-Compendium-2020.pdf.

NETL. n.d.(a). "9.2 Carbon Dioxide Capture Approaches." https://netl.doe.gov/research/carbon-management/energy-systems/gasification/gasifipedia/capture-approaches.

NETL. n.d.(b). "Point Source Carbon Capture." https://netl.doe.gov/carbon-management/carbon-capture.

NPC (National Petroleum Council). 2019. "Meeting the Dual Challenge: A Roadmap to At-Scale Deployment of Carbon Capture, Use, and Storage." https://dualchallenge.npc.org/downloads.php.

G

Key Features of Effective Siting and Permitting Processes

Siting policy and social acceptance of energy projects has been studied since the emergence of the anti-nuclear movement in the 1970s. Research has shown that "the character and quality of the process of engaging the public in the context of siting and permitting projects will affect the pace and scale" of a project's deployment (NASEM 2023, p. 212). While creative and robust public engagement is unlikely to change the stance of ardent opponents, shortcutting public engagement can lead to delays in a project's cycle due to driving publics to courts or other forms of protest. A review of the literature has identified the following key features of effective siting and permitting processes:

1. **Public engagement requires inclusive, expansive, and immersive communication.** This includes communication that is conducted in multiple languages and in diverse and accessible formats; begins early in the process and features continuous updates of project progress; and utilizes both low- and high-tech strategies.
2. **Public engagement professionals should treat local perspectives as constructive expertise in project design.** This means giving local communities the opportunity to participate in shaping the process and outcomes of important design decisions.
3. **Public engagement professionals should support communities in the development of local and regional visions prior to discussions of facility siting whenever possible.** By prioritizing community visions for development, the siting discussion can build on and incorporate local goals rather than the other way around.
4. **Public engagement needs to be place-based, customized to unique regions, demographics, politics, economics, and social values.** Flexibility in public engagement processes must be a priority for permitting practitioners to align with local circumstances.
5. **Public engagement needs to be respectful of a community's time and priorities.** Clustering review processes for projects and zone permitting have merit for equitable and rapid infrastructure development, acknowledge the risk of consultation fatigue, and facilitate effective environmental impact assessments.
6. **Public engagement must emphasize clarity, transparency, and accountability in all activities.** Every effort must be made to provide social learning opportunities focused on how projects will affect communities in terms of public health, local environments, and economics.

REFERENCE

NASEM (National Academies of Sciences, Engineering, and Medicine). 2023. "Public Engagement to Build a Strong Social Contract for Deep Decarbonization." Chapter 5 in *Accelerating Decarbonization in the United States: Technology, Policy, and Societal Dimensions*. Washington, DC: The National Academies Press. https://doi.org/10.17226/25931.

H

CO_2 Stream Impurities and CO_2 Purity Requirements for Transport and Utilization

This appendix collects information about impurities commonly present in gas streams and those of greatest concern for CO_2 conversion processes. Table H-1 lists various impurities and their concentrations in flue gas streams from different types of CO_2 capture facilities. It is provided for reference on the impurities that may be present in CO_2 destined to be transformed into products. Table H-2 describes trace impurities by CO_2 source. Table H-3 lists recommended maximum impurity limits for CO_2 transported in pipelines and by ship.

TABLE H-1 Overview of Impurity Concentrations of CO_2 Streams from Different Illustrative Facility Types

Component	Subcritical Pulverized Bituminous Coal (Illinois #6) Plant with Post-Combustion Capture[a]	Natural Gas with Carbon Capture[c]	Oxyfuel Combustion at Supercritical Pulverized Coal Plant[a,d]	Cement Plant[a]	Refinery Stack[a]	Bioethanol Plant[e]	Direct Air Capture[f]
	Gas leaving the carbon capture unit (post-combustion with MEA[b])	Gas leaving the carbon capture unit (post-combustion with MEA[b])	Gas leaving the boiler unit	Gas leaving the carbon capture unit (post-combustion with MEA[b])	Gas leaving the carbon capture unit (post-combustion with MEA[b])	Raw CO_2 gas from ethanol plant	Gas leaving the capture unit (KOH sorbent)
CO_2	99.7%	95%	96.65%	99.8%	99.6%	90%	97.11%
CO			750 ppmv	1.2 ppmv			
H_2O	640 ppmv		100 ppmv	640 ppmv	640 ppmv	1–5 ppmv	0.01%
CH_4		4%		0.026 ppmv		0–3 ppmv	
SO_2	<1 ppmv		50 ppmv	<0.1 ppmv	1.3 ppmv		
SO_3			20 ppmv				
NO_2	1.5 ppmv			0.86 ppmv	2.5 ppmv		
NO_x			100 ppmv				
O_2	61 ppmv		0.81 %	35ppmv	121 ppmv	10–100 ppmv	1.36%
H_2S		200 ppmv			7.9 ppmv		
N_2	0.18%	0.5%	1.96%	893 ppmv	0.29%	50–600 ppmv	1.51%
Ar	22 ppmv		0.57%	11 ppmv	38 ppmv		
Hg	0.0007 ppmv		0.011 ppmv	0.00073 ppmv			
As	0.0055 ppmv		0.026 ppmv	0.0029 ppmv			
Se	0.017 ppmv		0.08 ppmv	0.0088 ppmv			
Cl	0.85 ppmv			0.41 ppmv	0.4 ppmv		
Ethanol						25–950 ppmv	
Methanol						1–50 ppmv	
Acetaldehyde						3–75 ppmv	
Isoamyl acetate						0.6–3.0 ppmv	
Isobutanol						0–3 ppmv	
Ethylacetate						2–30 ppmv	

[a] Values from EC (2011).
[b] MEA = monoethanolamine.
[c] Values from SINTEF (2019).
[d] Values from Rütters et al. (2015).
[e] Values from McKaskle et al. (2018).
[f] Values from Keith et al. (2018).
NOTE: ppmv = parts per million by volume.
SOURCE: Reproduced from NASEM (2023).

TABLE H-2 Overview of Trace Impurities by CO_2 Source

	CO_2 Source									
Impurity	Combustion	Wells/Geothermal	Fermentation/Bioethanol Anaerobic Digestion (Purely Energy Crops)	Anaerobic Digestion (waste)	Hydrogen or Ammonia	Phosphate Rock	Coal Gasification	Ethylene Oxide	Acid Neutralization	Vinyl Acetate
Aldehydes	✓	✓	✓	✓	✓		✓	✓		✓
Amines	✓				✓					
Benzene	✓	✓	✓	✓	✓		✓	✓	✓	✓
Carbon monoxide	✓	✓	✓	✓	✓	✓	✓	✓	✓	✓
Carbonyl sulfide	✓	✓	✓	✓	✓		✓	✓		✓
Cyclic aliphatic hydrocarbons	✓	✓		✓	✓		✓	✓		✓
Dimethyl sulfide		✓	✓	✓		✓	✓		✓	
Ethanol	✓	✓	✓	✓	✓		✓	✓		✓
Ethers		✓	✓	✓	✓		✓	✓		✓
Ethyl acetate		✓	✓	✓			✓	✓		✓
Ethyl benzene		✓		✓	✓		✓	✓		✓
Ethylene oxide							✓	✓		
Halocarbons	✓			✓			✓	✓		✓
Hydrogen cyanide	✓						✓			
Hydrogen sulfide	✓	✓	✓	✓	✓	✓	✓	✓	✓	✓
Ketones	✓	✓	✓	✓	✓		✓	✓		✓
Mercaptans	✓	✓	✓	✓	✓	✓	✓	✓		✓
Mercury	✓	✓					✓			
Methanol	✓	✓	✓	✓	✓		✓	✓		✓
Nitrogen oxides	✓		✓	✓	✓		✓	✓	✓	
Phosphine						✓				
Radon		✓				✓			✓	
Sulfur dioxide	✓	✓	✓	✓	✓	✓	✓		✓	
Toluene		✓	✓	✓	✓		✓	✓		✓
Vinyl chloride	✓						✓	✓		✓
Volatile hydrocarbons	✓	✓	✓	✓	✓		✓	✓		✓
Xylene		✓	✓	✓	✓		✓	✓		✓

SOURCES: Adapted from EIGA (2016) and NASEM (2023).

TABLE H-3 Overview of Recommended Maximum Impurity Limits for CO_2 Transport in Pipelines and Shipping

	Pipelines				Shipping
	NETL (United States)[a]		National Grid Carbon	Northern Light	EU CCUS
Component	Conceptual Design	Range in Literature	(United Kingdom)[b]	Project (Norway)[c]	Projects Network[d]
CO_2 (minimum vol %)	95	90–99.8	≥91 (gaseous phase) ≥96 (dense phase)	99.81	>99.7
H_2O (ppmv)	500	20–650	50	≤30	<30
N_2	4 vol%	0–7 vol%		≤50 ppmv	
O_2	0.001 vol%	0.001–4 vol%	0.001 vol%	≤10 ppmv	<10 ppmv
Ar	4 vol%	0.01–4 vol%		≤100 ppmv	
CH_4	4 vol%	0.01–4 vol%		≤100 ppmv	
H_2	4 vol%	0–4 vol%	2	≤50 ppmv	<500 ppmv
CO ppmv	35	10–5000	200	≤100	<12,000
H_2S	0.01 vol%	0.002–1.3 vol%	0.002 vol% (for dense-phase 150 barg) 0.008 vol% (for gas-phase 38 barg)	≤9 ppmv	<5 ppmv
SO_2 ppmv	100	10–50000			
SO_x ppmv			100	≤10	<10
NO_x ppmv	100	20–2500	100	≤1.5	<1.5
NH_3 ppmv	50	0–50		≤10	<10
COS ppmv	trace	trace			
C_2H_6 ppmv	1	0–1		≤75	
C_{3+} ppmv	<1	0–1		≤1100	
Particulates	1 ppmv	0–1 ppmv		≤1 μm	
Hg ppmv				≤0.0003	<0.03
Glycol ppmv	46	0–174		≤0.005[e]	
Cd, Tl, ppm				≤0.03 (sum)	<0.03

[a] Values from NETL (2019).
[b] Values from Gibbins and Lucquiaud (2021).
[c] Values from Northern Lights (2024).
[d] Values from Aramis (2023).
[e] Concentration limit is for mono-ethylene glycol; tri-ethylene glycol is not allowed.
NOTES: The data in this table have been corrected to reflect accurate values. Please disregard previous versions of this table published in earlier editions. EU CCUS = European Union carbon capture, utilization, and storage.
SOURCE: Adapted from NASEM (2023).

REFERENCES

Aramis. 2023. "CO_2 Specifications for Aramis Transport Infrastructure." https://www.aramis-ccs.com/news/co2-specifications-for-aramis-transport-infrastructure.

EC (European Commission and Directorate-General for Climate Action). 2011. *Implementation of Directive 2009/31/EC on the Geological Storage of Carbon Dioxide: Guidance Document 2, Characterisation of the Storage Complex, CO2 Stream Composition, Monitoring and Corrective Measures*. Publications Office. https://doi.org/10.2834/98293.

EIGA (European Industrial Gases Association). 2016. "Carbon Dioxide Food and Beverages Grade, Source Qualification, Quality Standards and Verification." EIGA Doc 70/17, revision of Doc 70/08. https://www.eiga.eu/ct_documents/doc070.pdf.

Gibbins, J., and M. Lucquiaud. 2021. "BAT Review for New-Build and Retrofit Post-Combustion Carbon Dioxide Capture Using Amine-Based Technologies for Power and CHP Plants Fuelled by Gas and Biomass as an Emerging Technology Under the IED for the UK, UKCCSRC Report." UKCCSRC Report, Ver.1.0. Sheffield, UK: UK CCS Research Centre. https://ukccsrc.ac.uk/wp-content/uploads/2021/06/BAT-for-PCC_V1_0.pdf.

Keith, D.W., G. Holmes, D. St. Angelo, and K. Heidel. 2018. "A Process for Capturing CO_2 from the Atmosphere." *Joule* 2(8):1573–1594. https://doi.org/10.1016/j.joule.2018.05.006.

McKaskle, R., K. Fisher, P. Selz, and Y. Lu. 2018. "Evaluation of Carbon Dioxide Capture Options from Ethanol Plants." Circular 595. Champaign, IL: Illinois State Geological Survey Prairie Research Institute. https://library.isgs.illinois.edu/Pubs/pdfs/circulars/c595.pdf.

NASEM (National Academies of Sciences, Engineering, and Medicine). 2023. *Carbon Dioxide Utilization Markets and Infrastructure: Status and Opportunities: A First Report*. Washington, DC: The National Academies Press. https://doi.org/10.17226/26703.

NETL (National Energy Technology Laboratory). 2019. "CO_2 Impurity Design Parameters." Quality Guidelines for Energy System Studies. Pittsburgh, PA: National Energy Technology Laboratory. https://www.netl.doe.gov/projects/files/QGESSCO2ImpurityDesignParameters_010119.pdf.

Northern Lights. 2024. "Liquid CO_2 Quality Specifications." https://norlights.com/wp-content/uploads/2024/02/Northern-Lights-GS-co2-Spec2024.pdf.

Rütters, H., D. Bettge, R. Eggers, A. Kather, C. Lempp, U. Lubenau, and COORAL-Team. 2015. *CO_2-Reinheit Für Die Abscheidung Und Lagerung (COORAL) – Synthese*. Hannover: Bundesanstalt für Geowissenschaften und Rohstoffe. https://www.bgr.bund.de/DE/Themen/Nutzung_tieferer_Untergrund_CO2Speicherung/CO2Speicherung/COORAL/Downloads/Synthesebericht.pdf?__blob=publicationFile&v=6.

SINTEF. 2019. "CO_2 Impurities: What Else Is There in CO_2 Except CO_2?" *#SINTEFblog*, November 6. https://blog.sintef.com/sintefenergy/energy-efficiency/what-else-is-there-in-co2-except-co2.

I

Additional Information on Markets for CO_2 Utilization

Tables I-1 and I-2 provide context about market volumes for global chemical production, and for alternative carbon feedstocks that compete with CO_2. To better understand the current chemical industry, Table I-1 describes the major fossil-derived chemical products, excluding fuels, by global volume in 2007, and their production methods. Although the data are from 2007, they describe a baseline of fossil chemical production, which in the future will need to evolve into an industry producing a related-but-not-identical suite of products, with sustainable carbon feedstocks, and likely at larger volume overall, with projected increases in demand for chemicals production.

Table I-2 contains information on availability, conversion technologies, applications and markets, and barriers to adoption for alternative carbon feedstocks that represent competitors to CO_2 feedstocks. Issues associated with feedstock availability and suitability are discussed in Section 2.3.3 of this report.

TABLE I-1 Highest-Volume Products of the Chemical Industry, Global Product Volumes, and Global Fossil Production Method Share as of 2007

Chemical	Product Volume, Global (ktonne/year)	Fossil Production Method Share, Global, 2007
Ammonia[a]	134,330	Steam reforming of natural gas for hydrogen production, 83% Partial oxidation of oil for hydrogen production, 9% Partial oxidation of coal for hydrogen production, 9%
Urea	118,436	Reaction of ammonia with CO_2, 100%
Ethylene	91,000	Steam cracking of naphtha, 51% Steam cracking of gas oil, 7% Steam cracking of propane, 21% Steam cracking of ethane, 21%
Chlorine[a]	44,084	Electrolysis of sodium chloride (diaphragm), 60% Electrolysis of sodium chloride (mercury cathode), 20% Electrolysis of sodium chloride (membrane), 20%
Polyethylene	40,856	Addition polymerization of ethylene, 100%
Benzene from pyrolysis-gasoline (aromatics)	30,200	Benzene separation from pyrolysis-gasoline, 39%
Benzene from toluene (aromatics)	30,200	Hydrodealkylation of toluene from pyrolysis-gasoline, 5%
Polyethylene terephthalate	29,000	Esterification of terephthalic acid with ethylene glycol, 100%
Methanol	27,900	Steam reforming of natural gas, 88% Partial oxidation of residues, 9% Partial oxidation of coal, 3%
Polypropylene	27,833	Addition polymerization of propylene, 100%
Vinylchloride	26,746	Integrated chlorination and oxychlorination of ethylene, 100%
Polyvinylchloride	25,398	Addition polymerization of vinylchloride, 100%
Methyl tert-butyl ether	20,867	Reaction of isobutene and methanol, 100%
Ethylbenzene	20,351	Alkylation of benzene, 100%
Styrene	20,067	Dehydrogenation of ethylbenzene, 85%
Terephthalic acid	17,000	Oxidation of *p*-xylene, 100%
p-xylene from reformate (aromatics)	16,000	*p*-xylene from C8 aromatics cut, 100%
Ethylene oxide	13,410	Oxidation of ethylene, 100%
Polystyrene	13,244	Addition polymerization of styrene, 100%
Ethylene glycol	12,200	Hydration of ethylene oxide, 100%
Cumene	9631	Alkylation of propylene with benzene, 100%
Butadiene	7868	From steam cracking hydrocarbons, 100%
Polyurethane	7720	Reaction of toluene diisocyanate with polyols, 50% Reaction of methylene diphenyl diisocyanate with polyols, 50%
Acetic acid	7310	Carbonylation of methanol, 80% Oxidation of acetaldehyde, 20%
Formaldehyde	6450	Oxydehydration of methanol, 100%
Phenol	5586	Oxidation of cumene, 96% Oxidation of toluene, 4%
Cyclohexane	5100	Hydrogenation of benzene, 100%
Propylene oxide	4877	Indirect oxidation via chlorohydrin, 51% Indirect oxidation via *tert*-butyl hydroperoxide, 30% Indirect oxidation via ethylbenzene hydroperoxide, 19%

TABLE I-1 Continued

Chemical	Product Volume, Global (ktonne/year)	Fossil Production Method Share, Global, 2007
Polyetherpolyols	4816	Polyaddition of epoxies to an initiator, 100%
Acrylonitrile	4704	Ammoxidation of propylene, 100%
Caprolactam	4160	From cyclohexane, 54% From phenol, 46%
Acetone	3900	Dehydrogenation of isopropanol, 10%
Phthalic anhydride	3200	Oxidation of *o*-xylene, 85% Oxidation of naphthalene, 15%
Dimethyl terephthalate	3096	Oxidation of *p*-xylene, esterification with methanol, 100%
Aniline	3010	Hydrogenation of nitrobenzene, 100%
Dioctylphthalate	2880	Esterification of phthalic anhydride with 2-ethylhexanol, 100%
Acetaldehyde	2566	Oxidation of ethylene, 100%
Nitrobenzene	2468	Nitration of benzene, 100%
2-ethylhexanol	2408	Hydroformylation of propylene, 100%
Bisphenol-A	2300	Condensation of phenol with acetone, 100%
Polyamide 66	2237	Polycondensation of adipic acid with hexamethylenediamine, 100%
Polyamide 6	2237	Polymerization of caprolactam, 100%
Methylene diphenyl diisocyanate	2159	Condensation of aniline with formaldehyde, phosgenation to methylene diphenyl diisocyanate, 100%
Urea formaldehyde resin	2129	Condensation of urea with formaldehyde, 100%
Adipic acid	2100	Oxidation of cyclohexane, 100%
Isopropanol	1806	Hydration of propene, 100%
Polycarbonate	1500	Polycondensation of bisphenol-A with phosgene, 100%
Hexamethylenediamine	1346	Ammonia with adipic acid, 52% Hydrogen cyanide with butadiene, 25% Hydrogenation of acrylonitrile, 23%
Toluene diisocyanate	1213	Nitration of toluene, phosgenation to TDI, 100%
n-butanol	1019	Hydroformylation of propylene, hydrogenation of buteraldehyde, 100%

[a] Ammonia and chlorine are not carbon-based chemicals but are included in this table as they are major parts of the chemical industry.
SOURCE: Modified from Neelis et al. (2007).

TABLE I-2 Availability, Conversion Technologies, Relevant Application and Markets, and Barriers to Wider Adoption of Alternative Carbon Feedstocks Compared to CO_2 for Selected Applications or Markets

Carbon Feedstocks	Global Feedstock Availability (Data Year)[a]	Conversion Technologies	Relevant Application/ Markets	Barriers to Wider Adoption
Woody biomass	1100 Mt C/yr (2019–2020)	Pyrolysis Gasification[b,c,d]	Biodiesel and gasoline Sustainable aviation fuel Biochar—soil amendments Combined heat and power Renewable natural gas Biochemicals	• Geographic constraints and variation • Challenges with logistics and handling • Pre-processing/grinding • Low conversion efficiency • Land competition • Variable feedstock quality and consistency
Agricultural, forestry and livestock residues Municipal solid waste and food losses Crops	770 Mt C/yr (2019–2020) 870 Mt C/yr 2300 Mt C/yr (2019–2020)	Fermentation Anaerobic digestion Gasification Pyrolysis	Mixed alcohols Renewable natural gas Combined heat and power[e] Biodiesel and gasoline[b,d] Basic chemicals and intermediates Sustainable aviation fuel[f]	• Seasonal variability • Land use changes • Water availability • Variable feedstock quality and consistency • Presence of contaminants • Challenges with logistics and handling • Collection and sorting • Low conversion efficiency • Odor and emissions
Aquatic biomass, algae, etc.	25 Mt C/yr (2019–2020)	Fermentation Anaerobic digestion Photobioreactors Gasification Pyrolysis	Basic chemicals and intermediates Pharmaceuticals[e] Animal feed[e] Biodiesel and gasoline[e] Sustainable aviation fuel Renewable natural gas	• Life cycle impacts, including water, energy and land use • Risk of invasive species • Ecological risks • Relatively higher costs of cultivation and harvest
Coal waste	70–90 Mt/yr (United States, 2021–2022)[g]	Precipitation Compounding Pyrolysis Electrochemical Gasification Liquefaction Melt spinning Extraction	Pigments Agriculture Construction materials Energy storage materials Carbon fiber Carbon foam Three-dimensional (3D) printing materials Cement Concrete Critical minerals	• Variable feedstock composition • Locality • Separation of coal from mineral matter • Lack of property information to demonstrate code compliance • Lack of occupational and environmental safety studies • Impurities • Complex homogeneous chemistry • Limited life cycle assessment studies
Recycled plastics	360 Mt C/yr (2020–2022)[h]	Pyrolysis Gasification Hydrolysis Mechanical	Biodiesel and gasoline Basic chemicals and intermediates Combined heat and power Polymers and their precursors	• Feedstock purity, reliable composition, and quality • Reliable availability • Low conversion efficiency • Availability of hydrogen • Higher product cost

[a] Unless otherwise noted, data are from Kähler et al. (2023).
[b] From Hrbek (2021).
[c] From Mednikov (2018).
[d] From Molino et al. (2018).
[e] From Bacovsky et al. (2022).
[f] From Mesfun (2021).
[g] From Gassenheimer and Shaynak (2023). Includes impoundment waste, which is a mixture of water, coal fines (small particles of coal), and other substances generated during coal mining and processing activities. Does not include coal waste from acid mine drainage and coal combustion residuals.
[h] This value is based on the volume of embedded carbon in all global polymers. Current production of recycled plastics is at 24.3 Mt.
NOTE: This table is not exhaustive, and there may feedstocks, conversion technologies, applications, and barriers to adoption not mentioned.
SOURCES: Based on data from Al-Rumaihi et al. (2022); Bacovsky et al. (2022); Hrbek (2021); Kähler et al. (2023); Mednikov (2018); Mesfun (2021); Molino et al. (2018); Sorunmu et al. (2020).

REFERENCES

Al-Rumaihi, A., M. Shahbaz, G. Mckay, H. Mackey, and T. Al-Ansari. 2022. "A Review of Pyrolysis Technologies and Feedstock: A Blending Approach for Plastic and Biomass Towards Optimum Biochar Yield." *Renewable and Sustainable Energy Reviews* 167(October):112715. https://doi.org/10.1016/j.rser.2022.112715.

Bacovsky, D., C. DiBauer, B. Drosg, M. Kuba, D. Matschegg, C. Schmidl, E. Carlon, F. Schipfer, and F.F. Kraxner. 2022. "IEA Bioenergy Report 2023: How Bioenergy Contributes to a Sustainable Future." IEA Bioenergy. https://www.ieabioenergyreview.org/wp-content/uploads/2022/12/IEA_BIOENERGY_REPORT.pdf.

Gassenheimer, C., and C. Shaynak. 2023. "Coal Waste Recovery Presentation." Presentation to the committee. November 3. Washington, DC: National Academies of Sciences, Engineering, and Medicine.

Hrbek, J. 2021. "Status Report on Thermal Gasification of Biomass and Waste 2021." IEA Bioenergy.

Kähler, F., O. Porc, and M. Carus. 2023. "RCI Carbon Flows Report: Compilation of Supply and Demand of Fossil and Renewable Carbon on a Global and European Level." Renewable Carbon Initiative (RCI). https://doi.org/10.52548/KCTT1279.

Mednikov, A.S. 2018. "A Review of Technologies for Multistage Wood Biomass Gasification." *Thermal Engineering* 65(8):531–46. https://doi.org/10.1134/S0040601518080037.

Mesfun, S.A. 2021. "Biomass to Liquids (BtL) via Fischer-Tropsch—A Brief Review." ETIP Bioenergy. https://www.etipbioenergy.eu/images/ETIP_B_Factsheet_BtL_2021.pdf.

Molino, A., V. Larocca, S. Chianese, and D. Musmarra. 2018. "Biofuels Production by Biomass Gasification: A Review." *Energies* 11(4):811. https://doi.org/10.3390/en11040811.

Neelis, M., M. Patel, K. Blok, W. Haije, and P. Bach. 2007. "Approximation of Theoretical Energy-Saving Potentials for the Petrochemical Industry Using Energy Balances for 68 Key Processes." *Energy* 32(7):1104–1123. https://doi.org/10.1016/j.energy.2006.08.005.

Sorunmu, Y., P. Billen, and S. Spatari. 2020. "A Review of Thermochemical Upgrading of Pyrolysis Bio-oil: Techno-Economic Analysis, Life Cycle Assessment, and Technology Readiness." *GCB Bioenergy* 12(1):4–18. https://doi.org/10.1111/gcbb.12658.

J

Background Information About Life Cycle, Techno-Economic, and Societal/Equity Assessments

Material in this appendix provides additional information on techno-economic assessments (TEAs) and life cycle assessments (LCAs), including an example of resources for supporting TEAs and supplemental examples of the wide variation observed in a review of LCA results for production of CO_2-based chemicals.

Given the variety of techniques available, guidelines for selecting an uncertainty analysis method for TEAs are gaining traction in the literature. Figure J-1 shows an example decision tree recommending the type of uncertainty analysis based on purpose.

Tables J-1 and J-2 show compiled LCA results for CO_2 emissions released to produce dimethyl ether (DME) and dimethyl carbonate (DMC) from Garcia-Garcia et al. (2021), demonstrating the wide variety of technologies and processes that have been examined and how these technologies and processes for the same product may incur different environmental impacts.

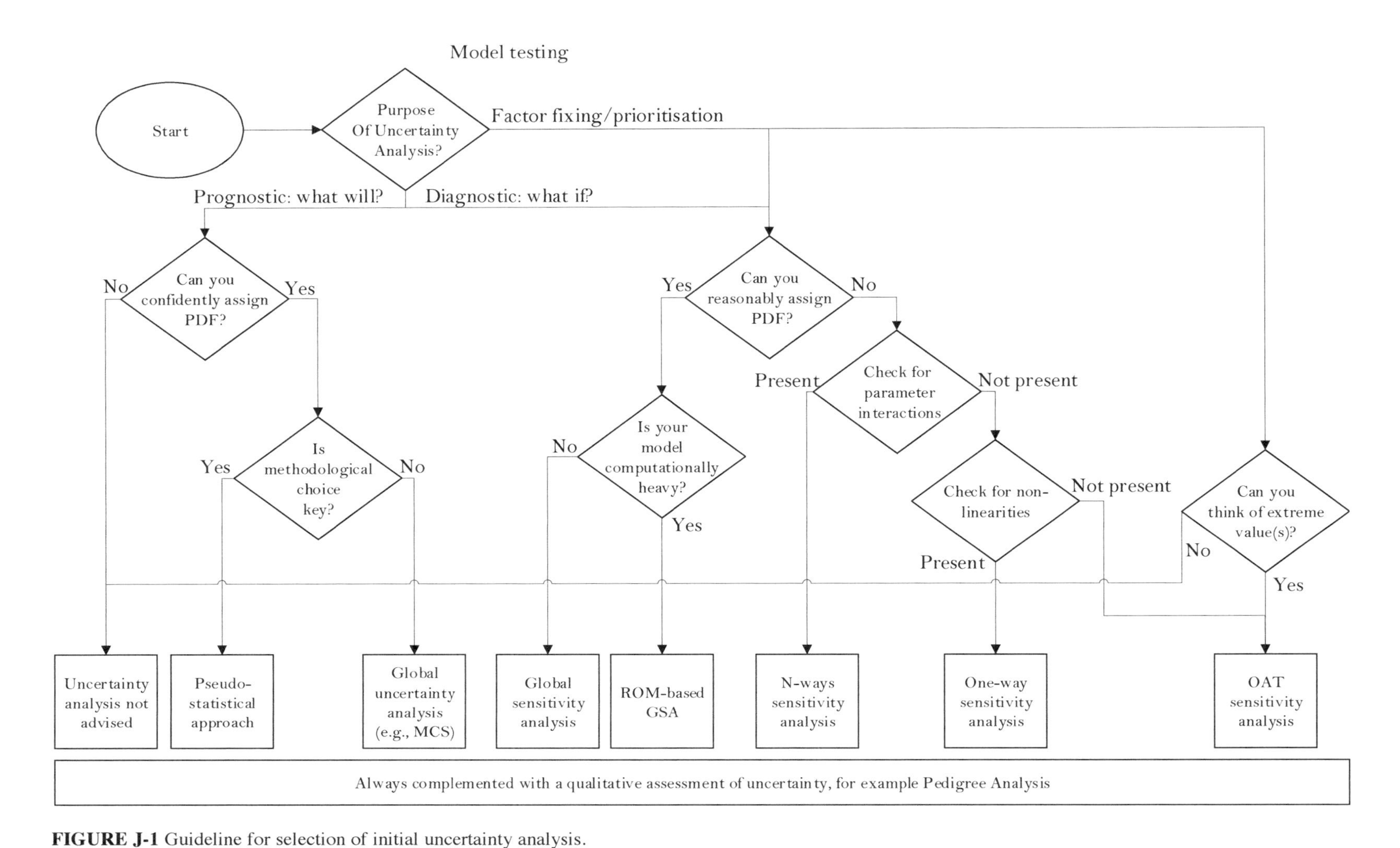

FIGURE J-1 Guideline for selection of initial uncertainty analysis.
NOTE: GSA = global sensitivity analysis; MCS = Monte Carlo simulation; OAT = one-at-a-time sensitivity analysis; PDF = probability distribution function; ROM = reduced order model.
SOURCE: Roussanaly et al. (2021), https://doi.org/10.2172/1779820. CC BY 4.0.

TABLE J-1 Compiled Life Cycle Assessment Results for CO_2 Emissions Using Different System Boundaries, Assumptions, and Processes for Dimethyl Ether (DME) Production from CO_2

Technology/Process	System Boundaries	CO_2 Emissions
Synthesis by dehydrogenation of methanol[a]	Cradle-to-gate	1.27 $kg_{CO_{2\ eq}}\ kg^{-1}_{DME}$
DME from natural gas[a]	Cradle-to-grave, including feedstock production and transport, fuel production, distribution and reforming, and vehicle fueling and combustion	91.1 $g_{CO_{2\ eq}}\ MJ^{-1}_{DME}$
CO_2 converted to syngas via dry reforming of methane ($Ni/Rh/Al_2O_3$ catalyst), then transformed into DME (γ-Al_2O_3 catalyst)	Cradle-to-gate plus combustion Cradle-to-gate	35.8 $g_{CO_{2\ eq}}\ MJ^{-1}_{DME}$ 0.12–0.15 $kg_{CO_{2\ eq}}\ MJ^{-1}_{DME}$[b] −1.07–0.48 $kg_{CO_{2\ eq}}\ kg^{-1}_{DME}$[b]
DME from high solid anaerobic digestion of food and yard waste	Cradle-to-grave, including feedstock production and transport, fuel production, distribution and reforming, and vehicle fueling and combustion	−5 $g_{CO_{2\ eq}}\ MJ^{-1}_{DME}$
CO_2 converted to methanol, then transformed to DME via a condensation reaction	Cradle-to-gate	0.5 $t_{CO_{2\ eq}}\ MJ^{-1}_{DME}$
CO_2 enhanced gasification of gumwood to produce DME	Cradle-to-gate including the pre-treatment process; production of DME; and utilization of DME as renewable fuel for diesel engines.	bio-DME emissions 46.2 $kg_{CO_{2\ eq}}$ per 100 km, and 162 $kg_{CO_{2\ eq}}$ per 100 km for diesel

[a] Standard production (non-CO_2 utilization) processes for comparison
[b] Range contingent on hydrogen and electricity sources and other assumptions
SOURCE: Adapted from Garcia-Garcia (2021), Table 7.

TABLE J-2 Compiled Life Cycle Assessment Results for CO_2 Emissions Using Different System Boundaries, Assumptions, and Processes for Dimethyl Carbonate Production from CO_2

Technology/Process	System Boundaries	CO_2 Emissions ($kg_{CO_{2\ eq}}\ kg^{-1}_{DMC}$)
Conventional production, via phosgene from CO and Cl_2, and the Bayer process[a]	Cradle-to-gate	0.52–132
Direct synthesis from CO_2 and methanol	Cradle-to-gate	7.26–7.33[b]
Electrochemical reaction of CO_2 with methanol in the presence of potassium methoxide and 1-butyl-3-methylimidazolium bromide	Cradle-to-gate	381–465[b]
Electrosynthesis from CO_2 and methanol	Cradle-to-gate	78.9
Oxidative carbonylation of methanol (Eni)	Cradle-to-gate	3.18
Transesterification of ethylene carbonate	Cradle-to-gate	0.45–0.77[b]
Transesterification of urea	Cradle-to-gate	2.94
Via ethylene oxide	Cradle-to-gate	0.86
Via urea from NH_3 and CO_2	Cradle-to-gate	30.6
Via urea methanolysis	Cradle-to-gate	0.34

[a] Standard production (non-CO_2 utilization) processes for comparison
[b] Range contingent on hydrogen and electricity sources and other assumptions
SOURCE: Adapted from Garcia-Garcia et al. (2021).

REFERENCES

Garcia-Garcia, G., M. Cruz Fernandez, K. Armstrong, S. Woolass, and P. Styring. 2021. "Analytical Review of Life-Cycle Environmental Impacts of Carbon Capture and Utilization Technologies." *ChemSusChem* 14(4):995–1015. https://chemistry-europe.onlinelibrary.wiley.com/doi/10.1002/cssc.202002126.

Roussanaly, S., E. Rubin, M. Der Spek, G. Booras, N. Berghout, T. Fout, M. Garcia, et al. 2021. "Towards Improved Guidelines for Cost Evaluation of Carbon Capture and Storage." OSTI ID:1779820. https://doi.org/10.2172/1779820.

K

Elemental Carbon Products Literature Review

A summary of carbon dioxide to elemental carbon (CTEC) research and development history is given in Table K-1. Table K-1 not only lists the major technology and the major reaction conditions used for CTEC but also the structure characteristics of CTEC products. The table shading helps to cluster rows by method; blue represents thermochemical, yellow represents electrochemical, orange represents photochemical, and green represents plasmachemical processes.

TABLE K-1 Relationship Among Carbon Dioxide to Elemental Carbon Methods, Reaction Conditions, and Generated Products

Method	Method Detail	No.	Year	Main Reaction Conditions	Product	Reference
Thermochemical	Cation-excess	1	1990	3 g cation-excess magnetite Reaction system: batch Reaction time: 1.7 h	Unknown-structure carbon	Tamaura and Tahata (1990)
	Cation-excess	2	1992	2.0 g active wustite ($Fe_{\delta}O$, with a δ value of 0.98) Reaction system: batch Reaction temperature: 300°C	Unknown-structure carbon	Kodama et al. (1992)
	Cation-excess	3	1992	Oxygen-deficient magnetite Reaction system: batch Reaction temperature: 300°C	Unknown-structure carbon	Tamaura and Nishizawa (1992)
	Cation-excess	4	1993	Rhodium-bearing magnetite Reaction temperature: 300°C	Unknown-structure carbon	Akanuma et al. (1993a)
	Cation-excess	5	1993	Oxygen-deficient Mn(II) ferrite Reaction temperature: 300°C	Unknown-structure carbon	Tabata et al. (1993a)
	Cation-excess	6	1993	Oxygen-deficient Mn(II)-bearing ferrites ($Mn_xFe_{3-x}O_{4-\delta}$, $0 \leqslant x \leqslant 1$, $\delta>0$) Reaction system: semi-batch Reaction temperature: 300°C	Unknown-structure carbon	Tabata et al. (1993b)
	Cation-excess	7	1993	Oxygen-deficient magnetite (ODM) Reaction system: batch Reaction temperature: 520°C	Mixture of carbon nanomaterials (CNMs)	Akanuma et al. (1993b)
	Cation-excess	8	1994	Hydrogen-activated Ni(II)-bearing ferrite Reaction temperature: 300°C	Unknown-structure carbon	Kato et al. (1994)
	Cation-excess	9	1994	Ni(II)- and Co(II)-bearing ferrites Reaction system: batch Reaction temperature: 300°C	Unknown-structure carbon	Kodama et al. (1994b)
	Cation-excess	10	1994	Oxygen-deficient Zn II-bearing ferrites ($Zn_xFe_{3-x}O_{4-\delta}$, $0 \leqslant x \leqslant 1$, $\delta>0$) Reaction system: batch Reaction temperature: 300°C	Unknown-structure carbon	Tabata et al. (1994a)
	Cation-excess	11	1994	Oxygen-deficient magnetite Reaction system: batch Reaction temperature: 350°C	Graphite	Tsuji et al. (1994)
	Cation-excess	12	1994	Oxygen-deficient Zn(II)-bearing ferrites ($Zn_xFe_{3-x}O_{4-\delta}$, $0 \leq x \leq 1$, $\delta>0$) Reaction system: batch Reaction temperature: 520°C	Unknown-structure carbon	Tabata et al. (1994b)
	Cation-excess	13	1994	Ni(II)- and Co(II)-bearing ferrites Reaction system: batch Reaction temperature: 300°C	Graphite	Kodama et al. (1994a)
	Cation-excess	14	1995	Ni(II)-bearing ferrite/magnetite Reaction system: batch Reaction temperature: 300°C	Unknown-structure carbon	Kodama et al. (1995c)
	Cation-excess	15	1995	Cation-excess magnetite Reaction temperature 80°C (358 K)	Unknown-structure carbon	Zhang et al. (1995)
	Cation-excess	16	1995	Oxygen-deficient Ni(II)-bearing ferrite (ODNF: $Ni_{0.39}Fe_{2.61}O_{4-\delta}$) Reaction temperature 300°C	Unknown-structure carbon	Kodama et al. (1995a)

continued

TABLE K-1 Continued

Method	Method Detail	No.	Year	Main Reaction Conditions	Product	Reference
	Cation-excess	17	1995	Ni(II)-bearing ferrite (UNF) $Ni^{2+}_{0.36}Fe^{2+}_{0.45}Fe^{3+}_{2.19}O_{4.10}$ Reaction system: batch Reaction temperature: 300°C	Unknown-structure carbon	Kodama et al. (1995b)
	Cation-excess	18	1995	Oxygen-deficient magnetite Reaction system: semi-batch Reaction temperature: 300°C	Unknown-structure carbon	Wada et al. (1995)
	Cation-excess	19	1996	Cation-excess magnetite Reaction system: batch Reaction temperature: 290°C (563 K)	Unknown-structure carbon	Zhang et al. (1996)
	Cation-excess	20	1996	1 g Ni(II)-bearing ferrite (NF) Reaction system: batch Reaction temperature: 300°C Reaction time: 60 min	Unknown-structure carbon	Tsuji et al. (1996a)
	Cation-excess	21	1996	Impregnated Rh, Pt, and Ce on Ni(II)-bearing ferrite (NF) Reaction system: batch Reaction temperature: 300°C	Unknown-structure carbon	Tsuji et al. (1996b)
	Cation-excess	22	1997	1 g Nanophase Zn ferrites Reaction system: batch Reaction temperature: 300°C Reaction time: 30 min	Unknown-structure carbon	Komarneni et al. (1997)
	Cation-excess	23	1997	0.3 kg Ni ferrite Reaction system: semi-batch Reaction temperature: 350°C	Unknown-structure carbon	Yoshida et al. (1997)
	Cation-excess	24	1997	Wurtzite ($Fe_{1-y}O$); 500°C (773 K) Reaction system: semi-batch	Unknown-structure carbon	Ehrensberger et al. (1997)
	Cation-excess	25	1998	Oxygen-deficient Ni(II)-bearing ferrite (ODNF)	Unknown-structure carbon	Sano et al. (1998)
	Cation-excess	26	1999	20 g active wustite ($Fe_{\delta}O$, with a δ value of 0.98) Reaction system: batch Reaction temperature: 300°C (573 K)	Unknown-structure carbon	Zhang et al. (1999)
	Cation-excess	27	2000	20 g oxygen-deficient magnetite Reaction system: batch Reaction temperature: 300°C Reaction time: 180 min	Unknown-structure carbon	Zhang et al. (2000a)
	Cation-excess	28	2000	20 g oxygen-deficient magnetite Reaction system: batch Reaction temperature: 300°C Reaction time: 180 min	Unknown-structure carbon	Zhang et al. (2000b)
	Cation-excess	29	2001	1 g ultra-fine (Ni,Zn)-ferrites Reaction system: semi-batch Reaction temperature: 300°C Reaction time: 7 min	Unknown-structure carbon	Kim et al. (2001)
	Cation-excess	30	2001	$(Ni_x, Zn_{1-x}) Fe_2O_{4-\delta}$ ferrites Reaction system: semi-batch Reaction temperature: 300°C	Unknown-structure carbon	Kim and Ahn (2001)
	Cation-excess	31	2001	Nano-size ferrites $(Ni_{0.5}Cu_{0.5}) Fe_2O4$ Reaction system: semi-batch Reaction temperature: 800°C	Unknown-structure carbon	Shin et al. (2004)

TABLE K-1 Continued

Method	Method Detail	No.	Year	Main Reaction Conditions	Product	Reference
	Cation-excess	32	2004	$(Mn_{0.67}Ni_{0.33})$ Fe_2O4 Reaction system: semi-batch Reaction temperature: 300°C	Unknown-structure carbon	Hwang and Wang (2004)
	Cation-excess	33	2005	Co-doped ferrite ($NiFe_2O_4$)	Unknown-structure carbon	Fu et al. (2005)
	Cation-excess	34	2006	$CoFe_2O_4$ nanoparticles Reaction temperature: 500°C	Carbon nanotubes (CNTs)	Khedr et al. (2006)
	Cation-excess	35	2006	Spinel structure $NiFe_{2-x}Cr_xO_4$ ($x = 0, 0.08$) Reaction system: batch	Unknown-structure carbon	Linshen et al. (2006)
	Cation-excess	36	2007	5 g mechanically milled magnetite Reaction system: batch Reaction temperature: 500°C (773 K) Reaction time: 3 hours	Graphite	Yamasue et al. (2007b)
	Cation-excess	37	2007	0.5 g nickel ferrite Ni $Fe_2O_{4-\delta}$ Reaction system: batch Reaction temperature: 320°C Reaction time: 120 min	Unknown-structure carbon	Fu et al. (2007)
	Cation-excess	38	2007	2 g Ni-ferrite doping different contents of Cr^{3+} reaction system: batch	Mixture of CNMs	Ma et al. (2007b)
	Cation-excess	39	2007	Nanocrystallines Fe_2O_3; 400–600/0°C; reaction system: semi-batch	CNTs	Khedr et al. (2007)
	Cation-excess	40	2007	$NiCr_{0.08}Fe_{1.92}O_4$; 310°C; reaction system: batch	CNMs	Ma et al. (2007a)
	Cation-excess	41	2007	Milled wustite powders; 500°C (773 K); reaction system: batch	Mixture of CNMs	Yamasue et al. (2007a)
	Cation-excess	42	2009	$Ni_{0.49}Cu_{0.24}Zn_{0.24}Fe_2O_4$; 310°C Reaction system: batch	Amorphous carbon	Ma et al. (2009b)
	Cation-excess	43	2009	$CoCr_{0.08}Fe_{1.92}O_4$ Reaction system: semi-batch Reaction temperature: 310°C	Unknown-structure carbon	Ma et al. (2009a)
	Cation-excess	44	2011	1.5–2 g nickel ferrite nanoparticles Reaction system: semi-batch Reaction temperature: 300°C Reaction time: 24 min O_2 detected	Unknown-structure carbon	Lin et al. (2011)
	Cation-excess	45	2011	MFe_2O_4 (M = Ni, Co, Cu, Zn) Reaction system: batch Reaction temperature: 350°C	Unknown-structure carbon	Ma et al. (2011)
	Cation-excess	46	2012	1 g zinc-modified zeolite Y material Reaction system: batch Reaction temperature: 300°C Reaction time: 8 h	Unknown-structure carbon	Wang et al. (2012)
	Cation-excess	47	2013	1.5-2 g nickel ferrite nanoparticles Reaction system: semi-batch Reaction temperature: 300°C Reaction time: 20 min	Unknown-structure carbon	Lin et al. (2013)
	Cation-excess	48	2015	H_2-reduced Fe_2O_3 and Fe_3O_4 Reaction system: batch Reaction temperature: 400°C	Unknown-structure carbon	Li et al. (2015)

continued

TABLE K-1 Continued

Method	Method Detail	No.	Year	Main Reaction Conditions	Product	Reference
	Cation-excess	49	2016	Spinel M-ferrites (M=Co, Ni, Cu, Zn) Reaction system: batch Reaction temperature: 310°C	Unknown-structure carbon	Jiaowen et al. (2016)
	Cation-excess	50	2017	$Ba_2Ca_{0.66}Nb_{1.34-x}Fe_xO_{6-\delta}$ (BCNF) Reaction system: semi-batch Reaction temperature: 300°C	Unknown-structure carbon	Mulmi et al. (2017)
	Cation-excess	51	2019	1.5 g $SrFeCo_{0.5}O_x$ Reaction system: semi-batch Reaction temperature: 300°C	Unknown-structure carbon	Kim et al. (2019)
	Cation-excess	52	2019	Fe_3O_4 Reaction temperature: 600°C	CNMs	Jo et al. (2019)
	Cation-excess	53	2020	1.0 g $SrFeO_{3-x}$ Reaction system: semi-batch Reaction time: 170 min	Unknown-structure carbon	Sim et al. (2020)
	Cation-excess	54	2021	0.1 g milled natural magnetite Reaction system: semi-batch Reaction time: 90 min	Amorphous carbon	Liu et al. (2021)
	Cation-excess	55	2021	Neat NaY zeolite (control) and Zn-NaY zeolite Reaction temperature: 300–500°C	Unknown-structure carbon	Bajaj et al. (2021)
	Cation-excess	56	2024	Spinel Nano-$Mn_xFe_{3-x}O_4$ Reaction system: semi-batch Reaction temperature: 340°C	Unknown-structure carbon	Wang et al. (2024)
	Reacting with metals	57	1978	Two blocks of dry ice with magnesium turnings	Unknown-structure carbon	Driscoll (1978)
	Reacting with metals	58	2001	2.6 g CO_2 + 0.3 g Mg Reaction system: closed cell Reaction temperature: 1000°C Reaction time: 3 h	Mixture of CNMs	Motiei et al. (2001)
	Reacting with metals	59	2003	CO_2: 8.0 G; metallic Li: 0.5 g Reaction pressure: 700 atm Reaction temperature: 550°C Reaction time:10 h	CNTs	Lou et al. (2003)
	Reacting with metals or metal oxides	60	2008	React with Zn/ZnO and FeO/Fe_3O_4	Unknown-structure carbon	Gálvez et al. (2008)
	Reacting with metals	61	2009	0.5 g metallic lithium; 8.0 g dry ice Reaction temperature: 700°C Reaction pressure: 100 MPa Reaction time: 10 h	C_{60}	Chen and Lou (2009)
	Reacting with metals	62	2011	3 g of Mg ribbon ignited inside a dry ice vessel, covered by another dry ice slab	Graphene	Chakrabarti et al. (2011)
	Reacting with metals	63	2013	2 g of Mg ribbon ignited inside a dry ice vessel at room temperature	Graphene	Moghaddam et al. (2013)
	Reacting with metals	64	2014	Lithium and dry ice, ignited with an oxygen–hydrogen torch	Graphene	Poh et al. (2014)
	Reacting with metals	65	2014	2.0 g Mg ribbon ignited inside a vessel containing dry ice at room temperature	Graphene	Samiee and Goharshadi (2014)

TABLE K-1 Continued

Method	Method Detail	No.	Year	Main Reaction Conditions	Product	Reference
	Reacting with metals	66	2014	Mg and Ca metals, ignited in a CO_2 atmosphere	Graphene	Zhang et al. (2014)
	Reacting with metals	67	2015	Mg powder: 1.5 g CO_2 flowrate: 70 mL/min Reaction temperature: 680°C Reaction time: 60 min	Graphene	Xing et al. (2015
	Reacting with metals	68	2015	CO_2 reacted with 1 g Mg ribbons Reaction system: semi-batch Reaction temperature: 800°C	CNTs	Wang et al. (2015)
	Reacting with metals	69	2016	React with liquid Na Reaction temperature: 600°C	Graphene	Wei et al. (2016)
	Reacting with metals	70	2016	React with liquid Li Reaction temperature: 550°C	Graphene	Smith et al. (2016)
	Reacting with metals	71	2017	React with liquid K Reaction temperature: 550°C	Graphene	Wei et al. (2017b)
	Reacting with metals	72	2017	React with liquid Na Reaction temperature: 550°C	Carbon nanowires (CNWs)	Wei et al. (2017a)
	Reacting with metals	73	2017	0.1 mol of potassium (from Aldrich) reacted with CO_2 in a batch ceramic-tube reactor at a temperature of 550°C and an initial pressure of 50 psi for a selected time (12, 24, or 48 h)	Graphene	Wei et al. (2017c)
	Reacting with metals	74	2018	Burning of Mg, Zn, and Ni metals in presence of CO_2 (dry ice)	Mixture of CNMs	Bagotia et al. (2018)
	Reacting with metals	75	2019	Reacting Ni and Mg with CO_2 Reaction temperature: 650°C	Mixture of CNMs	Baik et al. (2020)
	Reacting with metals	76	2020	React with Alkali metals, including lithium (Li), sodium (Na), and potassium (K)	Graphene	Sun and Hu (2020)
	Reacting with metals	77	2020	CO_2 reacted with Na liquid	Graphene	Wang et al. (2020c)
	Reacting with metals	78	2021	Zn/Mg M ratios: 0, 0.5, 1, 2, 3, 4, 5, and 6 CO_2 flowrate: 70 mL/min Reaction time: 180 min	Graphene	Luchetta et al. (2021)
	Reacting with metals	79	2021	Mg metal ribbon ignited in presence of CO_2 (dry ice, two blocks)	Mixture of CNMs	Sharma and Bagotia (2021)
	Reacting with metals	80	2022	Reduction agent: a eutectic of gallium and indium (EGaIn alloy) Reaction temperature: 25°C and 500°C	Unknown-structure carbon	Zuraiqi et al. (2022)
	Reacting with metals	81	2022	Mg molten temperature:720°C CO_2 flowrate: 900 mL/min	Graphene	Li et al. (2022)
	Reacting with metals	82	2022	Mg molten temperature:720°C CO2 flowrate: 995 mL/min Reaction time: 60 min	Graphene	Wei et al. (2022)
	Reacting with metals	83	2022	Mg and CO_2 ignition in reaction chamber	Graphene	Colson et al. (2022)

continued

TABLE K-1 Continued

Method	Method Detail	No.	Year	Main Reaction Conditions	Product	Reference
	Reacting with metals	84	2023	Reducing with Mg and Ga Reaction temperature: 40°C—near room temperature	Unknown-structure carbon	Ye et al. (2023)
	Reacting with metals	85	2023	Mg molten temperature:720°C CO2 flowrate: 500 mL/min Reaction time: 30 min	Graphene	Li et al. (2023)
	Reacting with H_2	86	1991	Catalyst: WO_3 (H_2) Reaction system: batch Reaction temperature: 700°C (973 K)	Unknown-structure carbon	Ishihara et al. (1991)
	Reacting with H_2	87	2008	Catalyst: 3%Ni-K/Al_2O_3 Reaction temperature: 500°C	Carbon nanofibers (CNFs)	Chen et al. (2011)
	Reacting with H_2	88	2009	Catalyst: Ni/Al_2O_3 Reaction temperature: 440–500°C	CNFs	Chen et al. (2009)
	Reacting with H_2	89	2010	Catalyst: Ni/Al_2O_3 Reaction temperature: 440–500°C	CNFs	Chen et al. (2010)
	Reacting with H_2	90	2022	Catalyst: Ni/Al_2O_3 Reaction pressure: 1 atm Reaction temperature: 500°C	CNFs	Lin et al. (2022)
	Reacting with LiH	91	2019	Reacting LiH with CO_2 Reaction pressure: 5, 15, 30 bar Reaction temperature: 210°C, 340°C, 470°C Reaction time: 30 s	CNMs	Liang et al. (2019)
	Reacting with $NaBH_4$	92	2006	Catalyst: 1.5 g $NaBH_4$ Reaction system: batch Reaction temperature: 700°C Reaction time: 8 h	CNTs	Lou et al. (2006)
	Reacting with $NaBH_4$	93	2020	Catalyst: $NiCl_2$; reducing agent: $NaBH_4$ Reaction pressure: 1 atm Reaction temperature: 500–700°C	CNTs	Kim et al. (2020b)
	Reacting with strong reducing agents	94	1991	Catalyst: WO_3 Reaction temperature: 700°C	Unknown-structure carbon	Ishihara et al. (1991)
	Reacting with strong reducing agents	95	2021	Reaction system: semi-batch CO_2 flowrate: 100 mL/min Reaction pressure: 1 atm Reaction temperature: 423°C (700 K) Reaction time: 4 h	Mixture of CNMs	Watanabe and Ohba (2021)
	Reacting with strong reducing agents (CVD)	96	2013	Ni/Al_2O_3 Reaction temperature: 1000°C	Graphene	Luo et al. (2013)
	Reacting with strong reducing agents (CVD)	97	2015	Monometallic FeNi0–$Al_2O_3$3 (FNi0) and bimetallic $FeNi_x$–Al_2O_3 (FNi_2, FNi_4, FNi_8, and FNi_{20}) Reaction temperature: 700°C	CNMs	Hu et al. (2015)
	Reacting with strong reducing agents (CVD)	98	2019	Cu–Pd alloy Reaction pressure: 1 atm Reaction temperature: 1000°C	Graphene	Molina-Jirón et al. (2019)

TABLE K-1 Continued

Method	Method Detail	No.	Year	Main Reaction Conditions	Product	Reference
	Reacting with strong reducing agents (CVD)	99	2020	30 mg Fe, Ni, Co Reaction temperature: 560°C Reaction time: 1 h	CNFs	Nakabayashi et al. (2020)
	Reacting with strong reducing agents (CVD)	100	2022	Ni/Al_2O_3 Reaction temperature: 1050°C	Graphene	Gong et al. (2022)
	Reacting with strong reducing agents (CVD)	101	2007	Catalyst: Fe/CaO Reaction system: semi-batch Reaction temperature: 790–810°C Reaction time: 45 min	CNTs	Xu and Huang (2007)
	Reacting with strong reducing agents (CVD)	102	2015	Reaction system: semi-batch Reaction temperature: 1060°C Reaction time: 60 min	Graphene	Strudwick et al. (2015)
	Reacting with strong reducing agents (CVD)	103	2015	Reaction system: semi-batch Reaction temperature: ~1000°C Reaction time: 30 min	Graphene	Seekaew et al. (2022)
	Reacting with strong reducing agents (CVD)	104	2015	Reaction system: semi-batch CO_2 flowrate: 900 mL/min Reaction temperature:1100°C Reaction time: 60 min	CNTs	Allaedini et al. (2015)
	Reacting with strong reducing agents (CVD)	105	2016	Reaction system: semi-batch; reaction temperature:1100°C; CO2 flowrate: 900 mL/min reaction time:1 h	Graphene	Allaedini et al. (2016a)
	Reacting with strong reducing agents (CVD)	106	2016	Ge/MgO Reaction system: semi-batch Reaction temperature: 1226°C	CNTs	Allaedini et al. (2016b)
Electrochemical	Electrochemical	107	2013	CO_2 9.7% or 90% (CO_2-Ar mixture); Electrolysis: 3.1 V (molten $CaCl_2$–CaO) or 3.2 V (molten LiCl–Li_2O) Reaction temperature: 654°C (923 K)	Mixture of CNMs	Otake et al. (2013)
		108	2013	Electrolysis current range: 0.2 mA–70 mA Reaction temperature: 750°C	Unknown-structure carbon	Guo et al. (2013)
		109	2015	Cathode: a coiled galvanized steel wire Anode: nickel Electrolyte: melt $LiCO_3$ Reaction temperature: ~800°C Electrolysis current density: 0.1 A/cm^2	CNFs	Ren et al. (2015a)
		110	2015	Electrolyte: Li_2CO_3/Na_2CO_3 or $Li_2CO_3/BaCO_3$ or $Na_2CO_3/BaCO_3$ Cathode: a Muntz brass Anode: iridium foil Reaction temperature: 750°C Electrolysis current density (A/cm^2): 0 ~ −1.2 Electrolysis voltage: <1 V	Unknown-structure carbon	Ren et al. (2015b)

continued

TABLE K-1 Continued

Method	Method Detail	No.	Year	Main Reaction Conditions	Product	Reference
		111	2016	CO_2 90% (CO_2-Ar mixture) Cathode: metallic Ca Anode: ZrO_2 Reaction temperature: 900°C (1173 K)	CNTs	Ozawa et al. (2016)
		112	2016	Cathode: a stainless steel Anode: RuO_2–TiO_2 Reaction temperature: 650–850°C	Graphene	Hu et al. (2016)
		113	2016	Cathode: a galvanized steel Anode: nickel Electrolyte: molten carbonate Reaction temperature: 725°C Reaction time: 1 h Electrolysis current density: 0.1 A/cm^2	CNTs	Licht et al. (2016)
		114	2016	Cathode: a Fe spiral Anode: a Ni-Cr spiral Electrolyte: Li_2CO_3-Na_2CO_3-K_2CO_3 (61:22:17 wt%, analytically pure) Electrolysis current densities: 200 mA/cm^2 and 400 mA/cm^2 Reaction temperature: 600°C	CNTs	Wu et al. (2016)
		115	2016	Cathode: galvanized steel Anode: nickel Electrolyte: lithiated molten carbonate	CNTs	Lau et al. (2016)
		116	2017	Cathode: three different steels (16 gauge galvanized steel wire, 316 stainless steel shim, and 1010 steel shim) Anode: untreated Ni wire, thermally oxidized Ni wire, and Ni wire coated with 500 cycles of Al_2O_3 Reaction temperature: 750°C Reaction time: 1 h Electrolysis current density: 0.1 A/cm^2	CNTs	Douglas et al. (2017a)
		117	2017	Cathode: Varieties of metals Anode: pure nickel or Nichrome wire Reaction temperature: 750°C Electrolysis current density; 0.1 A/cm^2	CNTs	Johnson et al. (2017a)
		118	2017	Cathode: a Ni sheet Anode: a graphite rod Electrolyte: 2 mol % $CaCO_3$-containing LiCl–KCl Reaction temperature: 450°C Reaction time: 1 h Electric voltage: 2.8 V	Hollow carbon sphere (HCS)	Deng et al. (2017)
		119	2017	Cathode: glassy carbon and graphite Anode: RuO_2–TiO_2 Electrolyte: molten $CaCl_2$–NaCl–CaO Electrolysis current densities: 200 mA/cm^2 Reaction temperature: 650–850°C	CNTs	Hu et al. (2017)

TABLE K-1 Continued

Method	Method Detail	No.	Year	Main Reaction Conditions	Product	Reference
		120	2017	Cathode: scrap metals including steel and brass; Anode: Al_2O_3 coated Ni wire Electrolyte: 40 g lithium carbonate Reaction temperature: 750°C	CNTs	Douglas et al. (2017b)
		121	2017	Cathode: a coiled galvanized steel wire Anode: nickel Electrolyte: molten Li_2CO_3	CNTs	Licht (2017a)
		122	2017	Cathode: steel Anode: nickel Electrolyte: 50/50 wt% of Na_2CO_3 mixed with Li_2CO_3	CNTs	Ren et al. (2017)
		123	2017	Cathode: nickel Anode: SnO_2 Electrolyte: mixed melt of Li_2CO_3–Na_2CO_3–K_2CO_3–Li_2SO_4 (40.02: 28.98: 23: 8 mol%) Reaction temperature: 450°C	CNMs	Chen et al. (2017c)
		124	2017	Cathode: a U-shape Ni sheet Anode: SnO2 or platinum plated titanium Electrolyte: mixed melt of Li_2CO_3–Na_2CO_3–K_2CO_3–Li_2SO_4 (40: 29: 23: 8 mol%) Reaction temperature: 475–825°C	Graphite	Chen et al. (2017a)
		125	2017	Cathode: U-shape Ni sheet Anode: SnO_2 Electrolyte: mixed melt of Li_2CO_3–Na_2CO_3–K_2CO_3 (43.5:31.5:25.0 mol%) Reaction temperature: 450°C	Amorphous carbon	Chen et al. (2017b)
		126	2017	Electrolyte: Li_2CO_3+ 0.1 wt% LiBO2 Cathode: Monel/Munz brass/(Ni+Cu alloy) Anode: iridium/Nichrome 60 Reaction temperature: 770°C Electrolysis current density (A/cm^2): 0.1–0.2	CNTs	Johnson et al. (2017b)
		127	2018	Cathode: a galvanized iron Anode: nickel Reaction pressure: 1 atm Reaction time: 4 h Electrolysis voltage: 0.5 ~ 2.5 V Current density: 0.2 A/cm^2	CNTs	Li et al. (2018)
		128	2018	Cathode: 316 stainless steel Anode: Al2O3-coated Ni wire Reaction pressure: 1 atm Reaction temperature: 750°C Reaction time: 1 h Electrolysis current density: 0.1 A/cm^2	CNTs	Douglas et al. (2018)

continued

TABLE K-1 Continued

Method	Method Detail	No.	Year	Main Reaction Conditions	Product	Reference
		129	2018	Cathode: a Ni wire Anode: a graphite rod Electrolyte: $CaCO_3$-containing LiCl–KCl Reaction temperature: 450°C, 550°C, 650°C Electric voltage: 2.8 V	HCS	Deng et al. (2018)
		130	2019	Electrolyte: calcium chloride anhydrous; calcium oxide; sodium carbonate; CO_2 100% (1 mL/min) Electrolysis: 3.0 V Current: 10 A Reaction temperature: 850°C	Graphite	Abbasloo et al. (2019)
		131	2019	Cathode: graphite rod; Anode: RuO_2–TiO_2 Reaction temperature: 625/725°C Reaction time: 4 h; Electrolysis current: 0.75 A	Graphite	Hu et al. (2019)
		132	2019	Cathode: galvanized iron wire Anode: nickel wire Electrolyte: Pure Li_2CO_3 (40 g), Li/Ca (40 g Li_2CO_3-4 g $CaCO_3$), Li/Sr (40 g Li_2CO_3-4 g $SrCO_3$), and Li/Ba (40 g Li_2CO_3-4 g $BaCO_3$) Electrolysis current densities: 200 mA/cm^2 Reaction temperature: 500–850°C	CNTs	Li et al. (2019)
		133	2019	Cathode: galvanized steel Anode: Ir/Pt Electrolyte: Li_2CO_3 Reaction temperature: 450°C	Carbon nano-onion	Liu et al. (2019)
		134	2019	Cathode: copper/galvanized steel/Monel Electrolyte: molten Li_2CO_3 Reaction temperature: 770°C	CNTs	Licht et al. (2019)
		135	2019	Cathode: brass sheet Anode: Inconel 718 Electrolyte: Li_2CO_3-Na_2CO_3-$LiBO_2$ or Li_2CO_3-K_2CO_3-$LiBO_2$ Reaction temperature: 740°C	CNTs	Wang et al. (2019)
		136	2019	Electrocatalyst: cerium oxide Electrolyte: liquid metal-containing cerium (LMCe)—a dimethylformamide (DMF)-based electrolyte Reaction temperature: room temperature	CNMs	Esrafilzadeh et al. (2019)
		137	2020	Cathode: 5 cm^2 galvanized (zinc coated) steel Anode: 5 cm^2 Pt Ir foil anode Electrolysis current: 0.05 A, 0.10 A, 0.2 A, 0.4 A, 1 A, 2 A Reaction temperature: 730°C	Graphene	Liu et al. (2020)

TABLE K-1 Continued

Method	Method Detail	No.	Year	Main Reaction Conditions	Product	Reference
		138	2020	Cathode: protonic ceramic fuel cell (PCFC) Anode: solid oxide fuel cell (SOFC) Reaction temperature: above 900°C	CNTs	Kim et al. (2020a)
		139	2020	Cathode: 316 stainless steel with Fe deposited Anode: Copper wire, Platinum wire, and Alumina coated Ni wire Reaction time: 1 h Reaction temperature: 750°C Electrolysis current density (A/cm^2): 0.05, 0.1, 0.2, 0.4	CNTs	Moyer et al. (2020)
		140	2020	Cathode: Muntz brass Anode: Inconel 718, Nichrome or Incoloy Reaction temperature: 770°C Electrolysis current density; 0.1 A/cm^2	CNTs	Wang et al. (2020d)
		141	2020	Cathode: 0.25-inch-thick Muntz brass sheet Anode: 0.04-inch-thick Nichrome sheet Electrolyte: molten lithium carbonate Electrolysis current densities: 200 mA/cm^2 Reaction temperature: 770°C	CNTs	Wang et al. (2020b)
		142	2020	Electrolyte: Na_2CO_3/Li_2CO_3 Cathode: a Muntz brass; anode: an Inconel Reaction temperature: 670°C Electrolysis current density (A/cm^2): 0.4	CNMs	Wang et al. (2020a)
		143	2021	Electrolyte: 20% Na_2CO_3 + 80% Li_2CO_3 Cathode: a brass sheet Anode: an Inconel 718 sheet Reaction temperature: 750°C Electrolysis current density (A/cm^2): 0.2	CNTs	Wang et al. (2021a)
		144	2021	Electrolyte: ionic liquid (0.5M LiTFSI in 1-butyl-1-methyl-pyrrolidinium bis(trifluoromethylsulfonyl)imide (Pyr14TFSI)) Cathode: 0.5M LiTFSI/Pyr14TFSI electrolyte and a porous carbon layer Anode: a stainless-steel coin cell current collector + a Li foil anode + a glass fiber separator Reaction temperature: room temperature	Amorphous carbon	Wang et al. (2021b)

continued

TABLE K-1 Continued

Method	Method Detail	No.	Year	Main Reaction Conditions	Product	Reference
		145	2022	Electrolyte: lithium carbonate (0.1 wt% Fe_2CO_3) Cathode: a Muntz brass Anode: high-surface-area Inconel 600 (screen) on Inconel 718 Reaction temperature: 770°C Electrolysis current density: 0.15 mA/cm^2	CNTs	Liu et al. (2022)
		146	2022	Electrolyte: Li_2CO_3 Cathode: Stainless Steel 304 or a Muntz brass Anode: Nichrome A/C or Inconel 600/625 or Monel 400 Reaction temperature: 670°C Electrolysis current density (A/cm^2): 0.01–0.4	Mixture of CNMs	Liu et al. (2022)
		147	2022	Electrolyte: Na_2CO_3 + $BaCO_3$ Cathode: a planar brass Anode: a planar Nichrome C Reaction temperature: 770°C. Electrolysis current density (A/cm^2): 0.05/0.1	CNTs	Wang et al. (2023)
		148	2022	Electrolyte: Li_2CO_3 + 0.1wt% Fe_2O_3 Cathode: Muntz brass Anode: Nichrome C Reaction temperature: 750°C Electrolysis current density (A/cm^2): 0.6	CNMs	Liu et al. (2021)
		149	2022	Metal electrocatalysts: Ag, Bi, Co, Zn, and Au Electrolyte: various ternary, binary, and aqueous electrolyte Applied potential: between −1.1 and −1.6 V versus Ag/AgCl Reaction temperature: room temperature	Mixture of CNMs	Nganglumpoon et al. (2022)
		150	2022	Electrolyte: electrodeposited Bi on Sn substrate Catholyte: mixture of PC:[BMIM] BF4:water Anolyte: $KHCO_3$ Reaction temperature: room temperature Applied potential: −1.5 V versus Ag/AgCl	Graphene	Pinthong et al. (2022)
		151	2023	Catalyst: vanadium-based EGaIn (V-EGaIn) Onset potential (−0.97 V versus Ag/ Ag+) Electrolyte: dimethylformamide (DMF) Electrolysis current density (mA/cm^2): −0.4~0	Unknown-structure carbon	Irfan et al. (2023)

TABLE K-1 Continued

Method	Method Detail	No.	Year	Main Reaction Conditions	Product	Reference
		152	2023	Electrolyte: Li_2CO_3 Cathode: nickel foam Anode: a glassy carbon rod Reaction temperature: 780°C Electrolysis current density (A/cm^2): 0.6	Graphite	Thapaliya et al. (2023)
		153	2023	Electrolyte: 0.01 M silver nitrate and 0.6 M of ammonium sulphate Cathode: copper substrate Anode: platinum rod Electrocatalyst: silver Reaction temperature: room temperature Applied potential: −1.6 V versus Ag/AgCl	Mixture of CNMs	Watmanee et al. (2024)
		154	2023	Metal electrocatalysts: Ag, Bi, Co, Zn Electrolyte: ternary electrolyte system containing [BMIM]+[BF_4]−/propylene carbonate/H_2O Cathode: copper substrate; anode: platinum rod Applied potential: between −1.1 and −1.6 V versus Ag/AgCl Reaction temperature: room temperature	Amorphous carbon	Watmanee et al. (2022)
	Electro-thermochemical	155	2016	Cathode: galvanized steel Anode: nickel Electrolyte: mixed ^{13}C lithium carbonate, ^{13}C carbon dioxide, lithium carbonate and lithium oxide Reaction temperature: 750°C	CNTs	Ren and Licht (2016)
		156	2017	Electrolyte: lithium carbonate Reaction temperature: 727°C Electrolysis current density (A/cm^2): 0.1	CNFs	Licht (2017b)
		157	2024	Cathode: stainless steel Anode: titanium Electrolyze: zero-gap MEA Catalyst loaded for the thermochemical reactor: Fe_3Co_6/CeO_2 200 mg Reaction temperature: 450°C Electrolysis current density (A/cm^2): −0.06, −0.1, −0.15, −0.2	CNFs	Xie et al. (2024)
Photochemical	Photochemical	—	—	—	—	—
	Photo-thermochemical	158	2013	Catalyst: 1 g reduced $NiFe_2O_4$ Light source: 300 W UV lamp (365 nm of wavelength)	Mixture of CNMs	Duan et al. (2013)

continued

TABLE K-1 Continued

Method	Method Detail	No.	Year	Main Reaction Conditions	Product	Reference
Plasmachemical	Plasmachemical	159	2006	Dielectric barrier discharge microplasma	Mixture of CNMs	Tomai (2007)
		160	2015	Plasma zone: a stainless-steel rod of inner electrode and a copper foil of outer electrode; plasma power supply: a monopolar pulsed electric generator and a AC high-voltage generator	Unknown-structure carbon	Yap et al. (2015)
		161	2023	Dielectric barrier discharge plasma Catalyst: dispersed liquid metal Ga	Amorphous carbon	Babikir et al. (2023)
	Plasma-thermochemical	—	—	—	—	—

REFERENCES

Abbasloo, S., M. Ojaghi-Ilkhchi, and M. Mozammel. 2019. "Reduction of Carbon Dioxide to Carbon Nanostructures in Molten Salt: The Effect of Electrolyte Composition." *JOM* 71:2103–2111.

Akanuma, K., K. Nishizawa, T. Kodama, M. Tabata, K. Mimori, T. Yoshida, M. Tsuji, and Y. Tamaura. 1993a. "Carbon Dioxide Decomposition into Carbon with the Rhodium-Bearing Magnetite Activated by H2-Reduction." *Journal of Materials Science* 28(4):860–864. https://doi.org/10.1007/BF00400865.

Akanuma, K., M. Tabata, N. Hasegawa, M. Tsuji, Y. Tamaura, Y. Nakahara, and S. Hoshino. 1993b. "Characterization of Carbon Deposited from Carbon Dioxide on Oxygen-Deficient Magnetites." *Journal of Materials Chemistry* 3(9):943–946. https://doi.org/10.1039/JM9930300943.

Allaedini, G., S.M. Tasirin, and P. Aminayi. 2015. "Synthesis of CNTs via Chemical Vapor Deposition of Carbon Dioxide as a Carbon Source in the Presence of NiMgO." *Journal of Alloys and Compounds* 647:809–814.

Allaedini, G., S.M. Tasirin, and P. Aminayi. 2016a. "Synthesis of Graphene Through Direct Decomposition of CO 2 with the Aid of Ni–Ce–Fe Trimetallic Catalyst." *Bulletin of Materials Science* 39:235–240.

Allaedini, G., S.M. Tasirin, and P. Aminayi. 2016b. "Yield Optimization of Nanocarbons Prepared via Chemical Vapor Decomposition of Carbon Dioxide Using Response Surface Methodology. " *Diamond and Related Materials* 66:196–205.

Babikir, A.H., M.G.M. Ekanayake, O. Oloye, J.D. Riches, K. Ostrikov, and A.P. O'Mullane. 2023. "Plasma-Assisted CO_2 Reduction into Nanocarbon in Water Using Sonochemically Dispersed Liquid Gallium." *Advanced Functional Materials* October:2307846. https://doi.org/10.1002/adfm.202307846.

Bagotia, N., H. Mohite, N. Tanaliya, and D.K. Sharma. 2018. "A Comparative Study of Electrical, EMI Shielding and Thermal Properties of Graphene and Multiwalled Carbon Nanotube Filled Polystyrene Nanocomposites." *Polymer Composites* 39(S2):E1041–E1051. https://doi.org/10.1002/pc.24465.

Baik, S., J.H. Park, and J.W. Lee. 2020. "One-Pot Conversion of Carbon Dioxide to CNT-Grafted Graphene Bifunctional for Sulfur Cathode and Thin Interlayer of Li–S Battery." *Electrochimica Acta* 330(January):135264. https://doi.org/10.1016/j.electacta.2019.135264.

Bajaj, N.K., S. Periasamy, R. Singh, Y. Tachibana, T. Daeneke, and K. Chiang. 2021. "The Catalytic Decomposition of Carbon Dioxide on Zinc-Exchanged Y-Zeolite at Low Temperatures." *Journal of Chemical Technology & Biotechnology* 96(9):2675–2680. https://doi.org/10.1002/jctb.6815.

Chakrabarti, A., J. Lu, J.C. Skrabutenas, T. Xu, Z. Xiao, J.A. Maguire, and N.S. Hosmane. 2011. "Conversion of Carbon Dioxide to Few-Layer Graphene." *Journal of Materials Chemistry* 21(26):9491–9493.

Chen, C., and Z. Lou. 2009. "Formation of C60 by Reduction of CO_2." *The Journal of Supercritical Fluids* 50(1):42–45. https://doi.org/10.1016/j.supflu.2009.04.008.

Chen, C.-S., J.-H. Lin, J.-H. Wu, and C.-Y. Chiang. 2009. "Growth of Carbon Nanofibers Synthesized from CO_2 Hydrogenation on a K/Ni/Al2O3 Catalyst." *Catalysis Communications* 11(3):220–224. https://doi.org/10.1016/j.catcom.2009.10.012.

Chen, C.-S., J.-H. Lin, J.H. You, and K.H. Yang. 2010. "Effects of Potassium on Ni-K/Al2O3 Catalysts in the Synthesis of Carbon Nanofibers by Catalytic Hydrogenation of CO_2." *The Journal of Physical Chemistry. A* 114(11):3773–3781. https://doi.org/10.1021/jp904434e.

Chen, C.-S., J.H. You, and C.C. Lin. 2011. "Carbon Nanofibers Synthesized from Carbon Dioxide by Catalytic Hydrogenation on Ni– Na/Al2O3 Catalysts." *The Journal of Physical Chemistry C* 115(5):1464–1473.

Chen, Z., Y. Gu, L. Hu, W. Xiao, X. Mao, H. Zhu, and D. Wang. 2017a. "Synthesis of Nanostructured Graphite via Molten Salt Reduction of CO_2 and SO_2 at a Relatively Low Temperature." *Journal of Materials Chemistry A* 5(39):20603–20607.

Chen, Z., Y. Gu, K. Du, X. Wang, W. Xiao, X. Mao, and D. Wang. 2017b. "Enhanced Electrocatalysis Performance of Amorphous Electrolytic Carbon from CO_2 for Oxygen Reduction by Surface Modification in Molten Salt." *Electrochimica Acta* 253(November):248–256. https://doi.org/10.1016/j.electacta.2017.09.053.

Chen, Z., B. Deng, K. Du, X. Mao, H. Zhu, W. Xiao, and D. Wang. 2017c. "Flue-Gas-Derived Sulfur-Doped Carbon with Enhanced Capacitance." *Advanced Sustainable Systems* 1(6):1700047. https://doi.org/10.1002/adsu.201700047.

Colson, M., L. Alvarez, S.M. Soto, S.H. Joo, K. Li, A. Lupini, K. Nawaz, et al. 2022. "A Novel Sustainable Process for Multilayer Graphene Synthesis Using CO_2 from Ambient Air." *Materials* 15(17):5894. https://doi.org/10.3390/ma15175894.

Deng, B., X. Mao, W. Xiao, and D. Wang. 2017. "Microbubble Effect-Assisted Electrolytic Synthesis of Hollow Carbon Spheres from CO_2." *Journal of Materials Chemistry A* 5(25):12822–12827.

Deng, B., J. Tang, M. Gao, X. Mao, H. Zhu, W. Xiao, and D. Wang. 2018. "Electrolytic Synthesis of Carbon from the Captured CO_2 in Molten LiCl–KCl–CaCO3: Critical Roles of Electrode Potential and Temperature for Hollow Structure and Lithium Storage Performance." *Electrochimica Acta* 259(January):975–985. https://doi.org/10.1016/j.electacta.2017.11.025.

Douglas, A., R. Carter, N. Muralidharan, L. Oakes, and C.L. Pint. 2017a. "Iron Catalyzed Growth of Crystalline Multi-Walled Carbon Nanotubes from Ambient Carbon Dioxide Mediated by Molten Carbonates." *Carbon* 116(May):572–578. https://doi.org/10.1016/j.carbon.2017.02.032.

Douglas, A., N. Muralidharan, R. Carter, and C.L. Pint. 2017b. "Sustainable Capture and Conversion of Carbon Dioxide into Valuable Multiwalled Carbon Nanotubes Using Metal Scrap Materials." *ACS Sustainable Chemistry & Engineering* 5(8):7104–7110. https://doi.org/10.1021/acssuschemeng.7b01314.

Douglas, A., R. Carter, M. Li, and C.L. Pint. 2018. "Toward Small-Diameter Carbon Nanotubes Synthesized from Captured Carbon Dioxide: Critical Role of Catalyst Coarsening." *ACS Applied Materials & Interfaces* 10(22):19010–19018.

Driscoll, J.A. 1978. "A Demonstration of Burning Magnesium and Dry Ice." *Journal of Chemical Education* 55(7):450.

Duan, Y.Q., T. Du, X.W. Wang, F.S. Cai, and Z.H. Yuan. 2013. "Photoassisted CO_2 Conversion to Carbon by Reduced $NiFe_2O_4$." *Advanced Materials Research* 726–731:420–424. https://doi.org/10.4028/www.scientific.net/AMR.726-731.420.

Ehrensberger, K., R. Palumbo, C. Larson, and A. Steinfeld. 1997. "Production of Carbon from Carbon Dioxide with Iron Oxides and High-Temperature Solar Energy." *Industrial & Engineering Chemistry Research* 36(3):645–648. https://doi.org/10.1021/ie950780y.

Esrafilzadeh, D., A. Zavabeti, R. Jalili, P. Atkin, J. Choi, B.J. Carey, R. Brkljača, et al. 2019. "Room Temperature CO_2 Reduction to Solid Carbon Species on Liquid Metals Featuring Atomically Thin Ceria Interfaces." *Nature Communications* 10(1):865. https://doi.org/10.1038/s41467-019-08824-8.

Fu, M.S., L.S. Chen, and S.Y. Chen. 2005. "Preparation, Structure of Doped Ferrite and Its Performance of Decomposition of Carbon Dioxide to Carbon." *Chemical Journal of Chinese Universities-Chinese* 26(12):2279–2283.

Fu, M., L. Chen, J. Li, and S. Chen. 2007. "Effect of Reduction Condition on Structure Stability of NiFe2O4–δ and Its Catalytic Performance of CO_2 Decomposition." *Journal of Fuel Chemistry and Technology* 35(4):431–435. https://doi.org/10.1016/S1872-5813(07)60027-9.

Gálvez, M.E., P.G. Loutzenhiser, I. Hischier, and A. Steinfeld. 2008. "CO $_2$ Splitting via Two-Step Solar Thermochemical Cycles with Zn/ZnO and FeO/Fe_3O_4 Redox Reactions: Thermodynamic Analysis." *Energy & Fuels* 22(5):3544–3550. https://doi.org/10.1021/ef800230b.

Gong, P., C. Tang, B. Wang, T. Xiao, H. Zhu, Q. Li, and Z. Sun. 2022. "Precise CO_2 Reduction for Bilayer Graphene." *ACS Central Science* 8(3):394–401. https://doi.org/10.1021/acscentsci.1c01578.

Guo, H., B. Kang, and A. Manivannan. 2013. "Carbon Dioxide Decomposition and Oxygen Generation Via SOEC." *ECS Transactions* 50(49):129

Hu, J., Z. Guo, W. Chu, L. Li, and T. Lin. 2015. "Carbon Dioxide Catalytic Conversion to Nano Carbon Material on the Iron–Nickel Catalysts Using CVD-IP Method." *Journal of Energy Chemistry* 24(5):620–625. https://doi.org/10.1016/j.jechem.2015.09.006.

Hu, L., Y. Song, S. Jiao, Y. Liu, J. Ge, H. Jiao, J. Zhu, J. Wang, H. Zhu, and D.J. Fray. 2016. "Direct Conversion of Greenhouse Gas CO_2 into Graphene via Molten Salts Electrolysis." *ChemSusChem* 9(6):588–594.

Hu, L., Y. Song, J. Ge, J. Zhu, Z. Han, and S. Jiao. 2017. "Electrochemical Deposition of Carbon Nanotubes from CO_2 in $CaCl_2$–NaCl-Based Melts." *Journal of Materials Chemistry A* 5(13):6219–6225. https://doi.org/10.1039/C7TA00258K.

Hu, L., Z. Yang, W. Yang, M. Hu, and S. Jiao. 2019. "The Synthesis of Sulfur-Doped Graphite Nanostructures by Direct Electrochemical Conversion of CO_2 in $CaCl_2NaCl$ CaO Li_2SO_4." *Carbon* 144(April):805–814. https://doi.org/10.1016/j.carbon.2018.12.049.

Hwang, C.-S., and N.-C. Wang. 2004. "Preparation and Characteristics of Ferrite Catalysts for Reduction of CO_2." *Materials Chemistry and Physics* 88(2–3):258–263.

Irfan, M., K. Zuraiqi, C.K. Nguyen, T.C. Le, F. Jabbar, M. Ameen, C.J. Parker, et al. 2023. "Liquid Metal-Based Catalysts for the Electroreduction of Carbon Dioxide into Solid Carbon." *Journal of Materials Chemistry A* 11(27):14990–14996. https://doi.org/10.1039/D3TA01379K.

Ishihara, T., T. Fujita, Y. Mizuhara, and Y. Takita. 1991. "Fixation of Carbon Dioxide to Carbon by Catalytic Reduction Over Metal Oxides." *Chemistry Letters* 20(12):2237–2240. https://doi.org/10.1246/cl.1991.2237.

Jiaowen, S., D.W. Kim, S.B. Kim, and Y.M. Jo. 2016. "CO_2 Decomposition Using Metal Ferrites Prepared by Co-Precipitation Method." *Korean Journal of Chemical Engineering* 33(11):3162–3168. https://doi.org/10.1007/s11814-016-0192-5.

Jo, C., Y. Mun, J. Lee, E. Lim, S. Kim, and J. Lee. 2019. "Carbon Dioxide to Solid Carbon at the Surface of Iron Nanoparticle: Hollow Nanocarbons for Sodium Ion Battery Anode Application." *Journal of CO_2 Utilization* 34:588–595.

Johnson, M., J. Ren, M. Lefler, G. Licht, J. Vicini, X. Liu, and S. Licht. 2017a. "Carbon Nanotube Wools Made Directly from CO_2 by Molten Electrolysis: Value Driven Pathways to Carbon Dioxide Greenhouse Gas Mitigation." *Materials Today Energy* 5(September):230–236. https://doi.org/10.1016/j.mtener.2017.07.003.

Johnson, M., J. Ren, M. Lefler, G. Licht, J. Vicini, and S. Licht. 2017b. "Data on SEM, TEM and Raman Spectra of Doped, and Wool Carbon Nanotubes Made Directly from CO_2 by Molten Electrolysis." *Data in Brief* 14(October):592–606. https://doi.org/10.1016/j.dib.2017.08.013.

Kato, H., T. Kodama, M. Tsuji, Y. Tamaura, and S.G. Chang. 1994. "Decomposition of Carbon Dioxide to Carbon by Hydrogen-Reduced Ni(II)-Bearing Ferrite." *Journal of Materials Science* 29(21):5689–5692. https://doi.org/10.1007/BF00349965.

Khedr, M.H., A.A. Omar, and S.A. Abdel-Moaty. 2006. "Reduction of Carbon Dioxide into Carbon by Freshly Reduced CoFe2O4 Nanoparticles." *Materials Science and Engineering: A* 1–2:26–33.

Khedr, M.H., M. Bahgat, M.I. Nasr, and E.K. Sedeek. 2007. "CO_2 Decomposition Over Freshly Reduced Nanocrystalline Fe2O3." *Colloids and Surfaces A: Physicochemical and Engineering Aspects* 302(1):517–524. https://doi.org/10.1016/j.colsurfa.2007.03.024.

Kim, J.-S., and J.-R. Ahn. 2001. "Characterization of Wet Processed (Ni, Zn)-Ferrites for CO_2 Decomposition." *Journal of Materials Science* 36:4813–4816.

Kim, J.-S., J.-R. Ahn, C.W. Lee, Y. Murakami, and D. Shindo. 2001. "Morphological Properties of Ultra-Fine (Ni, Zn)-Ferrites and Their Ability to Decompose CO_2." *Journal of Materials Chemistry* 11(12):3373–3376.

Kim, S.-H., J.T. Jang, J. Sim, J.-H. Lee, S.-C. Nam, and C.Y. Park. 2019. "Carbon Dioxide Decomposition Using $SrFeCo_{0.5}O_x$, a Nonperovskite-Type Metal Oxide." *Journal of CO_2 Utilization* 34:709–715.

Kim, G.M., W.Y. Choi, J.H. Park, S.J. Jeong, J.-E. Hong, W. Jung, and J.W. Lee. 2020a. "Electrically Conductive Oxidation-Resistant Boron-Coated Carbon Nanotubes Derived from Atmospheric CO_2 for Use at High Temperature." *ACS Applied Nano Materials* 3(9):8592–8597. https://doi.org/10.1021/acsanm.0c01909.

Kim, G.M., W.-G. Lim, D. Kang, J.H. Park, H. Lee, J. Lee, and J.W. Lee. 2020b. "Transformation of Carbon Dioxide into Carbon Nanotubes for Enhanced Ion Transport and Energy Storage." *Nanoscale* 12(14):7822–7833. https://doi.org/10.1039/C9NR10552B.

Kodama, T., K. Tominaga, M. Tabata, T. Yoshida, and Y. Tamaura. 1992. "Decomposition of Carbon Dioxide to Carbon with Active Wustite at 300°C." *Journal of the American Ceramic Society* 75(5):1287–1289. https://doi.org/10.1111/j.1151-2916.1992.tb05574.x.

Kodama, T., T. Sano, S.-G. Chang, M. Tsuji, and Y. Tamaura. 1994a. "Decomposition of O_2 to Carbon by H2-Activated Ni(II)- and Co(II)-Bearing Ferrites At 300°C." *MRS Online Proceedings Library* 344(1):63–68. https://doi.org/10.1557/PROC-344-63.

Kodama, T., H. Kato, S. G. Chang, N. Hasegawa, M. Tsuji, and Y. Tamaura. 1994b. "Decomposition of CO_2 to Carbon by H_2-Reduced Ni(II)- and Co(II)-Bearing Ferrites at 300°C." *Journal of Materials Research* 9(2):462–467. https://doi.org/10.1557/JMR.1994.0462.

Kodama, T., M. Tabata, T. Sano, M. Tsuji, and Y. Tamaura. 1995a. "XRD and Mössbauer Studies on Oxygen-Deficient Ni(II)-Bearing Ferrite with a High Reactivity for CO_2 Decomposition to Carbon." *Journal of Solid State Chemistry* 120(1):64–69. https://doi.org/10.1006/jssc.1995.1377.

Kodama, T., Y. Wada, T. Yamamoto, M. Tsuji, and Y. Tamaura. 1995b. "CO_2 Decomposition to Carbon by Ultrafine Ni(II)-Bearing Ferrite at 300°C." *Materials Research Bulletin* 30(8):1039–1048. https://doi.org/10.1016/0025-5408(95)00077-1.

Kodama, T., T. Sano, T. Yoshida, M. Tsuji, and Y. Tamaura. 1995c. "CO_2 Decomposition to Carbon with Ferrite-Derived Metallic Phase at 300°C." *Carbon* 33(10):1443–1447. https://doi.org/10.1016/0008-6223(95)00094-T.
Komarneni, S., M. Tsuji, Y. Wada, and Y. Tamaura. 1997. "Nanophase Ferrites for CO_2 Greenhouse Gas Decomposition." *Journal of Materials Chemistry* 7(12):2339–2340.
Lau, J., G. Dey, and S. Licht. 2016. "Thermodynamic Assessment of CO_2 to Carbon Nanofiber Transformation for Carbon Sequestration in a Combined Cycle Gas or a Coal Power Plant." *Energy Conversion and Management* 122:400–410.
Li, S., Z. He, Y. Zheng, and C. Chen. 2015. "Absorption and Decomposition of CO_2 by Active Ferrites Prepared by Atmospheric Plasma Spraying." *Journal of Thermal Spray Technology* 24(8):1574–1578. https://doi.org/10.1007/s11666-015-0333-0.
Li, X., H. Shi, X. Wang, X. Hu, C. Xu, and W. Shao. 2022. "Direct Synthesis and Modification of Graphene in Mg Melt by Converting CO_2: A Novel Route to Achieve High Strength and Stiffness in Graphene/Mg Composites." *Carbon* 186:632–643.
Li, X., X. Wang, X. Hu, C. Xu, W. Shao, and K. Wu. 2023. "Direct Conversion of CO_2 to Graphene via Vapor–Liquid Reaction for Magnesium Matrix Composites with Structural and Functional Properties." *Journal of Magnesium and Alloys* 11(4):1206–1212.
Li, Z., D. Yuan, H. Wu, W. Li, and D. Gu. 2018. "A Novel Route to Synthesize Carbon Spheres and Carbon Nanotubes from Carbon Dioxide in a Molten Carbonate Electrolyzer." *Inorganic Chemistry Frontiers* 5(1):208–216.
Li, Z., G. Wang, W. Zhang, Z. Qiao, and H. Wu. 2019. "Carbon Nanotubes Synthesis from CO_2 Based on the Molten Salts Electrochemistry: Effect of Alkaline Earth Carbonate Additives on the Diameter of the Carbon Nanotubes." *Journal of the Electrochemical Society* 166(10):D415. https://doi.org/10.1149/2.0861910jes.
Liang, C., L. Pan, S. Liang, Y. Xia, Z. Liang, Y. Gan, H. Huang, J. Zhang, and W. Zhang. 2019. "Ultraefficient Conversion of CO_2 into Morphology-Controlled Nanocarbons: A Sustainable Strategy Toward Greenhouse Gas Utilization." *Small* 15(33):1902249. https://doi.org/10.1002/smll.201902249.
Licht, S. 2017a. "Co-production of Cement and Carbon Nanotubes with a Carbon Negative Footprint." *Journal of CO_2 Utilization* 18:378–389.
Licht, S. 2017b. "Carbon Dioxide to Carbon Nanotube Scale-Up." arXiv preprint arXiv:1710.07246.
Licht, S., A. Douglas, J. Ren, R. Carter, M. Lefler, and C.L. Pint. 2016. "Carbon Nanotubes Produced from Ambient Carbon Dioxide for Environmentally Sustainable Lithium-Ion and Sodium-Ion Battery Anodes." *ACS Central Science* 2(3):162–168.
Licht, S., X. Liu, G. Licht, X. Wang, A. Swesi, and Y. Chan. 2019. "Amplified CO_2 Reduction of Greenhouse Gas Emissions with C2CNT Carbon Nanotube Composites." *Materials Today Sustainability* 6:100023. https://doi.org/10.1016/j.mtsust.2019.100023.
Lin, K.-S., A.K. Adhikari, Z.-Y. Tsai, Y.-P. Chen, T.-T. Chien, and H.-B. Tsai. 2011. "Synthesis and Characterization of Nickel Ferrite Nanocatalysts for CO_2 Decomposition." *Catalysis Today* 174(1):88–96.
Lin, K.-S., A.K. Adhikari, C.-Y. Wang, P.-J. Hsu, and H.Y. Chan. 2013. "Synthesis and Characterization of Nickel and Zinc Ferrite Nanocatalysts for Decomposition of CO_2 Greenhouse Effect Gas." *Journal of Nanoscience and Nanotechnology* 13(4):2538–2548. https://doi.org/10.1166/jnn.2013.7427.
Lin, K.-S., C.-Y. Tang, N.V. Mdlovu, C.J. Chang, C.-L. Chiang, and Z.-M. Cai. 2022. "Preparation and Characterization of Ni/Al_2O_3 for Carbon Nanofiber Fabrication from CO_2 Hydrogenation." *Catalysis Today, 4th International Conference on Catalysis and Chemical Engineering* 388–389(April):341–350. https://doi.org/10.1016/j.cattod.2020.06.008.
Linshen, C., C. Songying, and L. Guanglie. 2006. "Study the Structure Stability of NiFe 2–x Crx O4 (x= 0, 0.08) During H2/CO2 Cycle Reaction." *Journal of Materials Science* 41:6465–6469.
Liu, X., J. Ren, G. Licht, X. Wang, and S. Licht. 2019. "Carbon Nano-Onions Made Directly from CO_2 by Molten Electrolysis for Greenhouse Gas Mitigation." *Advanced Sustainable Systems* 3(10):1900056. https://doi.org/10.1002/adsu.201900056.
Liu, X., X. Wang, G. Licht, and S. Licht. 2020. "Transformation of the Greenhouse Gas Carbon Dioxide to Graphene." *Journal of CO_2 Utilization* 36:288–294.
Liu, X., G. Licht, and S. Licht. 2021. "The Green Synthesis of Exceptional Braided, Helical Carbon Nanotubes and Nanospiral Platelets Made Directly from CO_2." *Materials Today Chemistry* 22:100529.
Liu, X., G. Licht, X. Wang, and S. Licht. 2022. "Controlled Transition Metal Nucleated Growth of Carbon Nanotubes by Molten Electrolysis of CO_2." *Catalysts* 12(2):137. https://doi.org/10.3390/catal12020137.
Lou, Z., Q. Chen, W. Wang, and Y. Zhang. 2003. "Synthesis of Carbon Nanotubes by Reduction of Carbon Dioxide with Metallic Lithium." *Carbon* 41(15):3063–3067. https://doi.org/10.1016/S0008-6223(03)00335-X.
Lou, Z., C. Chen, H. Huang, and D. Zhao. 2006. "Fabrication of Y-Junction Carbon Nanotubes by Reduction of Carbon Dioxide with Sodium Borohydride." *Diamond and Related Materials* 15(10):1540–1543. https://doi.org/10.1016/j.diamond.2005.12.044.
Luchetta, C., E.C. Oliveira Munsignatti, and H.O. Pastore. 2021. "CO_2 Metallothermal Reduction to Graphene: The Influence of Zn." *Frontiers in Chemical Engineering* 3(August):707855. https://doi.org/10.3389/fceng.2021.707855.

Luo, B., H. Liu, L. Jiang, L. Jiang, D. Geng, B. Wu, W. Hu, Y. Liu, and G. Yu. 2013. "Synthesis and Morphology Transformation of Single-Crystal Graphene Domains Based on Activated Carbon Dioxide by Chemical Vapor Deposition." *Journal of Materials Chemistry C* 1(17):2990–2995. https://doi.org/10.1039/C3TC30124A.

Ma, L.J., L.S. Chen, and S.Y. Chen. 2007a. "Study on the Cycle Decomposition of CO_2 Over NiCr0. 08Fe1. 92O4 and the Microstructure of Products." *Materials Chemistry and Physics* 105(1):122–126.

Ma, L.J., L.S. Chen, and S.Y. Chen. 2007b. "Study of the CO_2 Decomposition Over Doped Ni-Ferrites." *Journal of Physics and Chemistry of Solids* 68(7):1330–1335.

Ma, L.J., L.S. Chen, and S.Y. Chen. 2009a. "Studies on Redox H_2–CO_2 Cycle on CoCrxFe2–xO4." *Solid State Sciences* 11(1):176–181. https://doi.org/10.1016/j.solidstatesciences.2008.05.008.

Ma, L.J., L.S. Chen, and S.Y. Chen. 2009b. "Study on the Characteristics and Activity of Ni–Cu–Zn Ferrite for Decomposition of CO_2." *Materials Chemistry and Physics* 114(2–3):692–696.

Ma, L.J., R. Wu, H. Liu, W. Xu, L.S. Chen, and S.Y. Chen. 2011. "Studies on CO_2 Decomposition Over H2-Reduced MFe2O4 (M = Ni, Cu, Co, Zn)." *Solid State Sciences* 13(12):2172–2176. https://doi.org/10.1016/j.solidstatesciences.2011.10.003.

Moghaddam, M.B., E.K. Goharshadi, M.H. Entezari, and P. Nancarrow. 2013. "Preparation, Characterization, and Rheological Properties of Graphene–Glycerol Nanofluids." *Chemical Engineering Journal* 231(September):365–372. https://doi.org/10.1016/j.cej.2013.07.006.

Molina-Jirón, C., M.R. Chellali, C.N.S. Kumar, C. Kübel, L. Velasco, H. Hahn, E. Moreno–Pineda, and M. Ruben. 2019. "Direct Conversion of CO_2 to Multi-Layer Graphene Using Cu–Pd Alloys." *ChemSusChem* 12(15):3509–3514. https://doi.org/10.1002/cssc.201901404.

Motiei, M., Y. Rosenfeld Hacohen, J. Calderon–Moreno, and A. Gedanken. 2001. "Preparing Carbon Nanotubes and Nested Fullerenes from Supercritical CO_2 by a Chemical Reaction." *Journal of the American Chemical Society* 123(35):8624–8625. https://doi.org/10.1021/ja015859a.

Moyer, K., M. Zohair, J. Eaves-Rathert, A. Douglas, and C.L. Pint. 2020. "Oxygen Evolution Activity Limits the Nucleation and Catalytic Growth of Carbon Nanotubes from Carbon Dioxide Electrolysis via Molten Carbonates." *Carbon* 165:90–99.

Mulmi, S., H. Chen, A. Hassan, J.F. Marco, F.J. Berry, F. Sharif, P.R. Slater, E.P.L. Roberts, S. Adams, and V. Thangadurai. 2017. "Thermochemical CO_2 Splitting Using Double Perovskite-Type Ba2Ca0.66Nb1.34-xFexO6-δ." *Journal of Materials Chemistry A* 5(15):6874–6883. https://doi.org/10.1039/C6TA10285A.

Nakabayashi, K., Y. Matsuo, K. Isomoto, K. Teshima, T. Ayukawa, H. Shimanoe, T. Mashio, I. Mochida, J. Miyawaki, and S.-H. Yoon. 2020. "Establishment of Innovative Carbon Nanofiber Synthesis Technology Utilizing Carbon Dioxide." *ACS Sustainable Chemistry & Engineering* 8(9):3844–3852. https://doi.org/10.1021/acssuschemeng.9b07253.

Nganglumpoon, R., S. Watmanee, T. Teerawatananond, P. Pinthong, K. Poolboon, N. Hongrutai, D.N. Tungasmita, et al. 2022. "Growing 3D-Nanostructured Carbon Allotropes from CO_2 at Room Temperature Under the Dynamic CO_2 Electrochemical Reduction Environment." *Carbon* 187(February):241–255. https://doi.org/10.1016/j.carbon.2021.11.011.

Otake, K., H. Kinoshita, T. Kikuchi, and R.O. Suzuki. 2013. "CO_2 Gas Decomposition to Carbon by Electro-Reduction in Molten Salts." *Electrochimica Acta* 100(June):293–299. https://doi.org/10.1016/j.electacta.2013.02.076.

Ozawa, S., H. Matsuno, A. Fujibayashi, T. Uchiyama, T. Wakamatsu, N. Sakaguchi, and R.O. Suzuki. 2016. "Influence of Gas Injection Pipe on CO_2 Decomposition by $CaCl_2$–CaO Molten Salt and ZrO_2 Solid Electrolysis." *ISIJ International* 56(11):2093–2099. https://doi.org/10.2355/isijinternational.ISIJINT-2016-179.

Pinthong, P., S. Phupaichitkun, S. Watmanee, R. Nganglumpoon, D.N. Tungasmita, S. Tungasmita, Y. Boonyongmaneerat, N. Promphet, N. Rodthongkum, and J. Panpranot. 2022. "Room Temperature Nanographene Production via CO_2 Electrochemical Reduction on the Electrodeposited Bi on Sn Substrate." *Nanomaterials* 12(19):3389. https://doi.org/10.3390/nano12193389.

Poh, H.L., Z. Sofer, J. Luxa, and M. Pumera. 2014. "Transition Metal-Depleted Graphenes for Electrochemical Applications via Reduction of CO_2 by Lithium." *Small* 10(8):1529–1535. https://doi.org/10.1002/smll.201303002.

Ren, J., and S. Licht. 2016. "Tracking Airborne CO_2 Mitigation and Low Cost Transformation into Valuable Carbon Nanotubes." *Scientific Reports* 6(1):27760. https://doi.org/10.1038/srep27760.

Ren, J., F.-F. Li, J. Lau, L. González-Urbina, and S. Licht. 2015a. "One-Pot Synthesis of Carbon Nanofibers from CO_2." *Nano Letters* 15(9):6142–6148.

Ren, J., J. Lau, M. Lefler, and S. Licht. 2015b. "The Minimum Electrolytic Energy Needed to Convert Carbon Dioxide to Carbon by Electrolysis in Carbonate Melts." *The Journal of Physical Chemistry C* 119(41):23342–23349. https://doi.org/10.1021/acs.jpcc.5b07026.

Ren, J., M. Johnson, R. Singhal, and S. Licht. 2017. "Transformation of the Greenhouse Gas CO_2 by Molten Electrolysis into a Wide Controlled Selection of Carbon Nanotubes." *Journal of CO_2 Utilization* 18(March):335–344. https://doi.org/10.1016/j.jcou.2017.02.005.

Samiee, S., and E.K. Goharshadi. 2014. "Graphene Nanosheets as Efficient Adsorbent for an Azo Dye Removal: Kinetic and Thermodynamic Studies." *Journal of Nanoparticle Research* 16:1–16.

Sano, T., H. Ono., H. Amano, M. Tsuji, and Y. Tamaura. 1998. "CO_2 Decomposition with Ferrite for Utilizing Carbon as Solar H2 Energy Carrier-CO_2 Decomposition Mechanism with Metal Substituted Ferrites." *Journal of the Magnetics Society of Japan* 22(S1):423–424.

Seekaew, Y., N. Tammanoon, A. Tuantranont, T. Lomas, A. Wisitsoraat, and C. Wongchoosuk. 2022. "Conversion of Carbon Dioxide into Chemical Vapor Deposited Graphene with Controllable Number of Layers via Hydrogen Plasma Pre-Treatment." *Membranes* 12(8):796. https://doi.org/10.3390/membranes12080796.

Sharma, D.K., and N. Bagotia. 2021. "Production of Graphene and Carbon Nanotubes Using Low Cost Carbon-Based Raw Materials and Their Utilization in the Production of Polycarbonate/Ethylene Methyl Acrylate-Based Nanocomposites." *Indian Journal of Engineering and Materials Sciences* 27(6):1127–1135.

Shin, H.C., J.H. Oh, J.C. Lee, K.D. Jung, and S.C. Choi. 2004. "CO_2 Decomposition with Nano-Size Ferrite Prepared by Irradiation of Ultrasonic Wave." *Journal of the Ceramic Society of Japan, Supplement 112-1* 112(5):S1373.

Sim, J., S.-H. Kim, J.-Y. Kim, K. Bong Lee, S.-C. Nam, and C.Y. Park. 2020. "Enhanced Carbon Dioxide Decomposition Using Activated SrFeO3–δ." *Catalysts* 10(11):1278.

Smith, K., R. Parrish, W. Wei, Y. Liu, T. Li, Y.H. Hu, and H. Xiong. 2016. "Disordered 3D Multi-Layer Graphene Anode Material from CO_2 for Sodium-Ion Batteries." *ChemSusChem* 9(12):1397–1402. https://doi.org/10.1002/cssc.201600117.

Strudwick, A.J., N.E. Weber, M.G. Schwab, M. Kettner, R.T. Weitz, J.R. Wünsch, K. Müllen, and H. Sachdev. 2015. "Chemical Vapor Deposition of High Quality Graphene Films from Carbon Dioxide Atmospheres." *ACS Nano* 9(1):31–42.

Sun, Z., and Y.H. Hu. 2020. "3D Graphene Materials from the Reduction of CO_2." *Accounts of Materials Research* 2(1):48–58. https://doi.org/10.1021/accountsmr.0c00069.

Tabata, M., Y. Nishida, T. Kodama, K. Mimori, T. Yoshida, and Y. Tamaura. 1993a. "CO_2 Decomposition with Oxygen-Deficient Mn(II) Ferrite." *Journal of Materials Science* 28(4):971–974. https://doi.org/10.1007/BF00400881.

Tabata, M., K. Akanuma, K. Nishizawa, K. Mimori, T. Yoshida, M. Tsuji, and Y. Tamaura. 1993b. "Reactivity of Oxygen-Deficient Mn(II)-Bearing Ferrites (MnxFe3-xO4-δ, O≤x≤1, Δ>0) Toward CO_2 Decomposition to Carbon." *Journal of Materials Science* 28(24):6753–6760. https://doi.org/10.1007/BF00356427.

Tabata, M., K. Akanuma, T. Togawa, M. Tsuji, and Y. Tamaura. 1994a. "Mössbauer Study of Oxygen-Deficient ZnII-Bearing Ferrites (ZnxFe3–xO4–δ, 0 ≤x≤ 1) and Their Reactivity Toward CO_2 Decomposition to Carbon." *Journal of the Chemical Society, Faraday Transactions* 90(8):1171–1175. https://doi.org/10.1039/FT9949001171.

Tabata, M., H. Kato, M. Tsuji, and Y. Tamaura. 1994b. "Decomposition of CO_2 to Carbon Using Oxygen-Deficient Zn(II)-Bearing Ferrite." *MRS Proceedings* 344:157. https://doi.org/10.1557/PROC-344-157.

Tamaura, Y., and K. Nishizawa. 1992. "CO2 Decomposition into C and Conversion into CH4 Using the H2-Reduced Magnetite." *Energy Conversion and Management* 33(5–8):573–577. https://doi.org/10.1016/0196-8904(92)90058-5.

Tamaura, Y., and M. Tahata. 1990. "Complete Reduction of Carbon Dioxide to Carbon Using Cation-Excess Magnetite." *Nature* 346(6281):255–256.

Thapaliya, B.P., A.S. Ivanov, H.-Y. Chao, M. Lamm, M. Chi, H.M. Meyer, X.-G. Sun, T. Aytug, S. Dai, and S.M. Mahurin. 2023. "Molten Salt Electrochemical Upcycling of CO_2 to Graphite for High Performance Battery Anodes." *Carbon* 212(August):118151. https://doi.org/10.1016/j.carbon.2023.118151.

Tomai, T., K. Katahira, H. Kubo, Y. Shimizu, T. Sasaki, N. Koshizaki, and K. Terashima. 2007. "Carbon Materials Syntheses Using Dielectric Barrier Discharge Microplasma in Supercritical Carbon Dioxide Environments." *Journal of Supercritical Fluids* 41(3):404–411. https://doi.org/10.1016/j.supflu.2006.12.003.

Tsuji, M., K. Nishizawa, T. Yoshida, and Y. Tamaura. 1994. "Methanation Reactivity of Carbon Deposited Directly from CO_2 on to the Oxygen-Deficient Magnetite." *Journal of Materials Science* 29(20):5481–584. https://doi.org/10.1007/BF01171565.

Tsuji, M., Y. Wada, T. Yamamoto, T. Sano, and Y. Tamaura. 1996a. "CO 2 Decomposition by Metallic Phase on Oxygen-Deficient Ni (II)-Bearing Ferrite." *Journal of Materials Science Letters* 15:156–158.

Tsuji, M., T. Kodama, T. Yoshida, Y. Kitayama, and Y. Tamaura. 1996b. "Preparation and CO_2 Methanation Activity of an Ultrafine Ni (II) Ferrite Catalyst." *Journal of Catalysis* 164(2):315–321.

Wada, Y., T. Yoshida, M. Tsuji, and Y. Tamaura. 1995. "CO_2-Decomposition Capacity of H_2-Reduced Ferrites." *Energy Conversion and Management, Proceedings of the Second International Conference on Carbon Dioxide Removal* 36(6):641–644. https://doi.org/10.1016/0196-8904(95)00087-T.

Wang, C., F. Li, H. Qu, Y. Wang, X. Yi, Y. Qiu, Z. Zou, Y. Luo, and B. Yu. 2015. "Fabrication of Three Dimensional Carbon Nanotube Foam by Direct Conversion Carbon Dioxide and Its Application in Supercapacitor." *Electrochimica Acta* 158(March):35–41. https://doi.org/10.1016/j.electacta.2015.01.112.

Wang, J.-F., K.-X. Wang, J.-Q. Wang, L. Li, and J.-S. Chen. 2012. "Decomposition of CO_2 to Carbon and Oxygen Under Mild Conditions Over a Zinc-Modified Zeolite." *Chemical Communications* 48(17):2325. https://doi.org/10.1039/c2cc17382d.

Wang, X., X. Liu, G. Licht, B. Wang, and S. Licht. 2019. "Exploration of Alkali Cation Variation on the Synthesis of Carbon Nanotubes by Electrolysis of CO_2 in Molten Carbonates." *Journal of CO_2 Utilization* 34(December):303–312. https://doi.org/10.1016/j.jcou.2019.07.007.

Wang, X., G. Licht, X. Liu, and S. Licht. 2020a. "One Pot Facile Transformation of CO_2 to an Unusual 3-D Nano-Scaffold Morphology of Carbon." *Scientific Reports* 10(1):21518. https://doi.org/10.1038/s41598-020-78258-6.

Wang, X., X. Liu, G. Licht, and S. Licht. 2020b. "Calcium Metaborate Induced Thin Walled Carbon Nanotube Syntheses from CO_2 by Molten Carbonate Electrolysis." *Scientific Reports* 10(1):15146.

Wang, L., J. Deng, J. Deng, Y. Fei, Y. Fang, and Y.H. Hu. 2020c. "Ultra-Fast and Ultra-Long-Life Li Ion Batteries with 3D Surface-Porous Graphene Anodes Synthesized from CO_2." *Journal of Materials Chemistry A* 8(26):13385–13392. https://doi.org/10.1039/D0TA03606D.

Wang, X., F. Sharif, X. Liu, G. Licht, M. Lefler, and S. Licht. 2020d. "Magnetic Carbon Nanotubes: Carbide Nucleated Electrochemical Growth of Ferromagnetic CNTs from CO_2." *Journal of CO_2 Utilization* 40. https://doi.org/10.1016/j.jcou.2020.101218.

Wang, X., G. Licht, and S. Licht. 2021a. "Green and Scalable Separation and Purification of Carbon Materials in Molten Salt by Efficient High-Temperature Press Filtration." *Separation and Purification Technology* 255:117719.

Wang, Y., W. Wang, J. Xie, C.-H. Wang, Y.-W. Yang, and Y.-C. Lu. 2021b. "Electrochemical Reduction of CO_2 in Ionic Liquid: Mechanistic Study of Li–CO_2 Batteries via *in Situ* Ambient Pressure X-Ray Photoelectron Spectroscopy." *Nano Energy* 83(May):105830. https://doi.org/10.1016/j.nanoen.2021.105830.

Wang, R., L. Sun, X. Zhu, W. Ge, H. Li, Z. Li, H. Zhang et al. 2023. "Carbon Nanotube-Based Strain Sensors: Structures, Fabrication, and Applications." *Advanced Materials Technologies* 8(1):2200855.

Wang, J., Z. Su, Y. Zhang, and T. Jiang. 2024. "Synthesis, Characterization, and Catalytic Reduction CO_2 Properties of Spinel Nano-Mn_XFe_3-XO_4: Effect of X on Mn^{3+}/Mn^{2+} Cation Occupancies and CO_2 Reduction Mechanism." *Chemical Engineering Journal* 479(January):147926. https://doi.org/10.1016/j.cej.2023.147926.

Watanabe, T., and T. Ohba. 2021. "Low-Temperature CO_2 Thermal Reduction to Graphitic and Diamond-Like Carbons Using Perovskite-Type Titanium Nanoceramics by Quasi-High-Pressure Reactions." *ACS Sustainable Chemistry & Engineering* 9.10:3860–3873.

Watmanee, S., R. Nganglumpoon, N. Hongrutai, P. Pinthong, P. Praserthdam, S. Wannapaiboon, P.Á. Szilágyi, Y. Morikawa, and J. Panpranot. 2022. "Formation and Growth Characteristics of Nanostructured Carbon Films on Nascent Ag Clusters During Room-Temperature Electrochemical CO_2 Reduction." *Nanoscale Advances* 4(10):2255–2267. https://doi.org/10.1039/D1NA00876E.

Watmanee, S., T. Klinaubol, R. Nganglumpoon, P. Pinthong, A. Christian Serraon, M. R. Chiong III, Y. Morikawa, and J. Panpranot. 2024. "Origin of Surface-Bonded Oxygen on Dendritic Ag Particles Towards 3D-Nanocrystalline Carbon Formation During CO_2 Electrochemical Reduction Reaction." *Chemical Engineering Journal* 480(January):148182. https://doi.org/10.1016/j.cej.2023.148182.

Wei, W., K. Sun, and Y.H. Hu. 2016. "Direct Conversion of CO 2 to 3D Graphene and Its Excellent Performance for Dye-Sensitized Solar Cells with 10% Efficiency." *Journal of Materials Chemistry A* 4(31):12054–12057.

Wei, W., K. Sun, and Y.H. Hu. 2017a. "Synthesis of Mesochannel Carbon Nanowall Material from CO_2 and Its Excellent Performance for Perovskite Solar Cells." *Industrial & Engineering Chemistry Research* 56.7:1803–1809.

Wei, W., D.J. Stacchiola, and Y.H. Hu. 2017b. "3D Graphene from CO_2 and K as an Excellent Counter Electrode for Dye-Sensitized Solar Cells." *International Journal of Energy Research* 41.15:2502–2508.

Wei, W., B. Hu, F. Jin, Z. Jing, Y. Li, A.A. García Blanco, D.J. Stacchiola, and Y.H. Hu. 2017c. "Potassium-Chemical Synthesis of 3D Graphene from CO_2 and Its Excellent Performance in HTM-Free Perovskite Solar Cells." *Journal of Materials Chemistry A* 5(17):7749–7752. https://doi.org/10.1039/C7TA01768E.

Wei, S., H. Shi, X. Li, X. Hu, C. Xu, and X. Wang. 2022. "A Green and Efficient Method for Preparing Graphene Using CO_2@ Mg in-Situ Reaction and Its Application in High-Performance Lithium-Ion Batteries." *Journal of Alloys and Compounds* 902:163700.

Wu, H., Z. Li, D. Ji, Y. Liu, L. Li, D. Yuan, Z. Zhang, et al. 2016. "One-Pot Synthesis of Nanostructured Carbon Materials from Carbon Dioxide via Electrolysis in Molten Carbonate Salts." *Carbon* 106(September):208–217. https://doi.org/10.1016/j.carbon.2016.05.031.

Xie, Z., E. Huang, S. Garg, S. Hwang, P. Liu, and J.G. Chen. 2024. "CO_2 Fixation into Carbon Nanofibres Using Electrochemical–Thermochemical Tandem Catalysis." *Nature Catalysis* 7(1):98–109. https://doi.org/10.1038/s41929-023-01085-1.

Xing, Z., B. Wang, W. Gao, C. Pan, J.K. Halsted, E.S. Chong, J. Lu, et al. 2015. "Reducing CO_2 to Dense Nanoporous Graphene by Mg/Zn for High Power Electrochemical Capacitors." *Nano Energy* 11(January):600–610. https://doi.org/10.1016/j.nanoen.2014.11.011.

Xu, X., and S. Huang. 2007. "Carbon Dioxide as a Carbon Source for Synthesis of Carbon Nanotubes by Chemical Vapor Deposition." *Materials Letters* 61(21):4235–4237.

Yamasue, E., H. Yamaguchi, H. Nakaoku, H. Okumura, and K.N. Ishihara. 2007a. "Carbon Dioxide Reduction into Carbon by Mechanically Milled Wustite." *Journal of Materials Science* 42(13):5196–5202. https://doi.org/10.1007/s10853-006-0458-0.

Yamasue, E., H. Yamaguchi, H. Okumura, and K.N. Ishihara. 2007b. "Decomposition of Carbon Dioxide Using Mechanically-Milled Magnetite." *Journal of Alloys and Compounds, Proceedings of the 12th International Symposium on Metastable and Nano-Materials (ISMANAM-2005)* 434–435(May):803–805. https://doi.org/10.1016/j.jallcom.2006.08.197.

Yap, D., J.-M. Tatibouët, and C. Batiot-Dupeyrat. 2015. "Carbon Dioxide Dissociation to Carbon Monoxide by Non-Thermal Plasma." *Journal of CO_2 Utilization* 12:54–61.

Ye, L., N. Syed, D. Wang, J. Guo, J. Yang, J. Buston, R. Singh, M.S. Alivand, G.K. Li, and A. Zavabeti. 2023. "Low-Temperature CO_2 Reduction Using Mg–Ga Liquid Metal Interface." *Advanced Materials Interfaces* 10(3):2201625. https://doi.org/10.1002/admi.202201625.

Yoshida, T., M. Tsuji, Y. Tamaura, T. Hurue, T. Hayashida, and K. Ogawa. 1997. "Carbon Recycling System Through Methanation of CO_2 in Flue Gas in LNG Power Plant." *Energy Conversion and Management* 38:S443–S448.

Zhang, C.L., T.-H. Wu, H.-M. Yang, Y.-Z. Jiang, and S.-Y. Peng. 1995. "Reduction of Carbon-Dioxide to Carbon with Active Cation Excess Magnetite." *Chemical Journal of Chinese Universities-Chinese* 16(6):955–957.

Zhang, C.L., Z. Liu, T.-H. Wu, H.-M. Yang, Y.-Z. Jiang, and S.-Y. Peng. 1996. "Complete Reduction of Carbon Dioxide to Carbon and Indirect Conversion to O2 Using Cation-Excess Magnetite." *Materials Chemistry and Physics* 44(2):194–198. https://doi.org/10.1016/0254-0584(95)01652-B.

Zhang, C.-L., S. Li, T.-H. Wu, and S.-Y. Peng. 1999. "Reduction of Carbon Dioxide into Carbon by the Active Wustite and the Mechanism of the Reaction." *Materials Chemistry and Physics* 58(2):139–145. https://doi.org/10.1016/S0254-0584(98)00267-3.

Zhang, C., S. Li, L. Wang, T. Wu, and S. Peng. 2000a. "Studies on the Decomposition of Carbon Dioxide into Carbon with Oxygen-Deficient Magnetite." *Materials Chemistry and Physics* 62(1):44–51. https://doi.org/10.1016/S0254-0584(99)00169-8.

Zhang, C., S. Li, L. Wang, T. Wu, and S. Peng. 2000b. "Studies on the Decomposing Carbon Dioxide into Carbon with Oxygen-Deficient Magnetite: II. The Effects of Properties of Magnetite on Activity of Decomposition CO_2 and Mechanism of the Reaction." *Materials Chemistry and Physics* 62(1):52–61. https://doi.org/10.1016/S0254-0584(99)00168-6.

Zhang, J., T. Tian, Y. Chen, Y. Niu, J. Tang, and L.-C. Qin. 2014. "Synthesis of Graphene from Dry Ice in Flames and Its Application in Supercapacitors." *Chemical Physics Letters* 591:78–81.

Zuraiqi, K., A. Zavabeti, J. Clarke-Hannaford, B.J. Murdoch, K. Shah, M.J.S. Spencer, C.F. McConville, T. Daeneke, and K. Chiang. 2022. "Direct Conversion of CO2 to Solid Carbon by Ga-Based Liquid Metals." *Energy & Environmental Science* 15(2):595–600.

L

Extraction of Select Critical Minerals from Coal Wastes: Literature Review

SUMMARY OF EXTRACTION METHODS, LEACHING AGENTS, AND LEACHING EFFICIENCY

As described in Chapter 9, the committee reviewed publications on the extraction of rare earth elements (REEs), lithium, and nickel from coal wastes since 2015. This review analyzed the extraction method and leaching agent employed in each publication, as well as the resultant leaching efficiencies of REEs, lithium, and nickel, as summarized in Table L-1.

TABLE L-1 Summary of Literature on Extraction of Rare Earth Elements (REEs), Lithium, and Nickel from Coal Wastes

Target Element(s)	Extraction Method	Leaching Agent	Leaching Efficiency	Citation
Journal Articles				
REEs	Acid leaching	H_2SO_4 or HCl or HNO_3	84.3% (maximum achieved)	Yang and Honaker (2020)
	Acid leaching	Citric acid	Y: 50% La and Ce: 40%	Prihutami et al. (2021)
	Acid leaching	HCl	La: 71.9% Ce: 66% Nd: 61.9%	Cao et al. (2018)
	Acid leaching	HCl	Y: 62.1% Nd: 55.5% Dy: 65.2	Tuan et al. (2019)
	Acid leaching	$HNO_3 + H_2SO_4$	No significant leachability	Lange et al. (2017)
	Acid leaching	HCl or HNO_3 or H_2SO_4 or H_3PO_4	La: 65.5% Ce: 64.4% Nd: 64.3%	Znamenáčková et al. (2021)
	Acid leaching	Methanesulphonic acid or p-toluenesulphonic acid	60–70%	Banerjee et al. (2022a)
	Acid leaching	HCl or HNO_3 or H_2SO_4 or acetic acid or formic acid	As high as 73%	Burgess et al. (2024)
	Acid leaching	HCl	Dy: 73.38% Er: 76.34% Eu: 88.02% Nd: 70.08% Tb: 90.01%	Dahan et al. (2022)
	Acid leaching	HF + HNO_3	Insufficient information	Hood et al. (2017)
	Acid leaching	HCl or HNO_3	59% (with HNO_3) 51% (with HCl)	Deng et al. (2022)
	Calcination → Acid leaching	HCl	TREEs: 72% (Western Kentucky No. 13 sample); 57% (Fire Clay sample)	Ji et al. (2022a)
	Roasting → acid leaching	NaOH, Na_2O_2, CaO, Na_2CO_3, $CaSO_4$, or $(NH_4)_2SO_4$ → HNO_3	>90% of total REE content (with NaOH or Na_2O_2) <50% of total REE content (with CaO, Na_2CO_3, $CaSO_4$, or $(NH_4)_2SO_4$)	Taggart et al. (2018)
	Roasting → acid leaching	NaOH, Na_2CO_3, $Ca(OH)_2$, $CaCl_2$, or $(NH_4)_2SO_4$ → HCl, H_2SO_4, or HNO_3	90% (maximum achieved)	Pan et al. (2021)
	Roasting → acid leaching	NaOH → HNO_3	Fe, Al, and REEs (except Ce): >90%	Wu et al. (2022)
	Alkali leaching	NaOH	REY: 30% (West Java coal sample); 24% (East Java coal sample)	Rosita et al. (2020c)
	Alkali-acid leaching	NaOH → HCl	Highest REE recovery: 95.5%	Wen et al. (2022b)
	Alkali-acid leaching	NaOH or $(NH_4)_2SO_4$ and H_2SO_4	REEs and Sc: 70–80% (after 5 h at 110°C and 5 M acid)	Shoppert et al. (2022)

continued

TABLE L-1 Continued

Target Element(s)	Extraction Method	Leaching Agent	Leaching Efficiency	Citation
	Alkali-acid leaching	NaOH → acetic acid	Maximum recovery of leaching: • Ce: 20.58% • Dy: 43.53% • La: 17.38% • Nd: 40.96% • Y: 18.45% • Yb: 32.74%	Manurung et al. (2020)
	Alkali-acid leaching	NaOH → HCl or HNO_3 or H_2SO_4	>90%	Trinh et al. (2022)
	Alkali-acid leaching	NaOH → HCl	>85%	Kuppusamy et al. (2019)
	Alkali-acid leaching	NaOH → HCl	LREEs: 71% (with 5 M NaOH at 90°C) HREEs: 41% (with 5 M NaOH at 90°C)	Li et al. (2022)
	Alkali fusion → acid leaching	Na_2CO_3 or NaCl or Na_2O_2 or NaOH or KOH or $Ca(OH)_2$ → HCl	49.25% (with Na_2O_2) 57.45% (with Na_2CO_3) 64.93% (with KOH) 74.23% (with NaOH)	Tang et al. (2022)
	Alkali-acid leaching	NaOH → HCl	64.9% (at 433K with 30 wt.-% NaOH)	Żelazny et al. (2023)
	Alkaline-acid leaching	NaOH + citric acid	77.6%	Rosita et al. (2023)
	Alkali treatment → acid leaching	NaOH → citric acid	REY recovery is 55%	Pan et al. (2023)
	Acid leaching	HNO_3	1.6–93.2% (via heated HNO_3 extraction)	Taggart et al. (2016)
	Na_2O_2 alkaline sintering → acid leaching	Na_2O_2 → HCl	The percentage recovery for total REEs for ashes was 80–90%	Middleton et al. (2020)
	Roasting → alkali-acid leaching	ZnO → NaOH → H_2SO_4	REEs: 87.1% • Ce: 70.7% • La: 82.5% • Gd: 83.2% • Nd: 87.1% • Dy: 62.3% • Y: 81.7%	Fan et al. (2022)
	Alkaline sintering-water immersion-acid leaching method	Na_2CO_3 à water à HCl	up to 85.81%	Zou et al. (2017)
	Acid leaching or alkali leaching	HNO_3 or HCl or H_2SO_4 or NaOH	98% (with HNO_3)	Penney and Alam (2023)
	Alkali fusion → acid leaching	Na_2CO_3 → HCl	~72.78%	Tang et al. (2019)
	Water leaching → acid leaching	Deionized water → HNO_3	>50%	Modi et al. (2023a)
	Calcination → water leaching → acidic/basic leaching	Deionized water → H_2SO_4 or NaOH	Insufficient information	Modi et al. (2023b)

TABLE L-1 Continued

Target Element(s)	Extraction Method	Leaching Agent	Leaching Efficiency	Citation
	Subcritical water + acid leaching	Subcritical water + HCl or HNO_3 or H_2SO_4	Maximum efficiencies achieved: • Y: 87.9% • Sm: 93.0% • Er: 86.2%	Liu and Lomanjaya (2022)
	Acid baking → water leaching	Sulfuric acid → water	80%	Kuppusamy and Holuszko (2022)
	Acid leaching or alkali leaching or water (Millipore Milli-Q) leaching	HCl or, NaOH, or doubly deionized water	~100% (for Powder River Basin coal samples)	King et al. (2018)
	Ionic liquid (IL) leaching → stripping	([Hbet][Tf_2N]) + $NaNO_3$ → HCl	>90% (with 1, 5, or 10 mg/g betaine)	Stoy et al. (2021)
	Microwave pretreatment → acid leaching Note: This study focused on using electron paramagnetic resonance to identify rare earth elements plus yttrium (REYs) in coal fly ash.	HNO_3 + HF + $HClO_4$ → H_3BO_3 → HNO_3	Insufficient information	Liu et al. (2021)
	Microwave-assisted pretreatment → acid leaching	Carbon lampblack powder → HNO_3	83.4%	Yakaboylu et al. (2019)
	Acid leaching → solvent extraction	HNO_3 → DEHPA	>80%	Honaker et al. (2017)
	Solvent extraction	$(NH_4)_2SO_4$, ionic liquid (1-butyl-3-methylimidazolium chloride) or deep eutectic solvent (2:1 molar ratio mixture of urea and choline chloride)	89% (with $(NH_4)_2SO_4$) 80% (with ionic liquid) 71% (with deep eutectic solvent)	Rozelle et al. (2016)
	Solvent extraction → stripping	[Hbet][Tf_2N] + $NaNO_3$/NaCl/ $Ca(NO_3)_2$/$CaCl_2$ → [Hbet] [Tf_2N] + HCl	>68.6%	Stoy et al. (2022a)
	Citrate and EDTA leaching	Citric acid + trisodium citrate or EDTA	11% (with citrate buffer) 33% (with EDTA)	Yang et al. (2021)
	Alkaline-acid leaching → stripping	NaOH → NaCl + ([Hbet] [Tf_2N]) → HCl	66%	Liu et al. (2023)
	Subcritical water acid leaching or microwave assisted acid leaching	Subcritical water → HCl	Y: 80.23% Sm: 68.19%	Lomanjaya and Liu (2023)
	Calcination extraction → acid leaching	Na_2CO_3 → HCl	95.8% (coal gangue sample) 93.2% (coal ash sample)	Zhang et al. (2022)
	Alkali calcination → supercritical CO_2 treatment → acid leaching	Na_2CO_3 → supercritical CO_2	>90%	Zhang et al. (2023)

continued

TABLE L-1 Continued

Target Element(s)	Extraction Method	Leaching Agent	Leaching Efficiency	Citation
	Solvent extraction	Diphosphate (2-ethylhexyl) (trade name: P_2O_4) + kerosene	La: 89.16% Ce: 94.11% Pr: 95.56% Nd: 96.33% Y: 99.80%	Pan et al. (2022a)
	Ionic liquid extraction	1-butyl-3-methylimidazolium tetrafluoroborate	26%	Thakare and Masud (2022)
	7-step sequential extraction	H_2O → $MgCl_2$ → NaOAc (pH=5) → NH_2OH•HCl (25% CH_3COOH) → HNO_3+H_2O_2 → NH_4OAc → microwave digestion	Insufficient information	Nie et al. (2022)
	6-step sequential extraction	MilliQ water → NH_4Ac or $MgCl_2$ → HCl or NaAc → HNO_3 or NH_2OH•HCl (25% CH_3COOH) → (HF + HCl), (HNO_3 + H_2O_2, H_2O_2 or NH_4OAc in HNO_3)	92.7–113.6% for individual REEs	Wu et al. (2020)
	5-step sequential extraction Or physical separation → acid leaching	$MgCl_2$ → NaOAc/HOAc → NH_2OH•HCl (25% CH_3COOH) → (HNO_3 + H_2O_2)/(NH_4OAc in HNO_3) → H_2SO_4+HF Or sieving and magnetic separation → HCl	79.85%	Pan et al. (2020)
	7-step sequential extraction Or acid leaching	Water → $MgCl_2$ → NaOAc (pH=5) → CH_3COOH + NH_2OH•HCl → HNO_3 + H_2O_2 → CH_3COONH_4 (pH=2) Or HCl	up to 98%	Pan et al. (2022b)
	4-step sequential extraction	NaOAc → NH_2OH•HCl in CH_3COOH → HNO_3+H_2O_2 → CH_3COONH_4 in HNO_3	~100% (Class C fly ash) 30–70% (Class F fly ash)	Liu et al. (2019)
	Precipitation → redissolution → complexation	$NaAlO_2$ → HNO_3 → tributyl phosphate	Ce: 41.8% La: 40.1% Nd: 58.2%	Song et al. (2021)
	Trap-extract-precipitate	$Na_2S_2O_4$ + $Na_3C_6H_5O_7$	>98%	Miranda et al. (2022)
	Selective precipitation → solvent extraction	NaOH → tributyl phosphate	97%	Talan and Huang (2020)
	Acid leaching → IL extraction → precipitation	HCl+HNO_3+HF → [N_{1888}]Cl / [$P_{6,6,6,14}$]Cl / [$P_{6,6,6,14}$][SOPAA] / [N_{1888}][SOPAA] → NH_4HCO_3/$Na_2C_2O_4$ solution	37.4%	Huang et al. (2019)
	Staged precipitation	NaOH	>80% Purity: 1.1%	Zhang and Honaker (2018)
	Acid leaching → solvent extraction → precipitation	HCl → tris-2-ethylhexyl amine → NH_3(aq)	30–90% for individual REEs	Kumari et al. (2019)

TABLE L-1 Continued

Target Element(s)	Extraction Method	Leaching Agent	Leaching Efficiency	Citation
	Citrate leaching → oxalate precipitation	Sodium citrate → sodium oxalate	10% (Class F fly ash sample) 60% (Class C fly ash sample)	Liu et al. (2023)
	Acid leaching → solvent extraction → stripping → precipitation	HNO_3 → tributyl phosphate or di-(2-ethylhexyl)-phosphoric acid in Elixore 205 → HNO_3 → oxalic acid	~100%	Wang et al. (2022)
	Acid leaching → biosorption	HCl → two microbe immobilization systems (polyethylene glycol diacrylate microbe beads and Si sol–gels) in immobilizing *Arthrobacter nicotianae*	82–90%	Alipanah et al. (2020)
	Bioleaching → precipitation	*Acidothiobacillus ferrooxidans* → H_2O_2 → NaOH → HNO_3 → oxalic acid	~40–60% Purity: 36.7%	Zhang et al. (2021)
	Bioleaching	*Candida bombicola*, *Phanerochaete chrysosporium*, or *Cryptococcus curvatus*	La, Ce, Pr, and Nd: 28.1–30.7% Yb: 67.7% Er: 64.6% Sc: 63.0% Y: 62.2%	Park and Liang (2019)
	Bioleaching	Mesophilic acidophilic chemolithotrophic microbial community	Sc: 52.0% Y: 52.6% La: 59.5%	Muravyov et al. (2015)
	Bioleaching	*Aspergillus niger*	30.91%	Ma et al. (2023)
	Bioweathering or acid leaching	*Shewanella oneidensis* or H_2SO_4	Total REEs: 98.4%	Sachan et al. (2023)
	Hydrothermal alkali treatment → bioleaching	NaOH → *Aspergillus niger*	Ti: 89.20% Ga: 32.00% Sr: 54.30% Zr: 74.50% Ba: 35.40%	Su et al. (2020a)
	Sieving → gravity separation → magnetic separation → flotation separation	Sieving → gravity separation → magnetic separation → flotation separation	65% (maximum achieved)	Abaka-Wood et al. (2022)
	Alkali fusion → TEHDGA resin extraction	$NaOH+NaNO_3$ → HNO_3 → TEHDGA (N,N,N′,N′-tetrakis-2-ethylhexyldiglycolamide) impregnated XAD-7 resin	Insufficient information	Mondal et al. (2019)
	Elution of ion exchange resins	Ion exchange resins	Insufficient information	Mostajeran et al. (2021)
	Flotation → mechanical grinding → acid leaching	HCl	25%	Wen et al. (2022a)
	Physical separation	Sieving and magnetic separation	REY: 71.21%	Rosita et al. (2020b)

continued

TABLE L-1 Continued

Target Element(s)	Extraction Method	Leaching Agent	Leaching Efficiency	Citation
	Absorption by high surface area carbon material	Absorbent: Microsphere Flower carbons	>85%	Brown and Balkus (2021)
	Acid leaching → precipitation → nanofiltration/ microfiltration	Microfiltration and nanofiltration membrane	92.8–99.3%	Kose Mutlu et al. (2018)
	Laser separation	A numerical study	Insufficient information	Phuoc et al. (2015)
	Electrodialytic remediation	Distilled water, $NaNO_3$, sodium acetate in acetic acid, or citric acid	40% (with citric acid)	Lima and Ottosen (2022)
	Insufficient information Acid leaching	Insufficient information HCOOH leaching → Removing Ca, Mg, and Fe with NH_4OH → Precipitating REEs with oxalic acid → Decomposing $RE_2(C_2O_4)_3$	Purity: 99.4% in REEOs	Huang et al. (2018)
	Bioleaching (helped by acid and ferric ions) → solvent extraction → precipitation	*Leptospirillum ferrooxidans*, *Acidithiobacillus ferrooxidans*, *Acidithiobacillus thiooxidans*, *Acidithiobacillus acidophilus*, and Sulfolobus-like bacteria	Insufficient information	Sarswat et al. (2020)
	A review	A review	A review	Arbuzov et al. (2019)
	A review	A review	A review	Das et al. (2018)
	A review	A review	A review	Rybak and Rybak (2021)
	A review	A review	A review	Bagdonas et al. (2022)
	A review	A review	A review	Zhang et al. (2020c)
	A review	A review	A review	Eterigho-Ikelegbe et al. (2021)
	A review	A review	A review	Fu et al. (2022)
	A review	A review	A review	Talan and Huang (2022)
	A review	A review	A review	Kursun Unver and Terzi (2018)
	A review	A review	A review	Wilfong et al. (2022)
	A review	A review	A review	Liu and Chen (2021)
	A review	A review	A review	Mwewa et al. (2022)
	A review	A review	A review	Ju et al. (2021)
	A review	A review	A review	Dai et al. (2016)
	A review	A review	A review	Zhang et al. (2015)
	A review	A review	A review	Dodbiba and Fujita (2023)
	A review	A review	A review	Royer-Lavallée et al. (2020)
	A review	A review	A review	Peiravi et al. (2021)

TABLE L-1 Continued

Target Element(s)	Extraction Method	Leaching Agent	Leaching Efficiency	Citation
HREE (heavy REE) + LREE (light REE)	Acid leaching	Carboxylic acid (tartaric acid, malonic acid, lactic acid, citric acid, or succinic acid)	62%	Banerjee et al. (2021)
	Acid leaching	HCl	~80%	Honaker et al. (2019)
	Calcination → acid leaching	H_2SO_4	TREE: 74%	Gupta et al. (2023)
	Calcination → acid leaching	HCl or $HClO_4$ or HNO_3	98.17%	Hamza et al. (2022)
	Calcination → acid baking	H_2SO_4	~80%	Nawab et al. (2022)
	Calcination → acid leaching	HCl or citric acid or maleic acid or D,L-malic acid or oxalic acid	~60% (with 0.05M HCl)	Ji et al. (2022b)
	Acid leaching → ion exchange leaching	H_2SO_4 → $(NH_4)_2SO_4$	TREE: 75–80% (with thermal activation or alkaline pretreatment)	Yang et al. (2019)
	Small-scale leaching or large-scale leaching → column separation → Precipitation and calcination	Small-scale or large-scale leaching: HCl, HNO_3, or H_2SO_4 → bisethylhexyl diethylenetriaminepentaacetic acid (bisethylhexyl DTPA) → oxalic acid	>70% (with mineral acid leaching) Purity: >10 wt.%	Dardona et al. (2023)
	Desilication → microwave-assisted acid leaching	NaOH → HNO_3, HCl, $HClO_4$, or HF	98.03% (with HNO_3 + HCl + HF)	Ju et al. (2023)
	Acid leaching → solvent extraction	HNO_3 → tributyl phosphate or Cyanex 572 or di-(2-ethylhexyl)phosphoric acid (DEHPA) or their combinations	~99%	Peiravi et al. (2017)
	Sequential leaching or single-step acid leaching or float-sink separations or humic acid extraction	sequential leaching; or single-step acid leaching agent: HCl or H_3PO_4 or H_2SO_4; or float-sink separations; or humic acid extraction via acetone-H_2O-HCl method	70-90%	Laudal et al. (2018)
	Tessier sequential extraction or BCR sequential extraction	Tessier sequential extraction: NaOAc → $NH_2OH{\cdot}HCl$ in CH_3COOH → HNO_3 + H_2O_2 Or BCR sequential extraction: CH_3COOH → $NH_2OH{\cdot}HCl$ + HNO_3 → H_2O_2 (pH 2–3) → HNO_3 + HCl	85% (with Tessier sequential extraction) 60–70% (with BCR sequential extraction)	Park et al. (2021)
	5-step sequential extraction	$MgCl_2$ → NaOAc →$NH_2OH{\cdot}HCl$ in CH_3COOH → HNO_3 → aqua regia + HF	45% (maximum achieved)	Zhang and Honaker (2019b)
	4-step sequential extraction	$MgCl_2$ → NaOAc → CH_3COOH + $NH_2OH{\cdot}HCl$ → HNO_3 + H_2O_2/ $NH_4CH_3CO_2$ + HNO_3	95.42% (Faer sample) 94.28% (Panbei sample)	Pan et al. (2019)

continued

TABLE L-1 Continued

Target Element(s)	Extraction Method	Leaching Agent	Leaching Efficiency	Citation
	4-step sequential extraction	CH_3COOH → $NH_2OH{\cdot}HCl$ → H_2O_2 → Ammonia acetate ($C_2H_7NO_2$) → HCl + HNO_3	45% (maximum achieved)	Okeme et al. (2022)
	Acid leaching → solvent extraction → stripping → selective precipitation	H_2SO_4 → DEHPA → HCl → oxalic acid	REE: 75% Efficiency of recovering leached REEs via solvent extraction: 95%	Honaker et al. (2020)
	Deep eutectic solvents leaching → precipitation	(choline chloride (ChCl) + lactic acid (LA)) or (ChCl + para toluene sulphonic acid monohydrate [pTSA])) → oxalic acid dihydrate or NaF or Na_2SO_4	85–95% Purity: 13.8% (REE-oxalate); 7.3% (REE-fluoride)	Karan et al. (2022)
	Desilication → solvent extraction → stripping → precipitation	Gelatin → DEHPA solvent → HCl →Na_2SO_4 or sulfamic acid + $NaNO_3$ or oxalic acid dihydrate	HREE: 94% LREE: 86% Purity: 17.6% (TREE)	Rao et al. (2022)
	Two-step staged precipitation	Na_2CO_3	TREE: 85%	Hassas et al. (2021)
	Acid leaching → biosorption	HCl → biosorbent (carbonized ginkgo leaves [GL450])	Er: 99.22%	Ponou et al. (2016)
	Roasting → acid leaching → Two liquid membrane separation (liquid emulsion membranes and supported liquid membranes)	NaOH → HNO_3 → (DEHPA in kerosene or mineral oil) + HNO_3	Y, Tb, Dy, Ho, Er, Tm, Yb, Lu: >75% La, Ce, Pr, Nd: <50%	Smith et al. (2019)
	Froth flotation → magnetic separation → acid leaching	HNO_3	>80%	Zhang et al. (2018)
	Acid leaching or electrodialytic separation	HNO_3 or electrodialytic recovery	>70% (maximum achieved)	Couto et al. (2020)
HREE + LREE (REE + Li + Ni)	Calcination → acid leaching or ion exchange leaching	HCl → $(NH_4)_2SO_4$	LREE: 80–90% HREE: 40–60%	Zhang and Honaker (2019a)
HREE + LREE + MREE	Alkali fusion-acid leaching	NaOH → HCl	32.624% (with 2 M HCl)	Mokoena et al. (2022)
	7-step sequential extraction Or magnetic separation → hydrothermal alkaline treatment	MilliQ water → $(NH_4)_2SO_4$ → CH_3COOH → hydroxylammonine chloride → ammonium oxalate + oxalic acid → ammonium oxalate, oxalic acid, and ascorbic acid → acidified H_2O_2 digestion + ammonium acetate extraction → $LiBO_2$ fusion Or magnetic separation → NaOH	Total REE: 97.8%	Lin et al. (2018)

TABLE L-1 Continued

Target Element(s)	Extraction Method	Leaching Agent	Leaching Efficiency	Citation
	7-step sequential extraction or alkali-acid leaching	Deionized water → Ascorbic acid → CH_3COOH → hydroxylammonium chloride → Ammonium oxalate + oxalic acid → Ammonium oxalate + oxalic acid + ascorbic acid → H_2O_2 or $(NH_4)_2SO_4$; or NaOH/KOH → HCl/oxalic acid	Insufficient information	Choudhary et al. (2024)
HREE + CREE (critical REE)	Acid leaching	H_2SO_4	80%	Honaker et al. (2018b)
HREE + LREE + CREE	5-step sequential	$MgCl_2$ → $NH_4CH_3CO_2$ + CH_3COOH → CH_3COOH + $NH_2OH{\cdot}HCl$ → HNO_3 + H_2O_2 → microwave digestion (NaOH + HNO_3)	45–75% bounded to Fe-Mn oxides	Wang et al. (2021)
	Acid leaching → Stagewise precipitation → solvent extraction → stripping → Oxalic acid precipitation	H_2SO_4 → NaOH → di-(2-ethylhexyl) phosphoric acid (DEHPA), DEHPA + tri-butyl phosphate, or DEHPA + H_2O_2 → HCl → oxalic acid	TREEs: 87.85% (average of 10 tests at pH 0.5) Purity: 80% in REO	Cicek et al. (2023)
REE + Li + Ni	Biomacromolecular extraction	Lanmodulin	La: 99.5% Sc: 96% Y: 96%	Deblonde et al. (2020)
	Sequential leaching	$NH_4CH_3CO_2$→ HCl → HF → HNO_3	Insufficient information	Finkelman et al. (2018)
REE + Li	Acid-alkali-based alternate extraction	HCl → NaOH → HCl → NaOH → HCl	REY: 65% Li: 84%	Ma et al. (2019)
	5-step sequential extraction Or calcination → acid leaching	$MgCl_2$ → NaOAc → $NH_2OH{\cdot}HCl$ in CH_3COOH → HNO_3 + H_2O_2 → $NH_4CH_3CO_2$ in HNO_3 Or calcination → HCl	LREEs: 80-90% Li: 70%	Zhang and Honaker (2020a)
	6-step sequential extraction	$MgCl_2$ → NaOAc → CH_3COOH + $NH_2OH{\cdot}HCl$ → HNO_3 +H_2O_2 + $NH_4CH_3CO_2$ → HF → HF + HNO_3	Insufficient information	Xu et al. (2022)
	A review	A review	A review	Sahoo et al. (2016)
	A review	A review	A review	Wang et al. (2020)
REE + Ni	Alkaline pretreatment → IL leaching → stripping	NaOH → [Hbet][Tf_2N] + $NaNO_3$ → [Hbet][Tf_2N] + HCl	LREEs: ~70–100%	Stoy et al. (2022b)
	7-step sequential extraction	Distilled water → $(NH_4)_2SO_4$ → sodium acetate trihydrate → $NH_2OH{\cdot}HCl$ → ammonium oxalate + oxalic acid + ascorbic acid → ammonium oxalate + oxalic acid → H_2O_2 / $NH_4CH_3CO_2$	Total REE: 2–21%	Bauer et al. (2022)

continued

TABLE L-1 Continued

Target Element(s)	Extraction Method	Leaching Agent	Leaching Efficiency	Citation
	Sequential precipitation → re-dissolution → oxalic acid precipitation	NaOH → HNO_3 → oxalic acid	REE: 95% Ni: insufficient information Purity: >98% in REO	Zhang and Honaker (2020b)
	One-step bioleaching or two-step bleaching (hydrothermal-alkali/acid treatment + bioleaching)	Bioleaching: *Acidithiobacillus thiooxidans*	La: 75.08% Ce: 87.08%	Su et al. (2020b)
Books				
REE	A review	A review	A review	Zhao et al. (2019)
	A review	A review	A review	Lai et al. (2021)
	A review	A review	A review	Sreenivas et al. (2021)
	Acid leaching by organic or mineral acids	Tartaric acid, lactic acid, HCl, or HNO_3	62% (tartaric acid) 56% (lactic acid) ~72% (HCl or HNO_3)	Banerjee et al. (2022b)
	A review	A review	A review	Rao and Sreenivas (2019)
	Acid leaching	HCl, HNO_3, or H_2SO_4	>90% (with HCl)	Kumari et al. (2020)
	Ionic liquid extraction	Review	A review	Danso et al. (2021)
	A review	A review	A review	Mahandra et al. (2021)
	A review	A review	A review	Arellano Ruiz et al. (2021)
REM	A review	A review	A review	Kumari et al. (2018)
REE + Li	A review	A review	A review	Vu et al. (2021)
Conference Papers				
REE	Alkaline digestion → acid leaching	Alkali → H_2SO_4	REY: 75.25% (maximum recovery)	Rosita et al. (2020a)
	Acid leaching	H_2SO_4	Ce: 37.5% La: 33.8% Nd: 40.6% Sc: 28.1% Y: 54.5%	Swinder et al. (2017)
	Alkaline treatments	Density separation → magnetic separation → size separation → NaOH treatment	REE enriching efficiency: 270%	Soong et al. (2019)
	Direct acid leaching	HCl or H_2SO_4	Insufficient information	Taggart (2015)
	Sintering → acid leaching	Na_2O_2 → HNO_3		
	Pressure-digestion acid leaching	NaOH → HCl		
	Alkali-acid leaching	NaOH → HCl	~90%	Roth et al. (2017)
	Acid leaching or roasting with chemical additives	Insufficient information	70–100%	Taggart et al. (2017)
	Acid leaching	H_2SO_4	85% Purity: 50%	Honaker et al. (2018a)

TABLE L-1 Continued

Target Element(s)	Extraction Method	Leaching Agent	Leaching Efficiency	Citation
	Magnetic separation → flotation	Talon 9400, sodium oleate and oleic acid (used for pH)	20%	Honaker et al. (2016)
	Acid leaching	Citric acid	Y: 83.35% (at 45°C); 51.00% (at 26°C)	Prihutami et al. (2020)
Government Reports				
REE	Two stage SX rougher and cleaner circuit or roasting → acid leaching → DEHPA/TBP → stripping	H_2SO_4 → DEHPA/TBP → oxalic acid	>97% (with SX rougher and cleaner circuit) Purity: >90% in REO (with SX rougher and cleaner circuit)	Honaker et al. (2021)
	Acid digestion process	HCl or HNO_3	REE + Y + Sc: 99.66% Purity: 1.04%	Peterson et al. (2017)
	Acid leaching → re-precipitation	HNO_3	~100% Purity: 6% in REE	Ziemkiewicz (2020)
	Alkaline pretreatment → acid leaching	NaOH → HCl	91%	Carlson (2018)
	Insufficient information	HF + HNO_3	Insufficient information	Hsu-Kim et al. (2020)
	Insufficient information	Citric acid	~30% (at pH ~ 2)	Yang et al. (2022)
	Insufficient information	HF + HCl + HNO_3	43% (maximum achieved) Purity: 54.4% (maximum achieved)	Mann et al. (2021)
	Insufficient information	Insufficient information	Insufficient information	Jayne et al. (2019)
	Insufficient information	HCl	Insufficient information	Sutterlin (2019)
	A review	A review	A review	Costis et al. (2019)
	Insufficient information	Insufficient information	Insufficient information	Bryan (2015)
	Insufficient information	Insufficient information	Insufficient information	Granite et al. (2016)
REE + Li + Ni	Milling and caustic leaching → acid leaching → solvent leaching → stripping → purification	NaOH + HNO_3	>90–95% in TREO	Argumedo et al. (2020)
REE + Li	Leaching → solvent extraction → precipitation	Bacterial leaching solution → CYANEX 272, CYANEX 923, D2EPHA, or Versatic 10 → oxalic acid	Extraction efficiency: 42.5% Purity of REE oxalate in final product: 36.7%	Free et al. (2020)

continued

TABLE L-1 Continued

Target Element(s)	Extraction Method	Leaching Agent	Leaching Efficiency	Citation
Theses/Dissertations				
REE	Alkali-acid leaching	NaOH → HCl	74%	Choi (2018)
	Alkaline sintering → acid leaching	Na_2O_2, NaOH, CaO, Na_2CO_3, $CaSO_4$, or $(NH_4)_2SO_4$ → HNO_3	TREE recovery efficiency: >90% (NaOH or Na_2O_2 sintering) <50% (CaO, Na_2CO_3, $CaSO_4$, or $(NH_4)_2SO_4$ sintering) ~100% for PRB samples regardless of sintering agent and additive:ash ratio	Taggart (2018)
	Acid baking and water leaching	H_2SO_4 and DI H_2O	TREE: 79.1% (maximum achieved)	Kuppusamy (2022)
	Acid leaching	H_2SO_4, HCl, H_3PO_4	70–90%	Laudal (2017)
	Dense medium circuit → flotation		60–76% (overall REE recovery)	Gupta (2016)
HREE + LREE	Acid leaching → precipitation → solvent extraction → stripping → precipitation → roasting	H_2SO_4 → H_2O_2 → DEHPA → HCl → oxalic acid	98.77% (with 0.5 M DEHPA) Purity: 80% by weight in REO	Cicek (2023)
	Solvent extraction	DEHPA	Purity: 4.63% in TREE in solid phase	Ren (2019)
	Acid leaching	H_2SO_4	49.6%	Yang (2019)
REE + Ni	Ionic liquid extraction	IL: [Hbet][Tf_2N]	~100%	Stoy (2021)

REFERENCES

Abaka-Wood, G.B., J. Addai-Mensah, and W. Skinner. 2022. "The Concentration of Rare Earth Elements from Coal Fly Ash." *Journal of the Southern African Institute of Mining and Metallurgy* 122(1):1–7. https://doi.org/10.17159/2411-9717/1654/2022.

Alipanah, M., D.M. Park, A. Middleton, Z. Dong, H. Hsu-Kim, Y. Jiao, and H. Jin. 2020. "Techno-Economic and Life Cycle Assessments for Sustainable Rare Earth Recovery from Coal Byproducts Using Biosorption." *ACS Sustainable Chemistry & Engineering* 8(49):17914–17922. https://doi.org/10.1021/acssuschemeng.0c04415.

Arbuzov, S.I., R.B. Finkelman, S.S. Il'enok, S.G. Maslov, A.M. Mezhibor, and M.G. Blokhin. 2019. "Modes of Occurrence of Rare-Earth Elements (La, Ce, Sm, Eu, Tb, Yb, Lu) in Coals of Northern Asia (Review)." *Solid Fuel Chemistry* 53(1):1–21. https://doi.org/10.3103/S0361521919010026.

Arellano Ruiz, V.C., P.K. Parhi, J.-Y. Lee, and R.K. Jyothi. 2021. "Investigation on Extraction and Recovery of Rare Earth Elements from Coal Combustion Products." Pp. 311–337 in *Clean Coal Technologies*, R.K. Jyothi and P.K. Parhi, eds. Cham: Springer International Publishing. https://doi.org/10.1007/978-3-030-68502-7_13.

Argumedo, D., K. Johnson, M. Heinrichs, R. Peterson, R. Winburn, and J. Brewer. 2020. "Recovery of High Purity Rare Earth Elements (REE) from Coal Ash via a Novel Electrowinning Process (Final Report)." DOE-Battelle-FE0031529. https://doi.org/10.2172/1631038.

Bagdonas, D.A., A.J. Enriquez, K.A. Coddington, D.C. Finnoff, J.F. McLaughlin, M.D. Bazilian, E.H. Phillips, and T.L. McLing. 2022. "Rare Earth Element Resource Evaluation of Coal Byproducts: A Case Study from the Powder River Basin, Wyoming." *Renewable and Sustainable Energy Reviews* 158(April):112148. https://doi.org/10.1016/j.rser.2022.112148.

Banerjee, R., A. Mohanty, S. Chakravarty, S. Chakladar, and P. Biswas. 2021. "A Single-Step Process to Leach out Rare Earth Elements from Coal Ash Using Organic Carboxylic Acids." *Hydrometallurgy* 201(May):105575. https://doi.org/10.1016/j.hydromet.2021.105575.

Banerjee, R., S. Chakladar, A. Mohanty, S. Kumar Chattopadhyay, and S. Chakravarty. 2022a. "Leaching Characteristics of Rare Earth Elements from Coal Ash Using Organosulphonic Acids." *Minerals Engineering* 185(July):107664. https://doi.org/10.1016/j.mineng.2022.107664.

Banerjee, R., S. Chakladar, and S. Chakravarty. 2022b. "Extraction of Rare Earth Metals from Coal Ash Using Mild Lixiviants in a Single Step Process." Pp. 63–70 in *Rare Metal Technology*. Springer International Publishing.

Bauer, S., J. Yang, M. Stuckman, and C. Verba. 2022. "Rare Earth Element (REE) and Critical Mineral Fractions of Central Appalachian Coal-Related Strata Determined by 7-Step Sequential Extraction." *Minerals* 12(11):1350. https://doi.org/10.3390/min12111350.

Brown, A.T., and K.J. Balkus. 2021. "Critical Rare Earth Element Recovery from Coal Ash Using Microsphere Flower Carbon." *ACS Applied Materials & Interfaces* 13(41):48492–48499. https://doi.org/10.1021/acsami.1c09298.

Bryan, R.C. 2015. "Assessment of Rare Earth Elemental Contents in Select United States Coal Basins." US DOE, National Energy Technology Laboratory.

Burgess, W., C.F. Chiu, T. Cain, E. Roth, M. Keller, and E. Granite. 2024. "Extractability Indices for Screening Coal Combustion Byproduct Feedstocks for Recovery of Rare Earth Elements." *International Journal of Coal Geology* 281(January):104401. https://doi.org/10.1016/j.coal.2023.104401.

Cao, S., C. Zhou, J. Pan, C. Liu, M. Tang, W. Ji, T. Hu, and N. Zhang. 2018. "Study on Influence Factors of Leaching of Rare Earth Elements from Coal Fly Ash." *Energy & Fuels* 32(7):8000–8005. https://doi.org/10.1021/acs.energyfuels.8b01316.

Carlson, G. 2018. "Economical and Environmentally Benign Extraction of Rare Earth Elements (REES) from Coal & Coal Byproducts." DOE-Tusaar–0027155. https://doi.org/10.2172/1430514.

Choi, H. 2018. "Development of Separation and Purification Methods for Producing Rare Earth Elements from Coal Fly Ash." Purdue University.

Choudhary, A.K.S., S. Kumar, and S. Maity. 2024. "A Study on Speciation and Enrichment of Rare Earth Elements (REE) by Sequential Extraction from a Potential Coal Fly Ash Resource and Its Role in REE Extractability." *Hydrometallurgy* 224(February):106256. https://doi.org/10.1016/j.hydromet.2023.106256.

Cicek, Z. 2023. "Selective Recovery of Rare Earth Elements from Acid Mine Drainage Treatment Byproduct." *Graduate Theses, Dissertations, and Problem Reports*. 11788. https://researchrepository.wvu.edu/etd/11788.

Cicek, Z., A.A. Mira, and Q. Huang. 2023. "Process Development for the Extraction of Rare Earth Elements from an Acid Mine Drainage Treatment Sludge." *Resources, Conservation and Recycling* 198(November):107147. https://doi.org/10.1016/j.resconrec.2023.107147.

Costis, S., K.K. Mueller, J.-F. Blais, A. Royer-Lavallee, L. Coudert, and C.M. Neculita. 2019. "Review of Recent Work on the Recovery of Rare Earth Elements from Secondary Sources." R1859. National Resources Canada.

Couto, N., A.R. Ferreira, V. Lopes, S.C. Peters, E.P. Mateus, A.B. Ribeiro, and S. Pamukcu. 2020. "Electrodialytic Recovery of Rare Earth Elements from Coal Ashes." *Electrochimica Acta* 359(November):136934. https://doi.org/10.1016/j.electacta.2020.136934.

Dahan, A.M.E., R.D. Alorro, M.L.C. Pacaña, R.M. Baute, L.C. Silva, C.B. Tabelin, and V.J.T. Resabal. 2022. "Hydrochloric Acid Leaching of Philippine Coal Fly Ash: Investigation and Optimisation of Leaching Parameters by Response Surface Methodology (RSM)." *Sustainable Chemistry* 3(1):76–90. https://doi.org/10.3390/suschem3010006.

Dai, S., I.T. Graham, and C.R. Ward. 2016. "A Review of Anomalous Rare Earth Elements and Yttrium in Coal." *International Journal of Coal Geology* 159(April):82–95. https://doi.org/10.1016/j.coal.2016.04.005.

Danso, I.K., A.B. Cueva-Sola, Z. Masaud, J.-Y. Lee, and R.K. Jyothi. 2021. "Ionic Liquids for the Recovery of Rare Earth Elements from Coal Combustion Products." Pp. 617–638 in *Clean Coal Technologies*, R.K. Jyothi and P.K. Parhi, eds. Cham: Springer International Publishing. https://doi.org/10.1007/978-3-030-68502-7_25.

Dardona, M., S.K. Mohanty, M.J. Allen, and T.M. Dittrich. 2023. "From Ash to Oxides: Recovery of Rare-Earth Elements as a Step Towards Valorization of Coal Fly Ash Waste." *Separation and Purification Technology* 314(June):123532. https://doi.org/10.1016/j.seppur.2023.123532.

Das, S., G. Gaustad, A. Sekar, and E. Williams. 2018. "Techno-Economic Analysis of Supercritical Extraction of Rare Earth Elements from Coal Ash." *Journal of Cleaner Production* 189(July):539–551. https://doi.org/10.1016/j.jclepro.2018.03.252.

Deblonde, G.J.-P., J.A. Mattocks, D.M. Park, D.W. Reed, J.A. Cotruvo, and Y. Jiao. 2020. "Selective and Efficient Biomacromolecular Extraction of Rare-Earth Elements Using Lanmodulin." *Inorganic Chemistry* 59(17):11855–11867. https://doi.org/10.1021/acs.inorgchem.0c01303.

Deng, B., X. Wang, D.X. Luong, R.A. Carter, Z. Wang, M.B. Tomson, and J.M. Tour. 2022. "Rare Earth Elements from Waste." *Science Advances* 8(6):eabm3132. https://doi.org/10.1126/sciadv.abm3132.

Dodbiba, G., and T. Fujita. 2023. "Trends in Extraction of Rare Earth Elements from Coal Ashes: A Review." *Recycling* 8(1):17. https://doi.org/10.3390/recycling8010017.

Eterigho-Ikelegbe, O., H. Harrar, and S. Bada. 2021. "Rare Earth Elements from Coal and Coal Discard—A Review." *Minerals Engineering* 173(November):107187. https://doi.org/10.1016/j.mineng.2021.107187.

Fan, X., J. Xia, D. Zhang, Z. Nie, Y. Liu, L. Zhang, and D. Zhang. 2022. "Highly-Efficient and Sequential Recovery of Rare Earth Elements, Alumina and Silica from Coal Fly Ash via a Novel Recyclable ZnO Sinter Method." *Journal of Hazardous Materials* 437(September):129308. https://doi.org/10.1016/j.jhazmat.2022.129308.

Finkelman, R.B., C.A. Palmer, and P. Wang. 2018. "Quantification of the Modes of Occurrence of 42 Elements in Coal." *International Journal of Coal Geology* 185(January):138–160. https://doi.org/10.1016/j.coal.2017.09.005.

Free, M., A. Noble, L. Allen, Z. Zhang, P. Sarswat, M. Leake, D. Kim, and G. Luttrell. 2020. "Economic Extraction, Recovery and Upgrading of Rare Earth Elements from Coal-Based Resources." DOE-UofU-31526. https://doi.org/10.2172/1634992.

Fu, B., J.C. Hower, W. Zhang, G. Luo, H. Hu, and H. Yao. 2022. "A Review of Rare Earth Elements and Yttrium in Coal Ash: Content, Modes of Occurrences, Combustion Behavior, and Extraction Methods." *Progress in Energy and Combustion Science* 88(January):100954. https://doi.org/10.1016/j.pecs.2021.100954.

Granite, E.J., E. Roth, and M.A. Alvin. 2016. "Characterization and Recovery of Rare Earths from Coal and By-Products." NETL-PUB–20414, 1245760. https://doi.org/10.2172/1245760.

Gupta, T. 2016. "Recovery of Rare Earth Elements from Alaskan Coal and Coal Combustion Products." University of Alaska Fairbanks.

Gupta, T., A. Nawab, and R. Honaker. 2023. "Optimizing Calcination of Coal By-Products for Maximizing REE Leaching Recovery and Minimizing Al, Ca, and Fe Contamination." *Journal of Rare Earths* 42(7):1354–1365. https://doi.org/10.1016/j.jre.2023.08.004.

Hamza, H., O. Eterigho-Ikelegbe, A. Jibril, and S.O. Bada. 2022. "Application of the Response Surface Methodology to Optimise the Leaching Process and Recovery of Rare Earth Elements from Discard and Run of Mine Coal." *Minerals* 12(8):938. https://doi.org/10.3390/min12080938.

Hassas, B.V., M. Rezaee, and S.V. Pisupati. 2021. "Effect of Various Ligands on the Selective Precipitation of Critical and Rare Earth Elements from Acid Mine Drainage." *Chemosphere* 280(October):130684. https://doi.org/10.1016/j.chemosphere.2021.130684.

Honaker, R., J. Groppo, A. Bhagavatula, M. Rezaee, and W. Zhang. 2016. "Recovery of Rare Earth Minerals and Elements from Coal and Coal Byproducts." Presented at the International Conference of Coal Preparation. April 2016. Louiseville, KY.

Honaker, R.Q., J. Groppo, R.-H. Yoon, G.H. Luttrell, A. Noble, and J. Herbst. 2017. "Process Evaluation and Flowsheet Development for the Recovery of Rare Earth Elements from Coal and Associated Byproducts." *Minerals & Metallurgical Processing* 34(3):107–115. https://doi.org/10.19150/mmp.7610.

Honaker, R., X. Yang, A. Chandra, W. Zhang, and J. Werner. 2018a. "Hydrometallurgical Extraction of Rare Earth Elements from Coal." Pp. 2309–2322 in *Extraction*, B.R. Davis, M.S. Moats, S. Wang, D. Gregurek, J. Kapusta, T.P. Battle, M.E. Schlesinger, et al., eds. The Minerals, Metals & Materials Series. Cham: Springer International Publishing. https://doi.org/10.1007/978-3-319-95022-8_193.

Honaker, Rick Q., W. Zhang, X. Yang, and M. Rezaee. 2018b. "Conception of an Integrated Flowsheet for Rare Earth Elements Recovery from Coal Coarse Refuse." *Minerals Engineering* 122(June):233–240. https://doi.org/10.1016/j.mineng.2018.04.005.

Honaker, R. Q., W. Zhang, and J. Werner. 2019. "Acid Leaching of Rare Earth Elements from Coal and Coal Ash: Implications for Using Fluidized Bed Combustion to Assist in the Recovery of Critical Materials." *Energy & Fuels* 33(7):5971–5980. https://doi.org/10.1021/acs.energyfuels.9b00295.

Honaker, R. Q., W. Zhang, J. Werner, A. Noble, G.H. Luttrell, and R.H. Yoon. 2020. "Enhancement of a Process Flowsheet for Recovering and Concentrating Critical Materials from Bituminous Coal Sources." *Mining, Metallurgy & Exploration* 37(1):3–20. https://doi.org/10.1007/s42461-019-00148-x.

Honaker, R., J. Werner, X. Yang, W. Zhang, A. Noble, R.H. Yoon, G.H. Luttrell, and Q. Huang. 2021. "Pilot-Scale Testing of an Integrated Circuit for the Extraction of Rare Earth Minerals and Elements from Coal and Coal Byproducts Using Advanced Separation Technologies." DOE-UKY-0463. https://www.osti.gov/servlets/purl/1798663.

Hood, M.M., R.K. Taggart, R.C. Smith, H. Hsu-Kim, K.R. Henke, U. Graham, J.G. Groppo, J.M. Unrine, and J.C. Hower. 2017. "Rare Earth Element Distribution in Fly Ash Derived from the Fire Clay Coal, Kentucky." *Coal Combustion and Gasification Products* 9(1):22–33. https://doi.org/10.4177/CCGP-D-17-00002.1.

Hsu-Kim, H., D. Plata, J. Hower, Z. Hendren, and M. Wiesner. 2020. "Novel Membrane and Electrodeposition-Based Separation and Recovery of Rare Earth Elements from Coal Combustion Residues (Final Report)." DE–FE0026952-Final. https://doi.org/10.2172/1526006.

Huang, C., Y. Wang, B. Huang, Y. Dong, and X. Sun. 2019. "The Recovery of Rare Earth Elements from Coal Combustion Products by Ionic Liquids." *Minerals Engineering* 130(January):142–147. https://doi.org/10.1016/j.mineng.2018.10.002.

Huang, Z., M. Fan, and H. Tian. 2018. "Coal and Coal Byproducts: A Large and Developable Unconventional Resource for Critical Materials—Rare Earth Elements." *Journal of Rare Earths* 36(4):337–338. https://doi.org/10.1016/j.jre.2018.01.002.

Jayne, K., D.R. Carr, J. Rowean, and M.C. Kimble. 2019. "Rare Earth Element Extraction from Coal Fly Ash." DOE-Skyhavensystems-18528.

Ji, B., Q. Li, H. Tang, and W. Zhang. 2022a. "Rare Earth Elements (REEs) Recovery from Coal Waste of the Western Kentucky No. 13 and Fire Clay Seams. Part II: Re-Investigation on the Effect of Calcination." *Fuel* 315(May):123145. https://doi.org/10.1016/j.fuel.2022.123145.

Ji, B., Q. Li, and W. Zhang. 2022b. "Leaching Recovery of Rare Earth Elements from the Calcination Product of a Coal Coarse Refuse Using Organic Acids." *Journal of Rare Earths* 40(2):318–327. https://doi.org/10.1016/j.jre.2020.11.021.

Ju, T., S. Han, Y. Meng, and J. Jiang. 2021. "High-End Reclamation of Coal Fly Ash Focusing on Elemental Extraction and Synthesis of Porous Materials." *ACS Sustainable Chemistry & Engineering* 9(20):6894–6911. https://doi.org/10.1021/acssuschemeng.1c00587.

Ju, T., Y. Meng, S. Han, F. Meng, L. Lin, J. Li, and J. Jiang. 2023. "Analysis of Enrichment, Correlation, and Leaching Patterns of Rare Earth Elements in Coal Fly Ash Assisted by Statistical Measures." *Science of The Total Environment* 902(December):166070. https://doi.org/10.1016/j.scitotenv.2023.166070.

Karan, R., T. Sreenivas, M.A. Kumar, and D.K. Singh. 2022. "Recovery of Rare Earth Elements from Coal Flyash Using Deep Eutectic Solvents as Leachants and Precipitating as Oxalate or Fluoride." *Hydrometallurgy* 214(October):105952. https://doi.org/10.1016/j.hydromet.2022.105952.

King, J.F., R.K. Taggart, R.C. Smith, J.C. Hower, and H. Hsu-Kim. 2018. "Aqueous Acid and Alkaline Extraction of Rare Earth Elements from Coal Combustion Ash." *International Journal of Coal Geology* 195(July):75–83. https://doi.org/10.1016/j.coal.2018.05.009.

Kose Mutlu, B., B. Cantoni, A. Turolla, M. Antonelli, H. Hsu-Kim, and M.R. Wiesner. 2018. "Application of Nanofiltration for Rare Earth Elements Recovery from Coal Fly Ash Leachate: Performance and Cost Evaluation." *Chemical Engineering Journal* 349(October):309–317. https://doi.org/10.1016/j.cej.2018.05.080.

Kumari, A., M.K. Jha, and D.D. Pathak. 2018. "Review on the Processes for the Recovery of Rare Earth Metals (REMs) from Secondary Resources." Pp. 53–65 in *Rare Metal Technology*, H. Kim, B. Wesstrom, S. Alam, T. Ouchi, G. Azimi, N.R. Neelameggham, S. Wang, and X. Guan, eds. The Minerals, Metals & Materials Series. Cham: Springer International Publishing. https://doi.org/10.1007/978-3-319-72350-1_5.

Kumari, A., R. Parween, S. Chakravarty, K. Parmar, D.D. Pathak, J. Lee, and M.K. Jha. 2019. "Novel Approach to Recover Rare Earth Metals (REMs) from Indian Coal Bottom Ash." *Hydrometallurgy* 187(August):1–7. https://doi.org/10.1016/j.hydromet.2019.04.024.

Kumari, A., M.K. Jha, S. Chakravarty, and D.D. Pathak. 2020. "Indian Coal Ash: A Potential Alternative Resource for Rare Earth Metals (REMs)." Pp. 265–273 in *Rare Metal Technology*, G. Azimi, K. Forsberg, T. Ouchi, H. Kim, S. Alam, and A. Abdullahi Baba, eds. The Minerals, Metals & Materials Series. Cham: Springer International Publishing. https://doi.org/10.1007/978-3-030-36758-9_25.

Kuppusamy, V.K. 2022. "Characterization and Extraction of Rare Earth Elements from Metallurgical Coal-Based Source." The University of British Columbia.

Kuppusamy, V.K., and M. Holuszko. 2022. "Sulfuric Acid Baking and Water Leaching of Rare Earth Elements from Coal Tailings." *Fuel* 319(July):123738. https://doi.org/10.1016/j.fuel.2022.123738.

Kuppusamy, V.K., A. Kumar, and M. Holuszko. 2019. "Simultaneous Extraction of Clean Coal and Rare Earth Elements from Coal Tailings Using Alkali-Acid Leaching Process." *Journal of Energy Resources Technology* 141(7):070708. https://doi.org/10.1115/1.4043328.

Kursun Unver, I., and M. Terzi. 2018. "Distribution of Trace Elements in Coal and Coal Fly Ash and Their Recovery with Mineral Processing Practices: A Review." *Journal of Mining and Environment* 9(3):641–655. https://doi.org/10.22044/jme.2018.6855.1518.

Lai, Q.T., T. Thenepalli, and J.W. Ahn. 2021. "Utilization of Circulating Fluidized Bed Combustion Fly Ash for Simultaneous Recovery of Rare Earth Elements and CO2 Capture." Pp. 403–430 in *Clean Coal Technologies*, R.K. Jyothi and P.K. Parhi, eds. Cham: Springer International Publishing. https://doi.org/10.1007/978-3-030-68502-7_16.

Lange, C.N., I.M.C. Camargo, A.M.G.M. Figueiredo, L. Castro, M.B.A. Vasconcellos, and R.B. Ticianelli. 2017. "A Brazilian Coal Fly Ash as a Potential Source of Rare Earth Elements." *Journal of Radioanalytical and Nuclear Chemistry* 311(2):1235–1241. https://doi.org/10.1007/s10967-016-5026-8.

Laudal, D.A. 2017. "Evaluation of Rare Earth Element Extraction from North Dakota Coal-Related Feed Stocks." University of North Dakota.

Laudal, D.A., S.A. Benson, R.S. Addleman, and D. Palo. 2018. "Leaching Behavior of Rare Earth Elements in Fort Union Lignite Coals of North America." *International Journal of Coal Geology* 191(April):112–124. https://doi.org/10.1016/j.coal.2018.03.010.

Li, Q., B. Ji, Z. Xiao, and W. Zhang. 2022. "Alkali Pretreatment Effects on Acid Leaching Recovery of Rare Earth Elements from Coal Waste of the Western Kentucky No. 13 and Fire Clay Seams." *Minerals and Mineral Materials* 1:7. https://doi.org/10.20517/mmm.2022.05.

Lima, A.T., and L.M. Ottosen. 2022. "Rare Earth Elements Partition and Recovery During Electrodialytic Treatment of Coal Fly Ash." *Journal of The Electrochemical Society* 169(3):033501. https://doi.org/10.1149/1945-7111/ac56a6.

Lin, R., M. Stuckman, B.H. Howard, T.L. Bank, E.A. Roth, M.K. Macala, C. Lopano, Y. Soong, and E.J. Granite. 2018. "Application of Sequential Extraction and Hydrothermal Treatment for Characterization and Enrichment of Rare Earth Elements from Coal Fly Ash." *Fuel* 232(November):124–133. https://doi.org/10.1016/j.fuel.2018.05.141.

Liu, C., G. Han, B. Hu, F. Geng, M. Liu, S. Dai, and Y. Yang. 2021. "Fast Screening of Coal Fly Ash with Potential for Rare Earth Element Recovery by Electron Paramagnetic Resonance Spectroscopy." *Environmental Science & Technology* 55(24):16716–16722. https://doi.org/10.1021/acs.est.1c06658.

Liu, J.-C., and F. Lomanjaya. 2022. "Subcritical Water Extraction of Rare Earth Elements from Coal Fly Ash." *SSRN Electronic Journal*. https://doi.org/10.2139/ssrn.4057928.

Liu, P., R. Huang, and Y. Tang. 2019. "Comprehensive Understandings of Rare Earth Element (REE) Speciation in Coal Fly Ashes and Implication for REE Extractability." *Environmental Science & Technology* 53(9):5369–5377. https://doi.org/10.1021/acs.est.9b00005.

Liu, P., S. Zhao, N. Xie, L. Yang, Q. Wang, Y. Wen, H. Chen, and Y. Tang. 2023. "Green Approach for Rare Earth Element (REE) Recovery from Coal Fly Ash." *Environmental Science & Technology* 57(13):5414–5423. https://doi.org/10.1021/acs.est.2c09273.

Liu, T., and J. Chen. 2021. "Extraction and Separation of Heavy Rare Earth Elements: A Review." *Separation and Purification Technology* 276(December):119263. https://doi.org/10.1016/j.seppur.2021.119263.

Liu, T., J.C. Hower, and C.-H. Huang. 2023. "Recovery of Rare Earth Elements from Coal Fly Ash with Betainium Bis(Trifluoromethylsulfonyl)Imide: Different Ash Types and Broad Elemental Survey." *Minerals* 13(7):952. https://doi.org/10.3390/min13070952.

Lomanjaya, F., and J. Liu. 2023. "Intensified Extraction of Rare Earth Elements from Coal Fly Ash." *Journal of Chemical Technology & Biotechnology* 98(9):2266–2273. https://doi.org/10.1002/jctb.7451.

Ma, J., S. Li, J. Wang, S. Jiang, B. Panchal, and Y. Sun. 2023. "Bioleaching Rare Earth Elements from Coal Fly Ash by *Aspergillus niger*." *Fuel* 354(December):129387. https://doi.org/10.1016/j.fuel.2023.129387.

Ma, Z., S. Zhang, H. Zhang, and F. Cheng. 2019. "Novel Extraction of Valuable Metals from Circulating Fluidized Bed-Derived High-Alumina Fly Ash by Acid–Alkali–Based Alternate Method." *Journal of Cleaner Production* 230(September):302–313. https://doi.org/10.1016/j.jclepro.2019.05.113.

Mahandra, H., B. Hubert, and A. Ghahreman. 2021. "Recovery of Rare Earth and Some Other Potential Elements from Coal Fly Ash for Sustainable Future." Pp. 339–380 in *Clean Coal Technologies*, R.K. Jyothi and P.K. Parhi, eds. Cham: Springer International Publishing. https://doi.org/10.1007/978-3-030-68502-7_14.

Mann, M., N. Theaker, B. Rew, S. Benson, A. Benson, D. Palo, C. Haugen, and D. Laudal. 2021. "Investigation of Rare Earth Element Extraction from North Dakota Coal-Related Feedstocks (Phase 2 Final Technical Report)." DUNS–10-228-0781-Rev.01. https://doi.org/10.2172/1785352.

Manurung, H., W. Rosita, I.M. Bendiyasa, A. Prasetya, F. Anggara, W. Astuti, D.R. Djuanda, and H.T.B.M. Petrus. 2020. "Recovery of Rare Earth Elements and Yitrium from Non-Magnetic Coal Fly Ash Using Acetic Acid Solution." *Metal Indonesia* 42(1):35. https://doi.org/10.32423/jmi.2020.v42.35-42.

Middleton, A., D.M. Park, Y. Jiao, and H. Hsu-Kim. 2020. "Major Element Composition Controls Rare Earth Element Solubility During Leaching of Coal Fly Ash and Coal By-Products." *International Journal of Coal Geology* 227(July):103532. https://doi.org/10.1016/j.coal.2020.103532.

Miranda, M.M., J.M. Bielicki, S. Chun, and C.-M. Cheng. 2022. "Recovering Rare Earth Elements from Coal Mine Drainage Using Industrial Byproducts: Environmental and Economic Consequences." *Environmental Engineering Science* 39(9):770–783. https://doi.org/10.1089/ees.2021.0378.

Modi, P., A. Jamal, R. Varshney, and I.C. Rahi. 2023a. "Occurrence, Mobility, Leaching, and Recovery of Rare Earth Elements and Trace Elements in Sohagpur Coalfield, Madhya Pradesh, India." *International Journal of Coal Preparation and Utilization* 43(1):103–118. https://doi.org/10.1080/19392699.2021.2014823.

Modi, P., A. Jamal, R. Varshney, I.C. Rahi, and M.A. Siddiqui. 2023b. "Rare Earth Elements Mobility, Leaching and Recovery by Different Chemicals Treatment on Coal Samples and Calcined Samples of Sohagpur Coalfield, Madhya Pradesh, India." *International Journal of Coal Preparation and Utilization* 43(1):148–168. https://doi.org/10.1080/19392699.2022.2031171.

Mokoena, B.K., L.S. Mokhahlane, and S. Clarke. 2022. "Effects of Acid Concentration on the Recovery of Rare Earth Elements from Coal Fly Ash." *International Journal of Coal Geology* 259(July):104037. https://doi.org/10.1016/j.coal.2022.104037.

Mondal, S., A. Ghar, A.K. Satpati, P. Sinharoy, D.K. Singh, J.N. Sharma, T. Sreenivas, and V. Kain. 2019. "Recovery of Rare Earth Elements from Coal Fly Ash Using TEHDGA Impregnated Resin." *Hydrometallurgy* 185(May):93–101. https://doi.org/10.1016/j.hydromet.2019.02.005.

Mostajeran, M., J.-M. Bondy, N. Reynier, and R. Cameron. 2021. "Mining Value from Waste: Scandium and Rare Earth Elements Selective Recovery from Coal Fly Ash Leach Solutions." *Minerals Engineering* 173(November):107091. https://doi.org/10.1016/j.mineng.2021.107091.

Muravyov, M.I., A.G. Bulaev, V.S. Melamud, and T.F. Kondrat'eva. 2015. "Leaching of Rare Earth Elements from Coal Ashes Using Acidophilic Chemolithotrophic Microbial Communities." *Microbiology* 84(2):194–201. https://doi.org/10.1134/S0026261715010087.

Mwewa, B., M. Tadie, S. Ndlovu, G.S. Simate, and E. Matinde. 2022. "Recovery of Rare Earth Elements from Acid Mine Drainage: A Review of the Extraction Methods." *Journal of Environmental Chemical Engineering* 10(3):107704. https://doi.org/10.1016/j.jece.2022.107704.

Nawab, A., X. Yang, and R. Honaker. 2022. "An Acid Baking Approach to Enhance Heavy Rare Earth Recovery from Bituminous Coal-Based Sources." *Minerals Engineering* 184(June):107610. https://doi.org/10.1016/j.mineng.2022.107610.

Nie, T., C. Zhou, J. Pan, Z. Wen, F. Yang, and R. Jia. 2022. "Study on the Occurrence of Rare Earth Elements in Coal Refuse Based on Sequential Chemical Extraction and Pearson Correlation Analysis." *Mining, Metallurgy & Exploration* 39(2):669–678. https://doi.org/10.1007/s42461-022-00542-y.

Okeme, I.C., R.A. Crane, W.M. Nash, T.I. Ojonimi, and T.B. Scott. 2022. "Characterisation of Rare Earth Elements and Toxic Heavy Metals in Coal and Coal Fly Ash." *RSC Advances* 12(30):19284–19296. https://doi.org/10.1039/D2RA02788G.

Pan, J., C. Zhou, M. Tang, S. Cao, C. Liu, N. Zhang, M. Wen, Y. Luo, T. Hu, and W. Ji. 2019. "Study on the Modes of Occurrence of Rare Earth Elements in Coal Fly Ash by Statistics and a Sequential Chemical Extraction Procedure." *Fuel* 237(February):555–565. https://doi.org/10.1016/j.fuel.2018.09.139.

Pan, J., T. Nie, B. Vaziri Hassas, M. Rezaee, Z. Wen, and C. Zhou. 2020. "Recovery of Rare Earth Elements from Coal Fly Ash by Integrated Physical Separation and Acid Leaching." *Chemosphere* 248:126112. https://doi.org/10.1016/j.chemosphere.2020.126112.

Pan, J., B. Vaziri Hassas, M. Rezaee, C. Zhou, and S.V. Pisupati. 2021. "Recovery of Rare Earth Elements from Coal Fly Ash Through Sequential Chemical Roasting, Water Leaching, and Acid Leaching Processes." *Journal of Cleaner Production* 284:124725. https://doi.org/10.1016/j.jclepro.2020.124725.

Pan, J., X. Zhao, C. Zhou, F. Yang, and W. Ji. 2022a. "Study on Solvent Extraction of Rare Earth Elements from Leaching Solution of Coal Fly Ash by P204." *Minerals* 12(12):1547. https://doi.org/10.3390/min12121547.

Pan, J., T. Nie, C. Zhou, F. Yang, R. Jia, L. Zhang, and H. Liu. 2022b. "The Effect of Calcination on the Occurrence and Leaching of Rare Earth Elements in Coal Refuse." *Journal of Environmental Chemical Engineering* 10(5):108355. https://doi.org/10.1016/j.jece.2022.108355.

Pan, J., L. Zhang, Z. Wen, T. Nie, N. Zhang, and C. Zhou. 2023. "The Mechanism Study on the Integrated Process of NaOH Treatment and Citric Acid Leaching for Rare Earth Elements Recovery from Coal Fly Ash." *Journal of Environmental Chemical Engineering* 11(3):109921. https://doi.org/10.1016/j.jece.2023.109921.

Park, S., and Y. Liang. 2019. "Bioleaching of Trace Elements and Rare Earth Elements from Coal Fly Ash." *International Journal of Coal Science & Technology* 6(1):74–83. https://doi.org/10.1007/s40789-019-0238-5.

Park, S., M. Kim, Y. Lim, J. Yu, S. Chen, S.W. Woo, S. Yoon, S. Bae, and H.S. Kim. 2021. "Characterization of Rare Earth Elements Present in Coal Ash by Sequential Extraction." *Journal of Hazardous Materials* 402(January):123760. https://doi.org/10.1016/j.jhazmat.2020.123760.

Peiravi, M., L. Ackah, R. Guru, M. Mohanty, J. Liu, B. Xu, X. Zhu, and L. Chen. 2017. "Chemical Extraction of Rare Earth Elements from Coal Ash." *Minerals & Metallurgical Processing* 34(4):170–177. https://doi.org/10.19150/mmp.7856.

Peiravi, M., F. Dehghani, L. Ackah, A. Baharlouei, J. Godbold, J. Liu, M. Mohanty, and T. Ghosh. 2021. "A Review of Rare-Earth Elements Extraction with Emphasis on Non-Conventional Sources: Coal and Coal Byproducts, Iron Ore Tailings, Apatite, and Phosphate Byproducts." *Mining, Metallurgy & Exploration* 38(1):1–26. https://doi.org/10.1007/s42461-020-00307-5.

Penney, S., and S. Alam. 2023. "Critical Metals for Clean Energy: Extraction of Rare Earth Elements from Coal Ash." Pp. 193–197 in *Energy Technology* 2023, S. Alam, D. Post Guillen, F. Tesfaye, L. Zhang, S.A.C. Hockaday, N.R. Neelameggham, H. Peng, N. Haque, and Y. Liu, eds. The Minerals, Metals & Materials Series. Cham: Springer Nature Switzerland. https://doi.org/10.1007/978-3-031-22638-0_19.

Peterson, R., M. Heinrichs, D. Argumedo, R. Taha, S. Winecki, K. Johnson, A. Lane, D. Riordan, and National Energy Technology Lab. 2017. "Recovery of Rare Earth Elements from Coal and Coal Byproducts via a Closed Loop Leaching Process: Final Report." DOE-BATTELLE–27012. https://doi.org/10.2172/1377818.

Phuoc, T.X., P. Wang, and D. McIntyre. 2015. "Discovering the Feasibility of Using the Radiation Forces for Recovering Rare Earth Elements from Coal Power Plant By-Products." *Advanced Powder Technology* 26(5):1465–1472. https://doi.org/10.1016/j.apt.2015.08.004.

Ponou, J., G. Dodbiba, J.-W. Anh, and T. Fujita. 2016. "Selective Recovery of Rare Earth Elements from Aqueous Solution Obtained from Coal Power Plant Ash." *Journal of Environmental Chemical Engineering* 4(4):3761–3766. https://doi.org/10.1016/j.jece.2016.08.019.

Prihutami, P., W.B. Sediawan, W. Astuti, and A. Prasetya. 2020. "Effect of Temperature on Rare Earth Elements Recovery from Coal Fly Ash Using Citric Acid." IOP Conference Series: Materials Science and Engineering 742(1):012040. https://doi.org/10.1088/1757-899X/742/1/012040.

Prihutami, P., A. Prasetya, W.B. Sediawan, H.T.B.M. Petrus, and F. Anggara. 2021. "Study on Rare Earth Elements Leaching from Magnetic Coal Fly Ash by Citric Acid." *Journal of Sustainable Metallurgy* 7(3):1241–1253. https://doi.org/10.1007/s40831-021-00414-7.

Rao, K.A., R. Karan, M. Babu J, R.D. G, and S. T. 2022. "Development of Process Scheme for Recovery of Rare Earths from Leachate of Coal Flyash." *Cleaner Chemical Engineering* 4(December 1):100078. https://doi.org/10.1016/j.clce.2022.100078.

Rao, K.A., and T. Sreenivas. 2019. "Recovery of Rare Earth Elements from Coal Fly Ash: A Review." Pp. 343–364 in *Critical and Rare Earth Elements*, A. Akcil, ed. 1st edition. CRC Press. https://doi.org/10.1201/9780429023545-18.

Ren, P. 2019. "Recovery of Rare Earth Elements (REEs) from Coal Mine Drainage Sludge Leachate."

Rosita, W., I.M. Bendiyasa, I. Perdana, and F. Anggara. 2020a. "Recovery of Rare Earth Elements and Yttrium from Indonesia Coal Fly Ash Using Sulphuric Acid Leaching." In, 050004. Yogyakarta, Indonesia. https://doi.org/10.1063/5.0000836.

Rosita, W., I.M. Bendiyasa, I. Perdana, and F. Anggara. 2020b. "Sequential Particle-Size and Magnetic Separation for Enrichment of Rare-Earth Elements and Yttrium in Indonesia Coal Fly Ash." *Journal of Environmental Chemical Engineering* 8(1):103575. https://doi.org/10.1016/j.jece.2019.103575.

Rosita, W., I.M. Bendiyasa, I. Perdana, and F. Anggara. 2020c. "Experimental Study of Rare Earth Element Enrichment from Indonesian Coal Fly Ash: Alkaline Leaching." *Key Engineering Materials* 840(April):514–519. https://doi.org/10.4028/www.scientific.net/KEM.840.514.

Rosita, W., I. Perdana, I.M. Bendiyasa, F. Anggara, H.T.B.M. Petrus, A. Prasetyo, and I. Rodliyah. 2023. "Sequential Alkaline-Organic Acid Leaching Process to Enhance the Recovery of Rare Earth Elements from Indonesian Coal Fly Ash." *Journal of Rare Earths* September 7. https://doi.org/10.1016/j.jre.2023.09.001.

Roth, E., M.K. Macala, R. Lin, T.L. Bank, R. Thompson, B. Howard, Y. Soong, and E. Granite. 2017. "Distributions and Extraction of Rare Earth Elements from Coal and Coal By-Products." Presented at 2017 World of Coal Ash Conference. May 9–11, 2017. Lexington, KY. https://www.osti.gov/servlets/purl/1812004.

Royer-Lavallée, A., C.M. Neculita, and L. Coudert. 2020. "Removal and Potential Recovery of Rare Earth Elements from Mine Water." *Journal of Industrial and Engineering Chemistry* 89(September):47–57. https://doi.org/10.1016/j.jiec.2020.06.010.

Rozelle, P.L., A.B. Khadilkar, N. Pulati, N. Soundarrajan, M.S. Klima, M.M. Mosser, C.E. Miller, and S.V. Pisupati. 2016. "A Study on Removal of Rare Earth Elements from U.S. Coal Byproducts by Ion Exchange." *Metallurgical and Materials Transactions E* 3(1):6–17. https://doi.org/10.1007/s40553-015-0064-7.

Rybak, A., and A. Rybak. 2021. "Characteristics of Some Selected Methods of Rare Earth Elements Recovery from Coal Fly Ashes." *Metals* 11(1):142. https://doi.org/10.3390/met11010142.

Sachan, A., S. Dev, T. Ghosh, S. Aggarwal, F. Dehghani, M. Martinez, and B.R. Briggs. 2023. "Bioweathering Using Shewanella Oneidensis MR-1 Enhances Recovery of Rare Earth Elements from Alaskan Coal Mines." *ACS ES&T Engineering* 3(11):1686–1693. https://doi.org/10.1021/acsestengg.3c00178.

Sahoo, P.K., K. Kim, M.A. Powell, and S.M. Equeenuddin. 2016. "Recovery of Metals and Other Beneficial Products from Coal Fly Ash: A Sustainable Approach for Fly Ash Management." *International Journal of Coal Science & Technology* 3(3):267–283. https://doi.org/10.1007/s40789-016-0141-2.

Sarswat, P.K., M. Leake, L. Allen, M.L. Free, X. Hu, D. Kim, A. Noble, and G.H. Luttrell. 2020. "Efficient Recovery of Rare Earth Elements from Coal Based Resources: A Bioleaching Approach." *Materials Today Chemistry* 16(June):100246. https://doi.org/10.1016/j.mtchem.2020.100246.

Shoppert, A., D. Valeev, J. Napol'skikh, I. Loginova, J. Pan, H. Chen, and L. Zhang. 2022. "Rare-Earth Elements Extraction from Low-Alkali Desilicated Coal Fly Ash by $(NH_4)_2SO_4$ + H_2SO_4." *Materials* 16(1):6. https://doi.org/10.3390/ma16010006.

Smith, R.C., R.K. Taggart, J.C. Hower, M.R. Wiesner, and H. Hsu-Kim. 2019. "Selective Recovery of Rare Earth Elements from Coal Fly Ash Leachates Using Liquid Membrane Processes." *Environmental Science & Technology* 53(8):4490–4499. https://doi.org/10.1021/acs.est.9b00539.

Song, G., X. Wang, C. Romero, H. Chen, Z. Yao, A. Kaziunas, R. Schlake, et al. 2021. "Extraction of Selected Rare Earth Elements from Anthracite Acid Mine Drainage Using Supercritical CO_2 via Coagulation and Complexation." *Journal of Rare Earths* 39(1):83–89. https://doi.org/10.1016/j.jre.2020.02.007.

Soong, Y., R. Lin, M. Stuckman, B. Howard, C. Lopano, and E. Granite. 2019. "Recovery of the Rare Earth Elements (REE) from Coal Fly Ash via the Combination of the Physical Separation and Chemical Extraction Methods." Presented at the 257th ACS National Meeting. March 31–April 4, 2019. Orlando, FL. https://www.osti.gov/servlets/purl/1811654.

Sreenivas, T., M. Serajuddin, R. Moudgil, and K. Anand Rao. 2021. "Developments in Characterization and Mineral Processing of Coal Fly Ash for Recovery of Rare Earth Elements." Pp. 431–71 in *Clean Coal Technologies*, R.K. Jyothi and P.K. Parhi, eds. Cham: Springer International Publishing. https://doi.org/10.1007/978-3-030-68502-7_17.

Stoy, L., V. Diaz, and C.-H. Huang. 2021. "Preferential Recovery of Rare-Earth Elements from Coal Fly Ash Using a Recyclable Ionic Liquid." *Environmental Science & Technology* 55(13):9209–9220. https://doi.org/10.1021/acs.est.1c00630.

Stoy, L., Y. Kulkarni, and C.-H. Huang. 2022a. "Optimization of Iron Removal in the Recovery of Rare-Earth Elements from Coal Fly Ash Using a Recyclable Ionic Liquid." *Environmental Science & Technology* 56(8):5150–5160. https://doi.org/10.1021/acs.est.1c08552.

Stoy, L., J. Xu, Y. Kulkarni, and C.-H. Huang. 2022b. "Ionic Liquid Recovery of Rare-Earth Elements from Coal Fly Ash: Process Efficiency and Sustainability Evaluations." *ACS Sustainable Chemistry & Engineering* 10(36):11824–11834. https://doi.org/10.1021/acssuschemeng.2c02459.

Su, H., H. Chen, and J. Lin. 2020a. "A Sequential Integration Approach Using Aspergillus Niger to Intensify Coal Fly Ash as a Rare Metal Pool." *Fuel* 270(June):117460. https://doi.org/10.1016/j.fuel.2020.117460.

Su, H., F. Tan, and J. Lin. 2020b. "An Integrated Approach Combines Hydrothermal Chemical and Biological Treatment to Enhance Recycle of Rare Metals from Coal Fly Ash." *Chemical Engineering Journal* 395(September):124640. https://doi.org/10.1016/j.cej.2020.124640.

Sutterlin, W. 2019. "Recovery of Rare Earth Elements from Coal Mining Waste Materials." DOE-Inventure–30146, 1560384. https://doi.org/10.2172/1560384.

Swinder, H., B. Bialecka, and A. Jarosinski. 2017. "Recovery of Rare Earth Elements from Coal Combustion Fly Ashes." Pp. 995–1002 in *International Multidisciplinary Scientific GeoConference: SGEM, Sofia*. https://doi.org/10.5593/sgem2017/11.

Taggart, R.K. 2015. "Recovering Rare Earth Metals from Coal Fly Ash." Presented at 2015 World of Coal Ash Conference. May 5–7, 2015. Nashville, TN. https://uknowledge.uky.edu/cgi/viewcontent.cgi?article=1634&context=woca.

Taggart, R.K. 2018. "Recovery of Rare Earth Elements from Coal Combustion Ash: Survey, Extraction, and Speciation." Duke University.

Taggart, R.K., J.C. Hower, G.S. Dwyer, and H. Hsu-Kim. 2016. "Trends in the Rare Earth Element Content of U.S.-Based Coal Combustion Fly Ashes." *Environmental Science & Technology* 50(11):5919–5926. https://doi.org/10.1021/acs.est.6b00085.

Taggart, R.K., J.F. King, J.C. Hower, and H. Hsu-Kim. 2017. "Rare Earth Element Recovery from Coal Fly Ash by Roasting and Leaching Methods." Presented at 2017 World of Coal Ash Conference, Lexington, KY. May 9–11, 2017. https://uknowledge.uky.edu/cgi/viewcontent.cgi?article=1291&context=woca.

Taggart, R.K., J.C. Hower, and H. Hsu-Kim. 2018. "Effects of Roasting Additives and Leaching Parameters on the Extraction of Rare Earth Elements from Coal Fly Ash." *International Journal of Coal Geology* 196(August):106–114. https://doi.org/10.1016/j.coal.2018.06.021.

Talan, D., and Q. Huang. 2020. "Separation of Thorium, Uranium, and Rare Earths from a Strip Solution Generated from Coarse Coal Refuse." *Hydrometallurgy* 197(November):105446. https://doi.org/10.1016/j.hydromet.2020.105446.

Talan, D., and Q. Huang. 2022. "A Review Study of Rare Earth, Cobalt, Lithium, and Manganese in Coal-Based Sources and Process Development for Their Recovery." *Minerals Engineering* 189(November):107897. https://doi.org/10.1016/j.mineng.2022.107897.

Tang, M., C. Zhou, J. Pan, N. Zhang, C. Liu, S. Cao, T. Hu, and W. Ji. 2019. "Study on Extraction of Rare Earth Elements from Coal Fly Ash Through Alkali Fusion—Acid Leaching." *Minerals Engineering* 136(June):36–42. https://doi.org/10.1016/j.mineng.2019.01.027.

Tang, M., C. Zhou, N. Zhang, J. Pan, S. Cao, T. Hu, W. Ji, Z. Wen, and T. Nie. 2022. "Extraction of Rare Earth Elements from Coal Fly Ash by Alkali Fusion–Acid Leaching: Mechanism Analysis." *International Journal of Coal Preparation and Utilization* 42(3):536–555. https://doi.org/10.1080/19392699.2019.1623206.

Thakare, J., and J. Masud. 2022. "Low Temperature Electrochemical Extraction of Rare Earth Metals from Lignite Coal: An Environmentally Benign and Energy Efficient Method." *Journal of the Electrochemical Society* 169(2):023503. https://doi.org/10.1149/1945-7111/ac4f77.

Trinh, H.B., S. Kim, and J. Lee. 2022. "Recovery of Rare Earth Elements from Coal Fly Ash Using Enrichment by Sodium Hydroxide Leaching and Dissolution by Hydrochloric Acid." *Geosystem Engineering* 25(1–2):53–62. https://doi.org/10.1080/12269328.2022.2120092.

Tuan, L., T. Thenepalli, R. Chilakala, H. Vu, J. Ahn, and J. Kim. 2019. "Leaching Characteristics of Low Concentration Rare Earth Elements in Korean (Samcheok) CFBC Bottom Ash Samples." *Sustainability* 11(9):2562. https://doi.org/10.3390/su11092562.

Vu, H., T. Frýdl, T. Bastl, P. Dvořák, E. Kristianová, and T. Tomáško. 2021. "Recent Development in Metal Extraction from Coal Fly Ash." Pp. 575–603 in *Clean Coal Technologies*, R.K. Jyothi and P.K. Parhi, eds. Cham: Springer International Publishing. https://doi.org/10.1007/978-3-030-68502-7_23.

Wang, N., X. Sun, Q. Zhao, Y. Yang, and P. Wang. 2020. "Leachability and Adverse Effects of Coal Fly Ash: A Review." *Journal of Hazardous Materials* 396(September):122725. https://doi.org/10.1016/j.jhazmat.2020.122725.

Wang, Y., A. Noble, C. Vass, and P. Ziemkiewicz. 2021. "Speciation of Rare Earth Elements in Acid Mine Drainage Precipitates by Sequential Extraction." *Minerals Engineering* 168(July):106827. https://doi.org/10.1016/j.mineng.2021.106827.

Wang, Y., P. Ziemkiewicz, and A. Noble. 2022. "A Hybrid Experimental and Theoretical Approach to Optimize Recovery of Rare Earth Elements from Acid Mine Drainage Precipitates by Oxalic Acid Precipitation." *Minerals* 12(2):236. https://doi.org/10.3390/min12020236.

Wen, Z., H. Chen, J. Pan, R. Jia, F. Yang, H. Liu, L. Zhang, N. Zhang, and C. Zhou. 2022a. "Grinding Activation Effect on the Flotation Recovery of Unburned Carbon and Leachability of Rare Earth Elements in Coal Fly Ash." *Powder Technology* 398(January):117045. https://doi.org/10.1016/j.powtec.2021.117045.

Wen, Z., C. Zhou, J. Pan, S. Cao, T. Hu, W. Ji, and T. Nie. 2022b. "Recovery of Rare-Earth Elements from Coal Fly Ash via Enhanced Leaching." *International Journal of Coal Preparation and Utilization* 42(7):2041–2055. https://doi.org/10.1080/19392699.2020.1790537.

Wilfong, W.C., T. Ji, Y. Duan, F. Shi, Q. Wang, and M.L. Gray. 2022. "Critical Review of Functionalized Silica Sorbent Strategies for Selective Extraction of Rare Earth Elements from Acid Mine Drainage." *Journal of Hazardous Materials* 424(February):127625. https://doi.org/10.1016/j.jhazmat.2021.127625.

Wu, G., T. Wang, J. Wang, Y. Zhang, and W. Pan. 2020. "Occurrence Forms of Rare Earth Elements in Coal and Coal Gangue and Their Combustion Products." *Journal of Fuel Chemistry and Technology* 48(12):1498–1505. https://doi.org/10.1016/S1872-5813(20)30094-3.

Wu, G., T. Wang, G. Chen, Z. Shen, and W.-P. Pan. 2022. "Coal Fly Ash Activated by NaOH Roasting: Rare Earth Elements Recovery and Harmful Trace Elements Migration." *Fuel* 324(September):124515. https://doi.org/10.1016/j.fuel.2022.124515.

Xu, F., S. Qin, S. Li, J. Wang, D. Qi, Q. Lu, and J. Xing. 2022. "Distribution, Occurrence Mode, and Extraction Potential of Critical Elements in Coal Ashes of the Chongqing Power Plant." *Journal of Cleaner Production* 342(March):130910. https://doi.org/10.1016/j.jclepro.2022.130910.

Yakaboylu, G.A., D. Baker, B. Wayda, K. Sabolsky, J.W. Zondlo, D. Shekhawat, C. Wildfire, and E.M. Sabolsky. 2019. "Microwave-Assisted Pretreatment of Coal Fly Ash for Enrichment and Enhanced Extraction of Rare-Earth Elements." *Energy & Fuels* 33(11):12083–12095. https://doi.org/10.1021/acs.energyfuels.9b02846.

Yang, J., S. Montross, and C. Verba. 2021. "Assessing the Extractability of Rare Earth Elements from Coal Preparation Fines Refuse Using an Organic Acid Lixiviant." *Mining, Metallurgy & Exploration* 38(4):1701–1709. https://doi.org/10.1007/s42461-021-00439-2.

Yang, J., S. Bauer, and C. Verba. 2022. "Strategies to Recover Easily-Extractable Rare Earth Elements and Other Critical Metals from Coal Waste Streams and Adjacent Rock Strata Using Citric Acid." DOE/NETL-2022/3732. https://doi.org/10.2172/1884275.

Yang, X. 2019. "Leaching Characteristics of Rare Earth Elements from Bituminous Coal-Based Sources." https://doi.org/10.13023/ETD.2019.229.

Yang, X., and R.Q. Honaker. 2020. "Leaching Kinetics of Rare Earth Elements from Fire Clay Seam Coal." *Minerals* 10(6):491. https://doi.org/10.3390/min10060491.

Yang, X., J. Werner, and R.Q. Honaker. 2019. "Leaching of Rare Earth Elements from an Illinois Basin Coal Source." *Journal of Rare Earths* 37(3):312–321. https://doi.org/10.1016/j.jre.2018.07.003.

Żelazny, S., H. Świnder, A. Jarosiński, and B. Białecka. 2023. "The Recovery of Rare-Earth Metals from Fly Ash Using Alkali Pre-treatment with Sodium Hydroxide." *Gospodarka Surowcami Mineralnymi - Mineral Resources Management* 36(3):127–144. https://doi.org/10.24425/gsm.2020.133930.

Zhang, R., C. Zhang, and Y. Cao. 2022. "The Enhanced Extraction of Rare Earth Elements from Coal Gangue and Coal Fly Ash." Preprint. In Review. https://doi.org/10.21203/rs.3.rs-1438617/v1.

Zhang, R., C. Zhang, and Y. Cao. 2023. "Effective Extraction of Rare Earth Elements from Coal Slurry." *Separation Science and Technology* 58(1):51–60. https://doi.org/10.1080/01496395.2022.2102999.

Zhang, W., and R.Q. Honaker. 2018. "Rare Earth Elements Recovery Using Staged Precipitation from a Leachate Generated from Coarse Coal Refuse." *International Journal of Coal Geology* 195(July):189–199. https://doi.org/10.1016/j.coal.2018.06.008.

Zhang, W., and R. Honaker. 2019a. "Calcination Pretreatment Effects on Acid Leaching Characteristics of Rare Earth Elements from Middlings and Coarse Refuse Material Associated with a Bituminous Coal Source." *Fuel* 249(August):130–145. https://doi.org/10.1016/j.fuel.2019.03.063.

Zhang, W., and R. Honaker. 2019b. "Enhanced Leachability of Rare Earth Elements from Calcined Products of Bituminous Coals." *Minerals Engineering* 142(October):105935. https://doi.org/10.1016/j.mineng.2019.105935.

Zhang, W., and R. Honaker. 2020a. "Characterization and Recovery of Rare Earth Elements and Other Critical Metals (Co, Cr, Li, Mn, Sr, and V) from the Calcination Products of a Coal Refuse Sample." *Fuel* 267(May):117236. https://doi.org/10.1016/j.fuel.2020.117236.

Zhang, W., and R. Honaker. 2020b. "Process Development for the Recovery of Rare Earth Elements and Critical Metals from an Acid Mine Leachate." *Minerals Engineering* 153(July):106382. https://doi.org/10.1016/j.mineng.2020.106382.

Zhang, W., A. Noble, X. Yang, and R. Honaker. 2020c. "A Comprehensive Review of Rare Earth Elements Recovery from Coal-Related Materials." *Minerals* 10(5):451. https://doi.org/10.3390/min10050451.

Zhang, W., M. Rezaee, A. Bhagavatula, Y. Li, J. Groppo, and R. Honaker. 2015. "A Review of the Occurrence and Promising Recovery Methods of Rare Earth Elements from Coal and Coal By-Products." *International Journal of Coal Preparation and Utilization* 35(6):295–330. https://doi.org/10.1080/19392699.2015.1033097.

Zhang, W., X. Yang, and R.Q. Honaker. 2018. "Association Characteristic Study and Preliminary Recovery Investigation of Rare Earth Elements from Fire Clay Seam Coal Middlings." *Fuel* 215(March):551–560. https://doi.org/10.1016/j.fuel.2017.11.075.

Zhang, Z., L. Allen, P. Podder, M.L. Free, and P.K. Sarswat. 2021. "Recovery and Enhanced Upgrading of Rare Earth Elements from Coal-Based Resources: Bioleaching and Precipitation." *Minerals* 11(5):484. https://doi.org/10.3390/min11050484.

Zhao, Y., Y. Zhou, J. Zhang, and C. Zheng. 2019. "Trace Element Resource Recovery from Coal and Coal Utilization By-Products." Pp. 375–399 in *Emission and Control of Trace Elements from Coal-Derived Gas Streams*. Elsevier. https://doi.org/10.1016/B978-0-08-102591-8.00009-X.

Ziemkiewicz, P. 2020. "Recovery of Rare Earth Elements (REEs) from Coal Mine Drainage, Phase 2." DOE-WVU-26927. https://doi.org/10.2172/1614906.

Znamenáčková, I., S. Dolinská, S. Hredzák, V. Čablík, M. Lovás, and D. Gešperová. 2021. "Study of Extraction of Rare Earth Elements from Hard Coal Fly Ash." *Inżynieria Mineralna* 2(1). https://doi.org/10.29227/IM-2020-01-71.

Zou, J., H. Tian, and Z. Wang. 2017. "Leaching Process of Rare Earth Elements, Gallium and Niobium in a Coal-Bearing Strata-Hosted Rare Metal Deposit—A Case Study from the Late Permian Tuff in the Zhongliangshan Mine, Chongqing." *Metals* 7(5):174. https://doi.org/10.3390/met7050174.